ENCYCLOPEDIA OF
EARTH AND SPACE SCIENCE
VOLUME II

ENCYCLOPEDIA OF
EARTH AND SPACE SCIENCE
VOLUME II

TIMOTHY KUSKY, Ph.D.

Katherine Cullen, Ph.D., Managing Editor

Facts On File
An imprint of Infobase Publishing

For Daniel and Shoshana

ENCYCLOPEDIA OF EARTH AND SPACE SCIENCE

Copyright © 2010 by Timothy Kusky, Ph.D.

Facts On File, Inc.
An imprint of Infobase Publishing
132 West 31st Street
New York NY 10001

Library of Congress Cataloging-in-Publication Data
Kusky, Timothy M.
Encyclopedia of Earth and space science / Timothy Kusky ; managing editor, Katherine Cullen.
p. cm.
Includes bibliographical references and index.
ISBN 978-0-8160-7005-3 (set : acid-free paper) 1. Earth sciences—Encyclopedias. 2. Space sciences—Encyclopedias. I. Cullen, Katherine E. II. Title.
QE5.K845 2010
550.3—dc22   2009015655

Facts On File books are available at special discounts when purchased in bulk quantities for businesses, associations, institutions, or sales promotions. Please call our Special Sales Department in New York at (212) 967-8800 or (800) 322-8755.

You can find Facts On File on the World Wide Web at http://www.factsonfile.com

Text design by Annie O'Donnell
Illustrations by Dale Williams
Photo research by Suzanne M. Tibor
Composition by Hermitage Publishing Services
Cover printed by Times Offset (M) Sdn Bhd, Shah Alam, Selangor
Book printed and bound by Times Offset (M) Sdn Bhd, Shah Alam, Selangor
Date printed: May 2010
Printed in Malaysia

10 9 8 7 6 5 4 3 2 1

This book is printed on acid-free paper.

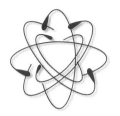

CONTENTS

Acknowledgments   vii

Introduction   viii

Entries Categorized by National Science Education Standards
    for Content (Grades 9–12)   x

Entries I–Z   429

Feature Essays:

    "Beauty and the Beach: Rethinking Coastal Living"   88

    "Diamonds Are Forever"   178

    "Drying of the American Southwest"   202

    "Love Canal Is Not for Honeymooners"   254

    "Liquefaction and Levees: Potential Double Disaster in the
        American Midwest"   321

    "Do Bay of Bengal Cyclones Have to Be So Deadly?"   407

    "The Search for Extraterrestrial Life"   492

    "The Sky Is on Fire"   544

    "The Large Hadron Collider"   604

    "Renewable Energy Options for the Future"   622

    "Predicting Future Earthquakes in the Western United States"   681

    "Tsunami Nightmare"   774

Appendixes:

    I. Chronology   817

    II. Glossary   827

    III. Further Resources   856

    IV. Geologic Timescale   863

    V. Periodic Table of the Elements   864

VI. SI Units and Derived Quantities   866

VII. Multipliers and Dividers for Use with SI Units   867

VIII. Astronomical Data   868

IX. Abbreviations and Symbols for Physical Units   869

X. The Greek Alphabet   870

XI. Common Conversions within the Metric System   871

XII. Common Conversions from U.S. Customary to Metric Unit Values   872

XIII. Temperature Conversions   873

Index   I-1

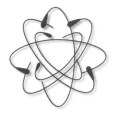

ACKNOWLEDGMENTS

I would like to express appreciation to Frank K. Darmstadt, executive editor, for his critical review of this manuscript, wise advice, patience, and professionalism, and Katherine E. Cullen for her expert editing. Thank you to Richard Garratt, Dale Williams, and the graphics department, who created the illustrations that accompany the entries in this work, and to Suzie Tibor for performing the photo research. I express deep thanks to Dr. Lu Wang for help in preparing this manuscript through its many drafts and stages. Many sections of the work draw from my own experiences doing scientific research in different parts of the world, and it is not possible to thank the hundreds of colleagues whose collaborations and work I have related in this book. Their contributions to the science that allowed the writing of this volume are greatly appreciated. I have tried to reference the most relevant works or, in some cases, more recent sources that have more extensive reference lists. Any omissions are unintentional. Finally, I would especially like to thank my wife and my children, Shoshana and Daniel, for their patience during the long hours spent at my desk preparing this book. Without their understanding this work would not have been possible.

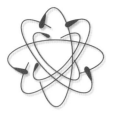

INTRODUCTION

Encyclopedia of Earth and Space Science is a two-volume reference intended to complement the material typically taught in high school Earth science and astronomy classes, and in introductory college geology, atmospheric sciences, and astrophysics courses. The substance reflects the fundamental concepts and principles that underlie the content standards for Earth and space science identified by the National Committee on Science Education Standards and Assessment of the National Research Council for grades 9–12. Within the category of Earth and space science, these include energy in the Earth system, geochemical cycles, origin and evolution of the Earth system, and origin and evolution of the universe. The National Science Education Standards (NSES) also place importance on student awareness of the nature of science and the process by which modern scientists gather information. To assist educators in achieving this goal, other subject matter discusses concepts that unify the Earth and space sciences with physical science and life science: science as inquiry, technology and other applications of scientific advances, science in personal and social perspectives including topics such as natural hazards and global challenges, and the history and nature of science. A listing of entry topics organized by the relevant NSES Content Standards and an extensive index will assist educators, students, and other readers in locating information or examples of topics that fulfill a particular aspect of their curriculum.

Encyclopedia of Earth and Space Science emphasizes physical processes involved in the formation and evolution of the Earth and universe, describes many examples of different types of geological and astrophysical phenomena, provides historical perspectives, and gives insight into the process of scientific inquiry by incorporating biographical profiles of people who have contributed significantly to the development of the sciences. The complex processes related to the expansion of the universe from the big bang are presented along with an evaluation of the physical principles and fundamental laws that describe these processes. The resulting structure of the universe, galaxies, solar system, planets, and places on the Earth are all discussed, covering many different scales of observation from the entire universe to the smallest subatomic particles. The geological characteristics and history of all of the continents and details of a few selected important areas are presented, along with maps, photographs, and anecdotal accounts of how the natural geologic history has influenced people. Other entries summarize the major branches and subdisciplines of Earth and space science or describe selected applications of the information and technology gleaned from Earth and space science research.

The majority of this encyclopedia comprises 250 entries covering NSES concepts and topics, theories, subdisciplines, biographies of people who have made significant contributions to the earth and space sciences, common methods, and techniques relevant to modern science. Entries average more than 2,000 words each (some are shorter, some longer), and most include a cross-referencing of related entries and a selection of recommended further readings. In addition, one dozen special essays covering a variety of subjects—especially how different aspects of earth and space sciences have affected people—are placed along with related entries. More than 300 color photographs and line art illustrations, including more than two dozen tables and charts, accompany the text, depicting difficult concepts, clarifying complex processes, and summarizing information for the reader. A glossary defines relevant scientific terminology. The back matter of *Encyclopedia of Earth and Space Science* contains a geological timescale, tables of conversion between different units used in the text, and the periodic table of the elements.

I have been involved in research and teaching for more than two decades. I am honored to be a Distinguished Professor and Yangtze Scholar at China's leading geological institution, China University of Geosciences, in Wuhan. I was formerly the P. C. Reinert Endowed Professor of Natural Sciences and am the founding director of the Center for Environmental Sciences at St. Louis University. I am actively involved in research, writing, teaching, and advising students.

My research and teaching focus on the fields of plate tectonics and the early history of the Earth, as well as on natural hazards and disasters, satellite imagery, mineral and water resources, and relationships between humans and the natural environment. I have worked extensively in North America, Asia, Africa, Europe, the Middle East, and the rims of the Indian and Pacific Oceans. During this time I have authored more than 25 books, 600 research papers, and numerous public interest articles, interviews with the media (newspapers, international, national and local television, radio, and international news magazines), and I regularly give public presentations on science and society. Some specific areas of current interest include the following:

- Precambrian crustal evolution
- tectonics of convergent margins
- natural disasters: hurricanes, earthquakes, volcanoes, tsunami, floods, etc.
- drought and desertification
- Africa, Madagascar, China
- Middle East geology, water, and tectonics

I received bachelor and master of science degrees from the Department of Geological Sciences at the State University of New York at Albany in 1982 and 1985, respectively, then continued my studies in earth and planetary sciences at the Johns Hopkins University in Baltimore. There I received a master of arts in 1988 and a Ph.D. in 1990. During this time I was also a graduate student researcher at the NASA Laboratory for Terrestrial Physics, Goddard Space Flight Center. After this I moved to the University of California at Santa Barbara where I did postdoctoral research in Earth-Sun-Moon dynamics in the Department of Mechanical Engineering. I then moved to the University of Houston for a visiting faculty position in the department of geosciences and allied geophysical laboratories at the University of Houston. In 1992 I moved to a research professor position in the Center for Remote Sensing at Boston University and also took a part-time appointment as a research geologist with the U.S. Geological Survey. In 2000 I moved to St. Louis University, then was appointed to a distinguished professor position at China University of Geoscience in 2009.

I have tried to translate as much of this experience and knowledge as possible into this two-volume encyclopedia. It is my hope that you can gain an appreciation for the complexity and beauty in the earth and space sciences from different entries in this book, and that you can feel the sense of exploring, learning, and discovery that I felt during the research related here, and that you enjoy reading the different entries as much as I enjoyed writing them for you.

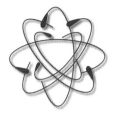

ENTRIES CATEGORIZED BY NATIONAL SCIENCE EDUCATION STANDARDS FOR CONTENT (GRADES 9–12)

When relevant an entry may be listed under more than one category. For example, Alfred Wegener, one of the founders of plate tectonic theory, is listed under both Earth and Space Science Content Standard D: Origin and Evolution of the Earth System, and Content Standard D: History and Nature of Science. Subdisciplines are listed separately under the category Subdisciplines, which is not a NSES category, but are also listed under the related content standard category.

Science as Inquiry (Content Standard A)
astronomy
astrophysics
biosphere
climate
climate change
Coriolis effect
cosmic microwave background radiation
cosmology
Darwin, Charles
ecosystem
Einstein, Albert
environmental geology
evolution
Gaia hypothesis
geological hazards
global warming
greenhouse effect
hydrocarbons and fossil fuels
ice ages
life's origins and early evolution
mass extinctions
origin and evolution of the Earth and solar system
origin and evolution of the universe
ozone hole
plate tectonics
radiation
sea-level rise

Earth and Space Science (Content Standard D): Energy in the Earth System
asthenosphere
atmosphere
aurora, aurora borealis, aurora australis
black smoker chimneys
climate
climate change
clouds
convection and the Earth's mantle
Coriolis effect
cosmic microwave background radiation
cosmic rays
Earth
earthquakes
Einstein, Albert
El Niño and the Southern Oscillation (ENSO)
electromagnetic spectrum
energy in the Earth system
Gaia hypothesis
geodynamics
geological hazards

geomagnetism, geomagnetic reversal
geyser
global warming
greenhouse effect
hot spot
hurricanes
ice ages
large igneous provinces, flood basalt
magnetic field, magnetosphere
mantle plumes
mass wasting
meteorology
Milankovitch cycles
monsoons, trade winds
ocean currents
paleomagnetism
photosynthesis
plate tectonics
precipitation
radiation
radioactive decay
subduction, subduction zone
Sun
thermodynamics
thermohaline circulation
thunderstorms, tornadoes
tsunami, generation mechanisms
volcano

Earth and Space Science (Content Standard D): Geochemical Cycles

asthenosphere
atmosphere
biosphere
black smoker chimneys
carbon cycle
climate change
clouds
continental crust
convection and the Earth's
    mantle
crust
diagenesis
Earth
economic geology
ecosystem
environmental geology
erosion
Gaia hypothesis
geochemical cycles
global warming
granite, granite batholith
groundwater
hydrocarbons and fossil fuels
hydrosphere
igneous rocks
large igneous provinces, flood
    basalt
lava
lithosphere
magma
mantle
mantle plumes
metamorphism and metamorphic
    rocks
metasomatic
meteoric
ocean currents
ophiolites
ozone hole
petroleum geology
photosynthesis
plate tectonics
precipitation
river system
seawater
sedimentary rock, sedimentation
soils
subduction, subduction zone
thermodynamics
thermohaline circulation
thunderstorms, tornadoes
volcano
weathering

Earth and Space Science (Content Standard D): Origin and Evolution of the Earth System

accretionary wedge
African geology
Andes Mountains
Antarctica
Arabian geology
Archean
Asian geology
asthenosphere
atmosphere
Australian geology
basin, sedimentary basin
beaches and shorelines
benthic, benthos
biosphere
Cambrian
Carboniferous
cave system, cave
Cenozoic
climate change
continental crust
continental drift
continental margin
convection and the Earth's
    mantle
convergent plate margin
    processes
coral
craton
Cretaceous
crust
crystal, crystal dislocations
deformation of rocks
deltas
deserts
Devonian
divergent plate margin processes
drainage basin (drainage system)
Earth
earthquakes
economic geology
Eocene
eolian
erosion
estuary
European geology
evolution
flood
fluvial
flysch
fossil
fracture
geoid
geomorphology

glacier, glacial systems
Gondwana, Gondwanaland
granite, granite batholith
greenstone belts
Grenville province and Rodinia
historical geology
hot spot
hydrocarbons and fossil fuels
hydrosphere
igneous rocks
impact crater structures
Indian geology
island arcs, historical eruptions
Japan
karst
large igneous provinces, flood
    basalt
lava
life's origins and early evolution
lithosphere
Madagascar
magma
mantle
mantle plumes
mass extinctions
mass wasting
mélange
Mesozoic
metamorphism and metamorphic
    rocks
meteor, meteorite
Milankovitch cycles
mineral, mineralogy
Neogene
Neolithic
North American geology
ocean basin
ocean currents
oceanic plateau
ophiolites
Ordovician
origin and evolution of the Earth
    and solar system
orogeny
ozone hole
paleoclimatology
Paleolithic
paleomagnetism
paleontology
Paleozoic
Pangaea
passive margin
pelagic, nektonic, planktonic
Permian
petroleum geology
petrology and petrography

Phanerozoic
photosynthesis
plate tectonics
Pleistocene
Precambrian
Proterozoic
Quaternary
radiation
radioactive decay
river system
Russian geology
seawater
sedimentary rock, sedimentation
seismology
sequence stratigraphy
Silurian
soils
South American geology
stratigraphy, stratification, cyclothem
structural geology
subduction, subduction zone
subsidence
Sun
supercontinent cycles
Tertiary
thermohaline circulation
transform plate margin processes
tsunami, generation mechanisms
unconformities
volcano
weathering
Wegener, Alfred

Earth and Space Science (Content Standard D): Origin and Evolution of the Universe
asteroid
astronomy
astrophysics
binary star systems
black holes
comet
cosmic microwave background radiation
cosmic rays
cosmology
dark matter
dwarfs (stars)
Einstein, Albert
galaxies
galaxy clusters
gravity wave
ice ages
interstellar medium
Jupiter

Mars
Mercury
meteor, meteorite
Neptune
nova
origin and evolution of the universe
planetary nebula
Pluto
pulsar
quasar
radiation
radio galaxies
Saturn
sea-level rise
solar system
star formation
stellar evolution
Sun
universe
Uranus
Venus

Science and Technology (Content Standard E)
astrophysics
cosmic microwave background radiation
electromagnetic spectrum
Galilei, Galileo
geochemistry
geochronology
geodesy
geodynamics
geographic information systems
geomagnetism, geomagnetic reversal
geophysics
gravity wave
gravity, gravity anomaly
Hubble, Edwin
magnetic field, magnetosphere
oceanography
paleomagnetism
radiation
remote sensing
seismology
telescopes
thermodynamics

Science in Personal and Social Perspectives (Content Standard F)
astronomy
aurora, aurora borealis, aurora australis
climate change

constellation
cosmology
Darwin, Charles
ecosystem
Einstein, Albert
El Niño and the Southern Oscillation (ENSO)
environmental geology
evolution
flood
Gaia hypothesis
geological hazards
global warming
greenhouse effect
hydrocarbons and fossil fuels
hydrosphere
island arcs, historical eruptions
life's origins and early evolution
mass extinctions
origin and evolution of the universe
sea-level rise
soils
subsidence
sun halos, sundogs, and sun pillars
supernova
tsunami, historical accounts

History and Nature of Science (Content Standard G)
astronomy
Bowen, Norman Levi
Brahe, Tycho
Cloud, Preston
Copernicus, Nicolas
Coriolis, Gustave
Dana, James Dwight
Darwin, Charles
Dewey, John F.
Du Toit, Alexander
Einstein, Albert
Eskola, Pentti
Galilei, Galileo
Gamow, George
Gilbert, Grove K.
Goldschmidt, Victor M.
Grabau, Amadeus William
Halley, Edmond
Hess, Harry
Hipparchus
Holmes, Arthur
Hubble, Edwin
Hutton, James
Huygens, Christian
Kepler, Johannes

Lawson, Andrew Cooper
Lemaître, Georges
Lyell, Sir Charles
Milankovitch, Milutin M.
Pettijohn, Francis John
Powell, John Wesley
Ptolemy, Claudius Ptolemaeus
Sedgwick, Adam
Smith, William H.
Sorby, Henry Clifton
Steno, Nicolaus
Stille, Wilhelm Hans
Wegener, Alfred
Werner, A. G.

Subdisciplines
astronomy
astrophysics

atmosphere
climate
cosmology
economic geology
evolution
geochemistry
geochronology
geodesy
geodynamics
geological hazards
geomorphology
geophysics
groundwater
historical geology
metamorphism and metamorphic
   rocks
meteorology
mineral, mineralogy

oceanography
paleoclimatology
paleomagnetism
paleontology
petroleum geology
petrology and petrography
plate tectonics
sedimentary rock, sedimentation
seismology
sequence stratigraphy
stratigraphy, stratification,
   cyclothem
structural geology
thermodynamics

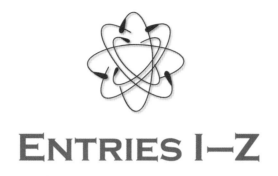

ENTRIES I–Z

ice ages Times when the global climate was colder and large masses of ice covered many continents are referred to as ice ages. At several times in Earth's history large portions of the Earth's surface have been covered with huge ice sheets. About 10,000 years ago all of Canada, much of the northern United States, and most of Europe were covered with ice sheets, as was about 30 percent of the world's remaining landmass. These ice sheets lowered sea level by about 320 feet (100 m), exposing the continental shelves and leaving present-day cities including New York, Washington, D.C., and Boston 100 miles (160 km) from the sea. In the last 2.5 billion years several periods of ice ages have been identified, separated by periods of mild climate similar to that of today. Ice ages seem to form through a combination of several different factors, including the following:

- the amount of incoming solar radiation, which changes in response to several astronomical effects
- the amount of heat retained by the atmosphere and ocean, or the balance between the incoming and outgoing heat
- the distribution of landmasses on the planet. Shifting continents influence the patterns of ocean circulation and heat distribution, and a large continent on one of the poles causes ice to build up on that continent, increasing the amount of heat reflected to space, lowering global temperatures in a positive feedback mechanism.

Glaciations have happened frequently in the past 55 million years and could occur again at almost any time. In the late 1700s and early 1800s Europe experienced a "little ice age" during which many glaciers advanced from the Alps and destroyed small villages in their path. Ice ages have occurred at several other times in the ancient geologic past, including in the Late Paleozoic (about 350–250 million years ago), Silurian (435 million years ago), and Late Proterozoic (about 800–600 million years ago). During parts of the Late Proterozoic glaciation, it is possible that the entire Earth surface temperature was below freezing and the planet was covered by ice.

In the Late Proterozoic the Earth experienced one of the most profound ice ages in the history of the planet. Isotopic records and geologic evidence suggests that the entire Earth's surface was frozen, though some scientists dispute the evidence and claim that there would be no way for the Earth to recover from such a frozen state. In any case it is clear that in the Late Proterozoic, during the formation of the supercontinent Gondwana, the Earth experienced one of the most intense glaciations ever, with the lowest average global temperatures in known Earth history.

One of the longest-lasting glacial periods was the Late Paleozoic ice age, which lasted about 100 million years, indicating a long-term underlying cause of global cooling. Of the variables that operate on these longtime scales, the distribution and orientation of continents seems to have caused the Late Paleozoic glaciation. The Late Paleozoic saw the amalgamation of the planet's landmasses into the supercontinent of Pangaea. The southern part of Pangaea, known as Gondwana, consisted of present-day Africa, South America, Antarctica, India, and Australia. During the drift of the continents in the Late Paleozoic, Gondwana slowly moved across the South Pole, and huge ice caps formed on these southern continents during their passage over the pole. The global climate was much colder overall, with the subtropical belts becoming very condensed and the polar and subpolar belts expanding to low latitudes.

During all major glaciations a continent was situated over one of the poles. Currently Antarctica is over the South Pole, and this continent has huge ice sheets. When continents rest over a polar region, they accumulate huge amounts of snow that gets converted into several-mile-thick ice sheets that reflect more solar radiation back to space and lower global seawater temperatures and sea levels.

Another arrangement that helps initiate glaciations is continents distributed in a roughly N-S orientation across equatorial regions. Equatorial waters receive more solar heating than polar waters. Continents block and modify the simple east to west circulation of the oceans induced by the spinning of the planet. When continents are present on or near the equator, they divert warm water currents to high latitudes, bringing warm water to higher latitudes. Since warm water evaporates much more effectively than cold water, having warm water move to high latitudes promotes evaporation, cloud formation, and precipitation. In cold, high-latitude regions the precipitation falls as snow, which persists and builds up glacial ice.

The Late Paleozoic glaciation ended when the supercontinent of Pangaea began breaking apart, suggesting a further link between tectonics and climate. The smaller landmasses might not have been able to divert the warm water to the poles anymore, or perhaps enhanced volcanism associated with the breakup caused additional greenhouse gases to build up in the atmosphere, raising global temperatures.

The planet began to enter a new glacial period about 55 million years ago, following a 10-million-year-long period of globally elevated temperatures and expansion of the warm subtropical belts into the subarctic. This Late Paleocene global hothouse saw the oceans and atmosphere holding more heat than at any other time in Earth history, but temperatures at the equator were not particularly elevated. Instead the heat was distributed more evenly around the planet, leading to fewer violent storms (with a small temperature gradient between low and high latitudes) and more moisture overall in the atmosphere. Several factors contributed to the abnormally warm temperatures on the planet during this time, including a distribution of continents that saw the equatorial region free of continents. This allowed the oceans to heat up more efficiently, raising global temperatures. The oceans warmed so much that the deep ocean circulation changed, and the normally cold deep currents became warm. These melted frozen gases (known as methane gas hydrates) that had

accumulated on the seafloor, releasing huge amounts of methane into the atmosphere. Methane is a greenhouse gas, and its increased abundance trapped solar radiation in the atmosphere, contributing to global warming. In addition volcanic eruptions released vast outpourings of mafic lavas in the North Atlantic Ocean realm, and the accompanying liberation of large amounts of CO_2 would have increased the greenhouse gases in the atmosphere and further warmed the planet. Global warming during the Late Paleocene was so extreme that about 50 percent of all single-celled organisms living in the deep ocean became extinct.

After the Late Paleocene hothouse, the Earth began a long-term cooling that continues today despite the present warming of the past century. This current ice age was marked by the growth of Antarctic glaciers, starting about 36 million years ago, until about 14 million years ago, when the Antarctic ice sheet covered most of the continent with several miles of ice. At this time global temperatures had cooled so much that many of the mountains in the Northern Hemisphere were covered with mountain and piedmont glaciers, similar to those in southern Alaska today. The ice age continued to intensify until 3 million years ago, when extensive ice sheets covered the Northern Hemisphere. North America was covered with an ice sheet that extended from northern Canada to the Rocky Mountains, across the Dakotas, Wisconsin, Pennsylvania, and New York, and on the continental shelf. At the peak of the glaciation (18,000–20,000 years ago) about 27 percent of the lands' surface was covered with ice. Midlatitude storm systems were displaced to the south, and desert basins of the southwest United States, Africa, and the Mediterranean received abundant rainfall and hosted many lakes. Sea level was lowered by 425 feet (130 m) to make the ice that covered the continents, so most of the world's continental shelves were exposed and eroded.

The causes of the Late Cenozoic glaciation are not well known but seem related to Antarctica coming to rest over the South Pole and other plate tectonic motions that have continued to separate the once contiguous landmasses of Gondwana, changing global circulation patterns in the process. Two of the important events seem to be the closing of the Mediterranean Ocean around 23 million years ago and the formation of the Panama isthmus 3 million years ago. These tectonic movements restricted the east-to-west flow of equatorial waters, causing the warm water to move to higher latitudes where evaporation promotes snowfall. An additional effect related to uplift of some high mountain ranges, including the Tibetan Plateau, has changed the pattern of the air circulation associated with the Indian monsoon.

The closure of the Panama isthmus correlates closely with the advance of Northern Hemisphere ice sheets, suggesting a causal link. This thin strip of land has drastically altered the global ocean circulation such that Pacific and Atlantic Ocean waters no longer communicate effectively, and it diverts warm currents to near-polar latitudes in the North Atlantic, enhancing snowfall and Northern Hemisphere glaciation. Since 3 million years ago the ice sheets in the Northern Hemisphere have alternately advanced and retreated, apparently in response to variations in the Earth's orbit around the Sun and other astronomical effects. These variations change the amount of incoming solar radiation on timescales of thousands to hundreds of thousands of years (Milankovitch Cycles). Together with the other longer-term effects of shifting continents, changing global circulation patterns, and abundance of greenhouse gases in the atmosphere, most variations in global climate can be approximately explained. This knowledge may help predict where the climate is heading and may help model and mitigate the effects of human-induced changes in the atmospheric greenhouse gases. If Earth is heading into another warm phase and the existing ice on the planet melts, sea level will quickly rise by 210 feet (64 m), inundating many of the world's cities and farmlands. Alternately, if the Earth enters a new ice sheet stage, sea levels will be lowered, and the planet's climate zones will be displaced to more equatorial regions.

See also ATMOSPHERE; GLACIER, GLACIAL SYSTEMS; GREENHOUSE EFFECT.

FURTHER READING

Erickson, J. *Glacial Geology: How Ice Shapes the Land.* New York: Facts On File, 1996.

Kusky, T. M. *Climate Change: Shifting Deserts, Glaciers, and Climate Belts.* New York: Facts On File, 2008.

igneous rock A rock that has crystallized from a melt or partially molten material (known as magma) is classified as igneous. Magma is a molten rock within the Earth; if it makes its way to the surface, it is referred to as lava. Different types of magma form in different tectonic settings, and many processes act on the magma as it crystallizes to produce a wide variety of igneous rocks.

Most magma solidifies below the surface, forming igneous rocks (*igneous* is Latin for fire). Igneous rocks that form below the surface are called intrusive (or plutonic) rocks, whereas those that crystallize on the surface are called extrusive (or volcanic) rocks. Rocks that crystallize at a very shallow depth are known as hypabyssal rocks. Intrusive igneous rocks crystallize slowly, giving crystals an extended time to grow, thus forming rocks with large mineral grains that are clearly distinguishable to the naked eye. These rocks are called phanerites. In contrast, magma that cools rapidly forms fine-grained rocks. Aphanites are igneous rocks in which the component grains cannot be distinguished readily without a microscope and are formed when magma from a volcano falls or flows across the surface and cools quickly. Some igneous rocks, known as porphyries, have two populations of grain size—a large group of crystals (phenocrysts) mixed with a uniform groundmass (matrix) that fills the space between the large crystals. This indicates two stages of cooling, as when magma has resided for a long time beneath a volcano, growing big crystals. When the volcano erupts, it spews out a mixture of large crystals and liquid magma that then cools quickly.

Once magmas are formed from melting rocks in the Earth, they intrude the crust and can take several forms. A pluton is a general name for a large, cooled, igneous, intrusive body in the Earth. The specific type of pluton is based on its geometry, size, and relations to the older rocks surrounding it, known as country rock. Concordant plutons have boundaries parallel to layering in the country rock, whereas discordant plutons have boundaries that cut across layering in the country rock. Dikes are generally thin with parallel sides exhibiting tabular shapes and cut across preexisting layers, and are therefore said to be discordant intrusions. In contrast, sills are tabular intrusions oriented parallel to layers and said to be concordant intrusives. Volcanic necks are conduits connecting a volcano with its underlying magma chamber (a famous example of a volcanic neck is Devils Tower, Wyoming). Some plutons are so large that they have special names. Batholiths, for example, have a surface area of more than 60 square miles (100 km^2).

The mechanisms by which large bodies of magma intrude into the crust are debated by geologists and may be different for different plutons. One mechanism, assimilation, involves the hot magma melting the surrounding rocks as it rises, causing them to become part of the magma. As the magma cools, its composition changes to reflect the added melted country rock. Magmas can rise only a limited distance by assimilation because they quickly cool before they can melt their way significant distances through the crust. If the magma is under high pressure, it may forcefully push into the crust. One variation of this forceful emplacement style is diapirism, where the weight of surrounding rocks pushes down on the melt layer, which then squeezes up through cracks that can expand and extend, forming volcanic vents at the surface. Yet another mechanism is stopping. During pluton emplacement by stopping large

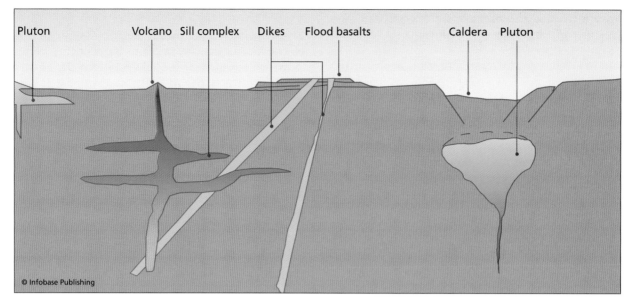

Pluton   Volcano Sill complex   Dikes   Flood basalts   Caldera   Pluton

© Infobase Publishing

Forms of plutons and intrusive rocks, showing names given to various types of intrusions

blocks of the surrounding country rock get thermally shattered and drop off the top of the magma chamber and fall into the chamber, eventually melting and becoming part of the magma.

NAMES OF IGNEOUS ROCKS

The first stage of naming an igneous rock is determining whether it is phaneritic or aphanitic. The next step involves the determination of its mineral constituents. The chemical composition of magma is closely related to how explosive and hazardous a volcanic eruption will be. The variation in the amount of silica (SiO_2) in igneous rocks is used to describe the variation in composition of igneous rocks and the magmas that formed them, as shown in the table "Names of Igneous Rocks Based on Silica Content and Texture." Rocks with low amounts of silica (basalt, gabbro) are known as mafic rocks, whereas rocks with high concentrations of silica (rhyolite, granite) are silicic or felsic rocks.

THE ORIGIN OF MAGMA

Some of the variation in the different types of volcanic eruptions can be understood by examining what causes magmas to have such a wide range in composition. Magmas come from deep within the Earth, but what conditions lead to the generation of melts in the interior of the Earth? The geothermal gradient reflects the relationship between increased temperature and increased depth in the Earth, and it provides information about the depths at which melting occurs and the depths at which magmas form. The differences in the composition of the oceanic and continental crusts lead to differing abilities to conduct the heat from the interior of the Earth, and thus different geothermal gradients. The geothermal gradients show that temperatures within the Earth quickly exceed 1,832°F (1,000°C) with increasing depth, so why are these rocks not molten? The answer is that pressures are very high, and pressure influences the ability of a rock to melt. As the pressure rises, the temperature at which the rock melts also rises. The presence of water significantly modifies this effect of pressure on melting, because wet minerals melt at lower temperatures than dry minerals. As the pressure rises, the amount of water that can be dissolved in a melt increases. Increasing the pressure on a wet mineral has the opposite effect

NAMES OF IGNEOUS ROCKS BASED ON SILICA CONTENT AND TEXTURE

Magma Types	Percent SiO$_2$	Volcanic Rock	Plutonic Rock
Mafic	45–52%	Basalt	Gabbro
Intermediate	53–65%	Andesite	Diorite
Felsic	> 65%	Rhyolite	Granite

from increasing the pressure on a dry mineral—it decreases the melting temperature.

PARTIAL MELTING

If a rock melts completely, the magma has the same composition as the rock. Rocks are made of many different minerals, all of which melt at different temperatures. Therefore if a rock is slowly heated, the resulting melt, or magma, will first have the composition of the first mineral that melts and then the first plus the second mineral that melts, and so on. If the rock continues to melt completely, the magma will eventually end up with the same composition as the starting rock, but this does not always happen. Often the rock only partially melts so that the minerals with low melting temperatures contribute to the magma, whereas the minerals with high melting temperatures did not melt and are left as a residue (or restite). In this way the end magma can have a composition different from the rock it came from.

The phrase "magmatic differentiation by partial melting" refers to the forming of magmas with differing compositions through the incomplete melting of rocks. For magmas formed in this way, the composition of the magma depends on both the composition of the parent rock and the percentage of melt.

BASALTIC MAGMA

Partial melting in the mantle leads to the production of basaltic magma, which forms most of the oceanic crust. Examination of the mineralogy of the oceanic crust, which is dominated by olivine, pyroxene, and feldspar, reveals that little water is involved in the production of the oceanic crust. These minerals are all anhydrous, that is, without water in their structure. Dry partial melting of the upper mantle must lead to the formation of oceanic crust. By collecting samples of the mantle that have been erupted through volcanoes, we know that it has a composition of garnet peridotite (olivine + garnet + orthopyroxene). Analyzing samples of this in the laboratory, by raising its temperature and pressure so that it is equal to 62 miles (100 km) depth, shows that 10 percent to 15 percent partial melt of this garnet peridotite yields a basaltic magma.

Magma that forms at 50 miles (80 km) depth is less dense than the surrounding solid rock, so it rises, sometimes quite rapidly (at rates of half a mile, or one kilometer, per day measured by earthquakes under Hawaii). In fact it may rise so fast that it does not cool appreciably, erupting at the surface at more than 1,832°F (1,000°C), generating basaltic magma.

GRANITIC MAGMA

Granitic magmas are very different from basaltic magmas. They have about 20 percent more silica, and the minerals in granite (mica, amphibole) have a lot of water in their crystal structures. Granitic magmas are mostly exclusive to regions of continental crust. Inference from these observations leads to the conclusion that the source of granitic magmas is within the continental crust. Laboratory experiments suggest that when rocks with the composition of continental crust start to melt at temperature and pressure conditions found in the lower crust, a granitic liquid is formed, with 30 percent partial melting. These rocks can begin to melt by either the addition of a heat source, such as basalt intruding the lower continental crust, or by burying water-bearing minerals and rocks to these depths. The geological processes responsible for bringing the water-bearing minerals down to the level in the crust where the water can be released from their crystal structure, and lower the melting

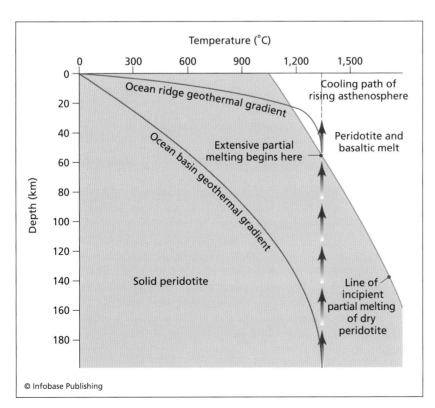

© Infobase Publishing

Typical geothermal gradient beneath the oceans, showing also the line of incipient partial melting of a dry peridotite

Red granite at the coast *(TTphoto, Shutterstock, Inc.)*

temperature of surrounding rocks, are limited. Deep burial by sedimentation can cause the release of water, which escapes to the surface but does not usually cause the rocks to melt and form granite. In subduction zones water-bearing minerals can be carried to great depths in the Earth, and when they release water, the fluid can rise up into the overlying mantle and crust, and cause the granite melts to form in those regions.

These granitic magmas rise slowly (because of their high SiO_2 and high viscosities), until they reach the level in the crust where the temperature and pressure conditions are consistent with freezing or solidification of magma with this composition. This occurs about three to six miles (5–10 km) beneath the surface, which explains why large portions of the continental crust are not molten lava lakes. In many region, crust lies above large magma bodies (called batholiths) that are heated by the cooling magma. An example is Yellowstone National Park, where hot springs, geysers, and many other features indicate the presence of a large hot magma body at depth. Much of Yellowstone Park is a giant valley called a caldera, formed when an ancient volcanic eruption emptied an older batholith of its magma, and the overlying crust collapsed into the empty hole formed by the eruption.

ANDESITIC MAGMA

The average composition of the continental crust is andesitic, a composition that falls between that of basalt and rhyolite. Laboratory experiments show that partial melting of wet oceanic crust yields an andesitic magma. Remember that oceanic crust is dry, but after it forms it interacts with seawater, which fills cracks to several miles' (kilometers') depth. Also the sediments on top of the oceanic crust are full of water, but these are for the most part nonsubductable. Andesite forms above places where water is released from the subducted slabs, and it migrates into the mantle wedge above the subducting slab, forming water-rich magmas. These magmas then intrude the continental crust above, some forming volcanic andesites, others crystallizing as plutons of diorite at depth.

SOLIDIFICATION OF MAGMA

Just as rocks partially melt to form different liquid compositions, magmas can solidify to different minerals at different times to form different solids (rocks). This process also results in the continuous change in the composition of the magma—if one mineral is removed, the resulting composition is different. If some process removes these solidified crystals from the system of melts, a new magma composition results.

Several processes can remove crystals from the melt system, including squeezing melt away from the crystals or sinking of dense crystals to the bottom of a magma chamber. These processes lead to magmatic differentiation by fractional crystallization, as first described by Canadian petrologist Norman L. Bowen (1887–1956). Bowen systematically documented how crystallization of the first minerals changes the composition of the magma and results in the formation of progressively more silicic rocks with decreasing temperature.

See also ISLAND ARCS, HISTORICAL ERUPTIONS; MINERAL, MINERALOGY; OPHIOLITES; PLATE TECTONICS; VOLCANO.

FURTHER READING

Mahoney, J. J., and M. F. Coffin, eds. *Large Igneous Provinces, Continental, Oceanic, and Planetary Flood Volcanism.* Washington, D.C.: American Geophysical Union, 1997.
McBirney, Alexander R. *Igneous Petrology.* 3rd ed. Boston: Jones and Bartlett, 2007.

impact crater structures   Ample evidence shows that many small and some large meteorites have hit Earth frequently throughout time. Several hundred impact craters resulting from these events have been recognized to be preserved on continents, and a few have been recognized on the ocean floor. These craters exhibit a wide range of appearances. Some are small, only a few yards (m) across, such as the small, approximately 5,000-year-old craters at Henbury in the Northern Territories of Australia, while others are up to hundreds of miles (hundreds of km) across, such as the Precambrian Vredefort dome in South Africa. Eyewitness accounts describe many events, such as fireballs in the sky, recording the entry of a meteorite into the Earth's atmosphere, to the huge explosion over Tunguska, Siberia, in 1980 that leveled thousands of square miles (km) of trees and created atmospheric shock waves that traveled around the world. Many theories have been proposed for the Tunguska event, the most favored of which is the impact of a comet fragment with Earth. Fragments of meteorites are regularly recovered from places like the Antarctic ice sheets, where rocky objects on the surface have no place to come from but space. Although meteorites may appear as flaming objects moving across the night skies, they are generally cold, icy bodies when they land on Earth, as only their outermost layers get heated from the deep freeze of space during their short transit through the atmosphere.

Most meteorites that hit Earth originate in the asteroid belt, situated between the orbits of Mars and Jupiter. There are at least a million asteroids in this belt with diameters greater than 0.6 miles (1 km), 1,000 with diameters greater than 19 miles (30 km), and 200 with diameters greater than 62 miles (100 km). These are thought to be either remnants of a small planet that was destroyed by a large impact event or perhaps fragments of rocky material that failed to coalesce into a planet, probably owing to the gravitational effects of the nearby massive planet Jupiter. Most scientists favor the second hypothesis but recognize that collisions between asteroids have fragmented a large body to expose a planetlike core and mantle now preserved in the asteroid belt.

Other objects from space may collide with Earth. Comets are masses of ice and carbonaceous material mixed with silicate minerals thought to originate in the outer parts of the solar system, in a region called the Oort Cloud. Other comets have a closer origin, in the Kuiper belt just beyond the orbit of Neptune. Small, icy Pluto, long considered the outermost planet, was recently reclassified from a planet to a dwarf planet, and may actually be a large Kuiper belt object. Comets may be less common near Earth than meteorites, but they still may hit Earth, with severe consequences. There are estimated to be more than a trillion comets in our solar system. Since they are lighter than asteroids and have water- and carbon-rich compositions, many scientists have speculated that cometary impact may have brought water, the atmosphere, and even life to Earth.

SURVEY OF IMPACT CRATERS ON EARTH AND THE MOON

Impact craters are known from every continent including Antarctica. Several hundred impact craters have been mapped and described in detail by geologists, and some patterns about the morphology, shape, and size of the craters have emerged from these studies. The most obvious variations in crater style and size are related to the size of the impacting meteorite, but other variations depend on the nature of the bedrock or cover, the angle and speed of the impact, and what the impactor was—rock or ice. Impacts are known from all ages and are preserved at various states of erosion and burial, allowing study of the many different levels of cratering and a better understanding of the types of structures and rocks produced during impacts.

The collision of meteorites with Earth produces impact craters, which are generally circular, bowl-shaped depressions. There are more than 200 known impact structures on Earth, although processes of weathering, erosion, volcanism, and tectonics have undoubtedly erased many thousands more. The Moon and other planets show much greater densities of impact craters, and since Earth has a greater gravi-

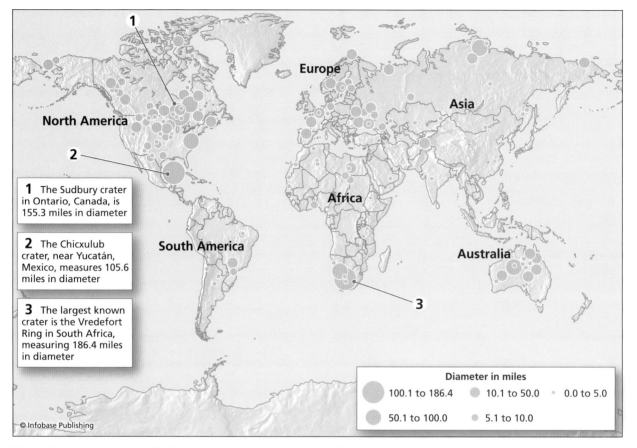

1 The Sudbury crater in Ontario, Canada, is 155.3 miles in diameter

2 The Chicxulub crater, near Yucatán, Mexico, measures 105.6 miles in diameter

3 The largest known crater is the Vredefort Ring in South Africa, measuring 186.4 miles in diameter

© Infobase Publishing

Diameter in miles
- 100.1 to 186.4
- 50.1 to 100.0
- 10.1 to 50.0
- 5.1 to 10.0
- 0.0 to 5.0

Map of impact craters on Earth—larger circles representing larger impact craters

tational pull than the Moon, it should have been hit by many more impacts than the Moon.

Meteorite impact craters have a variety of forms but are of two basic types. Simple craters are circular, bowl-shaped craters with overturned rocks around their edges, and are generally fewer than three miles (5 km) in diameter. They are thought to have been produced by impact with objects of fewer than 100 feet (30 m) in diameter. Examples of simple craters include the Barringer Meteor Crater in Arizona and Roter Kamm in Namibia. Complex craters are larger, generally greater than two miles (3 km) in diameter. They have an uplifted peak in the center of the crater and a series of concentric rings around the excavated core of the crater. Examples of complex craters include Manicougan, Clearwater Lakes, and Sudbury in Canada; Chicxulub in Mexico; and Gosses Bluff in Australia.

The style of impact crater depends on the size of the impacting meteor, the speed at which it strikes the surface, and, to a lesser extent, the underlying geology and angle at which the meteor strikes Earth. Most meteorites hit Earth with a velocity between 2.5 and 25 miles per second (4–40 km/sec), releasing tremendous energy when they hit. Meteor Crater in Arizona was produced about 50,000 years

ago by a meteorite approximately 100 feet (30 m) in diameter that hit the Arizona desert, releasing the equivalent of 4 megatons (3.6 megatonnes) of TNT. The meteorite body and a large section of the ground at the site were suddenly melted by shock waves from the impact, which released about twice as much energy as the eruption of Mount Saint Helens. Most impacts generate so much heat and shock pressure that the entire meteorite and a large amount of the rock it hits are melted and vaporized. Temperatures may exceed thousands of degrees within a fraction of a second as pressures increase a million times atmospheric pressure during passage of the shock wave. These conditions cause the rock at the site of the impact to accelerate downward and outward, and then the ground rebounds and tons of material are shot outward and upward into the atmosphere.

Impact cratering is a complex process. When the meteorite strikes, it explodes, vaporizes, and sends shock waves through the underlying rock, compressing the rock and crushing it into breccia, and ejecting material (conveniently known as ejecta) up into the atmosphere, from where it falls out as an ejecta blanket around the impact crater. Large impact events may melt the underlying rock forming an impact melt and may crystallize distinctive minerals that form only at exceedingly high pressures.

After the initial stages of the impact crater-forming process, the rocks surrounding the excavated crater slide and fall into the deep hole, enlarging the diameter of the crater, typically making it much wider than it is deep. Many of the rocks that slide into the crater are brecciated or otherwise affected by the passage of the shock wave, and may preserve these effects as brecciated rocks, high-pressure mineral phases, shatter cones, or other deformation features.

Impact cratering was probably a much more important process in the early history of Earth than it is at present. The flux of meteorites from most parts of the solar system was much greater in early times, and it is likely that impacts totally disrupted the surface in the early Precambrian. At present the meteorite flux is about a hundred tons (91 tonnes) per day (somewhere between 10^7–10^9 kg/yr), but most of this material burns up as it enters the atmosphere. Meteorites that are about one-tenth of an inch to several feet (mm–m) in diameter produce a flash of light (a shooting star) as they burn up in the atmosphere, and the remains fall to Earth as a tiny glassy sphere of rock. Smaller particles, known as cosmic dust, escape the effects of friction and slowly fall to Earth as a slow rain of extraterrestrial dust.

Meteorites must be greater than 3 feet (1 m) in diameter to make it through the atmosphere without burning up from friction. The Earth's surface is currently hit by about one small meteorite per year. Larger-impact events occur much less frequently, with meteorites 300 feet (90 m) in diameter hitting once every 10,000 years, 3,000 feet (900 m) in diameter hitting Earth once every million years, and six miles (10 km) in diameter hitting every 100 million years. Meteorites of only several hundred feet (hundreds of m) in diameter could create craters about one mile (1–2 km) in diameter, or if they hit in the ocean, they would generate tsunamis, more than 15 feet (5 m) tall over wide regions. The statistics of meteorite impact show that the larger events are the least frequent.

Barringer Meteor Impact Crater, Arizona
One of the most famous and visited impact craters in the United States is the Barringer meteor impact crater in Arizona, the first structure almost universally accepted by the scientific community as a meteorite impact structure. The crater is 0.75 miles (1.2 km) across, and has an age of 49,000 years before present. Its acceptance as an impact crater did not come easily. The leading proponent of the meteorite impact model was Daniel Barringer, a mining company executive who argued for years against the

Barringer meteor impact crater, Arizona *(François Gohier/Photo Researchers, Inc.)*

powerful head of the U.S. Geological Survey, Grove K. Gilbert, who maintained that the crater was a volcanic feature. For years Daniel Barringer lost the argument because in his model the crater should have been underlain by a massive iron-nickel deposit from the meteorite, which was never found since the meteorite was vaporized by the heat and energy of the impact. By 1930, however, enough other evidence for the origin of the crater had been accumulated to convince the scientific community of its origin by impact from space.

Barringer crater is a small crater with a roughly polygonal outline partly controlled by weak zones (fractures and joints) in the underlying rock. The bedrock of the area is simple, consisting of a cover of alluvium, underlain by the thin Triassic Moenkopi sandstone, about 300 feet (90 m) of the Permian Kaibab limestone, and 900 feet (270 m) of the Permian Coconino sandstone. The rim of the crater is 148 feet (45 m) higher than the surrounding desert surface, and the floor of the crater lies 328 feet (100 m) lower than the surrounding average desert elevation. The rim of the crater is composed of beds of material that originally filled the crater but was thrown or ejected during the impact. The beds in the rim rocks are fragmented and brecciated, and the whole sequence that was in the crater is now upside down lying on the rim. The underlying beds are turned upward as the crater is approached, reflecting this powerful bending and overturning that occurred when the interior rocks were thrown onto the rim during the impact.

Although the large mass of iron and nickel that Daniel Barringer sought at the base of the crater does not exist, thousands of small meteorite fragments have been collected from around the outer rim of the crater, from as far away as four miles (7 km) from the crater. The soil around the crater, for a distance of up to six miles (10 km) is pervaded by meteorite dust, suggesting that the impacting meteorite vaporized on impact, and the debris settled around the crater in a giant dust cloud. Estimates of the size of the meteorite that hit based on the amount of meteorite debris found are about 12,000 tons (10,884 tonnes).

Several lines of evidence indicate an impact origin for Meteor crater. The widespread brecciation, or fragmentation, of the rim and presence of iron-rich shale from weathering of the meteorite is consistent with an impact origin. More important, high-temperature glasses and minerals are preserved that form almost exclusively during the high pressures of meteorite impact. In some places the rock has been melted into impact glasses that require temperatures and shock pressures obtained only during impact of an object such as a meteor. Some rare minerals that form only at high pressures have been found at Barringer crater, including high-pressure phases of quartz known as coesite and stishovite, and also small diamonds formed by high pressures associated with the impact.

Chicxulub Impact and the Cretaceous/Tertiary Mass Extinction

The geologic record of life on Earth shows that there have been several sudden events that led to the extinction of large numbers of land and marine species within a short interval of time, and many of these are thought to have been caused by the impact of meteorites with Earth. Many of the boundaries between geologic time periods have been selected based on these mass extinction events. Some of the major mass extinctions include that between the Cretaceous and Tertiary Periods, marking the boundary between the Mesozoic and Cenozoic Eras. At this boundary occurring 66 million years ago dinosaurs, ammonites, many marine reptile species, and a large number of marine invertebrates suddenly died off, and the planet lost about 26 percent of all biological families and numerous species. At the boundary between the Permian and Triassic Periods (which is also the boundary between the Paleozoic and Mesozoic eras), 245 million years ago, 96 percent of all species became extinct. Many of the hallmark life-forms of the Paleozoic era were lost, such as the rugose corals, trilobites, many types of brachiopods, and marine organisms including many foraminifer species. Several other examples of mass extinctions have been documented from the geological record, including one at the boundary between the Cambrian and Ordovician Periods at 505 million years ago, when more than half of all families disappeared forever.

These mass extinctions have several common features that point to a common origin. Impacts have been implicated as the cause of many of the mass extinction events in Earth history. The mass extinctions seem to have occurred on a geologically instantaneous timescale, with many species present in the rock record below a thin clay-rich layer, and dramatically fewer species present immediately above the layer. In the case of the Cretaceous-Tertiary extinction, some organisms were dying off slowly before the dramatic die-off, but a clear, sharp event occurred at the end of this time of environmental stress and gradual extinction. Iridium anomalies have been found along most of the clay layers, considered by many to be the "smoking gun" indicating an impact origin as the cause of the extinctions. One-half million tons (454,000 tonnes) of iridium are estimated to be in the Cretaceous-Tertiary boundary clay, equivalent to the amount that would be contained in a meteorite with a six-mile (10-km) diameter. Other scientists argue that volcanic processes within Earth can produce iridium and that impacts are not necessary. Still other theories about the mass extinctions and loss of the dinosaurs exist, including that they died off from disease, insect bites, and genetic evolution that led to a great dominance of male over female species. Other rare elements and geochemical anomalies are present along the Cretaceous-Tertiary boundary, however, supporting the idea that a huge meteorite hit Earth at this time and was related in some way to the extinction of the dinosaurs, no matter what else may have been contributing to their decline at the time of the impact.

Other features have been found in the Chicxulub impact structure that support the impact origin for the mass extinctions. One of the most important is the presence of high-pressure minerals formed at pressures not reachable in the outer layers of Earth. The presence of the high-pressure mineral equivalents of quartz, including coesite, stishovite, and an extremely high-pressure phase known as diaplectic glass, strongly implicates an impacting meteorite, which can produce tremendous pressures during the passage of shock waves related to the force of the impact. Many of the clay layers associated with the iridium anomalies also have layers of tiny glass spherules thought to be remnants of melted rock produced during the impact that were thrown skyward, where they crystallized as tiny droplets that rained back on the planet's surface. They also have abundant microdiamonds similar to those produced during meteorite impact events. Layers of carbon-rich soot are also associated with some of the impact layers, and these are thought to represent remains of the global wildfires ignited by the impacts. Finally, some of the impact layers also record huge tsunamis that swept across coastal regions.

Many of these features are found around and associated with an impact crater recently discovered on Mexico's Yucatán Peninsula. The Chicxulub crater is about 66 million years old; half of it lies buried beneath the waters of the Gulf of Mexico, and half is on land. Tsunami deposits of the same age are found in inland Texas, much of the Gulf of Mexico, and the Caribbean, recording a huge tsunami, perhaps several hundred feet high (hundred m), generated by the impact. The crater is at the center of a huge field of scattered spherules that extends across Central America and through the southern United States. Chicxulub is a large structure, and is the right age to be the crater that records the impact at the Cretaceous-Tertiary boundary, thus fixing the extinction of the dinosaurs and other families.

The discovery and documentation of the meteorite crater on the Yucatán Peninsula, near the town of Chicxulub (meaning "tail of the devil") was a long process, but the crater is now widely regarded as the one that marks the site of the impact that caused the Cretaceous-Tertiary (K-T) mass extinction, including the loss of the dinosaurs. The crater was first discovered and thought to be an impact crater in the 1970s by Glen Penfield, a geologist working on oil exploration. He did not publish his results but presented ideas for the impact origin of the structure in scientific meetings in the 1980s. In 1981 University of Arizona geology graduate student Alan Hildebrand wrote about impact-related deposits around the Caribbean that had an age coincident with the Cretaceous-Tertiary boundary. These deposits include brown clay with an anomalous concentration of the metal iridium, thought to be from a meteorite, and some impact melts in the form of small beads called tektites. Hildebrand also wrote about evidence for a giant tsunami around the Caribbean at the K-T boundary but did not know the location of the crater. It was not until 1990 that Carlos Byars, a reporter for the *Houston Chronicle,* put the observations together and contacted Hildebrand, leading Hildebrand and Penfield to work together on realizing that they had located the crater and the deposits of the giant impact from the Cretaceous-Tertiary mass extinction.

Initially the Chixculub crater was thought to be about 110 miles (177 km) wide, but later studies have shown that it is a complex crater, and an additional ring was located outside the initial discovery. The crater is now regarded as being 190 miles (300 km) in diameter. The outermost known ring of the Chicxulub crater is marked by a line of sinkholes, where water moved along fractures and dissolved the underlying limestone.

The Chicxulub crater is buried under younger limestone, lying beneath 3,200 feet (1 km) of limestone that overlies 1,600 feet (500 m) of andesitic glass and breccia found only within the circular impact structure. These igneous rocks are thought to be impact melts, generated by the melting of the surrounding rocks during the impact event, and the melts rose to fill the crater immediately after the impact. Supporting this interpretation is the presence of unusual minerals and quartz grains that show evidence of being shocked at high pressures, forming distinctive bands through the mineral grains. The center depression of the crater is about 2,000 to 3,600 feet (609.6 to 1,097.3 m) deep compared with the same layers outside the crater rim, although the surface expression is now minimal since the crater is buried so deeply by younger rocks. The band of sinkholes that marks the outer rim of the crater suggests that the interior of the crater may have been filled with water after the impact, forming a circular lake.

The size of the meteorite that hit Chixculub is estimated to have been six miles (10 km) in diameter, releasing an amount of energy equal to 10^{14} tons of TNT. Recent studies suggest that the type of meteorite that hit Chicxulub was a carbonaceous chondrite, based on the chemistry and large amount of carbonaceous material found in pieces of the meteorite recovered from the impact.

The impact at Chicxulub was devastating for both the local and the global environment. The impact hit near the break between the continental shelf and the continental slope, so it ejected huge amounts of dust into the atmosphere from the shelf, and caused a huge mass of the continental shelf to collapse into the Gulf of Mexico and Caribbean. This in turn generated one of the largest tsunamis known in the history of the planet. This tsunami was thousands of feet (hundreds of m) tall on the Yucatán Peninsula and was still 165–330 feet (50–100 m) tall as it swept into the present-day Texas coastline, reaching far inland.

As the impact excavated the crater at Chicxulub, probably in less than a second, the meteorite was vaporized and ejected huge amounts of dust, steam, and ash into the atmosphere. As this material reentered the atmosphere around the planet it would have been heated to incandescent temperatures, igniting global wildfires and quickly heating surfaces and waters. At the same time tremendous earthquake waves were generated, estimated to have reached magnitudes of 12 or 13 on the open-ended Richter scale. This is larger than any known earthquake since then, and would have caused seismic waves that uplifted and dropped the ground surface by hundreds to a thousand feet (up to 300 m) at a distance of 600 miles (965 km) from the crater.

The impact generated huge amounts of dust and particles that would have been caught in the atmosphere for months after the impact, along with the ash from the global fires. This would block the sunlight and create in ice condition across the planet. Countering this effect is the release of huge amounts of carbon dioxide by the vaporization of carbonate rocks during the impact, which would have helped induce a greenhouse warming of climate that could have lasted decades. Together these effects wreaked havoc on the terrestrial fauna and flora that survived the fires and impact-related effects.

The Chicxulub crater and impact are widely held to have caused the mass extinction and death of the dinosaurs at the Cretaceous-Tertiary boundary. The global environment was considerably stressed before the impact, however, with marine planktonic organisms experiencing a dramatic decline before the impact (and a more dramatic one after the impact), and global temperatures falling before the crash. Some scientists argue that not all fauna, such as frogs, went extinct during the impact, and they should have

if the impact was the sole cause of mass extinction. Others have argued that the age of the impact is not exactly the same age as the extinction, and may actually predate the extinction by up to 300,000 years. Some models suggest that the global environment was already stressed, and the impact was the final blow to the environment that caused the global mass extinction.

Manicouigan Crater, Ontario, Canada

At 62 miles (100 km) across, the Manicouigan structure in Ontario is one of the largest known impact structures in Canada, and the fifth-largest known impact structure on Earth. This circular, partly exposed crater was formed by the impact of a meteorite with a three-mile (5-km) diameter with Earth 214 million years ago, hitting what is now the Precambrian shield. The Manicouigan River flows south out of the south side of the crater and drains into the St. Lawrence River. For some time it was thought that the Manicouigan crater resulted from the impact that caused a mass extinction that killed 60 percent of all species on Earth at the Permian-Triassic boundary, but dating the impact melt showed that the crater was 12 million years too old to be associated with that event.

The Manicouigan crater has been deeply eroded by the Pleistocene glaciers that scraped the loose sediments off the Canadian shield, pushing them south into the United States. The crater, exposed in bedrock, is currently delineated by two semicircular lakes that are part of a hydroelectric dam project.

The Manicouigan structure has a large central dome, rising 1,640 feet (500 m) above the surrounding surface, culminating at Mount de Babel peak. Manicouigan is characterized by huge amounts of broken, brecciated rocks and shatter cones that point toward the center of the dome. Shatter cones are cone-shaped fractures that form during impacts from the shock wave passing through the adjacent rock, and typically have tips or apexes that point toward the point of impact, but may have more complex patterns that form by the shock waves bouncing off other surfaces in the bedrock.

The Manicouigan crater also has a thick layer of an igneous rock called an impact melt, generated when the force of the impact causes the meteorite and the rock it crashes into to vaporize and melt, and then the melt can fill the resulting crater. The impact melt layer at Manicouigan is more than 325 feet (100 m) thick. There have also been high-pressure mineral phases discovered in the rocks from the Manicouigan structure—phases that form at pressures that could occur only as a result of a meteorite impact and are impossible to reach by other mechanisms at shallow levels of Earth's crust.

Vredefort Impact Structure, South Africa

The Vredefort dome of South Africa is one of the world's largest and oldest known, well-preserved impact structures. Located in the late Archean Witwatersrand basin, the Vredefort dome is a large, multiring structure with a diameter of 87 miles (140 km) formed by the impact of a meteorite 1.97 billion years ago.

The geology and structure of the Vredefort dome is complex, and its origin was the subject of debates for many years. The crater is characterized by a series of concentric rims of rock outcroppings of the different Precambrian rocks in the area, including the Proterozoic Transvaal Supergroup, and the Archean Ventersdorp and Witwatersrand Supergroups, and the underlying Dominion volcanics and Archean granitic basement. Before the impact these rocks formed a shallowly dipping sequence on top of the Archean granitic basement, but the impact was so strong that it excavated a crater. Immediately following, the rocks rebounded, with the Archean basement being uplifted in a steep dome in the center of the crater that rose all the way to the present surface, tilting and folding the surrounding rocks in the process.

The Vredefort dome is famous for hosting a number of geologic features that are diagnostic of meteorite impacts. First, the crater is surrounded by numerous shock metamorphic features, including many impact breccias, some of which are invaded by impact melt glasses called pseudotachylites. Many shatter cones have been identified at Vredefort, and the vast majority of these point inward to a point located just above the present-day surface, presumably pointing to the place of the initial shock and point of impact. Finally, the Vredefort dome is associated with a number of high-pressure mineral phases including coesite, stishovite, and diaplectic silica glass. These features require pressures of about 25, 75–100, and 120 kilobars (equivalent to depths of up 215 miles, or 350 km, depth in Earth), respectively, to form. Since the surrounding rocks did not reach these pressures, the only known way to form these mineral phases is through meteorite impact.

Lunar Impact Craters

The Earth's Moon is the closest celestial object, and it is covered by many impact craters, large and small. The lack of water, crustal recycling through plate tectonics, and weathering as on Earth has preserved craters that are billions of years old, providing scientists with a natural laboratory to observe and model impact craters of different sizes and styles. Thousands on thousands of photographs have revealed the great diversity in styles of lunar craters and have yielded insight into the cratering mechanisms responsible for cratering events on Earth. The large number of impact craters on the Moon also hints at the importance of impact cratering on the early Earth. The Moon has a much lower gravity field than Earth, so Earth has a greater chance of attracting and being hit by any nearby asteroids than the Moon. Since there are so many impact craters on the Moon, there should have been even more on the early Earth. Crustal recycling, erosion, and weathering have simply erased the surface traces of these impacts. Their effects, however, in terms of adding energy, elements, including organic and volatile molecules to Earth, is cumulative, and impacts have played a large role in the evolution of the planet.

NASA missions have also landed on and explored the lunar surface, aiding the understanding of the evolution of the Earth-Moon system, including the materials associated with impact events on the Moon. *Apollo 11*, the first manned lunar mission, launched from the Kennedy Space Center, Florida, via a *Saturn V* launch vehicle on July 16, 1969, and safely returned to Earth on July 24, 1969. The three-man crew aboard the flight consisted of Neil A. Armstrong, commander; Michael Collins, command module pilot; and Edwin E. (Buzz) Aldrin Jr., lunar module pilot. The lunar module (LM), named *Eagle*, carrying astronauts Armstrong and Aldrin, was the first crewed vehicle to land on the Moon. Meanwhile astronaut Collins piloted the command module in a parking orbit around the Moon. Armstrong was the first human ever to stand on the lunar surface, followed by Aldrin. The crew collected 47 pounds of lunar surface material that returned to Earth for analysis. The surface exploration was concluded in 2.5 hours. With the success of *Apollo 11* the national objective to land men on the Moon and return them safely to Earth had been accomplished. The samples collected were distributed to research groups around the country, greatly advancing the understanding of formation and evolution of the Moon.

SUMMARY

Hundreds of meteorite craters have been identified on Earth. Small craters typically have overturned and uplifted rims and a semicircular depression or crater marking the site of the impact. Larger impact craters collapse inward to fill the crater excavated by the impact and develop a characteristic medial high and many rims of uplifted and depressed crustal rocks. Impacts are commonly identified on the basis of impact breccias and shatter cones, by impact melts, and by the presence of high-pressure mineral phases such as coesite, stishovite, and diaplectic glass. Impact craters of all sizes and shapes have been identified on Earth and the Moon.

The most famous and perhaps most consequential impact crater is the one at Chicxulub, on the Yucatán Peninsula of Mexico. Here, a six-mile (10 km) wide meteorite hit Earth 66 million years ago, generating global fires and giant tsunamis, and ejecting tremendous amounts of dust and carbon dioxide into the atmosphere. The result was a mass extinction event at the Cretaceous-Tertiary boundary, and the loss of many species including the dinosaurs.

See also ASTEROID; COMET; MASS EXTINCTIONS; SOLAR SYSTEM.

FURTHER READING

Albritton, C. C., Jr. *Catastrophic Episodes in Earth History.* London: Chapman and Hale, 1989.

Alvarez, Walter. *T. Rex and the Crater of Doom.* Princeton, N.J.: Princeton University Press, 1997.

Blong, Russel J. *Volcanic Hazards: A Sourcebook on the Effects of Eruptions.* New York: Academic Press, 1984.

Cox, Donald, and James Chestek. *Doomsday Asteroid: Can We Survive?* New York: Prometheus Books, 1996.

Dressler, B. O., R. A. F. Grieve, and V. L. Sharpton, eds. *Large Meteorite Impacts and Planetary Evolution.* Boulder, Colo.: Geological Society of America Special Paper 293, 1994.

Eldredge, N. *Fossils: The Evolution and Extinction of Species.* Princeton, N.J.: Princeton University Press, 1997.

Fisher, R. V., G. Heiken, and J. B. Hulen. *Volcanoes: Crucibles of Change.* Princeton, N.J.: Princeton University Press, 1998.

Francis, Peter. *Volcanoes: A Planetary Perspective.* Oxford: Oxford University Press, 1993.

Gehrels, T., ed. *Hazards Due to Comets and Asteroids.* Tucson: University of Arizona Press, 1994.

Geological Survey of Canada, Earth Impact Database. Available online. URL: http://www.unb.ca/passc/ ImpactDatabase/ Accessed January 24, 2009.

Hodge, Paul. *Meteorite Craters and Impact Structures of the Earth.* Cambridge: Cambridge University Press, 1994.

Mark, Kathleen. *Meteorite Craters.* Tucson: University of Arizona Press, 1987.

Martin, P. S., and R. G. Klein, eds. *Quaternary Extinctions.* Tucson, Ariz.: University of Arizona Press, 1989.

Melosh, H. Jay. *Impact Cratering: A Geologic Process.* Oxford: Oxford University Press, 1988.

Robock, Alan, and Clive Oppenheimer, eds. *Volcanism and the Earth's Atmosphere.* Washington, D.C.: American Geophysical Union, 2003.

Scarth, Alwyn. *Vulcan's Fury, Man against the Volcano.* New Haven, Conn.: Yale University Press, 1999.

Sepkoski, J. J., Jr. *Mass Extinctions in the Phanerozoic Oceans: A Review, In Patterns and Processes in the History of Life.* Amsterdam, Netherlands: Springer-Verlaag, 1982.

Simkin, T., and R. S. Fiske. *Krakatau 1883: The Volcanic Eruption and Its Effects.* Washington, D.C.: Smithsonian Institution Press, 1993.

Spencer, John R., and Jacqueline Mitton. *The Great Comet Crash: The Impact of Comet Shoemaker-Levy 9 on Jupiter.* Cambridge: Cambridge University Press, 1995.

Stanley, Steven M. *Extinction.* New York: Scientific American Library, 1987.

———. *Earth and Life through Time.* New York: W. H. Freeman, 1986.

Indian geology   The subcontinent of India is divided into four main geologic provinces, including two on the Indian peninsula, the mountains in the north, and the Indo-Gangetic Plain between the peninsula and the mountains. Approximately two-thirds of peninsular India consists of Precambrian rocks, including assemblages as old as 3.8 billion years. The Precambrian rocks include granites, high-grade metamorphic granulite terranes, and belts of strongly deformed and metamorphosed sedimentary and volcanic rocks called schist belts (broadly equivalent to greenstone belts of other continents). These Precambrian rocks are covered by thick deposits of lavas known as the Deccan Plateau flood basalts, erupted in the Late Cretaceous and Early Tertiary. The Deccan flood plateau basalts cover large sections of western and central India, especially near Mumbai. The eruption of these flood basalts was a consequence of India's rifting away from other parts of Gondwana as it started its rapid journey to crash into Asia.

Peninsular India contains two converging mountain ranges that run along the east and west coasts and form the east and west boundaries of the Deccan Plateau. The two ranges are joined by the Nilgiri Hills in the south, and the highest point in the ranges is Anai Mudi, with an elevation of 8,841 feet (2,697 m). The Eastern Ghats have an average elevation of 2,000 feet (600 m) and generally lie at 50 to 150 miles (80–240 km) from the Coromandel coastline, and locally form steep cliffs along the coast. The Eastern Ghats are crossed by the Godavari, Krishna, and Kaveri Rivers, and are covered by many hardwood trees. The Western Ghats extend along the Malabar coast from the Tapi River to Cape Comorin at the southern tip of India and are generally very close to the coastline. Elevations in the northern part of the Western Ghats reach 4,000 feet (1,200 m) and 8,652 feet (2,637 m) at Doda Beta in the south. The western side of the Western Ghats receives heavy monsoonal rainfalls, but the eastern side of the Western Ghats is generally dry.

Geologically the Western Ghats extend from the Deccan flood basalt plateau in the north to the Pre-

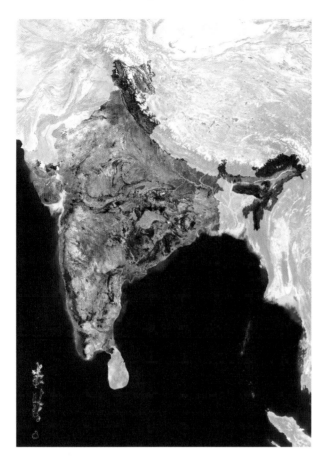

Satellite image of India *(M-Sat Ltd/Photo Researchers, Inc.)*

Sedimentary assemblages of peninsular India include many deposited during the Paleozoic and Mesozoic Eras when India was part of the supercontinent of Gondwana; such assemblages are known on the subcontinent as "Gondwanas." These are mostly thin sequences except in some narrow rift valleys and are mostly absent in interior India. In late Mesozoic and Tertiary times postrifting sedimentary assemblages were deposited on the continental margins and are several miles (km) thick in some isolated basins.

The Indo-Gangetic Plain forms a broad, several-hundred-mile (several-hundred-kilometer) wide plain between the Precambrian rocks exposed in southern India, from the active ranges of the Himalayan Mountains in the north. The rocks on the surface of the Indo-Gangetic Plain are very young, and with depth they record the progressive collision of India and Asia, containing the material eroded from the Himalayas as they were thrust upward during the collision. The Precambrian rocks of peninsular India warp downward and dip under the Indo-Gangetic Plain and form the basement for the sediments deposited in the Himalayan foreland basin.

The Himalaya Mountains contain a diverse suite of rocks ranging in age from Precambrian to Tertiary, and most are strongly deformed and metamorphosed from the Tertiary collision of India with Asia. The Main Boundary Thrust dips northward under the Himalayas and separates the deformed mountain ranges from the Indian shield.

PRECAMBRIAN SHIELD OF SOUTHERN INDIA

Southern India consists largely of Precambrian rocks, partly covered by the Deccan flood basalts and thin sedimentary sequences. The shield is made of seven main cratons of Archean age, including the Western Dharwar, Eastern Dharwar, Southern Granulite Terrane, Eastern Ghats, Bhandara, Singhbhum, and Aravalli cratons. Each of these cratons is somewhat different from the others, and most are now joined along intervening thrust belts or wide orogenic belts of intense deformation and metamorphism. In a few cases younger rifts separate the cratons, and the original relationships are unclear. Many of the orogenic belts between the cratons show strong deformation and metamorphism about 1.5 billion years ago, possibly representing the collision of the cratons to form a supercontinent.

Western Dharwar Craton

The Western Dharwar craton is bounded on the west by the Arabian Sea, covered in the north by the Deccan basalts, and separated on the east from the Eastern Dharwar craton by the Closepet granite and a major fault zone along the eastern margin of the Chitradurga schist belt.

cambrian basement shield including the Dharwar craton in the south. Isotopic ages of gneisses and greenstone belts in the Dharwar craton range from 2.6 to 3.4 billion years old. The Dharwar craton is well known for gold deposits associated with greenstone belts and banded-iron formations, with the best-known greenstone belt being the Chitradurga. The Dharwar craton is divided into eastern and western parts by the elongate north-northwest-striking 2.6-billion-year-old Closepet granite, probably of Andean arc affinity. Late Proterozoic metamorphism, locally to granulite grade and including large areas of charnockites, affects much of the southern part of peninsular India. Rocks of the Dharwar craton are overlain by Paleozoic sedimentary deposits of Gondwanan affinity. The Deccan flood basalts erupted at the end of the Cretaceous and overlie Gondwana and continental margin sequences that began developing with the breakup of Gondwana. The Eastern Ghats are entirely within the Precambrian basement rocks of the Indian subcontinent and the Aravalli craton in the north. The Aravalli craton is somewhat younger than the Dharwar craton, with isotopic ages falling in the range of 3.0 billion to 450 million years.

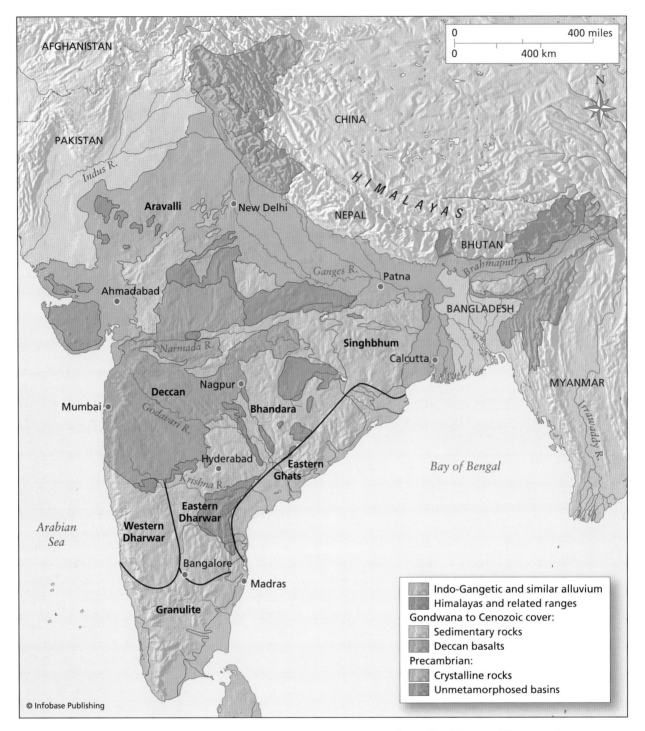

Simplified geologic-tectonic map of India. Main Archean terranes are Granulite, Western Dharwar, Eastern Dharwar, Bhandara, Singhbhum, and Aravalli. Other locations on map include Delhi, Patna, Ahmedabad, Calcutta, Nagpur, Mumbai, Hyderabad, Madras, and Bangalore. *(modeled after J. Rogers and S.M. Naqvi, 1987)*

The Western Dharwar craton is the best known of the Archean cratons of India, hosting several large gold deposits in the mafic volcanic/schist belts, especially in the Chitradurga belt. Structures in the Western Dharwar craton, including the main schist belts, strike generally northward, with a broad convex arc facing east toward the Eastern Dharwar craton. Greenstone belts in the northern part of the craton are generally larger and less metamorphosed than those in the south.

Rocks in the Western Dharwar craton are diverse in age and type. The so-called Peninsular Gneiss forms much of the craton and is made up of tonalitic-trondhjemitic gneiss with many inclusions of older sedimentary and igneous rocks. Several generations of igneous dikes and plutons intrude the Peninsular Gneiss, which has yielded isotopic ages of 3.4 to 3.0 billion years, with younger granites intruding the gneiss between 3.0 and 2.9 billion years ago. The term *Dharwar* has been used to describe mafic volcanic and sedimentary schist belts engulfed by quartz-feldspar gneisses of the Peninsular gneiss. There are three main types of mafic rocks included in this general classification: high-grade mafic rocks caught as enclaves in gneisses in the southern part of the craton, coherent belts of amphibolite facies metamorphic grade, and belts preserved at low metamorphic grade. The term *Sargurs* describes the highly deformed greenstone fragments that are generally older than 3.0 billion years. The high-grade schist belts of the Western Dharwar craton contain numerous fragments of ultramafic/mafic layered igneous complexes and include komatiites, basalts, and other magma types. Some of these may be examples of Archean ophiolites, and others may be intrusions into the continental crust. Other rock types include quartzites, conglomerates, greywacke and sandstones, banded-iron formations, mica schists, metamorphosed mafic volcanic rocks, chert, carbonates, and rare layers of evaporites. The 2.3–2.5-billion-year-old Closepet granite on the eastern side of the craton is rich in potassium feldspar, and its highly elongate shape (about 360 miles long by 30 miles wide [600 km by 50 km]) suggests that it may have intruded along a convergent continental margin type of tectonic setting, perhaps during the collision of the Eastern and Western Dharwar cratons.

Most rocks in the Western Dharwar craton are complexly deformed, with layered rocks typically preserving two or three folding generations. Older folds are generally tight to isoclinal, whereas younger folds are more open and upright, and fold interference patterns are common in the older rocks. The cause of the early deformation events in the schist and greenstone belts is not well known, but some of the younger events appear to be related to the collision of the Western Dharwar craton with other blocks, such as the Eastern Dharwar craton and the formation of supercontinents in the Proterozoic.

Eastern Dharwar Craton

The Eastern Dharwar craton is bordered on the west by the Closepet granite and parallel shear zone that separates it from the Western Dharwar craton. This craton has a gradational boundary with the high-grade Southern Granulite terrane in the south. The eastern boundary is a thrust fault that dips under the Eastern Ghats, and the northern boundary is marked by the Godavari rift in some places and is covered by the Deccan basalts in others.

The Eastern Dharwar craton is not as well known as the Western Dharwar and consists of four main rock associations including schist belts, gneiss, granite, and Late Proterozoic basins including the large Cuddapah basin. The most famous schist belt in the Eastern Dharwar craton is the Kolar belt in the south, which is well known for hosting major gold deposits in gold-quartz veins in shear zones. The schist belts consist of metamorphosed volcanic and sedimentary rocks and resemble the schist belts of the Western Dharwar craton. Granites belong to two main suites, including the arcuate bodies that are parallel and similar to the Closepet granite, and a more scattered and diffuse set especially concentrated near Hyderabad. The granites and schist belts are overlain by Proterozoic to Paleozoic sedimentary sequences, which are especially thick in the Cuddapah basin in the east, and in the Bhima basin in the northwest.

Ages of rocks in the Eastern Dharwar craton extend back to 3.01 billion years, with the oldest-known rocks being metasedimentary gneiss near the Closepet granite. Granite gneisses dated 2.95 billion years and Peninsular gneisses dated 2.6 billion years are exposed near Bangalore. The gneisses near the Kolar schist belt in the south are 2.9 billion years old. Mafic dikes with ages around 1.5–1.4 billion years intrude into the Eastern Dharwar craton.

Several kimberlite pipes in the Eastern Dharwar craton may have been the source for diamonds found as detrital grains in conglomerates in the lower part of the Cuddapah basin. The age of the kimberlites is not well established, but estimates span between 1.45 and 1.00 billion years.

The Cuddapah and related Bhima sedimentary basins, as well as the Kaladgi basin on the western Dharwar craton, are not strongly deformed or metamorphosed. The largest of these basins is the Cuddapah, which is preserved in a crescent-shaped structure on the eastern margin of the craton. The age of the lower-most stratum in the Cuddapah basin is 1.6 billion years, and the basin is cut by undeformed mafic dikes with ages of 980 million years. The oldest rock group in the basin is the Cuddapah Group, which is overlain by the Nallamalai Group, then by the Kurnool Group. Almost all of the rocks are mature sediments, including quartz sandstones and conglomerates, and carbonates (limestones and dolostones) that are free of clastic debris. The shales are clay-rich with only rare silty horizons. Most depositional environments are neritic, intertidal, and subaerial, with slightly deeper water facies recorded in eastern strata. Additionally there is a larger percentage of clastic debris in the eastern part of the basin, suggesting that the source was an uplifted orogen to the east, and the basin is a flexural-style foreland basin that records convergence between the Eastern Ghats and the Eastern Dharwar craton in middle Proterozoic times.

Strata in the Cuddapah basin are folded in increasingly tighter and larger folds toward the east, with the fold axes following the curved shape of the basin. Many west-directed thrust faults also cut the strata, and the overall shape and style of the basin are similar to many younger foreland basins and fold-thrust belts. Strata in the basin thicken and are more deformed toward the east. In this model the deformation would have been driven by orogenesis in the middle Proterozoic Eastern Ghats mobile belt that has been faulted against the eastern side of the basin.

Southern Granulite Terrane

The southern tip of the Indian subcontinent comprises the Southern Granulite terrane, a region of complexly and strongly metamorphosed granulite facies gneisses, as well as amphibolite facies gneisses and lower-grade sedimentary sequences. The Southern Granulite terrane is bounded to the north by the Eastern and Western Dharwar cratons.

Rock types in the Southern Granulite terrane are highly unusual and include a suite of metamorphic rocks derived from older sedimentary, volcanic, and gneissic sequences that were subject to very high temperatures and pressures in the presence of carbon dioxide–rich fluids. These conditions formed a suite of rocks that includes the following:

- charnockites, containing orthopyroxene, quartz, K-feldspar, plagioclase, and garnet
- enderbites, containing orthopyroxene, quartz, and plagioclase
- mafic granulite, containing dominantly pyroxene and plagioclase and commonly garnet
- khondalite, containing quartz, sillimanite, garnet, K-feldspar, plagioclase, and graphite
- leptynite, containing quartz, garnet, K-feldspar, plagioclase, amphibole, sillimanite, and biotite
- gneiss, containing quartz, K-feldspar, plagioclase, amphibole, and biotite

These unusual rocks are exposed well in some of the highland areas of southernmost India including the Nilgiri Hills, the Biligirirangan Hills, and the Kodaikanal massif. These exposures are cut by major Proterozoic shear zones including the Achankovil shear zone, the Palghat-Cauvery shear zone, and the Moyar-Bhavani-Attur shear zone. Many of these have been related to different phases of ocean clo-

sure and continental movements during the assembly of the Gondwana supercontinent at the end of the Precambrian.

Many if not most of the rocks in the Southern Granulite terrane are Archean, with most ages clustering around 2.6 billion years old, but some rocks are at least 3.0 billion years old. However, the age of the high-grade granulite facies metamorphism is much younger—falling in the range of 700–500 million years. Conditions of metamorphism are estimated to be in the general range of 1,470–1,560°F (800–850°C), at 11–19 miles (18–30 km) depth.

Eastern Ghats Province

The Eastern Ghats province, located east of the Eastern Dharwar craton, consists mainly of high-grade igneous and metamorphic rocks, but the ages of many of the rocks are not well established. The province is elongate in a north-northeast direction, and the main folds and shear zones are generally oriented in this direction as well. The western and northern margin of the Eastern Ghats province is a major thrust fault that places the Eastern Ghats over the Eastern Dharwar craton. This fault, known as the Sukinda thrust, corresponds to major changes in the density of the crust on either side, showing that it is a major structure. The Godavari rift cuts the center of the Eastern Ghats province, then extends across the coastline into the Bay of Bengal.

Rock types in the Eastern Ghats include several igneous suites and high-grade metamorphic rocks. These can be divided into six main rock associations:

- mafic schists, consisting of biotite, muscovite, and amphibole
- charnockites, containing quartz, feldspar, and hyperstene
- khondalites, including calc-silicates that represent strongly metamorphosed sedimentary rocks, and are now rich in garnet, sillimanite, cordierite, and sapphirine
- mafic granulites, containing plagioclase, clinopyroxene, orthopyroxene, and amphibole
- anorthosites, consisting of plagioclase and associated with gabbro, dunite, and chromite bearing serpentinite, some of which have been dated to be about 1.3 billion years old
- alkaline rocks, forming posttectonic plutons cutting the other rocks, some of which have yielded isotopic ages of around 1.3 billion years

All of the rocks in the Eastern Ghats were strongly deformed and metamorphosed in the mid-Proterozoic, around 1.4 billion years ago, and some evidence supports a magmatic event around 1.6 billion years ago. Even though there is a paucity of precisely determined isotopic Archean ages from the Eastern Ghats, many of the rocks, especially the schist belts, are quite similar to rocks in the Eastern and Western Dharwar cratons, and many Indian geologists have suggested that they may likewise be Archean.

Evolution of the Eastern Ghats began with deposition of silty and muddy sediments, carbonates, and basalts that later metamorphosed into khondalites and calc-silicate rocks. Volcano-sedimentary rocks that were deposited in several locations along the western side of the Eastern Ghats were later metamorphosed into schist belts. An orogenic event in the Middle Proterozoic caused east-west shortening and deformed the strata into northeast striking folds and shear zones and was associated with the high-grade granulite facies metamorphism. Later folds reorient some of the older folds, and intrusion of alkaline magmas occurred around 1.3–1.4 billion years ago. Structures along the western margin of the belt and the high-grade metamorphism suggest that the deformation and metamorphism formed during a continent-continent collision. Similarities between the Eastern Ghats and the Southern Granulite terrane suggest that the metamorphic events in the two may be related to the same major continent-continent collision event.

Bhandara Craton

The Bhandara craton is located in the north-central part of the Indian subcontinent, bounded on the east by the Eastern Ghats, the Godavari rift and Deccan basalts on the southwest, the Aravalli craton under the Deccan basalts and younger sediments to the northwest, and the Singhbhum craton to the northeast. The craton consists largely of granites and gneisses with many inclusions of older sedimentary and volcanic rocks, overlain by several Late Proterozoic basins, including the Chhattisgarh and Bastar basins. The groups of volcanic and sedimentary rocks include the Dongargarh, Sakoli, Sausar, Bengpal, Sukma, and Bailadila Groups.

The Satpura orogenic belt cuts east-west across the craton, disrupting the dominantly north-south strike of structures in other parts of the region. The Bhandara craton is known for its rich sedimentary manganese ores that are especially abundant in the Sausar Group.

Granites and gneisses of the Bhandara craton are abundant and intrude older continental shelf-type sediments, although no extensive older continental-type basement has been identified. Some of the gneisses are broadly similar to the peninsular gneisses of the Western Dharwar craton, but their ages have not been determined. Metamorphism of the gneisses occurred at 1.5 billion years ago.

The sedimentary assemblages engulfed in the granites and gneisses include mainly the Dongargarh, Sakoli, and Sausar Assemblages. The Dongargarh Supergroup includes quartz-feldspar-biotite gneisses with minor amounts of basalt, metamorphosed to amphibolite facies in the 2.3-billion-year-old Amgaon Orogeny. Younger rhyolites, sandstones, shales, and tuffaceous rocks are approximately 2.2 billion years old. The Dongargarh Supergroup is intruded by granites dated to be 2.27 billion years old, so the rocks are older than this and likely Early Proterozoic or Archean. Rocks of the Dongargarh Supergroup are deformed by three major fold sets, the first of which produced isoclinal folds in the Amgaon Orogeny, and the second and third of which produced tight folds during the Nandgaon and Khairagarh Orogenies.

The Sakoli Group includes metapelitic rocks disposed in a large, synclinal structure in the western part of the craton. The rocks are metamorphosed to lower amphibolite facies and deformed by an early generation of isoclinal folds and a later group of more open upright folds.

The Sausar Group forms a thin, elongate belt of sandy, shaly, and calcareous metamorphosed sedimentary rocks along the northern part of the craton and is one of the main manganese-producing units in India. The lack of volcanic material in the Sausar Group has made it difficult to determine the age of this sequence, but it is known that the rocks were deformed in the Satpura Orogeny at 1.53 billion years ago. The structural geology of the Sausar Group is interesting and unusual. The southern part of the belt is deformed into isoclinal folds, many of which are overturned and show nappe-style movements toward the north. These are bordered on the north by gneisses of the Tirodi suite and have many inclusions of the sedimentary rocks. This gneissic belt may represent the core of the orogen. To the north in the Satpura Ranges, the rocks are disposed in a series of south-directed nappes and thrust sheets.

Despite much work on the structural geology of the Sausar Group, details of the Satpura orogeny that affected these rocks are vague. The orogeny was a Middle Proterozoic event, with some metamorphic ages of circa 1.5 billion years. The main tectonic transport direction in the orogeny was likely toward the south, although the belt of northward-directed nappes south of the crystalline core of the orogen is enigmatic, and few other orogens show tectonic movement toward the center of the belt. Some fold interference or strike-slip motions in the core of the orogen may have eluded detection.

Singhbhum Craton

The Singhbhum craton is located in eastern India, bounded by the Mahamadi graben and Sukinda thrust fault in the south, the Narmada-Son lineament in the west, the Indo-Gangetic Plain in the north, and the Bay of Bengal in the east. The craton has three main parts: the old Archean Singhbhum nucleus in the south, the 2.2- to 1.0-billion-year-old Singhbhum-Dhalbhum mobile belt north of this, and the Chotanagpur-Satpura belt of gneisses and granites north and west of the mobile belt.

The Singhbum craton comprises many old rocks, including the older metamorphic group that is about 3.2 billion years old. Some evidence points to the possibility of rocks as old as 3.8 billion years in the Singhbum nucleus. Most magmatic activity ended in the Singhbum nucleus by 2.7 billion years ago. These rocks are intruded by granites 2.91–2.95 billion years old and then by diabasic intrusives between 1.5 and 1.0 billion years ago.

The Singhbum craton has three major thrust belts: the Dalma thrust in the northern part of the Singhbum-Dhalbhum mobile belt, the Singhbum thrust between the mobile belt and the Singhbum nucleus, and the Sukinda thrust along the southern margin of the craton. The center of the craton is cut by a major rift valley, filled with Gondwana sediments in the Damodar Valley.

The Singhbum thrust is a 120-mile (200-km) long, bow-shaped belt along the northern side of the Singhbum nucleus. It is more than 15 miles (25 km) wide, and contains at least three main thrust slices. Seismic evidence shows that the structure penetrates the thickness of the lithosphere and is therefore interpreted to be an ancient plate boundary. Blueschist facies rocks, which are high-pressure, low-temperature metamorphic rocks characteristic of younger subduction zone settings but exceedingly rare in Precambrian belts, have reportedly been found along the Singhbum thrust by Indian geologists S. N. Sarkar and A. K. Saha, but other geologists have disputed their finds.

A large part of the Singhbum nucleus consists of the Older Metamorphic Group, preserved as remnants in the intrusive Singhbum Granite complex. These rocks include mica schists, quartzites, calcsilicates, and amphibolites, along with gneissic remnants dated to be 3.8 and 3.2 billion years old. These older metasedimentary rocks are overlain by the Iron Ore Group of shales, hematitic jasper with iron ore layers, mafic lavas, sandstone, and conglomerate, but it is not clear which group is older. The iron ore deposits show three periods of folding, including F1 reclined folds, and F2 and F3 upright folds that interact to form fold interference patterns.

The Singhbum-Dhalbhum mobile belt includes rocks of the Singhbum Group, including mica schists, hornblende schists, quartzose schists, granulites, chloritic schists, and amphibolites. Rocks north of the Singhbum thrust are disposed in a large anticlinorium, then farther north they form a synclinorium containing ophiolitic-type volcanic and plutonic rocks of the Dalma Group in its uppermost sections. The rocks show up to four folding generations including early subhorizontal recumbent folds, followed by two generations of upright folds. The fourth generation of folds is associated with the shear zones in the south, related to the formation of the large Singhbum thrust. The Chotanagpur terrain north of the mobile belt contains a large area of gneissic and granitic rocks but also includes metasedimentary rocks, granulites, mafic/ultramafic schists, and anorthosites, and may represent a continental fragment or an island arc terrane.

Rocks in the Singhbum-Dhalbhum mobile belt are interpreted as a Proterozoic orogen deformed at 1.6 billion years ago. These rocks include shallow water sedimentary sequences overthrust by oceanic and ophiolitic assemblages preserved as structurally bounded mafic/ultramafic sequences during closure of a Proterozoic ocean. The ophiolite belt is succeeded southward by a flysch belt, a fold-thrust belt, then a molasse basin on the older Singhbum cratonic nucleus. The rocks were transported from north to south over the Singhbum granite, which represents the foreland to the orogen. The hinterland, or internal parts of the orogen, is in the Chotanagpur area, and the orogen represents the collision of the Chotanagpur block with the Singhbum nucleus about 1.6 billion years ago.

Aravalli Craton

The Aravalli craton is located in the northwestern part of peninsular India, bounded on the north by the Himalaya Mountain chain, the Cambay graben in the southwest, and the Narmada-Son lineament on the south and southeast. Young sediments cover the western boundary and may extend farther into Pakistan.

Rocks of the Aravalli craton are quite different from the Dharwar and other cratons of the Indian shield. They consist mostly of Proterozoic phyllites, graywackes, quartzites, and carbonates, with minor mafic and ultramafic schists. Stromatolites are common in the carbonates and are associated with phosphorite deposits. Banded-iron formations are almost totally absent from the Aravalli craton but common in other cratonic blocks of the Indian shield.

Major structures of the Aravalli craton include the Great Boundary fault on the eastern edge of the Aravalli-Delhi belt, the Delhi-Haridwar ridge, and the Faizabad ridge, which is an extension of the Bundelkhand massif under the Indo-Gangetic Plain. The Aravalli-Delhi belt contains a large number of granitic rocks emplaced over a wide range of time from 3.5 billion to 750 million years ago.

The oldest rocks of the Aravalli craton are found as metasedimentary/metavolcanic inclusions, named the Bhilwara suite, in the Banded Gneiss complex. The gneisses include metasedimentary units, migmatites, granitic gneisses and pegmatites, and metabasic layers. The Bhilwara suite is preserved in several elongate belts within the Banded Gneiss complex, especially between Karera and the Great Boundary fault. Rock types include shales, slates, quartzites, dolostones, marbles, cherts, graywackes, and hornblende and mica-schists. Sedimentary structures in these rocks suggest that they were deposited in a shallow shelf platformal environment.

The Aravalli Supergroup unconformably overlies this Banded Gneiss complex in some places but is intruded by the gneissic rocks in others. These conflicting relationships show that the Banded Gneiss complex consists of several different units that still need additional work to separate them from one another, as confirmed by a range of isotopic ages on the gneisses that extends from 3.5 to 2.0 billion years. Rock types in the Aravalli Supergroup, including quartzites, greywacke, and carbonates, have experienced greenschist facies metamorphism and are intruded by pre-, syn-, and posttectonic granitoids and mafic-ultramafic suites. Ages of the Aravalli Supergroup range between 2.5 and 2.0 billion years. The Aravalli Supergroup is complexly folded by as many as four generations, including early reclined, typically rootless folds, followed by later generations of folds distorting the outcrop belts into hook-shaped synclines, anticlines, and more complex shapes.

The Delhi Supergroup apparently overlies the Aravalli Supergroup, and parts of it are at least 1.6 billion years old, though other sections may be significantly younger. The Delhi Supergroup is the main rock suite outcropping in the Aravalli Mountains, over a distance of more than 350 miles (700 km) from Gujarat to Delhi. The Delhi Supergroup includes quartzites, conglomerates, arkoses, phyllites, slates, limestones, marbles, mafic volcanic rocks, and amphibolites. The rocks are complexly deformed and metamorphosed between greenschist and granulite grades. Two main deformation events are recognized from structures in the Delhi Supergroup. The first deformation event produced isoclinal folds with axial planar foliation parallel to bedding surfaces, whereas the second produced open to tight asymmetrical folds that plunge north to north-northwest.

Exposures of rock are few and far between in the Bundelkhand area, but the rocks there include granites and gneisses deposited over a considerable time spanning much of the Precambrian. The northeastern part of the craton is overlain by generally flat-lying rocks of the Late Proterozoic Vindhyan Supergroup, deformed only near the Great Boundary fault and in the south near the Satpura orogen. Vindhyan sedimentary rocks include basal conglomerates and quartzite, and grades up into quartz arenite, and higher in the section into limestones, shales, and other shallow-water sedimentary deposits. The upper part of the Vindhyan Supergroup includes sandstones and conglomerates, and some have diamonds probably eroded from nearby kimberlite pipes.

The Arvalli craton hosts a few alkaline igneous intrusions and kimberlite pipes. The alkaline rocks were emplaced around 1.5 billion years ago, and the kimberlites have yielded a variety of isotopic ages ranging from 1.63 to 1.12 billion years, all with large uncertainties.

HIMALAYA MOUNTAINS

The Himalaya Mountains were formed during the Tertiary continent-continent collision between India and Asia and contain the tallest mountains, as well as those exhibiting the greatest vertical relief over short distances, in the world. The range extends for more than 1,800 miles (3,000 km) from the Karakoram near Kabul, Afghanistan, past Lhasa, Tibet, to Arunachal Pradesh in the remote Assam Province of India. Ten of the world's 14 peaks that rise to more than 26,000 feet (8,000 m) are located in the Himalayas, including Mount Everest, 29,035 feet (8,850 m), Nanga Parbat, 26,650 feet (8,123 m), and Namche Barwa, 25,440 feet (7,754 m). The rivers that drain the Himalayas exhibit some of the highest sediment outputs in the world, including the Indus, Ganges, and Brahmaputra. The Indo-Gangetic Plain, on the southern side of the Himalayas, is a foreland basin filled by sediments eroded from the mountains and deposited on Precambrian and Gondwanan rocks of peninsular India. The northern margin of the Himalayas is marked by the world's highest and largest uplifted plateau, the Tibetan Plateau.

The Himalayas is one of the youngest mountain ranges in the world but has a long and complicated history best understood in the context of five main structural and tectonic units within the ranges. The Subhimalaya includes the Neogene Siwalik molasse, bounded on the south by the Main Frontal Thrust that places the Siwalik molasse over the Indo-Gangetic Plain. The Lower or Subhimalaya is thrust over the Subhimalaya along the Main Boundary Thrust, and consists mainly of deformed thrust sheets derived from the northern margin of the Indian shield. The High Himalaya is a large area of crystalline basement rocks, thrust over the Subhimalaya along the Main Central Thrust. Farther north, the High Himalaya sedimentary series or Tibetan Himalaya consists of sedimentary rocks deposited on the crystalline

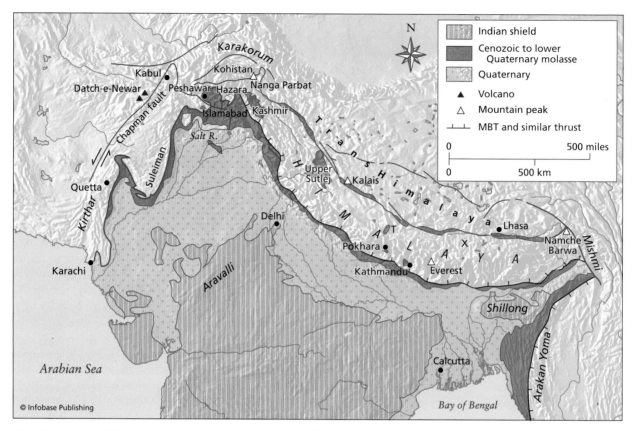

Simplified map of the Himalaya Mountains and surrounding areas, showing main tectonic zones and faults and Cenozoic-Quaternary basins. The Indus-Tsangpo suture is located along the boundary between the Transhimalaya and the Himalayas. Names are as follows: Ch F, Chaman faults; DeN, Dacht-e-Newar; E, Everest (29,015 feet [8,846 m]); H, Hazara; I, Islamabad; K, Kathmandu; Ka, Kashmir; Kb, Kabul; Ki, Karachi; Ko, Kohistan; Ks, Kailas; NB, Namche Barwa; NP, Nanga Parbat (26,650 feet or 8,125 m); P, Peshawar; Pk, Pokhra; Q, Quetta; T, Thakkhola; US, Upper Sutlej.

basement of the High Himalaya. Finally, the Indus-Tsangpo suture is the suture between the Himalayas and the Tibetan Plateau to the north.

Sedimentary rocks in the Himalayas record events on the Indian subcontinent, including a thick Cambrian-Ordovician through Late Carboniferous/Early Permian Gondwanan sequence, followed by rocks deposited during rifting and subsidence events on the margins of the Tethys and Neotethys Oceans. The collision of India with Asia was in progress by the Early Eocene. This collision exposed the diverse rocks in the Himalayas, revealing a rich geologic history that extends back to the Precambrian, where shield rocks of the Aravalli and Delhi cratons are intruded by 500-million-year-old granites. Subduction of Tethyan oceanic crust along the southern margin of Tibet formed an Andean-style arc represented by the Trans-Himalaya batholith that extends west into the Kohistan island arc sequence. The obduction of ophiolites and high-pressure (blueschist facies) metamorphism dated to have occurred around 100 million years ago is believed to be related to this subduction. Thrust stacks

began stacking up on the Indian subcontinent, and by the Miocene attempted deep intracrustal subduction of the Indian plate beneath Tibet along the Main Central Thrust formed high-grade metamorphism and generated a suite of granitic rocks in the Himalayas. After 15–10 million years ago movements were transferred to the south to the Main Frontal Thrust, which is still active.

INDO-GANGETIC PLAIN

The Indo-Gangetic Plain is the active foreland basin of the India-Asia collision, with sediments derived from erosion of the Himalaya Mountains and carried by numerous rivers that feed into the Indus and Ganges Rivers. Alluvial deposits of the Indo-Gangetic Plain stretch from the Indus River in Pakistan to the Punjab Plain in India and Pakistan, to the Haryana Plain and Ganges delta in Bangladesh. Sediments in the foreland basin extend up to 24,500 feet (7,500 m) thick over the basement rocks of the Indian shield, thinning toward the southern boundary of the basin plain. The plain has very little relief, with only occasional bluffs and terraces related to changes in river levels.

The northern boundary of the plain is marked by two narrow belts known as Terai, containing small hills formed by coarse remnant gravel deposits emerging from mountain streams. Many springs emanate from these gravel deposits forming large, swampy areas along the major rivers. In most places the Indo-Gangetic Plain is about 250 miles (400 km) wide. The southern boundary of the plains is marked by the front of the Great Indian Desert in Rajasthan, then continues eastward to the Bay of Bengal along the hills of the Central Highlands.

The Indo-Gangetic Plain can be divided into three geographically and hydrologically distinct sections. The Indus Valley in the west is fed by the Indus River, which flows out of Kashmir, the Hundu Kush, and the Karakoram range. The Punjab and Haryana Plains are fed by runoff from the Siwalik and Himalaya Mountains into the Ganges River, and fed by the Lower Ganga and Brahmaputra drainage systems in the east. The lower Ganga plains and Assam Valley are lush and heavily vegetated, and the waters flow into the deltaic regions of Bangladesh.

Clastic sediments of the foreland basin deposits under the Indo-Gangetic plain Eocene-Oligocene (about 50–30 million-year old) deposits, grading up to the Miocene to Pleistocene Siwalik clastic rocks, eroded from the Siwalik and Himalaya ranges. The basement of the Indian shield dips about 15° beneath the Great Boundary and other faults marking the deformation front at the toe of the Himalayas.

See also Archean; convergent plate margin processes; craton; Gondwana, Gondwanaland; greenstone belts; large igneous provinces, flood basalt.

FURTHER READING

Molnar, Peter. "The Geologic History and Structure of the Himalaya." *American Scientist* 74 (1986): 144–154.

Naqvi, S. Mahmood, and John J. W. Rogers. *Precambrian Geology of India*. Oxford: Oxford University Press, 1987.

Ramakrishnan, M., and R. Vaidyanadhan. *Geology of India*. 2 vols. Bangalore, India: Geological Society of India, 2008.

interstellar medium   The interstellar medium, which consists of the areas or voids between the stars and galaxies, represents a nearly perfect vacuum, with a density a trillion trillion times less than that of typical stars. The density of the interstellar medium is so low that there are only a couple of atoms per cubic inch (about 1 atom per cubic cm), or roughly a thimbleful of atoms in a volume the size of the Earth. Interstellar matter includes all of the gas, dust, dark matter, and other material in the universe not contained within stars, galaxies, or galaxy clusters. Despite the extremely low density of the interstellar medium, the vast size of the space is such that the amount of matter in the interstellar medium is about the same as that in all of the stars and galaxies.

Matter in the interstellar medium consists of two main components—gas and dust. In many places in the universe dust forms dark clouds of gigantic size that obscure distant light from passing through, making parts of the universe appear dark. This dust consists of clumps of atoms and molecules whose sizes are comparable to the wavelength of light (10^{-7} m), and much larger than the interstellar gas. The size of the dust particles explains why dust clouds appear dark. Electromagnetic radiation, such as visible light, can be effectively blocked only by particles of similar or greater size than the wavelength of the incident light (or other radiation), and since the size of the dust is similar to the wavelength of light, dust is an excellent blocker of light and appears dark, blocking light from sources behind the dust clouds. However, longer-wavelength radio waves can pass through dust clouds unimpeded. Gas in the interstellar medium is composed mainly of individual atoms of about 1 angstrom (10^{-10} m) in size, and a smaller amount of atoms combined into molecules. The size of the interstellar gas is less than the wavelengths of electromagnetic radiation in most of the visible and radio wavelengths, so this radiation, including light, can pass through the gas, with absorption occurring within a specific narrow range of wavelengths. Interstellar dust, however, blocks the shorter wavelength optical, ultraviolet, and X-ray radiation.

The gas and dust in the interstellar medium preferentially absorbs the longer wavelengths of light from the higher-frequency blue area of the spectrum, so stars and galaxies appear redder than they actually are. Despite this "reddening" of the appearance of stars, the spectral signature, and absorption lines in the stars' spectrums, are mostly unaffected by interstellar dust. With this relationship it is possible to measure the spectrum of a star, which will reveal its luminosity and color. Then by measuring the color on Earth, the amount of reddening by interaction with dust in the interstellar medium can be measured, and this in turn reveals information about the amount and type of interstellar dust the light encountered on its transit from the star to Earth.

Interstellar space is quite cold, with an average temperature of about 100 kelvin (which is -173°C, or -279°F), but ranging from a few kelvins (near absolute zero, where all motion of atoms stops) to several hundred kelvin near stars and other sources of radiation.

Cone Nebula pillar of gas and dust. Photo taken by Advanced Camera for Surveys aboard *Hubble* during space shuttle *STS-109* mission in March 2002 *(NASA, H. Ford (JHU), G. Illingworth (USCS/LO), M. Clampin (STScI), G. Hartig (STScI), the ACS Science Team, and ESA)*

Interstellar gas is made mostly (about 90 percent) of atomic and molecular hydrogen, followed by about 9 percent helium, and 1 percent heavier elements such as carbon, oxygen, silicon, and iron, but the heavier elements have a much lower concentration in the interstellar gas than in stars or in the Earth's solar system. This may be because the heavier elements were combined into molecules to form interstellar dust, whose composition is not well known. Dust is known to include these heavier elements such as silicon, iron, and graphite, as well as ices of water, ammonia, and methane, similar to the "dirty ice" found in comet nuclei. Interstellar dust polarizes light that passes through it; in other words it aligns the light so that the electromagnetic radiation vibrates in a single plane instead of randomly as in unpolarized light. The light becomes polarized because the interstellar dust is made of long, skinny needlelike particles aligned in a specific direction. The cause of this alignment of interstellar dust is the subject of much research and debate but is thought to be caused by the dust particles aligning themselves along weak magnetic field lines in interstellar space.

NEBULAE

Nebulae are areas of interstellar space that appear fuzzy yet are clearly distinguishable from surrounding areas of space, and many of those visible from Earth are concentrated in the plane of the Milky Way Galaxy. Many nebulae are clouds of interstellar gas and dust. In some cases these clouds block the light of stars that are located behind (from the observer's point of view) the nebula, and in other cases the nebulae appear bright and are lit up from the inside, typically by groups of young hot stars. Bright nebulae are known as emission nebulae, being hot glowing clouds of ionized interstellar gas. These nebulae have hot young stars in their centers that emit huge amounts of ultraviolet radiation, which in turn ionizes the surrounding gas, causing the spectacular glow of the clouds visible from Earth. Emission nebulae have red colors because the ionized hydrogen atoms in the gas clouds emit light in the red part of the visible spectrum, and hydrogen is the most abundant gas in the clouds. Other gases, such as helium, may lace the nebulae with other colors, and dark bands wisping across the nebulae are dust clouds that are integral parts of the nebulae. The gases in the nebulae emit spectra when ionized, and these spectra can reveal information about the composition of the gases in the nebulae. Most are made of 90 percent hydrogen, 9 percent helium, and 1 percent heavier elements, similar to the composition of interstellar gas. Nebulae have temperatures measured at around 8,000 kelvin (7,727°C, or 13,940°F), masses of about a couple of hundred to several thousand solar masses, and densities of several hundred to a thousand particles per cubic inch (a few hundred particles per cubic cm, or $8\text{-}10 \times 10^7$ particles per m^3).

Some emission nebulae have areas with green-colored glowing gases, known to be caused by doubly ionized oxygen atoms. These oxygen atoms contain electrons that exist in a raised level of energy and if undisturbed for several hours (without experiencing collisions with other particles) emit a photon, in the green wavelengths, when they drop down to the normal ionized state.

DUST CLOUDS

Dark areas of the sky can be voids, or alternatively, areas where the light is obscured by cold and relatively dense clouds of dust particles. They typically appear as dark, irregular areas in otherwise starlight areas of the sky. Dark dust clouds are typically about 100 kelvin (equivalent to -173°C, or -279°F), but can be considerably colder, range in size from bigger than Earth's solar system to many parsecs across, and have densities thousands to millions of times greater than surrounding voids in space. These dust clouds are composed predominantly of gas, but the property of absorbing light from stars is due primarily to the dust within the clouds.

Information about dust clouds can be learned from their characteristic absorption spectra, which can reveal details about the composition, temperature, and density of the dust cloud. As light leaves a star, the spectra will have the characteristic spectrum from the temperature and elements of that star, and if the light passes through a dust cloud, the spectrum of the dust will be added to that from the star. The cooler the temperature of the dust, the thinner the spectral absorption lines, so the spectral lines of the dust are easily distinguished from those of the original star from which the light arose. Spectral lines from dust clouds have revealed that most have compositions that are virtually identical to other parts of space, characterized by mostly (about 90 percent) atomic and molecular hydrogen, followed by about 9 percent helium, and followed by 1 percent heavier elements such as carbon, oxygen, silicon, and iron.

Although dark dust clouds do not emit light, since they are too cold and composed of neutral atoms and molecules, they do emit at longer wavelengths such as in the infrared. Interstellar matter also emits radiation with a wavelength of 8.2 inches (21 cm), caused by the emission of a low-energy photon when hydrogen changes from a relatively high-energy configuration, with the sense of spin of the atom's electron and proton being the same, to a lower energy configuration, in which the electron and proton have opposite senses of spin. Detection of the characteristic 8.2-inch (21-cm) radio wave radiation has enabled regions of the universe that contain enough hydrogen to emit these photons to be mapped and studied in terms of their temperature, density, composition, and velocity. The radio waves have a wavelength much larger than any interstellar dust particles, which means that the dust in interstellar space does not absorb the energy of or interfere with the 8.2-inch (21-cm) radiation, and it can be measured and studied undisturbed from Earth.

MOLECULAR CLOUDS

Molecular clouds are among the largest structures of interstellar space. They consist of cold and relatively dense (10^{12} particles/cm^3) collections of matter in molecular form. Molecules in these clouds can become excited by collision with other particles or by interacting with radiation. When either happens, the molecules reach a higher energy state when they are excited, and when they relax to a lower energy state, they emit a photon that can then be detected by astronomers. Molecules are more complex than atoms, so they can produce a greater variety of energy released during changes in rotation, electron transitions, and vibrations, each releasing a characteristic photon emission. Most molecular clouds are located in very dusty and dense areas in interstellar space, so energy released in these processes in the ultraviolet, optical, and most infrared wavelengths is absorbed by the local dust clouds, but photons and energy released at radio-wave frequencies moves through this medium and can be detected from Earth.

Spectra from molecular clouds reveal that they consist mostly of molecular hydrogen (H_2), but molecular hydrogen does not emit or absorb radio wave radiation, so it is not useful as a probe. But the spectra emitted from other molecules have proven useful for studying molecular clouds. Some of the most useful include carbon monoxide, hydrogen cyanide, ammonia, water, methyl alcohol, and formaldehyde, and many dozens of other complex molecules. These molecules are used as tracers of the physical and chemical makeup of the molecular clouds and are interpreted to have formed in the clouds. The spectral lines from molecular clouds can also be used to determine the composition of the clouds, their temperature, density, and distribution. One of the major discoveries about molecular clouds made in the past few decades is that the clouds are not isolated bodies, but form giant molecular cloud complexes, as large as 50 parsecs across, each containing millions of stars. The Milky Way Galaxy alone has more than 1,000 known molecular clouds.

See also ASTROPHYSICS; CONSTELLATION; COSMOLOGY; GALAXIES.

FURTHER READING

Chaisson, Eric, and Steve McMillan. *Astronomy Today.* 6th ed. Upper Saddle River, N.J.: Addison-Wesley, 2007.

Comins, Neil F. *Discovering the Universe.* 8th ed. New York: W. H. Freeman, 2008.

National Aeronautics and Space Administration. Goddard Space Flight Center Astronomical Data Center Quick Reference Page, Interstellar Medium (ISM) Web page. Available online. URL: http://adc.gsfc.nasa.gov/adc/quick_ref/ref_ism.html. Accessed April 30, 2002.

Snow, Theodore P. *Essentials of the Dynamic Universe: An Introduction to Astronomy.* 4th ed. St. Paul, Minn.: West, 1991.

island arcs, historical eruptions Island arcs are belts of high seismic activity and high heat flow with chains of active volcanoes, bordered by a submarine trench formed at a subduction zone. They form where plates of oceanic lithosphere are subducted beneath another oceanic plate, and the down-going oceanic lithosphere may be subducted to 500 miles (700 km) or more. A related type of volcanic arc, an Andean or continental-margin volcanic arc, forms on the edge of a continental plate where an oceanic slab is subducted beneath the edge of the continent. In both types of arcs fluids forced out of the subducting slab at 60–100 miles' depth (100–160 km) cause the mantle above the subducting slab to melt partially, and these magmas migrate upward to form the island or continental margin arc. In most cases these arcs are located 90–120 miles (150–200 km) from the trench, with the distance determined by the dip of the down-going slab.

Volcanic arcs developed above subduction zones have several different geomorphic zones defined largely on their topographic and structural expressions. The active arc is the topographic high with volcanoes, and the back arc region stretches from the active arc away from the trench; it may end in an older rifted arc or continent. The arc is succeeded seaward by the fore-arc basin, a generally flat topographic basin with shallow- to deep-water sediments, typically deposited over older accreted sediments and ophiolitic or continental basement. The accretionary prism includes uplifted, strongly deformed rocks scraped off the downgoing oceanic plate on a series of faults that branch off from the subduction zone thrust fault. Some accretionary prisms are 50–100 miles (80–160 km) wide, and can be thousands of miles long. The world's largest accretionary prisms currently extend around the Pacific Ocean rim, including the southern Alaska accretionary prism, the Franciscan complex in California, and prisms in Japan and the southwest Pacific. The trench may be several to five or more miles (10 km) deep below the average level of the seafloor in the region and marks the boundary between the overriding and under-thrusting plates. The outer trench slope is the region from the trench to the top of the flexed oceanic crust that forms a few hundred-meter-high topographic rise known as the forebulge on the down-going plate.

Trench depressions are triangular shaped in profile and typically partly to completely filled with graywacke-shale sediments derived from erosion of the accretionary wedge, and deposited by sediment-laden, fast-moving, down-slope flows known as turbidity currents. The resulting sedimentary rock types have a characteristic style of layering and an upward decrease in grain size, and are known as turbidites. Turbidites may also be transported by currents along the trench axis for large distances, up to hundreds or even thousands of miles from their ultimate source in uplifted mountains in the convergent plate boundary orogen. *Flysch* is a term that applies to rapidly deposited deep marine syn-orogenic clastic rocks that are generally turbidites. Chaotic deposits known as olistostromes that typically have clasts or blocks of one rock type, such as limestone or sandstone, mixed with a muddy or shaly matrix also characterize trenches. These are interpreted as slump or giant submarine landslide deposits. They are common in trenches because of the oversteepening of slopes in the wedge. Sediments that get accreted may also include pelagic (deep-water) sediments initially deposited on the subducting plate, such as red clay,

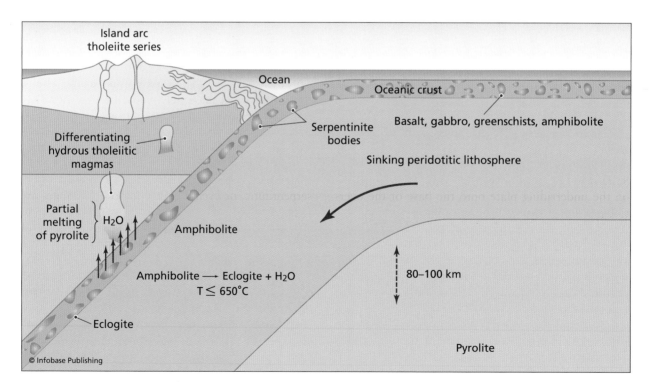

Cross section of island arc showing the physiography and geologic processes involved during subduction and formation of the arc

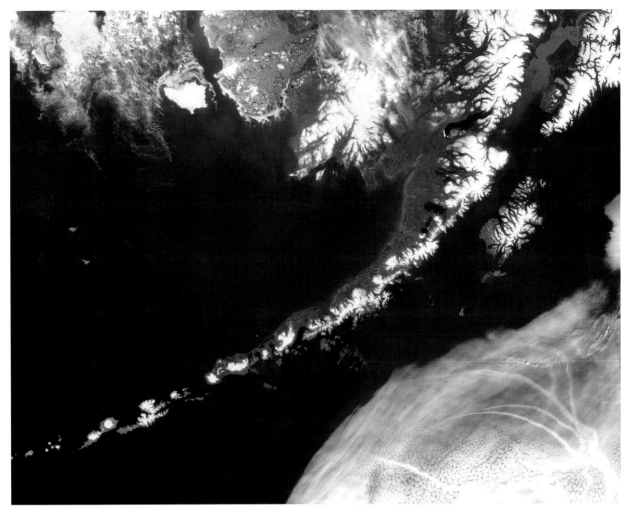

Terra MODIS satellite image of Aleutian Islands in Alaska, May 25, 2006 *(Jeff Schmaltz/NASA/Visible Earth)*

siliceous ooze, chert, manganiferous chert, calcareous ooze, and windblown dust.

The sediments are deposited as flat-lying turbidite packages, then are gradually incorporated into the accretionary wedge complex through folding and the propagation of faults through the trench sediments. Subduction accretion accretes sediments deposited on the underriding plate onto the base of the overriding plate. It causes the rotation and uplift of the accretionary prism, a broadly steady process that continues as long as sediment-laden trench deposits are thrust deeper into the trench. New faults will typically form and propagate beneath older ones, rotating the old faults and structures to steeper attitudes as new material is added to the toe and base of the accretionary wedge. This process increases the size of the overriding accretionary wedge and causes a seaward-younging in the age of deformation.

Parts of the oceanic basement to the subducting slab are sometimes scraped off and incorporated into the accretionary prisms. These tectonic slivers typi-cally consist of fault-bounded slices of basalt, gabbro, and ultramafic rocks; rarely, partial or even complete ophiolite sequences can be recognized. These ophiolitic slivers are often parts of highly deformed belts of rock known as mélanges. Mélanges are mixtures of many different rock types typically including blocks of oceanic basement or limestone in muddy, shaly, serpentinitic, or even a cherty matrix. Mélanges are formed by tectonic mixing of the many different types of rocks found in the fore arc, and are one of the hallmarks of convergent boundaries.

TYPES OF ISLAND AND CONVERGENT MARGIN ARCS

There are major differences in processes that occur at continental or Andean-style versus oceanic or island arc systems, also known as Marianas-type arcs. Andean-type arcs have shallow trenches, fewer than four miles (6 km) deep, whereas Marianas-type arcs typically have deep trenches reaching seven miles (11 km) in depth. Most Andean-type arcs subduct young oceanic crust and have very shallow-dipping subduction zones, whereas Marianas-type arcs subduct old oceanic crust and have steeply dipping zones of high earthquake activity, called Benioff zones, marking the top of the subducting slab. Andean arcs have back-arc regions dominated by foreland (retroarc) fold-thrust belts and sedimentary basins, whereas Marianas-type arcs typically have back-arc basins, often with active seafloor spreading. Andean arcs have thick crust, up to 45 miles (70 km) thick, and big earthquakes in the overriding plate, while Marianas-type arcs have thin crust, typically only 12 miles (20 km) thick, and have big earthquakes in the underriding plate. Andean arcs have only rare volcanoes, and these have magmas rich in SiO_2 such as rhyolites and andesites. Plutonic rocks are more common, and the basement is continental crust. Marianas-type arcs have many volcanoes that erupt lava low in silica content, typically basalt, and are built on oceanic crust.

Many arcs are transitional between the Andean or continental-margin types and the oceanic or Marianas-types, and some arcs have large amounts of strike-slip motion. The causes of these variations have been investigated, and it has been determined that the rate of convergence has little effect, but the relative motion directions and the age of the subducted oceanic crust seem to have the biggest effects. In particular, old oceanic crust tends to sink to the point where it has a near-vertical dip, rolling back through the viscous mantle, and dragging the arc and forearc regions of overlying Marianas-type arcs with it. This process contributes to the formation of back-arc basins.

As the cold subducting slab is pushed into the mantle, it cools the surrounding mantle and forearc. Therefore, the effects of the downgoing slab dominate the thermal and fluid structure of arcs. Fluids released from the slab as it descends past 70 miles (110 km) aid partial melting in the overlying mantle and form the magmas that form the arc on the overriding plate. This broad thermal structure of arcs results in the formation of paired metamorphic belts, where the metamorphism in the trench environment grades from cold and low-pressure at the surface to cold and high-pressure at depth, whereas the arc records low and high-pressure, high-temperature metamorphic facies series. One of the distinctive rock associations of trench environments is the formation of the unusual high-pressure, low-temperature blueschist facies rocks in subduction zones. The presence of index minerals glaucophane (a sodic amphibole), jadeite (a sodic pyroxene), and lawsonite (Ca-zeolite) indicate low temperatures extended to depths near 20 miles (20–30 km) (7–10 kbar). Since these minerals are unstable at high temperatures, their presence indicates they formed in a low-temperature environment, and the cooling effects of the subducting plate

offer the only known environment to maintain such cool temperatures at depth in the Earth.

Fore-arc basins may include several-mile (or kilometer) thick accumulations of sediments deposited in response to subsidence induced by tectonic loading or thermal cooling of fore arcs built on oceanic lithosphere. The Great Valley of California is a fore-arc basin formed on oceanic fore-arc crust preserved in ophiolitic fragments found in central California; Cook Inlet, Alaska, is an active fore-arc basin formed in front of the Aleutian and Alaska range volcanic arc.

The rocks in the active arcs typically include several different facies. Volcanic rocks may include subaerial flows, tuffs, welded tuffs, volcaniclastic conglomerate, sandstone, and pelagic rocks. Debris flows from volcanic flanks are common, and there may be abundant and thick accumulations of ash deposited by winds and dropped by Plinian and other eruption columns. Volcanic rocks in arcs include mainly calc-alkaline series, showing early iron enrichment in the melt, typically including basalts, andesites, dacites, and rhyolites. Immature island arcs are strongly biased toward eruption at the mafic end of the spectrum, and may also include tholeiitic basalts, picrites (a magnesium-rich basalt), and other volcanic and intrusive series. More mature continental arcs erupt more felsic rocks and may include large caldera complexes.

Back-arc or marginal basins form behind extensional arcs, or may include pieces of oceanic crust trapped by the formation of a new arc on the edge of an oceanic plate. Many extensional back arcs are found in the southwest Pacific, whereas the Bering Sea, between Alaska and Kamchatka, is thought to be a piece of oceanic crust trapped during the formation of the Aleutian chain. Extensional back-arc basins may have oceanic crust generated by seafloor spreading, and these systems very much resemble the spreading centers found at divergent plate boundaries. The geochemical signature of some of the lavas, however, shows some subtle and some not-so-subtle differences, with water and volatiles being more important in the generation of magmas in back-arc suprasubduction zone environments.

Compressional arcs such as the Andes have tall mountains, reaching heights of more than 24,000 feet (7,315 m) over broad areas. They have rare or no volcanism but much plutonism, and typically have shallow dipping slabs beneath them. They have thick continental crust with large compressional earthquakes, and show a foreland-style retro-arc basin in the back-arc region. Some compressional arc segments do not have accretionary forearcs but exhibit subduction erosion during which material is eroded and scraped off the overriding plate, and dragged down into the subduction zone. The Andes show remarkable along-strike variations in processes and tectonic style, with sharp boundaries between different segments. These variations seem to be related to what is being subducted and plate motion vectors. In areas where the down-going slab has steep dips the overriding plate has volcanic rocks; in areas of shallow subduction there is no volcanism.

TYPES OF VOLCANISM IN DIFFERENT ARCS

The most essential part of an island arc is the volcanic center, consisting of a line of volcanic islands comprising volcanic and pyroclastic debris, forming a linear chain about 60–70 miles (100–110 km) above the subducting slab. In island arcs the volcanic rocks are generally of several different types called volcanic series. These include a tholeiitic series, consisting of tholeiitic basalt, andesite, and less common dacite. The calc-alkaline series has basalts rich in alumina, abundant andesite, dacite, and some rhyolite. A third series, the alkali series, includes a sodic group dominated by alkali olivine basalt, alkalic andesite, trachyite, and alkalic rhyolite. A rarer assemblage of volcanic rocks consists of the shoshonite group, comprising of shoshonite, latite, and leucite-bearing magmatic rocks.

The magmatic rocks in island and continental-margin arcs are built on a basement or substrate made of older oceanic crust, or on the edge of a continent. Some arcs change along strike, being built on the edge of a continent in one location and extending off into the ocean, such as the Aleutians, which are built on North American basement in the east and extend to an oceanic island arc in the west. A similar transition exists in the arc system in Indonesia that extends from Bali to West Papua (Irian Jaya) to Sumatra. There is a general correlation between the type of magma that erupts in arc and the type of underlying crust. Arcs built on oceanic crust are dominated by tholeiitic-type volcanic series, whereas arcs built on thicker continental basement are more evolved and have more of the calc-alkaline and shoshonitic series erupted in their volcanic centers.

DESCRIPTIONS OF MAJOR ERUPTIONS FROM ISLAND AND CONTINENTAL-MARGIN ARC VOLCANOES

Volcanic arcs are important as the birthplace of continental crust, and they have also played a large role in the history of the world. Examination of some of the world's most significant volcanic eruptions helps understand the significance of arcs in natural and human history.

Thera, Greece, 3,650 Years before Present

One of the greatest volcanic eruptions in recorded history occurred approximately 3,650 years ago in the eastern Mediterranean region, then the cradle of civilization. Santorini is a small, elliptically shaped archipelago of islands, approximately 10 miles (16 km) across, located about 70 miles (110 km) north of the island of Crete. The islands are dark and ominous in stark contrast to the white limestone of the Greek islands, and they form ragged, 1,300-foot (390-m) peaks that seem to point up toward something that should be in the center of the ring-shaped archipelago but is no longer there. The peaks are pointing inward toward the center of a giant caldera complex that erupted in the late Bronze Age, approximately 3,650 years ago, devastating much of the eastern Mediterranean. The largest island on the rim of the caldera is Thera, and across two circular 900–1,000-foot (275–300-m) deep calderas rests the opposing island of Therasia, once part of the same volcano. In the center of the composite caldera complex are several smaller islands known as the Kameni Islands, which represent newer volcanic cones growing out of the old caldera.

Santorini and Thera (present-day Thíra) are part of the Cyclades islands that form part of the Hellenic island/continental-margin volcanic arc that stretches from western Turkey through Greece, lying above a subduction zone in the Mediterranean along which part of the African plate is being pushed beneath Europe and Asia. Volcanoes in the Hellenic arc are widely spaced, and numerous earthquakes also characterize the region. The area was apparently densely populated before the massive eruption 3,650 years ago, as remnants of Bronze Age and earlier Neolithic settlements and villages along the coastal Aegean are buried in ash from Thera. Settlers that arrived about 6,000–7,000 years ago and traded with Crete and Greece populated the island of Thera. In 1967 archaeologists discovered a Bronze Age city buried by ash from the eruption, and uncovered numerous paved streets and frescos, indicating that the city was at least the size of Pompeii, Italy, when it was buried by the eruption of Vesuvius more than 1,700 years later. At the time of the eruption the region was dominated by the Minoan culture, derived from Crete. The eruption occurred while a primitive form of writing was used by the Minoans, but has not been deciphered, and the Greek language had not yet been codified in writing. Thus no local texts record the eruption, although Hebrew text was in use in nearby Egypt and Israel. Some scholars have tried to link the eruption of Thera with biblical events such as the plagues, days of darkness, and parting of the Red Sea during the exodus of the Israelites from Egypt, although the timing of the eruption seems to be off by a couple of centuries for such a correlation to be made. The best current estimates for the age of the eruption are between 1690 and 1620 B.C.E.

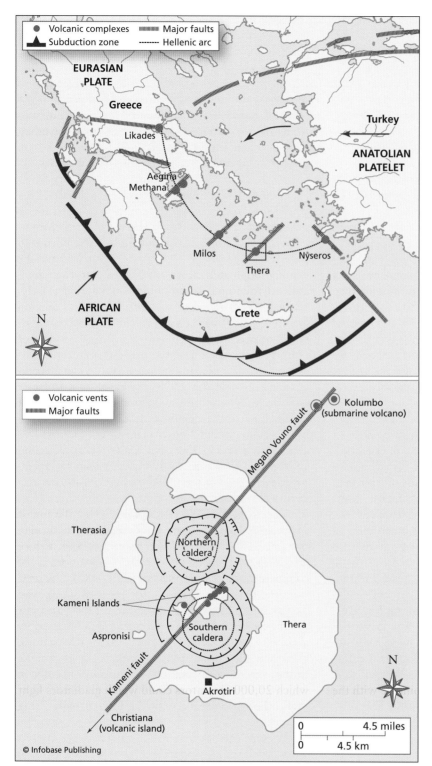

Map of the eastern Mediterranean (top) showing the tectonic setting of Thera in the Hellenic volcanic arc. Note that Thera is located along a major fault above a subduction zone. Detailed map (bottom) of Santorini archipelago, showing modern Thera and the large calderas that collapsed 3,650 years ago.

now referred to as Thíra. There are no known written firsthand accounts of the eruption of Thera, so the eruption history has been established by geological mapping and examination of the historical and archaeological records of devastation across the Mediterranean region. Volcanism on the island seems to have started 1 to 2 million years ago, and continues to this day. Large eruptions are known from 100,000 and 80,000 years ago, as well as 54,000, 37,000, and 16,000 years ago, then finally 3,500 years ago. The inside of Thera's caldera is marked by striking layers of black lava alternating with red and white ash layers, capped by a 200-foot (60-m) thick layer of pink to white ash and pumice that represents the deposits from the cataclysmic Bronze Age eruption. Ash from the eruption spread over the entire eastern Mediterranean and also fell on North Africa and across much of the Middle East. The most violent eruptions are thought to have occurred when the calderas collapsed and sea water rushed into the crater, forming a tremendous steam eruption and tsunami. The tsunami moved quickly across the Mediterranean, devastating coastal communities in Crete, Greece, Turkey, North Africa, and Israel. The tsunami was so powerful that it caused the Nile to run upstream for hundreds of miles (several hundred km).

Detailed reconstructions of the eruption sequence reveal four main phases. The first was a massive eruption of ash and pumice ejected high into the atmosphere, collapsing back on Thera and covering nearby oceans with 20 feet of pyroclastic deposits. This phase was probably a Plinian eruption column, and its devastating effects on Thera made the island uninhabitable. Approximately 20 years passed before activity continued, and some settlers tried to reinhabit the island. Huge fissures in the volcano began to open during the

Before the cataclysmic eruption, the Santorini islands existed as one giant volcano known to the Greeks as Stronghyle (Strongulè), or the round one,

second phase, and seawater entered these, initiating large steam eruptions and mudflows and leaving deposits up to 65 feet (20 m) thick. The third phase was the most cataclysmic and followed quickly after the second phase, as seawater began to enter deep into the magma chamber, initiating huge blasts that were heard across southern Europe, northern Africa, and the Middle East. Sonic blasts and pressure waves would have been felt for thousands of miles around. Huge amounts of ash and aerosols were ejected into the atmosphere, probably causing several days of virtual darkness over the eastern Mediterranean. The fourth phase of the eruption was marked by continued production of pyroclastic flows, settling many layers of ash, pumice, and other pyroclastic deposits around the island and nearby Aegean. Most estimates of the amount of material ejected during the eruption fall around 20 cubic miles (80 km^3), although some estimates are twice that amount. Ash layers from the eruption of Thera have been found in Egypt, Turkey, other Greek islands, and the Middle East.

Thera undoubtedly caused global atmospheric changes after ejecting so much material into the upper atmosphere. Data from Greenland ice cores indicate that a major volcanic eruption lowered Northern Hemisphere temperatures by ejecting aerosols and sulfuric acid droplets into the atmosphere in 1645 B.C.E. Additional evidence of an atmospheric cooling event caused by the eruption of Thera comes from tree-ring data from ancient bristle-cone pines in California, some of the oldest living plants on Earth. These trees, and other buried tree limbs from Ireland, indicate a pronounced cooling period from 1630 to 20 B.C.E. European and Turkish tree-ring data have shown cooling between 1637 and 28 B.C.E. Chinese records show that at this time there were unusual acidic fogs (probably sulfuric acid) and cold summers, followed by a period of drought and famine. The eruption of Thera therefore not only destroyed the Minoan civilization, but also changed atmospheric conditions globally, forming frosts in California and killing tea crops in China.

The eruption of Thera seems to coincide with the fall of Minoan civilization, certainly in the Santorini archipelago, and also on Crete and throughout the eastern Mediterranean. The cause of the collapse of Minoan society was probably multifold, including earthquakes that preceded the eruption, ash falls, and 30-foot (9-m) high tsunamis that swept the eastern Mediterranean from the eruption. Since the Minoans were sea merchants, tsunamis would have devastated their fleet, harbor facilities, and coastal towns, causing such widespread destruction that the entire structure of the society fell apart. Even vessels at sea would have been battered by the atmospheric pressure waves, covered in ash and pumice, and

stranded in floating pumice far from coastal ports. Crops were covered with ash, and palaces and homes were destroyed by earthquakes. The ash was acidic, so crops would have been ruined for years, leading to widespread famine and disease, driving people to seek relief by leaving Crete and the homeland of the Minoan culture. Many of the survivors are thought to have migrated to Greece and North Africa, including the Nile delta region, Tunisia, and to Levant, where the fleeing Minoans became known as Philistines.

Vesuvius, 79 C.E.

The most famous volcanic eruption of all time is probably that of Vesuvius in the year 79 C.E. Mount Vesuvius is the only active volcano on the European mainland, towering 4,195 feet (1,279 m) above the densely populated areas of Naples, Herculaneum, and surrounding communities in southern Italy. Vesuvius is an arc volcano related to the subduction of oceanic crust to the east of Italy beneath the Italian peninsula, and the volcano rises from the plain of Campania between the Apennine Mountains to the east and the Tyrrhenian Sea to the west. The volcano developed inside the collapsed caldera of an older volcano known as Monte Somma, only a small part of which remains along the northern rim of Vesuvius. Monte Somma and Vesuvius have had at least five major eruptions in the past 4,000 years, including 1550 and 217 B.C.E., and 79, 472, and 1631 C.E. There have been at least 50 minor eruptions of Vesuvius since 79 C.E.

Ash from the 79 C.E. eruption buried the towns of Pompeii, Herculaneum, and Stabiae in present-day Italy and killed tens of thousands of people. Before the eruption Pompeii was a well-known center of commerce, home to approximately 20,000. The area was known especially for its wines, cabbage, and fish sauce, and was also a popular resort area for wealthy Roman citizens. Pompeii became wealthy and homes were elaborately decorated with statues, patterned tile floors, and frescoes. The city built a forum, several theaters, and a huge amphitheater in which 20,000 spectators could watch gladiators fight each other or animals, with the loser usually being killed.

A large caldera north of Vesuvius presently has molten magma moving beneath the surface, causing a variety of volcanic phenomena that have inspired many legends. The Phlegraean Fields is the name given to a region where there are many steaming fumaroles spewing sulfurous gases and boiling mudpots, which may have inspired the Roman poet Virgil's description of the entrance to the underworld. The land surface in the caldera rises and falls with the movement of magma below the surface. This is most evident near the sea, where shorelines have moved up and down relative to coastal structures. At Pozzuoli the ancient Temple of Serapis lies partly submerged near the coast. The marble pillars of the temple show evidence of having been previously submerged, as they are partly bored through by marine organisms, leaving visible holes in the pillars. Charles Lyell, in his famous treatise *Principles of Geology,* used this observation to demonstrate that land can subside and be uplifted relative to sea level. A few miles north of Pozzuoli the ancient town of Port Julius is completely submerged beneath the sea, showing that subsidence has been ongoing for thousands of years.

Before the cataclysmic eruption in 79, the Campanians of southern Italy had forgotten that Vesuvius was a volcano and did not perceive any threat from the mountain, even though 50 years earlier the geographer Strabo had described many volcanic features of the mountain. The crater lake at the top of the mountain was used by the rebelling gladiator Sparticus and his cohorts to hide from the Roman army in 72 C.E., neither side of whom was aware of the danger of the crater. There were other signs that the volcano was returning to life. In 62 C.E. a powerful earthquake shook the region, damaging many structures and causing the water reservoir for Pompeii to collapse, flooding the town and killing and injuring many. It is likely that poisonous volcanic gases also escaped through newly opened fractures at this time, as Roman historians and philosophers write of a flock of hundreds of sheep dying mysteriously near Pompeii by a pestilence from within the Earth. The descriptions are reminiscent of the poison carbon dioxide gases emitted from some volcanoes, such as the disasters of Lake Nyos in Cameroon in 1984 and 1986. More earthquakes followed, including one in 63 C.E. that violently shook the theater while the emperor Nero was singing to a captive audience. Instead of evacuating the theater, it was said that Nero was convinced his voice was ever stronger and perhaps excited the gods of the underworld.

On August 24, 79 C.E., Vesuvius erupted after several years of earthquakes. The initial blast launched two and a half cubic miles (10 km³) of pumice, ash, and other volcanic material into the air, forming a mushroom cloud that expanded in the stratosphere. This first phase of the eruption lasted 12 hours, during which time Pompeii's terrified residents were pelted with blocks of pumice and a rain of volcanic ash that was falling on the city at a rate of 7–8 inches (15–20 cm) per hour. People were hiding in buildings and fleeing through the artificially darkened streets, many collapsing and dying from asphyxiation. Soon the weight of the ash caused roofs to collapse on structures throughout the city, killing thousands. The ash quickly buried Pompeii under 10 feet (3 m) of volcanic debris. About half a day after the eruption began, the eruption column began to collapse from decreasing pressure from the magma chamber, and the eruption entered a new phase. At this stage pyroclastic flows known as nuées ardentes began flowing down the sides of the volcano. These flows consisted of a mixture of hot gases, volcanic ash, pumice, and other particles; they raced downhill at hundreds of miles (km) per hour while maintaining temperatures of 1,800°F (1,000°C) or more. These hot pyroclastic flows ripped up and ignited anything in their path, and Herculaneum was first in that path.

Successive pyroclastic flows together killed about 4,000 people in Herculaneum and Pompeii, and neighboring towns suffered similar fates after the initial blast. During the second phase of the massive eruption huge quantities of ash were blown up to 20 miles (32 km) into the atmosphere, alternately surging upward and dropping tons of ash onto the surrounding region and killing most of the people who were not killed in the initial eruption. Daylight was quickly turned into a dark, impenetrable night, and the town of Pompeii was buried under another 6 to 7 feet (~2 m) of ash. Thick ash also accumulated on the slopes of the volcano and was quickly saturated with water from rains created by the volcanic eruption. Water-saturated mudflows called lahars moved swiftly down the slopes of Vesuvius, burying the town of Herculaneum under additional volcanic layers up to 65 feet (20 m) thick, and covering Pompeii by up to 20 feet (5 m) of mud. The towns of Herculaneum and Pompeii were not uncovered until archaeologists discovered their ruins nearly 2,000 years later when, in 1699 the Italian scientist Giuseppe Marcrini dug into an elevated mound and discovered parts of the buried city of Pompeii. The public was not interested in the early 1700s, however, and the region's history remained obscure. For the next century wealthy landowners discovered that if they dug tunnels through the solidified ash, they could fund ancient statues, bronze pieces, and other valuables that they used to decorate their homes. It was not until Italy became unified in 1860 and the archaeologist Giuseppe Fiorelli was put in charge of excavations in southern Italy that looting changed to systematic excavation and study.

This area has since been rebuilt, with farmland covering much of Pompeii, and the town of Ercolano now lies on top of the 20 feet (6 m) of ash that buried Herculaneum. Archaeological investigation at both sites, however, has led to a wealth of information about life in ancient Italy, and Pompeii in particular has proven to be a valuable time capsule preserved in pristine condition by the encapsulating volcanic ash that entrapped so many people in their homes or trying to escape into the streets or elsewhere. Vesuvius is still active, and has experienced many eruptions since the famous eruption in 79 C.E.

The 72 C.E. eruption of Vesuvius is the source of some terms commonly used to describe features of volcanic eruptions. Pliny the Elder, the Roman naturalist and naval officer in charge of a squadron of vessels in the Bay of Naples during the eruption, commanded one of his vessels to move toward the mountain during the initial eruption for a better view and to rescue a friend. Both efforts failed, as the eruption was too intense to approach, so he sailed to Stabiae to the south to attempt to save another friend. Pliny the Elder died there, apparently from a heart attack brought on by struggling through the thick ash in the city. In an account of the death of his uncle and of the eruption given to the Roman historian Tacitus, Pliny the Younger (the Elder's nephew) described the mushroom cloud associated with the initial eruption as being like an umbrella where the lower column rose up in a thin pipe then expanded outward in all directions at the top. The common term *Plinian eruption column* is taken from these descriptions.

Tambora, Indonesia, 1815

The largest volcanic eruption ever recorded is that of the Indonesian island arc volcano Tambora in 1815. This eruption initially killed an estimated 92,000 people and sent so much particulate matter into the atmosphere that it influenced the climate of the planet, cooling the surface and changing patterns of rainfall globally. The year after the eruption is known as "the year without a summer" in reference to the global cooling caused by the eruption, although people at the time did not know the reason for the cooling. In cooler climates the "year without

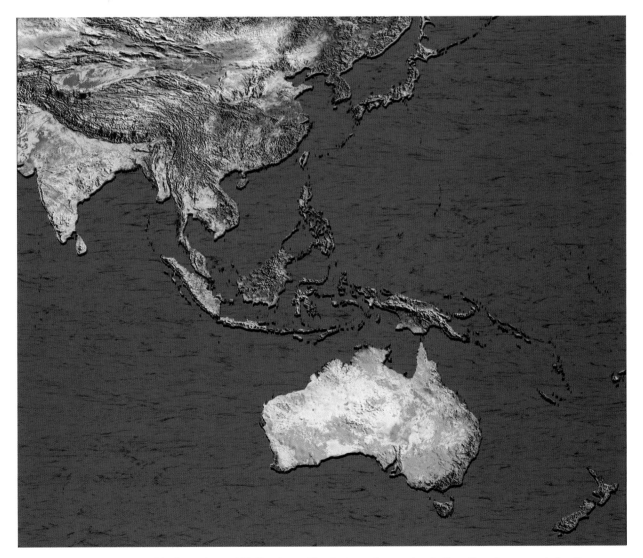

Colored satellite image of Indonesia and Philippines. The Indonesian region is the site of many subduction zones and island arc systems and has experienced some of the most dramatic and catastrophic volcanic eruptions in history. *(Dynamic Earth Imaging/Photo Researchers, Inc.)*

a summer" saw snow throughout the summer and crops could not grow. Great masses of U.S. farmers moved from New England to the Midwest and Central Plains seeking a better climate for growing crops. This mass migration and population of the American Midwest was all because of a volcanic eruption on the other side of the world.

Tambora is located in the Indonesian region, a chain of thousands of islands that stretch from Southeast Asia to Australia. The tectonic origins of these islands are complex and varied, but many of the islands along the southwest part of the chain are volcanic, formed above the Sumatra-Sunda trench system. Fertile soils host tropical rain forests, many of which have been deforested and replaced by tobacco, tea, coffee and spice plantations (hence their nickname, "spice islands"). This trench marks the edge of subduction of the Australian plate beneath the Philippine-Eurasian plates, and formed a chain of convergent-margin island arc volcanoes above the subduction zone. Tambora is one of these volcanoes, located on the island of Sumbawa, east of Java. Tambora is unique among the volcanoes of the Indonesian chain as it is located farther from the trench (210 miles; 340 km) and farther above the subduction zone (110 miles; 175 km) than other volcanoes in the chain. This is related to the fact that Tambora is located at the junction of subducting continental crust from the Australian plate and subducting oceanic crust from the Indian plate. A major fault cutting across the convergent boundary is related to this transition, and the magmas that feed Tambora seem to have risen through fractures along this fault.

Tambora has a history of volcanic eruptions extending back at least 50,000 years. The age difference between successive volcanic layers is large, and there appears to have been as many as 5,000 years between individual large eruptions. This is a large time interval for most volcanoes and may be related to Tambora's unusual tectonic setting far from the trench along a fault zone related to differences between the types of material being subducted on either side of the fault.

In 1812 Tambora started reawakening with a series of earthquakes plus small steam and ash eruptions. Inhabitants of the region did not pay much attention to these warnings, having not remembered the ancient eruptions of 5,000 years past. On April 5, 1815, Tambora erupted with an explosion heard 800 miles (1,300 km) away in Jakarta. Ash probably reached more than 15 miles (25 km) into the atmosphere, and this was only the beginning of what was to be one of history's greatest eruptions. Five days after the initial blast a series of huge explosions rocked the island, sending ash and pumice 25 miles (40 km) into the atmosphere, and sending hot pyroclastic flows (nuées ardentes) tumbling down the flanks of the volcano and into the sea. When the hot flows entered the cold water, steam eruptions sent additional material into the atmosphere, creating a scene of massive explosive volcanism and wreaking havoc on the surrounding land and marine ecosystems. More than 36 cubic miles (150 km³) of material were erupted during these explosions from Tambora, more than 100 times the volume of the Mount St. Helens eruption in Washington State of 1980.

Ash and other volcanic particles such as pumice from the April eruptions of Tambora covered huge areas that stretched many hundreds of miles across Indonesia. Towns located within a few tens of miles experienced strong, hurricane-force winds that carried rock fragments and ash, burying much in their path and causing widespread death and destruction. The ash caused a nightlike darkness that lasted for days even in locations 40 miles (65 km) from the eruption center, so dense was the ash. Roofs collapsed from the weight of the ash, and 15-foot (4.5-m) tall tsunamis were formed when the pyroclastic flows entered the sea. These tsunamis swept far inland in low-lying areas, killing and sweeping away many people and livestock. A solid layer of ash, lumber, and bodies formed on the sea extending several miles west from the island of Sumbawa, and pieces of this floating mass drifted off across the Java Sea. Although it is difficult to estimate, at least 92,000 people were killed in this eruption. Crops were incinerated or poisoned, and irrigation systems destroyed, causing additional famine and disease after the eruption ceased, killing tens of thousands of people who had survived the initial eruption, and forcing hundreds of thousands to migrate to neighboring islands.

The atmospheric effects of the eruption of Tambora were profound. During the eruptions Tambora shot huge amounts of sulfur dioxide and steam into the atmosphere and contaminated surface and groundwater systems. Much of the sulfur dioxide rained onto nearby lands and islands, causing illness including persistent diarrhea. Huge amounts of gas also entered the upper atmosphere, causing changes in weather patterns throughout the world. In the upper atmosphere sulfur dioxide combines with water molecules to form persistent sulfuric acid aerosols, which reflect large amounts of sunlight to space and cause a global cooling effect. Weather data show that the Northern Hemisphere experienced temperatures as many as 10 degrees Celsius cooler than normal for three years following the eruption, and much of that cooling is attributed to the aerosols in the upper atmosphere. The Indian Ocean monsoon was disrupted to such an extent that some regions expecting rain were plagued with drought instead, then when the rains were supposed to end, they finally came—but too late for the crops. Widespread famine, then an epidemic of cholera, engulfed north India and surrounding areas (present-day Pakistan and Bangladesh). In eastern China the Yangtze and Yellow Rivers experienced severe floods, destroying crops and killing many people. The cholera plague that started in India soon spread to Egypt, killing 12 percent of the population, then to Europe, where hundreds of thousands perished. The global outbreak soon spread to North America, hitting the immigrant cities of New York and Montreal with particular severity, where hundreds of people died each day. The cholera plague, lasting from 1817 through 1823, is one of the long-term secondary hazards of volcanic eruptions. The uncountable deaths from the secondary effects of the eruption thus far outnumber the deaths from the initial eruption.

The year 1816 is known as the "year without a summer," because of the atmospheric cooling from the sulfur dioxide released from Tambora. Snow fell in many areas across Europe and in some places was colored yellow and red from the volcanic particles in the atmosphere. Crops failed, people suffered, and

social and economic unrest resulted from the poor weather. Then the Napoleonic Wars soon erupted. Famine swept Europe, hitting France especially hard, with food and antitax riots breaking out in many places. The number of deaths from the famine in Europe is estimated at another 100,000.

North America also suffered from the global cooling from Tambora. The New England states were hardest hit, experiencing a cold drought, with deep frosts and snow storms even in the typically summer months of June, July, and August. Crops failed, livestock had nothing to eat and perished, and many farmers resorted to fishing in streams to get food for their animals. Others were unable to cope with loss of crops and rising food prices, and many became poor and dependent on charity. Others migrated to the fertile warmer plains of the American Midwest, expanding the country's wheat belt westward.

The eruption of Tambora illustrates the difficulty in estimating the numbers of deaths and the actual cost to the global population of a volcanic eruption. It is hard enough to determine the deaths in remote areas that have been affected by a catastrophe of this magnitude, but when secondary effects such as dis-

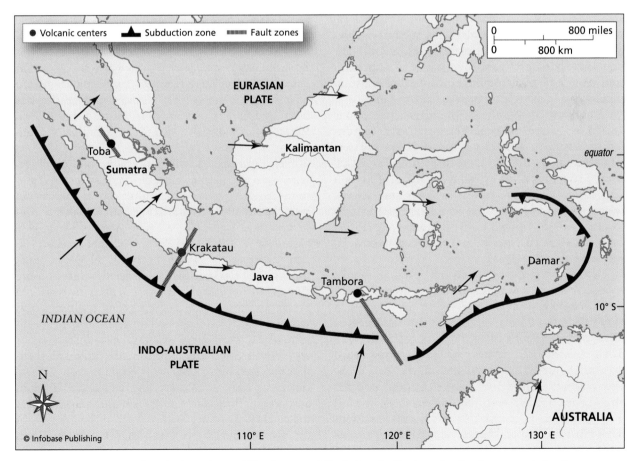

Map of Indonesia showing the locations of Tambora and Krakatau, both along faults above a subduction zone

ease, global epidemics, loss of crops, and changes of climate are taken into account, the numbers of those affected climbs by hundreds of thousands, millions, or more.

Krakatau, Indonesia, 1883

Indonesia has seen other catastrophic volcanic eruptions in addition to Tambora. The island nation of Indonesia has more volcanoes than any other country in the world, with more than 130 known active volcanoes in this island-continental margin-arc system. These volcanoes have been responsible for about one-third of all the deaths attributed to volcanic eruptions in the world. Indonesia stretches for more than 3,000 miles (5,000 km) between Southeast Asia and Australia and is characterized by fertile soils and warm climates. It is one of the most densely populated places on Earth. Its main islands include, from Northwest to Southeast, Sumatra, Java, Kalimantan (formerly Borneo), Sulawesi (formerly Celebes), and the Sunda Islands. The country averages one volcanic eruption per month; because of the dense population, Indonesia suffers from approximately one-third of the world's fatalities from volcanic eruptions.

One of the most spectacular and devastating eruptions of all time was that of 1883 from Krakatau, an uninhabited island in the Sunda Strait off the coast of the islands of Java and Sumatra. This eruption generated a sonic blast heard thousands of miles away, spewed enormous quantities of ash into the atmosphere, and initiated a huge tsunami that killed roughly 40,000 people and wiped out more than 160 towns. The main eruption lasted for three days, and the huge amounts of ash ejected into the atmosphere circled the globe, remained in the atmosphere for more than three years, forming spectacular sunsets and affecting global climate. Locally the ash covered nearby islands, killing crops, natural jungle vegetation, and wildlife, but most natural species returned within a few years.

Like Tambora, Krakatau is located at an anomalous location in the Indonesian arc. To the southeast of Krakatau, the volcanoes on the island of Java are aligned in an east-west direction, lying above the subducting Indo-Australian plate. To the northwest of Krakatau volcanoes on Sumatra are aligned in a northwest-southeast direction. Krakatau is thus located at a major bend in the Indonesian arc and lies along with a few other smaller volcanoes above the Krakatau fault zone that strikes through the Sunda Strait. This fault zone is accommodating differential motion between Java and Sumatra. Java is moving east at 1.5 inches (4 cm per year), whereas Sumatra is moving northeast at 1.5 inches (4 cm per year) and rotating in a clockwise sense, resulting in a zone of oblique extension along the Krakatau fault

zone in the Sunda Strait. The faults and fractures formed from this differential motion between the islands provide easy pathways for the magma and other fluids to migrate from great depths above the subduction zone to the surface. So like Tambora, Krakatau has had unusually large volcanic eruptions and is located at an anomalous structural setting in the Indonesian arc.

Legends in the Indonesian islands speak of several huge eruptions from the Sunda Strait area, and geological investigations confirm many deposits and calderas from ancient events. Before the 1883 eruption Krakatau consisted of several different islands including Perbuwatan in the north, and Danan and Rakata in the south. The 1883 eruption emptied a large underground magma chamber and formed a large caldera complex. During the 1883 eruption Perbuwatan, Danan, and half of Rakata collapsed into the caldera and sank below sea level. Since then a resurgent dome has grown out of the caldera, emerging above sea level as a new island in 1927. The new island, named Anak Krakatau (child of Krakatau), is growing to repeat the cycle of cataclysmic eruptions in the Sunda Strait.

Before the 1883 eruption the Sunda Strait was densely populated with many small villages built from bamboo with palm-thatched roofs, and other local materials. Krakatau is located in the middle of the strait, with many arms of the strait extending radially away into the islands of Sumatra and Java. Many villages, such as Telok Betong, lay at the ends of these progressively narrowing bays, pointed directly at Krakatau. These villages were popular with trading ships from the Indian Ocean which stopped to obtain supplies before heading through the Sunda Strait to the East Indies. The group of islands centered on Krakatau in the middle of the strait was a familiar landmark for these sailors.

Although not widely appreciated as such at the time, the first signs that Krakatau was not a dormant volcano but was about to become very active appeared in 1860 and 1861 with small eruptions, then a series of earthquakes between 1877 and 1880. On May 20, 1883, Krakatau entered a violent eruption phase, witnessed by ships sailing through the Sunda Strait. The initial eruption sent a seven-mile (11-km) high plume above the strait, with the eruption heard 100 miles (160 km) away in Jakarta. As the eruption expanded, ash covered villages in a 40-mile (60-km) radius. For several months the volcano continued to erupt sporadically, covering the straits and surrounding villages with ash and pumice, while the earthquakes continued.

On August 26, the form of the eruptions took a severe turn for the worse. A series of extremely explosive eruptions sent an ash column 15 miles (25 km) into the atmosphere, sending many pyroclastic flows and nuées ardentes spilling down the island slopes and into the sea. Tsunamis associated with the flows and earthquakes sent waves into the coastal areas surrounding the Sunda Strait, destroying or damaging many villages on Sumatra and Java. Ships passing through the straits were covered with ash, while others were washed ashore and shipwrecked by the many and increasingly large tsunamis.

On August 27, Krakatau put on its final show, exploding with a massive eruption that pulverized the island and sent an eruption column 25 miles (40 km) into the atmosphere. The blasts from the eruption were heard as far away as Australia, the Philippines, and Sri Lanka. Atmospheric pressure waves broke windows on surrounding islands and traveled around the world as many as seven times, reaching the antipode (area on the exactly opposite side of the Earth from the eruption) at Bogatá, Colombia, 19 hours after the eruption. The amount of lava and debris erupted is estimated at 18–20 cubic miles (75–80 km³), making this one of the largest eruptions known in the past several centuries. Many sections of the volcano collapsed into the sea, forming steep-walled escarpments cutting through the volcanic core, some of which are preserved to this day. These massive landslides were related to the collapse of the caldera beneath Krakatau and contributed to huge tsunamis that ravaged the shores of the Sunda Strait, with average heights of 50 feet (15 m), but reaching up to 140 feet (40 m) where the V-shaped bays amplified wave height. Many of the small villages were swept away with no trace, boats were swept miles inland or ripped from their moorings, and thousands of residents perished in isolated villages in the Sunda Strait.

Although it is uncertain how many people died in the volcanic eruption and associated tsunamis, the Dutch colonial government estimated in 1883 that 36,417 people died, with most of these deaths (perhaps 90 percent) from the tsunami. Several thousand people were also killed by extremely powerful nuées ardentes, or glowing clouds of hot ash that raced across the Sunda Strait on cushions of hot air and steam. These clouds burned and suffocated all who were unfortunate enough to be in their direct paths.

Tsunamis from the eruption spread out across the Indian Ocean and caused destruction along much of the coastal regions of the entire Indian Ocean, eventually moving around the world. Although documentation of this Indonesian tsunami is not nearly as good as that from the 2004 tsunami, many reports of the tsunami generated from Krakatau document this event. Residents of coastal India reported the sea suddenly receding to unprecedented levels, stranding fish that were quickly picked up by residents,

many of whom were then washed away by large waves. The waves spread into the Atlantic Ocean and were detected in France, and a seven-foot (2-m) high tsunami beached fishing vessels in Auckland, New Zealand.

Weeks after the eruption huge floating piles of debris and bodies were still floating in the Sunda Strait, Java Strait, and Indian Ocean, providing grim reminders of the disaster to sailors in the area. Some areas were so densely packed with debris that sailors reported those regions looked like solid ground, and one could walk across the surface. Fields of pumice from Krakatau reportedly washed up on the shores of Africa a year after the eruption, some mixed with human skeletal remains. Other pumice rafts carried live plant seeds and species to distant shores, introducing exotic species across oceans that normally acted as barriers to plant migration.

Ash from the eruption fell more than 1,500 miles (2,500 km) from the eruption for days after it, and many fine particles remained in the atmosphere for years, spreading across the globe by atmospheric currents. The ash and sulfur dioxide from the eruption caused a lowering of global temperatures by several degrees and created spectacular sunsets and atmospheric light phenomena by reflected and refracted sunlight through the particles and gas emitted into the atmosphere.

On western Java, one of the most densely populated regions in the world, destruction on the Ujong Kulon Peninsula was so intense that it was designated a national park, as a reminder of the power and continued potential for destruction from Krakatau. Such designations of hazardous coastal regions and other areas of potential destruction as national parks and monuments is good practice for decreasing the severity of future natural eruptions and processes.

Krakatau began rebuilding new cinder cones that emerged from beneath the waves in 1927 through 1929, when the new island, Anak Krakatau (child of Krakatau), went into a rapid growth phase. Several cinder cones have now risen to heights approaching 600 feet (190 m) above sea level. The cinder cones will undoubtedly continue to grow until Krakatau's next catastrophic caldera collapse eruption.

Mount Pelée, Martinique, 1902

Martinique was a quiet, West Indian island first discovered by Europeans in the person of Christopher Columbus in 1502. The native Carib people were killed off or assimilated into the black slave population brought by the French colonizers to operate the sugar, tobacco, and coffee plantations, and they exported sugar beginning in the mid-1600s. The city of St. Pierre, on the northwest side of the island, became the main seaport, as well as the cultural, educational, and commercial center of Martinique. The city became known as the "Paris of the West Indies," with many rum distilleries, red-roofed white masonry buildings, banks, schools, beautiful beaches, all framed by a picturesque volcano in the background.

The island is part of the Lesser Antilles arc, sitting above a west-dipping subduction zone built on the eastern margin of the Caribbean plate. Oceanic crust of the Atlantic Ocean basin that is part of the North American plate is being pushed beneath the Caribbean plate at about an inch (2 cm) per year. The oldest volcanoes on Martinique, including Morne Jacob, the Pitons du Carbet, and Mount Conil, emerged from the sea about 3–4 million years ago. They were built on a submarine island arc that had been active for approximately the last 16 million years. Pelée is a much younger volcano, first known to have erupted about 200,000 years ago. It rises to a height of 4,580 feet (1,397 m) on the north end of the island and has had major historic eruptions in 65 b.c.e., 280, and 1300, and smaller eruptions about every 50–150 years. Mount Pelée is located uphill and only six miles (10 km) from St. Pierre. It derives its name from the French word for bald (or peeled), after the eruptions of 1792 and 1851 that removed all vegetation from the top of the volcano.

Mount Pelée began to awaken slowly in 1898, when sulfurous gases were noticed coming out of a crater on top of the mountain and in the river Blanche, which flows through a gorge on the southwest flank of the volcano. Minor eruptions of steam became abundant in 1901, and additional gaseous emissions were common as well. In spring 1902 Mount Pelée began to show increased activity, with boiling lakes and intermittent minor pyroclastic flows and eruptions, associated with minor earthquake activity. All these phenomena were connected with the rise of magma beneath the volcano and served as warnings about the upcoming eruption. A dome of magma began growing in one of the craters on top of the volcano, then a huge, 650-foot (200-m) wide tower of solidified magma known as "the spine" formed a plug that rose to 375 feet (115 m) before the catastrophic May 8 eruption.

By April most people were becoming worried about the increasingly intense activity, and they congregated in St. Pierre to catch boats to leave the island. St. Pierre was located only six miles (~10 km) from the volcano. Landslides in the upper river Blanche triggered a series of massive mudflows that rushed down the river, while others crashed down the valley when water in the crater lake in L'Etang Sec, the main volcanic vent on Pelée, broke through a crack in the crater rim and escaped down the flank of the volcano. These landslides and mudflows were probably triggered by minor seismic activity associated with magma rising upward beneath Pelée. On May 3 groundwater began rushing out of fissures that opened in the ground and carried soil, trees, and carcasses of dead animals down through St. Pierre and out to sea. Flooding was widespread and destroyed many farms and villages, such as Le Prêcheur, carrying the wreckage downstream.

On May 4 a fissure opened in the ground in the village of Ajoupa Bouillon northeast of Pelée and caused a huge steam and mud eruption that killed several people. Floods and mudflows continued to flow down other rivers, including those that passed through St. Pierre. Submarine landslides ruptured communication cables and carried them to depths of a half mile (0.7 km). A larger eruption occurred on May 5, killing 40 people in a pyroclastic flow that raced down the river Blanche. Residents desperately wanted to leave the island for safety, but elections were five days away, and Governor Louis Mouttet did not want the inhabitants to leave for fear he might lose the election. He ordered the military to halt the exodus from the island. Governor Mouttet was running as the head of the békés, an ultraconservative white supremacist party opposed by a new mixed-race socialist party that was becoming increasingly powerful. A successful election by the socialists would have changed the balance of power in Martinique and other French colonies, so was resisted by any means (likely including election fraud) by the ruling békés. By this stage precursors to the eruption included strange behavior by animals and insects. A plague of armies of ants, venomous centipedes, poisonous snakes including pit vipers, and mammals began migrating in mass down the flanks of the volcano and invaded villages, plantations, and St. Pierre to the horror of residents. These insects and animals attacked people in factories, plantations, and homes, injuring and killing many. Poisonous gases killed birds in flight, which dropped on towns like ominous warnings from the sky. Mudflows continued down the volcano flanks and now included boiling lahars, one of which tore down the river Blanche, killing

(opposite page) (A) Map of the Caribbean and the Lesser Antilles arc, showing the location of Martinique and Mt. Pelée along a fault above a subduction zone; (B) map of the area around St. Pierre on Martinique devastated by the 1902 eruption; (C) map of the pyroclastic flows from the 1902 eruptions, in relation to the dome of Mount Pelée, and cities and settlements including St. Pierre, Carbet, Le Prêcheur

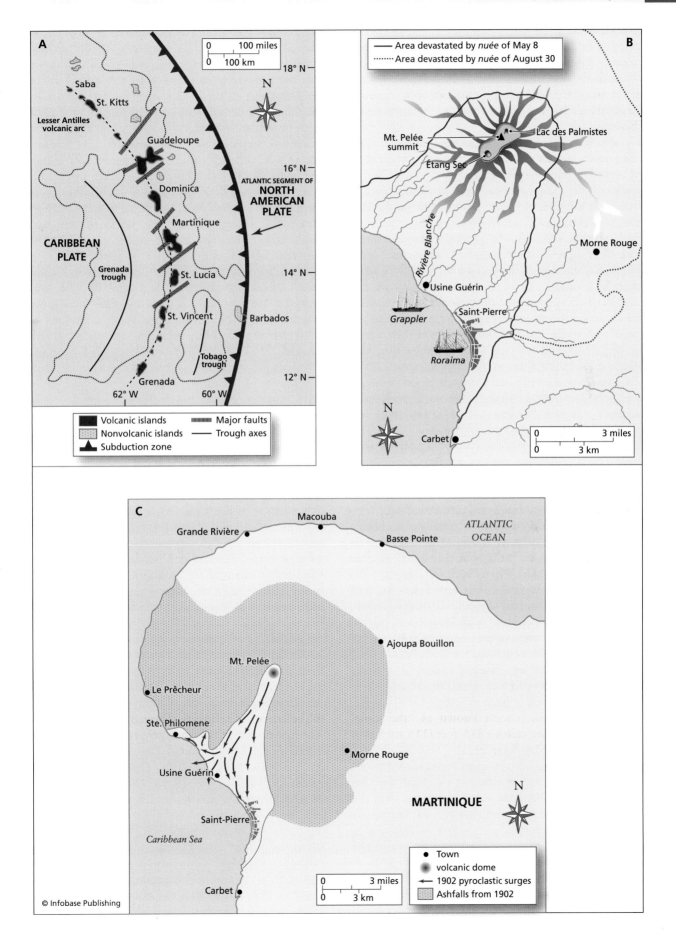

A

0   100 miles
0   100 km

18° N

Saba

St. Kitts

Lesser Antilles
volcanic arc

Guadeloupe

16° N

ATLANTIC SEGMENT OF
**NORTH
AMERICAN
PLATE**

Dominica

Martinique

**CARIBBEAN
PLATE**

Grenada
trough

14° N

St. Lucia

St. Vincent

Barbados

Tobago
trough

12° N

Grenada

62° W   60° W

Volcanic islands   Major faults
Nonvolcanic islands   Trough axes
Subduction zone

B

——— Area devastated by *nuée* of May 8
········ Area devastated by *nuée* of August 30

Mt. Pelée
summit   Lac des Palmistes

Étang Sec

Morne Rouge

Rivière Blanche

Usine Guérin

Saint-Pierre

Grappler

Roraima

N

Carbet

0   3 miles
0   3 km

C

Macouba

Grande Rivière   *ATLANTIC
OCEAN*

Basse Pointe

Ajoupa Bouillon

Mt. Pelée

Le Prêcheur

Ste. Philomene

Morne Rouge

Usine Guérin

MARTINIQUE

N

Saint-Pierre

Caribbean Sea

Town
volcanic dome
1902 pyroclastic surges
Ashfalls from 1902

Carbet

0   3 miles
0   3 km

© Infobase Publishing

many people in factories and homes in the boiling mixture of mud, ash, and water. Others in St. Pierre were dying rapidly of a contagious plague caused by drinking water contaminated by ash and sewage. Workers on the docks went on strike, and there was little chance for anybody to escape the condemned city.

On May 6, residents of the city, now in a state of turmoil, woke to a new layer of ash covering city streets and fields. This ash was not from Pelée but from another massive eruption that had occurred overnight 80 miles (130 km) to the south on the British colony island of St. Vincent, where the volcano La Soufrière, killed 1,650 residents of that island. The Soufrière eruption triggered submarine slides that broke all communication lines to and from Martinique, totally isolating that island from the rest of the world during the disaster that was about to occur. Ironically, communication lines to St. Vincent remained intact and surviving residents there could dispatch cables to England for help. Ships were dispatched but when they arrived they assumed all the ash and debris in the water was from St. Vincent, and help was not sent to St. Pierre until much later, when it was realized that two major eruptions had occurred from two different volcanoes.

On May 7 the volcano was thundering with loud explosions heard throughout the Lesser Antilles. Smoke, fires, and flashes of lightning made for a frightening spectacle, especially at night when the fires reflected off the low-lying clouds and ash. Volcanic bombs were being spewed from Pelée and landing on homes and fields on the outskirts of St. Pierre, starting many fires. On May 7 Governor Mouttet and his wife visited St. Pierre from the capital, Fort-de-France, to try to convince residents that it was safe to remain and to vote in the elections. However, at 7:50 A.M. on May 8 several huge sonic blasts rocked the island and a tall eruption column rose quickly from the top of the volcano. A few minutes later a huge nuée ardente or glowing avalanche erupted from Mount Pelée at 2,200°F (900°C), and moved over the six miles (~10 km) to St. Pierre at 115 miles (185 km) per hour (some estimates are as high as 310 miles [500 km] per hour), reaching the city 10 minutes after the eruption began. This began as a lateral blast or a collapse of the eruption column, aimed directly at St. Pierre below. As the pyroclastic flow rushed down the mountain, it expanded and scorched forests, sugarcane fields, and villages on the flanks of the volcano. The temperature of the flow cooled to an estimated 410°–780°F (200°–400°C) by the time it reached the sea, but this was enough to burn almost anything in its path. When the hot ash cloud reached St. Pierre, it demolished buildings and ignited flash fires throughout the city. Thousands of

casks of rum stored in the city ignited and flowed as rivers of fire to the sea, burning at temperatures high enough to melt glass. The eruption cloud continued past the city and screamed over the harbor on a cushion of hot steam, overturning and burning many ships at anchor, sending them to the seafloor and killing most sailors on board.

Many accounts of the eruption say that all but two of the city's 30,000 residents were killed. Although about 30 residents survived the initial eruption, more than 27,000 were killed in the first massive eruption. Most died in a matter of minutes during passage of the ash cloud, but others died slowly from suffocation from ash or volcanic gas. One of the survivors was a prisoner, Auguste Ciparis, who had been jailed for fighting and was spending his jail term in a deep, dungeonlike cell that sheltered him from the hot ash cloud. Ciparis spent the rest of his life paraded around freak shows in the United States by the Barnum and Bailey circus to show his badly burned body as "The Prisoner of St. Pierre." About 2,000 people from surrounding towns were killed in later eruptions. Governor Mouttet and his wife were buried in the flow and were never seen again.

After the catastrophic eruption magma continued to rise through the volcano and formed a new dome that rose above the crater's rim by the end of the month, and the "spine" rose to a height of 1,000 feet (300 m). Another pyroclastic eruption followed the same path on May 20, burning anything that had escaped the first eruption. Small eruptions continued to cover the region with ash until activity subsided in July 1905.

The 1902 eruption of Pelée was the first clearly documented example of a nuée ardente, or hot glowing avalanche cloud. Successive eruptions of Mount Pelée in 1904 provided additional documentation of this kind of eruption and its dangers to all in its path. Even though the amount of magma released in these flows may be relatively small, the destructiveness of these hot, fast-moving flows is dramatic.

Nevada del Ruiz, Colombia, 1985

The most deadly volcanic-induced disaster of modern times occurred in a relatively minor volcanic eruption in the Andes Mountains of South America. The Nevada del Ruiz volcano in Colombia entered an active phase in November 1984 and began to show rhythmically repeating harmonic earthquake tremors on November 10, 1985. At 9:37 P.M. that night a large Plinian eruption sent an ash cloud several miles into the atmosphere, and this ash settled on the ice cap on top of the mountain. This ash together with volcanic steam quickly melted large amounts of the ice, which mixed with the ash and formed giant lahars (mudflows) that swept down the east side of the mountain into the village of Chinchina, killing 1,800. The eruption continued and melted more ice that mixed with more ash, and sent additional, larger lahars westward. Some of these lahars moved nearly 30 miles (50 km) at nearly 30 miles per hour (50 km/hr), and under a thunderous roar buried the town of Armero beneath 26 feet (8 m) of mud. Twenty-two thousand people died in Armero that night. Many could have been saved, since warnings were issued before the mudflow, but the warnings went unheeded.

Nevada del Ruiz had experienced a year of intermittent precursory activity that indicated an eruption might occur, and the volcano was being studied by a group of Colombian geologists at the time of the eruption. At 3:05 P.M. on November 13, 1985, ranchers north of the volcano heard a low rumbling and observed a plume of black ash rise from the volcano and fall on the town of Armero, 45 miles (72 km) away about two hours later. By 4:00 P.M., local civil defense officials warned that an eruption was in progress and recommended that towns including Armero, Honda, and others be ready for immediate evacuation. After several hours of meetings the Red Cross ordered the evacuation of Armero at 7:30 P.M. Residents did not, however, hear the orders, and did not understand the danger moving their way.

At 9:08 P.M. two large explosions marked the start of a larger eruption, associated with a series of pyroclastic flows and surges that moved down the north flank of the volcano. The volcanic deposits moved across the ice cap on the mountain, scouring, melting, and covering it in various places. This released large amounts of melt water mixed with debris that moved down the slopes, quickly forming giant lahars that scoured the channels of the Nereidas, Molinos, Guali, Azufrado, and Lagunillas Rivers and picking up huge amounts of debris including rocks, soil, and vegetation in the process.

At 9:30 P.M. a Plinian eruption column was visible, rising to nearly seven miles (11 km), and it hurled blocks and bombs of andesitic pumice up to a couple of miles from the crater, with ash falling up to 250 miles (400 km) away. At 10:30 P.M., lahars began sweeping through the village of Chinchina, and additional warnings were sent to Armero. Later, survivors reported that electricity was out sporadically and many residents may not have heard the warnings. At 11:30 P.M. giant lahars surged into Armero in successive waves moving at 22–30 miles per hour (35–50 km/hr), sweeping away homes, cars, people, and livestock, and embedding all in 26 feet (8 m) of mud. Many survived the initial inundation but were trapped half-buried in the mud and died later of exposure.

Scientists have learned many lessons from Nevada del Ruiz that could be useful for saving lives in the future. First, even minor volcanic eruptions can trigger catastrophic mudflows under the right conditions, and geologic hazard maps should be made in areas of volcanism to understand the hazards and help emergency planning in times of eruption. Local topographic variations can focus lahars, enhancing their lethality in some places and spreading them out in others. Armero was located at the end of a canyon that focused the worst parts of the flow in the heart of the village. Understanding past hazards can help foreseeing what may happen in the future. If geologists had helped plan the location of Armero, they would have noticed that the town location was on top of an older lahar deposit that swept down the mountain in 1845, also killing all inhabitants more than a century earlier. Apparently the geologic record shows a number of repeated mudflows destroying villages at the site of Armero. A final lesson from Armero is that warning systems need to be in place, and even simple alarm systems can save thousands of lives. If residents of Armero had even an hour's warning, they could have fled to the valley slopes and survived. The mudflows traveled 45 miles (70 km), taking about one and a half hours to get to Armero, so even simple warnings could have saved lives.

Mount St. Helens, 1980, and the Cascades Today

The most significant eruption in the contiguous United States in the past 90 years is that of Mount St. Helens in 1980, a mountain that had lain dormant for 123 years. The volcano is part of the active Cascade volcanic arc, a continental-margin arc built on the western coast of North America above where the minor Juan de Fuca plate is being subducted beneath North America. The arc is relatively small (about 1,200 miles, or 2,000 km, long), and stretches from Lassen Peak in California to Mount Garibaldi in British Columbia. Cascade volcanoes in the United States include Lassen Peak, Mount Shasta, Crater Lake, the Three Sisters, Mounts Jefferson, Hood and Adams, Mount St. Helens, and Mount Rainier. Significant volcanic hazard threats remain from Mount St. Helens, Mount Rainer, and other Cascade volcanoes, especially in the densely populated Seattle area.

Mount St. Helens began to grow above the Cascadia subduction zone about 50,000 years ago, and the volcanic cone that blew up in 1980 formed about 2,500 years ago. Since its birth Mount St. Helens has been one of the most active Cascade volcanoes, erupting on average every 40–140 years. But the largest historical eruption from a Cascade volcano was from the present site of Crater Lake. The volcano Mount Mazama occupied this site about 6,000 years ago but exploded in a cataclysmic eruption that covered much of the Pacific Northwest with volcanic ash. As the crust above the emptied magma chamber collapsed, a giant caldera formed, now occupied by the 2,000-foot (610-m) deep Crater Lake, Oregon. If such a huge volcanic eruption were to occur today in the densely populated Pacific Northwest, the devastating effects would include many thousands of dead. The landscape would be covered with choking ash, rivers would be filled with mudflows, and the global climate would be adversely affected for years.

As the Pacific Northwest became densely settled, the natural areas around Mount St. Helens became popular recreation and tourist sites, attracting many to the beautiful scenery in the area. On March 20, 1980, the mountain rumbled with a magnitude 4.1 earthquake, prompting geologists from the U.S. Geological Survey to install a variety of volcano-monitoring equipment. Automatic cameras, seismographs, tilt meters, gravity meters, and gas collectors were installed to monitor the volcano for any new signs of impending eruption. A variety of precursory warnings of an impending eruption were observed, the most important of which included many swarms of closely spaced small earthquakes known as harmonic tremors and a seismic "humming" of the volcano. The U.S. Geological Survey and Forest Service began to consider volcanic hazards in the area and to propose evacuation routes in case of disaster.

On March 27, 1980, many small eruptions on Mount St. Helens were initiated when magma rose high enough to meet groundwater, which caused steam explosions to reach about two miles (3 km) into the sky. The volcano gradually bulged by about 300 feet (90 m), and harmonic tremors indicated an impending large eruption. By this time several small craters on the volcano's summit had merged to form a single, large crater 1,600 feet (500 m) across and 900 feet (1,600 feet) deep. The U.S. Geological Survey issued eruption warnings, the governor of Washington declared a state of emergency, and the area was evacuated.

The bulge on the volcano continued to grow and the Geological Survey established a volcano-monitoring station six miles (10 km) from the summit, while the state government declared an area around the summit off-limits to all unauthorized personnel.

At 8:32 A.M. on May 18, 1980, the upper 1,313 feet (394 m) of the volcano were blown away in an unusual lateral blast. A magnitude 5.1 earthquake initiated the lateral blast, causing parts of the bulge to collapse and slip away in three separate landslides. The rocks mixed with snow and debris, forming a huge debris avalanche that raced down the mountain at 150 miles per hour (200 km/h) in one of the largest debris avalanches ever recorded. More than a cubic mile (4 km³) of material moved downhill in

the debris avalanche, most of which remained in and around Spirit Lake in layers up to 300 feet (90 m) thick, forming a dam at the outlet of the lake that raised the lake level 200 feet (60 m), doubling its size. The blast from the eruption created a sonic boom heard up to 500 miles (800 km) away in Montana, yet those close to the volcano heard nothing. This is because the sound waves from the eruption initially moved vertically upward and bounced off a warm layer in the atmosphere, returning to spread across the surface at a distance of about 80 miles (130 km) wide around the volcano, leaving a "ring of silence" within the danger zone of the eruption. Some of the sound of the eruption near the blast may also have been absorbed by the huge amounts of ash in the air.

Loss of the weight from the bulge released pressure on the magma inside the volcano, and the side of the mountain exploded outward at 300 miles per hour (500 km/h). The preexisting rock mixed with magma and rose to temperatures of 200°F (93°C). This mass flew into Spirit Lake on the northern flank of the volcano and formed a wave 650 feet (200 m) high that destroyed much of the landscape. Lahars filled the north and south forks of the Toutle River, Pine Creek, and Muddy River. Some of these lahars included masses of ash, mud, and debris, filling much of the Toutle River to a depth of 150 feet (50 m) and locally to as deep as 600 feet (180 m). As the mudflows moved downhill and away from the volcano, they became mixed with more water from the streams and picked up speed as they moved downhill. This prompted a mass evacuation of areas downhill from the volcano, saving countless lives. Vehicles and more than 200 homes were swept away, bridges were knocked out, and the lumber industry was devastated.

A hot pyroclastic flow blasted out of the hole on the side of the volcano and moved at 250 miles per hour (400 km/h), knocking over and burying trees and everything else in its path for hundreds of square miles. The Geological Survey observation post was destroyed and the geologist monitoring the equipment was killed. The forests were a scene of utter devastation after the blast; trees up to 20 feet (6 m) in diameter were snapped at their bases and blown down for miles around, destroying an estimated 3.7 billion board feet (1.3 billion m) of timber. The blast moved so fast that it commonly jumped over ramps in the slope, knocking down trees on the side of a knob facing the volcano, while leaving standing charred remains of the trees that were in the blast shadow hidden behind cliffs on the downwind sides of hills. The wildlife in the forest was virtually wiped out but began to return within a couple of years of the eruption. Remarkably, some of the wildlife that lived underground in burrows or beneath lakes and rocks survived, so some of these resilient beavers, frogs, salamanders, crayfish, as well as some flowers, shrubs, and small trees helped regenerate the ecological systems in the eruption zone.

A huge Plinian ash cloud that erupted to heights of 12 miles (25 km) placed a half billion tons of volcanic ash and debris in the atmosphere. This ash was carried by winds and dropped over much of the western United States. Pyroclastic flows continued to move down the volcano at a rate of 60 miles per hour at temperatures of 550°–700°F (288°–370°C). Sixty-two people died in this relatively minor eruption, and damage to property is estimated at $1 billion. Volcanic ash fell like heavy black rain across much of eastern Washington, Idaho, and Montana. Finer-grained ash that made it into the upper atmosphere managed to circle the globe within 17 days and created spectacular sunsets for months. Even though the amount of ash that fell was relatively minor compared with some other historical eruptions, the ash caused major problems throughout the Pacific Northwest. Planes were grounded, and many electrical transformers short-circuited, while mechanical engines and motor vehicles became inoperable, stranding thousands of people. Breathing was difficult and many had to wear face masks. Rains came and turned the ash layers into a terrible mud that became like concrete, resting heavily on buildings and collapsing roofs.

Smaller eruptions and pyroclastic flows continued to move down the mountain less frequently over the next two years, as a new resurgent dome began to grow in the crater created by the blast that removed the top of the mountain. As the dome grows, residents await the next eruption of a Cascades volcano.

Mount Pinatubo, 1991

The eruption of Mount Pinatubo in the Philippines in 1991 was the second-largest volcanic eruption of the 20th century (after Katmai in Alaska in 1912). It offers many lessons in volcanic prediction, warning, evacuation, and resettlement that may serve as lessons for future eruptions around the world. Pinatubo offers reassuring evidence that careful volcanic monitoring can lead to public warnings and evacuations that can save thousands of lives during catastrophic volcanic eruptions.

Mount Pinatubo is located on the Philippine island of Luzon, about 50 miles (80 km) northwest of Manila and close to what was the U.S. Clark Air Force Base. Before 1991 Mount Pinatubo was known to be a volcano but was not thought to pose much of a threat since it had not erupted for approximately 500 years. About half a million people lived on or near the volcano, including some 16,000 American citizens stationed at Clark Air Force Base. As soon as Pinatubo began to show signs of activity, Philippine and U.S. Geological Survey personnel set up volcano-monitoring stations and equipment, providing detailed and constantly updated evaluations of the status of the volcano and the potential danger levels and likelihood of an impending eruption. Geologists coordinated efforts with local military, civil defense, and disaster-planning officials and quickly worked on determining the volcanic history and mapping areas the most at risk. Critically important was examining the past eruption history to determine how violent past eruptions had been, as an indicator about how bad any impending eruption could be. What the geologists found was frightening, as they determined that Pinatubo had a history of producing tremendous, extremely explosive eruptions, and that such an event might be in the making for the period of activity they were examining. They did not have long to realize their fears.

Precursors to the giant eruption of Pinatubo in June 1991 may have been initiated by a large earthquake on July 16, 1990, centered about 60 miles (100 km) northeast of Pinatubo. This earthquake may have somehow started a series of events that allowed magma to rise beneath the volcano, perhaps opening cracks and fissures beneath the volcano. Soon after this local villagers (known as the Aetas, a seminomadic people who had lived on the volcano for about 400 years, since they fled to the area to hide from conquering Spaniards) reported activity on the mountain, including low rumbling sounds, landslides near the summit, and steam eruptions from fissures. Seismologists measured five small earthquakes around Pinatubo in the next several weeks, but activity seemed to quiet for half a dozen months after the initial swarm of activity.

The first major steam eruption from Pinatubo was observed by villagers, who reported a mile-long (1.6-km) fissure exploding and emitting ash from the north side of the volcano on April 2, 1991. Ash covered surrounding villages, prompting the Philippine Institute of Volcanology and Seismology (PHIVOLCS) and military and civil defense authorities to set up a series of seismographs around the volcano, which recorded more than 200 earthquakes the next day. Authorities declared a volcanic emergency, recommending that villagers within six miles (10 km) of the summit be evacuated. Volcano experts from the U.S. Geological Survey joined the observation and monitoring team on April 23, bringing a plethora of monitoring equipment with them. The local and American teams set up a joint volcano-monitoring Observatory at Clark Air force base, where they also monitored seven seismic stations placed around the volcano.

By the middle of May the Pinatubo Volcano Observatory was recording 30–180 earthquakes a day—at the same time that geologists were scrambling to complete field work to understand the past eruptive behavior of the mountain. The geologists determined that the volcanic basement to the modern peak was about a million years old and that the mountain itself had been built by about six major violent eruptions in the past 35,000 years. They determined that the eruptions were becoming slightly less violent with time but more frequent, with the last major eruption about 500 years ago. The mountain was overdue for a large, violent eruption.

On May 13 volcanologists measured emissions of sulfur dioxide gas, an indicator that molten rock (magma) at depth was rising beneath the volcano. In consultation with the volcano observatory scientists, civil defense authorities issued a series of levels of alerts to warn the public in the event of a catastrophic eruption. These alert levels were updated several times per day, and geologists published new hazard maps showing locations where mudflows, lahars, ash falls, and nuées ardentes were the most likely to occur during an eruption.

Thousands of small earthquakes and greater amounts of sulfur dioxide indicated that magma was still rising beneath the volcano in May, with estimates showing that the magma had risen to within 1.2–4 miles (2–6 km) beneath the surface of the volcano. It was difficult to interpret the warning signs of the impending eruption and to determine whether and when it might occur. The geologists and government officials were torn between ordering immediate evacuations to save perhaps hundreds of thousands of lives, or letting villagers stay until the danger grew more imminent so they could harvest their fields.

On June 1 many of the earthquakes became concentrated in one area beneath a steam vent on the northwest side of the summit, which began to bulge outward. On June 3 the volcano began to spew a series of ash eruptions, which continued to increase such that the volcanic alert level was raised to level three (meaning an eruption was possible within two weeks). About 10,000 villagers were evacuated, and on June 7, the volcano had a minor eruption that sent ash and steam to a height of about five miles (8 km) above the summit. The volcano was bulging more, and there were more than 1,500 earthquakes per day, raising the alert level to four (eruption possible within 24 hours), and the evacuation zone was doubled in distance from the volcano.

June 7 saw a small magma dome oozing out of the volcano about half a mile northwest of its peak, the first sign that magma had reached the surface. Two days later large amounts of sulfur dioxide began

to escape again (after having stopped for several weeks), and small nuées ardentes began to tumble down the slopes of the volcano. The volcanic alert was raised to level five—eruption in progress—and massive evacuations began. The ash eruptions grew larger, and on June 10, all aircraft and 14,500 U.S. personnel from Clark Air Force Base left for safer ground at nearby Subic Bay Naval Base, leaving only 1,500 American personnel and three helicopters on the air force base.

At 8:51 A.M. on June 12 the mountain sent huge columns of hot ash surging to 12 miles (20 km) high in the atmosphere, spawning nuées ardentes that covered some now-evacuated villages. Skies became dark and everything for miles around began to be covered with ash. The evacuation radius was increased to 18 miles (30 km) from the volcano, with the numbers of evacuated now reaching approximately 73,000. Eruptions continued, and on June 13, another huge explosion sent ash past 15 miles (25 km) into the atmosphere. Then the volcano became ominously quiet.

Another large eruption broke the silence at 1:09 P.M., followed quickly by a series of 13 more blasts over the next day. Nuées ardentes roared through several more evacuated villages, burying them in hot ash, while typhoon Yunya pelted the area and added wind and rain to the ash, making a miserable mixture. Ash covered the entire region of a thousand square miles. Ash and pumice began falling heavily on Clark Air Force base, and the remaining staff from the volcano observatory fled to a nearby college.

As the scientists were leaving the observatory at Clark, the eruption style changed dramatically for the worse, rapidly moving into the realm of giant eruptions and becoming the second-largest eruption in the world in the 20th century. The cataclysmic eruption continued for more than 9 hours, and during this time more than 90 percent of the material erupted during the whole cycle was blasted into the air. Huge, billowing Plinian ash columns passed 21 miles (34 km) in height and spread across 250 square miles (1,000 km²). The top of the volcano began to collapse into the empty magma chamber, prompting fears of a truly catastrophic eruption that luckily proved unfounded. The magma was largely drained, and the top of the volcano collapsed into empty space, forming a large caldera whose summit lies 870 feet (265 m) below the former height of 5,724 feet (1,745 m) of the volcano. Ash, nuées ardentes, and lahar deposits hundreds of feet thick filled the valleys draining the flanks of Pinatubo, and thick ash covered buildings across an area of more than 210,000 square miles (340,000 km²). Approximately five to six cubic miles (20–25 km³) of volcanic material was blasted from Pinatubo, along with more than 17 megatons of sulfur dioxide 3–16 megatons of chlorine, and upwards of 420,234 megatons of carbon dioxide.

The devastation was remarkable, with buildings wiped away by nuées ardentes and mudflows, and others collapsed by the weight of wet ash on their roofs. Crops were destroyed and roads and canals impassable. Ash covered much of Luzon and fell across the South China Sea, while much ash remained in the atmosphere for more than a year afterward. Smaller eruptions continued, decreasing in frequency through July to about one per day by the end of August, stopping completely on September 4. A new lava dome rose in the caldera a year later in July 1992, but no large eruption ensued. Only between 200 and 300 people died in the initial eruption, although more were to die later in mudflows and other events, bringing the death toll to 1,202. Most of the initial deaths were people who took shelter in buildings whose roofs collapsed under the weight of the rain-soaked volcanic ash.

The well-documented atmospheric effects of the Pinatubo event clearly show the climatic effects of large volcanic eruptions. The gas cloud from Pinatubo, formed from the combination of ash, water, and sulfur dioxide, was the largest cloud of sulfuric acid aerosol since that produced by Krakatau in 1883. This sulfuric acid aerosol eventually reached the ozone layer, where it destroyed huge quantities of ozone and greatly increased the size of the ozone hole over Antarctica. In only three weeks the sulfuric acid cloud spread around the world between 10°S and 30°N latitudes, dropping global temperatures by up to one-half to one degree C and causing spectacular sunsets. The cloud remained detectable in the atmosphere until the end of 1993. Many unusual weather patterns have been attributed to the global lowering of temperatures by the Pinatubo cloud, including colder, wetter, and stormier winters in many locations.

Most of the evacuated villagers lost livestock, homes, and crops, but they survived because of the well-timed warnings and prompt, responsible evacuations by government officials. Conditions in the refugee camps were not ideal, however, and about 350 additional deaths occurred after the eruption from measles, diarrhea, and respiratory infections. The rainy season was approaching, and many mudflows began sweeping down the flanks of the volcano at 20 miles per hour (30 km/hr), then spread across once lush farmland. Hundreds of mudflows were recorded on the eastern flank of the volcano in the last few (rainy) months of 1993, killing another 100 people in 1993 and continued to do so every year after, though with fewer deaths. Mudflow warning systems and alert levels were set up, saving many additional lives in succeeding years.

Economic losses from the eruption of Pinatubo were tremendous, stunting the Philippine economy. Damage to crops and property amounted to $443 million by 1992, with $100 million more spent on refugees and another $150 million on mudflow controls. Eight thousand homes were destroyed, and 650,000 lost jobs for at least several months. The U.S. air and naval bases at Clark and Subic Bay both closed, causing additional job losses in the region.

Mount Pinatubo provided valuable information to geologists and atmospheric scientists about the amount and types of volcanic gases injected into the atmosphere during volcanic eruptions and the effects of these gases on climate and the environment. Most gases in the atmosphere are volcanic in origin, so volcanoes have had a direct link to climate and human activities over geologic time. The relative importance of the release of volcanic gases by volcanoes versus the emission of greenhouse gases by humans in climate change is currently a hotly debated topic. If volcanoes can produce more aerosols and gases in a few days or weeks than humans produce in years, then volcanic eruptions may drastically change the rates of climate change that are produced by natural cycles and human-related emissions.

Gases released from Mount Pinatubo produced an average global cooling around the planet, yet they were also associated with winter warming over the Northern Hemisphere continents for two years following the eruption. These unexpected effects resulted from complex differences in the way the aerosols were distributed in the stratosphere in different places. Aerosol heating in the lower stratosphere combined with a depletion of ozone to contribute to local warming in the Northern Hemisphere winters. The average atmospheric cooling has also been implicated in some biological responses to the volcanic eruption. Coral reefs are very sensitive to small variations in temperature, and after the eruption of Pinatubo coral reefs in the Red Sea saw a massive die-off that was probably related to the atmospheric cooling. The gases in the atmosphere also caused incoming solar radiation to be more diffuse, which led to greater vegetation growth. These additional plants in turn drew a greater amount of carbon dioxide from the atmosphere, further cooling the planet. It has also been hypothesized that Northern Hemisphere winter warming led to a spike in the number of polar bear cubs born the following spring. Most observations and models for atmospheric evolution following massive eruptions show that the aerosol and ozone levels recover to pre-eruption levels within five to 10 years after the eruption.

See also CONVERGENT PLATE MARGIN PROCESSES; MAGMA; PLATE TECTONICS; TSUNAMI, HISTORICAL ACCOUNTS; VOLCANO.

FURTHER READING

Blong, Russel J. *Volcanic Hazards: A Sourcebook on the Effects of Eruptions.* New York: Academic Press, 1984.

Fisher, R. V. *Out of the Crater: Chronicles of a Volcanologist.* Princeton, N.J.: Princeton University Press, 2000.

Fisher, R. V., G. Heiken, and J. B. Hulen. *Volcanoes: Crucibles of Change.* Princeton, N.J.: Princeton University Press, 1998.

Oregon Space Grant Consortium. "Volcanoworld." Available online. URL: http://volcano.oregonstate.edu/. Accessed October 10, 2008.

Scarpa, Roberto, and Robert I. Tilling. *Monitoring and Mitigation of Volcano Hazards.* New York: Springer, 1996.

Simkin, T., and R. S. Fiske. *Krakatau 1883: The Volcanic Eruption and Its Effects.* Washington, D.C.: Smithsonian Institution Press, 1993.

U.S. Geological Survey. Volcano Hazards Program home page. Available online. URL: http://volcanoes.usgs. gov/. Accessed September 11, 2008; data updated daily or more frequently.

Japan Japan is an island arc and subduction zone trench system that rests as a sliver of the North American plate above the Pacific plate outboard of the Eurasian plate. This tectonic scenario has existed since mid-Tertiary times. From the beginning of the Phanerozoic to the Tertiary, Japan was part of the Eurasian continental margin and was involved in interactions between Eurasia and the Tethys and Panthalassic Oceans. A few occurrences of middle Paleozoic rocks are known from Japan, but the vast majority of strata are younger than middle Paleozoic. Most of the rocks are aligned in strongly deformed structural belts that parallel the coast for 1,850 miles (3,000 km) and include fossiliferous marine strata, weakly to strongly metamorphosed pelitic to psammitic rocks, and granitic intrusions. Since most rocks are strongly deformed in fold-thrust belt structures, one can infer that the more strongly metamorphosed units have been uplifted from deeper in the arc-accretionary wedge system or metamorphosed near the plutons. The complexly deformed zones are overlain by little-deformed Mesozoic-Cenozoic nonmarine to shallow-marine basin deposits. In addition abundant Tertiary volcanic and volcaniclastic deposits are present along the western side of the islands and in the Fossa Magna in central Honshu Island.

In Japan the Sanbagawa belt represents a high-pressure, low-temperature metamorphic belt and the adjacent Ryoke-Abukuma belt represents a high-temperature, low-pressure metamorphic belt. Together these two contrasted metamorphic belts form Japan's paired metamorphic belt. In Japan Akiho Miyashiro (1920–2008) and others deduced in the 1960s that these adjacent belts with contrasted metamorphic histories formed during subduction of the oceanic plates beneath Japan. The low-temperature metamorphic series forms in the trench and immediately above the subduction zone where the cold subducting slab insulates overlying sediments from high mantle temperatures as they are brought down to locally deep high-pressure conditions. These rocks then get accreted to the overriding plate and may

Volcanic vents and sulfur deposits near Kushiro, Hokkaido, Japan, January 1999 *(© Wolfgang Kaehler/CORBIS)*

Mount Fuji and Motosuko Lake in Japan *(Gavin Hellier/Photographers' Choice/Getty Images)*

become exhumed and exposed at the surface in the high-pressure low-temperature belt. Blueschist facies rocks containing the diagnostic mineral galucophane, formed with pressures greater than four kilobars and temperatures between 390°–840°F (200°–450°C), are common.

The adjacent Ryoke-Abukuma belt contains low-pressure high-temperature, as well as medium-pressure, high-temperature metamorphic rocks. These metamorphic rocks form near the axis of the arc in close association with subduction-derived magmas. Since the rocks in this belt form over a considerable crustal thickness and are associated with many high-temperature magmas, they were metamorphosed at a range of pressures and generally high temperatures.

A number of other paired metamorphic belts have been recognized throughout the world. In the western United States the Franciscan complex contains rocks metamorphosed at high pressures and low temperatures, whereas rocks in the Sierra Nevada and Klamath Mountains contain high-temperature, low-pressure metamorphic facies. Other paired metamorphic belts are recognized in Alaska, New Zealand, Indonesia, Chile, Jamaica, and the European Alps.

See also CONVERGENT MARGIN PROCESSES; ISLAND ARCS, HISTORICAL ERUPTIONS; METAMORPHISM AND METAMORPHIC ROCKS.

FURTHER READING

Hashimoto, Mitsuo. *The Geology of Japan.* New York: Springer, 1991.

Miyashiro, Akiho. *Metamorphism and Metamorphic Belts.* London: Allen and Unwin, 1979.

Jupiter The fifth planet from the Sun, Jupiter is the gaseous giant of the solar system, named after the most powerful Roman god of the pantheon. It has more than twice the mass of all the other planets combined, estimated at 1.9×10^{27} kilograms, or 318 Earth masses, and a radius of 44,268 miles (71,400 km), or 11.2 Earth radii. Volumetrically it would take 1,400 Earths to fill the space occupied by Jupiter. Jupiter is the third-brightest object in the night sky, following the Moon and Venus. Four of its many moons are visible from the Earth. Orbiting the Sun at a distance of 483 million miles (778 million km, at its semimajor axis), Jupiter takes 11.9 Earth years to complete each orbit.

Visual observations of the surface of Jupiter indicate that the gaseous surface has a rapid differential rotation rate, with the equatorial zones rotating with a period of nine hours and 50 minutes, and higher latitudes rotating with a period of nine hours and 56 minutes. The interior of the planet is thought to be rotating with a period of nine hours 56 minutes, since the magnetic field rotates at this rapid rate. The rapid rotation has distorted the planet so that the equatorial radius (44,365 miles, or 71,400 km) is 6.5 percent greater than the polar radius (41,500 miles, or 66,800 km).

The outer layer of Jupiter is made of gases, with temperatures at the top of the cloud layers estimated to be -127°F (-88°C; 185 k) and a pressure of 10 bars. This is underlain by a layer of molecular hydrogen extending to 12,425 miles (20,000 km) below the surface. The temperature at this depth is estimated to be 19,340°F (11,000°K), with 4 Megabars pressure. Below this a layer of metallic hydrogen extends to 37,280 miles (60,000 km), with basal temperatures of 44,540°F (25,000°K), and 12 Megabars pressure. An internal rocky core extends another 6,215 miles (10,000 km).

Jupiter's surface and atmosphere is visibly dominated by constantly changing colorful bands extending parallel to the equator, and a great red spot, which is a huge, hurricane-like storm. The bands include yellows, blues, browns, tans, and reds, thought to be caused by chemical compounds at different levels of the atmosphere. The most abundant gas in the atmosphere is molecular hydrogen (86.1 percent), followed by helium (13.8 percent). Other chemical elements such as carbon, nitrogen, and oxygen are chemically mixed with helium. Hydrogen is so abundant on Jupiter because the gravitational attraction of the planet is sufficiently large to retain hydrogen, and most of the planet's original atmosphere has been retained.

Since Jupiter has no solid surface layer, the top of the troposphere is conventionally designated as the surface. A haze layer lies above the troposphere, then grades up into the stratosphere. The colorful bands on Jupiter are thought to reflect views deep into different layers of the atmosphere. Several to tens of miles (or kilometers) of white wispy clouds of ammonia ice overlie a layer of red ammonium hydrosulfide ice, then blue water ice extends to about 62 miles (100 km) below the troposphere. The cloud layers constantly change, reflecting different weather and convective systems in the atmosphere, creating the bands and the Great Red Spot. The leading hypothesis about the origin of the bands purports that the light or bright-colored bands are regions where the atmosphere is warm and upwelling, whereas the darker bands are places where the atmosphere is down-welling back to deeper levels. The rapid rotation of Jupiter causes these convective

bands to be wrapped around the planet in elongate fashion, unlike on Earth, where they tend to form isolated convective cells. The rotation of the planet causes a strong zonal flow, with most wind belts moving the atmosphere to the east at tens to several hundreds of miles (kilometers) per hour. Several belts are moving westward, with the largest and fastest being the 124-mile-per-hour (200-km/h) westward-moving belt associated with the northern edge of the great red spot. The southern edge of the great red spot is in an eastward-flowing zone (also about 124 miles per hour, or 200 km/h), and the great red spot rotates with the planet, caught between these two powerful belts. Many smaller, oval-shaped vortices that spin off the edges of the Great Red Spot are smaller storms that persist for several or several

Image of Jupiter showing swirling clouds, including the Great Red Spot. This true-color simulated view of Jupiter is composed of four images taken by NASA's *Cassini* spacecraft on December 7, 2000. To illustrate what Jupiter would have looked like if the cameras had a field-of-view large enough to capture the entire planet, the cylindrical map was projected onto a globe. The resolution is about 89 miles (144 km) per pixel. Jupiter's moon Europa is casting the shadow on the planet. Cassini is a cooperative mission of NASA, the European Space Agency, and the Italian Space Agency. JPL, a division of the California Institute of Technology in Pasadena, manages Cassini for NASA's Office of Space Science, Washington, D.C. *(NASA/JPL/University of Arizona)*

tens of years. Many similar features are found elsewhere on the planet.

Jupiter has many moons, with the Galilean satellites resembling a miniature solar system. The four largest moons include Io (1.22 Earth/Moon masses), Europa (0.65 Earth/Moon masses), Ganymede (2.02 Earth/Moon masses), and Callisto (1.47 Earth/Moon masses). Each moon is distinct and fascinating, showing different effects of the gravitational attraction of nearby Jupiter. Io and Europa are rocky, planetlike bodies, with Io exhibiting active sulfur-rich volcanism and very young surface material. The energy for the volcanism is thought to be the gravitational attraction of Jupiter, and the sulfur particles emitted from the volcanoes get entrained as charged ions in Jupiter's magnetosphere, forming a plasma torus ring around the planet. Europa has an icy surface with a rocky interior, criss-crossed by cracks on the surface that may be analogous to pressure ridges on terrestrial ice flows. Because the surface does not contain many craters, it must be relatively young. Ganymede and Callisto are icy satellites with low densities, and Ganymede's heavily cratered surface implies maturity. Callisto also has many craters, including two huge ones with multiple rings, reflecting cataclysmic impacts in its history.

See also EARTH; MARS; MERCURY; NEPTUNE; SATURN; SOLAR SYSTEM; URANUS; VENUS.

FURTHER READING

Chaisson, Eric, and Steve McMillan. *Astronomy Today.* 6th ed. Upper Saddle River, N.J.: Addison-Wesley, 2007.

Comins, Neil F. *Discovering the Universe.* 8th ed. New York: W. H. Freeman, 2008.

National Aeronautic and Space Administration. Solar System Exploration Page. "Jupiter." Available online. URL: http://solarsystem.nasa.gov/planets/profile.cfm?Object=Jupiter. Accessed June 25, 2008.

Snow, Theodore P. *Essentials of the Dynamic Universe: An Introduction to Astronomy.* 4th ed. St. Paul, Minn.: West, 1991.

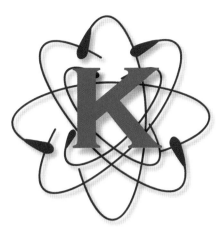

karst Areas affected by groundwater dissolution, cave complexes, and sinkhole development are called karst terrains. Globally, several regions are known for spectacular karst systems, including the cave systems of the Caucasus, southern Arabia including Oman and Yemen, Borneo, and the mature, highly eroded karst terrain of southern China's Guangxi Province. *Caves* are defined as underground openings and passageways in rock that are larger than individual spaces between the constituent grains of the rock. The term is often reserved for spaces large enough for humans to enter. Many caves are small pockets along enlarged or widened cavities, whereas others are huge open underground spaces. The largest cave in the world is the Sarawak Chamber in Borneo, with a volume of 65 million cubic feet (1.84 million m³). The Majlis Al Jinn (Khoshilat Maqandeli) Cave in Oman is the second-largest known cave and is big enough to hold several of the sultan of Oman's royal palaces, with a 747 flying overhead (for a few seconds). Its main chamber is more than 13 million cubic feet (370,000 m³), larger than the biggest pyramid at Giza. Other large caves include the world's third-, fourth-, and fifth-largest caves, the Belize Chamber, Salle de la Verna, and the largest "Big Room" of Carlsbad Cavern, a chamber 4,000 feet (1,200 m) long, 625 feet (190 m) wide, and 325 feet (100 m) high. Each of these has a volume of at least 3 million cubic feet (85,000 m³). Some caves form networks of linked passages that extend for many miles. Mammoth Cave, Kentucky, for instance, has at least 300 miles (485 km) of interconnected passageways. While the caves are forming, water flows through these passageways in underground stream networks.

The formation of karst topography begins with dissolution. Rainwater that filters through soil and rock works into natural fractures or breaks in the rock, and chemical reactions that remove ions from the limestone slowly dissolve and carry away parts of the limestone in solution. Fractures gradually enlarge, and groundwater flowing in underground stream networks through the rock creates new passageways. Dissolution of rocks is most effective if the rocks are limestone and if the water is slightly acidic (acid rain greatly helps cave formation). Carbonic acid (H_2CO_3) in rainwater reacts with the limestone, rapidly (at typical rates of a fraction of an inch, or a few millimeters per thousand years) creating open spaces, cave and tunnel systems, and interconnected underground stream networks.

When the initial openings become wider, they are known as caves. Many caves are small pockets along enlarged or widened cavities, whereas others are huge, open, underground spaces. In many parts of the world the formation of underground cave systems has led to parts of the surface collapsing into the caverns and tunnels, forming a distinctive type of topography known as karst topography. Karst is named after the Kars Limestone plateau region in Serbia, Bosnia, and Croatia (the southeast part of the former Yugoslavia), where it is especially well developed. Karst topography takes on many forms in different stages of landscape evolution but typically begins with the formation of circular pits on the surface known as sinkholes. These form when the roof of an underground cave or chamber suddenly collapses, bringing everything on the surface suddenly down into the depths of the cave. Striking examples of sinkhole formation surprised residents of Orlando, Florida, in 1981 when a series of sinkholes swallowed many businesses and homes with little warning. In this and many other examples sinkhole formation is initiated after a prolonged drought or drop in groundwater levels. This drains the water from underground cave networks, leaving the roofs of chambers unsupported and making them prone to collapse.

The sudden formation of sinkholes in the Orlando area is best illustrated by the formation of the Winter Park sinkhole on May 8, 1981. The first sign that trouble was brewing was the unusual spectacle of a tree suddenly disappearing into the ground at 7:00 P.M., as if being sucked in by an unseen force. Residents were rightfully worried. Within 10 hours a huge sinkhole nearly 100 feet (30 m) across and more than 100 feet deep had formed. It continued to grow, swallowing six commercial buildings, a home, two streets, six Porsches, and the municipal swimming pool, causing more than $2 million in damage. The sinkhole has since been converted into a municipal park and lake. More than 1,000 sinkholes have formed in parts of southern Florida in recent years, caused by the lowering of the groundwater level to accommodate residential and commercial growth in the region.

Many parts of the world exhibit sinkhole topography: Florida, Indiana, Missouri, Pennsylvania, and Tennessee in the United States, the karst regions of the Balkans, the Salalah region of Arabia, southern China, and many other places where the ground is underlain by limestone.

Sinkholes have many different forms. Some are funnel-shaped, with boulders and unconsolidated sediment along their bottoms; others are steep-walled, pipelike features that have dry or water-filled bottoms. Some sinkholes in southern Oman are up to 900 feet (247 m) deep pipes with caves at their bottoms, where residents obtained drinking water until recently, when wells were drilled. Villagers, mostly women, would have to climb down precarious vertical walls then back out carrying vessels of water. The bottoms of some of these sinkholes are littered with human bones, some dating back thousands of years, of water carriers who slipped on their route. Prehistoric cave art decorates some of the caves, showing that these sinkholes served as water sources for thousands or tens of thousands of years.

Sinkhole formation is intricately linked to the lowering of the water table, as exemplified by the Winter Park example. When water fills the underground caves and passages, it slowly dissolves the walls, floor, and roof of the chambers, carrying the limestone away in solution. When the water table is lowered by drought, by people overpumping the groundwater, or by other mechanisms, the roofs of the caves may no longer be supported, and they may

catastrophically collapse into the chambers, forming a sinkhole on the surface. In Florida many of the sinkholes formed because officials lowered the water-table level to drain parts of the Everglades in order to make more land available for development. This ill-fated decision was rethought, and attempts have been made to restore the water table, but in many cases it was too late and the damage was done.

Many sinkholes form suddenly and catastrophically, with the roof of an underground void suddenly collapsing, dropping all of the surface material into the hole. Other sinkholes form more gradually, with the slow movement of loose, unconsolidated material into the underground stream network, eventually leading to the formation of a surface depression that may continue to grow into a sinkhole.

The pattern of surface subsidence resulting from sinkhole collapse depends on the initial size of the cave that collapses, the depth of the cavity, and the strength of the overlying rock. Big caves that collapse can cause a greater surface effect. For a collapsed structure at depth to propagate to the surface, blocks must fall off the roof and into the cavern. The blocks fall by breaking along fractures and falling by the force of gravity. If the overlying material is weak, the fractures will propagate outward, forming a cone-shaped depression with its apex in the original collapse structure. In contrast, if the overlying material is strong, the fractures will propagate vertically upward, causing a pipelike collapse structure.

When the roof material collapses into the cavern, blocks of wall rock accumulate on the cavern floor. There is abundant pore space between these blocks, so the collapsed blocks occupy a larger volume than they did when they were attached to the walls. In this way the underground collapsed cavern can fill completely with blocks of the roof and walls before any effect migrates to the surface. If enough pore space is created, minimal subsidence will occur along the surface. In contrast, if the cavity collapses near the surface, a collapse pit will eventually form on the surface.

Migration of a deep-collapse structure from its initial depth to the surface can take years to decades. The first signs of a collapse structure migrating to the surface include tensional cracks in the soil, bedrock, and building foundations, formed as material pulls away from unaffected areas as it subsides. Circular areas of tensional cracks may enclose an area of contractional buckling in the center of the incipient collapse structure, as bending in the center of the collapsing zone forces material together.

After sinkholes form, they may take on several different morphological characteristics. Solution sinkholes are saucer-shaped depressions formed by the dissolution of surface limestone and have a thin cover of soil or loose sediment. These grow slowly and present few hazards, since they form on the surface and are not connected to underground stream or collapse structures. Cover-subsidence sinkholes form where the loose surface sediments move slowly downward to fill a growing, solution-type sinkhole. Cover-collapse sinkholes form where a thick section of sediment overlies a large solution cavity at depth, and the cavity is capped by an impermeable layer such as clay or shale. A perched water table develops over the aquiclude. Eventually, the collapse cavity becomes so large that the shale or clay aquiclude unit collapses into the cavern, and the remaining overburden rapidly sinks into the cavern, much like sand sinking in an hourglass. These are some of the most dangerous sinkholes, since they form rapidly and can be quite large. Collapse sinkholes are simpler but still dangerous. They form where the strong layers on the surface collapse directly into the cavity, forming steep-walled sinkholes.

Continued maturation of sinkhole topography can lead to the merging of many sinkholes into elongate valleys, and the former surface becomes flat areas on surrounding hills. Even this mature landscape may continue to evolve, until tall, steep-walled karst towers reach to the former land surface, and a new surface has formed at the level of the former cave floor. The Cantonese region of southern China's Guangxi Province best shows this type of karst tower terrain.

Detection of Incipient Sinkholes

Some of the damage from sinkhole formation could be avoided if the location and general time of sinkhole formation could be predicted. At present it may be possible to recognize places where sinkholes may be forming by monitoring for the formation of shallow depressions and extensional cracks on the surface, particularly circular depressions. Building foundations can be examined regularly for new cracks, and distances between slabs on bridges with expansion joints can be monitored to check for expansion related to collapse. Other remote sensing and geophysical methods may prove useful for monitoring sinkhole formation, particularly if the formation of a collapse structure is suspected. Shallow seismic waves can detect open spaces, and ground-penetrating radar can map the bedrock surface and look for collapse structures beneath soils. In some cases it may be worthwhile to drill shallow test holes to determine whether there is an open cavity at depth that is propagating toward the surface.

See also CAVE SYSTEMS, CAVE; SUBSIDENCE.

FURTHER READING

Beck, B. F. *Engineering and Environmental Implications of Sinkholes and Karst.* Rotterdam, Netherlands: Balkema, 1989.

Drew, D. *Karst Processes and Landforms.* New York: Macmillan Education Press, 1985.

Ford, D., and P. Williams. *Karst Geomorphology and Hydrology.* London: Unwin-Hyman, 1989.

Jennings, J. N. *Karst Geomorphology.* Oxford: Basil Blackwell, 1985.

Karst Waters Institute. Available online. URL: http://www. karstwaters.org/. Accessed December 10, 2007.

White, William B. *Geomorphology and Hydrology of Karst Terrains.* Oxford: Oxford University Press, 1988.

Kepler, Johannes (1571–1630) German *Mathematician, Astronomer, Astrologer* Johannes Kepler was a German mathematician and astronomer who became one of the most important and influential scientists of the 17th century for his derivation of the laws of planetary motion. His work contributed to Isaac Newton's theory of gravity, and he helped to confirm the observations of his contemporary Galileo Galilei. In the early 17th century there was no clear distinction between astronomy and astrology, and Kepler also worked with a strong religious conviction that he was exploring and trying to understand a universe created by God according to an intelligent plan that was accessible by reason. Kepler described his own work in astronomy as celestial physics.

EARLY YEARS

Johannes Kepler was born on December 27, 1571, in the imperial free city in what is now the Stuttgart region of Germany. Young Johannes had a rough childhood. His grandfather was mayor of the imperial free city, though the family's riches were declining when he was born, and his father, Heinrich, became a mercenary when Johannes was five. His troubles did not cease, and Heinrich became an alcoholic and abusive parent, leaving the family never to return when Johannes was 16 years old. Historians speculate that his father was killed in the Eighty Years' War in the Netherlands. Johannes was raised by his mother, Katharina Guldenmann, daughter of an innkeeper. Katharina was a healer who practiced herbalism and was by reputation "thin, garrulous, and bad-tempered." She was tried and imprisoned for 14 months for witchcraft and narrowly escaped death by torture. Johannes's grandmother helped with the family, though she was said to be "clever, deceitful, and blazing with hatred." Despite these shortcomings of his upbringing, Johannes's mother brought him outside at the age of six to observe the Great Comet of 1577 and again to witness the lunar eclipse of 1580 at the age of nine.

Johannes also had health problems from an early age. When he was three he contracted smallpox, which left him with impaired vision and crippled hands, so he was deemed unfit for most lines of work and sent to study for the ministry. In his early studies he quickly developed great abilities at mathematics and a keen interest in astronomy. In 1589 Kepler began studying at the University of Tübigen as a theology student, but specializing in astrology and the study of the stars, learning about both the Ptolemaic and the Copernican systems of planetary motion. During this time Kepler became convinced from both theoretical and theological perspectives that the Sun was the center of the universe, not the Earth. With this training and conviction he did not become a minister but was instead appointed to a position of teacher of mathematics and astronomy at the University of Graz (present-day Austria) in April 1594, at age 23. In 1595 Kepler met Barbara Muller, a widow (twice), whose late husbands had left her a fortune. The two were married on April 27, 1597, and the Keplers soon had two children, both of whom died in infancy. They later had other children: a daughter (Susanna) in 1602, and two sons (Friedrich and Ludwig) in 1604 and 1607, respectively.

SCIENTIFIC CONTRIBUTIONS

While lecturing in Graz, Kepler realized that the orbits of Mercury, Venus, Earth, Mars, Jupiter, and Saturn were geometrically regular and corresponded to the same geometric ratios as obtained by nesting the shapes of octahedron, isosahedron, dodedahedron, tetrahedron, and cube inside each other inside a circle. He reasoned that this geometrical regularity could explain the geometrical basis of the universe and used it to support the Copernican system of a Sun-centered solar system in his work *Mysterium Cosmographicum* (The cosmographic universe), published at Tübingen in 1596. Kepler thought he had found God's geometrical plan for the universe and dedicated one chapter of his book to reconciling this idea with biblical passages that most had interpreted to support geocentrism. He published a shorter version of the manuscript, *Mysterium*, later that same year. This was followed by a new edition in 1621, including 25 years of new calculations, notes, and improvements on his earlier work.

Kepler planned four additional books expanding on his ideas in *Mysterium*. These included works on the so-called stationary aspects of the universe (the Sun and fixed stars), one on the planets and their motions, one on the physical nature of planets (including the Earth) and their geographical features, and a fourth

on atmospheric optics, meteorology, and astrology. Kepler began corresponding with Danish astronomer Tycho Brahe in letters that discussed their scientific disagreements, but they remained professional and discussed the limits of accuracy of the measurements that Kepler used in his models. In 1600 Kepler went to Prague to work with Brahe, at a time when the financial burden and pressures of his teachings that seemed to go against the church in Graz were forcing him to seek employment elsewhere. Kepler arrived at Benatky nad Jizerou in what would be today the central Czech Republic, where Brahe was building a new observatory and castle, and the two, despite several intense arguments, negotiated a contract for Kepler to work for Brahe. Kepler returned to Graz to collect his family but ran into political difficulties. After refusing to convert to Catholicism, he was banished from Graz and moved to Prague to pursue his work with Brahe, who paid his salary for most of 1601 to work on observations of planetary motions in an attempt to discredit the models of Brahe's (deceased) rival, the mathematician Nicolas Reimers (Ursus). Tycho Brahe unexpectedly died (some rumors suggest that Kepler poisoned him) on October 24, 1601, however, and Kepler was then appointed Brahe's successor as imperial mathematician to Holy Roman Emperor Rudolph II. Kepler was first asked to complete Brahe's unfinished projects, and when Kepler was found to be appropriating Brahe's observations as his own, he encountered difficulties that delayed publication for several years. When this was solved Kepler continued with his productive yet troubled career.

Kepler was in charge of astrology for the emperor, as well as providing advice to him on political issues. But the emperor was incurring financial difficulty, and Kepler often did not get paid on time. Johannes Kepler continued his work on observations of Mars and solar eclipses, and in 1604 he published *Astronomiae Pars Optica* (The optical part of astronomy), including descriptions of the inverse square law of light, reflection of light by different types of mirrors, the principles of pinhole cameras, and observations on optical phenomena such as parallax of stars, where the apparent displacement or difference in orientation of an object when viewed along two different lines of sight is used to calculate the distance to the object. In this book Kepler became the first to recognize that images are projected inversely to the retina by the human eye and had to be corrected "in the hollows of the brain."

Kepler continued his studies of astronomy, including observations of the supernova of 1604, playing down astrological predictions based on the appearance of the new star. He continued to observe and record planetary motions, trying to find a formula that could explain his observations of the orbits and still fit his religious beliefs that the driving power from the Sun, as if a magnetic soul, would decrease with distance from the Sun so the speed of the orbits should decrease with increasing distance from the Sun. Drawing on this, he formulated what would become the second law of planetary motion, that planets sweep out equal areas in equal times. He then tried to fit the orbit of Mars better with his calculations, and after more than 40 failed attempts, he determined that Mars and the other planets followed elliptical orbits. This led to his first law of planetary motion stating that all planets move in ellipses with the Sun as their focus. Kepler wrote these conclusions in his treatise *Astronomia nova* (A new astronomy) in 1605, but legal arguments over his use of Tycho Brahe's measurements as his own prevented publication until 1609.

After Kepler finished *Astronomia nova*, he worked for many years on the *Rudolphine Tables,* which contained ephemerides, or predictions of when certain stars and planets would be found in specific locations. Kepler watched in 1610 as Galileo Galilei announced his discovery of four moons orbiting Jupiter, observed with his powerful new telescope. Kepler endorsed Galileo's observations and later that year published additional observations of the moons of Jupiter in *Narratio de Jovis Satellibus* (Narrative on the satellites of Jupiter).

The year of 1611 was one of misfortune for Kepler. Political tensions forced Emperor Rudolph to abdicate the throne as king of Bohemia to his brother Matthias, who did not favor Kepler. Barbara, Johannes's wife, contracted Hungarian spotted fever, then his three children fell sick with smallpox, and Friedrich, age six, died. Kepler sought new employment but his religious beliefs barred him from returning to the University of Tübingen, so he began to arrange a professorship in Austria. At this time Barbara relapsed into illness and died, then Emperor Rudolph died in 1812. Kepler was so distraught over the tragedies that he was unable to do research, but soon the new king, Matthias, reappointed him as imperial mathematician and allowed him to move to Linz. There Kepler worked on completing the *Rudolphine Tables* and teaching mathematics; he also performed astrological and astronomical services. In 1613 Kepler remarried the young Susanna Reuttinger, and the two had six children, three of whom survived childhood.

In 1615 Kepler completed the first three books of what would be his most influential works, his *Epitome astronomia Copernicanae* (Epitome of Copernican astronomy); the volumes were printed in 1617, 1620, and 1621. These books contained descriptions of the heliocentric model for the universe, the elliptical paths of planets, and all three laws of planetary motion. These laws include the following:

- The orbit of every planet is an ellipse with the Sun at its focus.
- A line joining a planet and the Sun sweeps out equal areas during equal intervals of time.
- The square of the orbital period of a planet is directly proportional to the third power of the semimajor axis of its orbit. Moreover, the constant of proportionality has the same value for all planets.

In addition to these works of science Kepler continued to publish astrological calenders, the payments for which served to pay his bills. These calenders were related to the work Kepler had done on the *Rudolphine Tables* and *Ephemerides* so did not take as much of his time as the *Epitome*. Many of Kepler's astrological predictions were vague, however, and people began to be suspicious, publicly burning his last astrological calender in Graz in 1624.

In one of Kepler's later works, *Harmonices Mundi* (The harmony of the worlds), he first described in detail his third law of planetary motion and attempted to explain the geometrical patterns of the world in terms of music. The *musica universalsi* (music of the spheres) had been described and studied previously by Pythagoras, Ptolemy, and others. Kepler sought out harmonic analysis of regular polygons (all sides of equal length) and regular solids, including the tetrahedron, cube, octahedron, dodecahedron, and icosahedron. The third law of planetary motion was one such harmony, where "the square of the periodic times are to each other as the cubes of the mean distances."

In 1623 Kepler completed his *Rudolphine Tables,* which was printed (after legal battles with Brahe's heir) in 1627. Religious tensions arose again in 1627, when the Catholic Counter Reformation sealed most of Kepler's library and besieged the city of Linz. Kepler and his family fled to Ulm in present-day Germany, where he published his *Rudolphine Tables* at personal expense. In 1628 he became an adviser to General Wallenstein, under Emperor Ferdinand, providing astronomical calculations for the general's astrologers. In this interval Kepler traveled widely, but he became ill and died on November 15, 1630, soon after his family moved to Regensburg, Germany.

See also ASTRONOMY; BRAHE, TYCHO; COPERNICUS, NICOLAUS; GALILEI, GALILEO; LEMAÎTRE, GEORGES.

FURTHER READING

Barker, Peter, and Bernard R. Goldstein. "Theological Foundations of Kepler's Astronomy." In *Osiris* 16, (2001): 88–113.

Ferguson, Kitty. *Tycho and Kepler: The Unlikely Partnership That Forever Changed Our Understanding of the Heavens.* New York: Walker, 2002.

Field, J. V. *Kepler's Geometrical Cosmology.* Chicago: University of Chicago Press, 1988.

Gingerich, Owen. "Kepler, Johannes." In *Dictionary of Scientific Biography.* Vol. 7, edited by Charles Coulston Gillispie, 302–304. New York: Scribners, 1973.

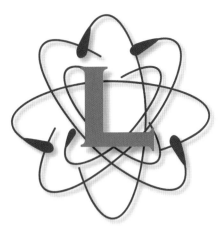

L

large igneous province, flood basalt A large igneous province, also known as a continental flood basalt, plateau basalt, and trap, is deposits that include vast plateaus of basalts, covering large areas of some continents. They have a tholeiitic basalt composition, but some show chemical evidence of minor contamination by continental crust. They are similar to anomalously thick and topographically high seafloor known as oceanic plateaus and to some volcanic rifted passive margins. In numerous instances over the past several hundred million years these vast outpourings of lava have accumulated, forming thick piles of basalt, representing the largest-known volcanic episodes on the planet. These piles of volcanic rock represent times when the Earth moved more material and energy from its interior than during intervals between the massive volcanic events. Such large amounts of volcanism also released large amounts of volcanic gases into the atmosphere, with serious implications for global temperatures and climate and may have contributed to some global mass extinctions.

The largest continental flood basalt province in the United States is the Columbia River flood basalt in Washington, Oregon, and Idaho. The Columbia River flood basalt province is 6–17 million years old and contains an estimated 1,250 cubic miles (4,900 km^3) of basalt. Individual lava flows erupted through fissures or cracks in the crust, then flowed laterally across the plain for up to 400 miles (645 km).

The 66-million-year-old Deccan flood basalts, also known as traps, cover a large part of western India and the Seychelles. They are associated with the breakup of India from the Seychelles during the opening of the Indian Ocean. Slightly older flood basalts (90–83 million years old) are associated with the breakaway of Madagascar from India. The volume of the Deccan traps is estimated to be 5 million cubic miles (20,840,000 km^3). This huge volume of volcanic rocks erupted over a period of about 1 million years, starting slightly before the great Cretaceous-Tertiary extinction. Most workers now agree that the gases released during the flood basalt volcanism stressed the global biosphere to such an extent that

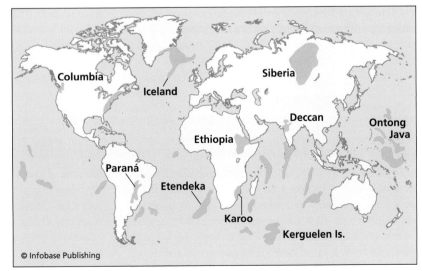

Map of the world showing distribution of flood basalts

Columbia River flood basalts along Snake River Birds of Prey National Conservation Area near Boise, Idaho *(David R. Frazier/Photo Researchers, Inc.)*

many marine organisms had gone extinct, and many others were stressed. Then the planet was hit by the massive Chicxulub impact, causing the massive extinction that included the end of the dinosaurs.

The breakup of east Africa along the East African rift system and the Red Sea is associated with large amounts of Cenozoic (fewer than 30 million years old) continental flood basalts. Some of the older volcanic fields are located in east Africa in the Afar region of Ethiopia, south into Kenya and Uganda, and north across the Red Sea and Gulf of Aden into Yemen and Saudi Arabia. These volcanic piles underlie younger (fewer than 15-million-year-old) flood basalts that extend both farther south into Tanzania and farther north through central Arabia, where they are known as Harrats, and into Syria, Israel, Lebanon, and Jordan.

An older volcanic province, the North Atlantic Igneous Province, also associated with the breakup of a continent, formed along with the breakup of the North Atlantic Ocean at 62–55 million years ago. The North Atlantic Igneous Province includes both onshore and offshore volcanic flows and intrusions in Greenland, Iceland, and the northern British Isles, including most of the Rockall Plateau and

Faeroe Islands. The opening of the ocean in the south Atlantic is similar to 129–134-million-year-old flood basalts, which, now split in half, comprise two parts. In Brazil the flood lavas are known as the Paraná basalts, and in Namibia and Angola of west Africa, as the Etendeka basalts.

These breakup basalts are transitional to submarine flood basalts that form oceanic plateaus. The Caribbean Ocean floor is one of the best examples of an oceanic plateau, with other major examples including the Ontong-Java Plateau, Manihiki Plateau, Hess Rise, Shatsky Rise, and Mid-Pacific Mountains. All of these oceanic plateaus contain between six- and 25-mile (10–40-km) thick piles of volcanic and subvolcanic rocks, representing huge outpourings of lava. The Caribbean seafloor preserves five- to 13-mile (8–21-km) thick oceanic crust formed before about 85 million years ago in the eastern Pacific Ocean. This unusually thick ocean floor was transported eastward by plate tectonics, where pieces of the seafloor collided with South America as it passed into the Atlantic Ocean. Pieces of the Caribbean oceanic crust are now preserved in Colombia, Ecuador, Panama, Hispaniola, and Cuba, and some scientists estimate that the Caribbean oceanic plateau was once twice its present size. In either case it represents a vast outpouring of lava that would have been associated with significant outgassing, with possible consequences for global climate and evolution.

The western Pacific Ocean basin contains several large oceanic plateaus, including the 20-mile (32-km) thick crust of the Alaskan-sized Ontong-Java Plateau, the largest outpouring of volcanic rocks on the planet. Having formed in two intervals, at 122 and 90 million years ago, respectively, entirely within the ocean, the Ontong-Java Plateau represents magma that rose in a plume from deep within the mantle and erupted on the seafloor. Estimates suggest that the volume of magma erupted in the first event was equivalent to that of all the magma being erupted at midocean ridges at the present time. Sea levels rose by more than 30 feet (9 m) in response to this volcanic outpouring. The gases released during these eruptions are estimated to have raised average global temperatures by 23°F (13°C).

ENVIRONMENTAL HAZARDS OF FLOOD BASALT VOLCANISM

The environmental impact of the eruption of large volumes of basalt in provinces including those described above can be severe. Huge volumes of sulfur dioxide, carbon dioxide, chlorine, and fluorine are released during large basaltic eruptions. Much of this gas may get injected into the upper troposphere and lower stratosphere during the eruption process, being released from eruption columns that reach two to eight miles (3–13 km) in height. Carbon dioxide is a greenhouse gas and can cause global warming, whereas sulfur dioxide and hydrogen sulfate have the opposite effect: they can cause short-term cooling. Many of the episodes of volcanism preserved in these large igneous provinces were rapid, repeatedly releasing enormous quantities of gases over periods of fewer than 1 million years, and releasing enough gas to change the climate significantly and more rapidly than organisms could adapt. For instance, one eruption of the Columbia River basalts is estimated to have released 9 billion tons of sulfur dioxide and thousands of millions of tons of other gases, whereas the eruption of Mount Pinatubo in 1991 released about 20 million tons of sulfur dioxide.

The Columbia River basalts of the Pacific Northwest continued erupting for years at a time, for approximately 1 million years. During this time the gases released would be equivalent to that of Mount Pinatubo every week, maintained for decades to thousands of years at a time. The atmospheric consequences are sobering. Sulfuric acid aerosols and acid from the fluorine and chlorine would form extensive poisonous acid rain, destroying habitats and making waters uninhabitable for some organisms. At the

very least the environmental consequences would be such that organisms were stressed to the point that they would be unable to handle an additional environmental stress, such as a global volcanic winter and subsequent warming caused by a giant impact.

Faunal extinctions have been correlated with the eruption of the Deccan flood basalts at the Cretaceous-Tertiary (K/T) boundary, and with the Siberian flood basalts at the Permian-Triassic boundary. There is still considerable debate about the relative significance of flood basalt volcanism and impacts of meteorites for extinction events, particularly at the Cretaceous-Tertiary boundary. Most scientists would now agree, however, that global environment was stressed shortly before the K/T boundary by volcanic-induced climate change, and then a huge meteorite hit the Yucatán Peninsula, forming the Chicxulub impact crater and causing the massive K/T boundary extinction and the death of the dinosaurs.

The Siberian flood basalts cover a large area of the Central Siberian Plateau northwest of Lake Baikal. They are more than half a mile thick over an area of 210,000 square miles (543,900 km²) but have been significantly eroded from an estimated volume of 1,240,000 cubic miles (3,211,600 km³). They were erupted over an extraordinarily short period of fewer than 1 million years, 250 million years ago, at the Permian-Triassic boundary. They are remarkably coincident in time with the major Permian-Triassic extinction, implying a causal link. The Permian-Triassic boundary at 250 million years ago marks the greatest extinction in Earth history, when 90 percent of marine species and 70 percent of terrestrial vertebrates became extinct. It has been postulated that the rapid volcanism and degassing released enough sulfur dioxide to cause a rapid global cooling, inducing a short ice age with associated rapid fall of sea level. Soon after the ice age took hold, the effects of the carbon dioxide took over and the atmosphere heated to cause global warming. The rapidly fluctuating climate postulated to have been caused by the volcanic gases is thought to have killed off many organisms, which were simply unable to cope with the wildly fluctuating climate extremes.

See also IGNEOUS ROCKS; MASS EXTINCTIONS; OCEANIC PLATEAU.

FURTHER READING

Albritton, C. C., Jr. *Catastrophic Episodes in Earth History.* London: Chapman and Hale, 1989.

MacDougall, J. D., ed. *Continental Flood Basalts,* Dordrecht, Germany: Kluwer Academic Publishers, 1988.

Mahoney, J. J., and M. F. Coffin, eds. *Large Igneous Provinces, Continental, Oceanic, and Planetary Flood*

Volcanism. Washington, D.C.: American Geophysical Union, 1997.

Robock, Alan, and Clive Oppenheimer, eds. *Volcanism and the Earth's Atmosphere.* Washington, D.C.: American Geophysical Union, 2003.

lava Molten rock or magma that flows on the surface of the Earth is known as lava. Lavas have a wide range in composition, texture, temperature, viscosity, and other physical properties, based on the composition of the melt and the amount of volatiles present. They tend to be very viscous (sticky, resistant to flow) when they are rich in silica and form slow-moving and steep-sided flows. The addition of a large amount of volatiles to silicic magma can cause explosive eruptions. Mafic, or low-silica, lavas are less viscous and tend to flow more easily, forming planar flows with gently sloping surfaces. Some basaltic flood lavas have flowed over hundreds or thousands of square miles (square kilometers), forming flat-lying layers of crystallized lava. Other mafic and intermediate lavas form shield volcanoes such as the Hawaiian Islands, with gently sloping sides built by numerous eruptions. If mafic lavas are rich in volatiles, they tend to become abundant in empty gas bubbles known as vesicles, forming pumice. Mafic lavas that flow on the surface often form ropey lava flows known as pahoehoes, or blocky flows known as aa lavas.

Some types of lava flows are extremely hazardous whereas others may be relatively harmless if treated with caution. In the most passive types of volcanic eruptions lava bubbles up or effuses from volcanic vents and cracks and flows like thick water across the land surface. During other eruptions lava oozes out more slowly, producing different types of flows with different hazards. Variations in magma composition, temperature, dissolved gas content, surface slope, and other factors lead to the formation of three main different types of lava flows. These include aa, pahoehoe, and block lava. Aa are characterized by a rough surface of spiny and angular fragments, whereas pahoehoe have smooth, ropeylike or billowing surfaces. Block lavas have larger fragments than aa flows and are typically formed by stickier, more silicic (quartz rich) lavas than aa and pahoehoe flows. Some flows are transitional between these main types, or may change from one type to another as surface slopes and flow rates change. Pahoehoe

Lava flow crossing Chain of Craters road from the west toward Hiiaka crater during eruption of Kilauea Volcano, Hawaii, May 5, 1973 *(R.L. Christiansen/USGS)*

flows commonly change into aa flows with increasing distance from the volcanic source.

Lava flows are most common around volcanoes that are characterized by eruptions of basalt with low contents of dissolved gasses. About 90 percent of all lava flows worldwide are made of magma with basaltic composition, followed by andesitic (8 percent) and rhyolitic (2 percent). Places with abundant basaltic flows include Hawaii, Iceland, and other exposures of oceanic islands and midoceanic ridges, all characterized by nonexplosive eruptions. Virtually the entire volume of all the islands of the Hawaiian chain are made of a series of lava flows piled high one on top of the other.

Lava flows generally follow topography, flowing from the volcanic vents downslope in valleys, much as streams or water from a flood would travel. Some lava flows move as fast as water, up to almost 40 miles per hour (65 km/hour) on steep slopes, but most lava flows move considerably more slowly. More typical rates of movement range from about 10 feet per hour (several meters per hour) to 10 feet (3 m) per day for slower flows. These rates of lava movement allow most people to move out of danger to higher ground, but lava flows are responsible for significant amounts of property damage in places like Hawaii. Lava flows have buried roads, farmlands, and other low-lying areas. One must keep in mind, however, that the entire Hawaiian Island chain was built by lava flows, and the real estate that is being damaged would not even exist if it were not for the lava flows. In general pahoehoe flows are the fastest, aa are intermediate, and blocky flows are the slowest.

Basaltic lava is extremely hot (typically about 1,830°–2,100°F, or 1,000°–1,150°C) when it flows across the surface, so when it encounters buildings, trees, and other flammable objects, they typically burst into flame and are destroyed. More silicic lavas are slightly cooler, in the range of 1,560°–1,920°F (850°–1050°C). Most lavas will become semisolid and stop flowing at temperatures approaching 1,380°F (750°C). Lavas cool quickly at first, until a crust or hard skin forms on the flow, then they cool more slowly. This property of cooling creates one of the greatest hazards of lava flows. A lava flow that appears hard, cool, and safe to walk on can hide an underlying thick layer of molten lava at temperatures of about 1,380°F (750°C) just below the thin surface. Many people have mistakenly thought it was safe to walk across a recent crusty lava flow, only to plunge through the crust to a fiery death. Thick flows take years to crystallize and cool, and residents of some volcanic areas have learned to use the heat from flows for heating water and piping it to nearby towns.

See also IGNEOUS ROCKS; VOLCANO.

FURTHER READING

Blong, Russel J. *Volcanic Hazards: A Sourcebook on the Effects of Eruptions.* New York: Academic Press, 1984.

Chester, D. *Volcanoes and Society.* London: Edward Arnold, 1993.

Decker, R. W., and B. B. Decker. *Volcanoes.* 3rd ed. New York: W. H. Freeman, 1997.

Fisher, R. V. *Out of the Crater.* Princeton, N.J.: Princeton University Press, 1999.

Fisher, R. V., G. Heiken, and J. B. Hulen. *Volcanoes: Crucibles of Change.* Princeton, N.J.: Princeton University Press, 1997.

Fisher, R. V., and H.-U. Schmincke. *Pyroclastic Rocks.* Berlin, Germany: Springer-Verlag, 1984.

Kusky, T. M. *Volcanoes: Eruptions and Other Volcanic Hazards.* New York: Facts On File, 2008.

Macdonald, G. A. *Volcanoes.* Englewood Cliffs, N.J.: Prentice-Hall, 1972.

Williams, H., and A. R. McBirney. *Volcanology.* San Francisco: Freeman, Cooper, 1979.

Lawson, Andrew Cooper (1861–1952) Scottish *Geologist* Andrew Cooper Lawson is known for his work in isostasy and the geology of western North America. He was born July 25, 1861, in Anstruther, Scotland, but moved with his family to Hamilton, Ontario, Canada, at age six. He studied in public schools in Hamilton and later received a bachelor's degree in natural sciences in 1883 from the University of Toronto and a master of arts degree in natural sciences from the same university in 1885. He then received another master's degree and a doctorate in geological sciences from the Johns Hopkins University in 1888. Lawson then spent seven years with the Geological Survey of Canada until he moved to the University of California at Berkeley, where he remained for 60 years until his death on June 16, 1952. During this time his studies of isostasy, the relative vertical movements of the crust and balance between crust and mantle rocks, continued over three decades, and these studies brought special attention to the importance of isostasy in deformation of the Earth's crust. He also developed the logical consequences of isostatic adjustment as an important factor in orogenesis. In some cases this was seen as the determinative agent in the elevation of mountains and the depression of deep troughs and basins. His research areas included the Sierra Nevada, the Great Valley of California, the Mississippi delta, the Cordillera, and the Canadian shield. Lawson also spent about 40 years studying the northwest region of Lake Superior, where he provided new information and revised the correlation of the pre-Cambrian rocks over a large part of North America.

His earlier studies of Lake of the Woods and Rainy Lake showed that the Laurentian granites were intrusive into metamorphosed volcanic and sedimentary rocks that he called the Keewatin Series. Under this series he found another sedimentary series that he called Coutchiching and saw this as the oldest rocks in the series. Later research on this area showed that two periods of batholithic invasion were followed by a great period of peneplaination. These surfaces, the Laurentian peneplain and the Eparchean peneplain, were used as references for the correlation of the invaded formations and for the subsequently deposited sedimentary beds over the areas. Lawson also worked as a consultant in a number of construction engineering projects including San Francisco's Golden Gate Bridge.

During his work in California Lawson became the first to identify and name the San Andreas Fault in 1895, and later, after the 1906 San Francisco earthquake, he was the first to map the entire length of the fault. Lawson is most famous for being the editor and coauthor of the 1908 report on the 1906 earthquake, a report that later became known as the Lawson Report.

See also GEOPHYSICS; LITHOSPHERE; NORTH AMERICAN GEOLOGY.

FURTHER READING

Lawson, Andrew C., ed. *The California Earthquake of April 18, 1906: Report of the State Earthquake Investigation Commission,* Andrew C. Lawson, chairman. Washington, D.C.: Carnegie Institution of Washington Publication 87, 2 vols. 1908. Available online. URL: http://earthquake.usgs.gov/regional/nca/1906/18april/references.php. Accessed December 3, 2008.

Lemaître, Georges (1894–1966) Belgian *Cosmologist* Georges Lemaître is most famous for proposing the theory of the big bang in 1933. He was born July 17, 1894, in Charleroi, Belgium, where he studied civil engineering and obtained a Ph.D. for his dissertation, "L'Approximation des fonctions de plusieurs variables réelles" (Approximation of functions of several real variables). He was ordained in 1923 as a Catholic priest before moving to Cambridge, United Kingdom, to study astrophysics at St. Edmund's College. He then moved to the United States, where at the Massachusetts Institute of Technology, he became fascinated by Edwin Hubble's observations of an expanding universe and discussed this with Harvard University astronomer Harlow Shapley, who also advocated an expanding-universe model. He extrapolated the consequences of an expanding universe back in time and proposed the model of the big bang for the origin of the universe.

Lemaître became a professor of astrophysics at the University of Louvain, Belgium, in 1927.

Georges Lemaître was a brilliant mathematician and cosmologist who pioneered the application of Albert Einstein's theory of general relativity to cosmology, publishing a precursor to Hubble's law in 1927 in the *Annales de la Société Scientifique de Bruxelles* (*Annals of the Scientific Society of Brussels*), entitled "Un Univers homogène de masse constante et de rayon croissant rendant compte de la vitesse radiale des nébuleuses extragalactiques" (A homogeneous universe of constant mass and growing radius accounting for the radial velocity of extragalactic nebulae). His next major paper hypothesized a big bang origin for the universe in a paper that was published in the prestigious journal *Nature* in 1931. He proposed that the universe started much like an incredibly dense egg or "primal atom" in which all the material for the entire universe was compressed into a sphere about 30 times larger than the Sun. Being an ordained priest, he noted the similarity of the big bang model to the "creation" of the universe in biblical accounts. In his models published in 1933 and 1946 he estimated that the primal atom exploded to create the universe, some 20 billion to 60 billion years ago. This model was widely criticized at first, in part for resembling too much the biblical account of creation, and proposed by a priest at that. Eventually, in 1933 Einstein endorsed Lemaître's theory, and at a meeting in California Einstein said of the theory, "This is the most beautiful and satisfactory explanation of creation to which I have ever listened." The scientific world listened, tested the theory, and soon the big bang theory became the leading model for the origin of the universe.

Lemaître soon became world renowned and was widely heralded as the leader of the new cosmological physics. In 1934 King Leopold III of Belgium awarded Lemaître the Francqui Prize, the highest Belgian scientific and scholarly prize, awarded to scientists under the age of 50 and named after Belgian diplomat and businessman Emile Francqui. In 1941 Lemaître was elected to the Royal Academy of Science and Arts of Belgium, and in 1953 he became the first recipient of the Eddington Medal, bestowed by the Royal Astronomical Society. In 1936 he was elected to the Pontifical Academy of Sciences and was president of this organization from 1960 until his death on June 20, 1966.

See also ASTRONOMY; ASTROPHYSICS; COSMIC MICROWAVE BACKGROUND RADIATION; DARK MATTER; HUBBLE, EDWIN; ORIGIN AND EVOLUTION OF THE UNIVERSE.

FURTHER READING

Lemaître, G. "The Beginning of the World from the Point of View of Quantum Theory." *Nature* 127, no. 3210 (1931): 706.

Murdin, Paul. *Encyclopedia of Astronomy and Astrophysics*, article 3804. Bristol, U.K.: Institute of Physics Publishing, 2001.

life's origins and early evolution The origin of life and its early evolution from simple, single-celled organisms to more complex forms has intrigued scientists, philosophers, theologians, and others from all over the world for much of recorded history. The question of the origin of life relates to where humans came from, why people are here, and what the future holds for the species. One of the most interdisciplinary of sciences, the study of the origin of life encompasses cosmology, chemistry, astrophysics, biology, geology, and mathematics.

The scientific community has supported several ideas about the location of the origin of life. Some scientists believe that life originated by chemical reactions in a warm little pond, whereas others suggest that it may have started in surface hot springs. Another model holds that the energy for life was first derived from deep within the Earth, at a hydrothermal vent on the seafloor. Still others hold that life may have fallen to Earth from outer reaches of the solar system, though this does not answer the question of where and how it began.

Life on the early Earth would have to have been compatible with conditions very different from what they are on the present-day Earth. The Earth's early atmosphere had very little if any oxygen, so the partial pressure of oxygen was lower and the partial pressure of carbon dioxide (CO_2) was higher in the early Archean. The Sun's luminosity was about 25 percent less than that of today, but since the early atmosphere was rich in CO_2, CH_4 (methane), NH_3 (ammonia), and N_2O (nitrous oxide), an early greenhouse effect warmed the surface of the planet. CO_2 was present at about 100 times its current levels, so the surface of the planet was probably hotter than today, despite the Sun's decreased luminosity. Evidence suggests that the surface was about 140°F (60°C), favoring thermophilic bacteria (heat adaptive) over organisms that could not tolerate such high temperatures. The lack of free oxygen and radiation-shielding ozone (O_3) in the early atmosphere led to a 30 percent higher ultraviolet flux from the Sun, which would have been deadly to most early life. The impact rate from meteorites was higher and heat flow from the interior of the Earth was about three times higher than at present. Early life would have to have been compatible with these conditions, so it would have to have been thermophilic and chemosynthetic (meaning early life-forms had the metabolic means for synthesizing organic compounds using energy extracted from reduced inorganic chemicals). The

best place for life under these extreme conditions would have been deep in the ocean. The surface would have been downright unpleasant.

Since the Earth is cool today, some process must have removed CO_2 from the atmosphere, otherwise it would have had a runaway greenhouse effect, similar to that on the planet Venus. Processes that remove CO_2 from the atmosphere include deposition of limestone ($CaCO_3$) and burial of organic matter. These processes are aided by chemical weathering of silicates (e.g., $CaSiO_3$) by CO_2-rich rainwater that produces dissolved Ca^{2+}, SiO_2, and bicarbonate (HCO_3^-), which is then deposited as limestone and silica. Life evolved in the early Precambrian and began to deposit organic carbon, removing CO_2 from the atmosphere. Limestones that formed as a result of organic processes acted as large CO_2 sinks and served to decrease global temperatures.

The present-day levels of CO_2 in the atmosphere are balanced by processes that remove CO_2 from the atmosphere and processes that return CO_2 to the atmosphere. Today sedimentary rocks store 78,000 billion tons of carbon, a quantity that would have required a few hundred million years to accumulate from the atmosphere. The return part of the carbon cycle is dominated by a few processes. The decomposition of organic matter releases CO_2. Limestone deposited on continental margins is eventually subducted, or metamorphosed, into calc-silicate ($CaSiO_3$) rocks, both processes that release CO_2. This system of CO_2 cycling regulates atmospheric CO_2, and thus global temperature on longtime scales. Changes in the rates of carbon cycling are intimately associated with changes in rates of plate tectonics, showing that tectonics, atmospheric composition and temperature, and the development of life are closely linked in many different ways.

Recognizing signs of life in very old, deformed rocks is often difficult. Searching for geochemical isotope fractionation is one method or detecting signs of previous life in rocks. Metabolism produces distinctive isotopic signatures in carbon (C)—organic and inorganic carbon 13/carbon 12 isotope ratios differ by about 5 percent. So the presence of isotopically light carbon in old rocks suggests the influence of life. Diverse forms of life—photosynthesizing, methanogenic, and methylotropic organisms may all have been present 3.5 or even 3.85 billion years ago. Early life, in a preoxygen-rich atmosphere, had to be adapted to the reducing environment.

Life 3.8 billion years ago consisted of primitive prokaryotic organisms (unicellular organisms that contain no nucleus and no other membrane-bound organelles). These organisms made their own organic compounds (carbohydrates, proteins, lipids, nucleic acids) from inorganic carbon derived from CO_2, water, and energy from the Sun by photosynthesis, but they did not release molecular oxygen (O_2) as a by-product. More familiar photosynthetic organisms use water (H_2O) as the reducing agent, oxidizing it to O_2 in a process called oxygenic photosynthesis. In contrast, many of these early prokaryotic organisms used hydrogen sulfide (H_2S) as their electron donors, producing elemental sulfur in a process called anoxygenic photosynthesis. The sulfur could then be further oxidized to form sulfate ions (SO_4^{2-}). Oxygen would have been toxic to these prokaryotes and the environment would have been devoid of oxygen. Because of this, they obtained their energy through anaerobic cellular respiration rather than aerobic respiration, reducing sulfate ions rather than O_2. By 3.5 Ga, cyanobacteria, a type of bacteria formerly called blue-green algae that are capable of oxygenic

photosynthesis, used CO_2 and emitted O_2 to the atmosphere. As a result the protective ozone (O_3) layer began to form, blocking ultraviolet (UV) radiation from the Sun and making the surface habitable for other organisms.

Ophiolites that are 2.5 billion years old with black smoker types of hydrothermal vents and evidence for primitive life-forms have been discovered in northern China. The physical conditions at these and even older midocean ridges permit the inorganic synthesis of amino acids and other prebiotic organic molecules, and this environment would have been sheltered from early high levels of UV radiation and its harmful physical effects on biomolecules. In this environment the locus of precipitation and synthesis for life might have been in small, iron-sulfide globules emitted by hydrothermal vents on

THE SEARCH FOR EXTRATERRESTRIAL LIFE

Ever since humans turned their eyes skyward and realized that the billions of stars out there are similar to the Earth's Sun, they wondered if there might be other forms of life in the universe. This speculation has run the range from deep philosophical and religious thought, to science fiction, through scientific investigation of the likelihood of other life existing beyond Earth. There are two main theories on the origin of extraterrestrial life, if any exists at all. One is that different life-forms may have risen independently in separate locations within the universe and be completely unrelated to each other. Another theory, called panspermia, suggests that life originated in one place, then spread to other habitable planets. Either or both theories could be true, although it is also possible that neither is true since no extraterrestrial life has ever been documented.

Spacecraft launched by NASA have contained messages intended as a welcome to any intelligent life-forms that may encounter them and wish to contact the Earth. The *Voyager 1* spacecraft launched September 5, 1977, to probe the outer solar system and beyond contained two

golden records with sound and images intended to portray the diversity of life and culture on Earth. The *Galileo* spacecraft, launched by NASA October 18, 1989, to study Jupiter and its moons also contained an apparatus to search for possible extraterrestrial life, particularly on the moon Europa, which contains a saltwater ocean beneath a layer of ice. The American astronomer Carl Sagan (1934–96) devised a set of experiments using Galileo's instruments to test for possible life using remote sensing. Sagan made a list of criteria needed to identify life on another planet system using remote sensing, and this became known as the "Sagan criteria for life." Included are the following:

- strong absorption of light at the red end of the visible spectrum (especially over continents), which was caused by absorption by chlorophyll in photosynthesizing plants
- absorption bands of molecular oxygen that is also a result of plant activity, infrared absorption bands caused by the ~1 micromole per mole (µmol/mol) of methane in the Earth's atmosphere (molecular oxygen is a gas that must

be replenished by either volcanic or biological activity)
- modulated narrowband radio wave transmissions uncharacteristic of any known natural source

The Mars *Rover* mission, launched by NASA in 2003, placed two mobile rovers on the surface of Mars with the primary aim of searching for evidence that liquid water once existed on the surface of Mars. Water is essential for life as we know it, and if evidence of water could be found then it would be possible that life once evolved there. The Mars *Rovers* did find abundant evidence for past episodes of water flowing across the surface of the red planet, but so far, no direct evidence of life has been documented.

Search for Extra-Terrestrial Intelligence (SETI) is the collective name for a number of activities to detect intelligent extraterrestrial life. The SETI Institute is a research institute dedicated to the search for extraterrestrial intelligence. Their mission is to explore, understand, and explain the origin, nature, and prevalence of life in the universe. The institute employs over 150 scientists in three separate cen-

the seafloor. Extant prokaryotic organisms, including both archaeans and bacteria, currently inhabit black smoker chimneys near the East Pacific Rise, at 230°F (110°C), the highest-known temperature at which life exists on the Earth. Life at these black smokers and other similar environments draws energy from the internal energy of the Earth (not the Sun) via oxidation in a reducing environment.

Life apparently remained relatively simple for more than a billion years. Roughly 2.5 billion years ago some prokaryotic life-forms evolved into eukaryotes, organisms whose cells contain nuclei and other membrane-bound organelles. Molecular biology yields some clues about life at 2.5 Ga. Molecular phylogenies compare genetic sequences and show that all living species cluster into three domains, Archea, Bacteria, and Eukarya (which includes plants, animals, protists, and fungi). They all have a common ancestor that is thermophilic, or heat-loving. The deepest branches of the "universal tree of life" are dominated by heat-loving species. This amazing fact suggests that hydrothermal systems provided the location for the origin and development of early life. The oldest thermophiles are all chemosynthetic organisms that use hydrogen (H) and sulfur (S) as major constituents in their metabolic processes. H and S are readily available at the black smoker chimneys, adding further support to the idea that submarine hydrothermal vents may have been the site of the development of Earth's earliest life.

Late Archean (2.5 Ga) banded-iron formations (BIFs) associated with the Dongwanzi, Zunhua, and Wutai Shan ophiolites in North China have black smoker chimneys associated with them, and some of

ters, the Center for SETI Research, the Carl Sagan Center for the Study of Life in the Universe, and the Center for Education and Public Outreach. Sponsors of research in the center include NASA, the U.S. Department of Energy, U.S. Geological Survey, Argonne National Laboratory, University Space Research Association, and many others.

The Center for SETI Research seeks evidence of intelligent life in the universe through signatures of its technologies. To this end scientists in the center have developed signal-processing technologies to search for signals of intelligent life in the cosmos. This includes the development of new signal-processing algorithms and the analysis of data collected from radio telescopes and other observational media. One of the most sophisticated instruments for collecting data and searching for signals from any possible extraterrestrial intelligence is the Allen telescope array, consisting of an array of 350 radio telescopes spread across one hectare, being constructed at the Hat Creek Radio Observatory 290 miles northeast of San Francisco. This radio interferometry telescope array will be dedicated 24 hours per day, seven days a week, to the search for signals from extraterrestrial technologies.

The Sagan Center for the Study of Life in the Universe is an astrobiology research center whose focus is to examine a wide variety of problems such as modeling the precursors of life on the Earth and in the depths of outer space and to investigate how life began and how its many forms evolved and survived different conditions. At the Sagan Center, scientists ask questions like How many planets are there that might support life? What is required for life to exist? How did life evolve, and what is the range of forms possible? and How many intelligent forms of life may exist in the universe? The Sagan Center obtains its funding from NASA, the National Science Foundation, and major universities.

One way of estimating the likelihood of the existence of extraterrestrial life in the universe is called the Drake equation, named after University of California Santa Cruz astronomer Frank Drake. The Drake equation multiplies estimates of the following terms together:

- the rate of formation of suitable stars
- the fraction of those stars that link with planets
- the number of Earth-like worlds per planetary system
- the fraction of planets where intelligent life develops
- the fraction of possible communicative planets
- the "lifetime" of possible communicative civilizations

Using these parameters, Drake estimated that there are about 10,000 planets in the Milky Way Galaxy that contain intelligent life and have the possibility of communicating with the Earth. Extrapolating this equation beyond the Milky Way Galaxy, scientists estimate that there are 125 billion galaxies in the universe. If 10 percent of all Sun-like stars have a planetary system, and if even only one thousandth of 1 percent of all stars are like the Sun, and there are about 500 billion stars in each galaxy, then there are about 6.25×10^{18} stars that have planets orbiting them in the universe. If one out of 1 billion of these stars has a planet that supports life, then there could be 6.25 billion planets with life-forms on them in the universe. Thus, intelligent life may be out there.

FURTHER READING

Dick, Steven J. *Life on Other Worlds: The 20th Century Extraterrestrial Life Debate.* Cambridge: Cambridge University Press, 2001.
SETI Institute homepage. Available online. URL: http://www.seti.org/Page.aspx?pid=1241. Accessed January 14, 2009.

Burgess Shale Paleozoic landscape showing opabinia, hallucigenia, wiwaxia, and pikaia *(Publiphoto/Photo Researchers, Inc.)*

these bear signs of early life. This was a time when the Earth's surface environment began a dramatic shift from reducing environments to highly oxidizing conditions. This may be when photosynthesis first developed in sulfur-reducing bacteria. Oxygenic photosynthesis first developed in cyanobacteria, later transferred to plants (eukaryotes) through an endosymbiotic association.

Life continued to have a major role in controlling atmospheric composition and temperature for the next couple of billion years. The first well-documented ice age occurred at the Archean-Proterozoic boundary, although some evidence points to other ice ages in the Archean. The Archean-Proterozoic ice age may have been related to decreasing tectonic activity and to less CO_2 in the atmosphere. Decreasing

plate tectonic activity resulted in less CO_2 released by metamorphism and volcanism. These trends resulted in global levels of atmospheric CO_2 falling, and this in turn caused a less effective greenhouse, enhancing cooling, and leading to the ice age.

Prolonged cool periods in the Earth history are called ice houses. Most result from decreased tectonic activity and the formation of supercontinents. Intervening warm periods are called hot houses, or greenhouses. In hot house periods, higher temperatures cause more water vapor to be evaporated and stored in the atmosphere, so more rain falls during hot houses than in normal times, increasing the rates of chemical weathering, especially of calcium silicates. These free Ca and Si ions in the ocean combine with atmospheric CO_2 and O_2, to form limestone and silica that gets deposited in the oceans. This increased removal of CO_2 from the atmosphere, in turn, cools the planet in a self-regulating mechanism. The cooling reduces the rate of chemical weathering. Previously deposited calc-silicates are buried and metamorphosed, and they release CO_2, which counters the cooling from a runaway ice age effect, warming the planet in another self-regulating step.

The earliest bacteria appear to have been sulfate-reducing thermophilic organisms that dissolved sulfate by reduction to produce sulfide. In this process the bacteria oxidize organic matter, transferring the electrons to sulfur and leading to the release of CO_2 into the atmosphere and the deposition of FeS_2 (pyrite), which became massive sulfide deposits in the ocean sediments.

BIFs are rocks rich in iron and silica that are common in 2.2 to 1.6-Ga old rock sequences. They are the source of 90 percent of the world's iron ore. BIFs were probably deposited during a hot house interval and require low oxygen in the atmosphere/hydrosphere system to form. Scientists have hypothesized that water with high concentrations of Fe^{2+} was derived from weathering of crust.

Eukaryotes with membrane-bound cell nuclei emerged at about 2 Ga. Aerobic photosynthetic cells evolved and very effectively generated oxygen. These organisms rapidly built up atmospheric oxygen, to high levels by 1.6 Ga. The eukaryotes evolved into plants and animals. For the next billion years oxygen increased and CO_2 fell in the atmosphere until the late Proterozoic, when the explosion of invertebrate metazoans (jellyfish) marked the emergence of complex Phanerozoic styles of life on the Earth. This transition occurred during the formation and breakup of the supercontinent of Gondwana, with associated climate changes from a 700-Ma global ice house (supercontinent), with worldwide glaciations, to equatorial regions. This was followed by warmer climates and rapid diversification of life.

See also ARCHEAN; BLACK SMOKER CHIMNEYS; CARBON CYCLE; CLOUD, PRESTON; COMET; FOSSIL; HISTORICAL GEOLOGY.

FURTHER READING

Farmer, Jack. "Hydrothermal Systems: Doorways to Early Biosphere Evolution." *GSA Today* 10, no. 7 (2000): 1–9.
Li, J. H., and T. M. Kusky. "World's Largest Known Precambrian Fossil Black Smoker Chimneys and Associated Microbial Vent Communities, North China: Implications for Early Life." In *Tectonic Evolution of China and Adjacent Crustal Fragments, Special Issue of Gondwana Research*, vol. 12, by M. G. Zhai, W. J. Xiao, T. M. Kusky, and M. Santosh, 84–100. Amsterdam: Elsevier, 2007.
Mojzsis, Stephen, and Mark Harrison. "Vestiges of a Beginning: Clues to the Emergent Biosphere Recorded in the Oldest Known Sedimentary Rocks." *GSA Today* 10, no. 4 (2000): 1–9.
Rasmussen, Birger. "Life at 3.25 Ga in Western Australia." *Nature* 405 (2000): 676–679.

lithosphere   The top of the mantle and the crust of the Earth is a relatively cold and rigid boundary layer called the lithosphere, which is typically about 60 miles (100 km) thick. Heat escapes through the lithosphere largely by conduction, transport of heat in igneous melts, and convection cells of water through midocean ridges. The lithosphere is about 75 miles (125 km) thick under most parts of continents, and 45 miles (75 km) thick under oceans, whereas the asthenosphere extends to about a 155-mile (250-km) depth. Lithospheric roots, also known as the tectosphere, extend to about 155 miles (250 km) beneath many Archean cratons.

The base of the crust, known as the Mohorovicic discontinuity (the Moho), is defined seismically and reflects the rapid increase in seismic velocities from basalt to peridotite at five miles per second (8 km/s). Petrologists distinguish between the seismic Moho, as defined above, and the petrologic Moho, reflecting the difference between the crustal cumulate ultramafics and the depleted mantle rocks from which the crustal rocks were extracted. This petrological Moho boundary is not recognizable seismically. In contrast, the base of the lithosphere is defined rheologically as where the same rock type on either side begins to melt, and it corresponds roughly to the 2,425°F (1,330°C) isotherm, or line in two dimensions and plane in three dimensions, along which temperatures have the same value.

Since the lithosphere is rigid, it cannot convect. It loses its heat by conduction and has a high temperature contrast (and geothermal gradient) across it compared with the upper mantle, which has a more uniform temperature profile. The lithosphere thus forms a rigid, conductively cooling thermal boundary layer riding on mantle convection cells, becoming convectively recycled into the mantle at convergent boundaries.

The elastic lithosphere is that part of the outer shell of the Earth that deforms elastically, and the thickness of the elastic lithosphere increases significantly with the time from the last heating and tectonic event. This thickening of the elastic lithosphere is most pronounced under the oceans, where the elastic thickness of the lithosphere is essentially zero to a few miles (or kilometers) at the ocean ridges. The lithosphere increases in thickness proportionally to the square root of age to about a 35-mile (60-km) thickness at an age of 160 million years.

One can also measure the thickness of the lithosphere by the wavelength and amplitude of the flexural response to an induced load. The lithosphere behaves in some ways like a thin beam or ruler on the edge of a table that bends and forms a flexural bulge inward from the main load. The wavelength is proportional to and the amplitude is inversely proportional to the thickness of the flexural lithosphere under an applied load, providing a framework to interpret the thickness of the lithosphere. Natural loads include volcanoes, sedimentary prisms, thrust belts, and nappes. The load of mountains on the edges of continents tends to deflect the underlying lithosphere into a bulge with an amplitude of typically 1,000 feet (300 meters) and a wavelength several hundred miles long. In contrast, oceanic crust exhibits shorter wavelength and higher amplitude bulges, because oceanic lithosphere is not as stiff or flexurally rigid as oceanic crust. Typically the thermal, seismic, elastic, and flexural thicknesses of the lithosphere are different because each method is measuring a different physical property, and also because elastic and other models of lithospheric behavior are overly simplistic.

See also ASTHENOSPHERE; CONTINENTAL CRUST; CRATON; OPHIOLITES; PLATE TECTONICS.

FURTHER READING

Kious, Jacquelyne, and Robert I. Tilling. "U.S. Geological Survey. This Dynamic Earth: The Story of Plate Tectonics." Available online. URL: http://pubs.usgs.gov/gip/dynamic/dynamic.html. Accessed March 27, 2007.

Moores, Eldridge M., and Robert Twiss. *Tectonics.* New York: W. H. Freeman, 1995.

Skinner, Brian, and B. J. Porter. *The Dynamic Earth: An Introduction to Physical Geology.* 5th ed. New York: John Wiley & Sons, 2004.

Lyell, Sir Charles (1797–1875) Scottish *Geologist* In the early 1830s Sir Charles Lyell authored the pioneering work *Principles of Geology*, a textbook that propelled uniformitarianism into the geological mainstream and is now considered a classic in the field. With this single influential work he firmly established geology as a science by convincing geologists to study the present in order to learn about the past. Since one cannot directly observe past processes, one must compare the results of those processes (such as fossils, mountains, and lavas) with modern geological phenomena currently forming by observable processes. Though the theme of *Principles* was not novel, Lyell reintroduced Scottish geologist James Hutton's ideas with a preponderance of supporting evidence that he gathered during numerous geological excursions across Europe and North America. He also dared to profess that humans were much older than creationists believed and named several geological eras: Eocene, Miocene, and older and newer Pliocene.

PREFERENCE OF GEOLOGY OVER LAW

Charles Lyell was the oldest of 10 siblings, born November 14, 1797, at the family estate, Kinnordy, at Kirriemuir, in the county of Angus, Scotland. His mother's maiden name was Frances Smith, and his father, Charles senior, was a wealthy lawyer who enjoyed collecting rare plants. His family moved to Hampshire, England, when Charles was an infant. At the age of seven he was sent to the first of several English schools and graduated at the top of his class in June 1815. When he was 11 he suffered a bout of pleurisy, and while recovering he began insect collecting, using his father's books to identify the various species. Entomology (the study of insects) spawned a more general interest in the natural sciences that persisted throughout his life.

Lyell entered Exeter College, Oxford University in January 1816 to study Greek, Latin, and the writings of Aristotle. Having already read Robert Bakewell's *Introduction to Geology* (1813), he was anxious to attend the mineralogy and geology lectures given by William Buckland. The English geology professor was a neptunist, meaning he supported the theories of the German geologist Abraham Gottlob Werner, who proposed the then commonly accepted idea that all rocks on the Earth were formed from a vast, ancient ocean that completely covered the planet and shaped the structure of its surface with its swirling, turbulent waters. While at Oxford Lyell began making geological excursions, a practice that continued throughout his lifetime. In 1817 he studied the column-shaped formations of basalt on the island of Staffa, Scotland. German geologist Leopold von Buch had proposed that Fingal's Cave on Staffa was formed by erosion of a dike of soft lava, but Lyell observed that the basalt columns of the cave's roof had broken ends, demonstrating Buch's theory to be false. While traveling to France, Switzerland, and Italy with his family in 1818, Lyell witnessed the effect of glaciers in the Alps and recognized an age sequence in the succession of rock exposures he observed.

In 1819 Lyell became a fellow of both the Geological Society and the Linnean Society of London. In December of that year he received a bachelor of arts degree in the classics from Oxford University. At his father's request Lyell entered Lincoln's Inn to study law, but he continued to study geology. He was admitted to the bar in 1822, but poor eyesight caused his eyes to swell and hurt frequently and made legal reading difficult for him. The Geological Society elected him joint secretary in 1823 and thus demonstrated that he was accepted as a geologist by his peers.

UNIFORMITARIANISM

Lyell visited Paris that year and met several famous scientists including Georges Cuvier, Alexander von Humboldt, and Constant Prévost. The alternating layers of marine and freshwater formations in the Paris basin intrigued Lyell, and he realized that minor changes in a geological barrier could explain the pattern. In 1824 Lyell accompanied Buckland on a trip through Scotland during which he pondered the Parallel Roads of Glen Roy, admired the granite veins of Glen Tilt, and studied limestone and marl deposits in small freshwater lakes in Bailie. He read his first paper to the Geological Society in December of that year, "On a Recent Formation of Freshwater Limestone in Forfarshire," followed by more on Tertiary exposures of the Hampshire coast of England. In 1826 an article he published in the *Quarterly Review*, "Transactions of the Geological Society of London," summarized the current knowledge and major areas of investigation in geology. While composing this review, he began to believe that ordinary geological forces such as earthquakes and volcanoes could explain unusual phenomena such as the presence of sedimentary strata formed on the ocean floor but found on mountain summits. The Royal Society of London elected Lyell a fellow in 1826.

Lyell continued to make geological excursions, including trips to France, Italy, and Scotland, collecting data wherever he visited. One journey in 1828 that affected his views about geological processes brought him to France, Germany, and Italy with Scottish geologist Roderick Murchison. In central France Lyell found analogies between the geological past and formations currently developing. Again, he thought modern processes must resemble those that had shaped ancient formations. The revelation struck him that the appearance of strata was determined by the conditions when it was laid, not just by age. Similar conditions in the present day could replicate a layer with characteristics in common with an ancient stratum; thus modern conditions and geological processes must resemble those of the past.

In Italy he observed layers of lava exposed in the mountain walls of Etna, fossils of living species buried at its base, and younger uplifted strata with large percentages of extant species. These observations led to his conclusion that the volcanic mountain had been formed relatively recently, layer by layer. Accruing evidence caused him to doubt the neptunist doctrine, and he observed more physical support for the vulcanists, who believed that volcanic activity was responsible for the major changes in the construction of the Earth's surface. Cuvier believed that life on the Earth was periodically destroyed through the violent actions of catastrophic events such as floods. Lyell rejected this idea of catastrophism. Instead he believed that geological changes were caused gradually by ordinary geological processes, a theory called uniformitarianism, proposed by Scottish geologist James Hutton in 1785. Lyell believed that the steady accumulation of changes from earthquakes and volcanic activity caused the elevation and disturbances found in the sedimentary rocks. He imagined that geological processes he directly observed also occurred in the past, forming analogous structures.

PRINCIPLES AND ELEMENTS

Geologists of the time were prepared to reject Werner's ideas for Hutton's, but they needed a push. They were ready to accept that basalt was of igneous origin but more hesitant to accept uniformity of geological processes of the past and the present and of uniform gradual rates of change. Lyell published *Principles of Geology: An Attempt to Explain the Former Changes in the Earth's Surface by Reference to Causes Now in Operation,* which appeared in three volumes between 1830 and 1833 and is now considered a classic in geology. The theme of the first volume was Hutton's uniformitarianism, for which Lyell clearly presented substantial geological reasoning. His arguments urged scientists to explain geological phenomena by comparison with modern processes and conditions. By studying modern occurrences, one could gain a better understanding of the past. He reviewed the processes of erosion, sediment accumulation, volcanic activity, and uplifting by earthquakes. Change was gradual, and even major changes in the Earth's surface structure could result from the buildup of relatively subtle changes over a sufficient period of time.

The second volume of *Principles of Geology* focused on organic evolution—the change over geological time in the populations of living species, a process that Lyell considered to be fixed. Lyell stated that as former species became extinct, new distinct species emerged to maintain a continuous, natural balance. Later he altered his views on organic evolution, agreeing with Darwin that life-forms have evolved from primitive into more complex forms over time. Extinction resulted from changes in an environment's physical characteristics as well as from dynamic relationships with other species in an ecosystem. Lyell made no claims concerning the mechanism by which new species emerged.

The beginning of the third volume addressed criticisms of his first two volumes. The remainder described the application of uniformitarianism and modern analogies to geological research and presented Lyell's classification scheme of the Tertiary formations that lay just below the most recent sedimentary deposits. (The Tertiary period spans the interval from 66 to 1.8 million years ago.) He identified the species of embedded fossil shells, figured out how many of the species were still living (extant), and declared the rock beds containing lower percentages of living species to be older than the ones containing more living species. This method was relative but served its purpose. He sorted the rock formations into epochs, Eocene (the oldest), Miocene, and the older and newer Pliocene (the most recent), and suggested that as time progressed, newer species replaced those driven to extinction by geological change. This volume included an appendix that contained tables of more than 3,000 Tertiary fossil shells.

King's College in London appointed Lyell professor of geology in 1831, and the general public attended his lectures in great numbers. He held this position for only two years, preferring not to have commitments outside his own research. Lyell married Mary Elizabeth Horner on July 12, 1832, and they set up a home in London. Fluent in German and French, Mary traveled with him and a translated for him. The couple had six daughters. As Lyell's eyes degenerated with age, his wife read to him and took dictation from him.

Revising and adding to his *Principles of Geology*, which had 12 editions in his lifetime, kept Lyell busy until his death. One significant change in its 10th edition (1867–68) was the modification of the entire text to incorporate Darwin's evolutionary theory that suggested natural selection as the mechanism of action. In 1838 Lyell published an introductory geology textbook for students, *Elements of Geology*, which had six editions. (Editions three, four, and five were titled *A Manual of Elementary Geology*).

EXPERTISE ABROAD

In 1841 the Lyells traveled to the United States for the first time. Lyell delivered a series of lectures at the Lowell Institute in Boston and explored the geology of the Atlantic coast. He was not a polished lecturer, but his engagements were always filled to capacity with those interested in his vision of the planet in ancient times. While touring North America he estimated the rate of recession of Niagara Falls toward Lake Erie, studied the Tertiary formations on the coasts of Virginia, the Carolinas, and Georgia, explored the Ohio Valley, Lake Erie, and Lake Ontario, examined coal in Nova Scotia, and visited an earthquake site in New Madrid, Missouri. In 1845 he published *Travels in North America*, then returned to deliver the Lowell lectures, explore the

South including the coalfields in Alabama, investigate the growth of the delta of the Mississippi River, and collect fossils. After publishing *A Second Visit to the United States of North America* (1849), he returned again in 1852 and 1853.

Lyell traveled to Madeira and the Canary Islands in the Atlantic Ocean from 1853 to 1854 to study volcanic geology. Buch had proposed the craters of elevation theory to explain the formation of volcanic islands such as Tenerife and Palma (of the Canary Islands). He thought that volcanoes were formed by the horizontal solidification of lava, followed by violent upheaval incomparable to any modern-day geological processes, and then the collapse of masses of Earth, forming tentlike roofs over large conical caverns. From his visit to France in 1828 Lyell recalled the intact cones and craters of extinct volcanoes and the unbroken sheets of lava extending from the cones in the Auvergne. In 1859 Lyell again visited the sheets of hardened rock on the slopes of Mount Etna and Mount Vesuvius in Italy and found no center of upheaval as would be predicted by Buch's proposed mechanism. In addition he had seen modern lavas solidifying on 15–20 degree angled slopes on Madeira and Palma. On Etna he witnessed lavas solidifying on slopes of up to 40 degrees. In 1858 he published "On the Structure of Lavas Which Have Consolidated on Steep Slopes; With Remarks on the Mode of Origin of Mount Etna, and on the Theory of Craters of Elevation," a paper that invalidated the theory of craters of elevation.

THE AGE OF MAN

Analysis of the flora and fauna of the Canary Islands in addition to the species's geographical distribution induced Lyell to ponder the question of the origin of species. In 1856 further discussions with Darwin prepared him to accept with certainty the process of species evolution. Acceptance that species could evolve into new species forced Lyell to consider the prehistory of humans. Scientists were uncovering paleontological evidence that suggested humans had been around much longer than believed at the time. A human skeleton with apelike features was discovered in Neanderthal, Germany, in 1857, and in 1859 a man-made tool was found embedded in ancient river gravel in France in a location previously believed to be much older than humans. These developments were too important simply to add into new editions of *Elements* or *Principles*, so Lyell composed a new work, *The Geological Evidences of the Antiquity of Man*. The book summarized substantial data for the evolution, or gradual change, in all species and provided evidence that humans had evolved from other animal species over a long period of time. Though Darwin had published *On the Origin of Species* in 1859, he had specifically omitted any discussion on the origin of humans, saving this discussion for his 1871 book, *The Descent of Man*. The former led to much controversy concerning the evolution of humans, however, and Lyell avoided stating a clear conclusion on the matter, leaving readers to draw their own conclusions based on the presented evidence. The blatant omission upset Darwin, who had developed a close friendship with Lyell. The following year Lyell publicly declared his full support for Darwin's theory of indefinite modification of species by means of natural selection and completely revised the 10th edition of *Principles* to reflect this.

KNIGHTHOOD AND BARONETCY

Lyell's wife, Mary, died of typhoid fever in 1873. Lyell's own health had begun to fail in 1869. He died on February 22, 1875, and was buried in Westminster Abbey. Considered a classic in geology today, Lyell's *Principles of Geology* was just as popular during the 19th century, evidenced by the fact that Lyell had just finished writing the 12th edition at the time of his death. Queen Victoria conferred knighthood on Lyell in 1848 and made him a baronet in 1864. He had served as president of the Geological Society from 1834 to 1836 and again in 1849, and president of the British Association for the Advancement of Science in 1864. The Royal Society of London awarded him both the Royal Medal (1834) and the Copley Medal (1858), and the Geological Society awarded him the Wollaston Medal (1866). At his request the Lyell Medal was established in 1875 and is awarded annually by the Geological Society. He also made provisions for the disbursement of money from the Lyell Geological Fund to support the geological sciences.

British geologist and biographer Edward Bailey sums up Lyell's enduring contribution to the field of geology in *Charles Lyell*: "He did more than anyone else to free geology from the authority of tradition. He steadfastly sought truth through deduction from observation." *Principles of Geology* exerted a profound influence on geologists of the time by shifting the focus away from catastrophism to uniformitarianism, but Lyell affected scientists in other areas as well. Biologist Charles Darwin was heavily influenced by the idea of gradual change over time and applied it to his proposed theory of evolution by means of natural selection.

See also DARWIN, CHARLES; EVOLUTION; HISTORICAL GEOLOGY; HUTTON, JAMES; STRATIGRAPHY, STRATIFICATION, CYCLOTHEM; WERNER, A. G.

FURTHER READING

Bailey, Edward. *Charles Lyell*. Garden City, N.Y.: Doubleday, 1963.

Carruthers, Margaret W., and Susan Clinton. *Pioneers of Geology: Discovering Earth's Secrets*. New York: Franklin Watts, 2001.

Chavez, Miguel. "SJG Archive: People: Charles Lyell." *The Unofficial Stephen Jay Gould Archive*. Available online. URL: http://www.stephenjaygould.org/people/charles_lyell.html. Accessed August 22, 2008.

Lyell, Charles. "On the Upright Fossil Trees Found at Different Levels in the Coal Strata of Cumberland, N. S." *Quarterly Journal of the Geological Society of London* (1843): 176–178.

———. *Principles of Geology*. New Haven, Conn.: Hezekiah Howe, 1830–33.

———. *Principles of Geology*. Chicago: University of Chicago Press, reprinted 1990–91.

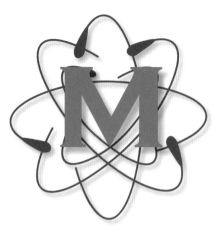

Madagascar Madagascar is the world's fourth largest island, covering 388,740 square miles (627,000 km²) in the western Indian Ocean off the coast of southeast Africa. Madagascar consists of a highland plateau fringed by a lowland coastal strip on the east, with a very steep escarpment dropping thousands of feet from the plateau to the coast over a distance of only 50–100 miles (80.5–161 km). Madagascar was separated from Africa by continental drift, and so is geologically part of the African mainland. The highest point in Madagascar is Mount Maromokotro in the north, which rises to 9,450 feet (2,882 m); the Ankaratra Mountains in the center of the plateau rise to 8,670 feet (2,645 m). The plateau dips gently to the western coast of the island toward the Mozambique Channel, where wide beaches are located. Several islands surround the main island, including Isle St. Marie in the northeast and Nosy-Be in the north. Most of the high plateau of Madagascar was once heavily forested, but intense logging over the last century has left most of the plateau a barren, rapidly eroding soil and bedrock-covered terrain. Red soil eroded from the plateau has filled many of the river estuaries along the coast.

Precambrian rocks underlie the eastern two-thirds of Madagascar, and the western third of the island is underlain by sedimentary and minor volcanic rocks that preserve a near complete record of sedimentation from the Devonian to Recent. The Ranotsara fault zone divides the Precambrian bedrock of Madagascar into two geologically different parts. The northern part is underlain by Middle and Late Archean orthogneisses, variably reworked in the Early and Late Neoproterozoic, whereas the southern part, known as the Bekily Block, consists dominantly of graphite-bearing paragneisses, bounded by

north-south trending shear zones that separate belts with prominent fold-interference patterns. All rocks south of the Ranotsara fault zone have been strongly reworked and metamorphosed to granulite conditions in the latest Neoproterozoic. Because the Ranotsara and other sinistral fault zones in Madagascar are subvertical, their intersections with Madagascar's continental margin provide ideal piercing points to match with neighboring continents in the East African Orogen. Thus, the Ranotsara fault zone is considered an extension of the Surma fault zone or the Ashwa fault zone in East Africa, or the Achankovil or Palghat-Cauvery fault zones in India. The Palghat-Cauvery fault zone changes strike (orientation relative to geographic north) to a north-south direction near the pre-breakup position of the India-Madagascar border and continues across northern Madagascar. The Precambrian rocks of northern Madagascar can be divided into three north-south trending tectonic belts defined, in part, by the regional metamorphic grade. These belts include the Bemarivo Block, the Antongil Block, and the Antananarivo Block.

The Bemarivo Block of northernmost Madagascar is underlain by calc-alkaline intrusive igneous rocks (Andriba Group) with geochemical compositions suggestive of rapid derivation from depleted mantle sources. These rocks are strikingly similar in age, chemistry, and isotopic characteristics to the granitoids of the Seychelles and Rajasthan (India). The Andriba granitoids are overlain by the Daraina-Milanoa Group (~750–714 million years old) in the north, and juxtaposed against the Sambirano Group in the south. A probable collision zone separates the Sambirano Group from the Andriba Group. The Daraina-Milanoa Group consists of two parts: a lower, largely clastic metasedimentary sequence and an upper volcanic sequence dominated by andesite with lesser basalt and rhyolite. Like the Andriba Group, volcanic rocks of the Daraina-Milanoa Group are calc-alkaline in chemistry and have geochemical signatures indicating that these rocks were derived directly from melting in the mantle. Copper (Cu) and

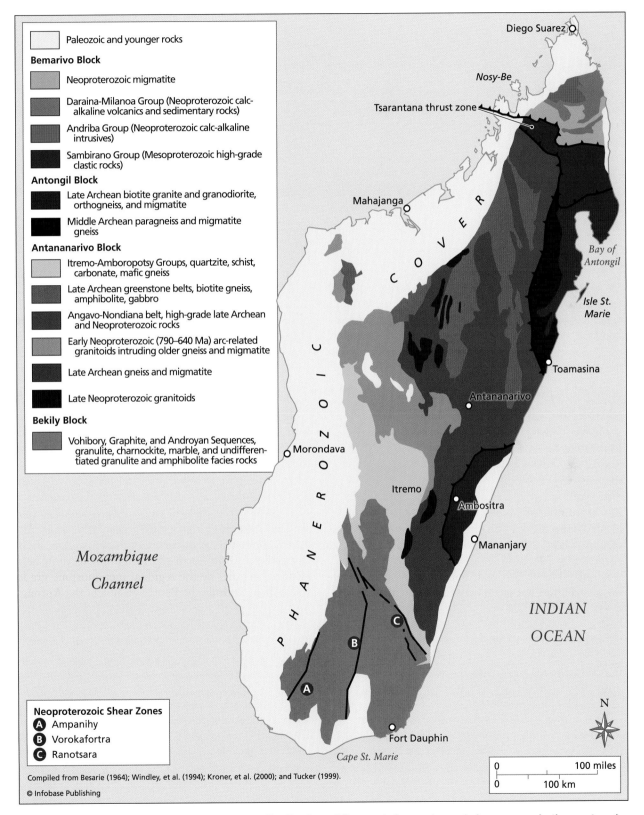

Geological map of Madagascar showing the distribution of Precambrian rocks and shear zones in the east and the Paleozoic basins in the west

gold (Au) mineralization occurs throughout the belt. The Sambirano Group consists of pelitic schist, and lesser quartzite and marble, which are variably metamorphosed to greenschist grade (in the northeast) and amphibolite grade (in the southwest). In its central part, the Sambirano Group is invaded by major massifs of migmatite gneiss and charnockite. The depositional age and provenance of the Sambirano Group is unknown.

The Antongil Block, surrounding the Bay of Antongil and Isle St. Marie, consists of late Archean biotite granite and granodiorite, migmatite, and tonalitic and amphibolitic gneiss, bounded on the west by a belt of Middle Archean metasedimentary gneiss and migmatite. These tonalitic gneisses of this region are the oldest rocks known on the island of Madagascar, dated using radiometric methods to be 3.2 billion years old. The older gneisses and migmatites are intruded by circa 2.5 billion-year-old epidote-bearing granite and granodiorite. Late Archean gneisses and migmatites near the coast in the Ambositra area may be equivalent to those near the bay of Antongil, although geochronological studies are sparse and have not yet identified middle Archean rocks in this area. Rocks of the Antongil Block have greenschist to lower-amphibolite metamorphic assemblages, in contrast to gneisses in the Antananarivo Block which tend to be metamorphosed to granulite facies. This suggests that the Antongil Block may have escaped high-grade Neoproterozoic events that affected most of the rest of the island. Gneisses in this block are broadly similar in age and lithology to the peninsular gneisses of southern India. High-grade psammites of the Sambirano Group unconformably overlie the northern part of the Antongil Block, and become increasingly deformed toward the north in the Tsarantana thrust zone, a Neoproterozoic or Cambrian collision zone between the Bemarivo Block and central Madagascar. The western margin of the Antongil Block is demarcated by a 30-mile (50-km) wide belt of pelitic metasediments with tectonic blocks of gabbro, harzburgite, and chromitites, with nickel and emerald deposits. This belt, named the Betsimisiraka suture, may mark the location of the closure of a strand of the Mozambique Ocean that separated the Antongil Block (and southern India?) from the Antananarivo Block within the Gondwanan supercontinent.

The Antananarivo Block is the largest Precambrian unit in Madagascar, consisting mainly of 2.55–2.49 billion-year-old granitoid gneisses, migmatites, and schist intruded by 1,000–640 million-year-old calc-alkaline granites, gabbro, and syenite. Rocks of the Antananarivo Block were strongly reworked by high-grade Neoproterozoic tectonism between 750 and 500 million years ago, and metamorphosed to granulite facies. Large, sheet-like granitoids of the stratoid series intruded the region, perhaps during a phase of extensional tectonism. Rocks of the Antananarivo Block were thrust to the east on the Betsimisiraka suture over the Antongil Block between 630–515 million years ago, then intruded by postcollisional granites (such as the 537–527 million-year-old Carion granite, and the Filarivo and Tomy granites) between 570–520 million years ago.

The Sèries Quartzo-Schisto-Calcaire or QSC (also known as the Itremo Group) consists of a thick sequence of Mesoproterozoic stratified rocks comprising, from presumed bottom to top, quartzite, pelite, and marble. Although strongly deformed in latest Neoproterozoic time (~570–540 million years ago), the QSC is presumed to rest unconformably on the Archean gneisses of central Madagascar because both the QSC and its basement are intruded by Early Neoproterozoic (~800 million-year-old) granitoids and no intervening period of tectonism is recognized. The minimum depositional age of the QSC is ~800 million years ago, and its maximum age, of ~1,850 million years ago, is defined by U-Pb detrital zircon geochronology. The QSC has been variably metamorphosed (~570–540 million years ago; greenschist grade in the east; amphibolite grade in the west) and repeatedly folded and faulted, but original sedimentary structures and facing-directions are well preserved. Quartzite displays features indicative of shallow subaqueous deposition, such as flat lamination, wave ripples, current ripple cross lamination, and dune cross-bedding, and carbonate rocks have preserved domal and pseudo-columnar stromatolites. To the west of the Itremo Group, rocks of the Amboropotsy and Malakialana Groups have been metamorphosed to higher grade, but include pelites, carbonates, and gabbro that may be deeper water equivalents of the Itremo Group. A few areas of gabbro/amphibolite-facies pillow lava/marble may represent strongly metamorphosed and dismembered ophiolite complexes.

Several large greenstone belts crop out in the northern part of the Antananarivo Block. These include the Maevatana, Andriamena, and Beforana-Alaotra greenstone belts, collectively called the Tsarantana sheet. Rocks in these belts include metamorphosed gabbro, mafic gneiss, tonalites, norite, and chromitites, along with pelites and minor magnetite-iron formation. Some early intrusions in these belts have been dated by Robert Tucker of Washington University in St. Louis to be between 2.75 and 2.49 billion years old, with some 3.26 billion-year-old zircon xenocrysts and Middle Archean neodymium (Nd) isotopic signatures. The chemistry, age, and nature of chromite mineralization all suggest an arc setting for the mafic rocks of the Tsarantana sheet, which is in thrust contact with underlying gneisses of the Antananarivo Block. The thrust zone is not yet well documented, but limited studies indicate east-directed thrusting. Gabbro intrusions that are 800–770 million years old cut early fabrics, but are deformed into east-vergent asymmetric folds cut by east-directed thrust faults.

The effects of Neoproterozoic orogenic processes are widespread throughout the Antananarivo Block. Archean gneisses and Mesoproterozoic stratified rocks are interpreted as the crystalline basement and platformal sedimentary cover, respectively, of a continental fragment of undetermined tectonic affinity (East or West Gondwanan, or neither). This continental fragment (both basement and cover) was extensively invaded by subduction-related plutons in the period from about 1,000 to ~720 million years ago, which were emplaced prior to the onset of regional metamorphism and deformation. Continental collision related to Gondwana's amalgamation began after ~720 million years ago and before ~570 million years ago and continued throughout the Neoproterozoic with thermal effects that lasted until about 520 million years ago. The oldest structures produced during this collision are kilometer-scale fold-and-thrust nappes with east or southeast-directed vergence (present-day direction). They resulted in the inversion and repetition of Archean and Proterozoic rocks throughout the region. During this early phase of convergence, warm rocks were thrust over cool rocks, thereby producing the present distribution of regional metamorphic isograds. The vergence of the nappes and the distribution of metamorphic rocks are consistent with their formation within a zone of west- or northwest-dipping continental convergence (present-day direction). Later upright folding of the nappes (and related folds and thrusts) produced kilometer-scale interference fold patterns. The geometry and orientation of these younger upright folds is consistent with east-west horizontal shortening (present-day direction) within a sinistral transpressive regime. This final phase of deformation may be related to motion along the Ranotsara and related shear zones of south Madagascar and to the initial phases of lower crustal exhumation and extensional tectonics within greater Gondwana.

South Madagascar, known as the Bekily Block, consists of upper amphibolite and higher-grade paragneiss bounded by north-south-striking shear zones that separate belts with prominent fold interference patterns. Archean rocks south of the Ranotsara shear zone have not been positively identified but certain orthogneisses have Archean ages (~2.9 billion years) that may represent continental basement to the paragneisses of the region. All rocks south of the Ranotsara shear zone have been strongly reworked and metamorphosed in the latest Neoproterozoic. The finite strain pattern of refolded folds results from the superimposition of at least two late Neoproterozoic deformation events characterized by early sub-horizontal foliations and a later network of kilometer-scale vertical shear zones bounding intensely folded domains. These latest upright shears are clearly related to late Neoproterozoic horizontal shortening in a transpressive regime under granulite facies conditions.

The western third of Madagascar is covered by Upper Carboniferous (300 million years old) to mid-Jurassic (180 million years old) basinal deposits that are equivalents to the Karoo and other Gondwanan sequences of Africa and India. From south to north, these include the Morondova, Majunga (or Mahajanga), and Diego (or Ambilobe) basins. Each has a similar three-fold stratigraphic division including the Sakoa, Sakamena, and Isalo Groups consisting of mainly sandstones, limestones, and basalts, overlain by unconsolidated sands in the south and along the western coast. These basins formed during rifting of Madagascar from Africa, and have conjugate margins along the east coast of southern and central Africa. The base of the Morondova basin, the oldest of the three, has spectacular glacial deposits including diamictites, tillites, and glacial outwash gravels. These are overlain by coals and arkoses, along with plant fossil (Glossopteris) rich mudstones thought to represent meandering stream deposits. Marine limestones cap the Sakoa Group. Fossiliferous deltaic and lake deposits of the Sakamena group prograde (from the east) over the Sakoa Group. The uppermost Isalo Group is 0.6–3.7 miles (1–6 km) thick, consisting of large-scale cross-bedded sandstones, overlain by red beds and fluvial deposits reflecting arid conditions. Mid-Jurassic limestones (Ankara and Kelifely Formations) mark a change to subaqueous conditions throughout the region.

EXTENSION OF THE EAST AFRICAN RIFT TO MADAGASCAR AND THE INDIAN RIDGE

Madagascar rifted from Gondwana in two stages, starting its separation from the Somali coast of Africa some 160 million years ago, and following up with a break from India and the Seychelles between 90 and 66 million years ago. The island has been tectonically isolated as an island in the Indian Ocean since the end of the Cretaceous.

The Davie Ridge runs north northwest through the Mozambique Channel to the west of Madagascar, and represents the now presumed-extinct transform along which Madagascar separated from Somalia between 160 and 117 Ma ago. Parts of the Davie Ridge are close to parallel with a prominent valley, the Alaotra-Ankay rift, that cuts the central plateau of Madagascar. Dredge samples from the Mozambique channel show that isolated blocks of Precambrian basement are preserved along this paleo-transform and that the region has seen volcanism and deformation since Madagascar reached its present location with respect to Africa. These volcanics include both Late Cretaceous fields and Cenozoic (Eocene-Miocene) flows.

With this simple post-Gondwana tectonic history in mind, Madagascar should by all rights have subdued topography, no volcanism, no earthquakes, and be a quiescent isolated continental block separated by rifting, passively weathering away as it waits to be incorporated into a collisional orogen as an exotic terrane in the future. However, Madagascar is anything but a passive isolated continental block, behaving by slowly subsiding and eroding. The thick tropical soils have in many places passed a critical threshold and are collapsing into steep canyonlands called lavakas, and the island-continent is beveled in many places by distinct erosion surfaces that have as yet escaped a unifying explanation. Some of these could be old, but others cut Cenozoic and even Neogene rocks, so much of the uplift of the island has occurred in the past 10–15 million years, perhaps spawning the current accelerated erosion. Madagascar has active earthquakes, active volcanoes, hot springs, and high juvenile topography that reaches 1.7 miles (2.8 km) above sea level within 250 miles (400 km) of the coast.

Perhaps the most striking features of Madagascar's active tectonics are the eastern coast, a straight, fault-controlled coastline that extends for more than 750 miles (1,200 km) of the 1,000-mile- (1,600-km-) long coastline, the high central Ankaratra plateau, and the oblique Ankay-Alaotra rift striking north-south through the island.

The east coast of Madagascar exhibits a juvenile, stepped topography that follows a staircase up from the coast to an elevation of more than 1.25 miles (2 km) over a cross strike distance of only 250 miles (400 km) at 22 degrees south latitude. Crustal extension and rotation of flat-topped fault blocks is demarcated by the juvenile topography along the southeast coast, and in the Ankay-Alaotra rift, similar in many ways to topography along the Gulf of Aqaba where Arabia is sliding north relative to Sinai.

At a slight angle to the coast is the Ankay-Alaotra rift, where topography falls 0.3–0.44 miles (0.5–0.7 km) across stepped fault scarps over a distance of several miles. Studies indicate tectonic activity in the Late Cretaceous, but indexes of active tectonics indicate active subsidence and relative uplift of rift shoulders as confirmed by the active seismicity, active volcanism, and the wonderfully warm hot springs of the high plateau.

Madagascar experiences thousands of earthquakes with magnitudes between 2.9–6.0 each year, and these earthquakes delineate several prominent areas of activity and northwest trends that parallel late structures on the surface. The most active area is beneath the Ankaratra plateau, where thousands of earthquakes occur annually at depths of 9–17 miles (15–28 km) and have included magnitude 5.2 and 5.5 events in 1985 and 1991. The rift valley of Lake Alaotra-Ankay graben is also seismically active, and the earthquakes show both a concentration parallel to the north-south strike of the rift, and a secondary northwest trend parallel to the swarm extending from the Ankaratra volcanic field.

Neogene to Quaternary volcanic rocks and active hot springs are found in several locations on the high central plateau of Madagascar. The Ankaratra volcanics form extensive flows and volcanic cones of basalt, basanite, and phonolite cover the Ankaratra plateau at elevations of 1.4–1.7 miles (2.3–2.7 km). Similar volcanics extend to the Itasy area to the northwest, and are found in the Neogene sediments of the Lake Alaotra rift basin. Ocean island basalts, basanites, and phonotephrites with ages of 7–10 Ma form prominent outcrops in the Ambohitra volcanic field of Nosy Be on the northern coast.

The volcanism on the northern coast of Madagascar is most likely related to young volcanism in the Comoros. Off the northwest corner of Madagascar a series of shallow marine platforms extends to the recently volcanically active Comoros Islands.

The Comoros consist of westward-younging volcanics that erupted through 55–60-mile- (~100-km-) thick, 135–142 Ma-old lithosphere. The similarity of the Comoros and Nosy Be volcanics suggests that they both formed either from migration of the Somali plate over a postulated Comoros hot spot, or northwest-propagation of rift-related faults from Madagascar formed deep faults that acted as conduits for the young magmas to reach the surface from the mantle. The progressive westward younging of the volcanics supports the hot-spot track model.

CRETACEOUS VOLCANICS AND THE MARION HOT SPOT

On the southern side of the island of Madagascar, the shallow-water Madagascar plateau extends 500 miles (800 km) south along the Marion hot spot track to Marion Island. The volcanics associated with this hot spot track correlate with the Late Cretaceous (90 Ma) volcanics abundant on the island, associated with the separation of Madagascar from the Seychelles and India. They become younger progressively to the south to zero-age volcanic on Marion Island. Madagascar was most likely flat and near sea level in the Late Cretaceous 90 million years ago, and may have been entirely covered by flood basalts from the large igneous province associated with the breakup of India, the Seychelles, and Madagascar. The flood basalts from different parts of the island had different sources, but still all formed between 92–84 Ma. In all of these cases, more than 80 million years has elapsed since Madagascar was influenced by the Marion plume. The active faulting, volcanism, and uplift are therefore not related to the passage of Madagascar over the Marion hot spot.

ACTIVE FAULTING OF LAKE ALAOTRA, CENTRAL MADAGASCAR

The Lake Alaotra-Ankay rift valley of central Madagascar forms a roughly northeast-southwest oriented depression that is filled with Neogene to Recent sediments, and is part of a more regional post-Miocene graben system that strikes north-south across much of the central part of the island. The region is characterized by a number of small earthquakes, steep fault-scarp bound valleys, several levels of terraces, and deeply incised topography related to intense tropical weathering. The origin and evolution of this extensional structure and its morphological expressions, however, are not clearly documented.

The rift valley is bordered by uplifted shoulders rising up to 0.4 miles (0.7 km) to the west and 0.3 miles (0.5 km) to the east. The biggest lake of the island, Lake Alaotra, is located adjacent to eastern termination of the Alaotra basin. The lake is 0.5 miles (0.75 km) above sea level with very shallow (average 16 feet or 5 meters) water level. It has lost 60 percent of its original size during the last 46 years due to deforestation-driven extreme erosion, and thus most of the modern basin is covered with recent alluvia, swamps, and rice fields. The eastern branch is morphologically more distinct in the field and in remotely sensed images, with several segments striking N-N35E with west-dipping fault planes, creating a series of escarpments up to 500 feet (150 m) high with well-developed triangular facets and short and steep fault scarp–bounded valleys. Terraces form prominent surfaces at several levels that correspond to successive stages in the local base level; higher, east-tilted surfaces between the highest and lowest surfaces correspond to fault displacements of a single surface. A number of faults cut basement units, forming fault zones and debris flow fans in intermountain domains of the eastern side of Lake Alaotra. These faults are mostly parallel to the structural fabric of basement rocks and form stair-step morphology. The western side of the Alaotra Basin is limited by north northeast-striking, east-dipping sub-parallel fault zones with less pronounced morphological signature.

The topographic gradient between the rift basin and shoulders is much steeper in the Ankay Basin, which is ~57 miles (92 km) long and ~20 miles (32 km) wide and bounded by a single west-dipping fault segment on the east and several east-dipping fault systems on the west. The western termination of the Ankay basin morphologically forms ~0.3-mile- (0.5-km-) high escarpments in some places. Although the border faults strike approximately north-south, they are cut by a set of northeast-southwest-striking faults. The Alaotra and Ankay basins are separated by a topographically high region, Andaingo Heights, which has dissected morphology by normal faulting but without any prominent basin fills.

POSSIBLE TECTONIC TRIGGERS OF THE ACTIVE TECTONICS OF MADAGASCAR

Several phenomena could potentially induce the active tectonic features of Madagascar described above. Passage of Madagascar and the Somali plate over a postulated Comoros hot spot is one mechanism that can explain the young volcanics and topography in northernmost Madagascar and in the Comoros, but it does not adequately explain the young topography in the center of the island, nor the active faulting in the Ankay-Alaotra rift, nor the straight eastern coastline.

Extension in the Ankay-Alaotra rift is oriented roughly the same as in the morphologically similar East African rift system, located only 300 miles (500 km) to the west. The eastern edge of the African continent is moving somewhat independently from the rest of Africa and the portion of the African plate east of the East African rift is regarded as a separate plate, the Somali plate. However, few if any plate configurations have clearly defined the southern extension of the Africa-Somali plate to where it must join the Southwest Indian Ocean ridge in order to be a true microplate bounded by plate boundaries. The plate boundary probably extends off the coast of Africa through the Comoros, then cuts through northern Madagascar and extends down the active Ankay-Alaotra rift to the fault-block dominated southeastern coast of Madagascar, then extends southeastward to the Southwest Indian Ocean ridge. If this suggestion is confirmed, then the southeastern part of the African plate is considerably fragmented by plume-related uplift and extensional events.

See also AFRICAN GEOLOGY; DIVERGENT PLATE MARGIN PROCESSES; ECONOMIC GEOLOGY; EROSION; GREENSTONE BELTS; OROGENY; PRECAMBRIAN; PROTEROZOIC.

FURTHER READING

Collins, Alan S., Ian C. W. Fitzsimons, Bregje Hulscher, and Theodore Razakamanana. "Structure of the Eastern East African Orogen in Central Madagascar." In *Evolution of the East African and Related Orogens, and the Assembly of Gondwana,* Special Issue of *Precambrian Research 123,* edited by Timothy M. Kusky, Mohamed Abdelsalam, Robert Tucker, and Robert Stern, 111–134. Amsterdam; Elsevier, 2003.

de Wit, Maarten J. "Madagascar: Heads It's a Continent, Tails It's an Island." *Annual Reviews of Earth and Planetary Sciences* 31 (2003): 213–248.

Handke, Michael J., Robert D. Tucker, and Lewis D. Ashwal. "Neoproterozoic Continental Arc Magmatism in West-Central Madagascar." *Geology* 27 (1999): 351–354.

Kusky, Timothy M., and Julian Vearncombe. "Structure of Archean Greenstone Belts." In *Tectonic Evolution of Greenstone Belts,* edited by Maarten J. de Wit and Lewis D. Ashwal, 95–128. Oxford: Oxford Monograph on Geology and Geophysics, 1997.

Tucker, Robert T., Timothy M. Kusky, Robert Buchwaldt, and Michael Handke. "Neoproterozoic Nappes and Superimposed Folding of the Itremo Group, West-Central Madagascar." *Gondwana Research* 12 (2007): 356–379.

Windley, Brian F., Adriantefison Razafiniparany, Theodore Razakamanana, and D. Ackemand. "Tectonic Framework of the Precambrian of Madagascar and Its Gondwanan Connections: A Review and Reappraisal." *Geologische Rundschau* 83 (1994): 642–659.

magma   Molten rock beneath the surface of the Earth is known as magma. When magma reaches the surface, it is known as lava, which may flow or explosively erupt from volcanoes. The wide variety of eruption styles and hazards associated with volcanoes around the world can be linked to variations in several factors, including different types of magma, different types of gases in the magma, different volumes of magma, different forces of eruption, and different areas that are affected by each eruption. Different types of magma form in distinct tectonic settings, explaining many of the differences above. Other distinctions between eruption styles are explained by the variations in processes that occur in the magma as it makes its way from deep within the Earth to the surface at a volcanic vent.

Most magma solidifies below the surface, forming igneous rocks (*igneous* is from the Latin word for fire). Igneous rocks that form below the surface are called intrusive or plutonic rocks, whereas those that crystallize on the surface are called extrusive or volcanic rocks. Rocks that crystallize at a very shallow depth are called hypabyssal rocks. Some common plutonic rock types include granites, and some of the most abundant volcanic rocks include basalts and rhyolites. Intrusive igneous rocks crystallize slowly, giving crystals an extended time to grow, thus forming rocks with large mineral grains that are clearly distinguishable with the naked eye. These rocks are called phanerites. In contrast, magma that cools rapidly forms fine-grained rocks. Aphanites are igneous rocks in which the component grains can not be distinguished readily without a microscope and are formed when magma from a volcano falls or flows across the surface and cools quickly. Some igneous rocks, known as porphyries, have two distinct populations of grain size. One group of very large crystals (called phenocrysts) is mixed with a uniform groundmass or matrix filling the space between the large crystals. This indicates two stages of cooling, as when magma has resided for a long time beneath a volcano, growing big crystals. When the volcano erupts it spews out a mixture of the large crystals and liquid magma that then cools quickly, forming the phenocrysts and the fine-grained groundmass.

MAGMA COMPOSITION AND NAMING IGNEOUS ROCKS

Determining whether an igneous rock is phaneritic or aphanitic is just the first stage in giving it a name. The second stage is determining its chemical com-

Fountaining and lava flow from Pu'u O eruption of Kilauea, Hawaii, January 31, 1984 *(J.D. Griggs, USGS)*

ponents. The composition of magma is controlled by the most abundant elements in the Earth, including silicon (Si), aluminum (Al), iron (Fe), calcium (Ca), magnesium (Mg), sodium (Na), potassium (K), hydrogen (H), and oxygen (O). Oxygen is the most abundant ion in the crust of the Earth, so petrologists usually express compositional variations of magmas in terms of oxides. For most magmas the largest constituent is represented by the combination of one silicon atom with two oxygen atoms, forming silicon dioxide, more commonly called silica. Three very narrow compositional variations in the silica content of magma are common. The first type has about 50 percent silica (SiO_2), the second 60 percent, and the third 70 percent. These volcanic rock types are called basalt, andesite, and rhyolite, and the corresponding plutonic rocks gabbro, diorite, and granite. The table "Classification of Igneous Rocks" discusses the different kinds of igneous rocks. Some of the variation in the nature of different types of volcanic eruptions can be understood by examining what causes magmas to have such a wide range in composition.

EXTRUSIVE IGNEOUS ROCKS

Magma that reaches the Earth's surface and flows as hot streams or is explosively blown out of a volcano is called lava. Lava has a range of compositions, a variety of high temperatures, and flows at various speeds.

The chemical composition of magma closely relates to how explosive and hazardous a volcanic eruption will be. The variation in the amount of silica (SiO_2) in igneous rocks is used to describe the variation in composition of igneous rocks—and the magmas that formed them. Rocks with low amounts of silica (basalt, gabbro) are known as mafic rocks, whereas rocks with high concentrations of silica (rhyolite, granite) are known as silicic or felsic rocks.

All magmas have a small amount of gas dissolved in them, usually comprising between 0.2–3 percent of the magma volume, and this is typically water vapor and carbon dioxide. The gases typically control such features as how explosive a volcanic eruption can be with greater abundances of gases leading to more explosive eruptions.

Magmas exhibit a wide range in temperatures. Measuring the temperature of an erupting volcano is difficult, since temperatures typically exceed 930°F (500°C) and melt most thermometers. Also, the volcano may explode, killing the people who try to measure its temperature. Therefore, temperature is measured from a distance using optical devices, yielding temperatures in the range of 1,900–2,200°F (1,040–1,200°C) for basaltic magma, and as low as 1,155°F (625°C) for some rhyolitic magmas.

Magmas can move downhill at very variable rates. For example, in Hawaii magma often flows downhill in magma streams at about 10 miles per hour (16 km/hr), destroying whole neighborhoods, whereas in other places it may move downhill so slowly as to be hardly detectable. At the other end of the spectrum some explosive volcanic ash clouds move downhill at speeds of several hundred miles (km) per hour, destroying all in their path. The measure of the resistance to flow of magma is called viscosity. The more viscous a magma, the less fluid it is. Honey is more viscous than water. The viscosity of magma depends on its temperature and composition. Higher-temperature magma such as basalt tends to have a higher fluidity (lower viscosity) than lower temperature magma such as rhyolite, explaining why basaltic flows tend to move over large distances, whereas rhyolitic magmas form large steep sided domes around the volcanic vent they erupted from. Magmas with more silica in them (like rhyolite) are more resistant to flow because the silica molecule forms bonds with other atoms (mostly oxygen), forming large chains and rings of molecules that offer more resistance to flow than magmas without these large interlocking molecules.

INTRUSIVE IGNEOUS BODIES

Once magmas are formed from melting rocks deep within the Earth, they rise to intrude the crust and may take several forms. A pluton is a general name for a large cooled igneous intrusive body in the Earth. The name of the specific type of pluton is based on its geometry, size, and relations to the older rocks surrounding the pluton, known as country rock. Concordant plutons have boundaries parallel to layering

CLASSIFICATION OF IGNEOUS ROCKS

Magma Types	% SiO₂	Volcanic Rock	Plutonic Rock
mafic	45–52%	basalt	gabbro
intermediate	53–65%	andesite	diorite
felsic	>65%	rhyolite	granite

in the country rock, whereas discordant plutons have boundaries that cut across layering in the country rock. Dikes are tabular but discordant intrusions, and sills are tabular and concordant intrusive rocks. Volcanic necks are conduits connecting a volcano with its underlying magma chamber. A famous example of a volcanic neck is Devils Tower in Wyoming. Some plutons are so large that they have special names. Batholiths have a surface area of more than 60 square miles (100 km²).

Batholiths contain hundreds to thousands of cubic miles of formerly molten magma, being more than 39 square miles (100 km²) on the surface and typically extending many miles deep. Scientists have long speculated on how such large volumes of magma intrude the crust and what relationships these magmas have to the style of volcanic eruption. One mechanism that may operate is assimilation, where the hot magma melts surrounding country rocks as it rises, causing it to become part of the magma. In doing this, the magma becomes cooler, and its composition changes to reflect the added melted country rock. Most geologists think that magmas may rise only a very limited distance by the process of assimilation. Some magmas may forcefully push their way into the crust if there are high pressures in the magma. One variation of this forceful emplacement style is diapirism, where the weight of surrounding rocks pushes down on the melt layer, which squeezes its way up through cracks that can expand and extend, forming volcanic vents at the surface. Stoping is a mechanism of igneous intrusion whereby big blocks of country rock above a magma body get thermally shattered, drop off the top of the magma chamber, and fall into the chamber, much like a glass ceiling breaking and falling into the space below; the magma then moves upward to take the place of the sunken blocks.

THE ORIGIN OF MAGMA

Magmas come from deep within the Earth. The processes of magma formation at depth and its movement to the surface have been the focus of research for hundreds of years. The temperature generally increases with depth in the Earth, since the surface is cool and the interior is hot. The geothermal gradient is a measure of how temperature increases with depth in the Earth, and it provides information about the depths at which melting occurs and the depths at which magmas form. The differences in the composition of the oceanic and continental crusts lead to differing abilities to conduct the heat from the interior of the Earth, and thus different geothermal gradients. The geothermal gradients show that temperatures within the Earth quickly exceed 1,830°F (1,000°C) with depth, so why are these rocks not molten? The

answer is that pressures are very high, and pressure influences the ability of a rock to melt. As the pressure rises, the temperature at which the rock melts also rises. However, this effect of pressure on melting is modified greatly by the presence of water, because wet minerals melt at lower temperatures than dry minerals. As the pressure rises, the amount of water that can be dissolved in a melt also increases. Therefore, increasing the pressure on a wet mineral has the opposite effect from increasing the pressure on a dry mineral—it decreases the melting temperature.

If a rock melts completely, the magma has the same composition as the rock. Rocks are made of many different minerals, all of which melt at different temperatures. Therefore, if a rock is slowly heated, the resulting melt or magma will first have the composition of the first mineral that melts. If the rock melts further, the melt will have the composition of the first plus the second minerals that melt, and so on. If the rock continues to completely melt, the magma will eventually end up with the same composition as the starting rock, but this does not always happen. What often occurs is that the rock only partially melts, so that the minerals with low melting temperatures contribute to the magma, whereas the minerals with high melting temperatures did not melt and are left as a residue (called a restite). In this way, the end magma can have a composition different from the rock it came from.

The phrase *magmatic differentiation by partial melting* refers to the process of forming magmas with differing compositions through the incomplete melting of rocks. For magmas formed in this way, the composition of the magma depends on both the composition of the parent rock and the percentage of melt.

BASALTIC MAGMA

Partial melting in the mantle leads to the production of basaltic magma, which forms most of the oceanic crust. By looking at the mineralogy of the oceanic crust, which is dominated by the minerals olivine (Mg_2SiO_4), pyroxene ($Mg,FeSi_2O_6$), and feldspar ($KAlSi_3O_8$), it is concluded that very little water is involved in the production of the oceanic crust. These minerals are all anhydrous, that is without water in their structure. Therefore partial melting of the upper mantle without the presence of water must lead to the formation of oceanic crust. By collecting samples of the mantle that have been erupted through volcanoes, geologists have determined that it has a composition of garnet peridotite (olivine + garnet + orthopyroxene). Experiments taking samples of this back to the laboratory and raising their temperature and pressure so that they reach the equivalent of the conditions at 60 miles (100 km) depth show that

10 percent to 15 percent partial melt of this garnet peridotite yields a basaltic magma.

Magma that forms at 60 miles (100 km) depth is less dense than the surrounding solid rock, so it rises, sometimes quite rapidly (at rates of half a mile [.8 km] per day measured by earthquakes under Hawaii). In fact, it may rise so fast that it does not cool off appreciably, erupting at the surface at more than $1,830°F$ ($1,000°C$). That is where basaltic magma comes from.

GRANITIC MAGMA

Granitic magmas are very different from basaltic magmas. They have about 20 percent more silica, and the minerals in granite include quartz (SiO_2) and the complex minerals mica (K,Na,Ca) $(Mg,Fe,Al)_2$ $AlSi_4$ O_{10} $(OH,F)_2$ and amphibole $((Mg,Fe,Ca)_2$ $(Mg,Fe,Al)_5$ $(Si,Al)_8$ $O_{22}(OH)_2)$, which both have a lot of water in their crystal structures. Also, granitic magmas are found almost exclusively in regions of continental crust. From these observations it is inferred that the source of granitic magmas is within the continental crust. Laboratory experiments suggest that when rocks with the composition of continental crust start to melt at temperature and pressure conditions found in the lower crust, a granitic liquid is formed, with 30 percent partial melting. These rocks can begin to melt either by the addition of a heat source, such as basalt intruding the lower continental crust, or by burying water-bearing minerals and rocks to these depths.

These granitic magmas rise slowly because of their high SiO_2 content and high viscosities until they reach the level in the crust where the temperature and pressure conditions are consistent with freezing or solidification of magma with this composition. This is about 3–6 miles (5–10 km) beneath the surface, which explains why large portions of the continental crust are not molten lava lakes. There are many regions with crust above large magma bodies (called batholiths) that are heated by the cooling magma. An example is Yellowstone National Park, where there are hot springs, geysers, and many features indicating that there is a large hot magma body at depth. Much of Yellowstone Park is a giant valley called a caldera, formed when an ancient volcanic eruption emptied an older batholith of its magma, and the overlying crust collapsed into the empty hole formed by the eruption.

ANDESITIC MAGMA

The average composition of the continental crust is andesitic, or somewhere between the composition of basalt and rhyolite. Laboratory experiments show that partial melting of wet oceanic crust yields an andesitic magma. Most andesites today are erupted along continental margin convergent boundaries where a slab of oceanic crust is subducted beneath the continent. Remember that oceanic crust is dry, but after it forms it interacts with seawater, which fills cracks to several miles (kilometers) depth. Also, the sediments on top of the oceanic crust are full of water, but these are for the most part nonsubductable. Andesite forms above places where water is released from the subducted slabs, and it migrates up into the mantle wedge above the subducting slab, forming water-rich magmas. These magmas then intrude the continental crust above, some forming volcanic andesites, others crystallizing as plutons of diorite at depth.

SOLIDIFICATION OF MAGMA

Just as rocks partially melt to form different liquid compositions, magmas may solidify to different minerals at different times to form different solids (rocks). This process also results in the continuous change in the composition of the magma—if one mineral is removed the resulting composition is different. If this occurs, a new magma composition results.

The removal of crystals from the melt system may occur by several processes, including the squeezing of melt away from the crystals or by the sinking of dense crystals to the bottom of a magma chamber. These processes lead to magmatic differentiation by fractional crystallization, as first described by Norman Levi Bowen. Bowen systematically documented how crystallization of the first minerals changes the composition of the magma, and results in the formation of progressively more silicic rocks with decreasing temperature.

See also CONTINENTAL CRUST; CONVERGENT PLATE MARGIN PROCESSES; DIVERGENT PLATE MARGIN PROCESSES; PETROLOGY AND PETROGRAPHY; VOLCANO.

FURTHER READING

Blong, Russel J. *Volcanic Hazards: A Sourcebook on the Effects of Eruptions.* New York: Academic Press, 1984.

Mahoney, J. J., and M. F. Coffin, eds. *Large Igneous Provinces, Continental, Oceanic, and Planetary Flood Volcanism.* Washington, D.C.: American Geophysical Union, 1997.

Volcano World home page. Available online. URL: http://volcano.oregonstate.edu/. Accessed October 10, 2008.

magnetic field, magnetosphere The Earth has a magnetic field that is generated within the core of the planet. The field is generally approximated as a dipole, with north and south poles, and magnetic field lines that emerge from the Earth at the magnetic south pole, and reenter at the magnetic north pole. The field is characterized at each place on the planet by an inclination and a declination. The inclination is a measure of how steeply inclined the field lines are with respect to the surface, with low inclinations near the equator, and steep inclinations near the poles. The declination measures the apparent angle between the rotational North Pole and the magnetic north pole.

The Earth's magnetic field orientation is commonly traced using a magnetic compass, an instrument that indicates the whole circle bearing from the magnetic meridian to a particular line of sight. It consists of a needle that aligns itself with the Earth's magnetic flux and with some type of index that allows for a numeric value for the calculation of bearing. A compass can be used for many things. The most common application is for navigation. People are able to navigate throughout the world by simply using a compass and map. The accuracy of a compass is dependent on other local magnetic influences such as man-made objects or natural abnormalities such as local geology. The compass needle does not really point true north, but is attracted and oriented by magnetic force lines that vary in different parts of the world and that are constantly changing. For example, north on a compass shows the direction toward the magnetic north pole. To offset this phenomenon a declination value is used to convert the compass reading to a usable map reading. Since the magnetic flux changes through time it is necessary to replace older maps with newer maps to insure accurate and precise up-to-date declination values.

The magnetic field originates in the liquid outer core of the Earth and is thought to result from electrical currents generated by convective motions of the iron-nickel alloy from which the outer core is made. The formation of the magnetic field by motion of the outer core is known as the geodynamo theory, pioneered by Walter M. Elsasser of Johns Hopkins University in the 1940s. The basic principle for the generation of the field is that the dynamo converts mechanical energy from the motion of the liquid outer core, which is an electrical conductor, into electromagnetic energy of the magnetic field. The convective motion of the outer core, maintained by thermal and gravitational forces, is necessary to maintain the field. If the convection stopped or if the outer core solidified, generation of the magnetic field would cease. Secular variations in the magnetic field have been well-documented by examination of the paleomagnetic record in the seafloor, lava flows, and sediments. Every few thousand years the magnetic field changes intensity and reverses, with the north and south poles abruptly flipping.

Studies of the record of ancient magnetism in rocks, called paleomagnetism, have revealed that the Earth's magnetic poles can flip suddenly, over a period of thousands or even hundreds of years. The magnetic poles also wander by about 10–20° around the rotational poles. On average, however, the magnetic poles coincide with the Earth's rotational poles. A researcher can use this coincidence to estimate the north-south directionality in ancient rocks that have drifted or rotated in response to plate tectonics. Determination of the natural remnant magnetism in rock samples can, under special circumstances, reveal the paleoinclination and paleodeclination, which can be used to estimate the direction and distance to the pole at the time the rock acquired the magnetism. If these parameters can be determined for a number of rocks of different ages on a tectonic plate, then an apparent polar wander path for that plate can be constructed. These show how the magnetic pole has apparently wandered with respect to (artificially) holding the plate fixed—when the reference frame is switched, and the pole is held fixed, the apparent polar wander curve shows how the plate has drifted on the spherical Earth.

MAGNETOSPHERE

The magnetosphere encompasses the limits of the Earth's magnetic field, as confined by the interaction of the solar wind with the planet's internal magnetic field. The natural undisturbed state of the Earth's magnetic field is broadly similar to a bar magnet, with magnetic flux lines (of equal magnetic intensity and direction) coming out of the south polar region, and returning back into the north magnetic pole. The solar wind, consisting of supersonic H^+ and $^4He^{2+}$ ions expanding away from the Sun, deforms this ideal state into a teardrop-shaped configuration known as the magnetosphere. The magnetosphere has a rounded compressed side with about 6–10 Earth radii facing the sun, and a long tail (magnetotail) on the opposite side that stretches past the orbit of the moon. The magnetotail is probably open, meaning that the magnetic flux lines never close but instead merge with the interplanetary magnetic field. The magnetosphere shields the Earth from many of the charged particles from the Sun by deflecting them around the edge of the magnetosphere, causing them to flow harmlessly into the outer solar system.

The Sun periodically experiences periods of high activity when many solar flares and sunspots form. During these periods the intensity of the solar wind emissions increases, and the solar plasma is emitted with greater velocity, in greater density, and with more energy than in its normal state. As a result of the high solar activity, the extra pressure of the solar wind distorts the magnetosphere and causes it to move around, also causing increased auroral activity.

See also AURORA, AURORA BOREALIS, AURORA AUSTRALIS; GEODYNAMICS; GEOPHYSICS.

FURTHER READING

Merrill, Ronald T., and Michael W. McElhinny. *The Earth's Magnetic Field: Its History, Origin and Planetary Perspective.* London: Academic Press, 1983.

Stern, David P., and Mauricio Peredo. *The Exploration of the Earth's Magnetosphere.* Available online. URL: http://www.istp.gsfc.nasa.gov/Education/Intro.html. Last updated November 20, 2003.

mantle The mantle forms about 80 percent of the Earth by volume, occupying the region between the crust and upper core, between about 22 and 1,802 miles (35–2,900 km) in depth. Divided into two regions, the upper and lower mantle, the crust is composed predominantly of silicate minerals in closely packed high-pressure crystal structures. Some of these are high-pressure forms of more common silicates, formed under high temperature and pressure conditions found in the deep Earth. The upper mantle extends to a depth of about 416 miles (670 km), and the lower mantle extends from there to the core-mantle boundary near 1,802 miles (2,900 km) depth.

Most knowledge about the mantle comes from seismological and experimental data as well as from rare samples of upper mantle material that has made its way to the surface in kimberlite pipes, volcanoes, and in some ophiolites and some other exhumed subcrustal rocks. Most of the mantle rock samples that have made their way to the surface are composed of peridotite, with a mixture of the minerals olivine, pyroxene, and some garnet. The velocities of seismic waves depend on the physical properties of the rocks through which they travel. Seismic experiments show that S-waves (shear waves) propagate through the mantle, and since S-waves do not propagate through liquids, the mantle must be a solid rocky layer. The temperature does not increase dramatically from the top to the bottom of the mantle, indicating that an effective heat transfer mechanism is operating in this region. Heat transfer by convection, in which the material of the mantle is flowing in large-scale rotating cells, is a very efficient mechanism that effectively keeps this region at nearly the same temperature throughout. In contrast, the lithosphere (occupying the top of the mantle and crust) cools by conduction rather than convection, so it shows a dramatic temperature increase from top to bottom.

There are several discontinuities in the mantle, where seismic velocities change across a discrete layer or zone. Between about 62 and 155 miles (100–250 km) depth, both P- and S-waves (compressional and shear waves) decrease in velocity, indicating that this zone probably contains a few percent partial melt. This low velocity zone is equated with the upper part of the asthenosphere, upon which the plates of the lithosphere move. The low velocity zone extends around most of the planet; however, it has not been detected beneath many Archean cratons that have thick roots, leading to uncertainty about the role that the low velocity zone plays in allowing these cratons to move with the tectonic plates. The asthenosphere extends to about 416–435 miles (670–700 km) depth. At 250 miles (400 km), an abrupt increase in seismic velocities occurs, associated with an isochemical phase change of the mineral olivine $((Mg,Fe)_2SiO_4)$ to a high-pressure mineral known as wadsleyite (or betaphase), and then at 323 miles (520 km) this converts to a high-pressure spinel known as ringwoodite. The base of the asthenosphere at 416 miles (670 km) is also associated with a phase change from spinel to the minerals perovskite $((Mg,Fe)SiO_3)$ and magnesiowüstite $((Mg,Fe)O)$, stable through the lower mantle or mesosphere that extends to 1,802 miles (2,900 km).

The nature of the seismic discontinuities in the mantle has been debated for decades. Professor Alfred E. "Ted" Ringwood, an Australian petrologist, proposed a model in which the composition of the mantle started off essentially homogeneous, consisting of the hypothetical composition pyrolite (82 percent harzburgite and 18 percent basalt). Extraction of basalt from the upper mantle has led to its depletion relative to the lower mantle. An alternative model, proposed by Professor Don Anderson from California Institute of Technology, poses that the mantle is compositionally layered, with the 415-mile (670-km) discontinuity representing a chemical as well as a phase boundary, with more silica-rich rocks at depth. As such, this second model requires that convection in the mantle be of a two-layer type, with little or no mixing between the upper and lower mantle to maintain the integrity of the chemical boundary. Recent variations on these themes include a two-layered convecting mantle with subducting slab penetration downward through the 415-mile (670-km) discontinuity, and mantle plumes that move up from the core mantle boundary through the 415-mile (670-km) discontinuity. In this model, the unusual region at the base of the mantle known as D″ (dee-double-prime) may be a place where many subducted slabs have accumulated.

The unusual basal D″ region of the mantle represents one of the most significant boundaries in the Earth. The viscosity contrast across the boundary is huge, being several times that of the rock/air interface. D″ is a boundary layer, so temperatures increase rapidly through the layer, and a huge seismic discontinuity exists at the boundary. P-waves drop in velocity from about 8.5 to 4 miles per second (14 to 8 km/s), and S-waves do not propagate across the boundary since the outer core is a liquid. Research into the nature of the D″ layer is active, and several ideas have emerged as possibilities for the nature of this region: the D″ layer could be a slab graveyard, where subducted slabs temporarily accumulate; a chemical reaction zone between the lower mantle and outer core; a remnant of chemical layering formed during the early accretion and differentiation of the Earth; or material that crystallized from the core and floated to accumulate at the core/mantle boundary. Whatever the case, an analog to plate tectonics may operate in this region, since it is a viscosity and thermal boundary layer subjected to basal traction forces by the rapidly convecting outer core.

See also CONVECTION AND THE EARTH'S MANTLE; ENERGY IN THE EARTH SYSTEM; LITHOSPHERE; MANTLE PLUMES; PLATE TECTONICS.

FURTHER READING

Anderson, Don L. *Theory of the Earth*. Oxford: Blackwell Scientific Publications, 1989.

Kious, Jacquelyne, and Robert I. Tilling. "U.S. Geological Survey. This Dynamic Earth: The Story of Plate Tectonics." Available online. URL: http://pubs.usgs.gov/gip/dynamic/dynamic.html. Last modified March 27, 2007.

Moores, Eldridge M., and Robert Twiss. *Tectonics*. New York: W. H. Freeman, 1995.

Ringwood, A. E. *Composition and Petrology of the Earth's Mantle*. New York: McGraw-Hill, 1975.

Skinner, Brian, and B. J. Porter. *The Dynamic Earth: An Introduction to Physical Geology*. 5th ed. New York: John Wiley & Sons, 2004.

mantle plumes The mantle of the Earth convects with large cells that generally upwell beneath the oceanic ridges and downwell with subduction zones. These convection cells are the main way that the mantle loses heat. In addition to these large cells, a number of linear plumes of hot material upwell from deep within the mantle, perhaps even from the core-mantle boundary. Heat and material in these plumes move at high velocities relative to the main mantle convection cells, and therefore burn their way through the moving mantle and reach the surface, forming thick sequences of generally basaltic lava. These lavas are chemically distinct from mid-ocean ridge and island arc basalts, and they form either as continental flood basalts, oceanic flood basalts (on oceanic plateaus), or shield volcanoes.

Mantle plumes are postulated to be upper mantle hot spots that were relatively stationary with respect to the moving plates, because a number of long linear chains of islands in the oceans were found to be parallel, and all old at one end and younger at the other end. In the 1960s when plate tectonics was first recognized, it was suggested that these hot spot tracks were formed when the plates moved over hot, partially molten spots in the upper mantle that burned their way, like a blow torch, through the lithosphere, and erupted basalts at the surface. As the plates moved, the hot spots remained stationary, so the plates had a series or chain of volcanic centers erupted through them, with the youngest volcano sitting above the active hot spot. The Hawaiian-Emperor island chain is one of the most exemplary of these hot spot tracks. Located in the north-central Pacific to the Pacific northwest near the Aleutian arc, the Hawaiian-Emperor island chain, which is about 70 million years old, shows a sharp bend in the middle of the chain where the volcanoes are 43 million years old, and then are progressively younger toward the island of Hawaii. Magmas beneath Hawaii are still molten, and are assigned an age of zero. The bend in the chain indicates a change in the plate motion direction and is reflected in a similar change in direction of many other hot spot tracks in the Pacific Ocean.

Geochemical data and seismic tomography has shown that the hot spots are produced by plumes of deep mantle material that probably rise from the D″ layer at the core-mantle boundary. These plumes may rise as a mechanism to release heat from the core or as a response to greater heat loss than is accommodated by convection. If heat is transferred from the core to D″, parts of this layer may become heated, become more buoyant, and rise as thin narrow plumes that rise buoyantly through the mantle. As they approach the base of the lithosphere the plumes expand outward, forming a mushroomlike plume head that may expand to more than 600 miles (1,000 km) in diameter. Flood basalts may rise from these plume heads, and large areas of uplift, doming, and volcanism may be located above many plume heads.

Geologists believe plumes exist beneath the African plate, such as beneath the Afar region, which has experienced uplift, rifting, and flood basalt volcanism. This region exemplifies a process whereby several (typically three) rifts propagate off a dome formed above a plume head, and several of these link up with rifts that propagated off other plumes formed over a large stationary plate. When several rifts link together, they can form a continental rift system that could become successful and expand into a young ocean basin, similar to the Red Sea. The linking of plume-related rifts has been suggested to be a mechanism to split supercontinents that have come to rest (in a geoid low) above a number of plumes. The heat from these plumes must eventually escape by burning through the lithosphere, forming linked rift systems that eventually rip apart the supercontinent.

Some areas of anomalous young volcanism may also be formed above mantle plume heads. For instance, the Yellowstone area has active volcanism and geothermal activity and is thought to rest above the Yellowstone hot spot, which has left a track extending northwest back across the flood basalts of the Snake River plain. Other flood basalt provinces probably formed in a similar way. For instance, the 65-million-year-old Deccan flood basalts of India formed when this region was over the Reunion hot spot that is presently in the Indian Ocean, and these may be related to a mantle plume.

Mantle plumes may also interact with mid-ocean ridge volcanism. For instance, the island of Iceland is located on the Reykjanes Ridge, part of the mid-Atlantic ridge system, but the height of the island is related to unusually thick oceanic crust produced in this region because a hot spot (plume) has risen directly beneath the ridge. Other examples of mantle plumes located directly beneath ridges are found in the South Atlantic Ocean, where the Walvis and Rio Grande Ridges both point back to an anomalously thick region on the present-day ridge where the plume head is located. As the South Atlantic opened, the thick crust produced at the ridge on the plume head split, half being accreted to the African plate, and half being accreted to the South American plate.

See also CONVECTION AND THE EARTH'S MANTLE; ENERGY IN THE EARTH SYSTEM; HOT SPOT; LARGE IGNEOUS PROVINCES, FLOOD BASALT; MANTLE.

FURTHER READING

Anderson, Don L. *Theory of the Earth*. Oxford: Blackwell Scientific Publications, 1989.

Kious, Jacquelyne, and Robert I. Tilling. "U.S. Geological Survey. This Dynamic Earth: The Story of Plate Tectonics." Available online. URL: http://pubs.usgs.gov/gip/dynamic/dynamic.html. Last modified March 27, 2007.

Moores, Eldridge M., and Robert Twiss. *Tectonics*. New York: W. H. Freeman, 1995.

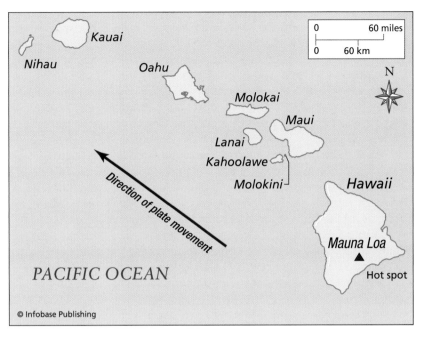

Map of Hawaii-Emperor chain formed by a hot spot track as the Pacific plate moved over a mantle plume. The youngest volcanoes are found on the island of Hawaii (Mauna Loa), and they get progressively older toward the northwest.

Ringwood, A. E. *Composition and Petrology of the Earth's Mantle*. New York: McGraw-Hill, 1975.

Skinner, Brian, and B. J. Porter. *The Dynamic Earth: An Introduction to Physical Geology*. 5th ed. New York: John Wiley & Sons, 2004.

Mars The fourth planet from the Sun, Mars is only 11 percent of the mass of Earth and has an average density of 3.9 grams per cubic centimeter and a diameter of 4,222 miles (6,794 km). Mars orbits the Sun every 687 days at a distance of 142 million miles (228 million km) and has a period of rotation about its axis of 24 hours, 37 minutes, and 23 seconds.

When viewed from Earth, Mars shows several striking surface features, including bright polar caps that consist mostly of frozen carbon dioxide (dry ice) that change in size with the seasons, almost disappearing in the Martian summer. Some spectacular canyons are also visible, including the 2,485-mile (4,000-km) long Valles Marineris. This canyon probably formed as a giant crack or fracture on the surface of the expanding bulge in the Tharsis region. Running water may have later modified its surface. Mars is prone to strong surface winds that kick up a lot of dust and generate dust storms that occasionally obscure the surface for long periods of time, an observation that led early observers to suggest that the planet may host vegetation and other life-forms.

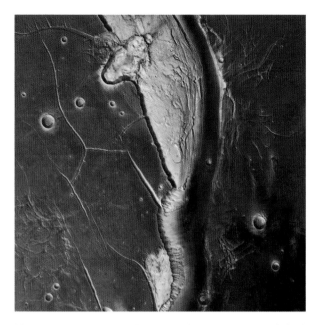

Northern main channel of Kasei Valles on Mars *(ESA/ DLR/FU Berlin, G. Neukum)*

Mars shows evidence for widespread volcanism in its past—its surface is covered with basaltic volcanic rocks, flows, and cones and hosts several large shield volcanoes in the Tharsis and Elysium regions. The Tharsis region is a huge, North America–sized bulge on the planet that rises on average 6.2 miles (10 km) above the elevation of surrounding regions. These volcanoes are huge compared to shield volcanoes on Earth. The largest volcano, Olympus Mons, is 435 miles (700 km) across, and several others are 220 to 250 miles (350–400 km) across and rise 12.5 miles (20 km) over the surrounding terrain. The northern hemisphere of Mars is made of rolling volcanic plains, similar to lunar maria (the dark flat areas on the Moon, shown to be lava flows), but formed by much larger volcanic flows than on Earth or the Moon. In contrast, the southern hemisphere consists of heavily cratered highlands, with a mean elevation several kilometers higher than the volcanic plains to the north. Estimates put the average age of the highland plains at 4 billion years, whereas the age of most of the volcanic plains in the north may be 3 billion years, with some volcanoes as young as 1 billion years old.

Recent high-resolution images of the surface of Mars have strengthened earlier views that water once ran across the surface of the planet. Outflow channels and runoff channels are common. Runoff channels form extensive systems in the southern hemisphere and resemble dried-up river systems. Outflow channels, which are most common in equatorial regions, may have formed during a catastrophic flooding epi-

sode about 3 billion years ago in the early history of the planet. Flow rates are estimated to have been at least one hundred times that of the Amazon River. The current absence of any water or visible water ice on the surface suggests that much of this water is frozen beneath the surface in a permafrost layer, reflecting a severe global cooling since 4 billion years ago.

The atmospheric pressure on Mars is only 1/150th that of Earth, and carbon dioxide makes up most of the gas in the atmosphere, with a few percent nitrogen, argon, oxygen, and carbon monoxide and less than one-tenth of a percent water vapor. The temperature rises from about -244°F at 62 miles (-153°C or 120 K at 100 km) above the surface, to about -10°F (-23°C or 250 K) at the surface, with surface temperatures on average about -82°F (-63°C or 50 K) cooler than on Earth.

Many early speculations centered on the possibility of life on Mars and several spectacular claims of evidence for life have been later found to be invalid. To date, no evidence for life, either present or ancient, has been found on Mars. The National Aeronautic and Space Administration (NASA) launched twin robot geologists to Mars in June and July of 2003, and these Rovers have been exploring the Martian surface since 2004. They have mapped a wide variety of rocks and surface structures on the planet, and returned numerous analyses and photographs. Some of the most exciting show evidence for past running water on the Martian surface and evidence that some of the landforms were carved by running water.

See also EARTH; JUPITER; MERCURY; NEPTUNE; SATURN; SOLAR SYSTEM; URANUS; VENUS.

FURTHER READING

Chaisson, Eric, and Steve McMillan. *Astronomy Today.* 4th ed. Upper Saddle River, N.J.: Prentice Hall, 2001.

Comins, Neil F. *Discovering the Universe.* 8th ed. New York: W. H. Freeman, 2008.

National Aeronautic and Space Administration. Mars Exploration Rover Mission page. Available online. URL: http://marsrover.nasa.gov/overview/. Last updated July 12, 2007.

National Aeronautic and Space Administration. Solar System Exploration page. Mars. Available online. URL: http://solarsystem.nasa.gov/planets/profile.cfm?Object=Mars. Last updated June 25, 2008.

Snow, Theodore P. *Essentials of the Dynamic Universe: An Introduction to Astronomy.* 4th ed. St. Paul, Minn.: West Publishing Company, 1991.

mass extinctions   Geologists and paleontologists study the history of life on Earth through detailed examination of that record as preserved in sedimentary rock layers laid down one upon the other. For hundreds of years paleontologists have recognized that many organisms are found in a series of layers, and then suddenly disappear at a certain horizon never to reappear in the succeeding progressively younger layers. These disappearances have been interpreted to mark extinctions of the organisms from the biosphere. After hundreds of years of work, many of these rock layers have been dated by using radioactive decay dating techniques on volcanic rocks in the sequences, and many of these sedimentary rock sequences have been correlated with each other on a global scale. Some are associated with other deposits along the layers, indicating unusual concentrations of the metal iridium, carbon from massive fires, and other impact-related features such as tektites and tsunami deposits.

One of the more important findings resulting from such detailed studies is that the rock record preserves evidence of several extinction events that have occurred simultaneously on a global scale. Furthermore, these events do not just affect thousands or hundreds of thousand of members of a species, but they have wiped out many species and families each containing millions or billions of individuals.

PROCESSES THAT DRIVE EVOLUTION AND EXTINCTION

The progression of life-forms, evolution, and extinction can be influenced or driven by many factors including impacts with space objects such as meteorites and comets, variations in the style of plate tectonics or the positions of the continents, the supercontinent cycle, and continental collisions. Plate tectonics also may cause glaciation and climate changes, which in turn influence evolution and extinction.

One of the primary mechanisms by which plate tectonics drives evolution and extinction is through tectonic-induced changes to sea level. Fluctuating sea levels cause the global climate to fluctuate between

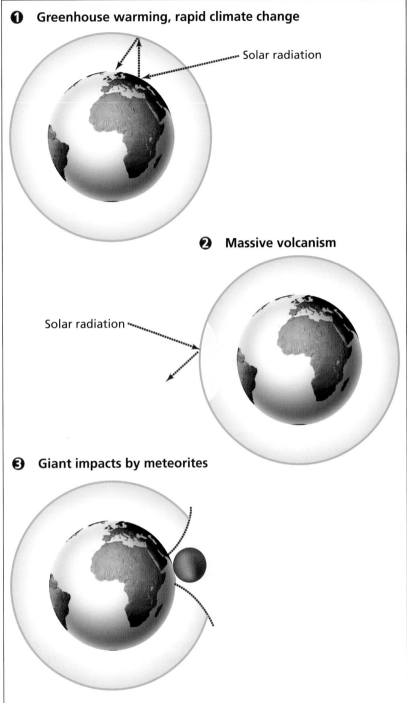

The main causes for mass extinctions, including (1) greenhouse warming and rapid climate change, followed by (2) massive volcanism, and (3) giant impact from space objects such as asteroids or comets

warm periods when shallow seas are easily heated, and cold periods when glaciation draws the water down to place shorelines along the steep continental slopes. Many species cannot tolerate such variations in temperature and drastic changes to their shallow shelf environments, and thus become extinct. After organisms from a specific environment die off, their environmental niches are available for other species to inhabit.

Sea levels have risen and fallen dramatically in Earth history, with water covering all but 5 percent of the land surface at times, and water falling so that continents occupy 40 percent or more of the planet's surface at other times. The most important plate tectonic mechanism of changing sea level is to change the average depth of the seafloor by changing the volume of the mid-ocean ridge system. If the undersea ridges take up more space in the ocean basins, then the water will be displaced higher onto the land, much like dropping pebbles into a birdbath may cause it to overflow.

The volume of the mid-ocean ridge system can be changed through several mechanisms, all of which have the same effect. Young oceanic crust is hotter, more buoyant, and topographically higher than older crust. Thus, if the average age of the oceanic crust is decreased, then more of the crust will be at shallow depths, displacing more water onto the continents. If seafloor spreading rates are increased then the average age of oceanic crust will be decreased, the volume of the ridges will be increased, the average age of the seafloor will be decreased, and sea levels will rise. This has happened at several times in Earth history, including during the mid-Cretaceous between 110–85 Ma when sea levels were 660 feet (200 m) higher than they are today, covering much of the central United States and other low-lying continents with water. This also warmed global climates, because the Sun easily warmed the abundant shallow seas. It has also been suggested that sea levels were consistently much higher in the Precambrian, when seafloor spreading rates were likely to have been generally faster.

Sea levels can also rise from additional magmatism on the seafloor. If the Earth goes through a period when seafloor volcanoes erupt more magma on the seafloor, then the space occupied by these volcanic deposits will be displaced onto the continents. The additional volcanic rocks may be erupted at hot spot volcanoes like Hawaii or along the mid-ocean ridge system; either way, the result is the same.

A third way for the mid-ocean ridge volume to increase sea level height is to simply have more ridges on the seafloor. At the present time the mid-ocean ridge system is 40,000 miles (65,000 km) long. If the Earth goes through a period where it needs to lose more heat, such as in the Precambrian, one of the ways it may do this is by increasing the length of the ridge system where magmas erupt and lose heat to the seawater. Ridge lengths were probably greater in the Precambrian, which together with faster sea floor spreading and increased magmatism may have kept sea levels high for millions of years.

Glaciation that may be induced by tectonic or astronomical causes may also change sea level. At present glaciers cover much of Antarctica, Greenland, and mountain ranges in several regions. Approximately 6 million cubic miles (25,000,000 km^3) of ice is currently locked up in glaciers. If this ice were to all melt, then sea levels would rise by 230 feet (70 m), covering many coastal regions, cities, and interior farmland with shallow seas. During the last glacial maximum in the Pleistocene ice ages (20,000 years ago) sea levels were 460 feet (140 m) lower than today, with shorelines up to hundred of miles (km) seaward of their present locations, along the continental slopes.

Continental collisions and especially the formation of supercontinents can cause glaciations. When continents collide, many of the carbonate rocks deposited on continental shelves are exposed to weathering. As the carbonates and other minerals weather, the products react with atmospheric elements and tend to combine with atmospheric carbon dioxide (CO_2). Carbon dioxide is a greenhouse gas that keeps the climate warm, and steady reductions of CO_2 in the atmosphere by weathering or other processes lowers global temperatures. Thus, times of continental collision and supercontinent formation tend to be times that draw CO_2 out of the atmosphere, plunging the Earth into a cold "icehouse" period. In the cases of supercontinent formation, this icehouse may remain in effect until the supercontinent breaks up, and massive amounts of seafloor volcanism associated with new rifts and ridges add new CO_2 back into the atmosphere.

The formation and dispersal of supercontinent fragments, and migrating landmasses in general, also strongly influence evolution and extinction. When supercontinents break up, a large amount of shallow continental shelf area is created. Life-forms tend to flourish in the diverse environments on the continental shelves, and many spurts in evolution have occurred in the shallow shelf areas. In contrast, when continental areas are isolated, such as Australia and Madagascar today, life-forms evolve independently on these continents. If plate tectonics brings these isolated continents into contact, the different species will compete for similar food and environments, and typically only the strongest survive.

The position of continents relative to the spin axes (or poles) of the Earth can also influence climate, evolution, and extinction. At times (like the present) when a continent is sitting on one or both of the poles, these continents tend to accumulate snow and ice, and to become heavily glaciated. This causes ocean currents to become colder, lowers global sea levels, and reflects more of the Sun's radiation back to space. Together, these effects can put a large amount of stress on species, inducing or aiding extinction.

THE HISTORY OF LIFE AND MASS EXTINCTIONS

Life on Earth has evolved from simple prokaryotic organisms such as Archaea that appeared on Earth by 3.85 billion years ago. Life may have been here earlier, but the record is not preserved, and the method by which life first appeared is also unknown and the subject of much thought and research by scientists, philosophers, and religious scholars.

The ancient Archaea derived energy from breaking down chemical bonds of carbon dioxide, water, and nitrogen, and anaerobic archaeans have survived to this day in environments where they are not poisoned by oxygen. They presently live around hot vents around mid-ocean spreading centers, deep in the ground in pore spaces between soil and mineral grains, and in hot springs. The Archaea represent one of the three main branches of life; the other two branches are the Bacteria and the Eukarya. The plant and animal kingdoms are members of the domain Eukarya.

Prokaryotic bacteria (single-celled organisms lacking a cell nucleus) were involved in photosynthesis by 3.5 billion years ago, gradually transforming atmospheric carbon dioxide to oxygen and setting the stage for the evolution of simple eukaryotes (organisms containing a cell nucleus and membrane-bound organelles) in the Proterozoic. Two and half billion years later, by one billion years ago, cells began reproducing sexually. This long-awaited step allowed cells to exchange and share genetic material, speeding up evolutionary changes by orders of magnitude.

Oxygen continued to build in the atmosphere, and some of this oxygen was combined into ozone (O_3). Ozone forms a layer in the atmosphere that blocks ultraviolet rays of the Sun, forming an effective shield against this harmful radiation. When the ozone shield became thick enough to block a large portion of the ultraviolet radiation, life began to migrate out of the deep parts of the ocean and deep in land soils, into shallow water and places exposed to the Sun. Multicellular life evolved around 670 million years ago, around the same time that the supercontinent of Gondwana was forming near the equator. Most of the planet's landmasses were joined together for a short while, and then began splitting up and drifting apart again by 550 million years ago. This breakup of the supercontinent of Gondwana is associated with the most remarkable diversification of life in the history of the planet. In an incredibly short period of no longer than 40 million years, life developed complex forms with hard shells, and an incredible number of species appeared for the first time. This period of change marked the transition from the Precambrian era to the Cambrian period, marking the beginning of the Paleozoic era. The remarkable development of life in this period is known as the Cambrian explosion. In the past 540 million years since the Cambrian explosion, life has continued to diversify with many new species appearing.

The evolution of life-forms is also punctuated with the disappearance or extinction of many species, some as isolated cases, and others that die off at the same time as many other species in the rock record. A number of distinct horizons represent times when hundreds, thousands, and even more species suddenly died, being abundant in the record immediately before the formation of one rock layer and absent immediately above that layer. Mass extinctions are typically followed, after several million years, by the appearance of many new species and the expansion and evolution of old species that did not go extinct. These rapid changes are probably a response to availability of environmental niches vacated by the extinct organisms. The new species rapidly populate these available spaces.

Mass extinction events are thought to represent major environmental catastrophes on a global scale. In some cases these mass extinction events can be tied to specific likely causes, such as meteorite impact or massive volcanism, but in others their cause is unknown. Understanding the triggers of mass extinctions has important and obvious implications for ensuring the survival of the human race.

EXAMPLES OF MASS EXTINCTIONS

Most species are present on Earth for about 4 million years. Many species come and go during background level extinctions and evolution of new species from old, but the majority of changes occur during the distinct mass dyings and repopulation of the environment. The Earth's biosphere has experienced five major and numerous less significant mass extinctions in the past 500 million years (in the Phanerozoic era). These events occurred at the end of the Ordovician, in the Late Devonian, at the Permian-Triassic boundary, the Triassic-Jurassic boundary, and at the Cretaceous-Tertiary (K-T) boundary.

The early Paleozoic saw many new life-forms emerge in new environments for the first time. The Cambrian explosion led to the development of trilobites, brachiopods, conodonts, mollusks, echinoderms, and ostracods. Bryozoans, crinoids, and rugose corals joined the biosphere in the Ordovician, and reef-building stromatoporoids flourished in shallow seas. The end-Ordovician extinction is one of the greatest of all Phanerozoic time. About half of all species of brachiopods and bryozoans died off, and more than 100 other families of marine organisms disappeared forever.

The cause of the mass extinction at the end of the Ordovician appears to have been largely tectonic, as no meteorite impacts or massive volcanic outpourings are known from this time. The major landmass of Gondwana had been resting in equatorial regions for much of the Middle Ordovician, but migrated toward the South Pole at the end of the Ordovician. This caused global cooling and glaciation, lowering sea levels from the high stand they had been resting at for most of the Cambrian and Ordovician. The combination of cold climates with lower sea levels, leading to a loss of shallow shelf environments for habitation, probably was enough to cause the mass extinction at the end of the Ordovician.

The largest mass extinction in Earth history occurred at the Permian-Triassic boundary, over a period of about 5 million years. The Permian world included abundant corals, crinoids, bryozoans, and bivalves in the oceans, and on land, amphibians wandered about amid lush plant life. At the end of the Permian, 90 percent of oceanic species were to become extinct, and 70 percent of land vertebrates died off. This great catastrophe in Earth history did not have a single cause, but reflects the combination of various elements.

First, plate tectonics was again bringing many of the planet's land masses together in a supercontinent (this time, Pangaea), causing greater competition for fewer environmental niches by Permian life-forms. The rich continental shelf areas were drastically reduced. As the continents collided, mountains were pushed up, reducing the effective volume of the continents available to displace the sea, so sea levels fell, putting additional stress on life by further limiting the availability of favorable environmental niches. The global climate became dry and dusty, and the supercontinent formation led to widespread glaciation. This lowered the sea level even more, lowered global temperatures, and put many life-forms on the planet in a very uncomfortable position, and many of these went extinct.

In the final million years of the Permian, northern Siberia witnessed massive volcanic outpouring that dealt life a devastating blow. The Siberian flood basalts began erupting at 250 million years ago, becoming the largest known outpouring of continental flood basalts ever. Carbon dioxide was released in hitherto unknown abundance, warming the atmosphere and melting the glaciers. Other gases were also released, perhaps also including methane, as the basalts probably melted permafrost and vaporized thick accumulations of organic matter in high latitudes like that at which Siberia was located 250 million years ago.

The global biosphere collapsed, and evidence suggests that the final collapse happened in less than 200,000 years, and perhaps in less than 30,000 years. Entirely internal processes may have caused the end-Permian extinction, although some scientists now argue that an impact may have dealt the final death blow. For some time it was argued that the Manicouagan impact crater in Canada may be the crater from an impact that caused the Permian-Triassic extinction, but radiometric dating of the Manicouagan structure showed it to be millions of years too old to be related to the mass extinction. After the Permian-Triassic extinction was over, new life-forms populated the seas and land, and these Mesozoic organisms tended to be more mobile and adept than their Paleozoic counterparts. The great Permian extinction created opportunities for new life-forms to occupy now empty niches, and the most adaptable and efficient organisms took control. The toughest of the marine organisms survived, and a new class of land animals grew to new proportions and occupied the land and skies. The Mesozoic, time of the great dinosaurs, had begun.

The Triassic-Jurassic extinction is not as significant as the Permian-Triassic extinction. Mollusks were abundant in the Triassic shallow marine realm, with fewer brachiopods, and ammonoids recovered from near total extinction at the Permian-Triassic boundary. Sea urchins became abundant, and new groups of hexacorals replaced the rugose corals. Many land plants survived the end-Permian extinction, including the ferns and seed ferns that became abundant in the Jurassic. Small mammals that survived the end-Permian extinction rediversified in the Triassic, many only to become extinct at the close of the Triassic. Dinosaurs evolved quickly in the late Triassic, starting off small, and attaining sizes approaching 20 feet (6 m) by the end of the Triassic. The giant pterosaurs were the first known flying vertebrates, appearing late in the Triassic. Crocodiles, frogs, and turtles lived along with the dinosaurs. The end of the Triassic is marked by a major extinction in the marine realm, including total extinction of the conodonts and a mass extinction of the mammal-like reptiles known as therapsids and the placodont marine reptiles. Although the causes of this major extinction event are poorly understood, the timing is coincident with the breakup of Pangaea and the formation of major evaporite and salt deposits. It is likely that this was a tectonic-induced extinction, with supercontinent breakup initiating new oceanic circulation patterns and new temperature and salinity distributions.

After the Triassic-Jurassic extinction, dinosaurs became extremely diverse and many grew quite large. Birds first appeared at the end of the Jurassic. The Jurassic was the time of the giant dinosaurs, which experienced a partial extinction affecting the largest varieties of stegosauroids, sauropods, and the marine ichthyosaurs and plesiosaurs. This major extinction is also poorly explained but may be related to global cooling. The other abundant varieties of dinosaurs continued to thrive through the Cretaceous.

The Cretaceous-Tertiary (K-T) extinction is perhaps the most famous of mass extinctions because the dinosaurs perished during this event. The Cretaceous land surface of North America was occupied by bountiful species, including herds of dinosaurs both large and small, some herbivores, and other carnivores. Other vertebrates included crocodiles, turtles, frogs, and several types of small mammals. The sky had flying dinosaurs, including the vulture-like pterosaurs, and insects, including giant dragonflies. The dinosaurs had dense vegetation to feed on, including the flowing angiosperm trees, tall grasses, and many other types of trees and flowers. Life in the ocean had evolved to include abundant bivalves, including clams and oysters, ammonoids, and corals that built large reef complexes.

Near the end of the Cretaceous, though the dinosaurs and other life-forms did not know it, things were about to change. High sea levels produced by mid-Cretaceous rapid seafloor spreading were falling, decreasing environmental diversity, cooling global climates, and creating environmental stress. Massive volcanic outpourings in the Deccan traps and the Seychelles formed as the Indian Ocean rifted apart and magma rose from an underlying mantle plume. Massive amounts of greenhouse gases were released, raising temperatures and stressing the environment. Many marine species were going extinct, and others became severely stressed. Then one day a visitor from space about six miles (10 km) across slammed into the Yucatán Peninsula of Mexico, instantly forming a fireball 1,200 miles (2,000 km) across, followed by giant tsunamis perhaps thousands of feet (hundreds of meters) tall. The dust from the fireball plunged the world into a dusty fiery darkness with months or years of freezing temperatures, followed by an intense global warming. Few species handled the environmental stress well, and more than a quarter of all the plant and animal kingdom families, including 65 percent of all species on the planet, became extinct forever. Dinosaurs, mighty rulers of the Triassic, Jurassic, and Cretaceous, were gone forever. Oceanic reptiles and ammonoids died off, and 60 percent of marine planktonic organisms went extinct. The great K-T extinctions affected not only the numbers of species, but also the living biomass—the death of so many marine plankton species alone amounted to 40 percent of all living matter on Earth at the time. Similar punches to land-based organisms decreased the overall living biomass on the planet to a small fraction of what it was before the K-T one-two-three knockout blows.

Some evidence suggests that the planet is undergoing the first stages of a new mass extinction. In the past 100,000 years, the ice ages have led to glacial advances and retreats, sea level rises and falls, the appearance and rapid explosion of human (*Homo sapiens sapiens*) populations, and the mass extinction of many large mammals. In Australia 86 percent of large (>100 pounds) animals have become extinct in the past 100,000 years, and in South America, North America, and Africa the extinction is an alarming 79 percent, 73 percent, and 14 percent. This ongoing mass extinction appears to be the result of cold climates and more importantly, predation and environmental destruction by humans. The loss of large-bodied species in many cases has immediately followed the arrival of humans in the region, with the clearest examples being found in Australia, Madagascar, and New Zealand. Similar loss of races through disease and famine has accompanied many invasions and explorations of new lands by humans throughout history.

The history of life on Earth shows that many species exist relatively unchanged for long periods of time and may diversify or become more specialized, in part in response to environmental changes. Occasionally, huge numbers of different species and individuals within species suddenly die off or are killed in mass extinction events. Some of these seem to result from a combination of severe environmental stresses, greatly enhanced volcanism, or a combination of different effects. Most mass extinction events are now known to also be associated with an impact event. However, not all large impact events are associated with a mass extinction, with a prime example being the Manicouagan impact structure, which formed from an impact occuring 214 million years ago, 12 million years older than the Permian-Triassic mass extinction. It seems that mass extinctions may be triggered by a combination of different forces all acting together, including environmental stresses from changes in climate associated with plate tectonics and orbital variations, from massive volcanism, and from impacts of meteorites or comets with Earth.

See also EVOLUTION; IMPACT CRATER STRUCTURES; METEOR, METEORITE.

FURTHER READING

Albritton, C. C., Jr. *Catastrophic Episodes in Earth History*. London: Chapman and Hale, 1989.

Alvarez, Walter. *T. Rex and the Crater of Doom*. Princeton, N.J.: Princeton University Press, 1997.

Blong, Russel J. *Volcanic Hazards: A Sourcebook on the Effects of Eruptions*. New York: Academic Press, 1984.

Eldredge, N. *Fossils: The Evolution and Extinction of Species*. Princeton, N.J.: Princeton University Press, 1997.

Fisher, R. V., G. Heiken, and J. B. Hulen. *Volcanoes: Crucibles of Change*. Princeton, N.J.: Princeton University Press, 1998.

Francis, Peter. *Volcanoes: A Planetary Perspective*. Oxford: Oxford University Press, 1993.

Martin, P. S., and R. G. Klein, eds. *Quaternary Extinctions*. Tucson: University of Arizona Press, 1989.

Melosh, H. Jay. *Impact Cratering: A Geologic Process*. Oxford: Oxford University Press, 1988.

Robock, Alan, and Clive Oppenheimer, eds. *Volcanism and the Earth's Atmosphere*. Washington D.C.: American Geophysical Union, 2003.

Scarth, Alwyn. *Vulcan's Fury, Man against the Volcano*. New Haven, Conn.: Yale University Press, 1999.

Sepkoski, J. J., Jr. *Mass Extinctions in the Phanerozoic Oceans: A Review*. In *Patterns and Processes in the History of Life*. Amsterdam: Springer-Verlag, 1982.

———. *Earth and Life through Time*. New York: W. H. Freeman, 1986.

Stanley, Steven M. *Extinction*. New York: Scientific American Library, 1987.

mass wasting   Mass wasting is the movement of soil, rock, and other Earth materials (together called regolith) downslope by gravity without the direct aid of a transporting medium such as ice, water, or wind. An estimated 2 million or more mass movements occur each year in the United States alone. Mass movements occur at various rates, from a few inches (few centimeters) per year to sudden catastrophic rock falls and avalanches that can bury entire towns under tons of rock and debris. In general, the faster the mass movement, the more hazardous it is to humans, although even slow movements of soil down hill slopes can be extremely destructive to buildings, pipelines, and other construction and infrastructure. In the United States alone, mass movements kill tens of people and cost more than 1.5 billion dollars a year. Other mass movement events overseas have killed tens to hundreds of thousands of people in a matter of seconds. Mass wasting occurs under a wide variety of environmental conditions and forms a continuum with weathering, as periods of intense rain reduce friction between regolith and bedrock, making movement easier. Mass movements also occur underwater, such as the giant submarine landslides associated with the 1964 Alaskan earthquake.

Mass movements are a serious concern and problem in hilly or mountainous terrain, especially for buildings, roadways, and other features engineered into hillsides, in addition to along riverbanks and in places with large submarine escarpments, such as along deltas (like the Mississippi Delta in Louisiana). Seismic shaking and severe storm-related flooding further compound the problem in some prone areas.

Imagine building a million-dollar mansion on a scenic hillside, only to find it tilting and sliding down the hill at a few inches (5–10 centimeters) per year. Less spectacular but common effects of slow downhill mass movements are the slow tilting of telephone poles along hillsides and the slumping of soil from oversteepened embankments onto roadways during storms.

Mass wasting is becoming more of a problem as the population moves from the overpopulated flat land to new developments in hilly terrain. In the past, small landslides in the mountains, hills, and canyons were not a serious threat to people, but now large numbers of people live in landslide-prone areas, thus landslide hazards and damage are rapidly increasing.

DRIVING FORCES OF MASS WASTING

Gravity is the main driving force behind mass-wasting processes, as it is constantly attempting to force material downhill. On a slope, gravity can be resolved into two components, one perpendicular to the slope, and one parallel to the slope. The steeper the angle of the slope, the greater the influence of gravity. The effect of gravity reaches a maximum along vertical or overhanging cliffs.

The tangential component of gravity tends to pull material downhill, resulting in mass wasting. When gt (the tangential component of gravity) is great enough to overcome the force of friction at the base of the loose mass, it falls downhill. The friction is really a measure of the resistance to gravity—the greater the friction, the greater the resistance to gravity's pull. Lubrication of surfaces in contact greatly reduces friction, allowing the two materials to slide past one another more easily. Water is a common lubricating agent, so mass-wasting events tend to occur more frequently during times of heavy or prolonged rain. For a mass-wasting event or a mass movement to occur, the lubricating forces must be strong enough to overcome the resisting forces that tend to hold the boulder (for example) in place against the wishes of gravity. Resisting forces include the cohesion between similar particles (like one clay molecule to another) and the adhesion between different or unlike particles (like the boulder to the clay beneath it). When the resisting forces are greater than the driving force (tangential component of gravity), the slope is steady and the boulder stays in place. When lubricating components reduce the resisting forces so much that the driving forces are greater than the resisting forces, slope failure occurs.

The process of the movement of regolith downslope (or under water) may occur rapidly, or it may proceed slowly. In any case, slopes on mountainsides typically evolve toward steady state angles, known as the angle of repose, balanced by material moving in from upslope, and out from downslope. The angle of repose is also a function of the grain size of the regolith.

Human activity can also increase driving forces for mass wasting. Excavation for buildings, roads, or other cultural features along the lower portions of slopes may actually remove parts of the slopes, causing them to become steeper than they were before construction and to exceed the angle of repose. This will cause the slopes to be unstable (or metastable) and susceptible to collapse. Building structures on the tops of slopes will also destabilize them, as the extra weight of the building adds extra stresses to the slope that could initiate the collapse of the slope.

PHYSICAL CONDITIONS THAT CONTROL MASS WASTING

Many factors control whether or not mass wasting occurs and, if it does, the type of mass wasting. These include characteristics of the regolith and bedrock, the presence or absence of water, overburden, angle of the slope, and the way that the particles are packed together.

Preexisting weaknesses in the rock that facilitate movement along them strongly influence mass wasting in solid bedrock terrain. For instance, bedding planes, joints, and fractures, if favorably oriented, can act as planes of weakness along which giant slabs of rock may slide downslope. Rock or regolith containing many pores, or open spaces between grains, is weaker than a rock without pores, because no material fills the pores, whereas if the open spaces were filled the material in the pore space could hold the rock together. Furthermore, pore spaces allow fluids to pass through the rock or regolith, and the fluids may further dissolve the rock, creating more pore space, and further weakening the material. Water in open pore space may also exert pressure on the surrounding rocks, pushing individual grains apart, making the rock weaker.

Water can either enhance or inhibit movement of regolith and rock downhill. Water inhibits downslope movement when the pore spaces are only partly filled with water and the surface tension (bonding of water molecules along the surface) acts as an additional force holding grains together. The surface tension bonds water grains to each other, water grains to rock particles, and rock particles to each other. An everyday example of how effective surface tension may be at holding particles together is found in sand castles at the beach—when the sand is wet, tall towers can be constructed, but when the sand is dry, only simple piles of sand can be made.

Water more typically acts to reduce the adhesion between grains, promoting downslope movements. When the pore spaces are filled, the water acts as a lubricant and exerts forces that push individual grains apart. The weight of the water in pore spaces also exerts additional pressure on underlying rocks and soils, known as loading. When the loading from water in pore spaces exceeds the strength of the underlying rocks and soil, the slope fails, resulting in a downslope movement.

Another important effect of water in pore spaces occurs when the water freezes; freezing causes the water to expand by a few percent, and this expansion exerts enormous pressures on surrounding rocks, in many cases pushing them apart. The freeze-thaw cycles found in many climates are responsible for many of the downslope movements.

Steep slopes are less stable than shallow slopes. Loose unconsolidated material tends to form slopes at specific angles that range from about 33–37 degrees, depending on the specific characteristics of the material. The arrangement or packing of the particles in the slope is also a factor; the denser the packing, the more stable the slope.

PROCESSES OF MASS WASTING

Mass movements are of three basic types, distinguished from each other by the way that the rock, soil, water, and debris move. Slides move over and in contact with the underlying surface, while flows include movements of regolith, rock, water, and air in which the moving mass breaks into many pieces that flow in a chaotic mass movement. Falls move freely through the air and land at the base of the slope or escarpment. A continuum exists between different processes of mass wasting, but many differ in terms of the velocity of downslope movement and also in the relative concentrations of sediment, water, and air. A landslide is a general name for any downslope movement of a mass of bedrock, regolith, or a mixture of rock and soil and indicates any mass wasting process.

A slump is a type of sliding slope failure in which a downward and outward rotational movement of rock or regolith occurs along a concave upslip surface. This produces either a singular or a series of rotated blocks, each with the original ground surface tilted in the same direction. Slumps are especially common after heavy rainfalls and earthquakes and are common along roadsides and other slopes that have been artificially steepened to make room for buildings or other structures. Slump blocks can continue to move after the initial sliding event, and in some cases this added slippage is enhanced by rainwater that falls on the back-tilted surfaces, infiltrates along the fault, and acts as a lubricant for added fault slippage.

A translational slide is a variation of a slump in which the sliding mass moves not on a curved surface, but moves downslope on a preexisting plane, such as a weak bedding plane or a joint. Translational slides may remain relatively coherent or break into small blocks forming a debris slide.

When mixtures of rock debris, water, and air begin to move under the force of gravity, they are said to flow. This is a type of deformation that is continuous and irreversible. The way in which this mixture flows depends on the relative amounts of solid, liquid, and air, the grain size distribution of the solid fraction, and the physical and chemical properties of the sediment. Mass-wasting processes that involve flow are transitional within themselves, and to stream-type flows in the amounts of sediment/water and in velocity. Many names for the different types of sediment flows include slurry flows, mudflows, debris flows, debris avalanches, earthflows, and loess flows. Many mass movements begin as one type of flow and evolve into another during the course of the mass-wasting event. For instance, flows commonly begin as rock falls or debris avalanches and evolve into debris flows or mudflows, as the flow picks up water and debris and flows over differing slopes along its length.

Creep is the slow, downslope-flowing movement of regolith; it involves the very slow plastic deformation of the regolith, as well as repeated microfracturing of bedrock at nearly imperceptible rates. Creep occurs throughout the upper parts of the regolith, and there is no single surface along which slip has occurred. The rates range from a fraction of an inch (about a centimeter) per year up to about two inches (five cm) per year on steep slopes. Creep accounts for leaning telephone poles, fences, and many of the cracks in sidewalks and roads. Although creep is slow and not very spectacular, it is one of the most important mechanisms of mass wasting, accounting for the greatest total volume of material moved downhill in any given year. One of the most common creep mechanisms is through frost heaving, an extremely effective means for moving rocks, soil, and regolith downhill. The ground freezes and ice crystals form and grow, pushing rocks upward perpendicular to the surface. As the ice melts in the freeze-thaw cycle, gravity takes over and the pebble or rock moves vertically downward, ending up a fraction of an inch (centimeter) downhill from where it started. Other mechanisms of surface expansion and contraction, such as warming and cooling, or the expansion and contraction of clay minerals with changes in moisture levels can also initiate creep. In a related phenomenon, the freeze-thaw cycle can push rocks upward through the soil profile, as revealed by farmers' fields in New England and other northern climates, where the fields seem to grow boulders. The fields are cleared of rocks, and years later, the

same fields are filled with numerous boulders at the surface. In these cases, the freezing forms ice crystals below the boulders that push them upward; during the thaw cycle, the ice around the edges of the boulder melts first, and mud and soil seep down into the crack, finding their way beneath the boulder. This process, repeated over years, is able to lift boulders to the surface, keeping the northern farmer busy.

The operation of the freeze-thaw cycle makes rates of creep faster on steep slopes than on gentle slopes, with more water, and with greater numbers of freeze-thaw cycles. Rates of creep of up to half an inch (1 cm) per year are common.

Solifluction, the slow viscous downslope movement of water-logged soil and debris, is most common in polar latitudes where the top layer of permafrost melts, resulting in a water-saturated mixture resting on a frozen base. This process is also common to very wet climates, as found in the Tropics. Rates of movement are typically an inch or two (2.5–5 cm) per year, slightly faster than downslope flow by creep. Solifluction results in distinctive surface features, such as lobes and sheets, carrying the overlying vegetation; sometimes the lobes override each other, forming complex structures. Solifluction lobes are relatively common sights on mountainous slopes in wet climates, especially in areas with permafrost. The frozen layer beneath the soil prevents drainage of water deep into the soil or into the bedrock, so the uppermost layers in permafrost terrains tend to be saturated with water, aiding solifluction.

A slurry flow is a moving mass of sediment saturated in water that is transported with the flowing mass. The mixture is so dense that it can suspend large boulders or roll them along the base. When slurry flows stop moving, the resulting deposit therefore consists of a nonsorted mass of mud, boulders, and finer sediment.

Debris flows involve the downslope movement of unconsolidated regolith, most of which is coarser than sand. Some debris flows begin as slumps, but then continue to flow downhill as debris flows. They fan out and come to rest when they emerge out of steeply sloping mountain valleys onto lower-sloping plains. Rates of movement in debris flows vary from several feet (1 m) per year to several hundred miles (kilometers) per hour. Debris flows are commonly shaped like a tongue with numerous ridges and depressions. Many form after heavy rainfalls in mountainous areas, and the number of debris flows is increasing with greater deforestation of mountain and hilly areas. This is particularly obvious on the island of Madagascar, where deforestation in places is occurring at an alarming rate, removing most of the island's trees. What was once a tropical rain forest is now a barren (but geologically spectacular) landscape, carved by numerous landslides and debris flows that bring the terra rossa soil to rivers, making them run red to the sea.

Most debris flows that begin as rock falls or avalanches move outward in relatively flat terrain less than twice the distance they fell. Internal friction (between particles in the flow) and external friction (especially along the base of the flow) slow them. Some of the largest debris flows that originated as avalanches or debris falls travel exceptionally large distances at high velocities—these are debris avalanches and sturzstrom deposits.

Mudflows resemble debris flows, except that they have higher concentrations of water (up to 30 percent), making them more fluid, with a consistency ranging from soup to wet concrete. Mudflows often start as a muddy stream in a dry mountain canyon. As it moves it picks up more and more mud and sand, until eventually the front of the stream is a wall of moving mud and rock. When this comes out of the canyon, the wall commonly breaks open, spilling the water behind it in a gushing flood, which moves the mud around on the valley floor. These types of deposits form many of the gentle slopes at the bases of mountains in the southwestern United States.

Mudflows have also become a hazard in highly urbanized areas such as Los Angeles, where most of the dry riverbeds have been paved over and development has moved into the mountains surrounding the basin. The rare rainfall events in these areas then have no place to infiltrate, and they rush rapidly into the city, picking up all kinds of street mud and debris and forming walls of moving mud that cover streets and low-lying homes in debris. Unfortunately, after the storm rains and waters recede, the mud remains and hardens in place. Mudflows are also common with the first heavy rains after prolonged droughts or fires, as many residents of California and other western states know. After the drought and fires of 1989 in Santa Barbara, California, heavy rains brought mudflows down out of the mountains, filling the riverbeds and inundating homes with many feet of mud. Similar mudflows followed the heavy rains in Malibu in 1994, which remobilized barren soil exposed by the fires of 1993. Three to four feet (more than a meter) of mud filled many homes and covered parts of the Pacific Coast highway. Mudflows are part of the natural geologic cycle in mountainous areas, and they serve to maintain equilibrium between the rate of uplift of the mountains and their erosion. Mudflows are catastrophic only when people have built homes, highways, and businesses in their paths.

Volcanoes, too, can produce mudflows. Rain or an eruption easily remobilizes layers of ash and volcanic debris, sometimes mixed with snow and ice, that can travel many tens of miles (kilometers). Volcanic mudflows are known as lahars. Mudflows have killed tens of thousands of people in single events and have been some of the most destructive of mass movements.

Granular flows are unlike slurry flows, in that, in granular flows, the full weight of the flowing sediment is supported by grain to grain contact between individual grains. Earthflows are relatively fast granular flows with velocities ranging from three feet per day to 1,180 feet per hour (1 m/day to 360 m/hour).

Rockfalls are the free falling of detached bodies of bedrock from a cliff or steep slope. They are common in areas of very steep slopes, where rockfall deposits form huge deposits of boulders at the base of the cliff. Rockfalls can involve a single boulder, or the entire face of a cliff. Debris falls are similar to rockfalls, but these consist of a mixture of rock and weathered debris and regolith.

Rockfalls have been responsible for the destruction of parts of many villages in the Alps and other steep mountain ranges, and rockfall deposits have dammed many a river valley, creating lakes behind the newly fallen mass. Some of these natural dams have been extended and heightened by engineers to make reservoirs, with examples including Lake Bonneville on the Columbia River and the Cheakamus Dam in British Columbia. Smaller examples abound in many mountainous terrains.

Rockslide is the term given to the sudden downslope movement of newly detached masses of bedrock (or debris slides, if the rocks are mixed with other material or regolith). These are common in glaciated mountains with steep slopes and also in places having planes of weakness, such as bedding planes or fracture planes that dip in the direction of the slope. Like rockfalls, rockslides can form fields of huge boulders coming off mountain slopes. The movement to this talus slope is by falling, rolling, and sliding, and the steepest angle at which the debris remains stable is known as the angle of repose, typically 33–37 degrees for most rocks.

Debris avalanches are granular flows moving at very high velocity and covering large distances. These rare, destructive (but spectacular) events have ruined entire towns, killing tens of thousands of people in them without warning. Some have been known to move as fast as 250 miles per hour (400 km/hr). These avalanches thus can move so fast that they move down one slope, then thunder right up and over the next slope and into the next valley. One theory of why these avalanches move so fast is that when the rocks first fall, they trap a cushion of air, and then travel on top of it like a hovercraft. Two of the worst debris avalanches in recent history originated from the same mountain, Nevados Huascaran, the highest peak in the Peruvian Andes. More than

22,000 people died in these two debris avalanches. Numerous debris avalanches were also triggered by the May 12, 2008, magnitude 7.9 earthquake in China that killed nearly 90,000 people.

SUBMARINE LANDSLIDES

Mass wasting is not confined to the land. Submarine mass movements are common and widespread on the continental shelves, slopes, and rises, and also in lakes. Mass movements under water, however, typically form turbidity currents, which leave large deposits of graded sand and shale. Under water, these slope failures can begin on very gentle slopes, even of < 1°. Other submarine slope failures are similar to slope failures on land.

Slides and slumps and debris flows are also common in the submarine realm. Submarine deltas, deep-sea trenches, and continental slopes are common sites of submarine slumps, slides, and debris flows. Some of these are huge, covering hundreds of square miles (<500 km²). Many of the mass-wasting events that produced these deposits must have produced large tsunamis. The continental slopes are cut by many canyons, produced by submarine mass-wasting events, which carried material eroded from the continents into the deep ocean basins.

Triggering Mechanisms for Submarine Mass Wasting

Submarine mass-wasting events are triggered by many phenomena, some similar to those in the subaerial realm and some different. Shaking by earthquakes and displacement by faulting can initiate submarine mass-wasting events, as can rapid release of water during sediment compaction. High sedimentation rates in deltas, continental shelves and slopes, and other depositional environments may create unstable slopes that can fail spontaneously or through triggers including agitation by storm waves. In some instances, sudden release of methane and other gases in the submarine realm may trigger mass-wasting events.

Submarine landslides tend to be larger than avalanches that originate above the water line. Many submarine landslides are earthquake-induced, whereas others are triggered by storm events and by increases in pressure induced by sea-level rise on the sediments on passive margins or continental shelves. A deeper water column above the sediments on shelf or slope environment can significantly increase the pressure in the pores of these sediments, causing them to become unstable and slide downslope. Since the last glacial retreat 6,000–10,000 years ago, sea levels have risen by 320–425 feet (98–130 m), which has greatly increased the pore pressure on continental slope sediments around the world. This increase in pressure is

thought to have initiated many submarine landslides, including the large Storegga slides from 7,950 years ago off the coast of Norway.

Many areas beneath the sea are characterized by steep slopes, including areas along most continental margins, around islands, and along convergent plate boundaries. Sediments near deep-sea trenches are often saturated in water and close to the point of failure, where the slope gives out and collapses, causing the pile of sediments to suddenly slide down to deeper water depths. When an earthquake strikes these areas large parts of the submarine slopes may give out simultaneously, often generating a tsunami.

Some steep submarine slopes that are not characterized by earthquakes may also be capable of generating huge tsunamis. Recent studies along the east coast of North America, off the coast of Atlantic City, New Jersey, have revealed that large sections of the continental shelf and slope are on the verge of failure. The submarine geology off the coast of eastern North America consists of a pile several thousands of feet (hundreds of m) thick of unconsolidated sediments on the continental slope. These sediments are so porous and saturated with water that the entire slope is near the point of collapsing under its own weight. A storm or minor earthquake may be enough to trigger a giant submarine landslide in this area, possibly generating a tsunami that could sweep across the beaches of Long Island, New Jersey, Delaware, and much of the east coast of the United States. With rising sea levels, the pressure on the continental shelves is increasing, and storms and other events may more easily trigger submarine landslides.

Storms are capable of generating submarine landslides even if the storm waves do not reach and disrupt the seafloor. Large storms are associated with storm surges that form a mound of water in front of the storm that may sometimes reach 20–32 feet (6–10 m) in height. As the storm surge moves onto the continental shelf it is often preceded by a drop in sea level caused by a drop in air pressure, so the storm surge may be associated with large pressure changes on the seafloor and in the pores of unconsolidated sediments. A famous example of a storm surge–induced submarine landslide and tsunami is the catastrophic event in Tokyo, Japan, on September 1, 1923. On this day, a powerful typhoon swept across Tokyo and was followed that evening by a huge submarine landslide and earthquake, which generated a 36-foot- (11-m-) tall tsunami that swept across Tokyo killing 143,000 people. Surveys of the seabed after the tsunami revealed that large sections had slid to sea, deepening the bay in many places by 300–650 feet (91–198 m), and locally by as much as 1,300 feet (396 m). Similar storm-induced slides are known from many continental slopes and delta environments, including the Mississippi delta in the Gulf of Mexico and the coasts of Central America.

Types of Submarine Landslides

Submarine slides are part of a larger group of processes that can move material downslope on the seafloor, including other related processes such as slumps, debris flows, grain flows, and turbidity currents. Submarine slumps are a type of sliding slope failure in which a downward and outward rotational movement of the slope occurs along a concave upslip surface. This produces either a singular or a series of rotated blocks, each with the original seafloor surface tilted in the same direction. Slumps can move large amounts of material short distances in short times and are capable of generating tsunamis. Debris flows involve the downslope movement of unconsolidated sediment and water, most of which is coarser than sand. Some debris flows begin as slumps, but then continue to flow downslope as debris flows. They fan out and come to rest when they emerge out of submarine canyons onto flat abyssal plains on the deep seafloor. Rates of movement in debris flows vary from several feet (1 m) per year to several hundred miles per hour (500 km/hr). Debris flows are commonly shaped like a tongue with numerous ridges and depressions. Large debris flows can suddenly move large volumes of sediment, so are also capable of generating tsunamis. Turbidity currents are sudden movements of water-saturated sediments that move downhill underwater under the force of gravity. These form when water-saturated sediment on a shelf or shallow water setting is disturbed by a storm, earthquake, or some other mechanism that triggers the sliding of the sediment downslope. The sediment-laden sediment/water mixture then moves rapidly downslope as a density current and may travel tens or even hundreds of miles at tens of miles (km) per hour until the slope decreases and the velocity of current decreases. As the velocity of the current decreases, the ability of the current to hold coarse material in suspension decreases. The current drops first its coarsest load, then progressively finer material as the current decreases further.

Many volcanic hot-spot islands in the middle of some oceans show evidence that they repeatedly generate submarine landslides. These islands include Hawaii in the Pacific Ocean, the Cape Verde Islands in the North Atlantic, and Réunion in the Indian Ocean. The shape of many of these islands bears the tell-tale starfish shape with cuspate scars indicating the locations of old curved landslide surfaces. These islands are volcanically active, and lava flows move across the surface, and then cool and crystallize quickly as the lava enters the water. This causes the islands to grow upward as very steep-sided columns, whose sides are prone to massive collapse and submarine sliding. Many volcanic islands are built up with a series of volcanic growth periods followed by massive submarine landslides, effectively widening the island as it grows. However, island growth by deposition of a series of volcanic flows over older landslide scars causes the island to be unstable—the old landslide scars are prone to later slip since they are weak surfaces, and the added stress of the new material piled on top of them makes them unstable. Other processes may also contribute to making these surfaces and the parts of the island above them unstable. For instance, on the Hawaiian Islands volcanic dikes have intruded along some old landslide scars, which can reduce the strength across the old surfaces by large amounts. Some parts of Hawaii are moving away from the main parts of the island by up to 0.5–4 inches (1–10 cm) per year by the intrusion of volcanic dikes along old slip surfaces. Also, many landslide surfaces are characterized by accumulations of weathered material and blocks of rubble, that under the additional weight of new volcanic flows can help to reduce the friction on the old slip surfaces, aiding the generation of new landslides. Therefore, as the islands grow, they are prone to additional large submarine slides that may generate tsunamis.

EXAMPLES OF LANDSLIDE DISASTERS

Mass wasting is one of the most costly of natural hazards, with the slow downslope creep of material causing billions of dollars in damage to properties every year in the United States. Earth movements do not kill many people in most years, but occasionally massive landslides take thousands or even hundreds of thousands of lives. Mass wasting is becoming more of a hazard in the United States as people move in great numbers from the plains into mountainous areas as population increases. This trend is expected to continue in the future, and more mass-wasting events like those described in this chapter may be expected every year. Good engineering practices and understanding of the driving forces of mass wasting will hopefully prevent many mass-wasting events, but it will be virtually impossible to stop the costly gradual downslope creep of material, especially in areas with freeze thaw cycles.

Examining the details of a few of the more significant mass-wasting events of different types, including a translational slide, a rockfall-debris avalanche, a mudflow, and whole scale collapse of an entire region is useful to help mitigate future landslide disasters. In this section lessons that can be learned from each different landslide are discussed with the aim that education can save lives in the future. The table "Significant Landslide Disasters" lists some of the more significant landslide and mass-wasting disasters with

SIGNIFICANT LANDSLIDE DISASTERS

Where	When	Trigger Process	How Many Deaths
Shaanxi Province, China	1556	m. 8 earthquake	830,000
Shaanxi Province, China	1920	earthquake	200,000
Sichuan Province, China	2008	m. 7.9 earthquake	87,587
Nevados Huascarán, Peru	1970	m. 7.7 earthquake	70,000
Nevado del Ruiz, Colombia	1985	volcanic eruption	20,000
Tadzhik Republic	1949	m.7.5 earthquake	12,000–20,000
Honduras, Nicaragua	1998	heavy rain	10,000
Venezuela	1999		10,000
Nevados Huascarán, Peru	1962		4,000–5,000
Vaiont, Italy	1963	heavy rain	3,000
Rio de Janeiro, Brazil	1966–67	heavy rain	2,700
Mount Coto, Switzerland	1618		2,430
Cauca, Colombia	1994	m. 6.4 earthquake	1,971
Serra das Araras, Brazil	1967	heavy rain	1,700
Leyte, Philippines	2006	heavy rain	1,450
Kure, Japan	1945		1,145
Shizuoka, Japan	1958	heavy rain	1,094
Rio de Janeiro, Brazil	1966	heavy rain	1,000
Napo, Ecuador	1987	m. 6.1 and 6.9 earthquakes	1,000

more than 1,000 deaths reported, with a bias toward events of the last 100 years.

Shaanxi, China, January 23, 1556

The deadliest earthquake and mass-wasting event on record occurred in 1556 in the central Chinese province of Shaanxi. Most of the 830,000 deaths from this earthquake resulted from landslides and the collapse of homes built into loess, a deposit of wind-blown dust that covers much of central China. The loess represents the fine-grained soil eroded from the Gobi desert to the north and west and deposited by wind on the great loess plateau of central China. Thus, this disaster was triggered by an earthquake but mass-wasting processes were actually responsible for most of the casualties.

The earthquake that triggered this disaster on the morning of January 23, 1556, leveled a 520-mile-wide area and caused significant damage across 97 counties in the provinces of Shaanxi, Shanxi, Henan, Hebei, Hubei, Shandong, Gansu, Jiangsu, and Anhui. Sixty percent of the population was killed in some counties. There were no modern seismic instruments at the time, but seismologists estimate that the earthquake had a magnitude of 8 on the Richter scale, with an epicenter near Mount Hua in Hua County in Shaanxi.

The reason for the unusually high death toll in this earthquake is that most people in the region at the time lived in homes carved out of the soft loess, or silty soil. People in the region would carve homes, called Yaodongs, out of the soft loess, benefit from the cool summer temperatures and moderate winter temperatures of the soil, and also have an escape from the sun and blowing dust that characterizes the loess plateau. The shaking from the magnitude 8 earthquake caused huge numbers of these Yaodongs to collapse, trapping the residents inside. Landslides raced down steep loess-covered slopes, and the long shaking caused the Yaodongs even in flat areas to collapse.

Time tends to make people forget about risks associated with natural hazards. For events that occur only every couple of hundred years, several generations may pass between catastrophic events, and each generation remembers less about the risks than the previous generation. This character of human nature was unfortunately illustrated by another earthquake in central China, nearly 400 years later. In 1920, a large earthquake in Haiyuan, in the Ningxia Authority of northern Shaanxi Province, caused about 675 major landslides in deposits of loess, killing another 100,000–200,000 people. Further south in 2008, the May 12 magnitude 7.9 earthquake in Sichuan Province similarly initiated massive landslides that killed an estimated 87,587 people.

Sichuan Province, China, May 12, 2008

The devastating magnitude 7.9 earthquake in the Longmenshan ranges of Sichuan Province, China, triggered thousands of landslides, including several gigantic landslides that buried several villages and blocked the Qingshui and Hongshi Rivers, forming two lakes, in Donghekou and Shibangou. Two of the first scientists to reach and study the Donghekou landslide after the earthquake were the British geologist Jian Guo Liu and the American geologist T. M. Kusky, who estimated the volume of material that collapsed and buried local villages. They mapped the thickness of the toe of the deposit, the length of the slide (>2 miles, or 3 km) and width (average >2,500 feet or 1 km), and estimated that at least 20–30 million cubic yards (15–25 million m³) of material raced down the slopes in this single landslide. The staggering reality is that during the earthquake, there were hundreds, if not thousands, of similar giant rock avalanches. The landslides and mass wasting were responsible for a large portion of the known 87,587 deaths from this earthquake.

A mountain on the outskirts of the buried village of Donghekou collapsed, and a large mass of earth slipped downhill at such a high speed it jumped a small mountain, plunged across a river, and then ran a thousand feet (300 m) up a mountain slope across from the village. The top of the mountain has many huge displaced blocks that slid away from the headwall scarp, suggesting that the event started as a massive rock slide/fall, and as the mass moved downhill, it quickly became a debris avalanche as the material in the fall was pulverized into a mixture of large boulders and finer material. Debris from the avalanche formed a natural dam in the river, blocking it and forming an avalanche lake. The scale of this massive landslide is impressive, covering several square miles (km²) of the surface, and being up to several hundred feet (hundred or more m) thick. The Donghekou landslide is about 30 percent larger than

the giant landslide/debris avalanche that buried the community of Barangay Guinsaugon in Leyte, Philippines, on February 17, 2006, killing 1,450 people.

Massive rockfall avalanches and rockflow rubble streams can travel at high speeds over large distances, particularly where a minimum of 650,000 cubic yards (500,000 m³) of rock falls at least 500 feet (150 m) onto a slope equal to or exceeding 25°. These landslides are able to travel so far because they must conserve mass and momentum. Mapping of the Donghekou landslide by geologists Liu and Kusky suggests that the momentum was maintained by a loss of friction along the base of the landslide, by having the falling mass ride on a cushion of compressed air. The Donghekou and Shibangou landslide avalanches were associated with air blasts that shot out from beneath the falling debris. Rapidly falling rock and soil compressed the air beneath the avalanches and allowed the material to move downhill quickly on a cushion of compressed air for great distances, and as the avalanches came to a halt the air was ejected from beneath the debris in strong jets that knocked down trees in radial patterns reaching a thousand feet (several hundred m) up opposing mountain slopes from the site of the avalanche tongues, in patterns reminiscent of bomb blasts.

The dam blocking the course of Qingshui River was channelled by the army using explosives to allow water to discharge gently, avoiding catastrophic dam collapse and flooding. The devastating consequence of landslides in the region is obvious, with several villages and schools buried by this slide. The landslide in Shibangou also buried a village and blocked the Qingshui River a second time, forming another lake upstream. The landslide at Shibangou is interesting in that the amphitheatre-shaped headscarp detachment surface at the top of the slide is visible, and it shows a typical slumplike crown with transverse cracks and fault zones, large slide blocks moving into the main slide, back rotation of slump blocks on the upper part of the slide, then a transition where the slide turns into an avalanche and eventually a debris avalanche grading into a debris field. This slide was also associated with powerful airblasts, as the trees are blown over in complex patterns going uphill and down valley away from the toe of the avalanche.

Vaiont, Italy, October 9, 1963

One of the worst translational slides ever occurred in Vaiont, Italy, in 1963. In 1960 a large dam was built in a deep valley in northern Italy. The valley occupies the core of a synclinal fold in which the limestone rock layers dip inward toward the valley center and form steeply dipping bedding surfaces along the valley walls. The valley floor was filled with glacial sediments left as the glaciers from the last ice age melted a few thousand years ago. The bottom of the valley was oversteepened by downcutting from streams, forming a steep V-shaped valley in the middle of a larger U-shaped glacial valley. The rocks on the sides of the valley are highly fractured, and broken into many individual blocks, and there was an extensive cave network carved in the limestones. The reservoir behind the dam held approximately 500 million cubic feet (14 million m³) of water.

After the dam was constructed the pores and caves in the limestone filled with water, exerting extra, unanticipated pressures on the valley walls and dam. Heavy rains in the fall of 1963 made the problem worse, and authorities predicted that sections of the valley might experience landslides. The rocks on the slopes surrounding the reservoir began creeping downhill, first at a quarter inch per day (0.3–0.65 cm/day), then accelerating to one and a half inches per day (4 cm per day) by October 6. Even though authorities were expecting landslides, they had no idea of the scale of what was about to unfold.

At 10:41 P.M., on October 9, 1963, a 1.1-mile- (1.8-km-) long and 1-mile- (1.6-km-) wide section of the south wall of the reservoir suddenly failed, and slid into the reservoir at more than 60 miles per hour (96 km/hr). Approximately 9.5 billion cubic feet (270,000,000 m³) of debris fell into the reservoir, creating an earthquake shock and an air blast that shattered windows and blew roofs off nearby houses. The debris that fell into the reservoir displaced a huge amount of water, and a series of monstrous waves was generated that raced out of the reservoir devastating nearby towns. A 780-foot- (238-m-) tall wave moved out of the north side of the reservoir, followed by a 328-foot- (100-m-) tall second wave. The waves combined and formed a 230-foot- (70-m-) tall wall of water that moved down the Vaiont Valley, inundating the town of Longarone, where more than 2,000 people were killed by the fast-moving floodwaters. Other waves bounced off the walls of the reservoir and emerged out the upper end of the reservoir, smashing into the town of San Martino, where another 1,000 people were killed in the raging waters.

Landslides in the Andes Mountains; Nevados Huascarán, Peru, 1962, 1970

The Andes, a steep mountain range in South America, are affected by frequent earthquakes and volcanic eruptions, are glaciated in places, and experience frequent storms from being close to the Pacific Ocean. All of these factors combine, resulting in many landslides and related mass-wasting disasters in the Andes. Some of the most catastrophic land-slides in the Andes have emanated from Nevados Huascarán, a tall peak on the slopes of the Cordillera Blanca in the Peruvian province of Ancash. In 1962, a large debris avalanche with an estimated volume of 16,900,000 cubic yards (13,000,000 m³) rushed down the slopes of Nevados Huascarán at an average velocity of 105 miles per hour (170 km/hr). The debris avalanche buried the village of Ranrahirca, killing 4,000–5,000 people. This scene of devastation was to be repeated eight years later. On May 31, 1970, a magnitude 7.7 earthquake was centered about 22 miles (35 km) offshore of Chimbote, a major Peruvian fishing port, causing widespread destruction and about 3,000 deaths in Chimbote. The worst destruction, however, was caused by a massive debris avalanche that rushed off Nevados Huascarán at 174 miles per hour (280 km/hr). This debris flow had a volume of 39–65,000,000 cubic yards (30–50,000,000 m³), and rushed through the Callejón de Huaylas, a steep valley that runs parallel to the coast. The debris avalanche covered the town of Yungay under thick masses of boulders, dirt, and regolith. Seventy percent of the buildings in the town were covered with tens of feet of debris. The death toll was enormous—most estimates place the deaths at 18,000, although local officials say that 20,000 died in Yungay alone, and as many as 70,000 people died in the region from the landslides associated with the May 31, 1970, earthquake.

There have been many other landslide disasters in the Andes Mountains. In 1974, a rock slide-debris avalanche in the Peruvian province of Huancavelica buried the village of Mayunmarca, killing 450 people. The debris avalanche raced down the mountain with an average velocity of 140 km/hr and caused the failure of a 150-meter-high older landslide dam, initiating major downstream flooding. Debris with a volume of 16,000,000,000 cubic meters from the 1974 avalanche blocked the Mantaro River, creating a new lake behind the deposit. In 1987, the Reventador landslides in Napo, Ecuador, were triggered by two earthquakes with magnitudes of 6.1 and 6.9. These earthquakes mobilized 98,000,000–143,000,000 cubic yards (75–110,000,000 cubic m) of soil that was saturated with water on steep slopes. These slides remobilized into major debris flows along tributary and main drainages, killing 1,000 people and destroying many miles (km) of the Trans-Ecuador oil pipeline, the main economic lifeline for the country. The magnitude 6.4 Paez earthquake in Cauca, Colombia, in 1994 also initiated thousands of thin soil slides that grouped together and were remobilized into catastrophic debris flows in the larger drainages. As these raced downstream, 1,971 people were killed, and more than 12,000 people were displaced from their destroyed homes.

Central America—Honduras, El Salvador, Nicaragua, October 1998

Hurricane Mitch devastated the Caribbean and Central America regions from October 24–31, 1998, striking as one of the strongest and most damaging storms to hit the region in more than 200 years. The storm reached hurricane status on October 24, and reached its peak on October 26–27 with sustained winds of 180 miles per hour (288 km/hr). The storm remained stationary off the coast of Honduras for more than 24 hours, then slowly moved inland across Honduras and Nicaragua, picking up additional moisture from the Pacific Ocean. The very slow forward movement of the storm produced unusually heavy precipitation even for a storm this size. Total amounts of rainfall have been estimated to range between 31–74 inches (80–190 cm) in different parts of the region. Most of the devastation from Mitch resulted from the torrential rains associated with the storm, which continued to fall at a rate of 4 inches (10 cm) per hour even long after the storm moved overland and its winds diminished. The rains from Mitch caused widespread catastrophic floods and landslides throughout the region, affecting Honduras, Guatemala, Nicaragua, and El Salvador. These landslides buried people, destroyed property, and clogged drainages with tons of fresh sediment deposited in the rivers. More than 10,000 people were killed by flooding, landslides, and debris flows across the region, with 6,600 of the deaths reported from along the northern seaboard of Honduras. More than 3 million people were displaced from their homes. The crater of Casitas volcano in Nicaragua filled up with water from the storm and ruptured, initiating large debris flows in this region.

The effects of Hurricane Mitch in Central America were unprecedented in the scope of damage from downed trees, floods, and landslides. Honduras suffered some of the heaviest rainfall, with three-day storm totals in the southern part of the country exceeding 36 inches (90 cm). The areas of heaviest rainfall experienced the greatest number of landslides. Most of the landslides were shallow debris flows with average thickness of only 3.3–6.6 feet (1–2 m), with runout distances of up to 1,000 feet (300 m). Landslide density (number of slides in specified area) in the southern part of Honduras reached about 750–800 per square mile (300 km²). In the capital of Tegucigalpa, most damage was caused by two large slump/earth flows and debris slides, with the largest being the Cerro El Berrinche slump/earthflow that destroyed Colonia Soto and parts of Colonias Catorce de Febrero and El Povenir. This landslide moved about 7,800,000 cubic yards (6 million m³) of material downhill, damming the Rio Choluteca, creating a reservoir of stagnant, polluted, sewage-filled water behind the newly created dam. This landslide initially moved slowly, so residents were able to be evacuated prior to the time it accelerated and dammed the river. The landslide consisted of three main parts, including a toe of buckled and folded rock and regolith, then upslope a tongue of a mass of regolith that slid across and dammed the Rio Choluteca, and an upper part consisting of a giant slump block. Other rock and debris slides were triggered by erosion of riverbanks by floods, destroying more homes.

In El Salvador flooding effects from the rain from Mitch did more damage than landslides, with landslides being more important in the north near the border with Honduras. Most of the storm-induced landslides in El Salvador were shallow features that displaced unconsolidated surface regolith derived from deep tropical weathering. Most of the landslides did not travel far but moved downhill as coherent masses that stopped at the bases of steep hills, but became highly fragmented by the time they stopped moving. Some landslides were fluid enough to evolve into debris flows that traveled up to several miles (several km) from their sources, moving quickly across low-gradient slopes and into drainage networks. The largest landslide in El Salvador resulting from the rains from Mitch was the Zompopera slide, a translational earth slide that evolved into a debris flow, traveling more than 3.7 miles (6 km) from the slide's source. This slide was a preexisting feature whose growth was accelerated by the rains of Mitch. Other preexisting landslides also were remobilized as earthslides and earthflows, with rounded headscarps and hummocky topography. Landslides in El Salvador did much less damage than landslides in Honduras.

In Nicaragua most of the landslides triggered by the rains from Hurricane Mitch were debris flows. These debris flows ranged from small flows that displaced only a few cubic meters of material to a depth of a few meters that moved a few tens of meters downslope, to larger debris flows covering 96,000 square yards (80,000 m²) that moved material up to two miles (three km) downslope in channels. The depth of landslide scars shows a clear relationship to the depth of weathered and altered material, with deeper weathering in deeper depths of landslides. In some areas landslides covered 80 percent of the terrain, but the disturbed areas were considerably smaller in most locations. Hurricane Mitch also initiated some slow-moving earthflows that continued to move more than a year after the storm. Mapping of the substrate in Nicaragua has shown that soils and rocks of certain types were more susceptible to landslides than other materials, so risk maps for future rain-induced landslide events can be made.

One of the worst individual mass movement events associated with the torrential rains from Hurricane Mitch was the collapse of the water-filled caldera of Casitas volcano. Water that rushed out of the volcanic caldera as it collapsed mixed with ash, soil, and debris that washed down rivers, destroying much in its path. Approximately 1,560–1,680 people were killed in this disaster, and many more were displaced. Several towns and settlements were completely destroyed and many bridges along the Pan-American Highway were destroyed. This was the worst disaster to affect Nicaragua since the 1972 Managua earthquake.

The Casitas volcano is a convergent margin volcano, part of the Cordillera Maribos volcanic chain that extends from the northern shore of Lake Managua to Chinandega in the south. Casitas is one of five main volcanic edifices and is a deeply dissected composite volcano that has a .6 mile (1 km) diameter crater at its summit. Heavy rainfall from Hurricane Mitch dropped 4 inches (10 cm) of rain per day on the summit starting on October 25, increasing to 8 inches (20 cm) per day by October 27, and reached a remarkable 20 inches (50 cm) per day on October 30, 1998, the day of the avalanche. The normal monthly average rainfall for October is 13 inches (33 cm), so the rainfall on Casitas was more than six times the normal level preceding the avalanche.

The avalanche began along an altered segment of rock along a fault that cuts the volcano about 66–88 yards (60–80 m) beneath the summit. The avalanche and release of water began when a slab of rock, measuring 137 yards (150 m) long, 55 yards (60 m) high, and 18 yards (20 m) thick, broke off this fault zone, and slid down the fault surface at a 45 degree angle toward the southeast. This initial rockslide released about 260,000 cubic yards (200,000 m³) of rock.

Residents who lived below the volcano and survived reported a sound like a helicopter along with minor ground shaking as the first avalanche was shattered into small pieces as the rock mass slid down the fault plane. For the first 1.3 miles (2 km) the avalanche deposit moved through a narrow valley, forming a mass of moving rock 500–825 feet (150–250 m) wide and 100–200 feet (30–60 m) deep. As the flow moved down the valley it sloshed back and forth from side to side of the valley, bouncing off the steep walls as it roared past at 50 feet (15 m) per second, sending large blocks of rock flying 6–10 feet (2–3 m) into the air as it passed, decapitating trees above the valley floor. As it continued to move down hill the blocks of rock in the avalanche scoured the clay-rich soil from the valley floor, excavating up to 33 feet (10 m) of material and incorporating this into the avalanche deposit. This process increased the volume of the downslope flow by about nine times its initial volume.

About three hours after the initial avalanche, a lahar was generated from the main accumulation of the avalanche rock, about 2 miles (3 km) from the summit and 2 miles (3 km) above the towns of El Porvenir and Rolando Rodriguez. It is likely that the rain and river flow continued to build up in the avalanche deposit until the pressure inside was great enough to break through the debris-blocked front of the deposit. The lahar formed a rapidly moving concentrated flow that was about 10 feet (3 m) thick, and spread across a width of almost a mile (about 1,500 m). As the lahar raced through the towns of El Porvenir and Rolando Rodriguez it destroyed all buildings and signs of human habitation, scouring the soil and moving this with the flow. Approximately 2,000 people are thought to have perished in the lahar that wiped the two towns away, but the death toll is inaccurate since the bodies were removed and burned for sanitary reasons before accurate counts were made. Several other smaller towns were also destroyed, and a large agricultural area, including many livestock, was wiped out.

Many warnings could have been heeded to avoid the Casitas disaster and the destruction of El Porvenir and Rolando Rodriguez. First, the towns were built in a low-lying area on deposits of old lahars, and the geologic risks should have been appreciated before the towns were settled. The incredibly high rainfall totals should have forewarned residents that the risks for landslides and lahars was high, and certainly after the avalanche occurred, stopping only 2 miles (3 km) upstream, the residents should have been evacuated before the lahar surged out of the avalanche deposit. These risks should be assessed, and applied to the other volcanoes in Central America, as similar situations exist in many places in this region.

Leyte, Philippines, February 17, 2006

A massive landslide buried the community of Barangay Guinsaugon in the central Philippines on February 17, 2006, killing an estimated 1,450 people. This landslide is classified as a rockslide-debris avalanche, having started as a rockslide and turning into a debris avalanche as it moved downslope and spread across the adjacent lowland. The debris avalanche apparently rode on a cushion of air trapped beneath the falling regolith, enabling the avalanche to reach speeds of 87 miles per hour (140 km/hr), and completely destroying the village within three to four minutes from the time of the start of the rockslide.

The Philippines are cut by a major fault that strikes through the entire length of the island chain. This fault is active and related to the convergence and left-lateral strike-slip motion between the Philippine plate and the Pacific plate, which is being subducted beneath the islands from the east. The Philippines are tectonically active, have many steep slopes, and experience heavy seasonal rainfall. All these factors can contribute to making slopes unstable, and undoubtedly helped initiate the 2006 landslide. In the region of the massive rockslide-debris avalanche, the Philippine fault strikes north-northwest, and a second fault branches off this toward the southeast. Movement along these faults has uplifted steep mountains, and these slopes show many horseshoe-shaped scars that probably represent a series of landslide scars.

The landslide covered an area of about 3,600,000 square yards (3,000,000 square m), and stretched 2.5 miles (4 km) from the head of the scarp to the toe of the slide. Considering that the thickness of the deposit ranges from about 100 feet (30 m) at the base of the slope of the mountain to 20–23 feet (6–7 m) near the toe of the slide, it is estimated that about 19.5–26 million cubic yards (15–20 million m³) of material collapsed from the mountain on February 17, 2006, burying the town of Barangay Guinsaugon. The velocity of the flow during the slide has been estimated at 60–87 miles per hour (100–140 km/hr), based on the time survivors remembered the event to take place in and the distance that the material traveled.

The rockslide originated on the fault surface of the Philippine fault zone and excavated a large, horseshoe-shaped amphitheater on the side of the mountain at a height of about 2,228 feet (675 m). Rock and regolith material consisting of a mixture of volcanic and sedimentary formations moved out of the head of the landslide, initially sliding along steeply dipping surfaces of the Philippine fault, leaving large striations known as slickensides parallel to the direction of movement of the slide. Below the head, or crown of the slide, material was deposited and deformed into terraces of regolith that slid along the slip surfaces. As the rockslide reached the base of the scarp, it spread out laterally and deposited a huge fan-shaped mass of regolith that has many ridges, radial cracks, and isolated hills known as hummocks. These characteristics show that the material initially slid along fracture surfaces as a rockslide, then exploded out of the mountain as a debris avalanche that may have ridden on a cushion of air trapped beneath the falling regolith, aiding the high velocity of the flow.

The origin of the Barangay Guinsaugon landslide is uncertain, but several factors seem to have played a role. First, the active tectonics uplifted the steep mountains along the Philippine fault, and the orientation of the fault surfaces were such that many rocks were perched above loose fracture surfaces that dipped toward the open face of the mountain, with the town of Barangay Guinsaugon directly below.

The landslide occurred during a period of very heavy rainfall, and the rain may have both lubricated the slip surfaces and filled open pore spaces in the regolith and rock above the slip surface, loading the slip planes beyond their ability to resist sliding. A small earthquake with a magnitude of 2.6 occurred about 15 miles (25 km) west of the head of the slide, at approximately the same time as the slide. The time of the slide is known from a telephone conversation, in which the speaker noted a loud noise, then screamed in fear, and then the line went dead. It is uncertain if such a small earthquake at this distance could initiate the landslide, but the apparent coincidence in time is remarkable. The most likely scenario is that the steep joint surfaces were already overloaded by the weight of the fresh water, and the small amount of shaking from the earthquake was just enough to change a metastable slope into a moving deadly landslide. Other observations suggest that the earthquake may have occurred slightly after the landslide, and may not have been a contributing cause to the disaster.

It appears that as the rockslide surged off the slopes and crashed onto the relatively flat plane below, where the town was located, that it may have trapped a cushion of air that became compressed and surged out beneath the moving debris avalanche. This surge of air moved many houses, even a three-story concrete building, 1650–2000 feet (550–600 m) from where they were built, and deposited them relatively intact, and in the same relative positions as the surrounding buildings. All the buildings that were moved in this fashion moved radially away from the source of the avalanche. Nearly all of the people that survived the debris avalanche were found along the edge of the deposit, where this cushion of air blasted structures and people from where they were, and deposited them 1,800–2,000 feet (550–600 m) away.

When the landslide occurred, it buried most of the town and its residents, including the school, which was full of 250 children and teachers. Cell phone signals and frantic text messages were received from teachers in the school, and there was tremendous hope that the students might be safe in the strong structure of the school building. The messages warned that cold waters were rapidly rising inside the school building, as the avalanche deposit became saturated with rain and river water. Rescue workers immediately focused their attention on trying to locate the site of the school beneath the rubble on the surface, consulting maps, using satellite global positioning systems (GPS), and other instruments such as ground-penetrating radar. Maps for the town were not very accurate, and it was very difficult to use the radar and other instruments, and rescue workers were constantly under the threat of additional landslides and had to be evacuated several times when the risk levels became too high. Still, after a couple of days, the rescue workers thought they had located the position of the school on the rubble on the surface, and called in heavy excavating equipment in the hope of finding survivors. Hopes were high, and high-tech sonic equipment brought in from the American and Malaysian military crews that happened to be in the area detected sounds from the site. Rhythmic beating and scratching was thought to be coming from people trapped in the school. After digging frantically, the workers found no school, and the scope of the tragedy set in, with little hope of finding any additional survivors. After several weeks or surveying, and analyzing data, geologists realized that the school had probably also been moved 1,650–2,000 feet (500–600 m) by the blast of compressed air from the debris avalanche, and like the other buildings that were moved, may have remained intact, and the rescue workers were simply digging in the wrong place as the students and teachers perished. If the nature of the blast of air was better understood, perhaps rescue workers would have focused their efforts looking in the direction the school would have been moved by the air blast preceding the debris avalanche.

REDUCTION OF LANDSLIDE HAZARDS AND DANGERS

To reduce the hazards to people and property from mass wasting, it is necessary to first recognize which areas may be most susceptible to mass wasting and then to recognize the early warning signs that a catastrophic mass-wasting event may be imminent. Some actions can be taken to protect people and valuable property that may be in the way of imminent downslope flows. As with many geological hazards, a past record of downslope flows is a good indicator that the area is prone to additional landslide hazards. Geological surveys and hazard assessments should be completed in mountainous and hilly terrain before construction of homes, roads, railways, power lines, and other features.

Hazards to Humans

From the descriptions of mass-wasting processes and specific events above, it should be apparent that mass wasting presents a significant hazard to humans. The greatest hazards are from building on mountain slopes, which when oversteepened may fail catastrophically. The fastest moving flows present the greatest threat to human life, with examples of the debris avalanches at Vaiont, Italy, in 1963, Nevados Huascarán, Peru, in 1962 and 1970, and the Leyte, Philippines, disaster of 2006 providing grim examples with tens of thousands of deaths. Gradual creep moves cultural and natural features downhill, which

accounts for the greatest cumulative amount of material moved through mass-wasting events. These slow flows do not usually hurt people but they do cause billions of dollars in damages every year. Occasionally slow flows will accelerate into fast-moving debris flows, so it is important to monitor areas that may experience accelerated creep. Human-built structures are not designed to move downhill or to be covered in debris, so mass wasting needs to be appreciated and accounted for when designing communities, homes, roads, pipelines, and other cultural features. The best planning involves not building in areas that pose a significant hazard, but if building is done, the hazards should be minimized through slope engineering, as described below.

Prediction of Downslope Flows

What can be done to reduce the damage and human suffering inflicted by mass movements? Greater understanding of the dangers and specific triggers of mass movements can help reduce casualties from individual catastrophes, but long-term planning is needed to reduce the costs from damage to structures and infrastructure inflicted by downslope movements of all types. One approach to reducing the hazards is to produce maps that show areas that have suffered or are likely to suffer from mass movements. These maps should clearly show hazard zones and areas of greatest risk from mass movements, and what types of events may be expected in any given area. These maps should be made publicly available and used for planning communities, roads, pipelines, and other constructions. It is the responsibility of community planners and engineers to determine and account for these risks when building homes, roads, communities and other parts of the nation's infrastructure.

Several factors need to be considered when making risk maps for areas prone to mass movements. First, slopes play a large role in mass movements, so anywhere there is a slope there is a potential for mass movement. In general, the steeper the slope, the greater the potential for mass movements. In addition, any undercutting or oversteepening of slopes (from coastal erosion or construction) increases the chances of downslope movements, and anything that loads the top of a slope (like a heavy building) also increases the chance of initiating a down slope flow. Slopes that are in areas prone to seismic shaking are particularly susceptible to mass flows, and the hazards are increased along these slopes. Slopes that are wet and have a buildup of water in the slope materials are well lubricated and exert extra pressure on the slope material, and are thus more susceptible to failure.

The underlying geology is also a strong factor that influences whether or not a slope may fail. The presence of joints, bedding planes, or other weak-

Slump of former State Highway 287 into Hebgen Lake, Montana, following an earthquake, August 1959 *(USGS)*

nesses increases the chance of slope failure. Additionally, rocks that are soluble in water may have large open spaces and are more susceptible to slope failure.

These features need to be considered when preparing landslide-potential maps, and once a significant landslide potential is determined for an area, it should be avoided for building. If this is not possible, several engineering projects can be undertaken to reduce the risk. The slope could be engineered to remove excess water, decreasing the potential for failure. This can be accomplished through the installation of drains at the top of the slope, and/or the installation of perforated pipes into the slope that help drain the excess water from the slope material, decreasing the chance of slope failure.

Prevention of Downslope Flows

Slopes can be reduced by removing material, reducing the potential for landslides. If this is not possible, the slope can be terraced, which decreases runoff and stops material from falling all the way to the base of the slope. Slopes can also be covered with stone, concrete, or other material that can reduce infiltration of water, and reduce erosion of the slope material. Retaining walls can be built to hold loose

material in place, and large masses of rocks can be placed along the base of the slope (called base loading), which serves to reduce the potential of the base of the slope slumping out by increasing the resistance to the movement. Unvegetated slopes can be planted, as plants and roots greatly reduce erosion, and may help soak up some of the excess water in the soil.

If a slope can not be modified, and people must use the area, there are several other steps that may be taken to help reduce the risk to people in the area. Cable nets and wire fences may be constructed around rocky slopes that are prone to rockfalls, and these wire meshes will serve to catch falling rocks before they hit passing cars or pedestrians. Large berms and ditches may be built to catch falling debris, or to redirect mudflows and other earthflows. Rock sheds and tunnels may be built for shelter in areas prone to avalanches, and people can seek shelter in these structures during snow and or rock avalanches.

Monitoring of Active Landslides

What are the signs that need to be watched for that may warn of an imminent mass-wasting event? Areas that have previously suffered mass-wasting events may be most prone to repeated events, so geomorphological evidence for ancient slumps and landslides should be viewed as a warning. It is recognized that seismic activity and periods of heavy rainfall destabilize slopes and are times of increased hazards. Activity of springs can be monitored to detect when the slopes may be saturated and unstable, and features such as wet areas or puddles oriented parallel to an escarpment should be viewed as potential warnings that the slope is saturated and perhaps ready to slide. In some cases, slopes or whole mountains have experienced accelerated rates of creep soon before large mass-wasting events, such as the Vaiont Dam disaster, described above.

The United States Geological Survey, along with other local agencies, has set up some real-time monitoring programs for a few areas in California, Colorado, Washington, and New Mexico that have active landslide features. These systems include a variety of sensors that collect data in all weather conditions, then transmit this information to geological survey computers where the information is automatically processed and made accessible online to local officials, engineers, and emergency managers.

These real-time monitoring systems operate on the principle that changes from slow flows to rapid flows can be fast and may occur during times of bad weather when residents may not be able to observe the outdoor conditions. Real-time monitoring can detect small changes in movement on ground that could be hazardous to be on, and having real-time data can be crucial in saving human lives and property. The continuous data also provides detailed information on the behavior of landslides over time that engineers can use to design controls to slow down or prevent the landslides from moving catastrophically.

Most of the systems involve monitoring of ground movements and water pressures, and how these change in time. The amount and rates of downslope movement can be recorded by extensometers that can detect stretching or shortening of the ground. Extensometers are basically very sensitive instruments that can measure the distance between two points, typically across an area that is moving and one that is not. One type of extensometer would have a tube inside a pipe, with one side anchored to either side of a moving landslide slip surface. As the landslide slips or creeps, the pipes would gradually move apart, and measurements of the amount over time would give the rate of slip. More sophisticated electronic extensometers are commonly used, but operate on the same principle.

Ground vibrations or micro-earthquakes are also commonly monitored along active landslide features. Increases in ground vibrations can be associated with enhanced slip and movement, and geophones buried in the slides are sensitive enough to detect these small vibrations. Groundwater pressure sensors within the slides monitor the groundwater conditions within the slides, and rain gauges record precipitation. High groundwater pressures or rapid changes in groundwater pressure can indicate that the landslide is on the verge of accelerated slip.

The real-time monitoring systems currently in use normally transmit data to the United States Geological Survey every 10 minutes. If there is strong ground motion or other indications of an imminent slide, the data is transmitted immediately, and warnings can be issued to areas at risk. Sites currently being monitored include several in the northern California Coast Ranges, in the Sierra Nevada, in Washington State near Seattle, in New Mexico, and in several places prone to slow landslides and rapid avalanches in Colorado. Active ground movement is occurring in every state in the country, and it is hoped that real-time monitoring programs can be extended to many other active and potentially hazardous sites.

Mitigation of Damages from Downslope Flows
Once a landslide has occurred it is difficult to recover property losses. Some areas in California were once very expensive ocean-view real estate, but once large slumps started moving whole neighborhoods downslope along systems of curved faults, the properties became worthless. The best way to avoid financial and property loss in places like this is to not try to rebuild, as once the land slips in these regions, it takes huge engineering efforts to prevent further movements. Sometimes the only way to prevent additional landslides is to completely reengineer the slopes, changing steep slopes into terraced low-angle hills. Even this type of engineering may not be enough to prevent additional slides, so the best protection is to avoid building on land that has a history of sliding.

Despite these cautions, many new developments in landslide mitigation techniques make living in mountainous or hilly terrain safer. Most early landslide repair and mitigation techniques involved the building and emplacement of buttresses along the toes of landslides to stop their forward advance and emplacement of pipes and other features to promote water drainage from within the slides. Later, in the middle part of the 20th century, as equipment for moving regolith became larger, it became more common to remove entire slide masses from the sides of hills, and to recompact the material to stabilize the slopes. Since the 1990s, new products such as geomembranes and geotextiles that can hold regolith in place and allow water to escape, have greatly increased the ability to stabilize slopes to make them safer from sliding.

The beginning of slope reengineering to prevent or mitigate landslides is thought to have started in the early 1830s, along railroad lines in England and France. With the industrial revolution in the late 1800s, engineers used steam engines to excavate slopes to 1:1 (horizontal to vertical), or a 45-degree slope. Steeper slopes were covered with masonry retaining walls, holding the slopes back with gravity. When slopes failed in downslope movements, they were typically repaired by cutting the slope back to more gentle slopes, or, if space in urban areas did not permit this, the slopes were reinforced with concrete or masonry walls.

After World War II, large earthwork projects were employed in the United States, particularly associated with construction started as a result of the Interstate Highway Act of 1955. At this time, a new style of landslide mitigation became common, that of excavating the entire slipped area, installing subdrainages, then refilling and compacting the slopes with the excavated material. These so-called buttress fills are still the most common form of landslide repair in the United States, and are moderately effective in most cases. Slopes can be modified and slip surfaces removed, and the subdrainages keep pore water pressures to a minimum.

Since the 1990s, new materials have been developed that help engineers mitigate the effects of landslides and reduce the risks of additional slope failures on repaired slopes. These materials are known as geosynthetics and geomembrane materials, and include many individual types of construction and materials. Pavement cloths are tack-coated membranes that are overlain on existing pavement, then paved over. They serve to hold the pavement together but allow water to escape through the membrane. Filter cloths are used beneath roads and railroad ballast and on hillsides to prevent settlement of gravels into the underlying soils. They help to stabilize slopes and prevent hillside drains and other embankments from settling. Liner membranes are impervious to water and can be used to isolate areas of contaminated or clean groundwater from regional groundwater systems. Drainage membranes are constructed as composites of the above materials, and can be used in the construction of retaining walls, combining different effects of not allowing water in some places, and forcing water to drain in others that are less hazardous. Other materials, known as geogrids, can be stretched across slopes, and these materials add strength and support to toes of slopes that might otherwise collapse.

Since the 1960s, soils on slopes have been engineered by mixing materials into the soil so that they have additional strength, much like the natural effect of having abundant roots in a soil. These reinforced earth walls have become common along highways and above retaining walls. In other cases, strong materials are partly buried along the toes of slides or bases of slopes that could potentially fail. These reinforcements are typically designed as grids, increasing the strength of the toe of the slopes so that they are less susceptible to failure. These grids are typically extended into the slope for a distance of about 1.5 times the slope height. In addition, the surfaces of faces of the slopes are wrapped with the reinforcement grids, and then the surface between the grids is planted, further promoting slope stability.

See also EARTHQUAKES; GEOLOGICAL HAZARDS; SOILS; VOLCANO; WEATHERING.

FURTHER READING

Armstrong, B. R., and K. Williams. *The Avalanche Book.* Armstrong, Colo.: Fulcrum Publishing, 1992.

Brabb, Earl E. "Landslides: Extent and Economic Significance." In *Proceedings of the 28th International Geological Congress: Symposium on Landslides, Washington, D.C., July 17, 1989,* edited by Earl E. Brabb and Betty L. Harrod, 25–50. Rotterdam, Netherlands: A. A. Balkema, 1989.

Bucknam, Robert C., Jeffrey A. Coe, Manuel Mota Chavarria, Jonathan W. Godt, Arthur C. Tarr, Lee-Ann Bradley, Sharon Rafferty, Dean Hancock, Richard L. Dart, and Margo L. Johnson. *Landslides Triggered by Hurricane Mitch in Guatemala—Inventory and Discussion.* Open-file report 01-0443, United States Geological Survey, 2001.

Coates, Donald R. *Landslides.* Vol. 3 of *Reviews in Engineering Geology.* Boulder, Colo.: Geological Society of America, 1977.

Hsu, K. J. "Catastrophic Debris Streams (Sturzstroms) Generated by Rockfalls." *Geological Society of America Bulletin* 86 (1989): 129–140.

Kusky, T. M. *Landslides: Mass Wasting, Soil, and Mineral Hazards.* New York: Facts On File, 2008.

Liu, Jian Guo, and Timothy M. Kusky. "After the 8.0 Mw Wenchuan Earthquake: A report on an International Field Excursion to Investigate the Earthquake Induced Geohazards (6–10 July 2008)." *Earth Magazine* (October 2008): 48–51.

Matthews, W. H., and K. C. McTaggert. "Hope Rockslides, British Columbia." In *Rockslides and Avalanches,* edited by B. Voight, 259–275. Amsterdam: Elsevier, 1978.

Natural Hazards Observer. Available online: http://www.colorado.edu/hazards/o/. Accessed May 23, 2007.

Norris, Robert M. "Sea Cliff Erosion." *Geotimes* 35 (1990): 16–17.

Pinter, Nicholas, and Mark Brandon. "How Erosion Builds Mountains." *Scientific American, Earth from the Inside Out* (1997): 74–79.

Plafker, George, and George E. Ericksen. "Nevados Huascaran Avalanches, Peru." In *Rockslides and*

Avalanches, edited by B. Voight, 277–314. Amsterdam: Elsevier, 1978.

Schultz, Arthur P., and C. Scott Southworth, eds. *Landslides in Eastern North America.* Circular 1008, United States Geological Survey. 1987.

Schuster, R. L., and R. W. Fleming. "Economic Losses and Fatalities Due to Landslides." *Bulletin of the Association of Engineering Geologists* 23 (1986): 11–28.

Shaefer, S. J., and S. N. Williams. "Landslide Hazards." *Geotimes* 36 (1991): 20–22.

Varnes, David J. "Slope Movement Types and Processes." In *Landslides, Analysis and Control,* edited by R. L. Schuster and Raymond J. Krizek, 11–33. Washington, D.C.: National Academy of Sciences, 1978.

mélange Mélanges are complex, typically chaotic tectonic mixtures of sedimentary, volcanic, and other types of rocks in a highly sheared sedimentary or serpentinitic matrix. Mélanges show inclusions of material of widely diverse origins at many different scales, showing that mélanges are fractal systems, with the same patterns appearing at multiple scales of observation. Some mélanges may be sedimentary in origin, formed by the slumping of sedimentary sequences down marine escarpments. These mélanges are more aptly termed olistostromes. Tectonic mélanges are formed by structural mixing between widely different units, typically in subduction zone settings.

Tectonic mélanges are one of the hallmarks of convergent margins, yet understanding their genesis and relationships of specific structures to plate kinematic parameters has proven elusive because of the complex and seemingly chaotic nature of these units. Many field workers regard mélanges as too deformed to yield useful information, and simply map the distribution of mélange-type rocks without further investigation. Other workers map clasts and matrix types, search for fossils or metamorphic index minerals in the mélange, and assess the origin and original nature of the highly disturbed rocks. Recent studies have made progress in being able to relate some of the structural features in mélanges to the kinematics of the shearing and plate motion directions responsible for the deformation at plate boundaries.

One of the most persistent questions raised in mélange studies relates to the relative roles of soft-sediment versus tectonic processes of disruption and mixing. Many mélanges have been interpreted as deformed olistostromes or giant submarine landslide deposits, whereas other models attribute disruption entirely to tectonic or diapiric processes. Detailed structural studies have the potential to differentiate between these three end-member models, in that soft-

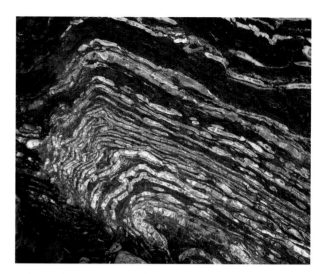

Mélange formed during Taconian Orogeny in the Cambrian Period at Lobster Cove Head, Gros Morne National Park, Newfoundland, Canada *(François Gohier/Photo Researchers, Inc.)*

sedimentary and some diapiric processes will produce clasts, which may then be subjected to later strains, whereas purely tectonic disruption will have a strain history beginning with continuous or semicontinuous layers that become extended parallel to initial layering. Detailed field, kinematic, and metamorphic studies may help further differentiate between mélanges of accretionary tectonic versus diapiric origin. Structural observations aimed at these questions should be completed at regional, outcrop, and hand-sample scales.

Analysis of deformational fabrics in tectonic mélange may also yield information about the kinematics of past plate interactions. Asymmetric fabrics generated during early stages of the mélange-forming process may relate to plate kinematic parameters such as the slip vector directions within an accretionary wedge setting. This information is useful for reconstructing the kinematic history of plate interactions along ancient plate boundaries, or how convergence was partitioned into belts of head-on and margin-parallel slip during oblique subduction.

See also ACCRETIONARY WEDGE; CONVERGENT PLATE MARGIN PROCESSES; STRUCTURAL GEOLOGY.

FURTHER READING

Kusky, Timothy M., and Dwight C. Bradley. "Kinematics of Mélange Fabrics: Examples and Applications from the McHugh Complex, Kenai Peninsula, Alaska." *Journal of Structural Geology* 21, no. 12 (1999): 1,773–1,796.

Raymond, Loren, ed. *Mélanges: Their Nature, Origin, and Significance.* Boulder, Colo.: Geological Society of America Special Paper 198, 1984.

Mercury The closest planet to the Sun, Mercury, is an astronomical midget. This planet has a mass of only 5.5 percent of the Earth's, a diameter of 3,031 miles (4,878 km), and an average density of 5.4 grams per cubic centimeter. Mercury rotates once on its axis every 59 Earth days and orbits the Sun once every 88 days at a distance of 36 million miles (58 million km). Because it is so close to the Sun, one cannot see it with only the naked eye unless the sun is blotted out, such as just before dawn, after sunset, or during total solar eclipses.

Mercury has such a weak gravitational field that it lacks an atmosphere, although bombardment by the solar wind releases some sodium and potassium atoms from surface rocks, and these may rest temporarily near the planet's surface. The magnetic field is very weak, approximately 1/100th as strong as Earth's. The surface of Mercury is heavily cratered and looks much like the Earth's Moon, showing no evidence for ever having sustained water, dust

Image mosaic of Mercury. After passing on the dark side of the planet, *Mariner 10* photographed the other, somewhat more illuminated hemisphere of Mercury. Note the rays of ejecta coming from some of the impact craters in the image. The *Mariner 10* spacecraft was launched in 1974. The spacecraft took images of Venus in February 1974 on the way to three encounters with Mercury in March and September 1974 and March 1975. The spacecraft took more than 7,000 images of Mercury, Venus, the Earth, and the Moon during its mission. The Mariner 10 mission was managed by the Jet Propulsion Laboratory for NASA's Office of Space Science in Washington, D.C. *(Image Note: Davies, M. E., S. E. Dwornik, D. E. Gault, and R. G. Strom, Atlas of Mercury, NASA SP-423, 1978; courtesy of NASA)*

storms, ice, plate tectonics, or life. The surface of Mercury is less densely cratered than the Moon's, however, and some planetary geologists suggest that the oldest craters may be filled in by volcanic deposits. Some surface scarps, estimated to be more than 4 billion years old, are thought to represent contraction of the surface associated with the core formation and shrinking of the planet in the first half-billion years of its history.

The density of Mercury and the presence of a weak magnetic field suggest that the planet has a differentiated iron-rich core with a radius of approximately 1,118 miles (1,800 km), but it is not known whether this is solid or liquid. A mantle probably exists between the crust and core, extending to 311 to 373 miles (500–600 km) depth. The small size of Mercury means it did not have enough internal energy to sustain plate tectonics or volcanism for long in its history, so the planet has been essentially dead for the past 4 billion years.

See also EARTH; JUPITER; MARS; NEPTUNE; SATURN; SOLAR SYSTEM; URANUS; VENUS.

FURTHER READING

Chaisson, Eric, and Steve McMillan. *Astronomy Today.* 6th ed. Upper Saddle River, N.J.: Addison-Wesley, 2007.

Comins, Neil F. *Discovering the Universe.* 8th ed. New York: W. H. Freeman, 2008.

National Aeronautic and Space Administration. Solar System Exploration page. Mercury. Available online. URL: http://solarsystem.nasa.gov/planets/profile. cfm?Object=Mercury. Last updated June 25, 2008.

Snow, Theodore P. *Essentials of the Dynamic Universe: An Introduction to Astronomy.* 4th ed. St. Paul, Minn.: West Publishing Company, 1991.

Mesozoic The fourth of five main geological eras, the Mesozoic falls between the Paleozoic and Cenozoic. The total stratigraphic record of rocks deposited in this era, the Mesozoic includes the Triassic, Jurassic, and Cretaceous periods. The era begins at 248 million years ago at the end of the Permian-Triassic extinction event and continues to 66.4 million years ago at the Cretaceous-Tertiary (K-T) extinction event. Named by the British geologist Charles Lyell in 1830, the term means middle life, recognizing the major differences in the fossil record between the preceding Paleozoic era and the succeeding Cenozoic era. Mesozoic life saw the development of reptiles and dinosaurs, mammals, birds, and many invertebrate species that are still flourishing, in addition to flowering plants and conifers inhabiting the land. The era is commonly referred to as the age of reptiles, since they dominated the terrestrial, marine, and aerial environments.

Pangaea continued to grow in the Early Mesozoic, with numerous collisions in eastern Asia, but as the supercontinent grew in some areas, it broke apart in others. As the fragments drifted apart especially in the later part of the Mesozoic, continental fragments became isolated and life-forms began to evolve separately in different places, allowing independent forms to develop, such as the marsupials of Australia. The Atlantic Ocean began opening as arcs and oceanic terranes collided with western North America. In the later part of the Mesozoic, in the Cretaceous, sea levels were high and shallow seas covered much of western North America and central Eurasia, depositing extensive shallow marine carbonates. Many marine organisms such as plankton rapidly diversified and bloomed, and thick organic-rich deposits formed source rocks for numerous coal and oil fields. With the high Cretaceous sea levels, Cretaceous rocks are abundant on many continents and are the most represented of the Mesozoic strata. These strata are rich in fossils that show both ancient and modern features, including dinosaurs and ammonoids, plus newly developed bony fishes and flowering plants. The dinosaurs and reptiles continued to rule the land, sea, and air until the devastating series of events culminating with the collision of an asteroid or comet with Earth at the end of the Cretaceous eliminated the dinosaurs, and caused the extinction of 45 percent of marine genera, including the ammonites, belemnites, inoceramid clams, and large marine reptiles.

See also CENOZOIC; CRETACEOUS; HISTORICAL GEOLOGY; PALEONTOLOGY; PALEOZOIC.

FURTHER READING

Kious, Jacquelyne, and Robert I. Tilling. U.S. Geological Survey. "This Dynamic Earth: The Story of Plate Tectonics." Available online. URL: http://pubs.usgs.gov/gip/dynamic/dynamic.html. Last modified March 27, 2007.

Pomerol, Charles. *The Cenozoic Era: Tertiary and Quaternary.* Chichester, England: Ellis Horwood, 1982.

Prothero, Donald, and Robert Dott. *Evolution of the Earth.* 6th ed. New York: McGraw-Hill, 2002.

Stanley, Steven M. *Earth and Life Through Time.* New York: Freeman, 1986.

metamorphism and metamorphic rocks
Metamorphism, a term derived from the Greek, means change of form or shape. Geologists use the term to describe changes in the minerals, chemistry, and texture within a rock. Metamorphism is typically induced by increases in pressure and temperature from burial, regional tectonics, or nearby igneous intrusions.

Any previously formed rocks may be deeply buried by sedimentary cover, affected by regional plate-boundary processes, or be heated close to an igneous intrusion, changing the temperature and pressure conditions from when and where they were formed. *Diagenesis* refers to early changes that occur to rocks, generally below 390°F (200°C). When temperatures rise above 390°F (200°C), the changes become more profound and are referred to as metamorphism.

When sedimentary rocks are deposited they contain many open spaces filled with water-rich fluids. When these rocks are deeply buried and subjected to very high temperatures and pressures, these fluids react with the mineral grains in the rock and play a vital role in the metamorphic changes that occur. These fluids act as a hot, reactive juice that transports chemical elements from mineral to fluids to new minerals. This is confirmed by observations of rocks heated to the same temperature and pressure without fluids, which hardly change at all.

When rocks are heated, certain minerals become unstable and others stabilize. Chemical reactions transform one assemblage of minerals into a new assemblage. Most temperature changes are accompanied by pressure changes, and it is the combined pressure-temperature (P-T) fluid composition that determines how the rock will change.

In liquids, pressures are equal in all directions, but in rocks pressures can be greater or lesser in one direction, and they are referred to as stresses. Textures in metamorphic rocks often reflect stresses that are greater in one direction than in another. Sheets of planar minerals become oriented with their flat surfaces perpendicular to the strongest or maximum stress. This planar arrangement of platy minerals is known as foliation.

Time is also an important factor in metamorphism. In general, the longer the reaction time, the larger the mineral grains and the more complete the metamorphic changes.

GRADES OF METAMORPHISM
Low-grade metamorphism refers to changes that occur at low temperatures and pressures, whereas high-grade metamorphism refers to changes that occur at high temperatures and pressures. At progressively higher grades of metamorphism the high temperature drives the water out of the pore spaces and eventually out of the hydrous mineral structures, so that at very high grades of metamorphism, the rocks contain fewer hydrous minerals (e.g., micas). Prograde metamorphism refers to changes that occur while the temperature and pressure are rising and pore fluids are abundant, whereas retrograde metamorphism refers to changes that occur when temperature and pressure are falling. At this stage, most

fluids have already been expelled and the retrograde changes are less pronounced. If this were not so, then all metamorphic rocks would revert back to clays stable at the surface.

METAMORPHIC CHANGES

Compressing a piece of paper would cause the flat dimensions to orient themselves perpendicular to the direction of compression. Likewise, when a metamorphic rock is compressed or stressed, the platy minerals, such as chlorite and micas, orient themselves so that their long dimensions are perpendicular to the maximum compressive stress. The planar fabric that results from this process is known as a foliation.

Slaty cleavage is a specific type of foliation in which the parallel arrangement of microscopic platy minerals causes the rock to break in parallel plate-like planes. Schistosity forms at higher metamorphic grades and is a foliation defined by a wavy or distorted plane containing large, visible, oriented minerals such as quartz, mica, and feldspar.

As rocks are progressively heated and subject to more pressure during metamorphism, different mineral assemblages are stable. Even though the rock may retain a stable overall composition, the minerals will become progressively recrystallized and the min-

eral assemblages (or parageneses) will change under different P-T conditions.

KINDS OF METAMORPHIC ROCKS

The names of metamorphic rocks are derived from their original rock type, their texture, and mineral assemblages. Shales and mudstones have an initial mineral assemblage of quartz, clays, calcite, and feldspar. Slate is the low grade metamorphic equivalent of shale and, with recrystallization, is made of quartz and micas. At intermediate grades of metamorphism, the mica grains grow larger so that individual grains become visible to the naked eye and the rock is called a phyllite. At high grades of metamorphism, the rock (shale, for example) now becomes a schist, which is coarse-grained, and the foliation becomes a bit irregular. Still higher grades of metamorphism separate the quartz and the mica into different layers; this rock is called a gneiss. For both schists and gneisses, a prefix is commonly added to the names to denote some of the minerals present in the rock. For instance, if garnet grows in a biotite schist, it could be named a garnet-biotite schist.

Fresh basalts contain olivine, pyroxene, and plagioclase, none of which contains abundant water. When metamorphosed, however, water typically

Marble (white) and schist (brown) on Charley River in Alaska *(USGS)*

enters the rock from outside the system. At low grades of metamorphism, the basalt is turned into a greenstone or greenschist, which has a distinctive color because of its mineral assemblage of chlorite (green) + albite (clear) + epidote (green) + calcite (clear).

At higher metamorphic grades, the greenschist mineral assemblage is replaced by one stable at higher temperature and pressure, typically plagioclase and amphibole, and the rock is known as amphibolite. Amphiboles have a chain structure which gives them an elongated shape. When they crystallize in a different stress field like that found in a metamorphic rock, the new minerals tend to align themselves so that their long axes are parallel to the least compressive stress, forming a lineation. At even higher metamorphic grades the amphiboles are replaced by pyroxenes and the rock is called a granulite.

Limestone metamorphoses into marble, which consists of a network of coarsely crystalline interlocking calcite grains. Most primary features, such as bedding, are destroyed during metamorphism and a new sugary texture appears.

When sandstone is metamorphosed, the silica remobilizes and fills in the pore spaces between the grains, making a very hard rock called a quartzite. Primary sedimentary structures may still be seen through the new mineral grains.

KINDS OF METAMORPHISM

Metamorphism is a combination of chemical reactions induced by changing pressure and temperature conditions and mechanical deformation caused by differential stresses. The relative importance of physical and chemical processes changes with metamorphism in different tectonic settings.

Near large plutons or hot igneous intrusions, rocks are heated to high temperatures without extensive mechanical deformation. These elevated temperatures cause rocks next to plutons to grow new minerals, but these kinds of contact metamorphic rocks may lack strong foliations like those formed during regional metamorphism. Rocks adjacent to these large plutons develop a contact metamorphic aureole of rocks, altered by heat from the intrusion. Large intrusions carry a lot of heat and typically have large contact aureoles, several miles (kilometers) wide.

The contact metamorphic aureole consists of several concentric zones, each with different mineral groups related to higher temperatures closer to the pluton. A hornfels is a hard, fine-grained rock composed of uniform interlocking grains, typically from metamorphosed and suddenly heated shale.

When rocks are buried by the weight of overlying sedimentary rocks, they undergo small changes called diagenesis, until they reach 390°F (200°C). At about 570°F (300°C), some recrystallization may begin, particularly the formation of a group of water-rich minerals known as zeolites.

The most common types of metamorphic rocks are the regional metamorphic rocks. Regional metamorphism involves a combination of chemical and mechanical effects, so these rocks tend to have a pronounced foliation (slate, schist). Most regional metamorphic rocks are found in mountain belts or old eroded mountain belts that formed by the collision of two tectonic plates. In regional metamorphic conditions, the rocks are compressed horizontally, resulting in large folds and faults, which place some rocks on top of other ones, burying them quickly and elevating their pressure and temperature conditions. In this type of environment there is a wide range of pressure/temperature conditions over which the rocks were metamorphosed, and geologists have defined a series of different metamorphic zones reflecting these conditions. These metamorphic zones are each defined by the appearance of a new metamorphic index mineral, which includes, in progressively higher grade order (for shale), chlorite-biotite-garnet-staurolite-kyanite-sillimanite. In the field, the geologist examines the rocks and looks for the first appearance of these different minerals and plots them on a map. By mapping out the distribution of the first appearance of these minerals on a regional scale, the geologists then defines isograds, which are lines on a map marking the first appearance of a given index mineral on the map. The regions between isograds are known as metamorphic zones.

METAMORPHIC FACIES

When rocks are metamorphosed their bulk chemistry remains about the same, except for water and CO_2, which are fairly mobile. The mineral assemblages constantly change but the chemistry remains the same. Thus, the temperature and pressure of metamorphism control the mineral assemblages in metamorphic rocks. In 1915, a famous petrologist named Pentti Eskola presented the concept of metamorphic facies, stating that different assemblages of metamorphic minerals that reach equilibrium during metamorphism within a specific range of physical conditions belong to the same metamorphic facies.

Eskola studied rocks of basaltic composition, so he named his facies according to the metamorphic names for basaltic rocks. His classification, shown in the upper figure on page 541, stands to this day.

METAMORPHISM AND TECTONICS

Regional metamorphism is a response to tectonic activity, and different metamorphic facies are found in different tectonic environments.

The lower figure on page 541 shows the distribution of metamorphic facies in relationship to the

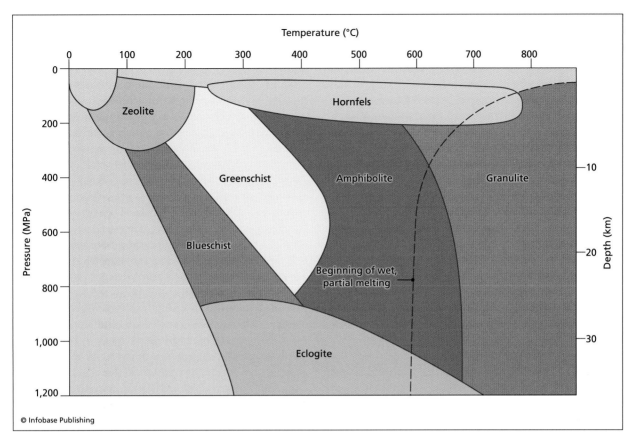

Metamorphic facies, showing the relationship between pressure, temperature, and different grades of metamorphism

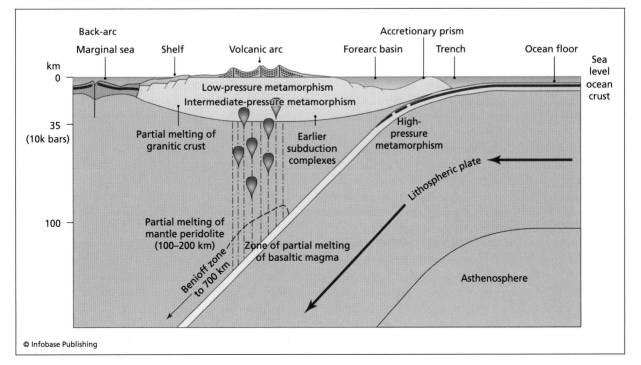

Relationship between metamorphism and tectonics in a convergent margin setting

structure of a subduction zone. Burial metamorphism occurs in the lower portion of the thick sedimentary piles that fill the trench, whereas deeper down the trench blue schist facies reflect the high pressures and low temperatures where magmas come up off the subduction slab and form an island arc; metamorphism is of greenschist to amphibolite facies. Closer to the plutons of the arc, the temperatures are high, but the pressures are low, so contact metamorphic rocks are found in this region.

See also MINERAL, MINERALOGY; PETROLOGY AND PETROGRAPHY; PLATE TECTONICS; STRUCTURAL GEOLOGY.

FURTHER READING

Cefrey, Holly. *Metamorphic Rocks*. New York: The Rosen Publishing Group, 2003.

Kornprobst, Jacques. *Metamorphic Rocks and Their Geodynamic Significance: A Petrological Handbook*. New York: Springer, 2002.

Moores, Eldridge M., and Robert Twiss. *Tectonics*. New York: W. H. Freeman, 1995.

Skinner, B. J., and B. J. Porter. *The Dynamic Earth: An Introduction to Physical Geology*. 5th ed. New York: John Wiley & Sons, 2004.

Stille, Darlene R., Lynn S. Fichter, Terrence E. Young, Jr., and Rosemary G. Palmer. *Metamorphic Rocks: Recycled Rock*. New York; Compass Point Books, 2008.

metasomatic The metamorphic process of changing a rock's composition or mineralogy by the gradual replacement of one component by another through the movement and reaction of fluids and gases in the pore spaces of a rock is called metasomatism, and these processes are metasomatic. Metasomatic processes are responsible for the formation of many ore deposits, which have extraordinarily concentrated abundances of some elements. They may also play a role in the replacement of some limestones by silica, and the formation of dolostones.

Many metasomatic rocks and ore deposits are formed in hydrothermal circulation systems that are set up around igneous intrusions. When magmas, particularly large batholiths, intrude country rocks, they set up a large thermal gradient between the hot magma and the cool country rock. Any water above the pluton gets heated and rises toward the surface, and water from the sides of the pluton moves in to replace that water. A hydrothermal circulation system is thus set up, and the continuous movement of hot waters in such systems often leach elements from some rocks and from the pluton and deposit them in other places, in metasomatic processes.

See also METAMORPHISM AND METAMORPHIC ROCKS.

meteor, meteorite Meteors are rocky objects from space, such as asteroids or smaller rocky objects called meteoroids, that enter the Earth's atmosphere. When meteors pass through Earth's atmosphere, they get heated and their surfaces become ionized, causing them to glow brightly, forming a streak moving across the atmosphere known as a shooting star or fireball. Most meteors burn up before they hit the surface of the Earth, but those that do make it to the surface are then called meteorites.

At certain times of the year, the Earth passes through parts of our solar system that are rich in meteors, and the night skies become filled with shooting stars and fireballs, sometimes as frequently as several per minute. These high-frequency meteor encounters, known as meteor showers, include the Perseid showers that appear around August 11 and the Leonid showers that appear about November 14 (both occur annually).

Many small meteorites have hit the Earth throughout time. Eyewitness accounts describe many meteors streaking across the sky with some landing on the surface. Fragments of meteorites are regularly recovered from the Antarctic ice sheets, where rocky objects on the surface could have come only from space. Although meteors may appear as flaming objects moving across the night skies, they are generally cold icy bodies when they land on Earth, and only their outermost layers get heated from the deep freeze of space during their short transit through the atmosphere.

Meteorites consist of several different main types. Stony meteorites include chondrites, which are very primitive, ancient meteorites made of silicate minerals, like those common in the Earth's crust and mantle, but chondrites contain small spherical object known as chondrules. These chondrules contain frozen droplets of material thought to be remnants of

Iron meteorite from Odessa, Ukraine *(Astrid & Hanns-Frieder Michler/Photo Researchers, Inc.)*

the early solar nebula from which the Earth and other planets initially condensed. Achondrites are similar to chondrites in mineralogy, except they do not contain the chondritic spheres. Iron meteorites are made of an iron-nickel alloy with textures that suggest they formed from slow crystallization inside a large asteroid or small planet that has since been broken into billions of small pieces, probably by an impact with another object. Stony-irons are meteorites that contain mixtures of stony and iron components and probably formed near the core-mantle boundary of the broken planet or asteroid. Almost all meteorites found on Earth are stony varieties.

ORIGIN OF METEORITES AND OTHER EARTH ORBIT-CROSSING OBJECTS

Most meteorites originate in the asteroid belt, situated between the orbits of Mars and Jupiter. At least 1 million asteroids reside in this belt with diameters greater than 0.6 mile (1 km), 1,000 with diameters greater than 18 miles (30 km), and 200 with diameters greater than 60 miles (100 km). Asteroids and meteoroids are distinguished only by their size, asteroids being greater than 328 feet (100 m) in diameter. Meteoroids are referred to as meteors only after they enter the Earth's atmosphere, and as meteorites after they land on the surface. These are thought to be either remnants of a small planet that was destroyed by a large impact event, or perhaps fragments of rocky material that failed to coalesce into a planet, probably due to the gravitational effects of the nearby massive planet of Jupiter. Most scientists favor the second hypothesis, but recognize that collisions between asteroids have fragmented a large body to expose a planetlike core and mantle now preserved in the asteroid belt.

HAZARDS OF METEORITE IMPACT WITH EARTH

The chances of the Earth's being hit by a meteorite are small at any given time, but they are greater than the chances of winning a lottery. The chances of dying from an impact are about the same as dying in a plane crash, for a person who takes one flight per year. These comparisons are statistical flukes, however, and reflect the fact that a large meteorite impact is likely to kill so many people that it raises the statistical chances of dying by impact. A globally catastrophic impact is generally thought of as one that kills more than 25 percent of the world's population, or currently about 1.6 billion people. The Earth has been hit by a number of small impacts and by some very large impacts that have had profound effects on the life on Earth at those times. Most geologists and astronomers now accept the evidence that an impact caused the extinction of the dinosaurs and caused many of the other mass extinctions in Earth history, so it is reasonable

to assume that a large impact would have serious consequences for life on Earth. Nations of the world need to consider more seriously the threat from meteorite impacts. Impacts that are the size of the blast that hit Siberia at Tunguska in 1908 happen about once every thousand years, and major impacts, that can seriously affect climate and life on Earth are thought to occur about every 300,000 years. Events like the impact at Chicxulub that killed the dinosaurs and resulted in a mass extinction are thought to occur once every 100 million years. The Chicxulub impact initiated devastating global wildfires that consumed much of the biomass on the planet, and sent trillions of tons of submicrometer dust into the stratosphere. After the flash fires, the planet became dark for many months, the atmosphere and ocean chemistry were changed, and the climate experienced a short-term but dramatic change. The global ecosystem was practically destroyed, and one of the greatest mass extinctions in geological time resulted. Events that approach or exceed this size place the entire population of the world at risk and threaten the survival of the human species. Much smaller events have the potential to destroy agricultural produce in fields around the world, leading to an instant global food shortage and mass starvation, collapse of global economies, and political strife.

There are thousands of near-Earth objects that have the potential to hit the Earth and form impact craters of various sizes. Most of these objects are asteroids diverted from the main asteroid belt and long-period comets. A couple hundred of these objects are in Earth orbit–crossing paths. They range widely in size, density, and composition, all of which can play a role in the type of hazard the body poses as it enters the Earth's atmosphere and falls to the surface. In most instances, for objects greater than several hundred feet (~100 m) in diameter, size is the most important factor, and the controlling factor on the style and hazard of the impact is related to the kinetic energy of the object.

Different hazards are associated with large and minor impacts. Passage of atmospheric shock waves, followed by huge solid Earth quakes, is described, followed by analysis of the tsunamis generated by ocean-hitting impacts. The atmospheric fires associated with the impact, followed by blocking of the Sun by particulate matter thrown up by the impact and fires, is considered to be one of the most hazardous elements of impacts through which many organisms would struggle to survive.

Mitigation strategies for avoiding large impacts of asteroids with Earth involve first locating and tracking objects in Earth orbit–crossing paths. If a collision appears imminent, then efforts should be made to try to deflect the asteroid out of the collision

THE SKY IS ON FIRE

When meteorites pass through Earth's atmosphere, they get heated, and their surfaces become ionized, causing them to glow brightly, forming a streak moving across the atmosphere known as a shooting star or fireball. If the meteorite is large enough, it may not burn up in the atmosphere and will then strike the Earth. Small meteorites may just crash on the surface, but the rare, large object can excavate a large impact crater or cause worse damage. At certain times of the year the Earth passes through parts of the solar system that are rich in meteorites, and the night skies become filled with shooting stars and fireballs, sometimes as frequently as several per minute. These times of high-frequency meteorite encounters are known as meteor showers, and include the Perseid showers that appear around August 11, and the Leonid showers that appear about November 14.

Eyewitness accounts describe many events such as fireballs in the sky recording the entry of a meteorite into the Earth's atmosphere, such as the massive fireball that streaked across western Canada on November 21, 2008, and the huge explosion over Tunguska, Siberia in 1908 that leveled thousands of square miles (km) of trees and created atmospheric shock waves that traveled around the world. Many theories have been proposed for the Tunguska event, the most favored of which is the impact of a comet fragment with Earth.

Few people lived in the core region of the Tunguska explosion in 1908. There were, however, many eyewitness accounts from places near the edges of the core damage zone and from other places as far as Australia and Scandinavia and the United Kingdom. The first published report of the explosion was in the Irkutsk City newspaper on July 2, 1908, published two days after the explosion, as related below:

> The peasants saw a body shining very brightly (too bright for the naked eye) with a bluish-white light.... The body was in the form of "a pipe," i.e., cylindrical. The sky was cloudless, except that low down on the horizon, in the direction in which this glowing body was observed, a small dark cloud was noticed. It was hot and dry and when the shining body approached the ground (which was covered with forest at this point) it seemed to be pulverized, and in its place a loud crash, not like thunder, but as if from the fall of large stones or from gunfire was heard. All the buildings shook and at the same time a forked tongue of flames broke through the cloud. All the inhabitants of the village ran out into the street in panic. The old women wept, everyone thought that the end of the world was approaching.

S. B. Semenov, an eyewitness from the village of Vanovara, located about 37 miles (60 km) south of the explosion site, described the event as follows:

> I was sitting in the porch of the house at the trading station of Vanovara at breakfast time . . . when suddenly in the north . . . the sky was split in two and high above the forest the whole northern part of the sky appeared to be covered with fire. At that moment I felt great heat as if my shirt had caught fire; this heat came from the north side. I wanted to pull off my shirt and throw it away, but at that moment there was a bang in the sky, and a mighty crash

course by blasting it with rockets, or mounting rockets on the asteroid to steer it away. Alternatively, if the asteroid is so big, a population from the Earth would need to escape the planet to start a new civilization on a new planet.

Atmospheric Shock Waves

The effect that an asteroid or meteor has on the Earth's atmosphere depends almost completely on the size of the object. Weak meteors that are up to about 30–90 feet (10–30 m) in diameter usually break up into fragments and completely burn up in the atmosphere before they hit the Earth's surface. The height in the atmosphere that these meteors break up depends on the strength of the meteor body, with most comets and carbonaceous chondrites of this size breaking up above 19 miles (30 km). Stronger meteors, such as irons, in this size range may make it to the surface.

Meteors and comets that enter the Earth's atmosphere typically are traveling with a velocity of about six miles/second (10 km/sec), or 21,600 miles per hour (37,754 km/hr). Small meteoroids enter the upper atmosphere every day; more rarely, large ones enter and compress and heat the air in front of them as they race toward the surface. This heat causes most of these bodies to burn up or explode before they reach the surface, with blasts the size of the Hiroshima or Nagasaki atomic bombs happening daily somewhere in the upper atmosphere, by meteors that are about 30 feet (10 m) in diameter. Larger meteors explode closer to the surface and can generate huge

was heard. I was thrown to the ground about three sajenes [about 23 feet, or 7 m] away from the porch and for a moment I lost consciousness.... The crash was followed by noise like stones falling from the sky, or guns firing. The earth trembled, and when I lay on the ground I covered my head because I was afraid that stones might hit it.

Local Inuit people reported the blast, as in the following testimony from Chuchan of the Shanyagir Tribe, recorded by ethnographer I. M. Suslov in 1926:

We had a hut by the river with my brother Chekaren. We were sleeping. Suddenly we both woke up at the same time. Somebody shoved us. We heard whistling and felt strong wind. Chekaren said, "Can you hear all those birds flying overhead?" We were both in the hut, couldn't see what was going on outside. Suddenly, I got shoved again, this time so hard I fell into the fire. I got scared. Chekaren got scared too. We started crying out for father, mother, brother, but no one answered. There was noise beyond the hut, we could hear trees falling down. Chekaren and I got out of our sleeping bags and wanted to run out,

but then the thunder struck. This was the first thunder. The Earth began to move and rock, wind hit our hut and knocked it over. My body was pushed down by sticks, but my head was in the clear. Then I saw a wonder: trees were falling, the branches were on fire, it became mighty bright, how can I say this, as if there was a second sun, my eyes were hurting, I even closed them. It was like what the Russians call lightning. And immediately there was a loud thunderclap. This was the second thunder. The morning was sunny, there were no clouds, our Sun was shining brightly as usual, and suddenly there came a second one!

Chekaren and I had some difficulty getting out from under the remains of our hut. Then we saw that above, but in a different place, there was another flash, and loud thunder came. This was the third thunder strike. Wind came again, knocked us off our feet, struck against the fallen trees.

We looked at the fallen trees, watched the tree tops get snapped off, watched the fires. Suddenly Chekaren yelled "Look up" and pointed with his hand. I looked there and saw another flash, and

it made another thunder. But the noise was less than before. This was the fourth strike, like normal thunder.

Now I remember well there was also one more thunder strike, but it was small, and somewhere far away, where the Sun goes to sleep.

All of these observers report a generally similar sequence of events from different perspectives. The bolide formed a giant, columnlike fireball that moved from southeast to northwest across the Siberian sky, then exploded in several pieces high above the ground surface, with audible to deafening sounds, and scorching to noticeable heat. The explosions were followed by the air blast that moved down in the center of the area, then outward toward the edges of the blast zone. These eyewitness accounts provide scientists with some of the best observation of airbursts of this magnitude and serve as a valuable lesson in the behavior of meteorites or asteroids that explode before hitting the surface.

FURTHER READING

Elkens-Tanton, Linda T. *Asteroids, Meteorites, and Comets.* New York: Facts On File, 2006.
Kusky, T. M. *Asteroids and Meteorites: Catastrophic Collisions with Earth.* New York: Facts On File, 2009.

air blasts like the explosion that leveled thousands of square miles (km) of trees in Siberia in 1908.

The flux of meteoroids of different sizes is calculated in part by comparing crater density on the Moon with the expected result in the higher gravity field of the Earth. Larger bodies that make it though the atmosphere hit with a greater frequency for the small objects, and less often for the larger bodies. Meteors in the 30-foot (10-m) diameter range release about as much energy as the nuclear bomb that was dropped on Hiroshima (0.01 megaton [9,070 tonnes]) of TNT equivalent) when they enter the atmosphere and burn before hitting the surface. Events of this size happen about one time per year on Earth, whereas larger events in the 1 megaton (1 megaton = 1,000,000 tons) range occur

about once a century, associated with the burning up of 100 foot (30 m) diameter bodies as they plunge through the atmosphere as shown by the table "Effects of Impacts as a Function of Energy and Crater Size."

A moderate-sized impact event, such as a collision with a meteorite with a 5–10 mile (8–16 km) diameter, moving at a moderate velocity of 7.5 miles/ sec (12 km/sec), would release energy equivalent to 100 megatons, or about 1,000 times the yield of all existing nuclear weapons on Earth. The meteoroid would begin to glow brightly as it approached Earth, encountering the outer atmosphere. As this body entered the atmosphere, it would create a huge fireball that would crash with the Earth after about 10 seconds. Events of this magnitude happen about

EFFECTS OF IMPACTS AS A FUNCTION OF ENERGY AND CRATER SIZE

Energy of Impact (Megatons)	Diameter of Meteorite or Comet	Crater Diameter Miles (km)	Consequences
< 10			Detonation of stones and comets in upper atmosphere. Irons penetrate to surface
10^1–10^2	245 feet (75 m)	1 (1.5)	Irons form craters (Meteor Crater), Stones produce airbursts (Tunguska). Land impacts destroy an area the size of a city (Washington, Paris)
10^2–10^3	525 feet (160 m)	1.9 (3)	Irons and stones produce ground-bursts, comets produce airbursts. Land impacts destroy an area the size of a large urban area (New York, Cairo)
10^3–10^4	1,150 feet (350 m)	3.7 (6)	Impacts on land produce craters. Ocean impacts produce significant tsunami. Land impacts destroy area size of small state (Delaware, Israel)
10^4–10^5	0.43 miles (0.7 km)	7.5 (12)	Tsunamis reach oceanic scales, exceeding damage from land impacts. Land impacts destroy area size of a moderate state (Virginia, Taiwan)
10^5–10^6	1.06 miles (1.7 km)	18.7 (30)	Land impacts raise enough dust to affect climate, and freeze crops. Ocean impacts generate hemispheric tsunamis. Global destruction of ozone. Land impacts destroy area size of large state (California, France)
10^6–10^7	1.9 miles (3 km)	37 (60)	Both land and ocean impacts raise dust, impact ejecta is global, changes global climate. Widespread fires. Land impacts destroy area size of a large nation (Mexico, India)
10^7–10^8	4.3 miles (7 km)	78 (125)	Global conflagration, prolonged climate effects, probably mass extinction. Direct destruction of continental scale area (United States, Australia)
10^8–10^9	10 miles (16 km)	155 (250)	Large mass extinction (K-T in scale)
10^9–10^{10}			Survival of all life threatened

* Note: table based on David Morrison, Clark Chapman, and Paul Slovic (*The Impact Hazard, 1994*).

once every 1,000 years on Earth, but obviously even in this energy range, impacts are not posing serious threats to the survival of life on Earth.

Objects that break up lower than 12 miles (20 km) above the surface cause much greater destruction. Objects about 165 feet (50 m) in diameter that break up at this height will generate significant airbursts that pose significant hazards. Larger objects will strike the ground, releasing energy in a manner similar to atomic bomb blasts (but without releasing radioactivity), with the amount of energy proportional to the size of the meteorite or comet. The impact that hit Tunguska in 1908 is estimated to have released about 10–20 megatons (9.1–18 megatonnes), with a radius of complete destruction of 15 miles (25 km), and a much larger affected area. The area of destruction increases to the two-thirds power of the magnitude of the blast.

Solid Earth Shock Wave and Earthquakes

When meteorites hit the surface of the Earth they generate seismic waves and cause earthquakes. The size of the earthquake is related to the energy released by the impactor, which is related to its mass and its velocity. The larger the energy released on the surface, the larger the earthquake. Meteors that explode in the atmosphere can also generate earthquakes, as the air blast transfers energy to the surface and also generates seismic waves.

Large impacts generate seismic waves that travel through the interior of the Earth and along the surface layers. For comparison, the Tunguska explosion and airburst generated a seismic event with a Richter magnitude that is estimated to be only about a magnitude 5 earthquake. In contrast, the Chicxulub impact at the Cretaceous-Tertiary boundary generated a magnitude 11, 12, or 13 earthquake, shaking the entire planet, and resulting in seismic waves that uplifted and dropped the ground surface by hundreds to a thousand feet (up to 300 m) at a distance of 600 miles (965 km) from the crater.

If an impact of this size and energy hit the Earth today, shock waves would be felt globally as earthquakes of unimaginable size, destroying much of the surface of the planet, and killing billions of people.

Tsunamis Generated by Meteorite Impacts in the Oceans

If a large meteorite struck the ocean, huge tsunamis would be formed, hundreds and perhaps thousands of feet tall (hundreds of m). These would run up on coastlines, washing away the debris from the earthquakes of a few moments or hours before. Impacts that hit water cause devastation over a larger area than impacts that hit land because of the far-traveling effects of the tsunamis. Impacts that have an energy release of 1,000 megatons (907 megatonnes) should generate tsunamis about 15 feet (5 m) tall and travel more than 600 miles (1,000 km). For impacts that are significantly larger than this (above 10,000 megatons, or 9,070 megatonnes) the damage from the tsunami is much greater, and covers a much larger area than the damage from the blast of the impactor itself. The tsunami associated with the Chicxulub impact on the Yucatán Peninsula may have initially been thousands of feet (hundreds of m) high, washing over much of the Gulf Coast of the United States and Mexico, and devastating the Caribbean.

Global Firestorm and Global Winter

The force of large- and medium-scale impacts ejects enormous quantities of superheated dust and gases into the atmosphere, some of which would fall back to Earth as flaming fireballs. Most of the dust would make it into the upper atmosphere, where it would encircle the entire planet. The energy from the impact would heat the atmosphere to such a degree that it would spontaneously ignite forests and much of the biomass, sending dark clouds of smoke into the atmosphere. This smoke and the dust from the impact would block out the Sun, leading to a rapid plunge into a dark mini-ice age with most of the Sun's energy blocked from reaching Earth, preventing photosynthesis and plant growth. This darkness and cold would last for several months as the dust slowly settled, forming a global layer of dust recording the chemical signature of the impact. The hallmark of such an impact includes a hallmark high concentration of the rare element iridium, produced by vaporization of the meteorite.

Gradually, rain would remove the sulfuric acid and dust from the atmosphere, but the chemical consequences include enhanced acid rain, which rapidly dissolves calcium carbonates from limestone and shells, releasing carbon dioxide to the atmosphere. The acid rain would wreak havoc on the ocean biosphere, changing the ocean chemistry to the point at which many life-forms become extinct. The carbon dioxide released to the atmosphere would heat the planet a few months after the impact, and the planet would enter an extended warm period caused by this greenhouse effect. Temperatures would be more than 10°C (18°F) hotter on average, shifting climate belts and leading to excessively long, hot summers.

MITIGATING THE DANGERS OF FUTURE IMPACTS

Collision of a meteorite about a mile (1.6 km) across with Earth would do enough damage to wipe out about a quarter of the human race, and events of this magnitude may occur approximately every 3 million years. Larger events happen less frequently, and smaller events occur more frequently. An estimated 50 objects with diameters of 50–100 feet (15–30 m) pass between the Earth and Moon every day, though these rarely collide with the Earth. Comets and stony meteorites of this size will typically break up upon entering the atmosphere, whereas iron meteorites would tend to make it all the way to the planet. Luckily, few of the near-Earth objects are iron meteorites.

To date, there have only been limited efforts by the nations of the world to monitor near-Earth objects and to try to prevent large meteorites from crashing into the Earth and wiping out much of the population and biosphere. NASA has estimated that there are about 2,000 near-Earth objects greater than one-half mile (1 km) in diameter and that about half of them may eventually hit the Earth. However, the time interval between individual impacts is greater than 100,000 years. If any of these objects hits the Earth, the death toll will be tremendous, particularly if any of them hit populated areas or a major city.

Collision of Earth with an asteroid only a mile or two (several km) in diameter would release as much energy as that released by the simultaneous explosion of several million nuclear bombs.

It is now technically feasible to map and track many of the large objects that could be on an Earth-impacting trajectory, and this is being done to some degree. Greater efforts would involve considerable expense to advanced societies, principally the taxpayers of the United States. NASA, working with the United States Air Force, has mounted a preliminary program for mapping and tracking objects in near-Earth orbit and has already identified many significant objects. Lawmakers and the public must decide if the calculated risk of the hazards of impacts hitting the Earth is worth greater expense. Risk assessment typically involves many variables, such as the likelihood of an event happening, how many deaths or injuries would result, and what can be done to reduce the risk. Also, other questions need to be asked, such as is it more realistic to try to stop the spread of disease, crime, poverty, and famine and prepare for other natural disasters than to spend resources looking for objects that might one day collide with the Earth. If an asteroid is determined to be on a collision course with Earth, some type of asteroid deflection strategy would need to be employed to attempt to prevent the collision.

Spaceguard is a term that refers to a number of different efforts to search for and monitor near-Earth objects. The United States Congress published a Spaceguard Survey Report, mandating that 90 percent of all large near-Earth objects be located by 2002, and some programs were funded at a level of several million dollars per year toward this goal. Present estimates are that the original goal will be met by the year 2020. One of these efforts is the Catalina Sky Survey, which discovered 310 near-Earth objects in 2005, 400 in 2006, and 450 in 2007. A loose organization of observers and astronomers in several countries meet to discuss strategies, progress, and ideas about asteroid detection through the International Astronomical Union. However, it is noteworthy that these efforts were not sufficient to detect two meteorite impacts with Earth: an explosion over the Mediterranean in 2002 and the crash of a meteorite in the Bodaybo area of Siberia on September 25, 2002. The meteorites were detected by United States military antimissile defense satellites only as they entered the Earth's atmosphere.

Societies have the technology to attempt to divert or blow up a meteorite using nuclear devices. Bombs could be exploded near the asteroid or meteorite in an attempt to move it out of Earth orbit or to break it up into pieces small enough to break up upon entering the atmosphere. Alternatively, given enough time, rockets could be installed on the meteorite and fired to try to steer it out of its impact trajectory. However, many asteroids rotate rapidly, and rockets mounted on these asteroids would not be so effective at changing their course. Other proposals have been made, including firing massive missiles at the asteroid, transferring kinetic energy to move it out of its collision course. However, if the object is very large it is likely that even all of the nuclear weapons or bombs on the planet would not have a significant effect on altering the trajectory of the meteorite or asteroid. Strategies for preventing catastrophic collisions of meteorites with Earth fall into two general categories—those that attempt to destroy or fragment the asteroid into small pieces that would burn up upon passing though the Earth's atmosphere, and those that attempt to divert the asteroid and move it out of its trajectory toward Earth. In some cases it may be enough to simply delay the arrival time of the asteroid with Earth's orbit so that the planet is no longer at the place where it would collide with the asteroid when it crosses Earth's orbit. Such strategies use less energy than blasting the asteroid out of the solar system.

Nuclear Attack

One of the most popular ideas for deflecting asteroids away from a potential collision with Earth is to fire many nuclear missiles at the asteroid, with the idea that the blast would vaporize the asteroid, eliminating the danger. However, the energy requirements may not be attainable with the world's current arsenal of nuclear weapons, as there are currently no nuclear weapons that release enough energy to destroy an asteroid only a half mile (1 km) in diameter. If enough blasts or a large enough blast could be directed at an incoming asteroid, it is likely that the blasts would simply fragment the asteroid into many pieces, which would then fall to Earth along with the radiation from the nuclear explosions.

Instead of aiming nuclear blasts directly at any incoming asteroid, it may be possible to divert the orbit of the asteroid by exploding many nuclear bombs near the asteroid, and the energy released from these blasts could effectively steer the asteroid away from its collision course with Earth. A group of researchers from Massachusetts Institute of Technology published a study called Project Icarus, showing such a strategy to be theoretically possible.

Kinetic Impact Strategies

One of the alternative strategies that may be effective in deflecting an asteroid from Earth's orbit is to send a massive spacecraft to collide with the asteroid, altering its momentum and removing it from the collision course. This strategy is currently the object of a major study and mission, called Don Quijote, by the European Space Agency. Early results from this mission have shown that it is possible, and a model of the deflection of near-Earth asteroid 99942 Apophis shows that it would only take a spacecraft with a mass of less than one ton (0.9 tonne) to deflect the asteroid out of its modeled collision course with Earth.

Gravitational Tractor Strategies

Many asteroids and comets are composed of piles of disconnected rubble. Deflection strategies that rely on kinetic impact or deflection by explosion would not necessarily work on these types of asteroids, since any impact would only deflect the fragment that it directly hit. One alternative type of deflection strategy involves slowly moving these asteroid rubble piles by moving a massive spacecraft near the asteroid and letting the gravitational attraction of the spacecraft slowly pull the asteroid out of threatening orbit. Since both the asteroid and the spacecraft would have a mutual gravitational attraction, the asteroid could be slowly pulled in one direction if small rockets on the spacecraft are used to counter the attraction toward the asteroid, and slowly pull it out of a path dangerous to Earth. This strategy would take several years to be effective.

Other Strategies

A number of other strategies have been proposed that could eventually be developed into working deflection missions for asteroids heading toward Earth. One proposal is to focus solar energy at the surface of the asteroid, vaporizing material from the surface, eventually deflecting it from its collision course. It may be possible to wrap incoming asteroids with reflective sheeting or to add reflective dust to the surface, so that part of the asteroid receives additional radiation from the Sun, and radiation pressure distributed unequally on the asteroid may be enough to deflect it from its orbit. Another idea is to attach to the asteroid a large solar sail, which would absorb solar energy and change its orbit.

RECENT NEAR-COLLISIONS OF ASTEROIDS WITH EARTH

Collisions between asteroids can alter their orbits and cause them to head into an Earth orbit–crossing path. At this point, the asteroid becomes hazardous to life on Earth and is known as an Apollo object. NASA and the United States Air Force estimate that approximately 20,000 objects in space could be on Earth orbit–crossing trajectories. Presently, about 150 Apollo asteroids with diameters of greater than half a mile (1 km) are known, and a couple of thousand objects this size are known in the entire

near-Earth object group. Objects larger than 460 feet (140 m) hit the Earth on average about once every 5,000 years. In 1996 an asteroid about one quarter mile (half km) across nearly missed hitting the Earth, speeding past at a distance about equal to the distance to the Moon. The sobering reality of this near collision is that the asteroid was not even spotted until a few days before it sped past Earth. If the object had been bigger or slightly closer, it might not have been stoppable, and its collision might have had major consequences for life on Earth. A similar near miss event was recorded again in 2001; Asteroid 2001 YB5 passed Earth at a distance of twice that to the Moon, and it too was not recognized until two weeks before its near miss. If YB5 hit Earth, it would have released energy equivalent to 350,000 times the energy released during the nuclear bomb blast in Hiroshima.

The objects that are in Earth orbit–crossing paths could not have been in this path for very long, because gravitational influences of the Earth, Mars, and Venus would cause them to hit one of the planets or be ejected from the solar system within about 100 million years. The abundance of asteroids in an Earth orbit–crossing path demonstrates that ongoing collisions in the asteroid belt are replenishing the source of potential impacts on Earth. A few rare meteorites found on Earth have chemical signatures that suggest they originated on Mars and on the Moon, probably being ejected toward the Earth from giant impacts on those bodies.

Other objects from space may collide with Earth. Comets are masses of ice and carbonaceous material mixed with silicate minerals that are thought to originate in the outer parts of the solar system, in a region called the Oort Cloud. Other comets have a closer origin, in the Kuiper Belt just beyond the orbit of Neptune, including the dwarf planet Pluto. Comets may be less common near Earth than meteoroids, but they still may hit the Earth with severe consequences. Astronomers estimate that there are more than a trillion comets in the solar system. Since they are lighter than asteroids and have water-rich and carbon-rich compositions, many scientists have speculated that cometary impact may have brought water, major components of the atmosphere, and even life to Earth. The relative risks of impact for different objects are described in the table "Torino Hazard Scale for Near-Earth Objects."

The importance of monitoring near-Earth objects is highlighted by a number of recent events where asteroids or comets nearly collided with Earth, or exploded in the planet's atmosphere. Several "Tunguska style" atmospheric explosions, where meteorites exploded in the atmosphere and formed air blasts have been noted, including the events in 1930

TORINO HAZARD SCALE FOR NEAR-EARTH OBJECTS

Scale	Description of Hazard
No Hazard	
0	The likelihood of a collision is zero, or is so low as to be effectively zero. Also applies to small objects such as meteors and bodies that burn up in the atmosphere as well as infrequent meteorite falls that rarely cause damage.
Normal	
1	A routine discovery in which a pass near the Earth is predicted but poses no unusual level of danger. Current calculations show the chance of collision is extremely unlikely with no cause for public attention or public concern. New telescopic observations very likely will lead to reassignment to level 0.
Meriting Attention by Astronomers	
2	A discovery, which may become routine with expanded searches, of an object making a somewhat close but not highly unusual pass near the Earth. While meriting attention by astronomers, there is no cause for public attention or public concern as an actual collision is very unlikely. New telescopic observations very likely will lead to reassignment to level 0.
3	A close encounter, meriting attention by astronomers. Current calculations give a 1 percent or greater chance of collision capable of localized destruction. Most likely, new telescopic observations will lead to reassignment to level 0. Attention by public and by public officials is merited if the encounter is less than a decade away.
4	A close encounter, meriting attention by astronomers. Current calculations give a 1 percent or greater chance of collision capable of regional devastation. Most likely, new telescopic observations will lead to reassignment to Level 0. Attention by public and by public officials is merited if the encounter is less than a decade away.
Threatening	
5	A close encounter posing a serious but still uncertain threat of regional devastation. Critical attention by astronomers is needed to determine conclusively whether a collision will occur. If the encounter is less than a decade away, governmental contingency planning may be warranted.
6	A close encounter by a large object posing a serious but still uncertain threat of a global catastrophe. Critical attention by astronomers is needed to determine conclusively whether a collision will occur. If the encounter is less than three decades away, governmental contingency planning may be warranted.
7	A very close encounter by a large object, which if occurring this century, poses an unprecedented but still uncertain threat of a global catastrophe. For such a threat in this century, international contingency planning is warranted, especially to determine urgently and conclusively whether a collision will occur.
Certain Collisions	
8	A collision is certain, capable of causing localized destruction for an impact over land or possibly a tsunami if close offshore. Such events occur on average between once per 50 years and once per several 1,000 years.
9	A collision is certain, capable of causing unprecedented regional devastation for a land impact or the threat of a major tsunami for an ocean impact. Such events occur on average between once per 10,000 years and once per 100,000 years.
10	A collision is certain, capable of causing global climatic catastrophe that may threaten the future of civilization as we know it, whether impacting land or ocean. Such events occur on average once per 100,000 years, or less often.

over the Amazon River, in 1965 over southeastern Canada, in 1965 over Lake Huron, in 1967 in
Alberta, Canada, in Russia in 1992, in Italy in 1993,
in Spain in 1994, in Russia and the Mediterranean
in 2002, and in Washington in 2004. Other asteroid
encounters luckily were "near-misses," where larger
asteroids narrowly escaped collision with Earth. In
1972, an asteroid estimated to be 6–30 feet (2–10
m) in diameter entered the Earth's atmosphere above
Salt Lake City, Utah, formed a huge fireball that
raced across the daytime sky, and exited the atmosphere near Calgary in Alberta, Canada. The geometry of the orbit was such that the meteorite just
grazed the outer parts of the atmosphere, getting
as close to the surface as 36 miles (58 km). In 1989
the 1,000-foot (300-m) diameter Apollo asteroid
4581 Asclepius crossed the exact place the Earth
had just passed through six hours earlier, missing
the planet by a mere 400,000 miles (700,000 km).
If an object that size had collided with Earth, the
results would have been catastrophic. On June 14,
2002, a 165–400 foot (50–120 m) diameter asteroid
named 2002 MN passed unnoticed at a distance of
75,000 miles (120,700 km), one-third the distance
to the moon. Remarkably, this asteroid was not recognized until three days after it passed that closely to
the Earth. On July 3, 2006, another asteroid, named
2004 XP14, passed at about 248,000 miles (400,000
km), moving at a velocity of 10.5 miles per second
(17 km/sec).

Several asteroids are known to be on near-colli-
sion courses with Earth. These include 99942 Apophis, which will pass within 20,000 miles (32,000
km) of Earth but will miss the planet. However, it
may come closer in 2036, with a possible impact on
that orbit. The chances of impact are estimated to
be one in 43,000, making 99942 Apophis a Level
O danger on the Torino impact hazard scale. On
March 16, 2880, asteroid 29075, with a diameter
of 0.7–0.9 miles (1.1–1.4 km) will pass close to
Earth. Some models suggest that this asteroid has
a one in 300 chance of hitting the planet, posing a
significant threat for a catastrophic collision, with
major changes to climate and possibly triggering
mass extinctions. This asteroid has the highest probability of any known large objects of hitting Earth.

SUMMARY

The chances of experiencing a large meteorite impact
on Earth are small, but the risks associated with
large impacts are extreme. Small objects hit the Earth
many times every day but burn up in the atmosphere. Events that release enough energy to destroy
a city happen about once every thousand years, while
major impacts that can significantly alter the Earth's
climate happen every 300,000 years. Truly catastrophic impacts that cause mass extinctions, death
of at least 25 percent of the world's population, and
could lead to the end of the human race occur about
every 100 million years.

Specific hazards from impacts include atmospheric shock waves and air blasts, major earthquakes, monstrous tsunamis, and global firestorms
that throw so much soot in the air the impact is followed by a global winter that could last years. Carbon dioxide can be released by impacts as well and
then can act as a greenhouse gas leading to global
warming.

More than 20,000 near-Earth objects are thought
to have a potential to collide with Earth, and more
than 150 of these are larger than half a mile (1 km)
across. A variety of programs to detect and track
these near Earth objects is under way, yet most meteorite impacts and near collisions in the past few years
have been complete surprises. If a large asteroid is
found to be on a collision course with Earth, several
strategies have been devised that may be able to move
the object out of its collision course with Earth. The
asteroid could be attacked with nuclear weapons that
could vaporize the object, removing the threat. However, this could also break up the asteroid and send
thousands of smaller, now radioactive, fragments
to Earth. If nuclear weapons are detonated near the
asteroid, the force of the explosions may be enough
to push it out of its collision course. A massive spacecraft could be crashed into the asteroid, changing its
momentum and moving it from orbit. It might be
possible to install rocket propulsion systems on the
asteroid and have it steer itself out of Earth orbit. A
variety of other techniques have been proposed to
deflect asteroids, including beaming solar radiation
at the body, or attaching thermal blankets or sails, to
have the solar radiation pressure move the asteroid
out of its collision course.

See also ASTEROID; COMET; SOLAR SYSTEM.

FURTHER READING

Alvarez, W. *T. Rex and the Crater of Doom.* Princeton, N.J.: Princeton University Press, 1997.

Angelo, Joseph A. *Encyclopedia of Space and Astronomy.* New York: Facts On File, 2006.

Chapman, C. R., and D. Morrison. "Impacts on the Earth by Asteroids and Comets: Assessing the Hazard." *Nature* 367 (1994): 33–39.

Cox, Donald, and James Chestek. *Doomsday Asteroid: Can We Survive?* New York: Prometheus Books, 1996.

Elkens-Tanton, Linda T. *Asteroids, Meteorites, and Comets.* New York: Chelsea House, 2006.

Kusky, T. M. *Asteroids and Meteorites: Catastrophic Collisions with Earth.* New York: Facts On File, 2009.

Martin, P. S., and R. G. Klein, eds., *Quaternary Extinctions*. Tucson: University of Arizona Press, 1989.

Melosh, H. Jay. *Impact Cratering: A Geologic Process*. New York: Oxford University Press, 1988.

National Aeronautic and Space Administration (NASA). NASA's Web site on Lunar and Planetary Science. Available online. URL: http://nssdc.gsfc.nasa.gov/planetary/planets/asteroidpage.html. Accessed February 17, 2008.

Poag, C. Wylie. *Chesapeake Invader, Discovering America's Giant Meteorite Crater*. Princeton, N.J.: Princeton University Press, 1999.

Sharpton, Virgil L., and P. D. Ward. "Global Catastrophes in Earth History." Special Paper 247, Geological Society of America. 1990.

Stanley, S. M. *Extinction*. New York: Scientific American Library, 1987.

meteoric   Water that has recently come from the Earth's atmosphere is called meteoric water. The term is usually used in studies of groundwater, to distinguish water that has resided in ground for extended periods of time versus water that has recently infiltrated the system from rain, snow melt, or stream infiltration. Measurements of oxygen isotopes and other elements are typically used to aid this differentiation, as water from different sources shows different isotopic compositions.

See also GROUNDWATER.

meteorology   Meteorology is the study of the Earth's atmosphere, along with its movements, energy, interactions with other systems, and weather forecasting. The main focus of meteorology is short-term weather patterns and data within a specific area, in contrast to climatology, which is the study of the average weather on longer timescales and often on a global basis. Different aspects of meteorology include the study of the structure of the atmosphere, such as its compositional and thermal layers, and how energy is distributed within these layers. It includes analysis of the composition of the atmosphere, how the relative and absolute abundance of elements have changed with time, and how different interactions of the atmosphere, biosphere, and lithosphere contribute to the atmosphere's chemical stability. A fundamental aspect of meteorology is relating how different factors, including energy from the Sun, contribute to cloud formation, movement of air masses, and weather patterns at specific locations. Meteorologists interpret these complex energy changes and moisture changes and try to predict the weather using this knowledge. Increasingly, meteorologists are able to use data collected from orbiting satellites to aid their interpretation of these complex phenomena. Satellites have immensely improved the ability to monitor and predict the strength and paths of severe storms, such as hurricanes, as well as monitor many aspects of the atmosphere, including moisture content, pollution, and wind patterns.

The Earth's atmosphere is rich in nitrogen and oxygen and has much lower abundances of water vapor, carbon dioxide, and other gases. Some gases, such as water vapor and carbon dioxide, have a tendency to trap heat in the atmosphere. Called greenhouse gases, such gases have varied in abundance throughout Earth history, causing large temperature changes of several to tens to even hundreds of degrees in the past 4.5 billion years.

The atmosphere is divided into several layers, including the lower troposphere (where most weather events take place), the stratosphere, the mesosphere, and finally the thermosphere, which is the hottest part of the atmosphere. The topmost layer of the atmosphere is called the exosphere, where many gas molecules escape from the gravitational pull of the Earth, and which grades into the highly charged ionosphere where many free electrons and ions exist.

Weather events in the atmosphere are driven by heat and energy transfer. Latent heat, the amount of energy in the form of heat that is absorbed or released by a substance during a change in state such as from a liquid to a solid, is an important source of atmospheric energy. Heat transfer by convection is also important in the atmosphere, as moving air transfers energy from one region to another. Radiation, or the transfer of energy by electromagnetic waves, is a third important source of energy in the atmosphere. The Sun emits energy as shortwave radiation that the Earth absorbs and subsequently emits as long wavelength infrared radiation. Water vapor and carbon dioxide can absorb energy at these wavelengths, warming the atmosphere. The atmosphere warms since it allows the Sun's short wavelength radiation through, but then traps the energy absorbed from the long wavelengths emitted from the Earth. The Earth then cools by radiation, which operates most efficiently on clear nights when the clouds do not trap the outgoing radiation.

Seasons on the Earth are caused by the Earth's tilt on its axis, which results in a seasonal variation in the amount of sunlight received in different hemispheres at different times of the year. Longer hours of more intense sunlight are associated with summer, and fewer hours of less intense sunlight are associated with winters in both hemispheres.

The daily variations in temperature in any place are controlled mainly by the balance between energy input from the Sun versus energy output by convection and radiation. With radiative cooling at night, the ground surface often cools more quickly than the overlying air, resulting in an inversion with the coldest air right next to the surface.

Water is an important element in the atmosphere. Absolute humidity is the density of water vapor in a given volume of air. Relative humidity is a measure of how close the air is to being saturated with water vapor, which also depends on temperature. The dew point is a measure of how much the air would have to be cooled for saturation to occur. When the air temperature and dew point are close the air feels much more humid and the relative humidity is high. Condensation occurs when the temperature reaches the dew point, and then small droplets of water form in the atmosphere or on surfaces, forming fog. If these small droplets of water freeze it produces small frozen droplets. Condensation above the surface produces clouds, which are classified according to their height and physical appearance and are commonly divided into high, middle, and low groups plus those that cut across many atmospheric levels.

Clouds tend to form horizontal layers in stable atmospheric conditions, but in unstable conditions parcels of air that get uplifted are warmer and lighter than surrounding air, so they continue to rise, forming large vertical clouds such as the towering cumulonimbus or thunderhead clouds. On warm days simple surface heating can cause cumulus clouds to form, at heights determined by the temperature and moisture content of the surface air. As droplets of moisture in clouds coalesce by moving in the convecting clouds and hitting each other, they gradually get large enough to form raindrops, or ice if the temperatures are low. Precipitation can have a variety of forms when it reaches the surface depending on the form it took in the cloud and on the near surface and surface temperatures. If the surface air is cold but the air aloft is warm, raindrops may fall and freeze on impact, a phenomenon called freezing rain. Snow can develop when both surface and higher level air is cold, and may fall as snowflakes, pellets, or grains. In situations where surface air is warm but cold air is aloft and there is a strong updraft (such as in a cumulonimbus cloud), hail stones may form in the cloud and hit the surface as balls of ice. Conditions in which there is warm air aloft and also on the ground cause precipitation to fall as rain.

Horizontal changes in temperature in the atmosphere produce areas with high and lower pressure. Plots of the height of equal air pressure show that low areas correspond to low pressure, and high areas to high pressure. The difference in the air pressure creates a force called the pressure gradient force that sets the air in motion in winds. This moving air is then acted on by the Coriolis force, which tends to move air to the right of its intended course in the Northern Hemisphere and to the left in the Southern Hemisphere. Winds in the Northern Hemisphere bend clockwise around high pressure and counterclockwise around low pressure centers. The Coriolis force causes the opposite pattern in the Southern Hemisphere.

There are many variations of microscale and mesoscale winds near the surface of the planet. The surface layer of air, extending to about half a mile (1 km) above the surface, is affected by surface friction, causing different types of winds to develop around different obstructions. Wind produces sand dunes and ripples in deserts and in snow fields and may deform vegetation near mountaintops where the winds are consistently strong. Mountains can produce strong rotations of the air downwind of the range, and frictional effects of fast-moving air aloft in jet streams can produce strong eddies in the surface layer, associated with strong turbulence. Local winds that blow uphill through mountain valleys during the day are called valley breezes, and those that flow downhill at night are called mountain breezes. Strong downslope winds are called katabatic winds. Larger scale mesocale wind systems often form near boundaries between the ocean and land, where differential heating of the land and water creates pressure differences that generate winds. Where winds blow across a large body of water, differential heating of the land and sea in different seasons may cause the winds to shift direction with the seasons, producing wind systems called monsoons.

There are many large-scale patterns of wind and pressure that persist around the world. Trade winds are those that blow toward the equator from the semipermanent high pressure zones located at 30° latitude. The trade winds from the Northern and Southern Hemispheres converge along the intertropical convergence zone. Poleward of the high pressure belts is a zone where the winds blow predominantly to the west (the westerlies). These meet a more poleward belt of east-flowing winds known as the easterlies along the polar front. Annual shifts in the positions of these belts produce the annual changes in patterns of precipitation that characterize many regions.

Jet streams form where strong winds aloft get concentrated into narrow bands, such as the polar jet stream that forms in response to temperature differences along the polar front, while subtropical jets form at high elevations above the subtropics along an upper level boundary called the subtropical front.

Interactions between the atmosphere and ocean are complex. Surface winds form ocean currents, and yet the oceans release energy that helps maintain atmospheric circulation. Atmospheric circulation patterns may change on seasonal or other timescales.

When warm air and water from the Austral-Indonesia region flows eastward toward South America it can form an El Niño, choking off the nutrient-rich upwelling and wreaking havoc on the environment and economics of South and Central America. The opposite effects, called La Niña, often dominate, and the alternating cycle of winds and currents is called the Southern Oscillation.

Air moves as coherent masses along boundaries called fronts. Stationary fronts have no movement, with cold air on one side and warm on the other. Winds usually blow parallel to fronts and in opposite directions on either side of the front. Fronts more typically move across continents and oceans, being driven by global atmospheric circulation. Leading edges of cold fronts are usually associated with showers as the cold air forces the warm air upward and replaces it, but in warm fronts the warm air rises over the colder surface air, producing cloudiness and widespread precipitation.

Mid-latitude cyclones form when an upper-level low-pressure trough forms west of a surface low-pressure area and when a shortwave disturbs this system, setting up surface and upper-level winds that enhance the development of the surface storm. The air converges at the surface level and rises in the center of the storm, forming precipitation. As the warm air rises and cool air sinks, energy is released, and the storm grows in strength as the potential energy is converted into kinetic energy. The storm may be steered by mid- or upper-level winds in the atmosphere.

Thunderstorms commonly develop when there is a humid layer of surface air, sunlight to heat the ground, and unstable air aloft. In these conditions the heated air may quickly rise, forming large cumulonimbus clouds that may drop locally heavy rains. When a strong vertical wind shear exists, severe thunderstorms may form. Supercells are large rotating thunderstorm systems that may persist for hours. Many thunderstorms form along frontal boundaries where cold air forces the warm air to rise, forming lines and clusters of storms called mesoscale convective complexes. Tornados, rapidly circulating columns of air that reach the ground, are often associated with supercells, and can have winds that reach a couple hundred miles per hour (few hundred km/hr) in the tornado core, typically less than several hundred yards (meters) wide.

Hurricanes are tropical cyclones with winds exceeding 74 miles per hour (119 km/hr) and include a well-organized mass of thunderstorms rotating about a central low pressure region in the storm's eye. Hurricanes form over warm tropical waters where surface winds converge along a tropical wave, initiating central airs to rise, forming a tropical depression. As the air continues to move into the storm system and rise in its center, much latent heat is released, causing more air to rise, and central pressures to reduce further, leading the storm to grow further. Energy is released by the storm in the cloud tops by radiational cooling so the strengthening of the storm depends on the balance between the energy gained by converting sensible and latent heat into kinetic energy in the storm eye and the energy lost in the cloud tops by radiation. Most hurricanes are steered to the west by the easterly winds in the tropics but may move westward when they move into mid-latitudes. Since the storms gain energy (and keep their energy balance) from the warm water, hurricanes rapidly lose strength when they move over cool water or land masses. Most damage from hurricanes is associated with the large storm surges that some generate, as well as the high winds and flooding rains.

See also ATMOSPHERE; CLIMATE; CLIMATE CHANGE; CLOUDS; EL NIÑO AND THE SOUTHERN OSCILLATION (ENSO); ENERGY IN THE EARTH SYSTEM; GREENHOUSE EFFECT; HURRICANES; MONSOONS, TRADE WINDS; PRECIPITATION; SUN; THERMODYNAMICS.

FURTHER READING

Ahrens, C. D. *Meteorology Today: An Introduction to Weather, Climate, and the Environment.* 8th ed. Pacific Grove, Calif.: Brooks/Cole, 2007.

Intergovernmental Panel on Climate Change. *Climate Change 2007: The Physical Science Basis. Contributions of Working Group I to the Fourth Assessment Report of the Intergovernmental Panel on Climate Change,* edited by S. Solomon, D. Qin, M. Manning, Z. Chen, M. Marquis, K. B. Averyt, M. Tignor, and H. L. Miller. Cambridge: Cambridge University Press, 2007. Also available online. URL: http://www.ipcc.ch/index.htm. Accessed October 10, 2008.

Thunderhead cloud (cumulonimbus) rising *(Greg F. Riegler, Shutterstock, Inc.)*

National Aeronautic and Space Administration (NASA). Earth Observatory. Available online. URL: http://earthobservatory.nasa.gov/. Accessed October 9, 2008; updated daily.

U.S. Environmental Protection Agency. Climate Change home page. Available online. URL: http://www.epa.gov/climatechange/. Updated September 9, 2008.

Milankovitch, Milutin M. (1879–1958) Serbian *Mathematician, Physicist* Milutin Milankovitch was born and educated in Serbia, and was appointed to a chair in the University of Belgrade in 1909, where he taught courses in mathematics, physics, mechanics, and celestial mechanics. He is well known for his research on the relationship between celestial mechanics and climate on the Earth, and he is responsible for developing the idea that rotational wobbles and orbital deviations combine in cyclic ways to produce the climatic changes on the Earth. He determined how the amount of incoming solar radiation changes in response to several astronomical effects such as orbital tilt, eccentricity, and wobble. These changes in the amount of incoming solar radiation in response to changes in orbital variations occur with different frequencies, and produce cyclical variations known as Milankovitch cycles. Milankovitch's main scientific work was published by the Royal Academy of Serbia in 1941, during World War II in Europe. He calculated that the effects of orbital eccentricity, wobble, and tilt combine every 40,000 years to change the amount of incoming solar radiation, lowering temperatures and causing increased snowfall at high latitudes. His results have been widely used to interpret climatic variations, especially in the Pleistocene record of ice ages, and also in the older rock record.

See also CLIMATE CHANGE; MILANKOVITCH CYCLES.

Milankovitch cycles Systematic changes in the amount of incoming solar radiation, caused by variations in Earth's orbital parameters around the Sun, are known as Milankovitch cycles. These changes can affect many Earth systems, causing glaciations, global warming, and changes in the patterns of climate and sedimentation.

Medium-term climate changes include those that alternate between warm and cold on time scales of 100,000 years or less. These medium-term climate changes include the semi-regular advance and retreat of the glaciers during the many individual ice ages in the past few million years. The last 2.8 million years have been marked by large global climate oscillations that have been recurring at approximately a 100,000-year periodicity at least for the past 800,000 years. The warm periods, called interglacial periods, appear to last approximately 15,000 to 20,000 years before regressing to a cold ice age climate. The last of these major glacial intervals began ending about 18,000 years ago, as the large continental ice sheets covering North America, Europe, and Asia began retreating. The main climate events related to the retreat of the glaciers, can be summarized as follows:

- 18,000 years ago: The climate begins to warm
- 15,000 years ago: Advance of glaciers halts and sea levels begin to rise
- 10,000 years ago: Ice Age megafauna goes extinct
- 8,000 years ago: Bering Strait land bridge becomes drowned, cutting off migration of people and animals.
- 6,000 years ago: The Holocene maximum warm period
- So far in the past 18,000 years, the Earth's temperature has risen approximately 16°F (10°C) and the sea level has risen 300 feet (91 m).

This past glacial retreat is but one of many in the past several million years, with an alternation of warm and cold periods apparently related to a 100,000 year periodicity in the amount of incoming solar radiation, causing the alternating warm and cold intervals. These systematic changes are known as Milankovitch cycles, after Milutin Milankovitch (1879–1958), a Serbian scientist who first clearly elucidated the relationships between the astronomical variations of the Earth orbiting the Sun and the climate cycles on Earth. Milankovitch's main scientific work was published by the Royal Academy of Serbia in 1941, during World War II in Europe. He was able to calculate that the effects of orbital eccentricity, wobble, and tilt combine every 40,000 years to change the amount of incoming solar radiation, lowering temperatures and causing increased snowfall at high latitudes. His results have been widely used to interpret the climatic variations, especially in the Pleistocene record of ice ages, and also in the older rock record.

Astronomical effects influence the amount of incoming solar radiation; minor variations in the path of the Earth in its orbit around the Sun and the inclination or tilt of its axis cause variations in the amount of solar energy reaching the top of the atmosphere. These variations are thought to be responsible for the advance and retreat of the Northern and Southern Hemisphere ice sheets in the past few million years. In the past two million years alone, the Earth

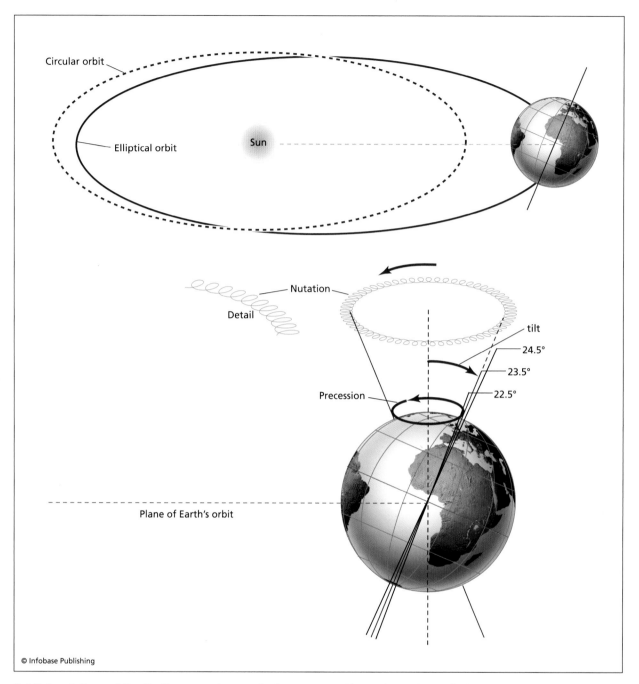

© Infobase Publishing

Orbital variations of the Earth cause changes in the amount of incoming solar radiation, known as Milankovitch cycles. Shown here are changes in the eccentricity of the orbit, the tilt of the spin axis (nutation), and precession of the equinoxes.

has seen the ice sheets advance and retreat approximately 20 times. The climate record as deduced from ice-core records from Greenland and isotopic tracer studies from deep ocean, lake, and cave sediments suggest that the ice builds up gradually over periods of about 100,000 years, then retreats rapidly over a period of decades to a few thousand years. These patterns result from the cumulative effects of different astronomical phenomena.

Several movements are involved in changing the amount of incoming solar radiation. The Earth rotates around the Sun following an elliptical orbit, and the shape of this elliptical orbit is known as its eccentricity. The eccentricity changes cyclically with time with a period of 100,000 years, alternately bringing the Earth closer to and farther from the Sun in summer and winter. This 100,000-year cycle is about the same as the general pattern of glaciers advancing and retreating every 100,000 years in the past two million years, suggesting that this is the main cause of variations within the present-day ice age. Presently, the Earth is in a period of low eccentricity (~3 percent) and this yields a seasonal change in solar energy of ~7 percent. When the eccentricity is at its peak (~9 percent), "seasonality" reaches ~20 percent. In addition a more eccentric orbit changes the length of seasons in each hemisphere by changing the length of time between the vernal and autumnal equinoxes.

The Earth's axis is presently tilting by 23.5°N/S away from the orbital plane, and the tilt varies between 21.5°N/S and 24.5°N/S. The tilt, also known as obliquity, changes by plus or minus 1.5°N/S from a tilt of 23°N/S every 41,000 years. When the tilt is greater, there is greater seasonal variation in temperature. For small tilts, the winters would tend to be milder and the summers cooler. This would lead to more glaciation.

Wobble of the rotation axis describes a motion much like a top rapidly spinning and rotating with a wobbling motion, such that the direction of tilt toward or away from the Sun changes, even though the tilt amount stays the same. This wobbling phenomenon is known as precession of the equinoxes, and it has the effect of placing different hemispheres closest to the Sun in different seasons. This precession changes with a double cycle, with periodicities of 23,000 years and 19,000 years. Presently the precession of the equinoxes is such that the Earth is closest to the Sun during the Northern Hemisphere winter. Due to precession, the reverse will be true in ~11,000 years. This will give the Northern Hemisphere more severe winters.

Because each of these astronomical factors acts on a different time scale, they interact in a complicated way, known as Milankovitch cycles. Using the power of understanding these cycles, we can make predictions of where the Earth's climate is heading, whether we are heading into a warming or cooling period and whether we need to plan for sea level rise, desertification, glaciation, sea level drops, floods, or droughts. When all the Milankovitch cycles (alone) are taken into account, the present trend should be toward a cooler climate in the Northern Hemisphere, with extensive glaciation. The Milankovitch cycles may help explain the advance and retreat of ice over periods of 10,000 to 100,000 years. They do not explain what caused the Ice Age in the first place.

Western aspect of the Pelmo massif in the Italian Dolomite Mountains. The cyclical layering recorded by the beds (horizontal) is interpreted as records of Milankovitch climate cycles. *(Gillian Price/Alamy)*

The pattern of climate cycles predicted by Milankovitch cycles is made more complex by other factors that change the climate of the Earth. These include changes in thermohaline circulation, changes in the amount of dust in the atmosphere, changes caused by reflectivity of ice sheets, changes in concentration of greenhouse gases, changing characteristics of clouds, and even the glacial rebound of land that was depressed below sea level by the weight of glaciers.

Milankovitch cycles have been invoked to explain the rhythmic repetitions of layers in some sedimentary rock sequences. The cyclical orbital variations cause cyclical climate variations, which in turn are reflected in the cyclical deposition of specific types of sedimentary layers in sensitive environments. There are numerous examples of sedimentary sequences where stratigraphic and age control are sufficient to be able to detect cyclical variation on the time scales of Milankovitch cycles, and studies of these layers have proven consistent with a control of sedimentation by the planet's orbital variations. Some examples of Milankovitch-forced sedimentation have been documented from the Dolomite Mountains of Italy, the Proterozoic Rocknest Formation of northern Canada, and from numerous coral reef environments.

Predicting the future climate on Earth involves very complex calculations, including inputs from the long- and medium-term effects described in this entry, and some short-term effects such as sudden changes caused by human inputs of greenhouse gases to the atmosphere, and effects such as unpredicted volcanic eruptions. Nonetheless, most climate experts expect that the planet will continue to warm on the hundreds-of-years time scale. However, based on the recent geological past, it seems reasonable that the planet could be suddenly plunged into another ice age, perhaps initiated by sudden changes in ocean circulation, following a period of warming. Climate is one of the major drivers of mass extinction, so the question remains if the human race will be able to cope with rapidly fluctuating temperatures, dramatic changes in sea level, and enormous shifts in climate and agriculture belts.

See also CLIMATE CHANGE; MILANKOVITCH, MILUTIN M.; SEQUENCE STRATIGRAPHY; STRATIGRAPHY, STRATIFICATION, CYCLOTHEM.

FURTHER READING

Ahrens, C. D. *Meteorology Today, An Introduction to Weather, Climate, and the Environment.* 6th ed. Pacific Grove, Calif.: Brooks/Cole, 2000.

Allen, John R. *Sedimentary Structures: Their Character and Physical Basis.* Amsterdam: Elsevier, 1982.

Allen, P. A., and J. R. Allen. *Basin Analysis, Principles and Applications.* Oxford: Blackwell Scientific Publications, 1990.

Dawson, A. G. *Ice Age Earth.* London: Routledge, 1992.

Erickson, J. *Glacial Geology: How Ice Shapes the Land.* New York: Facts On File, 1996.

Goldhammer, Robert K., Paul A. Dunn, and Lawrence A. Hardie. "High-Frequency Glacial-Eustatic Sea Level Oscillations with Milankovitch Characteristics Recorded in Middle Triassic Platform Carbonates in Northern Italy." *American Journal of Science* 287 (1987): 853–892.

Grotzinger, John P. "Upward Shallowing Platform Cycles: A Response to 2.2 Billion Years of Low-Amplitude, High-Frequency (Milankovitch Band) Sea Level Oscillations." *Paleoceanography* 1 (1986): 403–416.

Hayes, James D., John Imbrie, and Nicholas J. Shakelton. "Variations in the Earth's Orbit: Pacemaker of the Ice Ages." *Science* 194 (1976): 2,212–2,232.

Imbrie, John. "Astronomical Theory of the Pleistocene Ice Ages: A Brief Historical Review." *Icarus* 50 (1982): 408–422.

Intergovernmental Panel on Climate Change. *Climate Change 2007: The Physical Science Basis. Contributions of Working Group I to the Fourth Assessment Report of the Intergovernmental Panel on Climate Change,* edited by S. Solomon, D. Qin, M. Manning, Z. Chen, M. Marquis, K. B. Averyt, M. Tignor, and H. L. Miller. Cambridge: Cambridge University Press, 2007.

Intergovernmental Panel on Climate Change home page. Available online. URL: http://www.ipcc.ch/index.htm. Accessed January 29, 2009.

mineral, mineralogy The branch of geology that deals with the classification and properties of minerals is closely related to petrology, the branch of geology that deals with the occurrence, origin, and history of rocks. Minerals are the basic building blocks of rocks, soil, and sand. Most beaches are made of the mineral quartz, which is very resistant to weathering and erosion by the waves. Most minerals, like quartz or mica, are abundant and common, although some minerals like diamonds, rubies, sapphires, gold, and silver are rare and very valuable. Minerals contain information about the chemical and physical conditions in the regions of the Earth that they formed in. They can often help discriminate which tectonic environment a given rock formed in, and they can tell us information about the inaccessible portions of Earth. For example, mineral equilibrium studies on small inclusions in diamonds show that they must form below a depth of 90 miles (145 km). Economies of whole nations are based on exploitation of mineral wealth; for instance, South Africa is such a rich nation because of its abundant gold and diamond mineral resources.

The two most important characteristics of minerals are their composition and structure. The composition of minerals describes the kinds of chemical elements present and their proportions, whereas the structure of minerals describes the way in which the atoms of the chemical elements are packed together.

Mineralogists have identified nearly 4,000 minerals, most made out of the eight most common mineral-forming elements. These eight elements, listed in the table "The Eight Most Common Mineral-Forming Elements," make up greater than 98 percent of the mass of the continental crust. Most of the other 133 scarce elements do not occur by themselves, but occur with other elements in compounds by ionic substitution. For example, olivine may contain trace amounts of copper (Cu), nickel (Ni), cobalt (Co), manganese (Mn), and other elements.

The two elements oxygen and silicon make up more than 75 percent of the crust, with oxygen alone forming nearly half of the mass of the continental crust. Oxygen forms a simple anion (O^{2-}), and silicon forms a simple cation (Si^{4+}). Silicon and oxygen combine together to form a very stable complex anion that is the most important building block for minerals—the silicate anion ($SiO_4)^{4-}$. Minerals that contain this anion are known as the silicate minerals, and they are the most common naturally occurring inorganic compounds in the solar system. The other, less common building blocks of minerals (anions) are oxides (O^{2-}), sulfides (S^{2-}), chlorides (Cl^-), carbonates ($CO_3)^{2-}$, sulfates ($SO_4)^{2-}$, and phosphates ($PO_4)^{3-}$.

Minerals are classified into eight major groups based on the main type of cation present in the min-

Eight types of minerals: sulfur, sapphire, orpiment/realgar, cinnabar, malachite, olivine (peridot), copper, and beryl *(Charles D. Winters/Photo Researchers, Inc.)*

eral structure, a classification scheme championed by James Dana in the early 1800s. His classification recognized (1) native elements; (2) sulfides; (3) oxides and hydroxides; (4) halides; (5) carbonates, nitrates, borates, and iodates; (6) sulfates, chromates, molybdates, and tungstates; (7) phosphates, arsenates, and vanadates; and (8) silicates.

Approximately 20 minerals are so common that they account for greater than 95 percent of all the minerals in the continental and oceanic crust; these are called the rock-forming minerals. Most rock-forming minerals are silicates and they have some common features in the way their atoms are arranged.

THE SILICATE TETRAHEDRON

The silicate anion is made of four large oxygen atoms and one small silicon atom that pack themselves together to occupy the smallest possible space. This shape, with big oxygen atoms at four corners of the structure and the silicon atom at the center, is known as the silicate tetrahedron. Each silicate tetrahedron has four unsatisfied negative charges (Si has a charge of +4, whereas each oxygen has a charge of -2). To make a stable compound the silicate tetrahedron must therefore combine to neutralize this extra charge, which can happen in one of two ways:

(1) Oxygen can form bonds with cations (positively charged ions). For instance, Mg^{2+} has a charge of +2, and by combining with

THE EIGHT MOST COMMON MINERAL-FORMING ELEMENTS

Element	Abbreviation	Percentage of Continental Crust Mass
Oxygen	O	46.6
Silicon	Si	27.7
Aluminum	Al	8.1
Iron	Fe	5.0
Calcium	Ca	3.6
Sodium	Na	2.8
Potassium	K	2.6
Magnesium	Mg	2.1

Mg^{2+}, the silicate tetrahedron makes a mineral called olivine $(Mg_2)SiO_4$.

(2) Two adjacent tetrahedra can share an oxygen atom, making a complex anion with the formula $(Si_2O_7)^{6-}$. This process commonly forms long chains, so that the charge is balanced except at the ends of the structure. This process of linking silicate tetrahedra into large anion groups is called polymerization, and it is the most common way to build minerals, but in making the various possible combinations of tetrahedra, one rule must be followed, that is tetrahedra can only be linked at their apexes.

Olivine is one of the most important minerals on Earth, forming much of the oceanic crust and upper mantle. It has the formula $(Mg,Fe)_2SiO_4$ and forms the gem peridot.

Garnet is made of isolated silicate tetrahedra packed together without polymerizing with other tetrahedra. There are many different kinds of garnets, with almandine being one of the more common, deep red varieties that forms a common gemstone. Ionic substitution is common, with garnet having the chemical formula $A_3B_2(SiO_4)^3$, where:

$$A = Mg^{2+}, Fe^{2+}, Ca^{2+}, Mn^{2+}$$

$$B = Al^{3+}, Fe^{3+}$$

Pyroxene and amphibole both contain continuous chains of silicate tetrahedra. Polymerized chains of tetrahedra form pyroxenes, whereas amphiboles are built in double chains or linked rings. In both of these structures, the chains are bound together by cations such as Ca^{2+}, Mg^{2+}, and Fe^{2+}, which satisfy the negative charges of the polymerized tetrahedra. Pyroxenes are very common minerals in the oceanic crust and mantle and also occur in the continental crust. Amphiboles are very common in metamorphic rocks, have a complicated

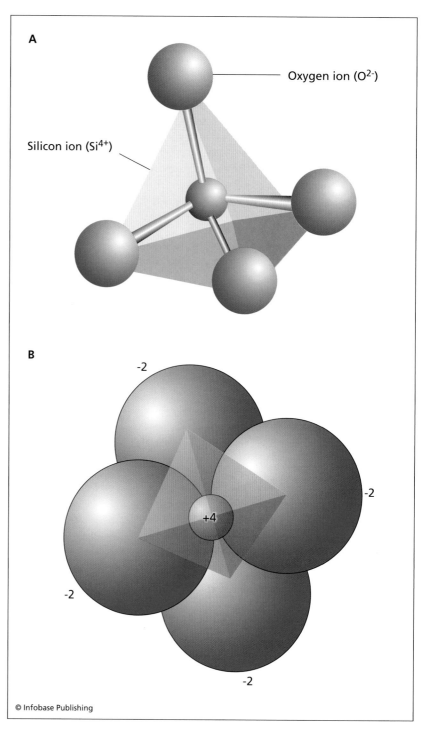

A — Oxygen ion (O^{2-})

Silicon ion (Si^{4+})

B — -2, -2, +4, -2, -2

© Infobase Publishing

Diagram showing silicate tetrahedra

chemical formula, and can hold a large variety of cations in their crystal structure.

Clays, micas, and chlorites are all closely related to sheet silicates, made of polymerized sheets of tetrahedra. By sharing three oxygen atoms with adjacent tetrahedra, each tetrahedron has only one oxygen atom remaining unbalanced, and it is typically balanced by Al^{3+} cations, which occupy spaces between the sheets. The sheet structure is why micas are easy to peel apart on cellophane-like surfaces.

Quartz, one of the most common minerals, also has one of the most common polymerizations. Its charges are satisfied by sharing all of its oxygen in a three-dimensional network. Quartz typically has six-sided crystals and has many other different forms and colors.

Feldspars are the most common minerals in the Earth's crust. They account for 60 percent of all minerals in the continental crust and 75 percent of the volume. Feldspars are also common in the oceanic crust. Like quartz, feldspars have a structure formed by polymerization of all the oxygen atoms, and some of the silicon atoms are replaced by Al^{3+}. Many different kinds of feldspar minerals form by the addition of different cations to the structure. For instance, potassium feldspar has the formula $K(Si_3Al)O_8$, albite has the formula $Na(Si_3Al)O_8$, and anorthite has the formula $Ca(Si_2Al_2)O_8$. A complete range of chemical compositions of feldspars is possible between the albite and anorthite varieties. These feldspar minerals are known as the plagioclase feldspars.

Silicates are the most abundant rock-forming minerals, but other types do occur in sufficient quantities to call them rock-forming minerals. Oxides use the oxygen anion and include ore minerals such as chromium, uranium, tin, and magnetite (FeO_4). Sulfides are minerals such as pyrite, copper, lead, zinc, cobalt, mercury, or silver that combine with the sulfur anion. For instance, FeS_2 is the formula for pyrite, commonly known as fool's gold. The carbonates calcite, aragonite, and dolomite form with the complex carbonate anion $(CO_3)^{2-}$. Phosphates are formed using the complex anion $(PO_4)^{3-}$. An example is the mineral apatite, used for fertilizers, and the same substance as that which forms teeth and bones. Sulfate minerals are formed using the complex sulfate

ion $(SO_4)^{2-}$. Gypsum and anhydrite, sulfate minerals formed by evaporation of salt water, are commonly used to make plaster.

One of the most important nonsilicate mineral groups is the carbonates, which are built using the carbonate ion. The most common carbonates are calcite ($CaCO_3$) and dolomite ($(Ca,MgCO_3)_2$). These minerals are common in sedimentary rock sequences deposited under marine conditions. Other nonsilicate minerals common in sedimentary rocks include the halide mineral halite (NaCl) and the sulfate gypsum ($CaSO_4$), both common in evaporate sequences formed when ocean waters evaporate and leave the dissolved minerals behind as sedimentary layers.

Native elements are not as common as many other mineral groups, and consist of free-occurring elements such as gold (Au), copper (Cu), silver (Ag), platinum (Pt), diamond (C), and graphite (C). Non-native-element economically important minerals include some oxides (hematite, magnetite, chromite, and ilmenite) and sulfides such as pyrite, chalcopyrite, galena, and sphalerite.

THE PROPERTIES OF MINERALS

Minerals have specific properties determined by their chemistry and crystal structure. Certain properties are characteristic of certain minerals, and one can identify minerals by learning these properties. The most common properties are crystal form, color, hardness, luster, cleavage, specific gravity, and taste.

When a mineral grows freely, it forms a characteristic geometric solid bounded by geometrically arranged plane surfaces (this is the crystal form). This symmetry is an external expression of the symmetric internal arrangement of atoms, such as in repeating tetrahedron arrays. Individual crystals of the same mineral may look somewhat different because the relative sizes of individual faces may vary, but the angles between faces are constant and diagnostic for each mineral.

Every mineral has a characteristic crystal form. Some minerals have such distinctive forms that they can be readily identified without measuring angles between crystal faces. For instance, pyrite is recognized as interlocking growth of cubes, whereas asbestos forms long silky fibers. These distinctive characteristics are known as growth habit.

Cleavage is the tendency of a mineral to break in preferred directions along bright reflective planar surfaces. External structure deters the planar surface along which cleavage occurs; cleavage occurs along planes where the bands between the atoms are relatively weak.

Luster is the quality and intensity of light reflected from a mineral. Typical lusters include metallic (like a polished metal), vitreous (like a polished glass),

Uncut emerald from Colombia *(Carl Frank/Photo Researchers, Inc.)*

MOH'S HARDNESS SCALE

10	Diamond
9	Corundum (ruby, sapphire)
8	Topaz
7	Quartz
6	Potassium feldspar (pocketknife, glass)
5	Apatite (teeth, bones)
4	Fluoride
3	Calcite (copper penny)
2	Gypsum (fingernail)
1	Talc

resinous (like resin), pearly (like a pearl), and greasy (oily).

Color is not reliable for identification of minerals, since it is typically determined by ionic substitution. For instance, sapphire and rubies are both varieties of the mineral corundum, but with different types of ionic substitution. However, the color of the streak a mineral leaves on a porcelain plate is often diagnostic for opaque minerals with metallic lusters.

The density of a mineral is a measure of mass per unit volume (g/cm^3). Density describes "how heavy the mineral feels." Specific gravity is an indirect measure of density; it is the ratio of weight of a substance to the weight of an equal volume of water (specific gravity has no units because it is a ratio).

Hardness is a measure of the mineral's relative resistance to scratching, as listed in the table "Moh's Hardness Scale." Hardness is governed by the strength of bonds between atoms and is very distinctive and useful for mineral identification. A mineral's hardness can be determined by the ease with which one mineral can scratch another. For instance, talc (used for talcum powder) is the softest mineral, whereas diamond is the hardest mineral.

See also CRYSTAL, CRYSTAL DISLOCATIONS; DANA, JAMES DWIGHT; ECONOMIC GEOLOGY; GEOCHEMISTRY; PETROLOGY AND PETROGRAPHY; THERMODYNAMICS.

FURTHER READING

Barthelmy, David. "Mineralogy Database." Available online. URL: http://www.webmineral.com/. Last updated August 31, 2008.

Skinner, Brian J., and Stephen C. Porter. *The Dynamic Earth, an Introduction to Physical Geology.* 5th ed. New York: John Wiley and Sons, 2004.

monsoons, trade winds A wind system that changes direction with the seasons is known as a monsoon, after the Arabic term *mausim*, meaning seasons. The Arabian Sea is characterized by monsoons, with the wind blowing from the northeast for six months, then from the southeast for the other half of the year. Seasonal reversal of winds is probably best known from India and southern Asia, where monsoons bring seasonal rains and floods.

In contrast to monsoons, trade winds are steady winds that blow between 0° and 30° latitude, from the northeast to southwest in the Northern Hemisphere and from southeast to northwest in the Southern Hemisphere. The trade winds are formed as the cool air from Hadley cell circulation returns to the surface at about 15–30° latitude, and then returns to the equatorial region. The Coriolis force deflects the moving air to the right in the Northern Hemisphere, causing the air to flow from northeast to southwest, and to the left in the Southern Hemisphere, causing a southeast to northwest flow. They are named trade winds because sailors used the reliability of the winds to aid their travels from Europe to the Americas. The doldrums, an area characterized by weak stagnant air currents, bound the trade winds, on high latitudes by the horse latitudes, characterized by weak winds, and toward the equator. The origin of the term *horse latitudes* is uncertain, but legend has it that it comes from ships traveling to the Americas that became stranded by the lack of winds in these regions, and were forced to kill onboard horses to conserve fresh water supplies and to eat their meat.

The Asian and Indian monsoon originates from differential heating of the air over the continent and ocean with the seasons. In the winter monsoon, the air over the continents becomes much cooler than the air over the ocean, and a large, shallow, high-pressure system develops over Siberia. This produces a clockwise rotation of air that rotates over the South China Sea and Indian Ocean, producing northeasterly winds and fair weather with clear skies over eastern and southern Asia. In contrast, in the summer monsoon, the air pattern reverses itself as the air over the continents becomes warmer than the air over the oceans. This produces a shallow, low-pressure system over the Indian subcontinent, within which the air rises. Air from the Indian Ocean and Arabian Sea rotates counterclockwise into the low-pressure area, bringing moisture-laden winds into the subcontinent. As the air rises due to convergence and orographic (mountain) effects, it cools below its saturation point, resulting in heavy rains and thunderstorms that characterize the summer monsoon of India from June through September. Some regions of India, especially the Cherrapunji area in the Khasi Hills of northeastern India, receive more than 40 inches (1,000 cm) of rain during a summer monsoon. A similar pattern develops over Southeast Asia. Other less intense monsoons are known from Australia, South America, Africa, and parts of the desert southwest, Pacific coast, and Mississippi Valley of the United States.

The strength of the Indian monsoon is related to the El Niño-Southern Oscillation. During the El Niño events, surface water near the equator in the central and eastern Pacific is warmer than normal, forming excessive rising air, thunderstorms, and rains in this region. This pattern causes air to sink over eastern Asia and India, leading to a summer monsoon with much lower than normal rainfall totals.

See also ATMOSPHERE; CLIMATE; CLIMATE CHANGE; EL NIÑO AND THE SOUTHERN OSCILLATION (ENSO); ENERGY IN THE EARTH SYSTEM.

FURTHER READING

Ahrens, C. D. *Meteorology Today, An Introduction to Weather, Climate, and the Environment.* 6th ed. Pacific Grove, Calif.: Brooks/Cole, 2000.

Intergovernmental Panel on Climate Change. *Climate Change 2007: The Physical Science Basis. Contributions of Working Group I to the Fourth Assessment Report of the Intergovernmental Panel on Climate Change,* edited by S. Solomon, D. Qin, M. Manning, Z. Chen, M. Marquis, K. B. Averyt, M. Tignor, and H. L. Miller. Cambridge: Cambridge University Press, 2007.

Intergovernmental Panel on Climate Change home page. Available online. URL: http://www.ipcc.ch/index.htm. Accessed January 29, 2009.

Neogene The Neogene is the second of three periods of the Cenozoic, including the Paleogene, Neogene, and Quaternary, and the second of two subperiods of the Tertiary, younger than Paleogene. Its base is at 23.8 million years ago and its top is at 1.8 million years ago, followed by the Quaternary period. Charles Lyell proposed the subdivision of the Neogene into the Miocene, Pliocene, Pleistocene, and Recent epochs in his book *Principles of Geology* in 1833. Austrian geologist Moriz Hörnes formally proposed the currently accepted division of the Neogene in 1835 and included only the older parts of Lyell's Neogene.

The Atlantic and Indian Oceans were open in the Neogene, and the Earth's plate mosaic resembled the modern configuration. The collision of India with Asia was well under way, and Australia had already rifted and was moving away from Antarctica, isolating Australia and leading to the development of the cold circumpolar current and the Antarctic ice cap. Subduction and accretion events were active along the Cordilleran margins of North and South America. Basin-and-range extension was active, and the Columbia River basalts were erupted in the northwestern United States. The San Andreas fault developed in California during subduction of the East Pacific Rise.

One of the more unusual events to mark the Neogene is the development of up to 1.2 miles (2 km) of salt deposits between 5.5 and 5.3 million years ago in the Mediterranean region. This event, known as the Messinian salt crisis, was caused by the isolation of the Mediterranean Sea by collisional tectonics and falling sea levels that caused the sea to at least partially evaporate several times during the 200,000-year-long crisis. Rising sea levels ended the Messinian crisis 5.3 million years ago, when waters of the Atlantic rose over the natural dam in the Strait of Gibraltar, probably forming a spectacular waterfall.

A meteorite impact event occurred about 15 million years ago, forming the 15-mile- (24-km-) wide Ries Crater near Nordlingen, Germany. The meteorite that hit the Earth in this event is estimated to have been half a mile (1 km) in diameter, releasing the equivalent of a 100,000 megaton explosion. About 55 cubic feet (155 m^3) of material was displaced from the crater, some of which formed fields of tektites, unusually shaped melted rock that flew through the air for up to 248.5 miles (400 km) from the crater.

The Neogene saw the spread of grasses and weedy plants across the continents and the development of modern vertebrates. Snakes, birds, frogs, and rats expanded their niches, whereas the marine invertebrates experienced few changes. Humans evolved from earlier apelike hominids. Continental glaciations in the Northern Hemisphere began in the Neogene, and continue to this day.

See also CENOZOIC; TERTIARY.

FURTHER READING

Prothero, Donald R., and Robert H. Dott. *Evolution of the Earth.* 6th ed. Boston: McGraw-Hill, 2002.
Stanley, Steven M. *Earth and Life through Time.* New York: Freeman, 1986.
Walsh, Stephen L. "The Neogene: Origin, Adoption, Evolution, and Controversy." *Earth Science Reviews* 89 (2008): 42–72.

Neolithic *Neolithic* is an archaeological term for the last division of the Stone Age, during which time humans developed agriculture and domesticated animals. The transition from hunter-gatherer and nomadic types of existence to the development of farming took place about 10,000–8,000 years ago in the Fertile Crescent, a broad stretch of land that extends from southern Israel through Lebanon, western Syria, Turkey, and through the Tigris-Euphrates Valley of Iraq and Iran. The Neolithic revolution and the development of stable agricultural practices led to an unprecedented explosion of the human population that continues to this day. About a million years ago, an estimated few thousand humans migrated on the Earth, and by about 10 thousand years ago this number had increased only to a mere 5–10 million. When humans began stable agricultural practices and domesticated some species of animals, the population rate increased substantially. The increased standards of living and nutrition caused the population growth to soar to about 20 million by 2,000 years ago, and 100 million by 1,000 years ago. By the 18th century, humans manipulated their environments to a greater degree, began public health services, and recognized and sought treatments for diseases that previously claimed many lives. The average life span soared, and world population surpassed 1 billion in the year

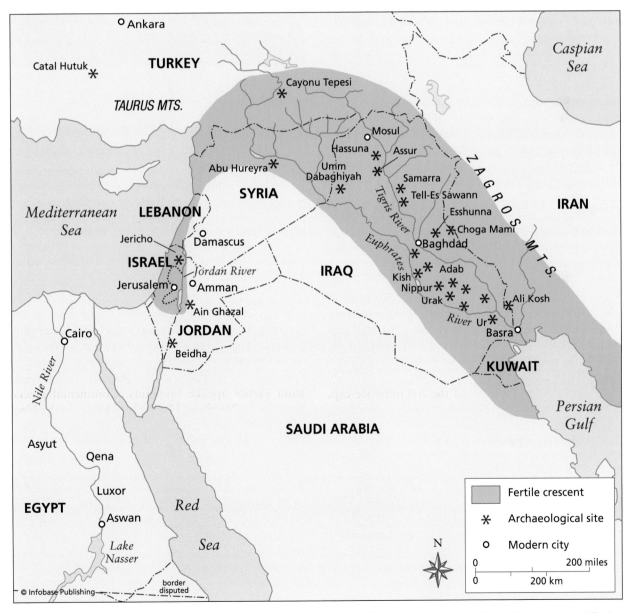

Map of the Fertile Crescent stretching from the Levant (Israel and Lebanon) through rolling hills in parts of Syria, southern Turkey, Iraq, and Iran. Ancient cities and agriculture arose in this area, with many early cities located in the Sumerian region between the Tigris and Euphrates Rivers, in what is now southern Iraq.

1810. A mere 100 years later, the world population doubled again to 2 billion, and had reached 4 billion by 1974. World population is now close to 7 billion and climbing more rapidly than at any time in history, doubling every 50 years.

FURTHER READING

Diamond, John. *Guns, Germs, and Steel: The Fates of Human Societies.* New York: W. W. Norton, 1999.
Leonard, Jonathan N. *The First Farmers: The Emergence of Man.* New York: Time-Life Books, 1973.

Neptune   The eighth and farthest planet from the center of the solar system, the giant Jovian planet Neptune orbits the Sun at a distance of 2.5 billion miles (4.1 billion km, or 30.1 astronomical units), completing each circuit every 165 years. Rotating about its axis every 16 hours, Neptune has a diameter of 31,400 miles (50,530 km) and a mass of more than 17.21 times that of Earth. Its density of 1.7 grams per cubic centimeter shows that the planet has a dense rocky core surrounded by metallic, molecular, and gaseous hydrogen, helium, and methane, giving the planet its blue color.

Neptune is unusual in that it generates its own heat, radiating 2.7 times more heat than it receives from the Sun. The source of this heat is uncertain, but it may be heat trapped from the planet's formation that is only slowly being released by the dense atmosphere. The cloud systems that trap this heat are visible from Earth-based telescopes and include some large hurricane-like storms such as the former Great Dark Spot, a storm about the size of the Earth, similar in many ways to the Great Red Spot on Jupiter, but that has dissipated.

Neptune has two large moons visible from Earth, Triton and Nereid, and six other smaller moons discovered by the *Voyager 2* spacecraft. Triton has a diameter of 1,740 miles (2,800 km) and orbits Neptune at a distance of 220,000 miles (354,000 km) from the planet. It is the only large moon in the solar system that has a retrograde orbit.

See also EARTH; JUPITER; MARS; MERCURY; SATURN; SOLAR SYSTEM; URANUS; VENUS.

FURTHER READING

Chaisson, Eric, and Steve McMillan. *Astronomy Today.* 6th ed. Upper Saddle River, N.J.: Addison-Wesley, 2007.
Comins, Neil F. *Discovering the Universe.* 8th ed. New York: W. H. Freeman, 2008.
National Aeronautic and Space Administration. Solar System Exploration page. Neptune. Available online. URL: http://solarsystem.nasa.gov/planets/profile.cfm?Object=Neptune. Last updated June 25, 2008.

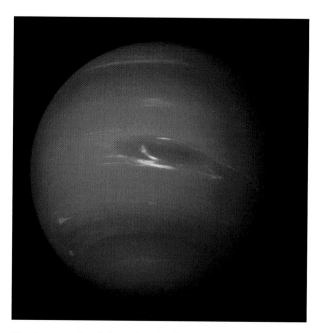

Neptune and swirling clouds. Neptune's blue-green atmosphere is shown in greater detail than ever before by the *Voyager 2* spacecraft as the craft rapidly approaches its encounter with the giant planet. This color image, produced from a distance of about 10 million miles (16 million km), shows several complex and puzzling atmospheric features. The Great Dark Spot (GDS) seen at the center is about 8,080 miles (13,000 km) by 4,100 miles (6,600 km) in size—as large along its longer dimension as the Earth. The bright, wispy "cirrus-type" clouds seen hovering in the vicinity of the GDS are higher in altitude than the dark material of unknown origin that defines its boundaries. A thin veil often fills part of the GDS interior, as seen on the image. The bright cloud at the southern (lower) edge of the GDS measures about 600 miles (1,000 km) in its north-south extent. The small, bright cloud below the GDS, dubbed the "scooter," rotates faster than the GDS, gaining about 30 degrees eastward (toward the right) in longitude every rotation. Bright streaks of cloud at the latitude of the GDS, the small clouds overlying it, and a dimly visible dark protrusion at its western end are examples of dynamic weather patterns on Neptune, which can change significantly on timescales of one rotation (about 16 hours). *(NASA Jet Propulsion Laboratory (NASA-JPL)*

Snow, Theodore P. *Essentials of the Dynamic Universe: An Introduction to Astronomy.* 4th ed. St. Paul, Minn.: West Publishing Company, 1991.

North American geology   The North American continent contains the oldest rocks known on the planet, and its core is made of a complex amalgam of some of the oldest Archean cratons on Earth. These cratons formed in complex accretionary orogenic events and then were brought together to form the cratonic core of North America in a series of collisional events in the Proterozoic. Embryonic North America formed an integral part of the supercontinents—Pangaea, and the progressively older Gondwana (1.0–0.54 billion years ago), Rodinia (1.6–1.0 billion years ago), Nuna (2.5–1.6 billion years ago), and Kenoraland (Archean)—and is the largest preserved fragment in the world of a continent assembled in the Proterozoic. Following the amalgamation of these continental blocks in the core of the continent, North America continued to evolve and grow by accretion and collision of exotic arc and other terranes in the Appalachian/Caledonian, Cordilleran, and Franklinian orogens. Understanding how the North American continent formed is instructive to understanding processes in continental formation and growth worldwide.

CRATONIC CORE OF NORTH AMERICA

There are four main Archean cratons that form the oldest core of North America. These are the Slave, Superior, Churchill, and Wyoming cratons. Each of these is reviewed briefly below, then their assembly together in the Proterozoic is discussed.

Slave Craton

The Slave craton is an Archean granite-greenstone terrane located in the northwestern part of the Canadian shield. The Archean history of the craton spans the interval from 4.03 billion years ago, the age of the world's oldest rocks, known as the Acasta Gneisses exposed in a basement culmination in the Wopmay orogen, to 2.6–2.5 billion years ago, the age of major granitic plutonism throughout the province. The margins of the craton were deformed and loaded by sediments during Proterozoic orogenies, and the craton is cut by several Proterozoic mafic dike swarms.

Most of the volcanic and sedimentary rocks of the Slave craton were formed in the interval between 2.7 and 2.65 billion years ago. Syntectonic to posttectonic plutons form about half of the map area of the province. The geology of the Slave Province shows some broad-scale tectonic zonations if the late granites are ignored. Greenstone belts are concentrated in a narrow northerly trending swath in the central part of the province, and the relative abundance of mafic volcanics, felsic volcanics, clastic rocks, and gneisses is different on either side of this line. The dividing line is coincident with a major Bouger gravity anomaly and with an isotopic anomaly indicating that older crust was involved in granitoid petrogenesis in the west, but not in the east. Greenstone belts west of the line comprise predominantly mafic volcanic and plutonic rocks, whereas volcanic belts to the east contain a much larger percentage of intermediate and felsic volcanic material. This is most evident in a large belt of northwest-trending felsic volcanics extending from south and east of Bathurst Inlet toward Artillery Lake. Quartzofeldspathic gneisses older than the greenstones are rare throughout the province and are confined to a line west of the central dividing line.

In the middle 1980s, American geologist T. M. Kusky proposed that these major differences in geology across the Slave Province reflect that it is divided into a number of different tectonic terranes. These ideas were initially debated but later largely accepted and modified by further mapping, seismic surveys, and geochemical analysis. An older gneissic terrane in the west, known as the Anton terrane, contains the world's oldest known rocks and is overlain by a platform-type sedimentary sequence. The Contwoyto terrane and Hackett River arc represent an accretionary prism and island arc that accreted to the Anton terrane in the late Archean, uplifting the Sleepy Dragon terrane in a basement culmination.

Gneissic rocks of the Anton terrane extend from Yellowknife to the Anialik River. The name is taken from the Anton complex exposed north of Yellowknife, which consists of metamorphosed granodiorite to quartz diorite, intruded by younger granitoids. The Anton terrane dips under the Wopmay orogen in the west, and its eastern contact is marked by a several-kilometer-thick, nearly vertical mylonite zone, best exposed in the vicinity of Point Lake. The Anton terrane includes the oldest rocks known in

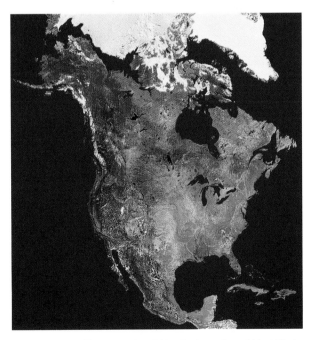

Landsat satellite mosaic of North America *(WorldSat International Inc./Photo Researchers, Inc.)*

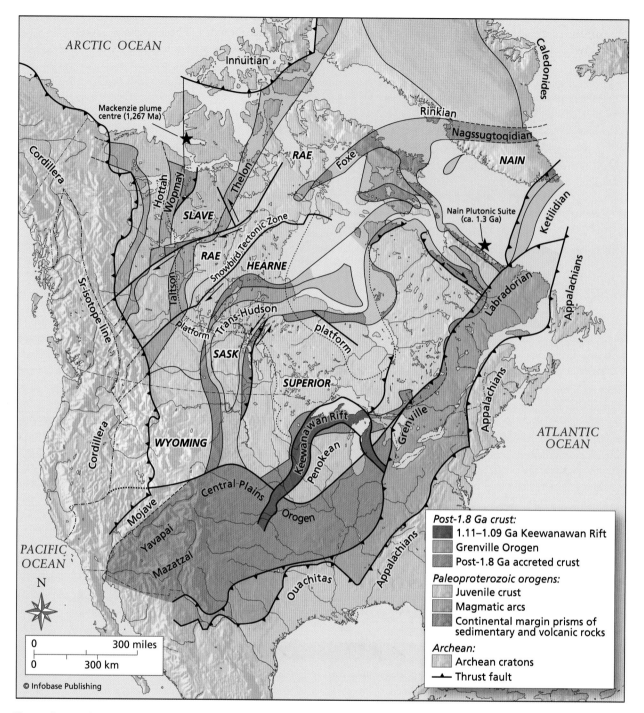

Tectonic province map of North America showing Archean cratons, including the Superior, Slave, and Wyoming Provinces, Paleoproterozoic orogens, Neoproterozoic orogens, including the Grenville orogen, and Phanerozoic orogens, including the Appalachians and Cordilleran belts *(Map inspired by Paul Hoffman)*

the world, the 4.03 billion-year-old Acasta gneisses exposed in a basement culmination along the border with the Wopmay orogen (slightly older rocks have recently been suggested to be present in Labrador, but their age is debated and the uncertainties on the dating methods are large). Also, 3.48 to 3.21 billion-year-old tonalitic gray gneisses are exposed in several locations, and similar undated old gray gneissic rocks are preserved as inclusions and small

outcrop belts within a sea of younger granites in the western part of the craton. Several different types of gneissic rocks are present in these areas, including a variety of metamorphosed igneous and sedimentary rocks. The oldest type of gneiss recognized in most places includes tonalitic to granodioritic layers with mafic amphibolite bands that are probably deformed dikes. Younger orthogneisses have tonalitic, granodioritic, and dioritic protoliths, and migmatization is common. Locally, especially near the eastern side of the Anton terrane, the older gneisses are overlain by a shallow-water sedimentary sequence that includes quartz-pebble conglomerate, quartzite, metapelite, and metacarbonates. These rocks are likely the remnants of a thin passive margin sequence.

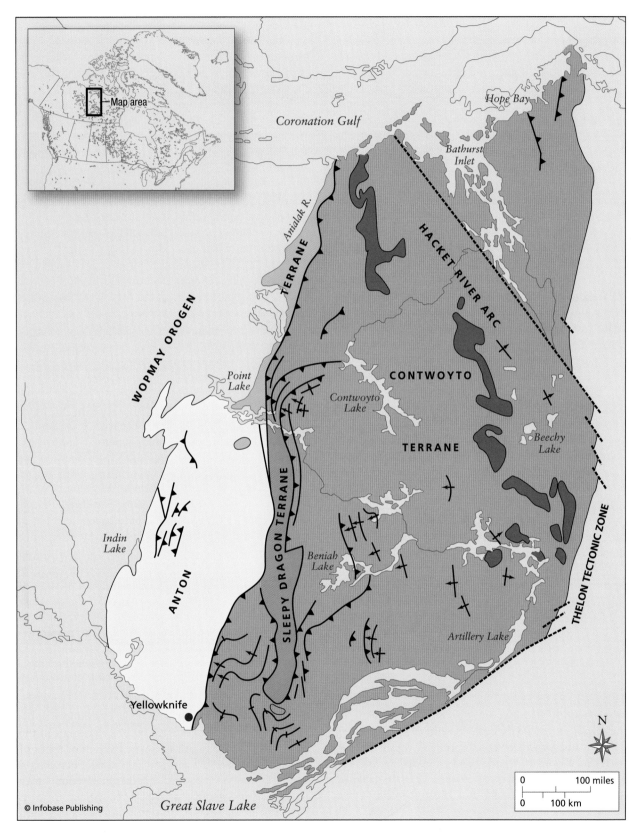

Tectonic zonation map of the Archean Slave Province showing ancient Gneissic terrane in the west (Anton terrane) and younger accreted arc in the east (Contwoyto terrane)

The Sleepy Dragon terrane extends from northeast of Yellowknife to the south shore of the south arm of Point Lake. This terrane includes intermediate to mafic quartzofeldspathic gneiss complexes such as the 2.8–2.7 billion-year-old Sleepy Dragon complex in the south, banded and migmatitic gneisses near Beniah Lake, and 3.15 billion-year-old chloritic granite on Point Lake. Isolated dioritic to gabbroic bodies are found as inclusions and enclaves. The most common protoliths to the gneisses are tonalites and granodiorites, and rock types in the Sleepy Dragon terrane are broadly similar to those in the Anton terrane. Sleepy Dragon gneisses are locally overlain unconformably by shallow water sedimentary sequences, notably along the southeastern margin of the complex near Detour Lake. Here, a basal tonalite, pebble-bearing conglomerate grades up into metaquartzose and calcareous sands, and then into a metacarbonate sequence consisting of marbles and calc-silicate minerals. From base to top this sequence is only 1,600 feet (500 m) thick, but it has been shortened considerably. Several tens of kilometers north at Beniah Lake in the Beaulieu River greenstone belt, up to 3,200 feet (1,000 m) of quartzite are recognized between shear zones. There are thus several locations where shallow water sediments appear to have been deposited on Sleepy Dragon gneisses. The similarities of the lithofacies successions in these rocks to those found in Phanerozoic rift and passive margin sequences are striking.

The Contwoyto terrane is composed of laterally continuous graywacke mudstone turbidites exposed in a series of westward-vergent folds and thrusts. Mapping in the Point Lake area revealed westward-directed thrusts placing high-grade metagraywackes over lower-grade equivalents. The graywackes are composed of matrix, rock fragments (felsic volcanics, mafic volcanics, chert, granite), and feldspars. Typically only the upper parts of the Bouma sequence are preserved. Black shales and iron formations form thin layers, especially near the structural base of the sequence. In many places greenstone belts conformably underlie the sediments, but the bases of the greenstone belts are either known to be truncated by faults or are poorly defined, suggesting that they are allochthonous. Ophiolite-like stratigraphy, including the presence of sheeted dikes and cumulate ultramafics, has been recognized in several greenstone belts. Other greenstone belts of the Contwoyto terrane are composed predominantly of basaltic pillow lavas and exhibit both tholeiitic and calc-alkaline differentiation trends.

Rocks of the Contwoyto terrane thus include tectonic slivers of ophiolite-like rocks, oceanic type sediments (shales, iron formations), and abundant graywackes exhibiting both volcanogenic and flysch-like characteristics. These rocks are contained in westward-verging folds and are disrupted by westward-directed thrusts. A series of granitoids intruded this package of rocks at various stages of deformation. These relationships are characteristic of an accretionary prism tectonic setting. In such an environment graywackes are eroded from a predominantly island arc source, as well as from any nearby continents, and are deposited over ophiolitic basement capped by abyssal muds and iron formations. Advance of the accretionary prism scrapes material off the oceanic basement and incorporates it in westward-vergent fold-and-thrust packages. This material is accreted to the front of the arc, and is intruded by arc-derived magmas during deformation. Metamorphism is of the low-pressure, high-temperature variety and is similar to that of accretionary prisms that have experienced subduction of young oceanic crust or subduction of a ridge segment.

The Hackett River arc consists of a series of northwest striking volcanic piles and synvolcanic granitoids, especially in the south. Felsic volcanics predominate but a spectrum of compositions including basalt, andesite, dacite, and rhyolite is present. Volcanic piles in the Hackett River arc therefore differ strongly from greenstone belts in the west, which consist predominantly of mafic volcanic and plutonic rocks. In the Back River area, cauldron subsidence features, rhyolitic ring intrusions, tuffs, breccias, flows, and domes, with well-preserved subaerial and subaquatic depositional environments, have been documented. Rhyolites from the Back River complex have been dated at 2.69 billion years old, and the volcanics are broadly contemporaneous with graywacke sedimentation because the flows overlie and interfinger with the sediments. Gneissic rocks in the area are not extensively intruded by mafic dike swarms like the gneisses of the Anton and Sleepy Dragon terranes, and they have yielded ages of 2.68 billion years, slightly younger than surrounding volcanics. Since none of the gneisses in the Hackett River arc have yielded ages significantly older than the volcanics, deformed plutonic rocks in this terrane are accordingly distinguished from gneisses in the western part of the Slave Province. These gneisses are suggested to represent subvolcanic plutons that fed the overlying volcanics. Another suite of tonalitic, dioritic, and granodioritic plutonic rocks with ages 60 to 100 million years younger than the volcanics also intrude the Hackett River arc. This suite of granitoids is equated with the late- to post-tectonic granitoids that cut all rocks of the Slave craton.

A belt of graywacke turbidites in the easternmost part of the Slave Province near Beechy Lake has dominantly eastward-vergent folds and possible thrusts, with some west-vergent structures. The change from regional west vergence to eastward vergence is consistent with a change from a forearc accretionary prism to a back-arc setting in the Beechy Lake domain.

The Hackett River terrane is interpreted as an island arc that formed above an east-dipping subduction zone at 2.7 to 2.67 billion years ago. The mafic to felsic volcanic suite, development of caldera complexes, and overall size of this belt are all similar to recent immature island arc systems. The Contwoyto terrane is structurally and lithologically similar to forearc accretionary complexes; west-vergent folds and thrusts in this terrane are compatible with eastward-dipping subduction, as suggested by the position of the accretionary complex to the west of the arc axis. The change from west to east vergence across the arc-axis into the Beechy Lake domain reflects the forearc and back-arc sides of the system. Mafic volcanic belts within the Contwoyto terrane are interpreted as ophiolitic slivers scraped off the subducting oceanic lithosphere.

The Anton terrane in the western part of the Slave Province contains remnants of an older Archean continent or microcontinent including the world's oldest known continental crust. Quartzofeldspathic gneisses here are as old as 4.03 billion years, with more abundant 3.5–3.1 billion-year-old crust. These gneissic rocks were deformed prior to the main orogenic event at 2.6 billion years ago. The origin of the gneisses in the Anton terrane remains unknown; many have igneous protoliths, but the derivation of the rocks is not yet clear. The Sleepy Dragon terrane might represent a microcontinent accreted to the Anton terrane prior to collision with the Hackett River arc, but more likely it represents an eastern part of the Anton terrane uplifted and transported westward during orogenesis. Studies at the southern end of the Sleepy Dragon terrane have shown that the gneisses occupy the core of a large fold or anticlinorium, consistent with the idea that the Sleepy Dragon terrane represents a basement-cored Alpine style nappe transported westward during the main orogenic event. The distribution of pregreenstone sediments lying unconformably over the gneisses is intriguing. At Point Lake, a few meters of conglomerates, shales, and quartzites lie with possible unconformity over the Anton terrane gneiss, whereas farther east, up to 1,600 feet (500 m) of sediments unconformably overlie Sleepy Dragon terrane gneisses. These include basal conglomerates and overlying sand, shale, and carbonate sequences, and thick quartzites with unknown relationships with surrounding rocks. These scattered bits of preserved older sediments in the Slave Province may represent remnants of an east-facing platform sequence developed on the Anton-Sleepy Dragon microcontinent.

In a simple sense, the tectonic evolution of the Slave craton can be understood in terms of a collision between an older continent with platformal cover in the west with a juvenile arc/accretionary prism in the east. The Anton terrane experienced a sequence of poorly understood tectonomagmatic events between 4.03 and 2.9 billion years ago, then was intruded by a set of mafic dikes probably related to lithospheric extension. After extension, the thermally subsiding Anton terrane was overlain by an eastward thickening shallow-water platform sequence. To the east, the Hackett River volcanic arc and Contwoyto terrane are formed as a paired accretionary prism and island arc above an east-dipping subduction zone. Numerous pieces of oceanic crust are sliced off the subducting lithosphere, and synvolcanic plutons intrude along the arc axis. Any significant rollback of the slab or progradation of the accretionary wedge will cause arc magmas to intrude the accreted sediments and volcanics. Graywacke sediments are also deposited on the back side of the system in the Beechy Lake domain.

As the arc and continent collided at about 2.65 billion years ago, large ophiolitic sheets were obducted, and a younger set of graywacke turbidites was deposited as conformable flysch. This is in contrast to other graywackes that were incorporated into the accretionary prism at an earlier stage and then thrust upon the Anton terrane. There are thus at least two ages of graywacke sedimentation in the Slave Province. Older graywackes were deposited contemporaneously with felsic volcanism in the Hackett River arc, whereas younger graywackes were deposited during obduction of the accretionary prism onto the Anton continent. Synvolcanic plutons along the arc axis became foliated as a result of the arc-continent collision, and back-thrusting shortened the Beechy Lake domain. Continued convergence caused the uplift and transportation of the Sleepy Dragon terrane as a basement nappe and strongly attenuated greenstone slivers. Numerous late- to post-tectonic granitoids represent postcollisional anatectic responses to crustal thickening, or pressure-release melts formed during postcollisional orogenic extension and collapse. These suites of granitoids have similar intrusion ages across the province.

Superior Craton

The Superior Province is the largest Archean craton preserved on the planet, with an area of 625,000 square miles (1.6 million km^2). Most rocks within the Superior Province resemble those found at younger subduction zones and in island and continental margin arc sequences, suggesting that the processes involved in forming the Superior Province were the result of plate tectonics. Contrary to popular geological myth that Archean cratons do not show linear tectonic zonations, the Superior Province exhibits strong linear tectonic zones with individual belts extending for thousands of miles (km) in a southwest-northeast arrangement. Some belts are dominated by metasedimentary rocks, others by mafic volcanic and plutonic rocks, high-grade gneisses, or granitoids. Most of these belts are separated from other belts by major fault zones, many of which have long tectonic histories. In general these different belts are older and were joined with each other at older ages in the north, extending back to 3.0 billion years ago, and get progressively younger to the south until the ages are about 2.7 billion years along the Canadian/United States border. The final belt to the south, however, is the Minnesota River Valley Province, with ages of old gneisses that extend back to 3.66 billion years. This old terrane probably represents an older continental fragment that the rest of the Superior Province was accreted to around 2.7 billion years ago.

The Minnesota River Valley Province occupies the southwest corner of the Superior Province and includes an assemblage of granitic gneisses and amphibolites with ages that go back to 3.66 billion years ago. These rocks experienced their first deformation event at 3.6 billion years ago, with additional deformation and metamorphic events at 3.05 and 2.7 billion years ago. This older continental block is separated from the Wawa terrane to the north by a major structure called the Great Lakes tectonic zone, interpreted to be a suture along which the terranes to the north were thrust southward over the Minnesota foreland.

The Wawa terrane consists of a thick assemblage of volcanic and sedimentary rocks, and plutons that largely represent the deeper level equivalent to the volcanic rocks. Most of the volcanic and plutonic rocks are 2.75 to 2.71 billion-year-old tholeiitic basalts intercalated with tholeiitic to calc-alkaline mixed mafic and felsic volcanics. These are intruded by a suite of tonalitic plutons, then a series of post-deformational plutons with ages of 2.68 billion years. The Wawa terrane is therefore interpreted as an island arc that formed at 2.75–2.71 billion years ago and was thrust over the Minnesota foreland by 2.68 billion years ago. In the eastern part of the province in Quebec, rocks of the Wawa terrane are tilted upward on a major crustal-scale shear zone known as the Kapuskasing structure that formed in the Early Proterozoic around 1.9 billion years ago, and uplifts from as deep as 12 miles (20 km) to the surface. The origin of the uplift is thought to be from a distant continent-continent collision such as along the Trans-Hudson orogen, forming an intracontinental uplift similar to the Tien Shan north of the contemporaneous India-Asia collision.

To the east of the Kapuskasing uplift, the Abitibi belt forms a wide area of island arc type rocks similar in aspect and age to the Wawa belt, with 2.73 to 2.70 billion year old basaltic to komatiitic lava complexes intruded by diorite-tonalite-granodiorite plutons and intruded by 2.7 to 2.67 billion-year-old suites of more silicic volcanic rocks. The Abitibi province is the site of many gold mines, with many of them hosted in quartz veins in shear zones. Other massive sulfide deposits are directly associated with the volcanic rock sequences. A group of strongly deformed turbidites called the Pontiac belt on the south side of the Abitibi belt is interpreted to be an accretionary prism associated with the Abitibi arc.

The Quetico Province is a large metasedimentary belt located north of the Wawa arc, consisting of strongly folded and sheared graywacke turbidites, conglomerates, and other rocks derived from a mixed mafic and felsic volcanic source. The belt is metamorphosed more strongly in the center of the belt than along its margins. The most reasonable interpretation of the Quetico is that it is a large accretionary prism terrane composed of trench-fill turbidites that formed the forearc to the Wabigoon arc to the north. Late stage deformation of the Quetico belt reflects dextral oblique subduction beneath the Wawa belt to the south.

The Wabigoon terrane probably formed as a south-facing island arc terrane. This arc is similar in terms of rock types to the other arcs in the south, but the age of deformation (2.71 to 2.70 billion years) is about 10 million years older than farther south in the Wawa and Abitibi belts and about 20 million years younger than in the arcs to the north. These relationships support a general north to south accretion of arc and accretionary prism terranes in the Superior Province during the Archean. The Wabigoon terrane is unusual though, in that it has some older gneissic fragments with ages of about 3.0 billion years old, and these are overlain by locally thick (~1,600 feet; 500 m) limestone and dolostone sequences with shallow-water stromatolites, interpreted to be remnants of a passive margin sequence later overthrust by the arc sequence.

The English River accretionary prism lies north of the Wabigoon arc and is similar in aspect to the Pontiac and Quetico belts. The English River belt is located between the Wabigoon arc and the Uchi-Sachigo arc to the north. Its northern boundary is a major dextral/south over north thrust, and deformation can be shown to have ended by 2.66 billion years ago based on the ages of cross-cutting undeformed granitic rocks. The Uchi-Sachigo island arc shows three distinct periods of magmatism at 2.93, 2.83, and 2.73 billion years ago, and deformation at 2.73 to 2.72 billion years ago, about 20–30 million

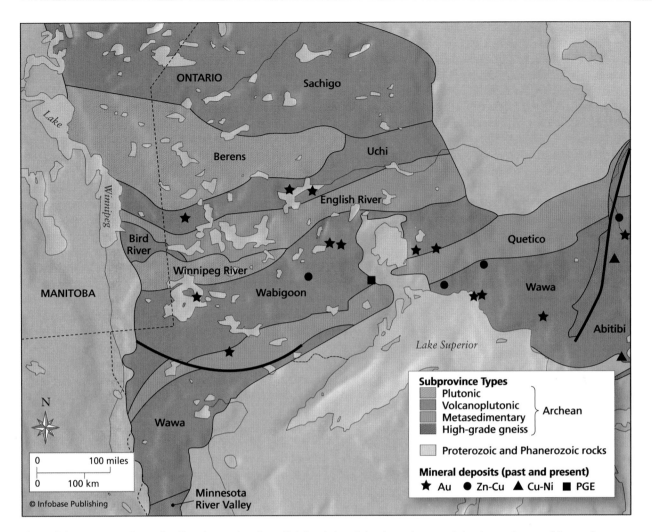

Map of the western Superior Province showing division into plutonic, volcano-plutonic, metamorphic, and metasedimentary subprovinces. The volcanic and volcano-plutonic subprovinces are interpreted to represent ancient island arc and continental arc terranes, whereas the metasedimentary subprovinces are interpreted to represent accretionary prisms that formed on the margins of these arcs. The metamorphic terranes include fragments of older continents. Collision of the arcs, accretionary prisms, and continental fragments formed the Superior Province. Large strike-slip faults currently separate many of the individual subprovinces. (Map modified after Ken Card and John Percival.)

years before deformation in the southern Superior Province.

The northwest part of the Superior Province consists of the Pikwitonei uplift and Thompson belt, consisting largely of high-grade granulites that are interpreted to represent a very deeply eroded section of the Sachigo arc terrane. The metamorphism and late deformation is related to the Proterozoic events in the 1.9 to 1.8 billion-year-old Trans-Hudson belt that borders the Superior Province just to the north and west of this location.

The northeastern part of the Superior Province is called the Minto block, and it was largely regarded as one of the world's largest granulite facies terranes until mapping in the late 1990s and 2000s by

Geological Survey of Canada geologists led by John Percival divided the block into a number of different terranes of different character. These include the Inukjuak terrane, Tikkerutuk terrane, and Lake Minto domains in the west that consist largely of granitic and granulitic gneisses, the Goudalie and Qalluviartuuq belts in the center of the block, consisting of oceanic assemblage ophiolitic and arc rocks tectonically mixed with metasediments, and in the east another group of granitoid and granulitic rocks in the Utsalik, Lepelle, and Douglas Harbour terranes. In a general way these are interpreted as two arc/continental fragments that collided across a closed ocean basin preserved in the Goudalie and Qalluviartuuq belts.

Wyoming and Churchill Cratons

The Wyoming craton is exposed discontinuously in the Laramide ranges of Wyoming and Montana. Rocks in the province include abundant shelf-type metasedimentary rocks, including quartzite, marble, and pelite, and older crust consisting of 3.6 to 3.1 billion-year-old granulites and gneisses. The rocks have been complexly folded and sheared, making correlations difficult, but most seem to have been deformed and assembled between 2.9 and 2.6 billion years ago. A late Archean suture has been interpreted to cut the southern part of the province in the Wind River Range and has a possible Archean ophiolite associated with the suture.

The Churchill Province consists of a number of different Archean and reworked Archean blocks separated by Proterozoic orogenic belts. Most important among these Archean blocks are the Hearne Province, which extends from the Wyoming craton in the south to Hudson Bay, and the Rae Province, which lies on the northwest side of the Hearne Province across the Snowbird tectonic line and extends to the Canadian Arctic islands. The Hearne Province is separated from the Superior craton by the Proterozoic Trans-Hudson orogen, and the Rae Province is separated from the Slave Province by the Thelon-Talston arc and orogen.

The Hearne Province is a Late Archean province consisting mostly of arc magmatic rocks but containing infolded remnants of platformal cover rocks and foreland basin sequences. The core of the Hearne Province consists of low-grade rocks in its core, with a progressive increase in metamorphic grade and the amount of strain outward toward the province boundaries until granulite grade is reached on its margins. The arc sequence in the core consists of submarine volcanic rocks and volcanogenic sediments, intruded by plutons of gabbro, diorite, tonalite and granite. Felsic volcanic rocks in the core of the Hearne Province have ages of 2.7 to 2.68 billion years.

Rocks in the Rae Province consist predominantly of Late Archean (~2.8 billion-year-old) quartzofeldspathic gneisses, enclosing belts of mafic volcanic and plutonic rocks, quartzites, iron formations, and graywacke-pelite assemblages. Some of the rocks enclosed in the gneiss have ages of 2.9–2.8 billion years, and a few are up to 3.1 billion years.

PROTEROZOIC ASSEMBLY OF THE CORE OF NORTH AMERICA

Most of the core of North America was assembled by the collision of Archean cratons in the Paleoproterozoic, between 2.5 and 1.6 billion years ago. Paleomagnetic data from the Slave, Superior, Churchill, and Wyoming cratons all show relative convergence of thousands of miles (up to 4,000 km) in the Late Paleoproterozoic, and wide collisional orogens developed between these cratons. At this time the Churchill Province consisted of the previously joined Rae and Hearne Provinces, and probably also the Wyoming craton in the south. These were joined with Baltica to form a Late Paleoproterozoic supercontinent called Nuna, after an Inuit name for the lands bordering the northern oceans.

During the formation, Nuna oceanic crust was subducted beneath the Churchill Province, forming extensive belts of Paleoproterozoic plutonic rocks that form large convergent margin arc sequences, now deeply eroded. The margins of the Slave and Superior cratons that collided with the Churchill show more passive margin–type sequences, and as the magmatically thickened and heated crust of the Churchill Province was caught between the colliding Slave and Superior cratons, it became strongly deformed and was extruded laterally in escape tectonics as the convergence continued. The extrusion was accompanied by thickening of the province, which formed a large collisionally thickened plateau, similar to deep levels of the Tibetan plateau forming in the present plate mosaic as a result of the India-Asia collision. The Churchill Province therefore preserves many large continental scale ductile shear zones formed during the Paleoproterozoic amalgamation of Nuna. After the collision was complete, the Churchill Province experienced extensional collapse and then intrusion of post-orogenic granitoid rocks, including rapakivi granites and other intraplate magmas.

The margins of the Slave and Superior cratons that collided with the Churchill Province in the Late Paleoproterozoic show large-scale fold-thrust belts and nappes that were directed away from the core of the orogen in the Churchill Province. The rocks in these fold-thrust belts include rifted margins formed between 2.1 and 2.0 billion years ago overlain by passive margin sediments that were caught in the thrusting at 1.6–1.9 billion years ago.

Some of the Paleoproterozoic suture zones surrounding the Superior craton contain remnants of closed oceans. On the Ungava Peninsula a 2.0 billion-year-old ophiolite, 1.92 billion-year-old oceanic plateau, and 1.87–1.83–billion-year-old oceanic island arc were thrust over a rifted margin of the Superior craton to the south, and similar sequences are known from the western side of Hudson's Bay. Overall, the structural style and types of rocks in these Paleoproterozoic orogens are very similar to orogenic belts formed in the Paleozoic and younger times.

During the formation of the Nuna supercontinent, the Slave, Superior, and other smaller blocks including the Nain craton collided with the Churchill Province, and these cratons all had passive margins on their trailing, or back sides. Soon after they collided with the Churchill Province, most of these passive margins experienced collision of arc terranes; then the direction of subduction along those margins reversed, so that after the collision, the margins were converted to Andean-style convergent margins. On the western side of the Slave Province the Wopmay orogen formed in this way, where a rift to passive margin formed on the Slave crust and was followed by the accretion of an arc and collision with the Great Bear magmatic arc at 1.88 billion years ago. Other belts experienced similar collisions at 1.85 and 1.84 billion years ago, showing that material continued to be accreted to Nuna, or ancestral North America, after the heartland formed.

The growth of North America continued at 1.8 to 1.6 billion years ago with the accretion of juvenile crust that underlies much of the Central Plains and the southwestern United States in the Central Plains, Yavapai, and Mazatzal orogens.

In the Mesoproterozoic from 1.6 to 1.0 billion years ago, the tectonics of North America were dominated by the formation of the Grenville orogen. The Grenville orogen forms a belt of high-grade metamorphic rocks that stretches from Newfoundland and Labrador through eastern Canada to the Great Lakes and Adirondack region of the United States, then through to Texas and the Sierra Madre of Mexico. The Grenville orogen is part of a worldwide belt of similar orogens with a total length of more than 6,000 miles (10,000 km) that formed as many continental nuclei and cratons similar in aspect to North America collided to form the supercontinent of Rodinia. These continental fragments included North America, Baltica, Amazonia, West Africa, the Kalahari craton, Congo craton, India, Australia, and parts of Siberia. Rodinia existed as a supercontinent from about 1.0 billion years ago until about 750 million years ago, when widespread rifting caused many of these fragments to be dispersed in a new phase of breakup in the supercontinent cycle. Some of the continents rifted off somewhat earlier, some even as early as 900 million years ago, not very long after Rodinia had just finished forming.

After the breakup of Rodinia in the Mesoproterozoic, many of the continental fragments reassembled on the opposite side of the globe in an "inside-out" configuration, where fragments formerly on the exterior parts of Rodinia found themselves located on the inside of the next supercontinent, Gondwana. This progressive breakup and reamalgamation also left formerly interior parts of the continents, including North America, outside the Gondwana supercontinent altogether or located on its outer fringes. Gondwana formed by the joining of the main blocks of the Congo and Kalahari cratons, West Africa, Amazon and Rio del Plata cratons, India, East Antarctica, and Australia at about 545 million years ago.

During the evolution of Gondwana, North America formed a northern supercontinent known as Laurentia. During the Paleozoic and Mesozoic, North America saw its most significant evolution along the western and eastern margins of the continent, in the Cordilleran or Rocky Mountain fold belts and the Applachian-Ouachita orogen. Other belts on the fringes of North America include the Caledonian belt and the Innuitian belt in the Canadian Arctic.

APPALACHIANS
The Appalachian Mountain belt extends for 1,600 miles (2,600 km) along the east coast of North America, stretching from the St. Lawrence Valley in Quebec, Canada, to Alabama. Many classifications consider the Appalachians to continue through Newfoundland in maritime Canada, and before the Atlantic Ocean opened, the Appalachians were continuous with the Caledonides of Europe. The Appalachians are one of the best-studied mountain ranges in the world, and understanding of their evolution was one of the factors that led to the development and refinement of the paradigm of plate tectonics in the early 1970s.

Rocks that form the Appalachians include those that were deposited on or adjacent to North America and thrust upon the continent during several orogenic events. For the length of the Appalachians, the older continental crust consists of Grenville Province gneisses, deformed and metamorphosed about 1.0 billion years ago during the Grenville orogeny. The Appalachians grew in several stages. After Late Precambrian rifting, the Iapetus Ocean evolved and hosted island arc growth, while a passive-margin sequence was deposited on the North American rifted margin in Cambrian-Ordovician times. In the Middle Ordovician, the collision of an island arc terrane with North America marks the Taconic orogeny, followed by the mid-Devonian Acadian orogeny, which probably represents the collision of North America with Avalonia, off the coast of Gondwana. This orogeny formed huge molassic fan delta complexes of the Catskill Mountains and was followed by strike-slip faulting. The Late Paleozoic Alleghenian orogeny formed striking folds and faults in the southern Appalachians, but was dominated by strike-slip faulting in the Northern Appalachians. This event appears to be related to the rotation of Africa to close the remaining part of the open ocean in the southern Appalachians. Late Triassic-Jurassic rifting reopened the Appalachians, forming the present Atlantic Ocean.

The history of the Appalachians begins with rifting of the 1-billion-year-old Grenville gneisses and the formation of an ocean basin known as Iapetus approximately 800–570 million years ago. Rifting was accompanied by the formation of normal-fault systems and grabens and by the intrusion of swarms of mafic dikes exposed in places in the Appalachians, such as in the Long Range dike swarm on Newfoundland's Long Range Peninsula. Rifting was also accompanied by the deposition of sediments, first in rift basins and then as a Cambrian transgressive sequence that prograded onto the North American craton. This unit is generally known as the Potsdam sandstone, and is well exposed around the Adirondack dome in northern New York State. Basal parts of the Potsdam sandstone typically consist of a quartz pebble conglomerate and a clean quartzite.

Overlying the basal Cambrian transgressive sandstone is a Cambrian-Ordovician sequence of carbonate rocks deposited on a stable carbonate platform or passive margin, known in the northern Appalachians as the Beekmantown Group. Deposition on the passive margin was abruptly terminated in the Middle Ordovician when the carbonate platform was progressively uplifted above sea level from the east, migrated to the west, and then suddenly dropped down to water depths too great to continue production of carbonates. In this period, black shales of the Trenton and Black River groups were deposited, first in the east and then in the west. During this time, a system of normal faults also migrated across the continental margin, active first in the east and then in the west. The next event in the history of the continental margin is deposition of coarser-grained clastic rocks of the Austin Glen and correlative formations, as a migrating clastic wedge, with older rocks in the east and younger ones in the west. Together, these diachronous events represent the first stages of the Taconic Orogeny, and they represent a response to the emplacement of the Taconic allochthons on the North American continental margin during Middle Ordovician arc-continent collision. In some places, an event that predates the Taconic orogeny, and postdates the late Precambrian is preserved and is known as the Penobscottian orogeny.

Penobscottian Orogeny

The northern Appalachians have been divided into a number of different terranes or tectonostratigraphic zones, reflecting different origins and accretionary histories of different parts of the orogen. The North American craton and overlying sedimentary sequences form the Humber zone, whereas fragments of peri-Gondwanan continents are preserved in the Avalon zone. The Dunnage terrane includes material accreted to the North American and Gondwanan continents during closure of the Paleozoic Iapetus Ocean. Gander terrane rocks were initially deposited adjacent to the Avalonian margin of Gondwana. A piece of northwest Africa left behind during Atlantic rifting is preserved in the Meguma terrane. For many years a pre-Middle Ordovician (Taconic age) deformation event has been recognized from parts of central Maine, New Brunswick, and a few other scattered parts of the Appalachian orogen. This Late Cambrian-Early Ordovician event probably occurred between two exotic terranes in the Iapetus Ocean, before they collided with North America, and is known as the Penobscottian orogeny in the northern Appalachians.

Because the Penobscottian orogenic deformation is less extensive than other orogenies in the Appalachians, interpretations have been based on information from limited areas; thus a comprehensive model is difficult to propose. The deformation is confined to the Gander terrane in the Northern Appalachians and the Piedmont in the Southern Appalachians. The timing of this orogenic event can be best determined in western Maine, where an exposure of sialic basement, an ophiolite, and associated accretionary complexes crop out. It has also been suggested that the Brunswick subduction complex in New Brunswick, Canada, represents the remains of the Penobscottian orogeny. In western Maine, the major units include the Chain Lakes massif, interpreted to be Grenville age; the Boil Mountain ophiolite, interpreted to be a relict of oceanic crust obducted during accretion of the Boundary Mountains terrane to the previously existing Gander terrane; and the Hurricane Mountain mélange, expressing the flysch deposits of the accretionary prism during amalgamation. These rock units together record a portion of the complex deformational history of the closing of the Iapetus Ocean. The Boil Mountain ophiolite occurs in a structurally complex belt of ophiolitic slivers, exotic microterranes (e.g., Chain Lakes massif), and mélange along the boundary between the Dunnage and Gander terranes in central Maine. Its geological evolution is critical to understanding the Late Cambrian-Early Ordovician Penobscottian orogeny.

The Penobscottian orogeny in the Northern Appalachians was initiated sometime during the Late Cambrian. Isotopic ages of the Boil Mountain ophiolite range from 500–477 million years old. Fossil evidence further supports this Middle Cambrian to Early Ordovician age range. Sponges of this age have been found in the black shales of the Hurricane Mountain mélange, constraining the maximum age for the mélange.

The Boil Mountain ophiolite and the associated Chain Lakes massif are instrumental in the interpretation of the Penobscottian orogeny. In order to better comprehend their relationship, it is important to understand their lithologies and regional context.

The Chain Lakes massif is characterized by meta diamictite that is composed of mostly metasandstone, minor amphibolite, granofels, and gneiss. The structure has been interpreted to be an elongate dome with an estimated exposed thickness of 9,840 feet (3,000 m). The unit is bounded by faults that strike northwest and comes in contact with several intrusive bodies, including the Attean batholith to the northeast and the Chain of Ponds pluton to the southwest. Along its eastern and western margin, Silurian and Devonian strata overlie the massif. The Boil Mountain ophiolite is in fault-bounded contact with the Chain Lakes massif on its southern and southeastern margin. Seismic reflection profiling has shown that the Chain Lakes massif is floored by a major fault dipping toward the southeast. This was interpreted to represent a thrust that originated to the southeast and formed during either the Acadian or the Taconic orogenies.

The massif may be divided into eight facies based on structural aspects and lithology of the complex. The structurally lowest sequence is first divided into three facies: (1) the Twin Bridges semipelitic gneiss, (2) the Appleton epidiorite, and (3) the Barrett Brook polycyclic epidiorite breccia. The next four facies represent the principal diamictite sequence and include (1) the McKenney Pond chaotic rheomorphic granofels, (2) the Coburn Gore semipelitic gneissic granofels, (3) the Kibby Mountain flecky gneiss, and (4) the Sarampus Falls massive to layered granofels. The structurally highest facies is the Bag Pond Mountain bimodal metavolcanic section and feldspathic meta-arenite. The highest facies are interpreted to represent the return to a passive margin sequence after metamorphism and deformation. However, the structural and metamorphic history of the area is complicated, and it is difficult to be certain about stratigraphic relationships across structural boundaries.

The metamorphic history of the Chain Lakes massif is one of repeated deformation over a long span of time. The metamorphism of this complex spans a period of 800 million years and has been interpreted to record a pressure-temperature time path of prograde and retrograde events that end with the emplacement of the Late Ordovician batholith. Isotopic ages of the Chain Lakes massif range from approximately 1,500 to 684 million years. The wide variance in the ages may be caused by deposition of material in the massif sometime between these ages as the deformational history was proceeding, or the zircons that produced the Precambrian dates must have been derived from a preexisting Precambrian unit that was eroding into the basin. The diamictite may have been deposited as an alluvial fan type of conglomerate along the rifted margin of the Iapetus

and subsequently overridden by the Boil Mountain ophiolite complex.

The Boil Mountain ophiolite lies in fault-bounded contact with the Chain Lakes massif along its southern boundary. This complex extends about 20 miles (30 km) along strike and has a maximum exposure across strike of about four miles (6 km). Units typical of ophiolites such as serpentinite, pyroxenite, metagabbro, and mafic volcanics and sediments consisting of metaquartzwacke, metapelite, slate, and metaconglomerate characterize the Boil Mountain ophiolite. The stratigraphic units consist of, from bottom to top: an ultramafic unit, a gabbroic unit, a tonalitic unit, mafic and felsic volcanics, and metasedimentary units. Thus, the ophiolite has all the components of an ophiolite sequence except the tectonite ultramafic unit and the sheeted dike unit. The basal contact is difficult to differentiate in the large scale, but any contacts that are present suggest that ductile faulting accompanied their emplacement. However, some of the units, including the serpentinite, are in sharp structural contact with the Chain Lakes massif along its southern boundary.

The complex may be divided based on chemistry into several units, including ultramafics, gabbros, two mafic volcanic units, and felsic volcanics. A felsic unit separates the two mafic volcanic units, termed upper and lower. Geochemical patterns reveal two distinct crystallization trends within the complex. The first magma to be erupted was the lower mafic volcanic unit. Geochemical trends suggest that the crystallization of olivine, clinopyroxene, and plagioclase resulted in this magma composition. This is accomplished by the formation of the ultramafics and gabbroic unit. The upper mafic unit has geochemical affinities to an island arc tholeiite (IAT) zone, while the upper mafic unit is similar to mid-ocean ridge basalts (MORB). Since there is such a large volume of felsic material associated with the ophiolite, it is likely that there were two phases of extrusion for the mafic volcanic unit. Because the presence of tonalites and other felsic volcanic rocks implies that there must be hydrous fluids, the first phase must include a subduction zone. The lower mafic unit and the felsics represent this phase. The upper mafic unit suggests that volcanism occurred in a marginal basin, because the mantle source for these units was not affected by the subducting slab.

The presence of tonalites in the Boil Mountain ophiolite, with small amounts of trondhjemite, is an unusual feature not found in normal ophiolite sequences. Other ophiolites that include a unit of tonalite are the Semail ophiolite in Oman and the Canyon Mountain ophiolite in California. Tonalites suggest the presence of fluids during crystallization and are instrumental in the interpretation of the ophiolites. The placement of these ophiolites with respect to the rest of the sequence is also of interest. The tonalite of the Boil Mountain complex appears in the sequence above the gabbros and below the volcanogenic units. It has the form of a sill, with abundant intrusive contacts evident in float and rare, poorly exposed outcrops in the northeast part of ophiolite. The tonalite was probably derived from partial melting of the lower mafic volcanics and gabbros, which then intruded as a sill.

The Boil Mountain ophiolite tonalite has yielded an isotopic age of 477 ± 1 million years. The age of 477 million years places a minimum plutonic age for the tonalites of the Boil Mountain ophiolite, which is significantly less than any previously determined age associated with the ophiolite. A Late Cambrian to Early Ordovician age for the Boil Mountain ophiolite has been previously suggested, based on several pieces of information. Felsic volcanics in the upper part of the ophiolite give an age of 500 Ma ± 10 million years. The age of 477 million years for the Boil Mountain tonalites is interpreted as a late-stage intrusive event, possibly related to partial melting of hydrated oceanic crust and the intrusion of a tonalitic extract.

A comparison of the age of the Boil Mountain ophiolite with nearby ophiolitic sequences in the Taconic allochthons of Quebec shows that ophiolite obduction was occurring on the Humber zone of the Appalachian margin of Laurentia, at similar times to the Boil Mountain ophiolite being emplaced over the Chain Lakes massif, interpreted as a piece of the Gander margin of Gondwana. Hornblendes from the metamorphic aureole to the Thetford Mines ophiolite have yielded isotopic ages of 477 ± 5 million years, with an initial detachment age of 479 ± 3 million years for the ophiolitic crust. Detachment of the circa 479 million-year-old sheet began at a ridge segment in a fore arc environment, in contrast to an older, circa 491 ± 11 million-year-old oceanic slab preserved as the Pennington sheet in the Flintkote mine that is interpreted as a piece of oceanic crust originally attached to Laurentia. Thus, there may be a protracted history of Taconian ophiolite obduction in the Quebec Appalachians. In Gaspé, a basal amphibolite tectonite gave an emplacement age of 456 ± 3 million years for the Mount Albert ophiolite.

The age and origin of the tonalites of the Boil Mountain ophiolite and their relationships to the Chain Lakes massif have considerable bearing on the Penobscottian orogeny. First, since tonalitic intrusives are confined to the allochthonous Boil Mountain ophiolite, it can be inferred that the ophiolite was not structurally emplaced in its final position over the Chain Lakes massif until after 477 mil-

lion years ago (Arenigian). This does not however preclude earlier emplacement (previous to tonalite intrusion) of the ophiolite in a different structural position. Dates for the Taconic orogeny range from 491 to 456 million years ago. Early ideas for the Penobscottian orogeny suggested that it took place prior to the Taconic orogeny. However, it seems more likely that these two orogenies were taking place at the same time, although not necessarily on the same margin of Iapetus.

Geochemical data suggests a two-stage evolution for the Boil Mountain ophiolite, including an early phase of arc or forearc volcanism, followed by a tholeiitic phase of spreading in an intra-arc or back-arc basin. The first phase includes felsic volcanics dated at 500 ± 10 million years, and the second phase, associated with partial melting of the older oceanic crust and the formation of the upper volcanic sequence, occurred at 477 ± 1 million years ago. This tectonic setting is compatible with the IAT characteristics of the lower volcanic unit, and the MORB characteristics of the upper mafic volcanic unit of the ophiolite. Sedimentary rocks intercalated with the upper volcanic unit include iron formation, graywacke, phyllite, and chert, consistent with deposition in a back-arc basin setting. The tonalites formed by hydrous melting of mafic crust of the lower volcanic unit during the initial stages of back-arc basin evolution.

The Boil Mountain ophiolite and Chain Lakes massif thus reveal critical insights about the Penobscottian orogeny. The lower volcanic group of the Boil Mountain ophiolite are island arc tholeiites formed within an immature arc setting 500 million years ago. This phase of volcanism is similar to other Exploits subzone (Dunnage zone) ophiolites, such as the 493 million-year-old Pipestone Pond complex, the 489 million-year-old Coy Pond complex, the South Lake ophiolite of Newfoundland, and 493 million-year-old rhyolites associated with serpentinites of the Annidale area of southern New Brunswick. The Pipestone Pond and Coy Pond complexes both occur as allochthons overlying domal metamorphic cores of Gander zone rocks (Mount Cormack and Meelpaeg Inliers) in a structural arrangement reminiscent of the Boil Mountain complex resting on the margin of the Chain Lakes massif. These ophiolites, including the Boil Mountain ophiolite, probably represent part of the Penobscot arc, which developed in the forearc over a west-dipping subduction zone near the Gander margin of Gondwana.

The 513 million-year-old Tally Pond volcanics of the Lake Ambrose volcanic belt of south-central Newfoundland may also be time-correlative with the lower volcanics of the Boil Mountain ophiolite; both preserve a mixed mafic-felsic volcanic section with arc-related geochemical affinities, and both are interpreted as parts of the Penobscot-Exploits arc accreted to Gondwana in the Early Ordovician. Thus, the Boil Mountain, Coy Pond, and Pipestone Pond ophiolites formed the forearc to the Penobscot-Exploits arc, and the Lake Ambrose volcanic belt represents volcanism in more central parts of the arc. Obduction of Exploits subzone forearc ophiolites over Gander zone rocks in Newfoundland and New Brunswick occurred in Tremadocian-Arenigian times (490–475 million years ago). This was followed by arc reversal in the Arenigian (475–465 million years ago), which formed a new arc (Popelogan arc in northern New Brunswick) and a back-arc basin (Fournier group ophiolites at 464 Ma) that widened rapidly to accommodate the collision of the Popelogan arc with the Notre Dame (Taconic) arc by 445 million years ago. This rifting event may have detached a fragment of Avalonian basement and Gander zone sediments, with overlying allochthonous ophiolitic slabs, now preserved as the Chain Lakes massif and the lower sections of the Boil Mountain ophiolite.

A second episode of magma generation in the Boil Mountain ophiolite is represented by the upper tholeiitic volcanic unit and by the 477 million-year-old tonalite sill, generated by partial melting of the lower volcanic unit. This phase of magma generation is associated with development of the Tetagouche back-arc basin behind the Popelogan arc during Late Arenigian-Llanvirnian. The timing of this event appears to be similar in southern Newfoundland, where several Ordovician granites, including the 477.6 ± 1.8 million-year-old Baggs Hill granite, and the 474 ± 3 million-year-old Partridgeberry Hills granite, intrude ophiolites obduced onto the Gondwanan margin during the early Ordovician.

Since the Boil Mountain tonalites are allochthonous and are not known to occur in the rocks beneath the ophiolite, final obduction of the Boil Mountain complex probably occurred after collision of the Popelogan and Notre Dame (Taconic) arcs, during the post-450 million-year-old collision of Gander margin of Avalon with the active margin of Laurentia. This event could be represented by the 445 million-year-old Attean pluton.

As an alternative model to that presented above, the entire Boil Mountain ophiolite complex may have formed in the forearc of the Popelogan arc, which formed after the Penobscot arc collided with the Avalonian margin of Gondwana. In this model, the Boil Mountain ophiolite would be correlative with other 480–475 million-year-old ophiolites of the Robert's Arm-Annieopsquotch belt that occur along the main Iapetus suture between the Notre Dame (Taconic) arc accreted to Laurentia, and the Penobscot-Exploits arcs accreted to Gondwana. This model would help explain why no evidence of pre-477 million-year-old obduction-related fabrics has been documented from the Boil Mountain complex-Chain Lakes massif contact, but it does not adequately account for the Cambrian ages of the lower volcanic unit of the Boil Mountain ophiolite.

The Penobscottian orogeny has presented difficulties to geologists for quite some time. However, recent studies of the exposure in Maine, including the Chain Lakes unit and the Boil Mountain ophiolite, have led to new models for the tectonic evolution of this complex terrane. New isotopic dates and geochemical evidence show that the Taconian and Penobscottian orogenies were most likely simultaneous events. The models presented in this entry take into account several factors, including the formation of the Chain Lakes unit, the timing of emplacement of the ophiolite, and the timing of the intrusion of tonalites that are thought to represent the suture of the Gander to the Boundary Mountains terrane.

Taconic Orogeny

The Taconic allochthons are a group of Cambrian through Middle Ordovician slates resting allochthonously on the Cambro-Ordovician carbonate platform. These allochthons are very different from the underlying rocks, implying that there have been substantial displacements on the thrust faults beneath the allochthons, probably on the order of 100 miles (160 km). The allochthons structurally overlie wild flysch breccias that are basically submarine slide breccias and mudflows derived from the allochthons.

Eastern sections of the Taconic-aged rocks in the Appalachians are more strongly deformed than those in the west. East of the Taconic foreland fold-thrust belts, a chain of uplifted basement with Grenville ages (about one billion years) extends discontinuously from Newfoundland to the Blue Ridge Mountains, and includes the Green Mountains of Vermont. These rocks generally mark the edge of the hinterland of the orogen and the transition into greenschist and higher metamorphic facies. Some of these uplifted basement gneisses are very strongly deformed and metamorphosed, and they contain domal structures known as gneiss domes, with gneisses at the core and strongly deformed and metamorphosed Cambro-Ordovician marbles around their rims. These rocks were deformed at great depths.

Also close to the western edge of the orogen is a discontinuous belt of mafic and ultramafic rocks comprising an ophiolite suite, interpreted to be remnants of the ocean floor of the Iapetus Ocean that closed during the Taconic orogeny. Spectacular examples of these ophiolites occur in Newfoundland, including the Bay of Islands ophiolite complex along Newfoundland's western shores.

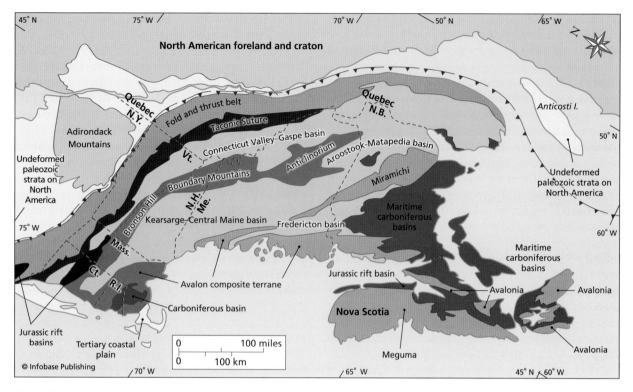

Tectonic zone map of northern Appalachians showing the Early Paleozoic tectonic terranes

Further east in the Taconic orogen are rocks of the Bronson Hill anticlinorium or terrane that are strongly deformed and metamorphosed and have been affected by both the Taconic and Acadian orogenies. These rocks have proven very difficult to map and have been of controversial significance for more than a century. Perhaps the best interpretation is that they represent rocks of the Taconic island arc that collided with North America to produce the Taconic orogeny.

The Piscataquis volcanic arc is a belt of Devonian volcanic rocks that extends from central Massachusetts to the Gaspé Peninsula. These rocks are roughly coextensive with the Ordovician arc of the Bronson Hill anticlinorium and include basalts, andesites, dacites, and rhyolites. Both subaerial volcanics and subaquatic pillow lavas are found in the belt. The Greenville plutonic belt of Maine (including Mount Kathadin) is included in the Piscataquis arc and is interpreted by some workers to be post-Acadian, but is more typical of syntectonic arc plutons. The Acadian orogeny also deformed the eastern part of the Taconic orogenic belt, which contains some younger rocks deposited on top of the eroded Taconic island arc.

The Taconic allochthons turn out to be continental rise sediments that were scraped off the North American continental margin and transported on thrusts for 60–120 miles (100–200 km) during the Taconic arc continent collision. A clastic wedge (Austin Glen and Normanskill formations) was deposited during emplacement of the allochthons by their erosion and spread out laterally in the foreland. As Taconic deformation proceeded, the clastic wedge, underlying carbonates, and the Grenville basement became involved in the deformation, rotating them, forming the Taconic angular unconformity.

Acadian Orogeny

The Acadian orogeny has historically been one of the most poorly understood aspects of the regional geology of the Appalachians. Some of the major problems in interpreting the Acadian orogeny include understanding the nature of pre-Acadian, post-Taconic basins such as the Kearsarge-Central Maine basin, Aroostook-Matapedia trough, and the Connecticut Valley–Gaspé trough. The existence and vergence of Acadian subduction zones is debated, and the relative amount of post-Acadian strike-slip movements is not well constrained.

Examining the regional geology of the northern Appalachians using only the rocks that are younger than the post-Taconic unconformity yields a picture of several distinctive tectonic belts including different rock types and structures. The North American craton includes Grenville gneisses and Paleozoic carbonates. The foreland basin includes a thick wedge of Devonian synorogenic clastic rocks, such as the

Catskill Mountains, that thicken toward the mountain belt. The Green Mountain anticlinorium is a basement thrust slice, and the Connecticut Valley–Gaspé trough is a post-Taconic basin with rapid Silurian subsidence and deposition. The Bronson Hill-Boundary Mountain anticlinorium (Piscataquis volcanic arc) is a Silurian–mid-Devonian volcanic belt formed along the North American continental margin. The Aroostook-Matapedia trough is a Silurian extensional basin, and the Miramichi massif represents remnants of a high-standing Ordovician (Taconic) arc. The Kearsarge-Central Maine basin (Merrimack trough) preserves Silurian deep-water sedimentary rocks, preserved in accretionary prisms, and is the most likely site where the Acadian Ocean closed. The Fredericton trough is a continuation of the Merrimack trough, and the Avalon Composite terrane (coastal volcanic arc) contains Silurian–Early Devonian shallow marine volcanics built upon Precambrian basement of Avalonia.

Synthesizing the geology of these complex belts, the tectonics of the Acadian orogeny in the Appalachian Mountains can be summarized as follows. The Grenville gneisses and some of the accreted Taconic orogen were overlain by a Paleozoic platform sequence, and by mid-Devonian times the region was buried beneath thick clastics of the Acadian foreland basin, best preserved in the Catskill Mountains. Nearly two miles (3 km) of fluvial sediments were deposited in 20 million years, derived from mountains to the east. Molasse and red beds of the Catskills once covered the Adirondack Mountains, as evidenced by pieces preserved in a diatreme in Montreal. These features are exposed along strike as the Old Red Sandstone in Scotland and on Spitzbergen Island.

The Connecticut Valley–Gaspé trough is a complex basin developed over the Taconic suture that was active from Silurian through Early Devonian. It is an extensional basin containing shallow marine sedimentary rocks and may have formed from oblique strike-slip after the Taconic collision, with subsidence in pull-apart basins. The Aroostook-Matapedia trough is an Ordovician-Silurian turbidite belt, probably a post-Taconic extensional basin, and perhaps a narrow oceanic basin.

The Miramichi massif contains Ordovician arc rocks intruded by Acadian plutons, and is part of the Taconic arc that persisted as a high area through Silurian times and became part of the Piscataquis volcanic arc in Silurian-Devonian times. The coastal volcanic arc (Avalon) is exposed in eastern Massachusetts though southern New Brunswick and includes about 5 miles (8 km) of basalt, andesites, rhyolite, and deep and shallow marine sediments. It is a volcanic arc that was built on Precambrian base-

ment that originated in the Avalonian or Gondwana side of the Iapetus Ocean.

The Kearsarge-Central Maine basin (Fredericton trough) is the location of a major post-Taconic, pre-Acadian ocean that closed to produce the Acadian orogeny. It contains polydeformed deepwater turbidites and black shales, mostly Silurian. The regional structural plunge results in low grades of metamorphism in Maine and high grades in New Hampshire, Massachusetts, and Connecticut. There are a few dismembered ophiolites present in the belt, structurally incorporated in about 3 miles (5 km) of turbidites.

Volcanic belts on either side of the Merrimack trough are interpreted to be arcs built over contemporaneous subduction zones. In the Late Silurian, the Acadian Ocean basin was subducting on both sides, forming accretionary wedges of opposite vergence, and forming the Coastal and Piscataquis volcanic arcs. The Connecticut Valley–Gaspé trough is a zone of active strike-slip faulting and pull-apart basin formation behind the Piscataquis arc. In the Devonian, the accretionary prism complexes collide, and west-directed overthrusting produces a migrating flexural basin of turbidite deposition, including the widespread Seboomook and Littleton formations. The collision continued until the Late Devonian, then more plutons intruded, and dextral strike-slip faulting continued.

Acadian plutons intrude all over the different tectonic zones and are poorly understood. Some are related to arc magmatism, some to crustal thickening during collision. Late transpression in the Carboniferous includes abundant dextral strike-slip faults, disrupted zones, and formed pull-apart basins with local accumulations of several miles of sediments. About 200 miles (300 km) of dextral strike-slip offsets are estimated to have occurred across the orogen.

The Late Paleozoic Alleghenian orogeny in the Carboniferous and Permian included strong folding and thrusting in the southern Appalachians, and formed a fold/thrust belt with a ramp/flat geometry. In the southern Appalachians the foreland was shortened by about 50 percent during this event, with an estimated 120 miles (200 km) of shortening. The rocks highest in the thrust belt have been transported the farthest and are the most allochthonous. At the same time, motions in the northern Appalachians were dominantly dextral strike-slip in nature.

In the Late Triassic-Jurassic, rifting and normal faulting were associated with the formation of many small basins and the intrusion of mafic dike swarms related to the opening of the present day Atlantic Ocean.

ROCKY MOUNTAINS
Extending 3,000 miles (4,800 km) from central New Mexico to northwest Alaska in the easternmost

Cordillera, the Rocky Mountains are one of the largest mountain belts in North America. The mountains are situated between the Great Plains on the east and a series of plateaus and broad basins on the west. Mount Elbert in Colorado is the highest mountain in the range, reaching 14,431 feet (4,399 m). The continental divide is located along the rim of the Rockies, separating waters that flow to the Pacific and to the Atlantic Oceans. The Rocky Mountains are divided into the Southern, Central, and Northern Rockies in the conterminous United States, Canadian Rockies in Canada, and the Brooks Range in Alaska. Several national parks are located in the system, including Rocky Mountain, Yellowstone, Grand Teton, and Glacier National Parks in the United States, and Banff, Glacier, Yoho, Kootenay, and Mount Revelstoke in Canada. The mountains were a major obstacle to traveling west during the expansion of the United States, but western regions opened up when the Oregon trail crossed the ranges through South Pass in Wyoming.

In New Mexico, Colorado, and southern Wyoming the Southern Rockies consist of two north-south ranges of folded mountains that have been eroded to expose Precambrian cores with overlying sequences of layered sedimentary rocks. Three basins are located between these ranges, known as the North, South, and Middle Parks. The southern Rockies are the highest section of the whole range, including many peaks more than 14,000 feet (4,250 m) high.

The Central Rockies in northeastern Utah and western Wyoming are lower and more discontinuous than the southern Rockies. Most are eroded down to their Precambrian cores, surrounded by Paleozoic-Mesozoic sedimentary rocks. Garnet Peak in the Wind River Range (13,785 feet; 4,202 m) and Grand Teton in the Teton Range (13,766 feet; 4,196 m) are the highest peaks in the Central Rockies.

The Northern Rockies in northeastern Washington, Idaho, and western Wyoming extend from Yellowstone National Park to the Canadian border. This section is dominated by north-south trending ranges separated by narrow valleys, including the Rocky Mountain trench, an especially deep and long valley that extends north from Flathead Lake. The highest peaks in the Northern Rockies include Borah Peak (12,655 feet; 3,857 m) and Leatherman Peak (12,230 feet; 3,728 m) in the Lost River Range.

The Canadian Rockies stretch along the British Columbia–Alberta border and reach their highest point in Canada on Mount Robson (12,972 feet; 3,954 m). The Rocky Mountain trench continues 800 miles (1,290 km) north-northwest from Montana, becomes more pronounced in Canada, and is joined by the Purcell trench in Alberta. In the Northwest Territories (Nunavet), the Rockies expand north-eastward in the Mackenzie and Franklin mountains and near the Beaufort Sea pick up as the Richardson Mountains that gain elevation westward into the Brooks Range of Alaska. Mount Chamberlin (9,020 feet; 2,749 m) is the highest peak in the Brooks Range.

The Rocky Mountains are rich in mineral deposits including gold, silver, lead, zinc, copper, and molybdenum. Principal mining areas include the Butte-Anaconda district of Montana, Leadville and Cripple Creek in Colorado, Coeur d'Alene in Idaho, and the Kootenay Trail region of British Colombia. Lumbering is an active industry in the mountains, but it is threatened by growing environmental concerns and tourism in the national park systems.

Mesozoic–Early Cenozoic contractional events produced the Rockies during uplift associated with the Cordilleran orogeny. Evidence for older events and uplifts are commonly referred to as belonging to the ancestral Rocky Mountain system. The Rocky Mountains are part of the larger Cordilleran orogenic belt that stretches from South America through Canada to Alaska, and it is best to understand the evolution of the Rockies through a wider discussion of events in this mountain belt. The Cordillera is presently active and has been active for the past 350 million years, making it one of the longest-lived orogenic belts on Earth. In the Cordillera, many of the structures are not controlled by continent/continent collisions as they are in many other mountain belts, since the Pacific Ocean is still open. In this orogen structures are controlled by the subduction/accretion process, collision of arcs, islands, and oceanic plateaus, and strike-slip motions parallel to the mountain belt. Present-day motions and deformation are controlled by complex plate boundaries between the North American, Pacific, Gorda, Cocos, and some completely subducted plates such as the Farallon. In this active tectonic setting the style, orientation, and intensity of deformation and magmatism depend largely on the relative convergence–strike-slip vectors of motion between different plates.

The geologic history of the North American Cordillera begins with rifting of the present western margin of North America at 750–800 million years ago, which is roughly the same age as rifting along the east coast in the Appalachian orogen. These rifting events reflect the breakup of the supercontinent of Rodinia at the end of the Proterozoic, and they left North America floating freely from the majority of the continental landmass on Earth. Rifting and the subsequent thermal subsidence of the rifted margin led to the deposition of Precambrian clastic rocks of the Windemere supergroup and carbonates of the Belt and Purcell supergroups, in belts stretching from southern California and Mexico to Canada. These are overlain by Cambrian-Devonian carbonates, Carboniferous clastic wedges, Carboniferous-Permian carbonates, and finally Mesozoic clastic rocks.

The Antler orogeny is a Late Devonian–Early Carboniferous (350–400 million-year-old) tectonic event formed during an arc-continent collision, in which deep-water clastic rocks of the Roberts Mountain allochthon in Nevada were thrust from west to east over the North American carbonate bank, forming a foreland basin that migrated onto the craton. This orogenic event, similar to the Taconic orogeny in the Appalachian Mountains, marks the end of passive margin sedimentation in the Cordillera, and the beginning of Cordilleran tectonism.

In the Late Carboniferous (about 300 million years ago), the zone of active deformation shifted to the east with a zone of strike-slip faulting, thrusts, and normal faults near Denver. Belts of deformation formed what is known as the ancestral Rocky Mountains, including the Front Ranges in Colorado, and the Uncompahgre uplift of western Colorado, Utah, and New Mexico. These uplifts are only parts of a larger system of strike-slip faults and related structures that cut through the entire North American craton in the Late Carboniferous, probably in response to compressional deformation that was simultaneously going on along three margins of the continent.

The Late Permian–Early Triassic Sonoma orogeny (260–240 million years ago) refers to events that led to the thrusting of deep-water Paleozoic rocks of the Golconda allochthon eastward over autochthonous shallow-water sediments just outboard (oceanward) of the Roberts Mountain allochthon. The Golconda allochthon in western Nevada includes deep-water oceanic pelagic rocks, an island-arc sequence, and a carbonate-shelf sequence and is interpreted to represent an arc/continent collision.

In the Late Jurassic (about 150 million years ago) a new, northwest-striking continental margin was established by cross cutting the old northeast-striking continental margin. This event, known as the early Mesozoic truncation event, reflects the start of continental margin volcanic and plutonic activity that continues to the present day. There is considerable uncertainty about what happened to the former extension of the old continental margin—it may have rifted and drifted away, or may have moved along the margin along large strike-slip faults.

Pacific margin magmatism has been active intermittently from the Late Triassic (220 million years ago) through the Late Cenozoic and in places continues to the present. This magmatism and deformation is a direct result of active subduction and arc magmatism. Since the Late Jurassic, there have been three main periods of especially prolific magmatism, including the Late Jurassic/Early Cretaceous Neva-

dan orogeny (150–130 million years ago), the Late Cretaceous Sevier orogeny (80–70 million years ago), and the Late Cretaceous/Early Cenozoic Laramide orogeny (66–50 million years ago).

Cretaceous events in the Cordillera resulted in the formation of a number of tectonic belts that are still relatively easy to discern. The Sierra Nevada ranges of California and Nevada represent the arc batholith, and contain high-temperature, low-pressure metamorphic rocks characteristic of arcs. The Sierra Nevada is separated from the Coast Ranges by flat-lying generally unmetamorphosed sedimentary rocks of the Great Valley, deposited over ophiolitic basement in a forearc basin. The Coast Ranges include high-pressure, low-temperature metamorphic rocks, including blueschists in the Franciscan complex. Together, the high-pressure, low-temperature metamorphism in the Franciscan complex with the high-temperature, low-pressure metamorphism in the Sierra Nevada, represent a paired metamorphic belt, diagnostic of a subduction zone setting.

Several Cretaceous foreland fold-thrust belts are preserved east of the magmatic belt in the Cordillera, stretching from Alaska to Central America. These belts include the Sevier fold-thrust belt in the United States, the Canadian Rockies fold-thrust belt, and the Mexican fold-thrust belt. They are all characterized by imbricate-style thrust faulting, with fault-related folds dominating the topographic expression of deformation.

The Late Cretaceous–Early Tertiary Laramide Orogeny (about 70–60 million years ago) is surprisingly poorly understood but generally interpreted as a period of plate reorganization that produced a series of basement uplifts from Montana to Mexico. Some models suggest that the Laramide Orogeny resulted from the subduction of a slab of oceanic lithosphere at an unusually shallow angle, perhaps related to its young age and thermal buoyancy.

The Late Mesozoic-Cenozoic tectonics of the Cordillera saw prolific strike-slip faulting, with relative northward displacements of terranes along the western margin of North America. The San Andreas fault system is one of the major transform faults formed in this interval, formed as a consequence of the subduction of the Farallon plate. Previous convergence between the Farallon and North American plates stopped when the Farallon plate was subducted, and new relative strike-slip motions between the Pacific and North American plates resulted in the formation of the San Andreas system. Remnants of the Farallon plate are still preserved as the Gorda and Cocos plates.

Approximately 15 million years ago the Basin and Range Province and the Colorado Plateau began uplifting and extending through the formation of rifts and normal faults. Much of the Colorado Plateau stands at more than a mile (1.5–2.0 km) above sea level but has a normal crustal thickness. The cause of the uplift is controversial but may be related to heating from below. The extension is related to the height of the mountains being too great for the strength of the rocks at depth to support, so gravitational forces are able to cause high parts of the crust to extend through the formation of normal faults and rift basins.

See also ACCRETIONARY WEDGE; ARCHEAN; CAMBRIAN; CONVERGENT PLATE MARGIN PROCESSES; CRATON; DEFORMATION OF ROCKS; DIVERGENT PLATE MARGIN PROCESSES; ECONOMIC GEOLOGY; FLYSCH; GONDWANA, GONDWANALAND; GRANITE, GRANITE BATHOLITH; OROGENY; PALEOZOIC; PANGAEA; PASSIVE MARGIN; PHANEROZOIC; PRECAMBRIAN; SILURIAN; STRUCTURAL GEOLOGY; SUPERCONTINENT CYCLES.

FURTHER READING

Ala Drake, A., A. K. Sinha, Jo Laird, and R. E. Guy. "The Taconic Orogen." Chapter 3 in *The Geology of North America*. Vol. A, *The Geology of North America, an Overview,* edited by A. W. Bally and A. R. Palmer, 10–78. Boulder, Colo: Geological Society of America, 1989.

Anderson, J. Lawford, ed. *The Nature and Origin of Cordilleran Magmatism.* Special Paper 174, Boulder, Colo.: Geological Society of America, 1990.

Bird, John M., and John F. Dewey. "Lithosphere Plate–Continental Margin Tectonics and the Evolution of the Appalachian Orogen." *Geological Society of America Bulletin* 81 (1970): 1,031–1,060.

Bleeker, Wouter, and William Davis, eds. "NATMAP Slave Province Project." *Canadian Journal of Earth Sciences* 36 (1999).

Bowring, Samuel A., Ian S. Williams, and William Compston. "3.96 Ga Gneisses from the Slave Province, Northwest Territories." *Geology* 17/11 (1989b): 969–1,064.

Bradley, Dwight C. "Tectonics of the Acadian Orogeny in New England and Adjacent Canada." *Journal of Geology* 91 (1983): 381–400.

Bradley, Dwight C., and Timothy M. Kusky. "Geologic Methods of Estimating Convergence Rates During Arc-Continent Collision." *Journal of Geology* 94 (1986): 667–681.

Burchfiel, Bert Clark, and George A. Davis. "Nature and Controls of Cordilleran Orogenesis, Western United States: Extensions of an Earlier Synthesis." *American Journal of Science* 275 (1975): 363–396.

———. "Structural Framework of the Cordilleran Orogen, Western United States." *American Journal of Science* 272 (1972): 97–118.

Corcoran, P. L., W. U. Mueller, and Timothy M. Kusky. "Inferred Ophiolites in the Archean Slave Province." In *Precambrian Ophiolites and Related Rocks, Developments in Precambrian Geology,* edited by Timothy M. Kusky, 363–404. Amsterdam: Elsevier, 2004.

Helmstaedt, Herart, William A. Padgham, and John A. Brophy. "Multiple Dikes in the Lower Kam Group, Yellowknife Greenstone Belt: Evidence for Archean Sea-floor Spreading?" *Geology* 14 (1986): 562–566.

Henderson, John B., "Archean Basin Evolution in the Slave Province, Canada." In *Precambrian Plate Tectonics,* edited by Alfred Kroner, 213–235. Amsterdam: Elsevier, 1981.

Kusky, Timothy M. "Structural Development of an Archean Orogen, Western Point Lake, Northwest Territories." *Tectonics* 10 (1991): 820–841.

———. "Evidence for Archean Ocean Opening and Closing in the Southern Slave Province." *Tectonics* 9 (1990): 1,533–1,563.

———. "Accretion of the Archean Slave Province." *Geology* 17 (1989): 63–67.

Kusky, Timothy M., J. Chow, and Samuel A. Bowring. "Age and Origin of the Boil Mountain Ophiolite and Chain Lakes Massif, Maine: Implications for the Penobscottian Orogeny." *Canadian Journal of Earth Sciences* 34, (1997): 646–654.

Kusky, Timothy M., and William S. F. Kidd. "Early Silurian Thrust Imbrication of the Northern Exploits Subzone, Central Newfoundland." *Journal of Geodynamics* 22 (1996): 229–265.

Kusky, Timothy M., William S. F. Kidd, and Dwight C. Bradley. "Displacement History of the Northern Arm Fault, and Its Bearing on the Post-Taconic Evolution of North-Central Newfoundland." *Journal of Geodynamics* 7 (1987): 105–133.

Oldow, John S., Albert W. Bally, Hans G. Avé Lallemant, and William P. Leeman. "Phanerozoic Evolution of the North American Cordillera; United States and Canada." In *The Geology of North America; The Appalachian-Ouchita Orogen in the United States,* Vol. F-2 of *Decade of North American Geology,* edited by Robert D. Hatcher, Jr., William A. Thomas, and George W. Viele, 179–232. Boulder, Colo.: Geological Society of America, 1989.

Osberg, Phil, James F. Tull, Peter Robinson, Rudolph Hon, and J. Robert Butler. "The Acadian Orogen." Chapter 4 in *The Geology of North America, an Overview.* Vol. A of *Decade of North American Geology,* edited by Albert W. Bally and Allison R. Palmer, 139–232. Boulder, Colo.: Geological Society of America, 1989.

Rankin, D. W., Avery Ala Drake, Jr., Lynn Glover III, Richard Goldsmith, Leo M. Hall, D. P. Murray, Nicholas M. Ratcliffe, J. F. Read, Donald T. Secor, Jr., and R. S. Stanley. "Pre-Orogenic Terranes." Chapter 2 in *The Geology of North America, The Appalachian-Ouchita Orogen in the United States,* Vol. F-2 of *Decade of North American Geology,* edited by Robert D. Hatcher, Jr., William A. Thomas, and George W.

Viele, 7–100, Boulder, Colo.: Geological Society of America, 1989.

Rast, Nick. "The Evolution of the Appalachian Chain." Chapter 12 in *The Geology of North America*. Vol. A *of The Geology of North America, an Overview*, edited by A. W. Bally and A. R. Palmer, 323–348. Boulder, Colo.: Geological Society of America, 1989.

Rowley, David B., and William S. F. Kidd. "Stratigraphic Relationships and Detrital Composition of the Medial Ordovician Flysch of Western New England: Implications for the Tectonic Evolution of the Taconic Orogeny." *Journal of Geology* 89 (1981): 199–218.

Roy, D. "The Acadian Orogeny: Recent Studies in New England, Maritime Canada, and the Autochthonous Foreland." In Geological Society of America Special Paper 275, edited by James W. Skehan, 1993.

Sisson, Virginia B., Sarah M. Roeske, and Terry L. Pavlis. *Geology of a Transpressional Orogen Developed During Ridge-Trench Interaction Along the North Pacific Margin*. Geological Society of America Special Paper 371, 2003.

Socci, Anthony D., James W. Skehan, and Geoffrey W. Smith. "Geology of the Composite Avalon Terrane of Southern New England." Special Paper 245 Geological Society of America, 1990.

Stanley, Rolfe S., and Ratcliffe, Nicholas M. "Tectonic Synthesis of the Taconian Orogeny in Western New England." *Geological Society of America Bulletin* 96 (1985): 1,227–1,250.

van Staal, Cees R., and Leslie R. Fyffe. "Dunnage Zone–New Brunswick." In *Geology of the Appalachian-Caledonian Orogen in Canada and Greenland*, edited by H. Williams, 166–178. *The Geology of North America* Vol. F-1, Geological Society of America, 1995.

nova A nova is a general name for a type of star that vastly increases in brightness (by up to 10,000 times) over very short periods of time, typically days or weeks. The word *nova* comes from the Latin for "new" and stems from early astronomers who thought that nova were new stars appearing in the sky, since the parent stars were too faint to be observed from the Earth before powerful telescopes were invented.

A nova forms when a white dwarf star has a companion, such as in a binary star system, and the companion contributes matter to the white dwarf after its initial death. In a simple system where a white dwarf exists alone, it will cool off indefinitely, approach absolute zero, and be invisible in space. However, in some binary star systems the large gravitational field of the white dwarf can pull material, predominantly hydrogen and helium, away from the companion main sequence star and accrete this material onto the white dwarf. As this gas builds up on the white dwarf surface it heats up and becomes denser until its temperature exceeds 100,000,000 K, at which point the hydrogen ignites and rapidly fuses into helium. This causes a sudden and dramatic flare-up of the surface of the white dwarf over a period of a few days, rapidly burning some of the fuel and expelling the rest of it to space in a nova event. After a few months the star's luminosity and surface temperature return to normal.

In white dwarf–binary star systems in which the companion star is a main sequence star, the material that is transferred off the main sequence star is affected by the rotation of the binary system and the gravitational field between the stars, and this material is forced to swirl around and orbits the white dwarf before it accretes to the surface. This forms what is known as an accretion disk. As the material in the accretion disk orbits the white dwarf, it is heated by friction and begins to glow and emit radiation in the visible, ultraviolet, and X-ray wavelengths and may become brighter than the star itself.

White dwarf–binary systems have the possibility of repeating the cycle of becoming novas many times, and some stars have done this hundreds of times. Such recurrent nova have been known since ancient times, and include systems such as RS Ophiuchi, located about 5,000 light years away in the constellation Ophiuchus; T Coronae Borealis, nicknamed the blaze star, in the constellation Coronos Borealis; and T Pyxidis, located about 6,000 light years from Earth in the constellation Pyxis.

See also ASTRONOMY; ASTROPHYSICS; BINARY STAR SYSTEMS; CONSTELLATION; DWARFS (STARS); STELLAR EVOLUTION; SUPERNOVA.

FURTHER READING

Chaisson, Eric, and Steve McMillan. *Astronomy Today*. 6th ed. Upper Saddle River, N.J.: Addison-Wesley, 2007.

Comins, Neil F. *Discovering the Universe*. 8th ed. New York: W. H. Freeman, 2008.

Dibon-Smith, Richard. The Constellations Web Page. Available online. URL: http://www.dibonsmith.com/index.htm. Last updated November 8, 2007.

Prialnik, Dina. "Novae." In *Encyclopedia of Astronomy and Astrophysics*, edited by Paul Murdin, 1,846–1,856. London: Institute of Physics Publishing Ltd and Nature Publishing Group, 2001.

Snow, Theodore P. *Essentials of the Dynamic Universe: An Introduction to Astronomy*. 4th ed. St. Paul, Minn.: West Publishing Company, 1991.

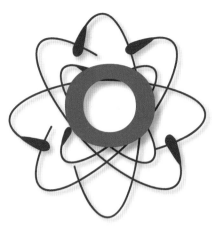

ocean basin The surface of the Earth is divided into two fundamentally different types of crust, including relatively light quartz and plagioclase-rich sial, forming the continental regions, and relatively dense olivine and pyroxene-rich sima underlying the ocean basins. The ocean basins are submarine topographic depressions underlain by oceanic (simatic) crust. Ocean basins are quite diverse in size, shape, depth, characteristics of the underlying seafloor topography, and types of sediments deposited on the oceanic crust. The largest ocean basins include the Pacific, Atlantic, Indian, and Arctic Oceans and the Mediterranean Sea, whereas dozens of smaller ocean basins are located around the globe.

The ocean basins' depths were first extensively explored by scientists aboard the H.M.S. *Challenger* in the 1800s, using depth reading from a weight attached to a several-mile (kilometer) long cable that was dropped to the ocean floor. Results from these studies suggested that the oceans were generally about three to four miles (five to six km) in depth. Later, with the development of echo-sounding technologies and war-induced mapping efforts, the variety of sea floor topography became appreciated. Giant submarine mountain chains were recognized where the depth is reduced to 1.7 miles (2.7 km), and these were later interpreted to be oceanic ridges where new oceanic crust is created. Deep-sea trenches with depths exceeding five miles (eight km) were delineated, and later recognized to be subduction zones where oceanic crust sinks back into the mantle. Other anomalous regions of thick oceanic crust (and reduced depths) were recognized, including large oceanic plateaus where excessive volcanism produced thick crust over large regions, and smaller seamounts (or guyots) where smaller, off-ridge volca-

nism produced isolated submarine mountains. Some of these rose above sea level, were eroded by waves, and grew thick reef complexes as they subsided with the cooling of the oceanic crust. Charles Darwin made such guyots and coral atolls famous in his study of coral reefs of the Pacific Ocean basin.

Pelagic sediments are deposited in the ocean basins and generally form a blanket of sediments draping over preexisting topography. Carbonate rocks produced mainly by the tests of foraminifera and nannofossils may be deposited on the ocean ridges and guyots that are above the carbonate compensation depth (CCD), above which the sea water is saturated with $CaCO_3$, and below which it dissolves in the water. Below this, sediments comprise red clays and radiolarian and diatomaceous ooze. Manganese nodules are scattered about on some parts of the ocean floor.

The abyssal plains are relatively flat, generally featureless parts of the ocean basins where the deep parts of the seafloor topography have been filled in with sediments, forming flat plains, broken occasionally by hills and volcanic islands such as the Bermuda platform, Cape Verde Islands, and the Azores. Some of these submarine plains are quite large, such as the vast 386,100-square mile (1 million-km²) Angolan abyssal plain in the South Atlantic, and the 1,428,578-square mile (3.7 million-km²) abyssal plain in the Antarctic Ocean basin. Other abyssal plains are much smaller, such as the 1,003-square mile (2,600-km²) Alboran Sea in the Mediterranean. Different deep-sea plains may also be characterized and distinguished on the basis of their sediment composition, their geometry, depth, and volume and thickness of the sediments they contain. The deep abyssal areas in the Pacific Ocean are characterized by the presence of more abundant hills or seamounts, which rise up to 0.6 mile (1 km) above the seafloor. For this reason the deep abyssal region of the Pacific is generally referred to as the abyssal hills instead of the abyssal plains. Approximately 80–85 percent of the Pacific Ocean floor lies close to areas with hills and seamounts, making the abyssal hills the most common landform on the surface of the Earth.

Many of the sediments on the deep seafloor (the abyssal plain) are derived from erosion of the continents and carried to the deep sea by turbidity currents or by wind (e.g., volcanic ash) or released from floating ice. Other sediments, known as deep-sea oozes, include pelagic sediments derived from marine organic activity. When small organisms die, such as diatoms in the ocean, their shells sink to the bottom and over time create significant accumulations. Calcareous ooze occurs at low to middle latitudes where warm water favors the growth of carbonate-secreting organisms. Calcareous oozes are not found in water that is more than 2.5–3 miles (4–5 km) deep, because this water is under such high pressure that it contains dissolved CO_2 that dissolves carbonate shells. Siliceous ooze is produced by organisms that use silicon to make their shell structure.

See also OCEANIC PLATEAU; OPHIOLITES; PASSIVE MARGIN; PLATE TECTONICS.

FURTHER READING

Erickson, Jon. *Marine Geology; Exploring the New Frontiers of the Ocean.* Rev. ed. New York: Facts On File, 2003.
Moores, Eldridge M., and Robert Twiss. *Tectonics.* New York: W. H. Freeman, 1995.

ocean currents   Like the atmosphere, the ocean is constantly in motion. Ocean currents are defined by the movement paths of water in regular courses, driven by the wind and thermohaline forces across the ocean basins. The wind primarily drives shallow currents, but the Coriolis force systematically deflects them to the right of the atmospheric wind directions in the Northern Hemisphere, and to the left of the prevailing winds in the Southern Hemisphere. Therefore, shallow water currents tend to be oriented about 45° from the predominant wind directions.

Thermohaline effects, the movement of water driven by differences in temperature and salinity, primarily drive deep-water currents. The Atlantic and Pacific Ocean basins both show a general clockwise rotation in the Northern Hemisphere, and a counterclockwise spin in the Southern Hemisphere, with the strongest currents in the midlatitude sectors. The pattern in the Indian Ocean is broadly similar but seasonally different and more complex because of the effects of the monsoon. Antarctica is bound on all sides by deep water, and has a major clockwise current surrounding it known as the Antarctic Circumpolar Current, lying between 40° and 60° south. This is a strong current, moving at 1.6–5 feet per second (0.5–1.5 m/s) and has a couple of major gyres in it at the Ross Ice Shelf and near the Antarctic Peninsula. The Arctic Ocean has a complex pattern, because it is sometimes ice-covered and is nearly completely surrounded by land with only one major entry and escape route east of Greenland, called Fram Strait. Circulation patterns in the Arctic Ocean are dominated by a slow, 0.4–1.6-inch-per-second (1–4-cm/s) transpolar drift from Siberia to the Fram Strait and by a thermohaline-induced anticyclonic spin known as the Beaufort Gyre that causes ice to pile up on the Greenland and Canadian coasts. Together the two effects in the Arctic Ocean bring numerous icebergs into North Atlantic shipping lanes and send much of the cold deep water around Greenland into the North Atlantic ocean basin.

EKMAN SPIRALS

Ekman spirals are differences in current directions with depth, and form through the turning of water with depth as a result of the Coriolis force. They form because each (infinitesimally thin) layer of the ocean water exerts a frictional drag on the layer below, so that as the top layer moves, the layers below move slightly less with each depth increment. Because the Coriolis force causes moving objects to deflect to the right in the Northern Hemisphere and to the left in the Southern Hemisphere, each successively deeper layer will also be slightly deflected to the right or left of the moving layer above. These effects cause moving water on the surface to be succeeded with depth by progressively slowing and turning particle paths. The Ekman spirals typically extend to about 325 feet (100 m), where the water moves in the direction opposite to that of the surface water that caused the initial flow. The movement of water by Eckman spirals causes a net transport of water to the right of the direction of surface water in the Northern Hemisphere and to the left of the direction of surface water in the Southern Hemisphere. This phenomenon is known as Ekman transport.

GEOSTROPHIC CURRENTS

Some currents in the oceans follow specific horizons on the topographic contours of the ocean basin, staying at the same depth for long distances. These currents in the ocean or atmosphere, in which the horizontal pressure is balanced by the equal but opposite Coriolis force, are known as geostrophic currents. Friction does not affect these currents, which flow to the right of the pressure gradient force along pressure

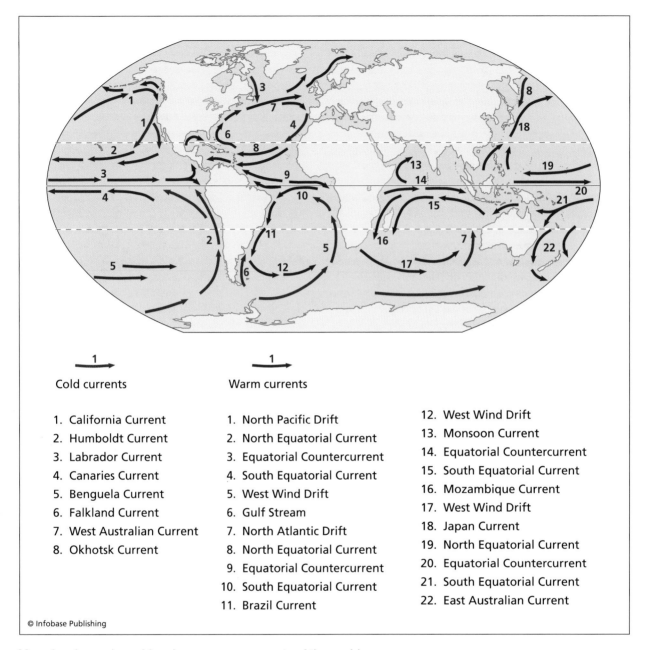

Map showing major cold and warm ocean currents of the world

Cold currents

1. California Current
2. Humboldt Current
3. Labrador Current
4. Canaries Current
5. Benguela Current
6. Falkland Current
7. West Australian Current
8. Okhotsk Current

Warm currents

1. North Pacific Drift
2. North Equatorial Current
3. Equatorial Countercurrent
4. South Equatorial Current
5. West Wind Drift
6. Gulf Stream
7. North Atlantic Drift
8. North Equatorial Current
9. Equatorial Countercurrent
10. South Equatorial Current
11. Brazil Current
12. West Wind Drift
13. Monsoon Current
14. Equatorial Countercurrent
15. South Equatorial Current
16. Mozambique Current
17. West Wind Drift
18. Japan Current
19. North Equatorial Current
20. Equatorial Countercurrent
21. South Equatorial Current
22. East Australian Current

© Infobase Publishing

isobars in the Northern Hemisphere atmosphere and oceans and to the left in the Southern Hemisphere. In the oceans, geostrophic currents are also known as contour currents, since they follow the bathymetric contours on the seafloor, flowing clockwise in the Northern Hemisphere and counterclockwise in the Southern Hemisphere. Downslope currents such as turbidity currents deposit most of the sediments on continental slopes. Since geostrophic or contour currents flow along the bathymetric contours, they rework bottom sediments at right angles to the currents that deposited the sediments. Their work is therefore detectable by examination of paleocurrent

indicators that swing from downslope to slope-parallel movement vectors at the top of turbidite and other slope deposits.

COASTAL UPWELLING AND DOWNWELLING CURRENTS

Coastal upwelling is a phenomenon caused by surface winds that blow parallel to the coast, forming ocean surface movements at 90 degrees to the direction of surface winds in Eckman spirals. In many cases, the upper hundred feet or so (few tens of meters) of surface waters move away from the shoreline, forcing a corresponding upwelling of water from depth to replace the water that has moved offshore. This is known as coastal upwelling. Upwelling is most common on the eastern sides of ocean basins, where the surface layer is thin, and near capes and other irregularities in the coastline. Upwelling also occurs away from the coasts along the equator, where surface waters diverge because of the change in sign of the Coriolis force across the equator. Water from depth upwells to replace the displaced surface water.

Zones of coastal and other upwelling, where the water comes from more than 325 feet (100 m) depth, are typically very productive organically, with abundant marine organisms including plants and fish. This is because upwelling coastal waters are rich in nutrients that suddenly become available to benthic and planktonic photic zone organisms.

In contrast to coastal upwelling, coastal downwelling is a phenomenon where winds moving parallel to the coast cause the ocean surface waters to move toward the shoreline, necessitating a corresponding deeper flow from the coast below the shoreward-moving coastal water. Ocean surface currents generally move at right angles to the dominant wind, and move at a velocity of about 2 percent of the wind's. Therefore, onshore winds cause currents to move parallel to the coasts, whereas along shore winds set up currents that move toward or away from the shore.

RAPID CHANGES IN OCEAN CIRCULATION PATTERNS AND CLIMATE CHANGE

Some models of climate change show that patterns of ocean circulation can suddenly change and cause global climate conditions to switch from warm to cold, or cold to warm, over periods of a few decades. Understanding how fast climate can shift from a warm period to a cold, or cold to a warm, is controversial. The record of climate indicators is incomplete and difficult to interpret. Only 18,000 years ago the planet was in the midst of a major glacial interval, and since then global average temperatures have risen 16°F (10°C) and are still rising, perhaps at a recently accelerated rate from human contributions to the atmosphere. Still, recent climate work is revealing that there are some abrupt transitions in the slow warming, in which there are major shifts in some component of the climate, where the shift may happen on scales of 10 years or less.

One of these abrupt transitions seems to affect the ocean circulation pattern in the North Atlantic Ocean, where the ocean currents formed one of two different stable patterns or modes, with abrupt transitions occurring when one mode switches to the other. In the present pattern the warm waters of the Gulf Stream come out of the Gulf of Mexico and flow along the eastern seaboard of the United States, past

the British Isles, to the Norwegian Sea. This warm current is largely responsible for the mild climate of the British Isles and northern Europe. In the second mode, the northern extension of the Gulf Stream is weakened by a reduction in salinity of surface waters from sources at high latitudes in the North Atlantic. The fresher water has a source in increased melting from the polar ice shelf, Greenland, and northern glaciers. With less salt, seawater is less dense, and is less able to sink during normal wintertime cooling.

Studies of past switches in the circulation modes of the North Atlantic reveal that the transition from mode 1 to mode 2 can occur over a period of only five to 10 years. These abrupt transitions are apparently linked to increase in the release of icebergs and freshwater from continental glaciers, which upon melting contribute large volumes of freshwater into the North Atlantic, systematically reducing the salinity. The Gulf Stream presently seems on the verge of failure, or of switching from mode 1 to mode 2, and historical records show that this switch can be very rapid. If this predicted switch occurs, northern Europe and the United Kingdom may experience a significant and dramatic cooling of their climate, instead of the warming many fear.

See also EL NIÑO AND THE SOUTHERN OSCILLATION (ENSO); ENERGY IN THE EARTH SYSTEM; HYDROSPHERE; OCEAN BASIN; OCEANOGRAPHY; PELAGIC, NEKTONIC, PLANKTONIC; THERMOHALINE CIRCULATION.

FURTHER READING

Erickson, Jon. *Marine Geology; Exploring the New Frontiers of the Ocean.* Rev. ed. New York: Facts On File, 2003.

Intergovernmental Panel on Climate Change. *Climate Change 2007: The Physical Science Basis. Contributions of Working Group I to the Fourth Assessment Report of the Intergovernmental Panel on Climate Change,* edited by S. Solomon, D. Qin, M. Manning, Z. Chen, M. Marquis, K. B. Averyt, M. Tignor, and H. L. Miller. Cambridge: Cambridge University Press, 2007.

Intergovernmental Panel on Climate Change home page. Available online. URL: http://www.ipcc.ch/index.htm. Accessed January 29, 2009.

Kusky, T. M. *Climate Change: Shifting Deserts, Glaciers, and Climate Belts.* New York: Facts On File, 2008.

———. *The Coast: Hazardous Interactions within the Coastal Zone.* New York: Facts on File, 2008.

oceanic plateau Regions of anomalously thick oceanic crust and topographically high seafloor are known as oceanic plateaus. Many have oceanic crust that is 12.5–25 miles (20–40 km) thick, and rise thousands of meters above surrounding oceanic crust of normal thickness. The Caribbean ocean floor represents one of the best examples of an oceanic plateau, with other major examples including the Ontong-Java Plateau, Manihiki Plateau, Hess Rise, Shatsky Rise, and Mid-Pacific Mountains. All of these oceanic plateaus contain thick piles of volcanic and subvolcanic rocks representing huge outpourings of lava; most erupted in a few million years. They typically do not show the magnetic stripes that characterize normal oceanic crust produced at oceanic ridges, and are thought to have formed when mantle plume heads reached the base of the lithosphere, releasing huge amounts of magma. Some oceanic plateaus have such large volumes of magma that the total magmatic flux in the plateaus would have been similar to or larger than all of the magma erupted at the mid-ocean ridges during the same interval. The Caribbean seafloor preserves 5–7.5-mile- (8–21-km-) thick oceanic crust formed before about 85 million years ago in the eastern Pacific Ocean. Plate tectonics transported this unusually thick ocean floor eastward, where pieces of the seafloor collided with South America as it passed into the Atlantic Ocean. Pieces of the Caribbean oceanic crust are now preserved in Colombia, Ecuador, Panama, Hispaniola, and Cuba, and some scientists estimate that the Caribbean oceanic plateau may have once been twice its present size. An accompanying vast outpouring of lava would have been associated with significant outgassing, with possible consequences for global climate and evolution.

The western Pacific Ocean basin contains several large oceanic plateaus, including the 20-mile- (32-km-) thick crust of the Ontong-Java Plateau, which is the largest outpouring of volcanic rocks on the planet. The Ontong-Java Plateau is the largest igneous province in the world not associated with the oceanic-ridge spreading-center network, covering an area roughly the size of Alaska (9,300,000 square miles, or 15,000,000 km²). The plateau is located northeast of Papua New Guinea and the Solomon Islands in the southwest Pacific Ocean, centered on the equator at 160°E longitude. Most of the plateau formed about 122 million years ago in the Cretaceous period, probably as a result of a mantle plume rising to the surface and causing massive amounts of volcanism over a geologically short interval, likely lasting only about a million years. Smaller amounts of volcanic material erupted later, at about 90 million years ago. Together, these events formed a lava plateau that is 20 miles (32 km) thick. The amount of volcanic material produced to form the plateau is estimated to be approximately the same as that erupted from the entire global ocean-ridge spreading-center system in the same period. Such massive amounts of volcanism cause worldwide changes in climate and ocean temperatures and typically have great impacts on the biosphere. Sea levels rose by more than 30 feet (9 m) in response to this volcanic outpouring. The gases released during these eruptions are estimated to have raised average global temperatures by 23°F (13°C). Perhaps more remarkably, the Ontong-Java Plateau is but one of many Cretaceous oceanic plateaus in the Pacific, suggesting that the Cretaceous was characterized by a long-standing eruption of massive amounts of deeply derived magma. Some geologists have suggested that events like this relate to major mantle overturn events, when plumes dominate heat loss from the Earth instead of oceanic-ridge spreading, as in the present plate mosaic.

The plateau is thought to be composed largely of basalt, based on limited sampling, deep-sea drilling, and seismic velocities. A covering by a thick veneer of sediments, exceeding thousands of feet (a kilometer or more) in most places, presents great difficulties in trying to sample the plateau. The plateau is colliding with the Solomon trench, but thick oceanic plateaus like the Ontong-Java are generally unsubductable. When oceanic plateaus are attempted to be subducted,

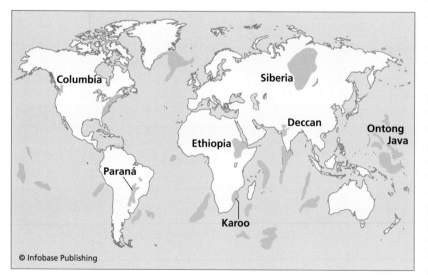

Map of the world showing major flood basalt regions, including oceanic plateaus

they typically get accreted to the continents, leading to continental growth.

See also AFRICAN GEOLOGY; LARGE IGNEOUS PROVINCES, FLOOD BASALT.

FURTHER READING

Kious, Jacquelyne, and Robert I. Tilling. U.S. Geological Survey. This Dynamic Earth: The Story of Plate Tectonics. Available online. URL: http://pubs.usgs.gov/gip/dynamic/dynamic.html. Last modified March 27, 2007.

Kusky, T. M., ed. *Precambrian Ophiolites and Related Rocks. Developments in Precambrian Geology* Vol. 13. Amsterdam: Elsevier Publishers, 2003.

Kusky, Timothy M., and William S. F. Kidd. "Remnants of an Archean Oceanic Plateau, Belingwe Greenstone Belt, Zimbabwe." *Geology* 20, no.1 (1992): 43–46.

Moores, Eldridge M. "Origin and Emplacement of Ophiolites." *Review Geophysics* 20 (1982): 735–750.

Moores, Eldridge M., and Robert Twiss. *Tectonics*. New York: W. H. Freeman, 1995.

oceanography The study of the physical, chemical, biological, and geological aspects of the ocean basins is oceanography. Oceanographers have begun using an Earth system science approach to study the oceans, with the appreciation that many of the different systems are related, and changes in the biological, chemical, physical, or geological conditions will result in changes in the other systems and also influence other Earth systems such as the atmosphere and climate. The oceans contain important geological systems, since the ocean basins are the places where oceanic crust is both created at mid-ocean ridges and destroyed at deep-sea trenches. Being topographic depressions, they are repositories for many of the sediments eroded from the continents and carried by rivers and the wind to be deposited in submarine settings. Seawater is the host of much of the life on Earth and also holds huge quantities of dissolved gases and chemicals that buffer the atmosphere, keeping global temperatures and climate hospitable for humans. Energy is transferred around the planet in ocean currents and waves, which interact with land, eroding or depositing shoreline environments. Being host to some of the planet's largest and most diverse biota, the oceans may hold the key to feeding the planet. Mineral resources are also abundant on the seafloor, many formed at the interface between hot volcanic fluids and cold seawater, forming potentially economically important reserves of many minerals.

The oceans cover two-thirds of the Earth's surface, yet scientists have explored less of the ocean's depths and mysteries than the surfaces of several nearby planets. The oceans have hindered migration of peoples and biota between distant continents, yet paradoxically now serve as a principal means of transportation. Oceans provide us with incredible mineral wealth and renewable food and energy sources, yet also breed devastating hurricanes. Life may have begun on Earth in environments simulated by hot volcanic events on the seafloor, and marine biologists are exploring the diverse and unique fauna that can still be found living in deep dark waters around similar vents today.

Ocean basins have continually opened and closed on Earth, and the continents have alternately been swept into large single supercontinents and then broken apart by the formation of new ocean

USNS *Bowditch*, a U.S. Navy T-AGS 60 Class oceanographic survey ship *(U.S. Navy photo)*

Japanese research vessel *Natsushima,* support ship for the Shinkai subs *(Science Source/Photo Researchers, Inc.)*

basins. The appearance, evolution, and extinction of different life-forms is inextricably linked to the opening and closing of ocean basins, partly through the changing environmental conditions associated with the changing distribution of oceans and continents.

Early explorers slowly learned about ocean currents and routes to distant lands, and some dredging operations revealed huge deposits of metals on the seafloor. Tremendous leaps in our understanding of the structure of the ocean basin seafloor were acquired during surveying for the navigation of submarines and detection of enemy submarines during World War II. Magnetometers towed behind ships and accurate depth measurements provided data that led to the formulation of the hypothesis of seafloor spreading, which added the oceanic counterpart to the idea of continental drift. The plate tectonic paradigm later unified these two theories.

Ocean circulation is responsible for much of the world's climate. For instance, mild foggy winters in London are caused by warm waters from the Gulf of Mexico flowing across the Atlantic via the Gulf Stream to the coast of the British Isles. Large variations in ocean and atmospheric circulation patterns in the Pacific lead to alternating wet and dry climate conditions known as El Niño and La Niña. These variations affect Pacific regions most strongly, but are felt throughout the world. Other more dramatic movements of water include the sometimes devastating tsunamis initiated by earthquakes, volcanic eruptions, and giant submarine landslides.

One of the most tragic tsunamis in recent history occurred on December 26, 2004, following a magnitude 9.0 earthquake off the coast of northern Sumatra in the Indian Ocean. The earthquake was the largest since the 1964 magnitude 9.2 event in southern Alaska and released more energy than all the earthquakes on the planet in the last 25 years combined. During this catastrophic earthquake, a segment of the seafloor the size of the state of California, lying above the Sumatra subduction zone trench, suddenly moved upward and seaward by more than 30 feet (9 m). The sudden displacement of this volume of undersea floor displaced a huge amount of water and generated the most destructive tsunami known in recorded history. Within minutes of the initial earthquake a mountain of water more than 100 feet (30 m) high was ravaging northern Sumatra, sweeping into coastal villages and resort communities with a fury that crushed all in its path, removing buildings and vegetation, and in many cases eroding shoreline areas down to bedrock, leaving no traces of the previous inhabitants or structures. Similar scenes of destruction and devastation rapidly moved up the coast of nearby Indonesia, where residents and tourists were enjoying a holiday weekend. Firsthand accounts of the catastrophe reveal similar scenes of horror where unsuspecting tourists and residents were enjoying themselves in beachfront playgrounds, resorts, and villages, and reacted as large breaking waves appeared off the coast. Many moved toward the shore to watch the high surf with interest, then ran in panic as the sea rapidly rose beyond expectations, and walls of water engulfed entire beachfronts, rising above hotel lobbies, and washing through towns with the force of Niagara Falls. In some cases the sea retreated to unprecedented low levels before the waves struck, causing many people to move to the shore to investigate the phenomenon; in other cases, the sea waves simply came crashing inland without warning. Buildings, vehicles, trees, boats and other debris were washed along with the ocean waters, forming projectiles that smashed at speeds of up to 30 miles per hour (50 km/hr) into other structures, leveling all in their paths, and killing more than a quarter million people.

Another tragic tsunami in history was generated by the eruption of the Indonesian volcano Krakatau in 1883. When Krakatau erupted, it blasted out a large part of the center of the volcano, and seawater rushed in to fill the hole. This seawater was immediately heated and it exploded outward in a steam eruption and a huge wave of hot water. The tsunami generated by this eruption reached more than 120 feet (36.5 m) in height and killed an estimated 36,500 people in nearby coastal regions. In 1998 a catastrophic 50-foot- (15-m-) high wave unexpectedly struck Papua New Guinea, killing more than 2,000 people and leaving more than 10,000 homeless.

Rich mineral deposits fill the oceans—oil and gas are on the continental shelves and slopes, and metal-liferous deposits form near mid-ocean ridge vents. Much of the world's wealth of manganese, copper, and gold lies on the seafloor. The oceans also yield rich harvests of fish, and care must be taken to not deplete this source. Sea vegetables are growing in popularity, and their use may help alleviate the growing demand for space in fertile farmland. The oceans may offer the world a solution to growing energy and food demands resulting from a rapidly growing population. New life-forms are constantly being discovered in the depth of the oceans, and precautions must be taken to understand these creatures before any changes people make to their environment cause them to perish forever.

HISTORY OF EXPLORATION OF THE WORLD'S OCEAN BASINS

The earliest human exploration of the oceans is poorly known, but pictures of boats on early cave drawings in Norway illustrate Viking-style ocean vessels known to be used by the Vikings centuries later. Other rock drawings around the world show dugout canoes, boats made of reeds, bark, and animal hides. Early migrations of humans must have utilized boats to move from place to place. For instance, analysis of languages and of genetics shows that the Polynesians moved south from China into southeast Asia and Polynesia, then somehow made it, by sea, all the way to Madagascar off the east coast of Africa. Other oceanic migrations include the colonization of Europe by Africans about 10,000 years ago, explorations and trade around and out of the Mediterranean by the Phoenicians about 3,000 years ago, and the colonization of North America by the Siberians and Vikings. Ming Dynasty ocean explorations in the early 1400s were massive, involving tens of thousands of sailors on 317 ships. The Chinese ships were huge, including as many as nine masts more than 444 feet (135 m) in length and 180 feet (55 m) in width. The Chinese mounted these expeditions to promote Chinese culture, society, and technology but did not contribute significantly to understanding the oceans.

The first European to reach North America was probably Leif Eriksson, who, in the year 1,000, landed at L'Anse-aux Meadows in the Long Range Peninsula of Newfoundland, after becoming lost on his way from Greenland to Norway. The Vikings established a temporary settlement in Newfoundland, and there are some speculations of further explorations by the Vikings to places as far south as New England and Narragansett Bay in Rhode Island. Their colonies disappeared during the Dark Ages, probably as a result of a global climate cooling trend that turned previously arable lands into arctic tundra.

Ptolemy (in the year 140) published maps of Europe's coastline that were largely inaccurate and took many years of ocean exploration to correct. The Greeks and Islamic explorers had made great strides in understanding the geography of the world centered on the Mediterranean Sea and Arabian Peninsula, and the records of these explorations eventually made it into European hands, where this knowledge was used for further explorations. The Portuguese, most notably Prince Henry the Navigator (1392–1460), were the most avid explorers of the Atlantic, exploring northwest Africa and the Azores in the early 1400s. In the late 1400s, Vasco da Gama (1460–1524) made it to southern Africa and eventually around the Cape of Good Hope, past Madagascar, and all the way to India in 1498. These efforts initialized economically important trade routes between Portugal and India, building the powerful Portuguese Empire. The timing was perfect for establishment of ocean trade routes, as the long-used overland Silk Roads had become untenable and dangerous with the collapse of the Mongol Empire and the Turk conquest of Constantinople (Istanbul) in 1453.

In the late 15th and early 16th centuries, many ocean exploration expeditions were mounted as a precursor to more widespread use of the oceans for transportation. In 1492, Christopher Columbus sailed from Spain to the east coast of North America, and from the late 1400s to 1521 Ferdinand Magellan sailed around the world, including a crossing of the Pacific Ocean, followed by Sir Francis Drake of England. Later, Henry Hudson explored North American waters, including attempts to find a northwest passage between the Atlantic and Pacific. During the 1700s, Captain James Cook made several voyages in the Pacific and coastal waters of western North America, improving maps of coastal and island regions.

The early explorations of the oceans were largely concerned with navigation and determining the positions of trade routes, coastlines, and islands. Later, sea-going expeditions aimed at understanding the physical, chemical, biological, and geological conditions in the ocean were mounted. In the late 1800s the British Royal Society sponsored the world's most ambitious scientific exploration of the oceans ever, the voyage of the H.M.S. *Challenger*. The voyage of the *Challenger* in 1872–76 established for the first time many of the basic properties of the oceans and set the standard for the many later expeditions.

Ocean exploration today is led by American teams based at several universities, Scripps Institute of Oceanography, and Woods Hole Oceanographic Institute, where the deep submersible Alvin is based and from where many oceanographic cruises are coordinated. The Ocean Drilling Program (formerly the Deep-Sea Drilling Project) has amassed huge quantities of data on the sediments and volcanic rocks deposited on the ocean floor, as well as information about biology, climate, chemistry, and ocean circulation. Many other nations, including Japan, China, France, and Russia have mounted ocean exploration campaigns, with a trend toward international cooperation in understanding the evolution of the ocean basins.

See also OCEAN BASIN; OCEAN CURRENTS; OPHIO- LITES; PLATE TECTONICS.

FURTHER READING

Erickson, Jon. *Marine Geology: Exploring the New Fron- tiers of the Ocean.* Rev. ed. New York: Facts On File, 2003.

Kusky, T. M. *The Coast: Hazardous Interactions within the Coastal Zone.* New York: Facts On File, 2008.

———. *Tsunamis: Giant Waves from the Sea.* New York: Facts On File, 2008.

Simkin, T., and R. S. Fiske. *Krakatau 1883: The Volcanic Eruption and Its Effects.* Washington, D.C.: Smithso- nian Institution Press, 1993.

ophiolites A distinctive group of rocks that includes basalt, diabase, gabbro, and peridotite, the ophiolites may also be associated with chert, metallif- erous sediments (umbers), trondhjemite, diorite, and serpentinite. Many ophiolites are altered to serpenti- nite, chlorite, albite, and epidote-rich rocks, possibly by hydrothermal seafloor metamorphism. German geologist G. Steinman introduced the term in 1905 for a tripartite assemblage of rocks, including basalt, chert, and serpentinite, that he recognized as a com- mon rock association in the Alps. Most ophiolites form in an ocean-floor environment, including at mid-ocean ridges, in back-arc basins, in extensional forearcs, or within arcs. Ophiolites are detached from the oceanic mantle and have been thrust upon conti- nental margins during the closure of ocean basins. Lines of ophiolites decorate many sutures around the world, marking places where oceans have closed. In the 1960s and 1970s much research focused on defining a type of ophiolite succession that became known as the Penrose-type of ophiolite. More recent research has revealed that the variations between individual ophiolites are as significant as any broad similarities between them.

A classic Penrose-type of ophiolite is typically three to nine miles (5–15 km) thick and if complete, consists of the following sequence from base to top, with a fault marking the base of the ophiolite. The lowest unit in some ophiolites is an ultramafic rock called lherzolite, consisting of olivine + clinopyrox- ene + orthopyroxene, generally interpreted to be fer- tile, undepleted mantle. The base of most ophiolites

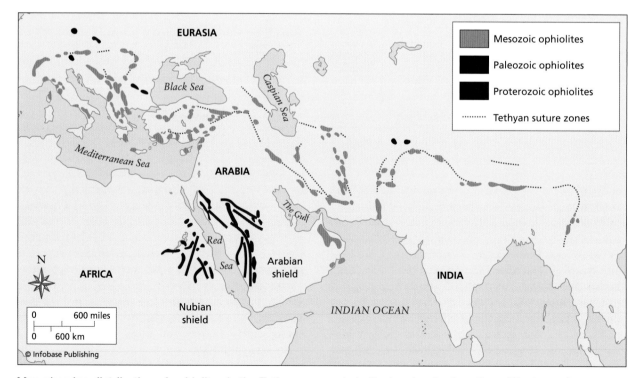

Map showing distribution of ophiolites in the Tethyan orogenic belt, showing the location of Proterozoic, Paleozoic, and Mesozoic ophiolites

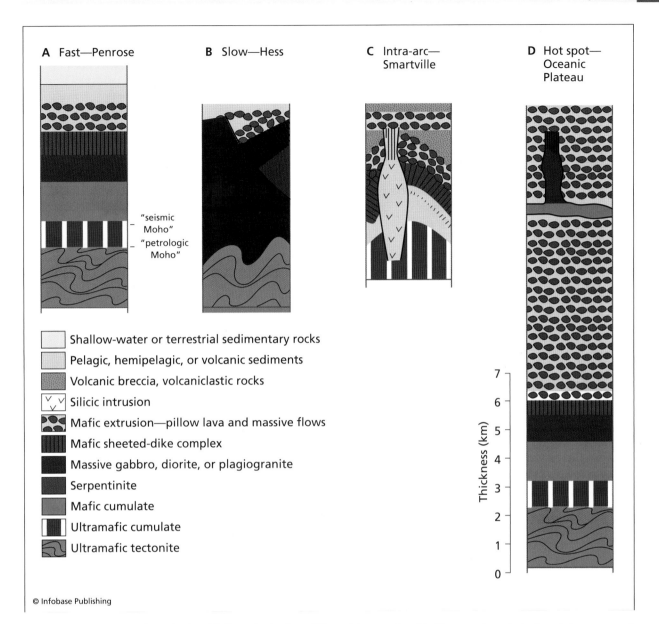

A Fast—Penrose

B Slow—Hess

C Intra-arc—Smartville

D Hot spot—Oceanic Plateau

"seismic Moho"
"petrologic Moho"

Thickness (km)

□ Shallow-water or terrestrial sedimentary rocks
□ Pelagic, hemipelagic, or volcanic sediments
▨ Volcanic breccia, volcaniclastic rocks
▨ Silicic intrusion
▨ Mafic extrusion—pillow lava and massive flows
▮ Mafic sheeted-dike complex
■ Massive gabbro, diorite, or plagiogranite
▨ Serpentinite
▨ Mafic cumulate
▮ Ultramafic cumulate
▨ Ultramafic tectonite

© Infobase Publishing

Cross sections through typical ophiolites, including different types of ophiolites produced at slow, intra-arc, and hotspot types of tectonic settings

consists of an ultramafic rock known as harzburgite, consisting of olivine + orthopyroxene (± chromite), often forming strongly deformed or transposed compositional layering, forming a distinctive rock known as harzburgite tectonite. In some ophiolites, harzburgite overlies lherzolite. The harzburgite is generally interpreted to be the depleted mantle from which overlying mafic rocks were derived, and the deformation is related to the overlying lithospheric sequence flowing away from the ridge along a shear zone within the harzburgite. The harzburgite sequence may be six miles (10 km) or more thick in some ophiolites, such as the Semail ophiolite in Oman, and the Bay of Islands ophiolite in Newfoundland.

Resting above the harzburgite is a group of rocks that were crystallized from a magma derived by partial melting of the harzburgite. The lowest unit of these crustal rocks includes crystal cumulates of pyroxene and olivine, forming distinctive layers of pyroxenite, dunite, and other olivine + clinopyroxene + orthopyroxene peridotites, including wehrlite, websterite, and pods of chromite + olivine. The boundary between these rocks (derived by partial melting and crystal fractionation) and those below from which melts were extracted is one of the most fundamental boundaries in the crust, known as the Moho, or base of the crust, named after the Yugoslavian seismologist Andrija Mohorovičić, who noted a fundamental seismic boundary beneath the continental crust. In this case, the Moho is a chemical boundary, without a sharp seismic discontinuity. A seismic discontinuity occurs about 1,500–1,600 feet (half a kilometer) higher than the chemical Moho in ophiolites.

The layered ultramafic cumulates grade upward into a transition zone of interlayered pyroxenite and plagioclase-rich cumulates, then into an approximately half-mile (1 km) thick unit of strongly layered gabbro. Individual layers within this thin unit may include gabbro, pyroxenite, and anorthosite. The layered gabbro is succeeded upward by one to three miles (2–5 km) of isotropic gabbro, which is generally structureless but may have a faint layering. The layers within the isotropic gabbro in some ophiolites define a curving trajectory, interpreted to represent crystallization along the walls of a paleomagma chamber. The upper part of the gabbro may contain many xenoliths of diabase and pods of trondhjemite (plagioclase plus quartz) and may be cut by diabase dikes.

The next highest unit in a complete, Penrose-style ophiolite is typically a sheeted dike complex, consisting of a 0.3–1.25-mile- (0.5–2 km-) thick complex of diabasic, gabbroic, to silicic dikes that show mutually intrusive relationships with the underlying gabbro. In ideal cases, each diabase dike intrudes into the center of the previously intruded dike, forming a sequence of dikes that have chilled margins developed only on one side. These dikes are said to exhibit one-way chilling. In most real ophiolites, examples of one-way chilling may be found, but statistically the one-way chilling may only show directional preference in 50–60 percent of cases.

The sheeted dikes represent magma conduits that fed basaltic flows at the surface. These flows are typically pillowed, with lobes and tubes of basalt forming bulbous shapes distinctive of underwater basaltic volcanism. The pillow basalt section is typically 0.3–0.6 mile (0.5–1 km) thick. Chert and sulfide minerals commonly fill the interstices between pillows.

Deep-sea sediments, including chert, red clay, in some cases carbonates, or sulfide layers, overlie many ophiolites. Many variations are possible, depending on tectonic setting (e.g., conglomerates may form in some settings) and age (e.g., siliceous biogenic oozes and limestones would not form in Archean ophiolites, before the life-forms that contribute their bodies developed).

PROCESSES OF OPHIOLITE AND OCEANIC-CRUST FORMATION

The sequences of rock types described above result from a specific set of processes that occurred along the oceanic-spreading centers where the ophiolites formed. As the mantle convects and the asthenosphere upwells beneath mid-ocean ridges, the mantle harzburgites undergo partial melting of 10–15 percent in response to the decreasing pressure. The melts derived from the harzburgites rise to form a magma chamber beneath the ridge, forming the crustal section of the oceanic crust. As the magma crystallizes, the densest crystals gravitationally settle to the bottom of the magma chamber, forming layers of ultramafic and higher mafic cumulate rocks. Above the cumulate a gabbroic fossil magma chamber forms, typically with layers defined by varying amounts of pyroxene and feldspar crystals. In many examples the layering in ophiolites is parallel to the fossil margins of the magma chamber. An interesting aspect of the magma chamber is that periodically, new magma is injected into the chamber, changing the chemical and physical dynamics. These new magmas are injected during extension of the crust so the magma chamber may effectively expand infinitely if the magma supply is continuous, as in fast-spreading ridges. In slow-spreading ridges the magma chamber may completely crystallize before new batches of melt are injected.

As extension occurs in the oceanic crust, dikes of magma shoot out of the gabbroic magma chamber, forming a diabasic (fine-grained, rapidly cooled magma with the same composition as gabbro) sheeted dike complex. The dikes tend to intrude along the weakest, least crystallized part of the previous dike, which is usually in the center of the last dike to intrude. In this way, each dike intrudes the center of the previous dike, forming a sheeted dike complex characterized by dikes that have only one chill margin, most of which face in the same direction.

Many of the dikes reach the surface of the seafloor where they feed basaltic lava flows. Basaltic lava flows on the seafloor are typically in the form of bulbous pillows that stretch out of magma tubes, forming the distinctive pillow lava section of ophiolites. Seafloor metamorphism typically alters the top of the pillow lava section, including having deposits of black smoker-type hydrothermal vents. Sediments deposited on the seafloor overlie the pillow lavas. If the oceanic crust forms above the calcium carbonate compensation depth, the lowermost sediments may be calcareous. These would be succeeded by siliceous oozes, pelagic shales, and other sediments as the seafloor cools, subsides, and moves away from the mid-ocean ridge. A third sequence of sediments may be found on the ophiolites. These would include sediments shed during detachment of the ophiolite from the seafloor basement and its thrusting or emplacement onto the continental margin.

The type of sediments deposited on ophiolites may have been very different in some of the oldest ophiolites that formed in the Precambrian.

For instance, in the Proterozoic and especially the Archean, organisms that produce the carbonate and siliceous oozes would not be present, as the organisms that produced these sediments had not yet evolved.

There is considerable variation in the classical ophiolite sequence described above, as first formally defined by the participants of a Penrose conference on ophiolites in 1972. First, because most ophiolite sequences are deformed and metamorphosed, recognizing many of the primary magmatic units, especially sheeted dikes, is difficult. Deformation associated with emplacement typically causes omission of some or several sections of the complete sequence, and repetition of others along thrust faults. Therefore the adjectives *metamorphosed, partial,* and *dismembered* often serve as prefixes to descriptions of individual ophiolites. The thickness of individual units also varies considerably—some may be totally absent, and different units may be present in specific examples. Similar variations are noted from the modern seafloor and island arc systems, likely settings for the formation of ophiolites. Most ophiolites are interpreted to be fragments of the ocean floor generated at mid-ocean ridges, but the thickness of the modern oceanic crustal section is about 4 miles (7 km), whereas the equivalent units in ophiolites average about 1.8–3.1 miles (3–5 km).

Some of the variations relate to the variety of tectonic environments in which ophiolites form. Results from the Ocean Drilling Program, in which the oceanic crust has been drilled in a number of locations, have helped geologists to determine which units in what thickness are present in different sections of oceanic crust. Fast-spreading centers such as the East Pacific Rise typically show the complete ophiolite sequence, whereas slow-spreading centers such as the Mid-Atlantic Ridge may be incomplete, in some cases entirely lacking the magmatic section. Other ophiolites may form at or near transform faults, in island arcs, back-arc basins, forearcs, or above plumes.

THE WORLD'S OLDEST OPHIOLITE

Prior to 2001, no complete Phanerozoic-like ophiolite sequences had been recognized from Archean rock sequences around the world, leading some workers to the conclusion that no Archean ophiolites or oceanic crustal fragments are preserved. These ideas were challenged by the discovery of a complete 2.5 billion-year-old ophiolite sequence in the North China craton. This remarkable rock sequence includes chert and pillow lava, a sheeted dike complex, gabbro and layered gabbro, cumulate ultramafic rocks, and a suite of strongly deformed mantle harzburgite tectonites, all complexly deformed in a series of fault blocks. The mantle rocks include a distinctive type of intrusion with metallic chrome nodules called a podiform chromite deposit, known to form only in oceanic crust.

Well-preserved black smoker chimney structures in metallic sulfide deposits have also been discovered in some sections of the Dongwanzi ophiolite belt, and these ancient seafloor hydrothermal vents are among the oldest known. Deep-sea hydrothermal vents host the most primitive thermophyllic, chemosynthetic, sulfate-reducing organisms known, believed to be the closest relatives of the oldest life on Earth, with similar vents having possibly provided nutrients and protected environments for the first organisms. These vents are associated with some unusual microscale textures that may be remnants of early life-forms, most likely bacteria. These ancient fossils provide tantalizing suggestions that early life may have developed and remained sheltered in deep-sea hydrothermal vents until surface conditions became favorable for organisms to inhabit the land.

Archean oceanic crust was possibly thicker than Proterozoic and Phanerozoic counterparts, resulting in accretion predominantly of the upper basaltic section of oceanic crust. The crustal thickness of Archean oceanic crust may in fact have resembled modern oceanic plateaus. If this were the case, complete Phanerozoic-like ophiolite sequences would have been very unlikely to be accreted or obducted during Archean orogenies. In contrast, only the upper, pillow lava–dominated sections would likely be accreted. Remarkably, Archean greenstone belts contain an abundance of tectonic slivers of pillow lavas, gabbros, and associated deep-water sedimentary rocks. The observation that Archean greenstone belts have such an abundance of accreted ophiolitic fragments compared to Phanerozoic orogens suggests that thick, relatively buoyant, young Archean oceanic lithosphere may have had a rheological structure favoring delamination of the uppermost parts during subduction and collisional events.

See also AFRICAN GEOLOGY; ARABIAN GEOLOGY; ASIAN GEOLOGY; CONVERGENT PLATE MARGIN PROCESSES; DIVERGENT PLATE MARGIN PROCESSES; OROGENY.

FURTHER READING

Anonymous. "Ophiolites." *Geotimes* 17 (1972): 24–25.
Dewey, John F., and John M. Bird. "Origin and Emplacement of the Ophiolite Suite: Appalachian Ophiolites in Newfoundland, in Plate Tectonics." *Journal of Geophysical Research* 76 (1971): 3,179–3,206.
Kusky, T. M., ed. *Precambrian Ophiolites and Related Rocks. Developments in Precambrian Geology* Vol. 13. Amsterdam: Elsevier Publishers, 2003.
Kusky, Timothy M., Jianghai Li, and Robert T. Tucker. "The Archean Dongwanzi Ophiolite Complex, North

China Craton: 2.505 Billion Year Old Oceanic Crust and Mantle." *Science* 292 (2001): 1,142–1,145.

Moores, Eldridge M. "Origin and Emplacement of Ophiolites." *Review Geophysics* 20 (1982): 735–750.

U.S. Geological Survey. "This Dynamic Earth: The Story of Plate Tectonics." Available online. URL: http://pubs.usgs.gov/gip/dynamic/dynamic.html. Last modified March 27, 2007.

Ordovician   The Ordovician is the second period of the Paleozoic Era and refers to the corresponding rock series, falling between the Cambrian and the Silurian. Commonly referred to as the age of marine invertebrates, the base of the Ordovician is defined on the Geological Society of America time scale (1999) as 490 million years ago, and the top or end of the Ordovician is defined at 444 million years ago. Charles Lapworth named the period, in 1879, after the Ordovices, a Celtic tribe that inhabited the Arenig-Bala area of northern Wales, where rocks of this series are well exposed.

By the Early Ordovician, North America had broken away from the supercontinent of Gondwana that amalgamated during the latest Precambrian and early Cambrian Period. It was surrounded by shallow water passive margins, and being at equatorial latitudes, these shallow seas were well suited for the proliferation of marine life-forms. The Iapetus Ocean separated what is now the east coast of North America from the African and South American segments of the remaining parts of Gondwana. By the Middle Ordovician, convergent tectonics brought an island arc system to the North American margin, initiating the Taconic orogeny as an arc/continent collision. This was followed by a sideways sweep of parts of Gondwana past the North American margin, leaving fragments of Gondwana attached to the modified eastern margin of North America.

During much of the Ordovician, carbonate sediments produced by intense organic productivity covered shallow epeiric seas in the tropical regions, including most of North America. This dramatic increase in carbonate sedimentation reflects a combination of tectonic activities that brought many low-lying continental fragments into the Tropics, high–sea level stands related to the breakup of Gondwana, and a sudden increase in the number of different organisms that started to use calcium carbonate to build their skeleton or shell structures.

Marine life included diverse forms of articulate brachiopods, communities of echinoderms such as the crinoids or sea lilies, and reef-building stro-

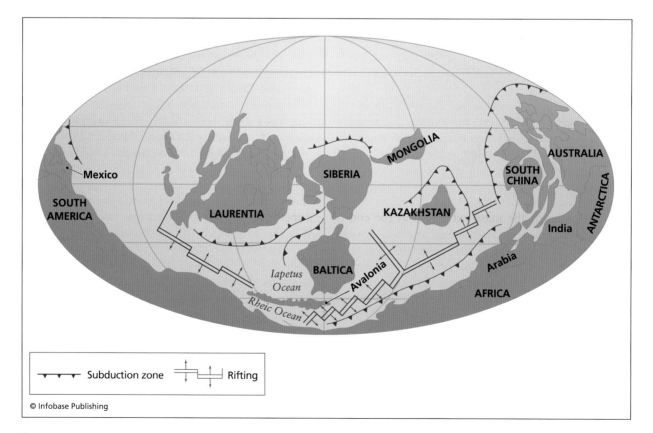

Ordovician paleogeography showing the distribution of continents approximately 500 million years ago

Two Ordovician trilobite fossils from the Wolchow River in Russia. Left is *Cerauinella ingrica* (12.5 cm long), right is *Pseudobasilica lawrowi* *(François Gohier/Photo Researchers, Inc.)*

matoporoids, rugose and tabulate corals. Trilobites roamed the shallow seafloors, and many forms emerged. The Ordovician saw rapid diversification and wide distribution of several planktonic and pelagic faunas, especially the graptolites and conodonts, which form useful index fossils for this period. Nautiloids floated across the oceans and some attained remarkably large sizes, reaching up to more than 10 feet (several meters) across. Fish fossils are not common from Ordovician deposits, but some primitive armored types may have been present. The end of the Ordovician is marked by a marine extinction event, apparently caused by rapid cooling of the shallow seas, perhaps related to continental glaciation induced by tectonic plate movements. The end-Ordovician extinction is one of the greatest of all Phanerozoic time. About half of all species of brachiopods and bryozoans died off, and more than 100 other families of marine organisms disappeared forever.

The cause of the mass extinction at the end of the Ordovician appears to have been largely tectonic. The major landmass of Gondwana had been resting in equatorial regions for much of the Middle Ordovician, but migrated toward the South Pole at the end of the Ordovician. This caused global cooling and glaciation, lowering sea levels from the high stand they had been resting at for most of the Cambrian and Ordovician. The combination of cold climates with lower sea levels, leading to a loss of shallow shelf environments for habitation, probably was enough to cause the mass extinction at the end of the Ordovician.

See also NORTH AMERICAN GEOLOGY; PALEOZOIC.

FURTHER READING
Condie, Kent C., and Robert E. Sloan. *Origin and Evolution of Earth: Principles of Historical Geology.* Upper Saddle River, N.J.: Prentice Hall, 1997.
Geological Society of America. Geologic Time Scale. Available online, URL: http://www.geosociety.org/science/timescale/timescl.htm. Accessed January 25, 2009.

origin and evolution of the Earth and solar system Understanding the origin of the Earth, planets, Sun, and other bodies in the solar system is a fundamental yet complex problem that has intrigued scientists and philosophers for centuries. Most of the records from the earliest history of the Earth have been lost to tectonic reworking and erosion, so most information about the formation of the Earth and solar system comes from the study of meteorites, the Earth's Moon, and observations of the other planets and interstellar gas clouds. In addition, isotope geochemistry can be used to understand some of the conditions on the early Earth.

The solar system displays many general trends with increasing distance from the Sun, and systematic changes like these imply that the Sun did not gravitationally capture planets, but rather the Sun and planets formed from a single event that occurred about 4.6 billion years ago. The nebular theory for the origin of the solar system suggests that a large spinning cloud of dust and gas formed and began to collapse under its own gravitational attraction. As it collapsed, it began to spin faster to conserve angular momentum (much like ice skaters spin faster when they pull their arms in to their chests) and eventually formed a disk. Collisions between particles in the disk formed protoplanets and a protosun, which then had larger gravitational fields than surrounding particles and began to sweep up and accrete loose particles.

The condensation theory states that particles of interstellar dust (many of which formed in older supernovas) act as condensation nucleii that grow through accretion of other particles to form small planetesimals that then have a greater gravitational field that attracts and accretes other planetesimals and dust. Some collisions cause accretion, other collisions are hard and cause fragmentation and breaking up of the colliding bodies. The Jovian planets became so large that their gravitational fields were able to attract and accrete even free hydrogen and helium in the solar nebula.

This condensation theory explains the main differences between the planets due to distance from the Sun, since the temperature of the solar nebula would have decreased away from the center where the Sun formed. The temperature determines which materials condense out of the nebula, so the composition of the planets was determined by the temperature at their position of formation in the nebula. The inner terrestrial planets are made of rocky and metallic material because high temperatures near the center of the nebula allowed only the rocky and metallic material to condense from the nebula. Farther out, water and ammonia ices also condensed out of the nebula, because temperatures were cooler at greater distances from the early Sun.

Early in the evolution of the solar system, the Sun was in a T-Tauri stage and possessed a strong solar wind that blew away most gases from the solar nebula, including the early atmospheres of the inner planets. Gravitational dynamics moved many of the early planetesimals into orbits in the Oort Cloud, where most comets and many meteorites are found. Some of these bodies have eccentric orbits that occasionally bring them into the inner solar system, and collisions with comets and smaller molecules likely brought the present atmospheres and oceans to Earth and the other terrestrial planets. Thus air and water, some of the basic building blocks of life, were added to the planet after it formed, being thrown in from deep space of the Oort Cloud.

The *Hadean* is the term used for the first of the four major eons of geological time: the Hadean, Archean, Proterozoic, and Phanerozoic. Some time classification schemes use an alternative division of early time, in which the Hadean is considered the earliest part of the Archean. As the earliest phase of Earth's evolution, ranging from accretion until approximately the age of first rocks [4.55 to 4.0 Ga (Ga = giga annee, or 10^9 years)], the Hadean is the most poorly understood interval of geologic time. Only a few mineral grains and rocks have been recognized from this Eon, so most of what is thought to be known about the Hadean is based on indirect geochemical evidence, meteorites, and models.

Between 4.55 and 3.8 Ga, meteorites bombarded the Earth; some were large enough to severely disrupt the surface, vaporize the atmosphere and ocean, and even melt parts of the mantle. By about 4.5 Ga, it appears as if a giant impactor, about the size of Mars, hit the protoearth. This impact ejected a huge amount of material into orbit around the protoearth, and some undoubtedly escaped. The impact probably also formed a new magma ocean, vaporized the early atmosphere and ocean (if present), and changed the angular momentum of the Earth as it spins and orbits

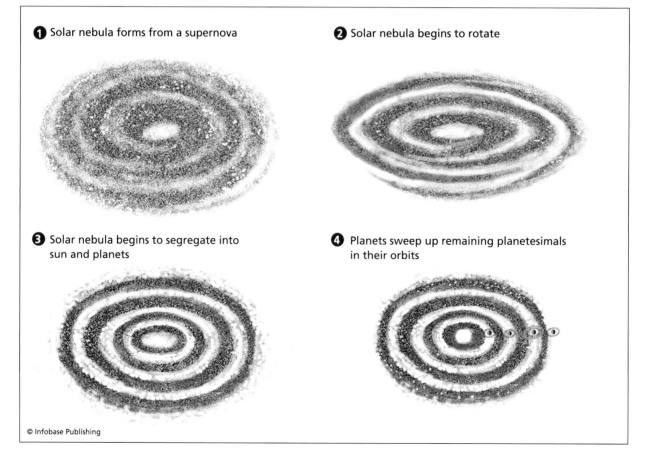

❶ Solar nebula forms from a supernova

❷ Solar nebula begins to rotate

❸ Solar nebula begins to segregate into sun and planets

❹ Planets sweep up remaining planetesimals in their orbits

© Infobase Publishing

Formation of the solar system from condensation and collapse of a solar nebula

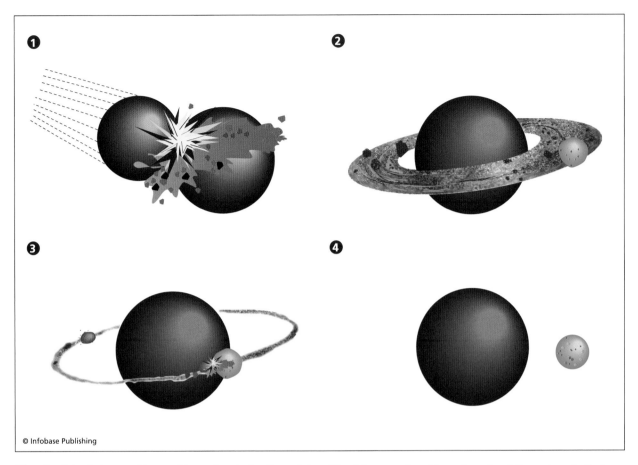

Sketch of the late great impact hypothesis for the origin of the Moon and melting of parts of the outer shell of the Earth early in its evolution

the Sun. The material in orbit coalesced to form the Moon, and the Earth-Moon system was born. Although not certain, this impact model for the origin of the Moon is the most widely accepted hypothesis, and it explains many divergent observations.

- The Moon's orbit is inclined by 5.1° from the ecliptic plane, whereas the Earth's orbit is inclined 23.4° from the ecliptic, suggesting that some force, such as a collision, disrupted the angular momentum and rotational parameters of the Earth-Moon system.
- The Moon is retreating from the Earth, resulting in a lengthening of the day by 15 seconds per year, but the Moon has not been closer to the Earth than 149,129 miles (240,000 km).
- The Moon is significantly less dense than the Earth and other terrestrial planets, being depleted in iron and enriched in aluminum, titanium, and other related elements.
- The oxygen isotopes of igneous rocks from the Moon are the same as from the Earth's mantle, suggesting a common origin.

These relationships suggest that the Moon did not form by accretion from the solar nebula at its present location in the solar system. The age of the Moon rocks shows that it formed at 4.5 Ga, with some magmatism continuing until 3.1 Ga, consistent with the impactor hypothesis.

The atmosphere and oceans of the Earth probably formed from early degassing of the interior by volcanism within the first 50 million years of Earth history. It is likely that the present atmosphere is secondary, in that the first or primary atmosphere would have been vaporized by the late great impact that formed the Moon, if it survived being blown away by an intense solar wind when the Sun was in a T-Tauri stage of evolution. The primary atmosphere would have been composed of gases left over from accretion, including primarily hydrogen, helium, methane, and ammonia, along with nitrogen, argon, and neon. The fact that the atmosphere has much less than the expected amount of these elements and is quite depleted in these volatile elements relative to the Sun suggests that the primary atmosphere has been lost to space.

Gases are presently escaping from the Earth during volcanic eruptions and also being released by weathering of surface rocks. Degassing of the mantle by volcanic eruptions, and perhaps also cometary impact, produced the secondary atmosphere. Gases released from volcanic eruptions include nitrogen (N), sulfur (S), carbon dioxide (CO_2), and water vapor (H_2O), closely matching the suite of volatiles that compose the present atmosphere and oceans. Most models show that little or no free oxygen was present in the early atmosphere, as oxygen was not produced until later, by photosynthetic life.

The early atmosphere was dense, with water vapor (H_2O), carbon dioxide (CO_2), sulfur (S), nitrogen (N), and hydrochloric acid (HCl). The mixture of gases in the early atmosphere would have caused greenhouse conditions similar to those presently existing on Venus. Since the early Sun during the Hadean era was approximately 25 percent less luminous than today, the atmospheric greenhouse served to keep temperatures close to their present range, where water is stable, and life can form and exist. As the Earth cooled, water vapor condensed to make rain that chemically weathered igneous crust, making sediments. Gases dissolved in the rain produced acids, including carbonic acid (H_2CO_3), nitric acid (HNO_3), sulfuric acid (H_2SO_4), and hydrochloric acid (HCl). These acids were neutralized by minerals (which are bases) that became sediments, and chemical cycling began. These waters plus dissolved components became the early hydrosphere, and chemical reactions gradually began changing the composition of the atmosphere, getting close to the dawn of life.

Speculation about the origin of life on Earth is of great intellectual interest. In the context of the Hadean, when life most likely arose, scientists are forced to consider different options for the initial trigger of life. Life could have come to Earth on late accreting planetesimals (comets) as complex organic compounds, or perhaps it came from interplanetary dust. If true, this would show how life got to Earth, but not how, when, where, or why it originated. Biological evidence supports the origination of life on Earth, in the deep sea near a hydrothermal vent or in shallow pools with the right chemical mixture. To start, life probably needed an energy source, such as lightning, or perhaps submarine hydrothermal vents, to convert simple organic compounds into building blocks of life—ribonucleic acids (RNA) and amino acids.

See also ARCHEAN; EARTH; LIFE'S ORIGINS AND EARLY EVOLUTION; METEOR, METEORITE; PLATE TECTONICS; SOLAR SYSTEM.

FURTHER READING

Chaisson, Eric, and Steve McMillan. *Astronomy Today.* 6th ed. Upper Saddle River, N.J.: Addison-Wesley, 2007.

Cloud, Preston. *Oasis in Space.* New York: W.W. Norton, 1988.

Comins, Neil F. *Discovering the Universe.* 8th ed. New York: W. H. Freeman, 2008.

Condie, Kent C., and Robert E. Sloan. *Origin and Evolution of Earth, Principles of Historical Geology.* Upper Saddle River, N.J.: Prentice Hall, 1997.

National Aeronautics and Space Administration. Goddard Space Flight Center Astronomical Data Center Quick Reference Page, Interstellar Medium (ISM) Web page. Available online. URL: http://adc.gsfc.nasa.gov/adc/quick_ref/ref_ism.html. Last updated April 30, 2002.

Schopf, J. William. *Cradle of Life: The Discovery of Earth's Earliest Fossils.* Princeton, N.J.: Princeton University Press, 1999.

Snow, Theodore P. *Essentials of the Dynamic Universe: An Introduction to Astronomy.* 4th ed. St. Paul, Minn.: West Publishing Company, 1991.

origin and evolution of the universe   One of the deepest questions in cosmology, the science that deals with the study and origin of the universe and everything in it, relates to how the universe came into existence. Cosmologists estimate that the universe is 10–20 billion years old and consists of a huge number of stars grouped in galaxies, clusters of galaxies, and superclusters of galaxies, surrounded by vast distances of open space. The universe is thought to be expanding because measurements show that the most distant galaxies, quasars, and other objects in the universe are moving away from each other and from the center of the universe.

THE BIG BANG

The big bang is one of several theoretical beginning moments of the universe. The big bang theory states that the expanding universe originated 10–20 billion years ago in a single explosive event in which the entire universe suddenly exploded out of a single, minuscule, infinitely dense and hot particle, reaching a pea-sized supercondensed state with a temperature of 10 billion million million degrees Celsius in 1 million-million-million-million-million-millionth (10^{-36}) of a second after the big bang. Some of the fundamental contributions of the expanding universe models come from the German-born American physicist Albert Einstein, who in 1915 proposed the general theory of relativity, which described how matter and energy warp space-time to produce gravity. When Einstein applied his theory to the universe in 1917, he discovered that gravity would cause the universe to be unstable and collapse, so he proposed adding a cosmological constant as a "fudge factor" to his equations. The cosmological constant added a repulsive force to the general theory, and this force

Milky Way Galaxy. Astronomers and cosmologists must estimate the amount of matter, dark matter, and energy in the galaxies and the interstellar medium to test different models for the evolution of the universe *(Sander van Sinttruye, 2008, used under license from Shutterstock, Inc.)*

counterbalanced gravity, enabling the universe to continue expanding in his equations. Dutch mathematician, physicist, and astronomer William de Sitter (1872–1934) further applied Einstein's theory of general relativity to predict that the universe is expanding. In 1927 Belgian priest and astronomer Georges Lemaître (1894–1966) proposed that the universe originated in a giant explosion of a primeval atom, an event now called the big bang. In 1929 American astronomer Edwin Hubble (1889–1953) measured the movement of distant galaxies and discovered that galaxies are moving away from each other, expanding the universe as if the universe is being propelled from a big bang. This idea of expansion from an explosion negated the need for Einstein's cosmological constant, which he retracted, referring to it as his biggest blunder. This retraction, however, would later come back to haunt cosmologists.

Also in the 1920s, the Russian physicist and cosmologist from Odessa, George Gamow (1904–68), worked with a group of scientists and suggested that elements heavier than hydrogen, specifically helium and lithium, could be produced in thermonuclear reactions during the big bang. Later, in 1957, English astronomer Fred Hoyle (1915–2001) and his colleagues, American astrophysicist William Fowler (1911–55) and American physicists Geoff Burbidge (1925–) and British-born Margaret Burbidge (1919–), showed how hydrogen and helium could be processed in stars to produce heavier elements necessary to life, such as carbon, oxygen, and iron.

The inflationary theory is a modification of the big bang theory, and suggests that the universe underwent a period of rapid expansion immediately after the big bang. This theory, proposed in 1980 by American physicist Alan Guth (1947–), attempts to explain the present distribution of galaxies, as well as the 3 K cosmic background radiation discovered by American physicists and Nobel laureates Arno Penzias (1933–) and Robert Woodrow Wilson (1936–) in 1965. This uniformly distributed radiation is thought to be a relic left over from the initial explosion of the big bang. For many years after the discovery of the cosmic background radiation, astronomers searched for answers to the amount of mass in the universe and to determine how fast the universe was expanding and to what degree the gravitational attraction of bodies in the universe was

THE LARGE HADRON COLLIDER

Physicists go to extreme lengths to solve some of the deepest mysteries of the universe. The Large Hadron Collider (LHC) is a huge, 16.7-mile- (27-km-) long ring containing 9,300 superconducting magnets buried 109 yards (100 m) underground near Geneva, crossing the border between Switzerland and France. It is the world's largest particle accelerator (and the largest machine of any type in the world), designed to study the smallest known particles that are the building blocks of all things. The LHC has the potential to revolutionize the scientific understanding of subatomic particle physics and the large-scale structure of the universe. Each magnet must be cooled to 1.9 K (-456.3°F; -271.3°C) using liquid nitrogen and liquid helium before the accelerator is operated.

The collider works by taking two beams of particles, called hadrons, made either of proton or lead ions, and accelerates them toward each other in opposite directions around the collider in an ultra-high vacuum. The particles can be made to travel around the hadron collider's loop many times to gain velocity, then they are aimed at each other so they collide. The two hadrons collide at extremely high energy, recreating conditions similar to those occurring just after the big bang. The experiments will be designed and the results monitored by scientific teams from around the world. When the particle accelerator is run at full power, trillions of protons will race around the track 11,245 times each second, travelling at 99.99 percent of the speed of light. Nearly 600 million collisions will take place every second, with energy of 14 Tev (teraelectronvolt). Temperatures generated at the collision points will be more than 100,000 times hotter than the center of the Sun, whereas the cooling rings surrounding the accelerator ring will be colder than outer space.

The particle detection system on the LHC contains the largest and most sophisticated detectors ever built, able to record the results of over 600 million proton collisions per second with micron (1-millionth of a meter) precision. The electronic trigger systems measure time for the passage of a particle to within a few billionths of a second, and the location of the particles is known to within a few millionths of a yard (m). These detectors feed the information to the most powerful scientific supercomputer in the world, including tens of thousands of computers located around the world in a distributed computing network called the Grid. The amount of data produced by experiments on the LHC will amount to about 1 percent of the world's information production rate, so a vast computer network is required to analyze and store the data.

The LHC ran into some problems that delayed its initial testing for many months, but it should shed new light on many unanswered questions in cosmology and the early history of the universe, as well as on fundamental particle physics. Some experiments are designed to determine why some particles have mass, while others apparently do not.

---

inhibiting the expansion. A relatively high density of matter in the universe would cause it to decelerate and eventually collapse back upon itself, forming a "big crunch," perhaps to be followed by a new big bang. Cosmologists called this the closed universe model. A low-density universe would expand forever, forming what cosmologists called an open universe. In between these end member models was a "flat" universe that would expand ever more slowly until it froze in place.

An alternative theory to the big bang is known as the steady state theory, in which the universe is thought to exist in a perpetual state with no beginning or end, with matter continuously being created and destroyed. The steady state theory does not adequately account for the cosmic background radiation. For many years cosmologists argued, almost religiously, whether the big bang theory or the steady state theory better explained the origin and fate of the universe. More recently, with the introduction of new high-powered instruments such as the Hubble space telescope, the Keck Mirror Array, and supercomputers, many cosmology theories have seen a convergence of opinion. A new, so-called standard model of the universe has been advanced and is currently being refined to reflect this convergence of opinion.

STANDARD MODEL OF THE UNIVERSE
In the standard model for the universe, the big bang occurred 14 billion years ago and marked the beginning of the universe. The cause and reasons for the big bang are not part of the theory, but left for the fields of religion and philosophy. Interestingly, ancient Jewish kabbalistic writings ranging from 700–1,500 years old concluded that the universe expanded from an object the size of a pea, roughly 15.34 billion years ago. Dr. William Percival of the University of Edinburgh leads a group of standard model cosmologists, and they calculate that the big bang occurred 13.89 billion years ago, plus or minus half a billion years. Most of the matter of

Data collected using the LHC may also shed light on why particles and atoms have the mass that they do. The LHC scientists plan to perform experiments designed to shed light on the distribution of mass in the universe—perhaps most importantly, that of dark matter. Contemporary models of cosmology suggest that the visible universe is only about 4 percent of the total matter in the universe, and the other 96 percent is made up of dark matter and dark energy, both extremely difficult to detect and study. When some dark matter is captured LHC scientists hope to be among the first to examine its properties, and consider the implications it has for the way the universe behaves.

Antimatter is similar to matter but has the opposite charge. A positron is similar to an electron but has a positive instead of a negative charge. Models for the early history of the universe suggest that at the big bang, equal amounts of matter and antimatter were created. When matter and antimatter meet they annihilate each other, releasing energy but losing matter. Scientists believe that most matter-antimatter pairs annihilated each other during early collisions, leaving only a small amount of matter in the entire universe, that forms all the matter visible (and invisible) today. Some experiments at the LHC will be geared to examine the differences between matter and antimatter and to try to understand why a small preference for one to be preserved led to such an imbalance in the universe.

The first second of the universe was the most dramatic second in the history of time. In this moment the universe expanded from essentially nothing, from a cocktail of fundamental particles about the size of a pea, into a rapidly expanding universe that would have the dimensions of time and space and the properties of mass and velocity. In the present universe ordinary matter consists of atoms, containing a nucleus made of protons and neutrons, in turn made of quarks that are bound together by gluons. Presently the bond between quarks and gluons is very strong, but in the first seconds of the early universe conditions would have been too hot and energetic for the gluons to hold the quarks together. The LHC will be able to reproduce the conditions in the first microseconds of the universe, so that scientists can investigate the physical environment of this hot, high-energy mixture of quarks and gluons, called a quark-gluon plasma.

There are complicated relationships between space and time. As the German-born American physicist Albert Einstein theorized, the three dimensions of space are related to time, and intense gravity fields can warp the space-time continuum. Further theoretical work has proposed that there may be even more hidden dimensions of space. String theory suggests that there may be numerous other dimensions to space, but they are difficult to observe. However, it is possible that these other dimensions may become detectable at the high energy conditions that will be created in the Large Hadron Collider.

FURTHER READING

CERN. LHC faq: The Guide. Available online. URL: http://cdsmedia.cern.ch/img/ CERN-Brochure-2008-001-Eng.pdf. Accessed January 14, 2009.

European Organization for Nuclear Research. The Large Hadron Collider. Available online. URL: http://public.web.cern. ch/public/en/LHC/WhyLHC-en.html. Accessed January 14, 2009.

the universe is proposed to reside in huge invisible clouds of dark matter, thought to contain elementary particles left over from the big bang. Galaxies and stars reside in these huge clouds of matter and comprise a mere 4.8 percent of the matter in the universe. The dark matter forms 22.7 percent of the universe, leaving another 72.5 percent of the universe as nonmatter. At the time of the proposal of the standard model, this ambiguous dark matter had yet to be conclusively detected or identified. In 2002 the first-ever atoms of antimatter were captured and analyzed by scientific teams from CERN, the European Laboratory for Particle Physics. In October of 2008 CERN opened a new generation of instruments designed to test theories of particle physics, the Large Hadron Collider, which is the largest particle accelerator (and largest scientific instrument) in the world. Unfortunately, the instrument suffered some serious malfunctions on its initial opening, and experiments were not carried out until 2009.

Detailed observations of the cosmic background radiation by space-borne platforms such as NASA's *Cosmic Background Explorer (COBE)* that in 1992 revealed faint variations and structure in the background radiation, are consistent with an inflationary, expanding universe. Blotches and patterns in the background radiation reveal areas that may have been the seeds or spawning grounds for the origin of galaxies and clusters. Detailed measurements of this background radiation have revealed that the universe is best thought of as flat—however, the lack of sufficient observable matter to have a flat universe requires the existence of some invisible dark matter. These observations were further expanded in 2002 and 2004, when teams working with the Degree Angular Scale Interferometer (DASI) experiment reported directional differences (called polarizations) in the cosmic microwave background radiation dating from 450,000 years after the big bang. The astronomers were able to relate these directional differences to forces that led to the formation of galaxies and the overall structure of the universe today. These density differences are quantum effects that effectively seeded the early universe with proto-galaxies during the early inflation period, and their observation provides strong support for the standard model for the universe.

Recent measurements have shown that the rate of expansion of the universe seems to be increasing, which has led cosmologists to propose the presence of a dark energy that is presently largely unknown. This dark energy is thought to comprise the remaining 72.5 percent of the universe, and it is analogous to a repulsive force or antimatter. Recognition in 1998 that the universe is expanding at ever-increasing rates has toppled questions about open versus closed universe models and has drastically changed perceptions of the fate of the universe. Amazingly, the rate of acceleration of expansion is remarkably consistent with Einstein's abandoned cosmological constant. The expansion seems to be accelerating so rapidly that eventually the galaxies will be moving apart so fast that they will not be able to see each other and the universe will become dark. Other cosmologists argue that so little is known of dark matter and dark energy that it is difficult to predict how it will act in the future, and the fate of the universe is not determinable from present observations and knowledge.

Alan Guth and coworkers recently proposed modifications of the inflationary universe model. They suggested that the initial inflation of the universe, in its first few microseconds, can happen over and over again, forming an endless chain of universes, called multiverses by Dr. Martin Rees of Cambridge University. With these ideas, our 14 billion-year-old universe may be just one of many, with big bangs causing inflations of the perhaps infinite other universes. According to the theories of particle physics, it takes only about one ounce of primordial starting material to inflate to a universe like our own. The process of growing chains of bubble-like universes through multiple big bangs and inflationary events has been termed eternal inflation by Dr. Andrei Linde of Stanford University.

Cosmologists, astronomers, and physicists are searching for a grand unifying theory that is able to link Einstein's general relativity with quantum mechanics and new observations of our universe. One attempt at a grand unifying theory is the string theory, in which elementary particles are thought to be analogous to notes being played on strings vibrating in 10- or 11-dimensional space. A newer theory emerging is called M-theory (for membrane theory or matrix theory), in which various dimensional membranes including universes can interact and collide, setting off big bangs and expansions that could continue or alternate indefinitely.

Cosmology and the fate of theories like the big bang are undergoing rapid and fundamental changes in understanding, induced by new technologies, computing abilities, and philosophy and from the asking of new questions about creation of the universe. Although it is tempting to think of current theories as complete, perhaps with a few unanswered questions, history tells us that much can change with a few new observations, questions, or understanding.

See also ASTRONOMY; ASTROPHYSICS; BINARY STAR SYSTEMS; BLACK HOLES; COSMIC MICROWAVE BACKGROUND RADIATION; COSMOLOGY; GALAXIES; GALAXY CLUSTERS; PULSAR; QUASAR; RADIO GALAXIES; SUPERNOVA.

FURTHER READING

Chaisson, Eric, and Steve McMillan. *Astronomy Today.* 6th ed. Upper Saddle River, N.J.: Addison-Wesley, 2007.

Comins, Neil F. *Discovering the Universe.* 8th ed. New York: W. H. Freeman, 2008.

Encyclopedia of Astronomy and Astrophysics. CRC /Taylor and Francis Press. Available online. URL: http://eaa.crcpress.com/. Accessed October 24, 2008.

Krauskopf, Konrad, and Arthur Beiser. *The Physical Universe.* 12th ed. Boston, Mass.: WCB-McGraw Hill, 2008.

ScienceDaily: Astrophysics News. ScienceDaily LLC. Available online. URL: http://www.sciencedaily.com/news/space_time/astrophysics/. Accessed October 24, 2008.

orogeny The process of building mountains is known as orogeny. For several or tens of centuries, many early philosophers, theologians, and scientists going back at least as far as Francis Bacon (1561–1626) were formulating theories about the forces involved in uplifting and deforming mountains. By the middle 1800s, the processes involved in the formation of mountains became known as orogeny. Early ideas suggested that mountains were deformed and uplifted by magmatic intrusions, or reflected an overall contraction of the Earth with mountain belts representing cooling wrinkles as on a shriveled prune. In a classical work in 1875, Eduard Suess published *Die Entstehung der Alpen* (The origin of mountains) in which he argued that the pattern of mountain belts on the planet did not follow any regular pattern that would indicate global contraction, and he suggested that the mountain belts represented contraction between rigid blocks (now called cratons) and surrounding rocks on the margins of these massifs. However, he still believed that the main driving force was global contraction induced by cooling of the Earth. In Suess's last volume of *Das Antlitz* *der Erde* (The Face of the Earth, 1909), he admitted that the amount of shortening observed in mountain belts was greater than global cooling and contraction could explain, and he suggested that perhaps other translations, or movements of crustal blocks, have occurred in response to tidal forces and the rotation of the planet.

In 1915 and 1929 Alfred Wegener published *Die Entstehung der Kontinente* and *Die Entstehung der Kontinente und Ozeane* (The origin of the continents and the oceans). Wegener argued strongly for large horizontal motions between cratons made of sial (light continental crust), and using such data as the match of restored coastlines and paleontological data, he founded the theory of continental drift. Several geologists, including Alex du Toit, Reginald Daly, and Arthur Holmes documented geological ties between different continents supporting the idea of continental drift. In the 1940s–60s, geophysical exploration of the seafloor led to the recognition of seafloor spreading, and provided the data that J. Tuzo Wilson needed to propose the modern theory of plate tectonics in 1965.

With the development of the ideas of plate tectonics, geologists now recognize that mountain belts are of three basic types: fold-and-thrust belts, volcanic mountain ranges, and fault block ranges. Fold-and-thrust belts are contractional mountain belts, formed where two tectonic plates collided, forming great thrust faults, folds, metamorphic rocks, and volcanic rocks. Detailed mapping of the structure in the belt enables geologists to reconstruct their history, and essentially pull them back apart. Many of the rocks in fold-and-thrust belt types of mountain ranges were deposited on the bottom of the ocean, or continental rises, slopes, or shelves, or on ocean margin deltas. When the two plates collide, many of the sediments get scraped off and deformed, forming the mountain belts; thus, these belts mark places where oceans have closed. Volcanic mountain ranges include places such as Japan's Fuji and Mount St. Helens in the Cascades of the western United States. Volcanism associated with subduction and plate tectonics, rather than deformation, primarily forms these mountain ranges. Fault-block mountains, such as the Basin and Range Province of the western United States, are formed by the extension or pulling apart of the continental crust, forming elongate valleys separated by tilted fault-bounded mountain ranges.

See also CONTINENTAL DRIFT; CONVERGENT PLATE MARGIN PROCESSES; DIVERGENT PLATE MARGIN PROCESSES; PLATE TECTONICS.

FURTHER READING

Bacon, Francis. *Novum Organum* (The New Organon, or True Directions Concerning the Interpretation of Nature, translated by Basil Montague.) Philadelphia, Pa.: Parry and MacMillan, 1854.

Miyashiro, Akiho, Keiti Aki, and A. M. Celal Sengor. *Orogeny*. Chichester, England: John Wiley & Sons, 1982.

Suess, Eduard. *Das Antlitz der Erde* (The Face of the Earth, translated by Hertha B. C. Sollas.) Oxford; Clarendon Press, 1909.

———. *Die Entstehung der Alpen* (The origin of mountains). Ann. Acad. Sci. Fennicae, Ser. A, III. Geol.-Geogr., No. 97, 28 p., 1875.

Wegener, Alfred, *Die Entstehung der Kontinente* and *Die Entstehung der Kontinente und Ozeane* (The Origin of the Continents and the Oceans), 1915; 1929.

Wilson, J. Tuzo. "A New Class of Faults and Their Bearing on Continental Drift." *Nature* 207 (1965): 343–347.

ozone hole Ozone (O_3) is a poisonous gas that is present in trace amounts in much of the atmosphere, but reaches a maximum concentration in a stratospheric layer between 9 and 25 miles (15–40 km) above the Earth, with a peak at 15.5 miles (25 km). The presence of ozone in the stratosphere is essential for most life on Earth, since it absorbs the most carcinogenic part of the solar spectrum with wavelengths between 0.000011 and 0.0000124 inches (280 and 315 nm). If these ultraviolet rays reached the Earth they would cause many skin cancers, and possibly depress the human immune system. These harmful rays would greatly reduce photosynthesis in plants and reduce plant growth to such an extent that the global ecosystems would crash.

Ozone naturally changes its concentration in the stratosphere, and is also strongly affected by human or anthropogenically produced chemicals that make their way into the atmosphere and stratosphere. Ozone is produced by photochemical reactions above 15 miles (25 km), mostly near the equator, and moves toward the poles, where it is most abundant and where it is gradually destroyed. The concentration of ozone does not vary greatly in equatorial regions, but at the poles tends to be the greatest in the winter and early spring. The formation of a strong vortex that isolates the stratospheric air over the pole characterizes stratospheric circulation during the night in the Antarctic winter.

Atmospheric and stratospheric flow dynamics can change the distribution of ozone, solar flare and sunspot activity can enhance ozone, and volcanic eruptions can add sulfates to the stratosphere that destroy ozone. In the 1970s scientists realized that some aerosol chemicals and refrigerants, such as chlorofluorocarbons (CFCs), could make their way into the stratosphere, where ultraviolet light broke them down, releasing ozone-destroying chlorine.

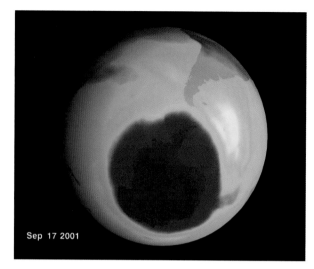

Satellite data shows the Antarctic ozone hole, 9/17/01. The hole is roughly the size of North America *(NASA Goddard Space Flight Center [NASA-GSFC])*

Sep 17 2001

Environmental regulations subsequently curtailed the use of CFCs, but the aerosols and chlorine have long residence times in the stratosphere, and each chlorine ion is capable of destroying large amounts of ozone. Since the middle 1980s, a large hole marked by large depletions of ozone in the stratosphere has been observed above Antarctica every spring, its growth aided by the polar vortex. The hole has continued to grow, but the relative contributions to the destruction of ozone by CFCs, other chemicals (such as supersonic jet and space shuttle fuel), volcanic gases, and natural fluctuations is uncertain. In 1999 the size of the Antarctic ozone hole was measured at more than 9,650,000 square miles (25 million km²), more than two and half times the size of Europe. The appearance of ozone depletion above Arctic regions has added credence to models that show the ozone depletion being largely caused by CFCs. Many models suggest that the CFCs may lead to a 5–20 percent reduction in global ozone, with consequent increases in cancers and disease and loss of crop and plant yield.

See also ATMOSPHERE; CLIMATE CHANGE; ELECTROMAGNETIC SPECTRUM; GLOBAL WARMING; METEOROLOGY.

FURTHER READING
Ahrens, C. Donald. *Meteorology Today: An Introduction to Weather, Climate, and the Environment.* 7th ed. Pacific Grove, Calif.: Thomson Brooks/Cole, 2003.

National Weather Service, National Oceanic and Atmospheric Administration, home page. Available online. URL: http://www.nws.noaa.gov/. Last modified September 15, 2008. Data updated continuously.

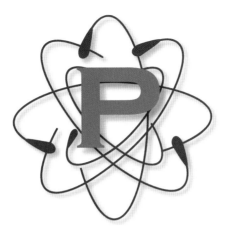

paleoclimatology Paleoclimatology is the study of past and ancient climates, their distribution and variation in space and time, and the mechanisms of long-term climate variations. A wide variety of different types of data are used to determine past climates, such as the distribution of certain plant and animal species that are climate sensitive and the distribution of certain rock types that form in restricted climate conditions. Other types of data serve as paleoclimate indicators, including tree-ring studies (dendrochronology), ice-core data, cave deposits (speleothems), and lake sediment studies. Increasingly, studies are using isotopic data, such as ratios between light and heavy oxygen isotopes, as paleoclimate indicators since these ratios are very sensitive to past global climates, glaciations, and elevations at which rainwater fell.

Most paleoclimate studies reveal that there have been major climate shifts on the planet throughout Earth history, with periods of near global glaciation, periods of intense heat and humidity, or hot and dry weather, and more temperate periods such as the current interglacial stage. Many factors play a role in climate change, including orbital and astronomical variations described by Milankovitch cycles, plate tectonics and the distribution of continental land masses, and volcanic productivity.

Paleoclimate studies have been used widely by scientists who study the past distribution of continents in supercontinents and continental drift. For instance, if a continental land mass moves equatorial regions to more polar regions, it will experience a progressive shift in the surface climates. Any rocks deposited during these different climates will reflect the climatic conditions that prevailed during that time period, and any plant or animal fossils in the rocks will reflect species that were able to survive under the prevailing climate conditions. Thus, studying the rock record reveals information not only about the past climates for a continental block, but also the history of climate zones that a plate or continental block moved through during its drift across the surface of the Earth.

See also CLIMATE; CLIMATE CHANGE; MILANKOVITCH CYCLES; PLATE TECTONICS; SUPERCONTINENT CYCLES.

FURTHER READING
Condie, Kent C., and Robert Sloan. *Origin and Evolution of Earth, Principles of Historical Geology.* Upper Saddle River, N.J.: Prentice Hall, 1997.

Windley, Brian F. *The Evolving Continents.* 3rd ed. Chichester, England: John Wiley & Sons, 1995.

Paleolithic The Paleolithic is the first division of the Stone Age in archaeological time, marked by the first appearance of humans and their associated tools and workings. The time of the Paleolithic corresponds generally with the Pleistocene (from 1.8 million years ago until 10,000 years ago) of the geological time scale, Earth's most recent period of repeated glaciations, but varies somewhat from place to place. The Paleolithic began with the first use of stone tools about 2.5 or 2.6 million years ago and ended with the introduction of agriculture at the end of the Pleistocene about 10,000 years ago. Artifacts that have survived from the Paleolithic are known as paleoliths.

Human technological prehistory is divided into three periods; the Stone Age, the Bronze Age, and the

Iron Age. Modern divisions of the Stone Age stretch from the Paleolithic to Neolithic according to the following scheme:

- Pleistocene epoch, a time of heavy glaciation
- Holocene epoch, a time of modern climate, divided into the Mesolithic, Epipaleolithic, and Neolithic ages, Copper Age, Bronze Age, and Iron Age, and
- the Historical period, from when modern records began.

During the Paleolithic a race of ancient hominids known as Neanderthals (also spelled Neandertals) inhabited much of Central Asia, the Middle East, Near East, western Siberia, and Europe, especially throughout the past 200,000 years during the Pleistocene ice ages. Neanderthals were heavily built and had large brains, and were probably adapted to the cold climate conditions in the periglacial environments they inhabited (i.e., they were probably hairy and fatty). These premodern humans were few in number, dwindled in numbers around 40,000 years ago, and disappeared into extinction by around 27,000 years ago.

The first remains of modern humans (*Homo sapiens sapiens*) are dated at around 30,000 to 40,000 years old and show that modern humans inhabited some of the same areas as the Neanderthals. The overlapping time and space ranges of Neanderthals and modern humans raises questions about the origins of modern humans and relationships between the two groups of hominids. Archaeologists debate whether the Neanderthals went extinct because they were hunted and killed by the modern humans, or whether perhaps climate conditions became unbearable to the Neanderthals and they simply died off. There is very little evidence about whether modern humans and Neanderthals lived in peace side by side, possibly interbreeding, or if they fought or avoided each other. Some questions relate to whether or not modern humans evolved from Neanderthals, if the two races of hominids are unrelated, or if they share common distant ancestors. A possible transitional form between Neanderthals and *Homo sapiens sapiens* has been described from Mount Carmel in Israel, although most genetic evidence so far suggests that the modern humans did not evolve from Neanderthals. Modern humans seem to have descended from a single African female that lived 200,000 years ago, although some theories suggest that humans evolved in different parts of the world at virtually the same time.

Understanding of the evolution of humans is a controversial and constantly changing field. Genetic studies show that humans and chimpanzees had a common ancestor that lived about 5–10 million years ago. The earliest known human ancestor is australopithecine, from 3.9–4.4-million-year-old hominids found in Ethiopia and Kenya. It is thought that australopithecine evolved into *Homo habilis* by 2 million years ago. *Homo habilis* was larger than australopithecine, walked upright, and was the first hominid to use stone tools. By 1.7 million years ago, *Homo erectus* appeared in Africa, probably evolving from *Homo habilis*. *Homo erectus* had prominent brow ridges, a flattened cranium and a rounded jawbone and was the first hominid to use fire and migrate out of Africa as far as China, Europe, and the British Isles. Modern humans (*Homo sapiens sapiens*) and Neanderthals (*Homo sapiens neanderthalensis*) are both probably descendants from *Homo erectus*.

Neanderthals were hunters and gatherers who roamed the plains, forests, and mountains of Europe and Eurasia. They left many stone tools, clothing, and possibly some art including cave drawings and sculptures. Around 40,000–30,000 years ago, the Neanderthals found their environments increasingly inhabited by modern humans, who had smaller, less robust skeletons and smaller brains and lacked many of the primitive traits that characterized earlier humans. Some early modern humans had some Neanderthal traits, but there is considerable debate in the anthropological community about whether this indicates an evolutionary trend, or more likely, that the two races interbred, producing mixed offspring. Whether the interbreeding was peaceful or a consequence of war and raids, there is no evidence that the mixed offspring successfully developed into a separate mixed race. The genetic and most archaeological evidence suggests that modern humans evolved separately, from a single African female, whose descendants came out of Africa and inhabited the Near East, Europe, and Asia.

These debates in the scientific community highlight two competing hypotheses for the origin of modern humans. The out-of-Africa theory follows the genetic evidence that modern humans arose about 200,000 years ago in Africa and spread outward, replacing older indigenous populations of Neanderthals and other hominids by 27,000 years ago. An opposing theory, called the multiregional evolution theory, argues that all modern humans are not descended from a single 200,000-year-old African ancestor. This model supposes that modern humans have older ancestors, such as *Homo erectus*, that spread out to Europe and Asia by 1 or 2 million years ago, then evolved into separate races of *Homo sapiens sapiens* independently in different parts of the world. There are many arguments against this theory of multiregional parallel evolution, the strongest of which notes the unlikelihood of the same evolutionary path being followed independently in several different places at the same time. However, the multiregionalists argue that the evolutionary advances were driven by similar technological and lifestyle advances and that many adjacent groups may have been interbreeding and thereby exchanging genetic material. The multiregionalists need to allow enough genetic exchange between regional groups to form an early worldwide-web dating or genetic exchange system, whereby enough traits are transmitted between groups to keep *Homo sapiens sapiens* the same species globally, but to keep enough isolation so that individual groups maintain certain distinctive traits.

The multiregionalists see a common ancient ancestor 1–2 million years ago, with different groups evolving to different degrees toward what we call modern humans. In contrast, the out-of-Africans see different branches from *Homo erectus* 200,000 years ago, with Neanderthals first moving into Eurasia and the Middle East, to be later replaced by migrating early modern humans that followed the migrating climate zones north with the retreat of the Pleistocene glaciers.

See also EVOLUTION; ICE AGES; NEOLITHIC; PLEISTOCENE.

FURTHER READING

Leonard, Jonathan N. *The First Farmers: The Emergence of Man.* New York: Time-Life Books, 1973.

Prothero, Donald R. *Bringing Fossils to Life: An Introduction to Paleobiology.* New York: McGraw-Hill, 2004.

paleomagnetism Paleomagnetism is the study of natural remnant magnetism in rocks with the goal of understanding the intensity and direction of the Earth's magnetic field in the geologic past and understanding the history of plate motion. The Earth's magnetic field can be divided into two components at any location—the declination and the inclination. The declination measures the angular difference between the Earth's rotational north pole and the magnetic north pole. The inclination measures the angle at which the magnetic field lines plunge into the Earth. The inclination is 90° at the magnetic poles and 0° halfway between the poles.

Studies of paleomagnetism in young rocks have revealed that the Earth's magnetic poles can flip suddenly, over a period of thousands or even hundreds of years. The magnetic poles also wander by about 10–20° around the rotational poles. On average, however, the magnetic poles coincide with the Earth's rotational poles. A researcher can use this coincidence to estimate the north-south directionality in ancient rocks that have drifted or rotated in response to plate tectonics. Determination of the natural remnant magnetism in rock samples can, under special circumstances, reveal the paleo-inclination and paleo-declination, which can be used to estimate the direction and distance to the pole at the time the rock acquired the magnetism. If these parameters can be determined for a number of rocks of different ages on a tectonic plate, then an apparent polar wander path for that plate can be constructed. These show how the magnetic pole has apparently wandered with respect to (artificially) holding the plate fixed—when the reference frame is switched, and the pole is held fixed, the apparent polar wander curve shows how the plate has drifted on the spherical Earth.

Paleomagnetism played an enormous role in the confirmation of seafloor spreading, through the discovery and understanding of seafloor magnetic anomalies. In the 1960s geophysicists surveyed the magnetic properties of the ocean floor and began to discover some amazing properties. The seafloor has a system of linear magnetic anomalies where one "stripe" has its magnetic minerals all oriented the same way as the present magnetic field, and the magnetic minerals in alternate stripes are oriented in the opposite direction. These stripes are oriented parallel to the mid-ocean ridge system; where transform faults offset the ridges, the anomalies are also offset. The anomalies are symmetric on either side of the ridge, and the same symmetry is found across ridges worldwide.

Understanding the origin of seafloor magnetic stripes was paramount in acceptance of the plate tectonic paradigm. The magnetic stripes form in the following way. As oceanic crust is continuously formed at mid-ocean ridges, as if it were being extruded from a conveyor belt, all the magnetic minerals tend to align with the present magnetic field when the new crust forms. The oceanic crust thus contains a record of when and for how long the Earth's magnetic field has been in the "normal" position, and when and for how long it has been "reversed." Terrestrial rock sequences exhibit similar reversals of the Earth's magnetic field, and many of these have been dated. Using these data, geologists have established a magnetic polarity reversal time scale. The last reversal was about 700,000 years ago, and the one before that, about 2.2 million years ago. Oceanic crust is as old as Jurassic, and documentation of the age of seafloor magnetic stripes has led to the construction

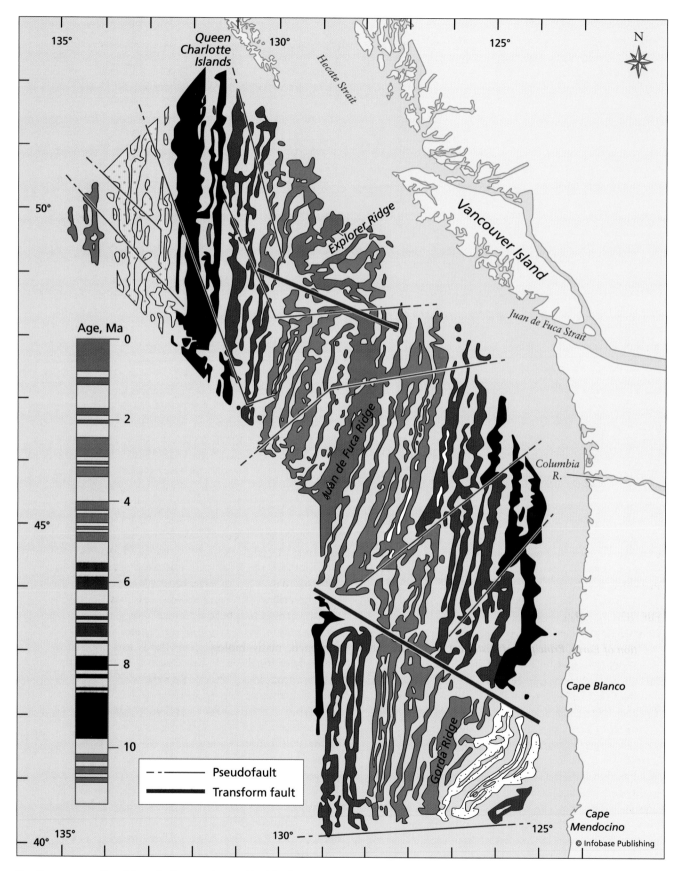

Seafloor magnetic stripes in the northeast Pacific Ocean produced by seafloor spreading on the Juan de Fuca, Gorda, and Explorer ridges

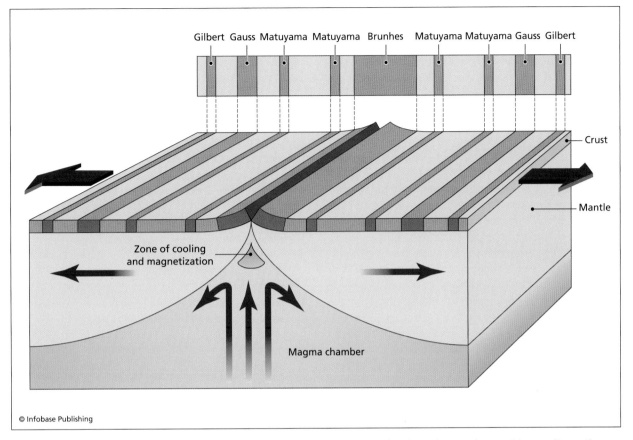

Gilbert Gauss Matuyama Matuyama Brunhes Matuyama Matuyama Gauss Gilbert

Crust

Mantle

Zone of cooling
and magnetization

Magma chamber

© Infobase Publishing

Symmetric magnetic anomalies produced by conveyor-belt style production of oceanic crust in an alternating magnetic field. The magnetic polarity timescale, above, is constructed by dating rocks with different magnetizations.

of the magnetic polarity time scale back to 170 million years ago.

See also MAGNETIC FIELD, MAGNETOSPHERE; PLATE TECTONICS.

FURTHER READING

Condie, Kent C., and Robert Sloan. *Origin and Evolution of Earth, Principles of Historical Geology.* Upper Saddle River, N.J.: Prentice Hall, 1997.

Moores, Eldridge M., and Robert Twiss. *Tectonics.* New York: W. H. Freeman, 1995.

Windley, Brian F. *The Evolving Continents.* 3rd ed. Chichester, England: John Wiley & Sons, 1995.

paleontology Paleontology is the study of past life based on fossil evidence, with a focus on the lines of descent of organisms and the relationships between life and other geological phenomena. Information from fossil distributions is used to understand ancient environments and climates and to determine the boundaries of former ties between landmasses that are now separated. Many paleontologists are also concerned with mechanisms of extinction and the appearance of new organisms, as well as the mode of life of organisms and their evolution. In the past, paleontology was mostly a descriptive science describing the morphology of fossils, but in recent years it has become much closer to biology, with a new science of geobiology emerging. In this approach, many biological methods are applied to the study of fossils, including cladistic methods, functional morphology, and even paleogenetic studies.

The preservation of the remains of organisms as fossils is a rare event and therefore represents unusual conditions, not necessarily representative of life at the time the organism lived. Despite this limitation, fossils are the best way to understand the past history of life on Earth. To correctly interpret the fossil record paleontologists must use statistical techniques to try to interpret the bias in the record and to understand the variations in the environment that may have led to some organisms being preserved while others were destroyed.

Organisms may leave traces of their former existence in body fossils, in which the body of the organism itself is preserved or replaced with other elements and preserved in shape, or as trace fossils, in which footprints, burrows, or other vestiges of the organism's having influenced the environment are left behind in the geological record. An organism may also leave evidence of its influence through geochemical tracers that can yield clues about the life-forms and how they interacted with their environment.

The science of paleontology experienced major changes in the 1970s and 1980s. In past centuries paleontology was mostly concerned with the taxonomy and anatomy of fossils, with tremendous efforts spent on describing and classifying small details of fossil morphology. The revolution in paleontology resulted from the mixture of taxonomy with biology, to form a new field of paleobiology in which the paleontologists asked new questions, such as how much of a preservational bias is there in the geological record, and what does the fossil record reveal about evolution. Some paleontologists began investigating the causes of mass extinctions, whereas others study the interactions of plate tectonics and the development and evolution of life in the field of paleobiogeography.

MAIN LIFE GROUPS

Metazoa are complex multicellular animals in which the cells are arranged in two layers in the embryonic gastrula stage. The Metazoa are extremely diverse and include 29 phyla, most of which are invertebrates. The phylum Chordata is an exception.

The Metazoa appeared about 620 million years ago and experienced a rapid explosion around the Precambrian-Cambrian boundary, probably associated with the formation and breakup of the supercontinents Rodinia and Gondwana and the rapidly changing environments associated with the supercontinent cycle. They evolved from eukaryotes, single-celled organisms with a nucleus, that appeared around 1,600 million years ago. Prokaryotes are older, probably extending back past 3,800 million years ago.

Some of the oldest soft-bodied Metazoa are remarkably well preserved in the Ediacarian fauna from southeast Australia and other locations around the world. These fauna include a remarkable group of very unusual shallow marine forms, including some giants up to a meter in length. This explosion from simple, small single-celled organisms that existed on Earth for the previous three billion years (or more) is truly remarkable. The Ediacarian fauna (and related fauna, collectively called the Vendoza fauna) died off after the period between 620–550 million years ago, as these organisms show no affinity with modern invertebrates.

After the Ediacarian and Vendoza fauna died off, other marine invertebrates saw a remarkable explosion through the Cambrian. These organisms in the Cambrian included shelly fossils, trilobites, brachiopods, mollusks, archeocyathids, and echinoderms, and eventually in the Ordovician were joined by crinoids and bryozoans. Modern Metazoa include a variety of organisms including corals, gastropods, bivalves, and echinoids.

See also CLOUD, PRESTON; EVOLUTION; FOSSIL; GRABAU, AMADEUS WILLIAM; HISTORICAL GEOLOGY; LIFE'S ORIGINS AND EARLY EVOLUTION; PHANEROZOIC.

FURTHER READING

McKinney, Michael L. *Evolution of Life, Processes, Patterns, and Prospects.* Englewood Cliffs, N.J.: Prentice Hall, 1993.

Prothero, Donald R. *Bringing Fossils to Life: An Introduction to Paleobiology.* New York: McGraw-Hill, 2004.

Raup, David M., and Steven M. Stanley. *Principles of Paleontology.* 3rd ed. San Francisco: W. H. Freeman, 2007.

Stanley, Steven M. *Earth and Life through Time.* New York: W. H. Freeman, 1986.

Paleozoic   The era of geologic time that includes the interval between 544 and 250 million years ago and the erathem of rocks deposited in this interval, called the Paleozoic, includes seven geological periods and systems of rocks: the Cambrian, Ordovician, Silurian, Devonian, Carboniferous (Mississippian and Pennsylvanian), and Permian. The name *Paleozoic*, meaning ancient life, was coined by British geologist Adam Sedgwick in 1838 for the deformed rocks underlying the Old Red Sandstone in Wales. The base of the Paleozoic is defined as the base of the Cambrian period, conventionally taken as the lowest occurrence of trilobites. Recently however, with the recognition of the advanced Vendian and Ediacaran fauna, the base of the Cambrian was reexamined and has been defined using fossiliferous sections in eastern Newfoundland and Siberia to be the base of an ash bed dated at 544 million years ago.

At the beginning of the Paleozoic, the recently formed supercontinent of Gondwana was breaking apart, but later regrouped as Pangaea by the Carboniferous. This supercontinent included the southern continents in the Gondwanan landmass and the northern continents in Laurasia, separated by the Pleionic and Tethys Oceans, and surrounded by the Panthalassa Ocean. With the breakup of the late Precambrian supercontinent, climates changed from icehouse to hothouse conditions. The volume of carbon dioxide (CO_2) emitted to the atmosphere by the mid-ocean ridge system caused this dramatic change. During supercontinent periods, the length of the ridge system is small, and relatively small amounts of CO_2 are emitted to the atmosphere. During supercontinent breakup however, much more CO_2 is released during enhanced volcanism associated with the formation of new ridge systems. Since CO_2 is a greenhouse gas, supercontinent breakup is associated with increasing temperatures and the establishment of hothouse conditions. The Pangaean supercontinent then experienced continental climates ranging from hot and dry to icehouse conditions, with huge continental ice sheets covering large parts of the southern continents. Many collisional and rifting events occurred, especially along the active margins of Pangaea, necessitating the subduction of huge tracts of oceanic crust in order to accommodate these collisional events.

The dramatic changes in continental configurations, the arrangement of ecological niches, and the huge climatic fluctuations at the base of the Paleozoic are associated with the most dramatic explosion of life in the history of the planet. Hard-shelled organisms first appeared in the lower Cambrian and are abundant in the fossil record by the mid-Cambrian. Fish first appeared in the Ordovician. All of the modern animal phyla and most of the plant kingdom are represented in the Paleozoic record, with fauna and flora inhabiting land, shallow seas, and deep-sea environments. There are several mass extinction events in the Paleozoic in which large numbers of species suddenly died off and were replaced by new species in similar ecological niches.

In addition to the development of hard-shelled organisms and skeletons, the Paleozoic saw the dramatic habitation of the terrestrial environment. Bacteria and algae crept into different environments such as soils before the Paleozoic, with land plants appearing in the Silurian. Dense terrestrial flora expanded by the Devonian and culminated in the dense forests of the Carboniferous. This profoundly changed the weathering, erosion, and sedimentation patterns from that of the Precambrian, and also significantly affected the atmosphere-ocean composition. Terrestrial fauna rapidly followed the plants onto land, with tetrapods roaming the continents by the Middle or Late Devonian. By the Devonian, invertebrates including spiders, scorpions, and cockroaches had invaded the land, and fish became abundant in the oceans.

In the Carboniferous much organic carbon got buried, reducing atmospheric CO_2 levels and ending the hothouse conditions. With the formation of Pangaea in the Carboniferous and Permian, global ridge lengths were reduced, and less CO_2 was released to the atmosphere. Together with the burial of organic carbon, new icehouse conditions were established, stressing the global fauna and flora. The largest mass extinction in geological history marks the end of the Paleozoic (end-Permian mass extinction) and the start of the Mesozoic. The causes of this dramatic event seem to be multifold. Conditions on the planet included the formation of a supercontinent (Pangaea), falling sea levels, evaporite formation, and rapidly fluctuating climatic conditions. At the boundary between the Paleozoic and Mesozoic Periods (245 million years ago), 96 percent of all species became extinct, including marine organisms such as the rugose corals, trilobites, many types of brachiopods, and many foraminifera species.

The Siberian flood basalts were erupted over a period of less than 1 million years 250 million years ago, at the end of the Permian at the Permian-Triassic boundary. They are remarkably coincident in time with the major Permian-Triassic extinction, implying a causal link. They cover a large area of the Central Siberian Plateau northwest of Lake Baikal and are more than half a mile thick over an area of 210,000 square miles (544,000 km^2), but they have been significantly eroded from an estimated volume of 1,240,000 cubic miles (5,168,500 km^3). The rapid volcanism and degassing could have released enough sulfur dioxide to cause a rapid global cooling, inducing a short ice age with an associated rapid fall of sea level. Soon after the ice age took hold the effects of the carbon dioxide took over and the atmosphere heated, resulting in a global warming. The rapidly fluctuating climate postulated to have been caused by the volcanic gases is thought to have killed off many organisms, which were simply unable to cope with the wildly fluctuating climate extremes.

Another possibility is that the impact of a meteorite or asteroid with the Earth aided the end-Permian extinction, adding environmental stresses to an already extremely stressed ecosystem. If additional research proves this to be correct, it will be shown that a one-two-three punch, including changes in plate configurations and environmental niches, dramatic climate changes, and extraterrestrial impacts together caused history's greatest calamity.

See also CAMBRIAN; CARBONIFEROUS; DEVONIAN; EVOLUTION; FOSSIL: GONDWANA, GONDWANALAND; PANGAEA; PERMIAN; SILURIAN.

FURTHER READING
Prothero, Donald R., and Robert H. Dott. *Evolution of the Earth.* 6th ed. Boston, Mass.: McGraw-Hill, 2002.

Windley, Brian F. *The Evolving Continents.* 3rd ed. Chichester, England: John Wiley & Sons, 1995.

Pangaea Pangaea was the supercontinent that formed in the Late Paleozoic, lasting from about 300–200 million years ago, and included most of the planet's continental masses. The former existence of Pangaea, meaning all land, was first postulated by Alfred Wegener in 1924, when he added the Australian and Antarctic landmasses to an 1885 supercontinent reconstruction of Gondwana by Eduard Suess that included Africa, India, Madagascar, and South America. He used the fit of the shapes of the coastlines of the now dispersed continental fragments, together with features such as mineral belts, faunal and floral belts, mountain ranges, and paleoclimate zones that matched across his reconstructed Pangaean landmass to support the hypothesis that the continents were formerly together. Wegener proposed that the supercontinent first broke up into two large fragments including Laurasia in the north and Gondwana in the south, and then continued breaking up, leading to the present distribution of continents and oceans. The scientific community did not generally accept Wegener's ideas at first, but since the discovery of seafloor magnetic anomalies and the plate tectonic revolution, the general framework of his Pangaea model has proved to be generally valid.

The Pangaean supercontinent began amalgamating from different continental fragments with the collision of Gondwana and Laurentia and Baltica in the Middle Carboniferous, resulting in the Alleghenian, Mauritanide, and Variscan orogenies. Final assembly of Pangaea involved the collision of the South China and Cimmerian blocks to the Paleo-Tethyan margin, resulting in the early Yanshanian and Indonesinian orogenies in the Middle to Late Triassic.

The formation of Pangaea is associated with global climate change and rapid biological evolution. The numerous collisions caused an overall thickening of the continental crust that decreased continental land area and resulted in a lowering of sea level. The uplift and rapid erosion of many carbonate rocks that had been deposited on trailing or passive margins caused a decrease in the carbonate strontium 87/strontium 86 ratios in the ocean. During the final stages of the coalescence of Pangaea, drainage systems were largely internal, erosion rates were high, and the climate, with large parts of the supercontinent lying between 15° and 30° latitude, became arid, with widespread redbed deposition. Soon however, the effects of the erosion and burial of large amounts of carbonate and the associated drawdown of atmospheric CO_2 caused climates to rapidly cool, resulting in high-latitude glaciations.

The main glaciations of Pangaea started in the Late Devonian and Early Carboniferous, began escalating in intensity by 333 million years ago, peaked in the Late Carboniferous by 292 million years ago, and ended in the Early Permian by 272 million years ago. These glaciations resulted in major global regressions as the continental ice sheets used much of the water on the planet. Wegener, and many geologists since, used the distribution of Pangaean glacial deposits as one of the main lines of reasoning to support the idea of continental drift. If the glacial deposits of similar age are plotted on a map of the present distribution of the continents, the ice flow patterns indicate that the oceans too must have been covered. However, the planet does not hold enough water to make ice sheets

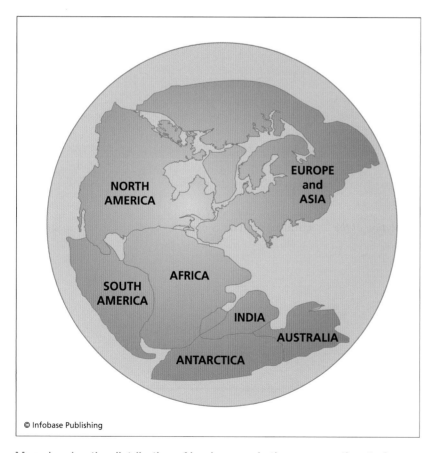

Map showing the distribution of landmasses in the supercontinent of Pangaea

so large that they can cover the entire area required if the continents have not moved. If the glacial deposits are plotted on a map of Pangaea however, they cover a much smaller area, the ice flow directions are seen to radiate outward from depocenters, and the amount of water on Earth can accommodate the total volume of ice.

Pangaea began rifting and breaking apart about 230 million years ago, with numerous continental rifts, flood basalts, and mafic dikes intruding into the continental crust. Major breakup and seafloor spreading began about 175 million years ago in the central Atlantic, when North and South America broke away from Pangaea, 165 million years ago off Somalia, and 160 million years ago off the coast of northwest Australia. Sea levels began to rise with breakup because of the increase in volume of the mid-ocean ridges that displaced seawater onto the continents, forming marine transgressions. Episodic transgressions and evaporation of seawater from restricted basins led to the deposition of thick salts in parts of the Atlantic, with some salt deposits reaching 1.2 miles (2 km) thick off the east coast of North America, Spain, and northwest Africa. Several rifts along these margins have up to 6 miles (10 kilometers) of nonmarine sandstones, shales, red beds, and volcanics associated with the breakup of Pangaea. Many are very fossiliferous, including plants, mudcracks, and even dinosaur footprints attesting to the shallow water nature of these deposits.

Breakup of the supercontinent was also associated with a dramatic climate change and high sea levels. The increased volcanism at the oceanic ridges released many gases to the atmosphere, inducing global warming, leading to a global greenhouse, ideal for carbonate production on passive or trailing continental margins.

See also PALEOZOIC; SUPERCONTINENT CYCLES.

FURTHER READING
Prothero, Donald R., and Robert H. Dott. *Evolution of the Earth.* 6th ed. Boston, Mass.: McGraw-Hill, 2002.
Rogers, J. J. W., and M. Santosh. *Continents and Supercontinents.* Oxford: Oxford University Press, 2004.
Windley, Brian F. *The Evolving Continents.* 3rd ed. Chichester, England: John Wiley & Sons, 1995.

passive margin Continental margins that are attached to the adjacent oceanic crust and do not have a plate boundary along the margin are known as passive or trailing margins. Most parts of the east coasts of North and South America, the west coasts of Europe and Africa, and most of the coastlines of India, Antarctica, and Australia are passive margins, many of which are characterized by thick accumula-

tions of marine carbonates, shales, and sandstones. Conditions for the formation, accumulation, and preservation of hydrocarbons are met along many passive margins, so contemporaneous and ancient passive margin sequences are the focus of intense petroleum exploration.

Trailing or passive margins typically develop from a continental rift and first form an immature passive margin, with the Red Sea being the main example present on the planet at this time. Rifting along the Red Sea began in earnest by 30 million years ago, separating Arabia from Africa. The Red Sea is characterized by uplifted rift shoulders that slope generally away from the interior of the sea, but have narrow down-dropped coastal plains where steep mountain fronts are drained by wadis (dry stream beds) with alluvial fans that form a typically narrow coastal plain. These have formed over stretched continental crust, forming many rotated fault blocks and grabens, intruded by mafic dike swarms that in some places feed extensive young volcanic fields, especially in Saudi Arabia. The center of the Red Sea, which is located in tropical to subtropical latitudes, has developed thick carbonate platforms along the stretched continental crust. As rifting and associated stretching of the continental crust proceeded, areas that were once above sea level subsided below sea level, but different parts of the Red Sea basin reached this point at different times. Together with global rises and falls of sea level, this led to episodic spilling of salty seawater into restricted basins which would then evaporate, leaving thick deposits of salt behind. As rifting continued these salts became buried beneath the carbonates, shales, and sandstones, but when salt gets buried deeply it rises buoyantly, forming salt domes that pierce overlying sediments. The movement of the salt forms broad open folds that in some places forms exceptionally good petroleum traps. Parts of the passive margins along the Red Sea are several kilometers thick and currently exhibit some of the world's best coral reefs. The center of the Red Sea is marked by steep slopes off the carbonate platforms, leading to the embryonic spreading center that is present only in southern parts of the sea. Abundant volcanism, hot black smoker vents, and active metalliferous and brine mineralization on the seafloor characterize this spreading center.

As spreading continues on passive margins, the embryonic or Red Sea stage gradually evolves into a young oceanic or mature passive margin stage where the topographic relief on the margins decreases and the ocean-to-passive-margin transition becomes very flat, forming wide coastal plains such as those along the east coast of North America. This transition is an important point in the evolution of passive margins, as it marks the change from rifting and heating of

618 pelagic, nektonic, planktonic

the lithosphere to drifting and cooling of the lithosphere. The cooling of the lithosphere beneath the passive margin leads to gradual subsidence, typically without the dramatic faulting that characterized the rifting and Red Sea stages of the margin's evolution. Volcanism wanes, and sedimentation on the margins evolves to exclude evaporites, favoring carbonates, mudstones, sandstones, and deltaic deposits. The overall thickness of passive margin sedimentary sequences can grow to 9 or even 12.5 miles (15–20 km), making passive margin deposits among the thickest found on Earth.

Most ancient passive margins have gone through stages of evolution similar to those described for the Red Sea, so they form a distinctive assemblage of rocks in the geological record. In general ancient passive margins can be recognized first, in that they are located on the flanks of cratons, continents, or microcontinents. The rocks of the passive margin typically overlie older continental crusts although some overlie other rock sequences such as rift deposits that record a geologic history of the margin prior to its development as a passive margin. The passive margin sequence has a geometry where initial rift phase deposits are overlain by a seaward thickening and seaward deepening wedge of sedimentary rocks, typically grading from a sandy shore facies, to an offshore muddy facies, and in cases where the climate permitted, to a carbonate platform. The passive margin sequence is flanked by deep water facies rocks that, during collisional orogenesis, may be thrust on top of the shallow water sediments of the passive margin. In some cases these thrust sheets contain slivers or large thrust sheets of oceanic crust and lithosphere known as ophiolites.

The oldest known well-preserved passive margin is the Steep Rock Lake belt, which formed between 3.0 and 2.7 billion years ago in the Superior Province of Canada, although many smaller and disrupted candidates exist elsewhere in the world including a 2.9–2.7 billion-year-old passive margin in the North China craton, and a 2.7 billion-year-old margin in Zimbabwe. Passive margins show a cyclic or episodic distribution of peak abundances through geological time, with the most abundant extant passive margins found at 1950, 550, and 0 million years ago, corresponding to times of supercontinent dispersal. The opposite is also true—times where there were relatively few passive margins, such as 1750 and 300 million years ago, correspond to times of supercontinent formation.

The present day total length of passive margins on the Earth is 64,500 miles (104,000 km), and these have an average age of 104 million years and a maximum age of 180 million years. Studies of ancient passive margins by U.S. Geological Survey geologist

Dwight Bradley show that they have a mean life span of 187 million years and a maximum life span of 550 million years. Divided by age, the mean life span changes from 182 million years for the Archean and Paleoproterozoic, 211 million years for the Neoproterozoic, 145 million years for the Cambrian to Carboniferous, and 142 million years for the Carboniferous to present. Overall there is a general trend for passive margins to have a longer life span with younger geological ages, suggesting that the rate of plate tectonic processes may have been faster in Precambrian times.

See also AFRICAN GEOLOGY; ASIAN GEOLOGY; CONTINENTAL MARGIN; DIVERGENT PLATE MARGIN PROCESSES; HYDROCARBONS AND FOSSIL FUELS; PLATE TECTONICS; SUBSIDENCE.

FURTHER READING

Bradley, Dwight C. "Passive Margins through Earth History." *Earth Science Reviews* 91 (2008): 1–26.
Kusky, Timothy M., and Peter J. Hudleston. "Growth and Demise of an Archean Carbonate Platform, Steep Rock Lake, Ontario Canada." *Canadian Journal of Earth Sciences* 36 (1999): 1–20.
Kusky, Timothy M., and Jianghai Li. "Paleoproterozoic Tectonic Evolution of the North China Craton." *Journal of Asian Earth Sciences* 22, no. 4 (2003): 383–397.
Kusky, Timothy M., and Pamela A. Winsky. "Structural Relationships along a Greenstone/Shallow Water Shelf Contact, Belingwe Greenstone Belt, Zimbabwe." *Tectonics* 14, no. 2 (1995): 448–471.

pelagic, nektonic, planktonic The pelagic environment includes the open ocean, inhabited by free-floating planktonic organisms and free-swimming nektonic organisms. Sediments that are deposited in an open ocean environment are said to be pelagic and largely consist of the remains of free-floating plankton that sink to the seafloor upon death.

Plankton, the bodies of aquatic organisms that float, drift freely, or swim weakly, includes a large variety of species in the marine realm: bacteria, phytoplankton (one-celled plantlike organisms), and zooplankton that are tiny animals such as jellyfish and invertebrates, as well as numerous nonmarine aquatic species. Planktonic species are contrasted with nektonic organisms, which are strong swimmers, and benthic organisms, which are bottom dwellers.

Planktonic species tend to be small and without strong skeletons, and they utilize the density of surrounding water to support their dominantly water-filled bodies. Many types sink or float to specific depths, where light and salinity characteristics meet their needs. They move vertically by changing the

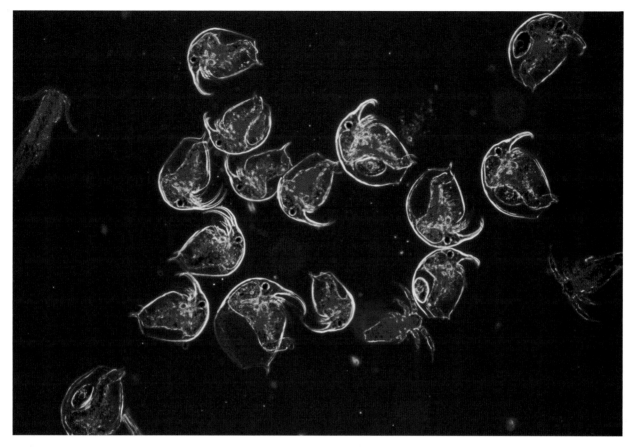

Color-enhanced micrograph of the crustacean zooplankton *Bosmina longicornis* *(Christian Gautier/Photo Researchers, Inc.)*

amount of air in their bodies, thus getting the nutrients they require and avoiding becoming food for predators. Other plankton utilize their transparency or live in large schools of similar organisms to avoid being eaten.

Phytoplankton are microscopic floating photosynthetic organisms that form an extremely important part of the biomass and food chain. Examples include diatoms, a type of algae that secretes walls of silica; cyanobacteria, photosynthetic bacteria that have been on the Earth for at least 3.8 billion years; and dinoflagellates, flagellate protists that exhibit characteristics of both plants and animals. Coccolithophores are one-celled, floating plants covered with an armor of small calcareous plates. Silicoflagellates are similar but have plates made of silica.

Zooplankton comprises a huge variety of protozoans and small metazoans that exhibit a wide range of temperature and salinity tolerances. Some zooplankton are holoplanktonic, meaning they remain free-floating throughout their lives. In addition to phytoplankton, this group includes zooplankton such as the extremely important foraminifera that produce calcium carbonate tests and radiolaria that produce silica tests, as well as tunicates, tiny jellyfish, and copepods. Other zooplankton are considered meroplankton, meaning they spend only part of their lives as plankton, then they join either the benthic or nektonic realm. Meroplankton are common in coastal waters and include most fish eggs and the eggs and larvae of other marine animals such as marine worms or crustaceans (arthropods with stiff, chitinous outer shells or skeletons including lobsters, shrimp, crabs, and krill).

Gelatinous plankton such as jellyfish include the siphonophores that paralyze prey with stinging cells made of barbs attached to poison sacs. The siphonophores are colonies of animals that live together but function as a single animal. Ctenophores resemble jellyfish and have trailing tentacles, used to trap prey. They are carnivorous and may occur in large swarms, greatly reducing local populations of crustaceans and small or young fish. Tunicates are primitive planktonic creatures with backbones inside a barrel-shaped gelatinous structure.

Nekton are pelagic animals that move through the water primarily by swimming. Nektons are distinguished from other pelagic organisms (plankton) that float in the water. The most important nektons in the water today are the fish, whereas in the Paleozoic several other forms were common. The ammonoids of the Devonian were coiled cephalopod mollusks that evolved from earlier nautilids, and these existed with the free-swimming, scorpion-like eurypterids. Fish first appeared in the marine record in the Cambrian-Ordovician, and included the early bony-skinned fish known as ostracoderms, followed in the Late Silurian by the finned acanthodians. Heavily armored large-jawed fish known as placoderms are found in many Late Devonian deposits, as are lung fish, ray-finned fish, and lobe-finned fish that include the coelacanths, one species that survives to this day. Lobe-finned fish are the ancestors of all terrestrial vertebrates. Sharks were very common in the marine realm by the Late Paleozoic.

See also BENTHIC, BENTHOS; BIOSPHERE; FOSSIL; OCEAN BASIN; PASSIVE MARGIN.

FURTHER READING

Prothero, Donald R. *Bringing Fossils to Life: An Introduction to Paleobiology.* New York: McGraw-Hill, 2004.

Permian Permian refers to the last period in the Paleozoic era, lasting from 290–248 million years ago, and to the corresponding system of rocks. Sir Roderick Murchison named it in 1841 after the Perm region of northern Russia, where rocks of this age were first studied in detail. The supercontinent of Pangaea included most of the planet's landmasses during the Permian. This continental landmass extended from the South Pole, across the equator to high northern latitudes, with a wide Tethys Sea forming an open wedge of water near the equator. The Siberian continental block collided with Laurasia in the Permian, forming the Ural Mountains. Most of Pangaea was influenced by hot and dry climate conditions and saw the formation of continental red-bed deposits and large-scale, cross-bedded sandstones, such as the Coconino sandstone of the southwestern United States and the New Red sandstone of the United Kingdom. Ice sheets covered the south-polar region, amplifying already low sea levels so they fell below the continental shelves, causing widespread mass extinctions. The glaciations continued to grow in intensity through the Permian, and together with weathering of continental calsilicates, were able to draw enough CO_2 out of the atmosphere to drastically lower global temperatures. This dramatic climate change enhanced the already widespread extinctions, killing off many species of corals, brachiopods, ammonoids, and forams in one

Fossil reptile *Orobates pabsti* Diadectomorpha Diadectidea from early Permian period. Found in the Bromacker, Germany, area *(Phil Degginger/Carnegie Museum/Alamy)*

of history's greatest mass extinctions, in which about 70–90 percent of all marine invertebrate species perished, as did large numbers of land mammals. This greatest catastrophe of Earth history did not have a single cause, but reflects the combination of various elements.

The complex multi-factor cause of the end-Permian mass extinction is explained as follows. Plate tectonics was bringing many of the planet's landmasses together in a supercontinent (Pangaea), causing greater competition for fewer environmental niches by Permian life-forms. Drastically reduced were the rich continental shelf areas. As the continents collided mountains were pushed up, reducing the effective volume of the continents available to displace the sea, so sea levels fell, putting additional stress on life by further limiting the availability of favorable environmental niches. The global climate became dry and dusty, and the supercontinent formation led to widespread glaciation. This lowered sea level even more, lowered global temperatures, and put many life-forms on the planet in a very uncomfortable position, and many perished.

In the final million years of the Permian, the northern Siberian plains became the site of one of the largest volcanic effusions in Earth history. The Siberian flood basalts began erupting at 250 million years ago, becoming the largest known outpouring of continental flood basalts ever. Carbon dioxide was released in hitherto unknown abundance, warming the atmosphere and melting the glaciers. Other gases were also released, perhaps including methane, as the basalts probably melted permafrost and vaporized thick deposits of organic matter that accumulated in high latitudes like that at which Siberia was located 250 million years ago.

The global biosphere collapsed, and evidence suggests that the final collapse happened in less than 200,000 years, and perhaps in less than 30,000 years. According to this scenario, entirely internal processes may have caused the end-Permian extinction, although some scientists now argue that an impact may have dealt the final death blow. After it was over, new life-forms populated the seas and land, and these Mesozoic organisms tended to be more mobile and adept than their Paleozoic counterparts. The great Permian extinction created opportunities for new life-forms to occupy now empty niches, and the most adaptable and efficient organisms took control. The toughest of the marine organisms survived, and a new class of land animals grew to new proportions and occupied the land and skies. The Mesozoic, time of the great dinosaurs, had begun.

See also EVOLUTION; FOSSIL; MASS EXTINCTIONS; PANGAEA; STRATIGRAPHY, STRATIFICATION, CYCLOTHEM; SUPERCONTINENT CYCLES.

FURTHER READING

Albritton, C. C., Jr. *Catastrophic Episodes in Earth History.* London: Chapman and Hale, 1989.

Eldredge, N. *Fossils: The Evolution and Extinction of Species.* Princeton, N.J.: Princeton University Press, 1997.

MacDougall, J. D., ed. *Continental Flood Basalts.* Dordrecht, Germany: Kluwer Academic Publishers, 1988.

Mahoney, J. J., and M. F. Coffin, eds. *Large Igneous Provinces, Continental, Oceanic, and Planetary Flood Volcanism.* Washington, D.C.: American Geophysical Union, 1997.

Prothero, Donald R., and Robert H. Dott. *Evolution of the Earth.* 6th ed. Boston, Mass.: McGraw-Hill, 2002.

Stanley, Steven M. *Earth and Life through Time.* New York: W. H. Freeman, 1986.

———. *Extinction.* New York: Scientific American Library, 1987.

Windley, Brian F. *The Evolving Continents.* 3rd ed. Chichester, England: John Wiley & Sons, 1995.

petroleum geology   Petroleum is a mixture of different types of hydrocarbons (fossil fuels) derived from the decomposed remains of plants and animals that are trapped in sediment, and can be used as fuel. The petroleum group of hydrocarbons includes crude oil, natural gas, and gas condensate. When plants and animals are alive, they absorb energy from the Sun (directly through photosynthesis in plants, and indirectly through consumption in animals) to make complex organic molecules. After these plants and other organisms die, if they become and remain buried before they decay, they may be converted into hydrocarbons and other fossil fuels.

Crude oil and natural gas may become concentrated in some regions and become minable for use under some special conditions. First, for oil and gas to form, more organic matter must be produced than is destroyed by scavengers and organic decay, conditions that are met in relatively few places. One of the best places for oil and gas to form is on offshore continental shelves, passive margins, or carbonate platforms, where organic productivity is high and the oxygen content of bottom waters is low, so organic decay is low and inadequate to break down the amount of organic material produced. Organic material may also be buried before it decays in sufficient quantities to make petroleum in some deltaic and continental rise environments.

Once the organic material is buried, it must reach a narrow window of specific pressure and temperature conditions to make petroleum. If these temperatures and pressures are not met or are exceeded, petroleum will not form or will be destroyed. When organic-rich rocks are in this petroleum window of

RENEWABLE ENERGY OPTIONS FOR THE FUTURE

Advanced societies require sources of external energy for transportation, heating, ventilation, production of goods, and comfort. As the world's population rapidly increases and the available supplies of economically, technologically, and environmentally extractable hydrocarbons diminishes, it is imperative that nations begin to seriously invest in developing alternative energy sources and strategies. Increased efficiency and conservation of presently known hydrocarbon reserves will help postpone the depletion of this resource, but it will eventually run out or become exceedingly expensive. To maintain levels of comfort and production in the world's civilizations, acceleration of the research and development of sustainable energy-harvesting techniques is necessary.

Primary energy sources are those that contain energy in a form (high potential) that enables them to be converted directly to lower forms of energy that are directly usable by people. These include fossil fuels, nuclear energy, and renewable resources such as biofuels, geothermal energy, hydroelectricity, solar power, tidal power, and wind power. Fossil fuels will eventually be depleted, cause significant pollution, disrupt the environment in their extraction, and emit greenhouse gases. Nuclear fuels are limited in abundance since the known resources of uranium will be depleted in dozens (uranium 235) to thousands (uranium 238) of years, and the generation of nuclear power, albeit efficient and clean, is plagued with the problem of how to dispose of the radioactive wastes. There are alternative energy sources that are renewable, clean, and efficient.

Biomass fuels involve using garbage, corn, or other vegetables to generate electricity. When garbage decomposes it generates methane that can be captured and then burned to produce electricity. Burning garbage can generate energy and alleviate land-use stress from landfills, but it also generates air pollution similar to that of burning fossil fuels. Biofuel such as vegetable oil is produced from sunlight and carbon dioxide (CO_2) by plants. It can be modified to burn like diesel fuel and is safer than gasoline since it has a higher flash point. However, the amount of plant matter required to generate biofuels is large and competes with land use for farming to feed people and livestock, driving up food prices. It also takes a significant amount of fuel to plant, fertilize, and grow biofuel stock, so some argue that there is no net gain in the biofuel economy.

Geothermal energy taps heat resources that are close to the surface in some regions of Earth, such as near volcanic centers. Less efficient geothermal sources are present in most places. Two wells are drilled in the ground; water is injected into one, and it becomes heated by the hot rocks at depth. The water is then extracted, or naturally shoots up as steam, from the other well. The energy from the steam can drive turbines to create electricity, and the heat from the water can be used as an energy source for things like home heating. Geothermal energy is cheap and generally clean, but is most efficient in specific locations where the underlying geology places hot material close to the surface.

Hydroelectricity uses the gravitational potential energy of a river by means of a dam, flume, or tunnel, using the pressure from water at high elevation to pass through a turbine or waterwheel to drive

---

specific temperature and pressure, organic-rich beds known as source rocks become compacted, and the organic material undergoes chemical reactions to form hydrocarbons including oil and gas. These fluids and gases have a lower density than surrounding rocks and a lower density than water, so they tend to migrate upward until they escape at the surface or are trapped between impermeable layers, where they may form a petroleum reservoir.

Oil traps are of many varieties, divided into mainly structural and stratigraphic types. Structural traps include anticlines, where the beds of rocks are folded into an upward arching dome. In these types of traps, petroleum in a permeable layer that is confined between impermeable layers (such as a sandstone bed between shale layers) may migrate up to the top of the anticlinal dome, where it becomes trapped. If a fault cuts across beds, it may form a barrier or it may act as a conduit for oil to escape, depending on the physical properties of the rock in the fault zone. In many cases faults juxtapose an oil-bearing permeable unit against an impermeable horizon, forming a structural trap. Salt domes in many places form diapirs that pierce through oil-bearing stratigraphic horizons. They typically cause an upwarping of the rock beds around the dome, forming a sort of anticlinal trap that in many regions has yielded large volumes of petroleum. Stratigraphic traps are found mainly where two impermeable layers such as shales are found above and below a lens-shaped sandstone unit that pinches out laterally to form a wedge-shaped trap. These conditions are commonly met along passive margins, where transgressions and regressions of the sea cause sand

electric generators or mills. Hydroelectric plants are clean, in many cases water can be stored to be used at peak demand, and there are many undeveloped places with high potential to generate new hydroelectric power. However, construction of dams on rivers seriously affects the river dynamics and ecology, so the future of hydroelectricity may lie mostly in situations where underground tunnels can be made that utilize the gravitational potential energy, but do not seriously disrupt the river environment.

Solar power uses solar cells to convert sunlight into electricity, utilizing the most steady source of energy in the solar system. Sunlight can heat water or air in solar panels, create steam using parabolic mirrors, or be used in a more passive way, utilizing the light entering windows to heat buildings. Solar power is most efficient in places where the solar radiation is the highest. Places like the sunny desert southwestern United States are much more suited for solar power than the Pacific Northwest, which is shrouded in clouds many days of the year. Solar panels operate at different efficiencies, depending on the sunlight conditions and their construction. Assuming a 20 percent efficiency, all of the energy needs of the United States could be met if an area the size of Arizona were covered

with solar panels. Solar energy requires a complementary energy storage system, since it is not available at night or during cloudy weather.

Tidal power can be harnessed by building impoundment dams that capture water at high tide and slowly release the water through turbines during low tide. Water turbines can also be located in areas with strong tidal currents, with most of the infrastructure located underwater. These turbines can be connected to electrical generators or to gas compressors, which then store the energy as compressed gas that can be slowly released to drive turbines as needed to generate electricity.

Wind power captures the energy of the wind by placing large wind turbines in persistently windy locations. The wind turns the blades of the turbines that cause the rotation of magnets that in turn generate electricity. Some locations are prone to steady and strong winds, and these locations can be the sites of wind farms where many tall towers with large blades are set up to harness the wind's power. This source of energy is clean, produces no chemical or air pollution, and is renewable. The blades of the turbines are high off the ground, so wind farms on land can function along with the primary use of the land, such as grazing or farm-

ing. Wind farms can also be placed offshore in appropriate locations; however, they can interfere with radar systems so can pose a risk to national security if placed inappropriately. Additionally, the wind is not always predictable or reliable, so wind power needs to have a storage system or be built in conjunction with other energy systems. Some communities consider wind farms to be an eyesore, and some are said to generate low-frequency noise that affects some people and animals.

The world's supply of cheap hydrocarbon reserves is running out. Examining the future trends in energy use, it is clear that many different types of renewable energy sources need to be integrated in an intelligent system that responds to local and national needs and continues to provide the energy needed for civilization's comforts and development.

FURTHER READING

Bilgen, S., and K. Kaygusuz. "Renewable Energy for a Clean and Sustainable Future." *Energy Sources* 26 (2004): 1,119–1,129.

Serra, J. "Alternative Fuel Resource Development." Clean and Green Fuels Fund. Available online. URL: http://cleanandgreenfuels.org/. Accessed February 5, 2009.

and mud facies to migrate laterally. When combined with continuous subsidence, passive margin sequences typically develop many sandstone wedges caught between shale layers. River systems and sandstone channels in muddy overbank delta deposits also form good trap and reservoir systems, since the porous sandstone channels are trapped between impermeable shales.

Most of the world's industrialized nations get the majority of their energy needs from petroleum and other fossil fuels, so exploration for and exploitation of petroleum is a major national and industrial endeavor. Huge resources are spent in petroleum exploration, and thousands of geologists are employed in the oil industry. In the early days of exploration the oil industry gained a reputation of being environmentally degrading, but increased regu-

Two geologists studying GeoProbe model. They are wearing glasses to make the model appear three-dimensional. *(Chris Sattiberger/Photo Researchers, Inc.)*

lations and awareness by these companies has greatly alleviated these problems, and most petroleum is now explored and extracted with minimal environmental consequences. The burning of fossil fuels however, continues to release huge amounts of carbon dioxide and other chemicals into the atmosphere, contributing to global warming.

See also GLOBAL WARMING; HYDROCARBONS AND FOSSIL FUELS; PASSIVE MARGIN.

FURTHER READING

Hunt, John M. *Petroleum Geochemistry and Geology*. San Francisco: W. H. Freeman, 1979.

North, F. K. *Petroleum Geology*. Dordrecht, Germany: Kluwer Academic Publishers, 1986.

Seeley, Richard. *Elements of Petroleum Geology*. New York: Academic Press, 1998.

petrology and petrography Petrology is the branch of geology that attempts to describe and understand the origin, occurrence, structure, and evolution of rocks. Petrography describes the minerals and textures in rock bodies. The two fields are related but differ in that petrography is largely a descriptive science, whereas petrology uses petrographic and other data to deduce the origin and history of rocks.

The petrologic classification of rocks recognizes three main categories with different modes of origin and histories. Igneous rocks crystallized from magma, and include plutonic and volcanic varieties that cooled below and at the surface, respectively. Metamorphic rocks are those that have been changed in some way, such as by the growth of new minerals or structures during heating and pressure from being subjected to tectonic forces. Sedimentary rocks include clastic varieties that represent the broken down, transported, deposited, and cemented fragments of older rocks, as well as chemical and biochemical varieties that represent chemicals that precipitated from a solution.

See also IGNEOUS ROCKS; METAMORPHISM AND METAMORPHIC ROCKS; MINERAL, MINERALOGY; SEDIMENTARY ROCK, SEDIMENTATION.

Polarized light micrograph of eucrite, a type of coarse-grained gabbro from an achondritic stony meteorite, recovered from the Frankenstein-range, Hessia, Germany. Magnification: ×8 at 6×7 cm size (Alfred Pasieka/Photo Researchers, Inc.)

Pettijohn, Francis John (1904–1999) American *Sedimentologist, Field Geologist* Francis Pettijohn was born in Waterford, Wisconsin, on June 20, 1904, and is widely known as the "father of modern sedimentology." Francis became interested in geology while growing up in Bloomington, Indiana, where he often explored the many caves of the region and collected fossils from outcrops near his home. He graduated from high school in Indianapolis in 1921, then entered the University of Minnesota where he received a bachelor of arts in geology in 1924 and a master of arts, also in geology, in 1925. In 1927 he entered graduate school at the University of California at Berkeley, and then transferred to the University of Minnesota, where he received a Ph.D. in Precambrian geology in 1930. He was a professor of geology at Johns Hopkins University in Baltimore from 1952 until his retirement in 1973, and he served as chair of the department there from 1963 to 1968.

Pettijohn is most famous for his studies on the sedimentology and geological evolution of the rocks in the Appalachian Mountains and for the 24 books that he authored or coauthored. Perhaps his most famous book is *Sedimentary Rocks*, in which the techniques of modern sedimentology were clearly described, and which has been reprinted many times since its first publication in 1949. This book has remained a standard in the field for more than 50 years. Pettijohn published his own autobiography in 1984, a humorous and anecdotal work titled *Memoirs of an Unrepentant Field Geologist*.

Pettijohn received numerous awards for his work, including the Sorby Medal of the International

Francis Pettijohn in the field using magnifying lens to examine rock specimen, 1985 *(The Ferdinand Hamburger Archives, Johns Hopkins University)*

Association of Sedimentologists in 1983, the Twenhofel Medal from the Society of Economic Paleontologists and Mineralogists, the Wollaston Medal from the Geological Society of London, the Penrose Medal from the Geological Society of America, the Francis J. Pettijohn Medal from the Society for Sedimentary Geology, and an honorary doctor of science degree from the University of Minnesota. He was professionally active, serving as president of the Society of Economic Paleontologists and Mineralogists and Councilor of the Geological Society of America. Pettijohn was elected a member of the National Academy of Sciences and a Fellow of the American Academy of Arts and Sciences.

See also NORTH AMERICAN GEOLOGY; SEDIMENTARY ROCK, SEDIMENTATION.

FURTHER READING

Pettijohn, Francis J. "In Defense of Outdoor Geology." *Bulletin of the American Association of Petroleum Geologists* 40 (1956): 1,455–1,461.

———. *Memoirs of an Unrepentant Field Geologist.* Chicago: University of Chicago Press, 1984.

———. *Sedimentary Rocks.* New York: Harper and Brothers, 1949.

Pettijohn, Francis J., and Paul E. Potter. *Atlas and Glossary of Primary Sedimentary Structures.* New York: Springer-Verlag, 1964.

Phanerozoic   The eon of geological time since the base of the Cambrian at 544 million years ago and extending to the present is called the Phanerozoic. Introduced by George H. Chadwick in 1930, the eon is characterized by the appearance of abundant visible life in the geological record, in contrast to the earlier eon referred to by Chadwick as the Cryptozoic, but now generally referred to as the Precambrian. Although paleontologists now recognize that many forms of life existed prior to the Phanerozoic, the first appearance of shelly fossils corresponds to the base of the Phanerozoic.

Phanerozoic time is divided mainly on the basis of fossil correlations, and new geochronological studies of important fossil-bearing units continuously force the revision of the absolute ages for the divisions. Its three main fundamental time divisions, called eras, include the Paleozoic, Mesozoic, and Cenozoic. These eras are in turn divided into smaller divisions known as periods, epochs, and ages.

See also CENOZOIC; MESOZOIC; PALEOZOIC.

FURTHER READING

Prothero, Donald R., and Robert H. Dott. *Evolution of the Earth.* 6th ed. Boston: McGraw-Hill, 2002.

Windley, Brian F. *The Evolving Continents.* 3rd ed. Chichester, England: John Wiley & Sons, 1995.

photosynthesis   Green plants, algae, and some bacteria trap solar energy and use it to drive a series of chemical reactions that results in the production of sugars, such as glucose, in a process called photosynthesis. The resulting sugars form the basic food for the plant as well as for insects and animals that eat these plants. Photosynthesis is therefore one of the most important processes for life on Earth. In order for photosynthesis to occur, the photosynthetic organisms must contain pigments such as chlorophyll. In plant cells, structures called chloroplasts contain the chlorophyll and the other necessary components. The most common type of photosynthesis also requires carbon dioxide and water and produces oxygen, also necessary for most life to exist on the planet. Before simple single-celled organisms developed the ability to carry out photosynthesis in the Precambrian, the atmosphere probably contained very little oxygen. Therefore, the process has also been largely responsible for changing the conditions on the planet's surface to make it more hospitable for the development of new, more complex life-forms and for conditions to evolve so that they are suitable for humans.

See also BIOSPHERE; ENVIRONMENTAL GEOLOGY.

planetary nebula   A planetary nebula is the name for a type of emission nebula consisting of a glowing shell of gas and plasma surrounding a central core of a star that went through its hydrogen- and helium-burning stages, ejecting the outer layers of the dying giant star in one of the last stages of its evolution. The core consists of dense carbon ash, formed as a product of nuclear fusion of hydrogen and then helium, which accumulated in the core, then blew away the outer hydrogen- and helium-rich parts of the outer layers of the star as it died. Planetary nebulae have nothing to do with planets, but were named as such in the 1700s by astronomers who could observe faint blurry areas that were not as sharp as stars, and that resembled planets at the time when only crude telescopes were available. Planetary nebulae last only a few tens of thousands of years, compared to billions of years for the earlier life times of the star. Famous examples of planetary nebulae include the Cat's Eye Nebula and the Dumbbell Nebula.

See also STELLAR EVOLUTION.

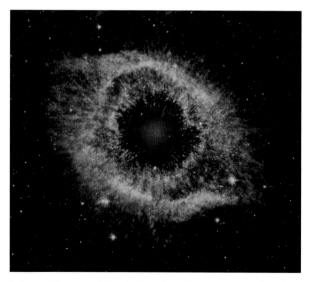

Infrared image of the Helix planetary nebula taken by *Spitzer Space Telescope,* 2/12/07. The Helix nebula is located 700 light-years away in the constellation Aquarius. Infrared light from the outer gaseous layers is represented in blues and greens, and the white dwarf star in the center is represented as a tiny white dot. The red color in the interior of the nebula is from the final layers of gas blown out as the white dwarf died. *(NASA Jet Propulsion Laboratory)*

plate tectonics   The theory of plate tectonics guides the study of the large-scale evolution of the lithosphere of the Earth. In the 1960s, the Earth sciences experienced a scientific revolution, when the paradigm of plate tectonics was formulated from a number of previous hypotheses that attempted to explain different aspects about the evolution of continents, oceans, and mountain belts. New plate material is created at mid-ocean ridges and destroyed when it sinks back into the mantle in deep-sea trenches. Scientists had known for some time that the Earth is divided into many layers defined mostly by chemical characteristics, including the inner core, outer core, mantle, and crust. The plate-tectonic paradigm led to the understanding that the Earth is also divided mechanically and includes a rigid outer layer, called the lithosphere, sitting upon a very weak layer containing a small amount of partial melt of peridotite, termed the asthenosphere. The lithosphere is about 78 miles (125 km) thick under continents and 47 miles (75 km) thick under oceans, whereas the asthenosphere extends to about 155 miles (250 km) depth. The basic tenet of plate tectonics is that the outer shell or lithosphere of the Earth is broken into about twelve large rigid blocks or plates that are all moving relative to one another. These plates are torsionally rigid, meaning that they can rotate about on the surface and not deform significantly internally. Most deformation of plates occurs along their edges, where they interact with other plates.

Plate tectonics has unified the Earth sciences, bringing together diverse fields such as structural geology, geophysics, sedimentology and stratigraphy, paleontology, geochronology, and geomorphology, especially with respect to active tectonics (also known as neotectonics). Plate motion almost always involves the melting of rocks, so other fields are also important, including igneous petrology, metamorphic petrology, and geochemistry (including isotope geochemistry).

The base of the crust, known as the Mohorovicic discontinuity, is defined seismically and reflects the difference in seismic velocities of basalt and peridotite. However, the base of the lithosphere is defined on the basis of the rock's mechanical properties, as where the same rock type on either side begins to melt, and it corresponds roughly to the 2,426°F (1,330°C) isotherm. The main rock types of interest to tectonics include granodiorite, basalt, and peridotite. The average continental crustal composition is equivalent to granodiorite. (The density of granodiorite is 2.6 g/cm^3; its mineralogy includes quartz, plagioclase, biotite, and some potassium feldspar.) The average oceanic crustal composition is equivalent to that of basalt. (The density of basalt is 3.0 g/cm^3; its mineralogy includes plagioclase, clinopyroxene, and

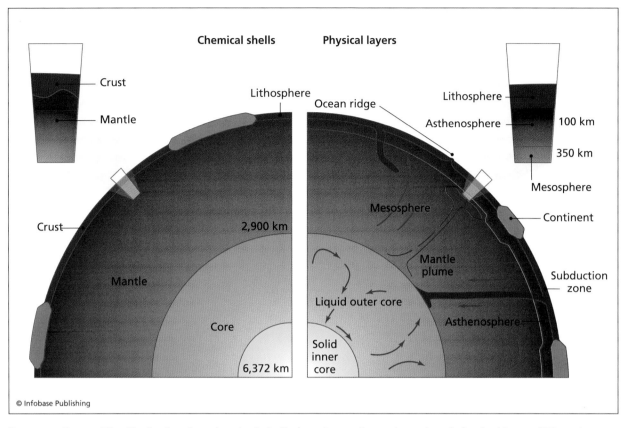

Chemical shells   Physical layers

Cross sections of the Earth showing chemical shells (crust, mantle, and core) and physical layers (lithosphere, asthenosphere, mesosphere, outer core, inner core)

olivine.) The average upper mantle composition is equivalent to peridotite. (The density of peridotite is 3.3 g/cm^3; its mineralogy includes olivine, clinopyroxene, and orthopyroxene.) Considering the densities of these rock types, the crust can be thought of as floating on the mantle; rheologically, the lithosphere floats on the asthenosphere.

The plate tectonic paradigm states that the Earth's outer shell, or lithosphere, is broken into 12 large and about 20 smaller blocks, called plates, that are all moving with respect to each other. The plates are rigid, and they deform along their edges but not internally when they move. The edges of plates therefore serve as home for most of the Earth's mountain ranges and active volcanoes and are where most of the world's earthquakes occur. The plates move in response to heating of the mantle by radioactive decay and somewhat resemble lumps floating in a pot of boiling stew.

The movement of plates on the spherical Earth can be described as a rotation about a pole of rotation, using a theorem first described by Euler in 1776. Euler's theorem states that any movement of a spherical plate over a spherical surface can be described as a rotation about an axis that passes through the center of the sphere. The place where the axis of rotation passes through the surface of the Earth is referred to as the pole of rotation. The pole of rotation can be thought of as analogous to a pair of scissors opening and closing. The motions of one side of the scissors can be described as a rotation of the other side about the pin in a pair of scissors, either opening or closing the blades of the scissors. The motion of plates about a pole of rotation is described using an angular velocity. As the plates rotate, locations near the pole of rotation experience low angular velocities, whereas points on the same plates that are far from the pole of rotation experience much greater angular velocities. Therefore, oceanic-spreading rates or convergence rates along subduction zones may vary greatly along a single plate boundary. This type of relationship is similar to a marching band going around a corner. The musicians near the corner have to march in place and pivot (acting as a pole of rotation) while the musicians on the outside of the corner need to march quickly to keep the lines in the band formation straight as they go around the corner.

Rotations of plates on the Earth lead to some interesting geometrical consequences for plate tectonics. We find that mid-ocean ridges are oriented

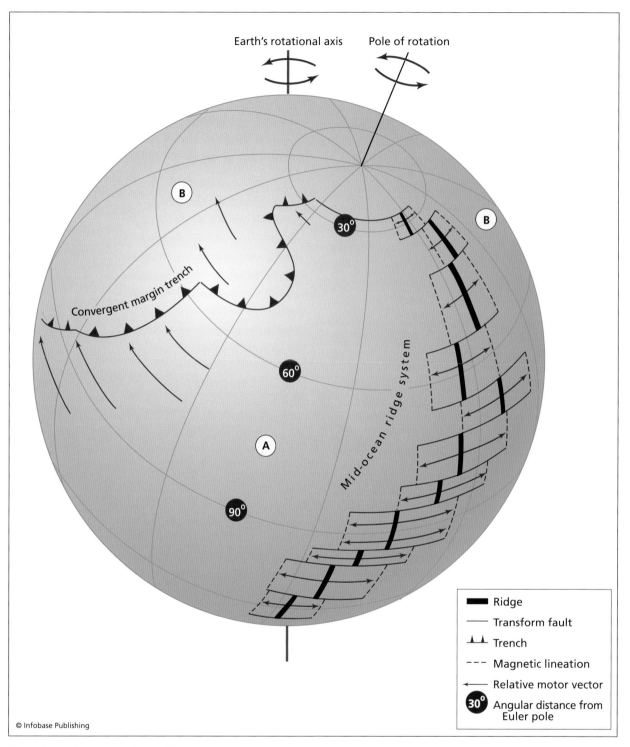

Pole of rotation on a sphere. Plate A rotates away from Plate B, with ridge axes falling on great circles intersecting at pole of rotation and oceanic transform faults falling along small circles that are concentric about the pole of rotation. The angular velocity of the plates increases with increasing distance from the pole of rotation.

so that the ridge axes all point toward the pole of rotation and are aligned on great circles about the pole of rotation. Transform faults lie on small circles that are concentric about the pole of rotation. In con-trast, convergent boundaries can lie at any angle with respect to poles of rotation.

Since plates do not deform internally, all the action happens along their edges. The type of action

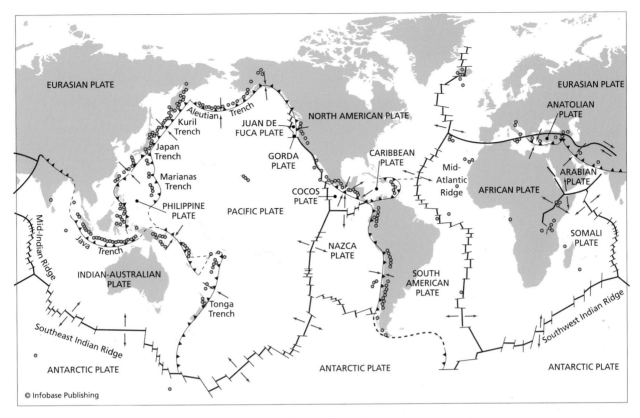

Plate tectonic map of Earth showing convergent, divergent, and transform boundaries

where two plates meet forms the basis for defining three fundamental types of plate boundaries. Divergent boundaries are where two plates move apart, creating a void that typically becomes filled by new oceanic crust that wells up to fill the progressively opening hole. Convergent boundaries are where two plates move toward each other, resulting in one plate sliding beneath the other (when a dense oceanic plate is involved), or collision and deformation (when two continental plates are involved). Transform boundaries form where two plates slide past each other, moving in opposite but parallel directions, such as along the San Andreas fault in California.

Since all plates are moving with respect to each other, the surface of the Earth is made up of a mosaic of various plate boundaries, providing geologists with an amazing diversity of different geological environments to study. Every time one plate moves, the others must move to accommodate this motion, creating a never-ending saga of different plate configurations.

DIVERGENT PLATE BOUNDARIES AND THE CREATION OF OCEANIC CRUST

Where plates diverge, seafloor spreading produces new oceanic crust. As the plates move apart, the pressure on deep underlying rocks decreases, which causes them to rise and partially melt by 15–25 per-

cent. Basaltic magma is produced by partially melting the peridotitic mantle, leaving a "residue" type of rock in the mantle known as harzburgite. The magma produced in this way upwells from deep within the mantle to fill the gap opened by the diverging plates. This magma forms a chamber of molten or partially molten rock that slowly crystallizes to form a coarse-grained igneous rock known as gabbro, which has the same composition as basalt. Before crystallization, some of the magma moves up to the surface through a series of dikes and forms the crustal sheeted-dike complex and basaltic flows. Many of the basaltic flows have distinctive forms with the magma forming bulbous lobes known as pillow lavas. Lava tubes are also common, as are fragmented pillows formed by the implosion of the lava tubes and pillows. Back in the magma chamber, other crystals grow in the gabbroic magma, including olivine and pyroxene, which are heavier than the magma and sink to the bottom of the chamber. These crystals form layers of dense minerals known as cumulates. Beneath the cumulates, the mantle material from which the magma was derived becomes progressively more deformed as the plates diverge, forming a highly deformed ultramafic rock known as a harzburgite or mantle tectonite. This process can be seen on the surface in Iceland along the Reykjanes Ridge.

TRANSFORM PLATE BOUNDARIES
AND TRANSFORM FAULTS

In many places in the oceanic basins, the mid-ocean ridges are apparently offset along great escarpments or faults, which fragment the oceanic crust into many different segments. In 1965 J. Tuzo Wilson correctly interpreted these not as offsets, but as a new class of faults, known as transform faults. The actual sense of displacement on these faults is opposite to the apparent offset, so the offset is apparent, not real. The solution to the real vs. apparent offsets along the transform faults is a primary feature of Wilson's model, proven correct by earthquake studies.

These transform faults are steps in the plate boundary where one plate is sliding past the other plate. Transform faults are also found on some continents, with the most famous examples being the San Andreas fault, the Dead Sea Transform, the North Anatolian fault, and the Alpine fault of New Zealand. All of these are large strike-slip faults with horizontal displacements and separate two different plates.

CONVERGENT PLATE BOUNDARIES

Oceanic lithosphere is being destroyed by sinking back into the mantle at the deep ocean trenches in a process called subduction. As the oceanic slabs sink downward, they experience higher temperatures that cause the release of water and other volatiles from the subducting slab, generating melts in the mantle wedge overlying the subducting slab. These melts then move upward to intrude the overlying plate, where the magma may become contaminated by melting through and incorporating minerals and elements from the overlying crust. Since subduction zones are long narrow zones where large plates are being subducted into the mantle, the melting produces a long line of volcanoes above the down-going plate. These volcanoes form a volcanic arc, either on a continent or over an oceanic plate, depending on which type of crust the overlying plate is composed of.

Island arcs are extremely important for understanding the origin of the continental crust because the magmas and sediments produced here have the same composition as the average continental crust. A simple model for the origin of the continental crust is that it represents a bunch of island arcs which formed at different times and which collided during plate collisions.

Since the plates are in constant motion, island arcs, continents, and other terranes often collide with each other. Mountain belts or orogens typically mark the places where lithospheric plates have collided, and the zone that they collided along is referred to as a suture. Suture zones are complex and include folded and faulted sequences of rocks that form on the two colliding terranes and in any intervening ocean basin.

Often, slices of the old ocean floor are caught in these collision zones (these are called ophiolites), and the process by which they are emplaced over the continents is called obduction (opposite of subduction).

In some cases, subduction brings two continental plates together and they collide, forming huge mountain belts like the Himalayan Mountain chain. In continent-continent collisions, deformation may be very diffuse and extend beyond the normal limit of plate boundary deformation that characterizes other types of plate interactions. For instance, the India-Asia collision has formed the huge uplifted Tibetan Plateau, a series of mountain ranges to the north including the Tien Shan and Karakoram, and deformation of the continents extends far into Asia, as far as Lake Baikal.

HISTORICAL DEVELOPMENT
OF THE PLATE TECTONIC PARADIGM

The plate tectonic paradigm was developed from a number of different models, ideas and observations that were advanced over the prior century by a number of scientists on different continents. Between 1912 and 1925, Alfred Wegener, a German meteorologist, published a series of papers and books outlining his ideas for the evolution of continents and oceans. Wegener was an early proponent of continental drift. He looked for a driving mechanism to move continents through the mantle, and invoked an imaginary force (which he called *Pohlfluicht*) that he proposed caused the plates to drift toward the equator because of the rotation of the Earth. Geophysicists showed that this force was unrealistic, and since Wegener's idea of continental drift lacked a driving mechanism, it was largely disregarded.

In 1929 British geologist Arthur Holmes proposed that the Earth produces heat by radioactive decay and that there are not enough volcanoes to remove all this heat. He proposed that a combination of volcanic heat loss and mantle convection can disperse the heat, and that the mantle convection drives continental drift. Holmes wrote a textbook on this subject, which became widely used and respected. Holmes proposed that the upwelling convection cells were in the ocean basins, and that downwelling areas could be found under Andean-type volcano chains.

Alex du Toit was a South African geologist who worked on Gondwana stratigraphy and published a series of important papers between 1920 and 1940. Du Toit compared stratigraphic sections on the various landmasses that he thought were once connected to form the supercontinent of Gondwana (Africa, South America, Australia, India, Antarctica, Arabia). He showed that the stratigraphic columns of these places were very similar for the periods he proposed the continents were linked, supporting his idea of an older, large, linked landmass. Du Toit also dem-

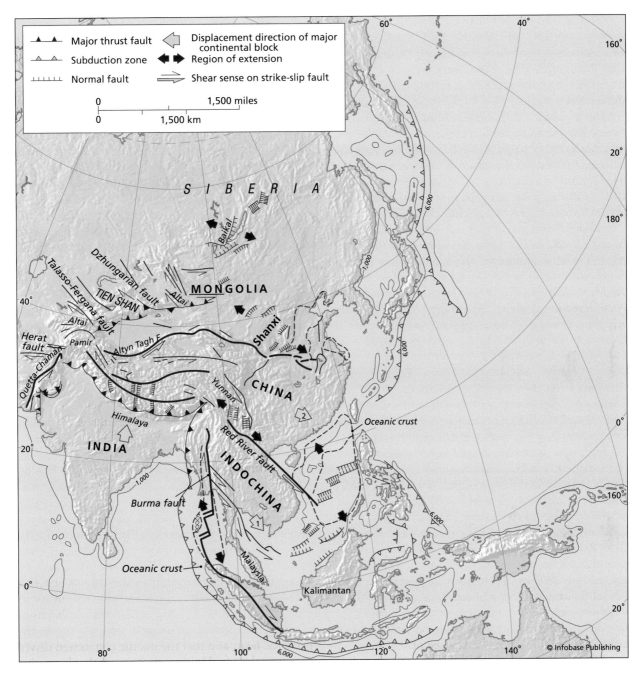

Map of central and eastern Asia collision showing the wide area affected by the collision of India with Asia

onstrated that the floral distributions had belts that matched when the continents were reconstructed, but that appeared disjointed in the continents' present distribution.

In the 1950s, paleomagnetism began developing as a science. The Earth has a dipolar magnetic field, with magnetic field lines plunging into and out of the Earth at the north and south magnetic poles. Field or flux lines are parallel or inclined to the surface at intermediate locations, and the magnetic field can be defined by the inclination of the field lines and their deviation from true north (declination) at any loca-

tion. When igneous rocks solidify, they pass through a temperature at which any magnetic minerals will preserve the ambient magnetic field at that time. In this way, some rocks acquire a magnetism when they solidify. The best rocks for preserving the ambient magnetic field are basalts, which contain 1–2 percent magnetite; they acquire a remnant inclination and declination as they crystallize. Other rocks, including sedimentary red beds with iron oxide and hematite cements, shales, limestones, and plutonic rocks, also may preserve the magnetic field, but they are plagued with other problems hindering interpretation.

In the late 1950s, British geophysicist Stanley K. Runcorn and Canadian geologist Earl Irving first worked out the paleomagnetism of European rocks and discovered a phenomenon they called apparent polar wandering (APW). The Tertiary rocks showed very little deviation from the present pole, but rocks older than Tertiary showed a progressive deviation from the expected results. They initially interpreted this to mean that the magnetic poles wandered around the planet, and the paleomagnetic rock record reflected this wandering. Runcorn and Irving made an APW path for Europe by plotting the apparent position of the pole while holding Europe stationary. However, they found that they could also interpret their results to mean that the poles were stationary and the continents were drifting around the globe. Additionally, they found that their results agreed with some previously hard to interpret paleolatitude indicators from the stratigraphy.

Next, Runcorn and Irving determined the APW curve for North America. They found it similar to Europe's from the late Paleozoic to the Cretaceous, implying that the two continents were connected for that time period and moved together, and later (in the Cretaceous) separated, as the APW curves diverged. This remarkable data set converted Runcorn from a strong disbeliever of continental drift into a drifter.

In 1954 Hugo Benioff, an American seismologist, studied worldwide, deep-focus earthquakes to about a 435 mile (700 km) depth. He plotted earthquakes on cross sections of island arcs and found earthquake foci were concentrated in a narrow zone beneath each arc, extending to about a 435 mile (700 km) depth. He noted that volcanoes of the island-arc systems were located about 62 miles (100 km) above this zone. He also noted compression in island-arc geology and proposed that island arcs are overthrusting oceanic crust. Geologists now recognize that this narrow zone of seismicity is the plate boundary between the subducting oceanic crust and the overriding island arc, and they have named this area the Benioff Zone.

Development of technologies associated with World War II led to remarkable advancements in understanding some basic properties of the ocean basins. In the 1950s, the ocean basin bathymetry, gravity, and magnetic fields were mapped for the U.S. Navy submarine fleet. After this, research scientists from oceanographic institutes such as Scripps, Woods Hole, and Lamont-Doherty Geological Observatory studied the immense sets of oceanographic data. As the raw data was acquired in the 1950s, the extent of the mid-ocean ridge system was recognized and documented by American geologists including Bruce Heezen, Maurice Ewing, and Harry Hess. They also documented the thickness of the sedimentary cover overlying igneous basement and showed that the sedimentary veneer is thin along the ridge system and thickens away from the ridges. Walter Pitman from Lamont-Doherty Geological Observatory in New York happened to cross the mid-oceanic ridge in the South Pacific perpendicular to the ridge and noticed the symmetry of the magnetic anomalies on either side of the ridge. In 1962 Harry Hess from Princeton proposed that the mid-ocean ridges were the site of seafloor spreading and the creation of new oceanic crust, and that Benioff Zones were sites where oceanic crust was returned to the mantle. In 1963 American geologists Fred Vine and Drummond Matthews combined Hess's idea and magnetic anomaly symmetry with the concept of geomagnetic reversals. They suggested that the symmetry of the magnetic field on either side of the ridge could be explained by conveyor-belt style formation of oceanic crust, forming and crystallizing in an alternating magnetic field, such that the basalts of similar ages on either side of the ridge would preserve the same magnetic field properties. Their model was based on earlier discoveries by a Japanese scientist, Motonori Matuyama, who in 1910 discovered recent basalts in Japan that were magnetized in a reversed field and proposed that the magnetic field of the Earth experiences reversals. Allen Cox (Stanford University) had constructed a geomagnetic reversal time scale in 1962, so it was possible to correlate the reversals with specific time periods and deduce the rate of seafloor spreading.

With additional mapping of the seafloor and the mid-ocean ridge system, the abundance of fracture zones on the seafloor became apparent with mapping of magnetic anomalies. In 1965, Canadian geologist J. Tuzo Wilson wrote a classic paper, "A New Class of Faults and Their Bearing on Continental Drift," published in *Nature*. This paper connected previous ideas, noted the real sense of offset of transform faults, and represented the final piece in the first basic understanding of the kinematics or motions of the plates. Lynn Sykes and other seismologists provided support for Wilson's model about one year later by using earthquake studies of the mid-ocean ridges. They noted that the ridge system divided the Earth into areas of few earthquakes, and that 95 percent of the earthquakes occur in narrow belts. They interpreted these belts of earthquakes to define the edges of the plates. They showed that about 12 major plates are all in relative motion to each other. Sykes and others confirmed Wilson's model, and showed that transform faults are a necessary consequence of spreading and subduction on a sphere.

See also CONVERGENT PLATE MARGIN PROCESSES; DIVERGENT PLATE MARGIN PROCESSES; GEODYNAMICS; TRANSFORM PLATE MARGIN PROCESSES.

FURTHER READING

Condie, Kent C., and Robert Sloan. *Origin and Evolution of Earth, Principles of Historical Geology.* Upper Saddle River, N.J.: Prentice Hall, 1997.

Erikson, Jon. *Plate Tectonics: Unraveling the Mysteries of the Earth.* New York: Facts On File, 2001.

Miyashiro, Akiho, Keiti Aki, and A. M. Celal Sengor. *Orogeny.* Chichester, England: John Wiley & Sons, 1982.

Moores, Eldridge M., and Robert Twiss. *Tectonics.* New York: W. H. Freeman, 1995.

Strahler, Arthur. *Plate Tectonics.* Cambridge, Mass.: Geo Books Publishing, 1998.

Windley, Brian F. *The Evolving Continents.* 3rd ed. Chichester, England: John Wiley & Sons, 1995.

Pleistocene The Pleistocene is the older of two epochs of the Quaternary Period, lasting from 1.8 million years ago until 10,000 years ago, at the beginning of the Holocene epoch. Charles Lyell formally proposed the name Pleistocene in 1839, after earlier informal proposals, based on the appearance of species of North Sea mollusks in Mediterranean strata.

The Pleistocene is recognized as an epoch of widespread glaciation, with glaciers advancing through much of Europe and North America and across the southern continents. Glaciers covered about 30 percent of the northern continents, most as huge ice sheets that advanced across Canada, the northern United States, and Eurasia. Smaller alpine glaciers dissected the mountain ranges, forming the glacial landforms visible today, including horns, arêtes, U-shaped valleys, and giant eskers and moraines. Some of the ice sheets were up to two miles (three kilometers) thick, acting as huge bulldozers that removed much of the soil from Canada and scraped the bedrock clean, depositing giant outwash plains in lower latitudes.

The continental ice sheets are known to have advanced and retreated several times during the Pleistocene, based on correlations of moraines, sea surface temperatures deduced from oxygen-isotope analysis of deep-sea cores, and magnetic stratigraphy. Eighteen major glacial expansions and retreats are now recognized from the past 2.4 million years, including four major glacial stages in North America. The Nebraskan glacial maximum peaked at 700,000

Protalus ramparts, dating from a late part of the last major glaciation, along the north base of Sunrise Ridge, northeast of Mount Rainier, Washington. The ramparts form when gravel and other debris fills cracks between the glacier and the canyon wall, leaving the debris behind as an elongate ridge when the glacier melts. *(USGS)*

years ago, followed by the Kansan, Illinoian, and the Wisconsin maximums. Ice from the Wisconsin glacial maximum retreated from the northern United States and Canada only 11,000 years ago, and it may return in a short amount of geological time.

Many species became extinct or otherwise changed in response to the rapid climate changes during the Pleistocene. Many species lived in the climate zone close to the glacier front, including the woolly mammoth, giant versions of mammals now living in the arctic, rhinoceros, and caribou. Farther from the ice, giant deer, mastodons, dogs, cats, ground sloths, and other mammals were common. Both humans (Homo sapiens) and Neanderthals roamed through Eurasia, but anthropologists do not know the nature of the interaction between these two hominid species. Many of the giant mammals became extinct in the latter part of the Pleistocene, especially between 18,000 and 10,000 years ago. Currently considerable debate exists about the relative roles of climate change and predation by hominids in these extinctions.

See also NEOGENE; QUATERNARY; TERTIARY.

FURTHER READING

Dawson, A. G. *Ice Age Earth*. London: Routledge, 1992.
Erickson, Jon. *Glacial Geology: How Ice Shapes the Land*. New York: Facts On File, 1996.
Intergovernmental Panel on Climate Change. Available online. URL: http://www.ipcc.ch/index.htm. Accessed January 30, 2008.
Kusky, T. M. *Climate Change: Shifting Deserts, Glaciers, and Climate Belts, The Hazardous Earth Set*. New York: Facts On File. 2008.

Pluto The solar system has long been considered to have nine planets, including Mercury, Venus, Earth, Mars, Jupiter, Saturn, Uranus, Neptune, and Pluto. Pluto has been thought to be the most distant planet, and a relatively small body at roughly one-fifth the mass of the Earth's Moon, or 0.66 percent of the Earth's volume. Pluto is considerably smaller than many moons in the solar system, including Ganymede, Titan, Callisto, Io, Earth's Moon, Europa, and Triton. In 2006 the International Astronomical Union met, decided that Pluto does not meet the formal criteria of being a planet, and demoted the status of the object to that of a "dwarf planet." What happened?

Pluto has a colorful history of discovery. In the early to mid-1800s scientists noticed that the orbit of Uranus showed some unusual perturbations, hypothesized that these were due to the gravitational attraction of a more distant planet, and were able to predict where a new planet, Neptune, should be. French mathematician Urbain Le Verrier (1811–77)

performed these calculations, then sent his calculation of where this planet should be to German astronomer Johann Gottfried Galle on September 23, 1846; Galle looked in this position and identified Neptune the following day, September 24, 1846. In the late 1800s further calculation showed that the orbit of Neptune was also being disturbed by something, so a search was mounted for another distant planet, dubbed planet "X." The search for planet "X" was pioneered by Percival Lowell, who founded the Lowell Astronomical Observatory in Arizona, who unsuccessfully searched the skies for this hypothesized planet from 1905 until his death in 1916. Clyde Tombaugh resumed the search in 1929 and then discovered a planet "X" in the right location on February 18, 1930. The name Pluto was suggested for "planet X" by an 11-year-old girl, Venetia Burney, and the name was adopted by a vote at the Lowell Observatory on March 24, 1930. Subsequent studies of Pluto revealed that it was a very small object, and many astronomers argued that it should not be a planet and could not have such an effect on the orbit of Neptune. In a twist of fate, later observations of the mass of Neptune by the *Voyager 2* spacecraft flyby in 1989 revealed that the mass of Neptune was overestimated by 0.5 percent, and this change was enough to explain the discrepancies in the orbits of Neptune and Uranus that initially led to the search for planet "X" (Pluto). Thus, the very reason for searching for a ninth planet was false.

Pluto is quite different from other planets. Not only is it very small compared to all other planets, but also it orbits far from the plane of the ecliptic, and it resides in the Kuiper Belt, along with objects that are about the same order of magnitude in size and mass as Pluto. Enis, an orbiting body further out than the Kuiper Belt in the scattered disk, is 27 percent larger than Pluto. Pluto is thought to be composed of 98 percent nitrogen ice, along with methane and carbon monoxide, similar to other objects, including comets, in the Kuiper belt and in the scattered disk that overlaps with the outer edge of the Kuiper belt. If Pluto were to orbit close to the Sun, it would develop a long cometary tail, and would be classified as a comet, not a planet.

With the advent of more powerful telescopes and data from space missions in the 1980s, 1990s, and 21st century, many objects with masses approaching that of Pluto have been discovered. After much debate and discussion, the International Astronomical Union proposed that planets need to be defined on the basis of three main criteria, and on August 24, 2006, adopted the following definition of a planet:

- A planet is a celestial body that (a) is in orbit around the Sun, (b) has sufficient mass for its self-gravity to overcome rigid body forces so that it assumes a hydrostatic equilibrium (nearly round) shape, and (c) has cleared the neighborhood around its orbit.

- A *"dwarf planet"* is a celestial body that (a) is in orbit around the Sun, (b) has sufficient mass for its self-gravity to overcome rigid body forces so that it assumes a hydrostatic equilibrium (nearly round) shape, (c) has not cleared the neighborhood around its orbit, and (d) is not a satellite.
- All other objects, except satellites, orbiting the Sun shall be referred to collectively as *"Small Solar System Bodies."*

The International Astronomical Union made some further footnotes to their revised definition of a planet.

- The eight planets are: Mercury, Venus, Earth, Mars, Jupiter, Saturn, Uranus, and Neptune.
- An IAU process will be established to assign borderline objects into dwarf planet and other categories.
- These currently include most of the Solar System asteroids, most Trans-Neptunian Objects (TNOs), comets, and other small bodies.

The IAU further resolved that "Pluto is a 'dwarf planet' by the above definition and is recognized as the prototype of a new category of Trans-Neptunian Objects." Thus, after 100 years of searching for planet "X," and 76 years of considering it to be a planet since Pluto's discovery in 1930, the formerly most distant planet is now regarded as the second largest known dwarf planet in the solar system, and as just another large object in the Kuiper belt of the outer solar system.

The dwarf planet Pluto has a variable orbital distance of about 30–40 astronomical units (2.7–3.7 billion miles, or 4.4–6 billion km) from the Sun, circling once every 249 Earth years, and has a retrograde rotation period of 6.4 Earth days. It is a small body with a mass of 0.003 Earth masses, a diameter of 1,400 miles (2,250 km; only 20 percent that of Earth), and a density of 2.3 grams/cubic centimeter. It has one moon known as Charon, and it closely resembles other asteroids of the outer solar system. The large 17.2° inclination of its orbital plane with respect to the ecliptic plane supports the contention that Pluto is a captured asteroid.

The physical properties of the Pluton-Charon system suggest that it is an icy dual-asteroid system similar to some of the Jovian moons, being most similar to Neptune's moon Triton. Models for the origin of Pluto range from its being a captured icy asteroid or an escaped moon to being a remnant of material left over from the formation of the solar system. The great distance and small size of the system make it difficult to observe, and certainly as deep planetary probes explore the outer reaches of the solar system, new theories and models for the origin and evolution of this system will emerge.

See also EARTH; JUPITER; MARS; MERCURY; NEPTUNE; SATURN; SOLAR SYSTEM; URANUS; VENUS.

FURTHER READING
Chaisson, Eric, and Steve McMillan. *Astronomy Today.* 6th ed. Upper Saddle River, N.J.: Addison-Wesley, 2007.
Comins, Neil F. *Discovering the Universe.* 8th ed. New York: W. H. Freeman, 2008.
National Aeronautic and Space Administration. Solar System Exploration page. Pluto. Available online. URL: http://solarsystem.nasa.gov/planets/profile.cfm?Object=Pluto. Last updated June 25, 2008.
Snow, Theodore P. *Essentials of the Dynamic Universe: An Introduction to Astronomy.* 4th ed. St. Paul, Minn.: West Publishing Company, 1991.

Powell, John Wesley (1834–1902) American *Geologist, Explorer* John Wesley Powell is most famous for his early explorations and geologic descriptions of the Grand Canyon and Colorado River in 1869. His expedition was the first passage through the Grand Canyon, and his explorations opened up the West to many later explorers. Powell became the second director of the U.S. Geological Survey, serving from 1881–94.

EARLY LIFE AND PERSONAL ACHIEVEMENTS
John Wesley Powell was born on March 24, 1834, in Mount Morris, New York, to Joseph and Mary Powell. His father was an impoverished preacher from England who moved his young family to Jackson, Ohio, then to Walworth County, Wisconsin, then to Boone County, Illinois. Powell attended Illinois College, then Wheaton College, and finally Oberlin College but never completed a degree. He developed deep knowledge of ancient Greek, Latin, and the natural sciences, particularly geology. This interest led him to explore the Mississippi River valley, including a trip in 1856, when he rowed from St. Anthony northeast of Minneapolis all the way to the Gulf of Mexico, south of New Orleans. The following year, in 1857, he rowed the Ohio River from Pittsburgh to St. Louis, followed by a trip in 1858 down the Illinois River and up the Mississippi River to central Iowa. These trips inspired him for his later explorations of the West and along the Colorado River and helped him be elected to the Illinois Natural History Society in 1859.

The Civil War broke out in 1861, and Powell enlisted with the Union Army as a topographer and military engineer to help abolish slavery. At the Battle of Shiloh on April 6 and 7, 1862, Powell was hit by a musket ball and lost most of one of his arms. He was later promoted to major and chief of artillery of the 17th Army Corps.

John Wesley Powell married Emma Dean in 1862. After the war has took a position as professor of geology at Illinois Wesleyan University and helped found the Illinois Museum of Natural History.

SCIENTIFIC CONTRIBUTIONS

In 1867 and 1868 Powell led a series of expeditions into the Rocky Mountains and along the Green and Colorado Rivers. His most famous expedition was in 1869, when he took nine men in four boats to explore the Colorado River and Grand Canyon. His geological observations led to the understanding that the canyon was formed by the river's gradually cutting down through the rocks of the region as the plateau was slowly uplifted. The team left Green River, Wyoming, on May 24 and entered the wild Colorado River near Moab, Utah. Powell took notes concerning the scenery and geology along the way, naming and traversing Glen Canyon, then reaching the Virgin River on August 30, 1869. One man had quit after the first month, and three more quit the expedition just before the junction with the Virgin River. Tragically, the three who abandoned the expedition at this point were killed in an ambush in the Wild West, but the perpetrators were never identified. On a second expedition along the same route, Powell completed an accurate map of the rivers, along with collecting many photographs and producing several scientific papers about the region.

Between 1874 and 1879, Powell was the director of the United States Geological and Geographical Survey of the Territories, and he led explorations into the Rocky Mountains of the southwestern regions. During these field excursions, Powell became convinced of the limits on development posed by the paucity of water in the desert southwest, and he completed many surveys of the region's water resources. In 1881 Powell became director of the U.S. Geological Survey, but he resigned in 1894 to pursue studies of the native peoples of the land.

See also FLUVIAL; GEOMORPHOLOGY; NORTH AMERICAN GEOLOGY.

FURTHER READING

Powell, John W. *The Exploration of the Colorado River.* New York: Doubleday and Company, 1961.
———. "The Laws of Hydraulic Degradation." *Science 16* (1888): 229–233.
———. "Remarks on the Structural Geology of the Valley of the Colorado of the West." *Bulletin—Philosophical Society of Washington* 1 (1874): 48–51.
Stegner, Wallace. *Beyond the Hundredth Meridian: John Wesley Powell and the Second Opening of the West.* Omaha: University of Nebraska Press, 1954.

Precambrian Comprising nearly 90 percent of geologic time, the Precambrian eon includes the time interval in which all rocks older than 544 million years formed. The Precambrian is preceded by the Hadean Eon, representing the time interval during which the Earth and other planets were accreting and from which no rocks are preserved, and is succeeded by the Cambrian, the dawn of advanced life on Earth. The Precambrian consists of two eras: the Archean, ranging in age from the oldest known rocks at about 4.0 billion years old to 2.5 billion years ago, and the Proterozoic, ranging from 2.5 billion years ago until 544 million years ago. The Archean is further divided into the Early (4.0 Ga–3.0 Ga) and Late (3.0–2.5 Ga), and the Proterozoic is divided into the Early or Paleoproterozoic (2.5 Ga–1.6 Ga), Middle or Mesoproterozoic (1.6 Ga–1.3 Ga), and Late or Neoproterozoic (1.3 Ga–0.54 Ga).

Most Precambrian rocks are found in cratons, areas of generally thick crust that have been stable since the Precambrian and that exhibit low heat flow, subdued topography, and few earthquakes. Many also preserve a thick lithospheric keel known as the tectosphere. Exposed parts of Precambrian cratons are known as shields. Many of the rocks in cratons are preserved in granite-greenstone terrains, fewer are preserved as linear high-grade gneiss complexes, and still fewer form relatively undeformed sedimentary and volcanic sequences deposited in shallow water or platform basins, resting on older Precambrian rocks. Platformal sequences form a thin veneer over many older Precambrian terrains, so geologic maps of cratons and continents show many essentially flat-lying platformal units, but these are volumetrically less significant than the underlying sections of the crust. Many other areas of Precambrian rocks are found as linear tectonic blocks within younger orogenic belts. These probably represent fragments of older cratons that have been rifted, dispersed, and accreted to younger orogens by plate tectonic processes, some traveling huge distances from where they initially formed in their primary tectonic settings.

The Precambrian is the most dramatic of all geological eons, as it marks the transition from the accretion of the Earth to a planet that has plate tectonics, a stable atmosphere-ocean system, and a temperature range all delicately balanced in such a

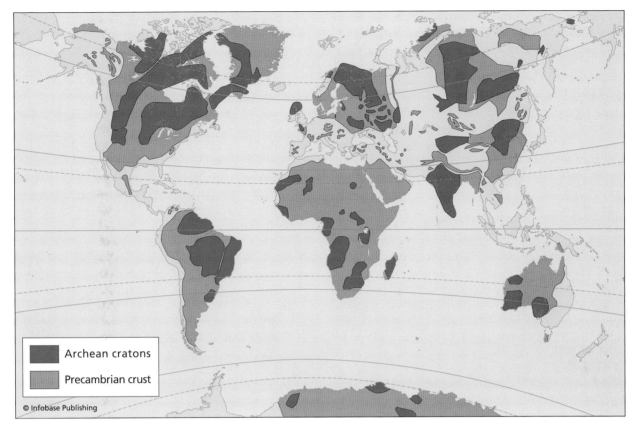

Map of the world, showing the distribution of Precambrian rocks in Archean cratons and other undifferentiated areas of Precambrian crust

way as to allow advanced life to develop and persist on the planet. The planet has been cooling steadily since accretion and was producing more heat by radioactive decay in the Precambrian than it has been since. However, scientists do not know if this greater amount of heat significantly heated the mantle and crust, or if this additional heat was simply lost faster than it is by the present style of plate tectonics. It is likely that more rapid seafloor spreading or a greater total length of oceanic ridges with active volcanism was able to accommodate this higher heat flow, keeping mantle and crustal temperatures close to what they have been in the Phanerozoic.

Understanding of the development of life in the Precambrian has been undergoing rapid advancement, and the close links between life, atmospheric chemistry, plate tectonics, and global heat loss are only recently being explored. Many mysteries remain about the events that led to the initial creation of life, its evolution to more complex forms, and the eventual development of multi-celled complex organisms at the end of the Precambrian.

BANDED-IRON FORMATION

Banded-iron formations are a distinctive type of sedimentary rock that formed predominantly dur-

ing the Precambrian and is the major source of the world's iron reserves. Banded-iron formations (BIFs) are a thinly bedded, chemically precipitated, iron-rich rock, with layers of iron ore minerals typically interbedded with thin layers of chert or microcrystalline silica. Many are completely devoid of detrital or clastic sedimentary input. Most banded-iron formations formed between 2.6 and 1.8 billion years ago, and only a few very small similar types of deposits have been discovered in younger mountain belts. This observation suggests that the conditions necessary to form the BIFs were present on Earth in early (Precambrian) time, but largely disappeared by 1.8 billion years ago. The chemical composition and reduced state of much of the iron of BIFs suggest that they may have formed in an oxygen-poor atmosphere/ocean system, explaining their disappearance around the time that atmospheric oxygen was on the rise. BIFs may also be intimately associated with early biological activity and may preserve the record of the development of life on Earth. The world's oldest BIF is located in the 3.8 billion-year-old Isua belt in southwestern Greenland, and some geologists have suggested that this formation contains chemical signatures that indicate biological activity was involved in its formation.

Two main types of banded-iron formations are classified based on the geometric and mineralogical characteristics of the deposits. Algoma-type BIFs are lens-shaped bodies that are closely associated with volcanic rocks, typically basalts. Most are around a thousand feet to several miles (several hundred meters to kilometers) in scale. In contrast, Superior-type BIFs are very large in scale, many initially covering tens of thousands of square miles (kilometers). Superior-type BIFs are closely associated with shallow marine shelf types of sedimentary rocks including carbonates, quartzites, and shales.

Banded-iron formations are also divisible into four types based on their mineralogy. Oxide-iron formations contain layers of hematite, magnetite, and chert (or cryptocrystalline silica). Silicate-iron formations contain hydrous silicate minerals including chlorite, amphibole, greenalite, stilpnomelande, and minnesotaite. Carbonate-iron formations contain siderite, ferrodolomite, and calcite. Sulfide-iron formations contain pyrite.

In addition to being rich in iron, BIFs are ubiquitously silica-rich, indicating that the water from which they precipitated was saturated in silica as well as iron. Other chemical characteristics of BIFs include low aluminum and titanium, elements that are generally increased by erosion of the continents. Therefore, BIFs are thought to have been deposited in environments away from any detrital sediment input. Some BIFs, especially the sulfide-facies Algoma-type iron formation, have chemical signatures compatible with formation near black smoker types of seafloor hydrothermal vents, whereas others may have been deposited on quiet marine platforms. In particular, many of the Superior-types of deposits have characteristics of deposition on a shallow shelf, including their association with shallow water sediments, their chemical and mineralogical constituency, and the very thin and laterally continuous nature of their layering. For instance, in the Archean Hammersley Basin of Western Australia, millimeter-thick layers in the BIF can be traced for hundreds of miles.

The environments where BIFs formed and the mechanism responsible for the deposition of the iron and silica in BIFs prior to 1.8 Ga is under current debate. Any model must explain the large-scale transport and deposition of iron and silica in thin layers, in some cases over large areas, for a limited time period of Earth's history. Some observations are pertinent. First, to form such thin layers, the iron and silica must have been dissolved in solution. For iron to be in solution, it needs to be in the ferrous (reduced) state, in turn suggesting that the Earth's early oceans and atmosphere had little if any free oxygen, and were reducing. The source of the iron and silica is also problematic; it may have come from

weathering of continents or from hydrothermal vents on the seafloor. Evidence supports both ideas for individual and different kinds of BIFs, although the scales seem to be tipped in favor of hydrothermal origins for Algoma-types of deposits, and weathering of continents for Superior-type deposits.

The mechanisms responsible for causing dissolved iron to precipitate from the seawater to form the layers in banded-iron formations have also proven elusive and problematic. Changes in pH and acidity of seawater may have induced the iron precipitation, with periods of heavy iron deposition occurring during a steady background rate of silica deposition. Periods of nondeposition of iron would then be marked by deposition of silica layers. Prior to 1.8 Ga the oceans did not have organisms (e.g., diatoms) that removed silica from the oceans to make their shells, so the oceans would have been close to saturated in silica at this time, easing its deposition.

Several models have attempted to bring together the observations and requirements for the formation of banded-iron formations, but none are completely satisfactory at present. Perhaps there is no unifying model or environment of deposition, and multiple origins are possible. One model calls on alternating periods of evaporation and recharge to a restricted basin (such as a lake or playa), with changes in pH

A banded iron formation—sample is about 1 inch (2.5 cm) across *(Dirk Wiersma/Photo Researchers, Inc.)*

and acidity being induced by the evaporation. This would cause deposition of alternating layers of silica and iron. Most BIFs do not appear to have been deposited in lakes. Another model calls on biological activity to induce the precipitation of iron, but fossils and other traces of life are generally rare in BIFs, although present in some. In this model, the layers would represent daily or seasonal variations in biological activity. Another model suggests that the layering was induced by periodic mixing of an early stratified ocean, where a shallow surface layer may have had some free oxygen resulting from near-surface photosynthesis, and a deeper layer would be made of reducing waters, containing dissolved elements produced at hydrothermal seafloor vents. In this model, precipitation and deposition of iron would occur when deep reducing water upwelled onto continental shelves and mixed with oxidized surface waters. The layers in this model would then represent the seasonal (or other cycle) variation in the strength of the coastal upwelling. This last model seems most capable of explaining features of the Superior-types of deposits, such as those of the Hamersley Basin in western Australia. Variations in the exhalations of deep-sea vents may be responsible for the layering in the Algoma-type deposits. Other variations in these environments, such as oxidation, acidity, and amount of organic material, may explain the mineralogical differences between different banded-iron formations. For instance, sulfide-facies iron formations have high amounts of organic carbon (especially in associated black shales and cherts) and were therefore probably deposited in shallow basins with enhanced biological activity. Carbonate-facies BIFs have lower amounts of organic carbon and sedimentary structures indicative of shallow water deposition, so these probably were deposited on shallow shelves but farther from the sites of major biological activity than the sulfide-facies BIFs. Oxide-facies BIFs have low contents of organic carbon but have a range of sedimentary structures indicating deposition in a variety of environments.

The virtual disappearance of banded-iron formation from the geological record at 1.8 billion years ago is thought to represent a major transition on the planet from an essentially reducing atmosphere to an oxygenated atmosphere. The exact amounts and rate of change of oxygen dissolved in the atmosphere and oceans would have changed gradually, but the sudden disappearance of BIFs at 1.8 Ga seems to mark the time when the rate of supply of biologically produced oxygen overwhelmed the ability of chemical reactions in the oceans to oxidize and consume the free oxygen. The end of BIFs therefore marks the new dominance of photosynthesis as one of the main factors controlling the composition of the atmosphere and oceans.

KOMATIITE

A komatiite is a high-magnesium, ultramafic lava exhibiting spinifex (a bladed quench pattern, like ice on a window pane) textures as shown by bladed olivine or pyroxene crystals. The composition of komatiite may range from peridotite, with 30 percent MgO and 44 percent SiO_2, to basalt, with 8 percent MgO and 52 percent SiO_2. The name is from the type section on the Komati River in Barberton, South Africa. Komatiites are very rare in Phanerozoic orogenic belts and have been recovered from few places, such as fracture zones, on the modern sea floor. They are more abundant but still rare in Archean greenstone belts. Early work on komatiites suggested that they reflected high degrees of partial melting of a high temperature mantle, with mantle melting temperatures estimated to be as high as 2,912–3,272°F (1,600–1,800°C). Since these temperatures are much higher than those in the melting region of the mantle today, and since komatiites are more abundant in Archean greenstone belts than younger orogenic belts, some workers used komatiites as evidence that the Archean mantle was much hotter than the mantle is today. More recent petrological work has shown that the earlier estimates were based on dry melting experiments, and komatiites have water in their structure. Adding water to the melting calculations, new estimates of komatiite source region melting temperatures fall in the range of 2,192–2,552°F (1,200–1,400°C), much more similar to present-day mantle temperatures.

See also ARCHEAN; ATMOSPHERE; CONTINENTAL CRUST; CRATON; GREENSTONE BELTS; LIFE'S ORIGINS AND EARLY EVOLUTION; PROTEROZOIC.

FURTHER READING

Condie, Kent C., and Robert Sloan. *Origin and Evolution of Earth: Principles of Historical Geology.* Upper Saddle River, N.J.: Prentice Hall, 1997.

Goodwin, Alan M. *Precambrian Geology.* London: Academic Press, 1991.

Kusky, Timothy M. *Precambrian Ophiolites and Related Rocks.* Amsterdam: Elsevier, 2004.

Morris, R. C. "Genetic Modeling for Banded Iron Formation of the Hammersley Group, Pilbara Craton, Western Australia." *Precambrian Research* 60 (1993): 243–286.

Simonson, Bruce M. "Sedimentological Constraints on the Origins of Precambrian Banded Iron Formations." *Geological Society of America Bulletin* 96 (1985): 244–252.

Windley, Brian F. *The Evolving Continents.* 3rd ed. Chichester, England: John Wiley & Sons, 1995.

precipitation Water that falls to the surface from the atmosphere in liquid, solid, or fluid form is called precipitation. Whether it falls as rain, drizzle, fog, snow, sleet, freezing rain, or hail, it is measured as a liquid-water equivalent. The types and amounts of precipitation in different parts of the world vary greatly, from places that have never had any measurable precipitation to places that regularly receive several to more than 10 feet (hundreds of centimeters) of rain per year. Precipitation is strongly seasonal in some places, with dry and wet seasons, and distributed more regularly in other climates.

Rain is liquid precipitation with droplets greater than 0.02 inches (0.5 mm) in diameter, whereas drizzle has droplets between 0.008–0.02 inches (0.2–0.5 mm) in diameter. Fog is a cloud whose base is at the surface and has smaller particles that only truly become precipitation when wind drives them against surfaces or the ground. Freezing rain and drizzle both fall in liquid form but freeze upon hitting cold surfaces on the ground, creating a frozen coating known as glaze. Sleet consists of frozen ice pellets less than 0.2 inches (5 mm) in diameter, and hail consists of larger transparent to opaque particles that typically have diameters of 0.2–0.8 inches (5–20 mm) but sometimes are as large as golf balls or rarely even grapefruit. Snow is frozen precipitation consisting of complex hexagonal ice crystals that fall to the ground.

In tropical regions and temperate climates in warmer parts of the year, most precipitation falls as rain and drizzle. Heavy rain is defined as more than 0.16 inches (4 mm) of precipitation per hour, moderate rain falls between 0.16–0.02 inches (4 mm and 0.5 mm) per hour, and light rain (commonly called drizzle) is less than 0.02 inches (0.5 mm) per hour. Frequent and steady rains characterize some regions; others are characterized by infrequent but intense downpours, including thunderstorms that may shed hailstones. At high elevation, high latitudes, and in midlatitudes during colder months, most precipitation falls as frozen solid particles. Most frozen precipitation falls as snow that typically has a water equivalent of one-tenth the amount of snow that falls (i.e., 10 cm of snow equals 10 mm of rain). More freezing rain and sleet than snow characterize some regions, particularly coastal regions influenced by warm ocean currents.

Uplift within clouds or larger-scale systems are generally necessary to initiate the formation of water droplets that become precipitation. Convection cells in thunderheads, air forced over mountains, zones of convergence along fronts, and cyclonic systems can all produce dramatic uplift and induce precipitation. In order to form precipitation, the small water (or ice) droplets that are separated by very wide

Dark storm clouds at beach *(Javarman, Shutterstock, Inc.)*

spaces must coalesce into particles large enough to fall as precipitation. Additionally, the particles must overcome the forces of evaporation as they rise or fall through unsaturated air in order to make it to the ground. Rapid lateral and vertical motions in clouds, leading to collisions between particles, aid the coalescence of particles, and gravity then accelerates particles to the ground with larger particles initially falling faster than smaller ones since they are less affected by updrafts. Large particles therefore tend to collide with and incorporate smaller particles. After frozen particles form in the upper levels of vertically extensive cloud systems, they alternately fall into and rise out of lower levels, where they partially melt, grow, and rise on updrafts. Such cycling produces relatively large particles that may fall as precipitation.

See also CLIMATE; CLIMATE CHANGE; CLOUDS; HURRICANES.

FURTHER READING

Ahrens, C. Donald. *Meteorology Today.* 7th ed. Pacific Grove, Calif.: Brooks/Cole, 2002.

Proterozoic The Proterozoic refers to the younger of the two Precambrian eras and the erathem of rocks deposited in this era. Divisions of the Proterozoic include the Early or Paleoproterozoic (2.5 Ga–1.6 Ga), Middle or Mesoproterozoic (1.6 Ga–1.3 Ga), and Late or Neoproterozoic (1.3 Ga–0.54 Ga). Proterozoic rocks are widespread on many continents, with large areas preserved especially well in North America, Africa and Saudi Arabia, South America, China, and Antarctica.

Like the Archean, Proterozoic terrains are of three basic types: rocks preserved in cratonic associations, orogens (often called mobile belts in Proterozoic literature), and cratonic cover associations. Wide shear zones, extensive mafic dike swarms, and layered mafic-ultramafic intrusions cut many Proterozoic terrains. Proterozoic orogens have long linear belts of arc-like associations, metasedimentary belts, and widespread, well-developed ophiolites.

Many geologists believe that clear records of plate tectonics first appeared in the Proterozoic, although many others have challenged this view, placing the operation of plate tectonics earlier, in the Archean. This later view is supported by the recent recognition of Archean ophiolites (including the Dongwanzi ophiolite) in northern China.

The Proterozoic saw the development of many continental-scale orogenic belts, many of which have been recently recognized to be parts of global-scale systems that reflect the formation, breakup, and reassembly of several supercontinents. Paleoproterozoic orogens include the Wopmay in northern Canada, interpreted to be a continental margin arc that rifted from North America and then collided soon afterwards, closing the young back-arc basin. There are many 1.9–1.6 Ga orogens in many parts of the world, including the Cheyenne belt in the western United States, interpreted as a suture that marks the accretion

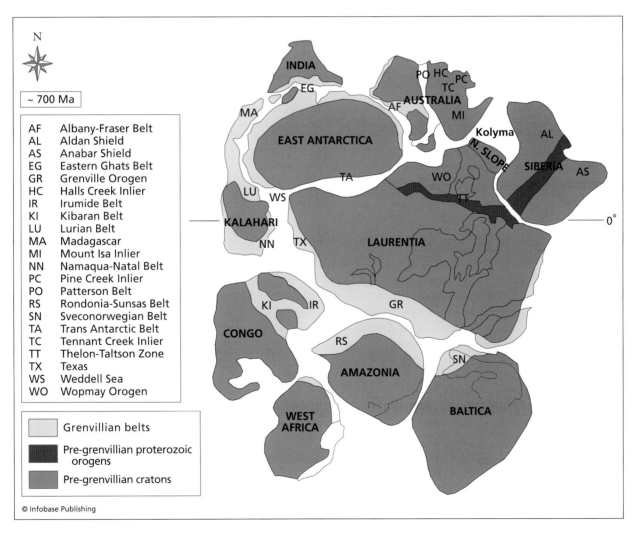

AF	Albany-Fraser Belt
AL	Aldan Shield
AS	Anabar Shield
EG	Eastern Ghats Belt
GR	Grenville Orogen
HC	Halls Creek Inlier
IR	Irumide Belt
KI	Kibaran Belt
LU	Lurian Belt
MA	Madagascar
MI	Mount Isa Inlier
NN	Namaqua-Natal Belt
PC	Pine Creek Inlier
PO	Patterson Belt
RS	Rondonia-Sunsas Belt
SN	Sveconorwegian Belt
TA	Trans Antarctic Belt
TC	Tennant Creek Inlier
TT	Thelon-Taltson Zone
TX	Texas
WS	Weddell Sea
WO	Wopmay Orogen

Grenvillian belts

Pre-grenvillian proterozoic orogens

Pre-grenvillian cratons

© Infobase Publishing

Plate reconstruction of the continents at 700 million years ago showing the supercontinent of Rodinia, with North America situated in the center of the supercontinent

of the Proterozoic arc terrains of the southwestern United States with the Archean Wyoming Province.

The supercontinent Rodinia formed in Meso-proterozoic times by the amalgamation of Laurentia, Siberia, Baltica, Australia, India, Antarctica, and the Congo, Kalahari, West Africa, and Amazonia cratons between 1.1 and 1.0 Ga. The joining of these cratons resulted in the terminal collisional events at convergent margins on many of these cratons, including the ca. 1.1–1.0-Ga Ottawan and Rigolet orogenies in the Grenville Province of Laurentia's southern margin. Globally, these events have become known as the Grenville orogenic period, named after the Grenville orogen of eastern North America. Grenville-age orogens are preserved along eastern North America, as the Rodonia-Sunsas belt in Amazonia, the Irumide and Kibaran belts of the Congo craton, the Namaqua-Natal and Lurian belts of the Kalahari craton, the Eastern Ghats of India, and the Albany-Fraser belt of Australia. Many of these belts now preserve deep-crustal metamorphic rocks (granulites) that were tectonically buried to 19–25-mile- (30–40-km-) depth; then the overlying crust was removed by erosion, forcing the deeply buried rocks to the surface. Since 30–40 kilometers of crust still underlies these regions, they might have had double crustal thickness during the peak of metamorphism. Today such thick crust is produced in regions of continent-continent collision and locally in Andean arc settings. The linear quality and wide distribution of the Grenville-aged orogens suggest they delineate the sites of continent-continent collisions where the various cratonic components of Rodinia collided between 1.1 and 1.0 Ga.

The Neoproterozoic breakup of Rodinia and the formation of Gondwana at the end of the Precambrian and the dawn of the Phanerozoic represents one of the most fundamental problems being studied in earth sciences today. There have been numerous and rapid changes in the understanding of events related to the assembly of Gondwana. One of the most fundamental and most poorly understood aspects of the formation of Gondwana is the timing and geometry of closure of the oceanic basins, separating the continental fragments that amassed to form the Late Neoproterozoic supercontinent. The final collision between East and West Gondwana most likely followed the closure of the Mozambique Ocean, forming the East African Orogen, which encompasses the Arabian-Nubian shield in the north and the Mozambique Belt in the south. These and several other orogenic belts are commonly referred to as Pan-African belts, recognizing that many distinct belts in Africa and other continents experienced deformation, metamorphism, and magmatic activity spanning the period of 800–450 Ma. Pan-African tectonothermal activity in the Mozambique Belt was broadly contemporaneous with magmatism, metamorphism, and deformation in the Arabian-Nubian shield. Geologists attribute the difference in lithology and metamorphic grade between the two belts to the difference in the level of exposure, with the Mozambican rocks interpreted as lower crustal equivalents of the juvenile rocks in the Arabian-Nubian shield. Recent geochronologic data indicate the presence of two major "Pan-African" tectonic events in East Africa. The East African Orogeny (800–650 Ma) represents a distinct series of events within the Pan-African of central Gondwana, responsible for the assembly of greater Gondwana. Collectively, paleomagnetic and age data indicate that another later event at 550 Ma (Kuunga orogeny) represents the final suturing of the Australian and Antarctic segments of the Gondwana continent. The Arabian-Nubian shield in the northern part of the East African orogen preserves many complete ophiolite complexes, making it one of the oldest orogens with abundant Penrose-style ophiolites, with crustal thicknesses similar to those of Phanerozoic orogens.

The Proterozoic record preserves several continental-scale rift systems. Rift systems with associated mafic dike swarms cut across the North China craton at 2.4 and 1.8 billion years ago, as well as in many other cratons. One of the best-known of Proterozoic rifts is the 1.2–1.0-Ga Keweenawan rift, a 950-mile- (1,500-km-) long 95-mile- (150-km-) wide trough that stretches from Lake Superior to Kansas in North America. This trough, like many Proterozoic rifts, is filled with a mixture of basalts, rhyolites, arkose, conglomerate, and other, locally red, immature sedimentary rocks, all intruded by granite and syenite. Some of the basalt flows in the Keweenawan rift are 1–4 miles (2–7 km) thick.

Massive Proterozoic diabase dike swarms cut straight across many continents and may be related to some of the Proterozoic rift systems or to mantle plume activity. Some of the dike swarms are more than 1,865 miles (3,000 km) long, hundreds of kilometers wide, and consist of thousands of individual dikes ranging from less than three feet to more than 1,640 feet (1–500 m). The 1.267 Ga Mackenzie swarm of North America and others show radial patterns and point to a source near the Coppermine River basalts in northern Canada. Other dike swarms are more linear and parallel failed or successful rift arms. The direction of magma flow in the dikes is generally parallel to the surface, except in the central 300–650 miles (500–1,000 km) of the swarms, suggesting that magma may have fed upward from a plume that initiated a triple-armed rift system, and then the magma flowed away from the plume head. In some cases, such as the Mackenzie swarm, one of the rift arms succeeded in forming an ocean basin.

Arabian-Nubian shield in eastern Sudan *(Earth Sciences and Image Analysis Laboratory, NASA Johnson Space Center)*

Cratonic cover sequences are well preserved from the Proterozoic in many parts of the world. In China, the Mesoproterozoic Changcheng Series consists of several-kilometer-thick accumulations of quartzite, conglomerate, carbonate, and shale. In North America, the Paleoproterozoic Huronian Supergroup of southern Canada consists of up to 7.5 miles (12 km) of coarse clastic rocks dominated by clean beach and fluvial sandstones, interbedded with carbonates and shales. Thick sequences of continentally derived clastic rocks interbedded with marine carbonates and shales represent deposition on passive continental margins, rifted margins of back arc basins, and cratonic cover sequences from epicontinental seas. Many parts of the world have similar cratonic cover sequences, showing that continents were stable by the Proterozoic, that they were at a similar height with respect to sea level (freeboard), and that the volume of continental crust at the beginning of the Proterozoic was at least 60 percent of the present volume of continental crust.

One of the more unusual rock associations from the Proterozoic record is the 1.75–1.00 Ga granite-anorthosite association. The anorthosites (rocks consisting essentially of all plagioclase) have chemi-cal characteristics indicating that they were derived as cumulate rocks from fractional crystallization of a basaltic magma extracted from the mantle, whereas partial melting of the lower crustal rocks produced the granites. The origin of these rocks is not clearly understood—some geologists suggest they were produced on the continental side of a convergent margin, others suggest an extensional origin, and still others suggest an anorogenic association.

Proterozoic life began with very simple organisms similar to those of the Archean, and 2.0 Ga planktonic algae and stromatolitic mounds with prokaryotic filaments and spherical forms are found in many cherts and carbonates. The stromatolites formed by cyanobacteria exhibit a wide variety of morphologies, including columns, branching columns, mounds, cones, and cauliflower type forms. In the 1960s, many geologists, particularly from the Russian academies, attempted to correlate different Precambrian strata based on the morphology of the stromatolites they contained, but this line of research proved futile as all forms are found in rocks of all ages. The diversity and abundance of stromatolites peaked about 750 million years ago, and declined rapidly after that time period, probably due to the

Stromatolites in the Helena Formation along Highline Trail, Glacier National Park, Montana *(USGS)*

sudden appearance of grazing multicellular metazoans, such as worms, at this same time. Eukaryotic cells (with membrane-bound nuclei and other distinct organelles) are preserved in sedimentary rocks from as early as 1.8 Ga, reflecting increased oxygen in the atmosphere and ocean. The acritarchs are spherical fossils of single-celled, photosynthetic marine plankton found in a wide variety of rock types. Around 750 million years ago some of the prokaryotes experienced a sudden decline, as eukaryotic life-forms adapted to fill their niches. This dramatic change is not understood, but its timing is coincident with the breakup of Rodinia and the formation of Gondwana is notable; thus tectonic changes could have induced atmospheric and environmental changes that favored one type of organism over the other.

A wide range of metazoans, complex multicellular organisms, are recognized from the geological record by 1.0 Ga, and probably evolved along several different lines before the record was well established. A few metazoans up to 1.7 Ga have been recognized from North China, but the fossil record from this interval is poorly preserved since most animals were soft-bodied. The transition from the Proterozoic fauna to the Paleozoic is marked by a remarkable group of fossils known as the Ediacaran fauna,

first described from the Ediacara Hills in the Flinders Ranges of southern Australia. These 550–540 million-year-old fauna represent an extremely diverse group of multicellular, complex metazoans including jellyfishlike forms, flatwormlike forms, soft-bodied arthropods, echinoderms, and many other species. The ages of these fauna overlap slightly with the sudden appearance and explosion of shelly fauna in Cambrian strata at 540 million years ago, showing the remarkable change in life coincident with the formation of Gondwana at the end of the Proterozoic.

See also ARCHEAN; GONDWANA, GONDWANA-LAND; PRECAMBRIAN; SUPERCONTINENT CYCLES.

FURTHER READING

Condie, Kent C., and Robert Sloan. *Origin and Evolution of Earth: Principles of Historical Geology.* Upper Saddle River, N.J.: Prentice Hall, 1997.

Dalziel, Ian W. D. "Neoproterozoic-Paleozoic Geography and Tectonics: Review, Hypothesis, Environmental Speculation." *Geological Society of America Bulletin* 109 (1997): 16–42.

Grotzinger, John P., Samuel A. Bowring, Beverly Z. Saylor, and Alan J. Kaufman. "Biostratigraphic and Geochronologic Constraints on Early Animal Evolution." *Science* 270 (1995): 598–604.

Hoffman, Paul F. "Did the Breakout of Laurentia Turn Gondwanaland Inside-Out?" *Science* 252 (1991): 1,409–1,411.

Kaufman, Alan J., Andrew J. Knoll, and Guy M. Narbonne. "Isotopes, Ice Ages, and Terminal Proterozoic Earth History." *National Academy of Sciences Proceedings* 94 (1997): 6,600–6,605.

Kusky, Timothy M., Mohamed Abdelsalam, Robert Tucker, and Robert Stern, eds., *Evolution of The East African and Related Orogens, and the Assembly of Gondwana.* Amsterdam; Elsevier, Special Issue of Precambrian Research, Vol. 123. Amsterdam: Elsevier, 2003.

Kusky, Timothy M., and Jianghai H. Li. "Paleoproterozoic Tectonic Evolution of the North China Craton." *Journal of Asian Earth Sciences* 22 (2003): 383–397.

Moores, E. M. "Southwest United States-East Antarctic (SWEAT) Connection: A Hypothesis." *Geology* 19 (1991): 425–428.

Stern, Robert J. "Arc Assembly and Continental Collision in the Neoproterozoic East African Orogen: Implications for Consolidation of Gondwanaland." *Annual Review of Earth and Planetary Science* 22 (1994): 319–351.

Windley, Brian F. *The Evolving Continents.* 3rd ed. Chichester, England: John Wiley & Sons, 1995.

Ptolemy, Claudius Ptolemaeus (83–168)

Greek *Mathematician, Astronomer, Astrologer, Geographer* Claudius Ptolemaeus, known in English as Ptolemy, was an important ancient Greek mathematician, astronomer, and geographer whose works contributed to building the basis for European and Islamic sciences. Ethnically Ptolemy was Greek, though his name Claudius shows he had Roman citizenship. Egyptians knew him as "The Upper Egyptian," suggesting he was from southern Egypt, but this is not certain. His works show that he had access to the older Babylonian astronomical observations and data. He is most famous for *Almagest* (The Great Treatise, or The Mathematical Treatise). His second main work was *Geographia,* which discussed the geography of the Greco-Roman world. The third major treatise that bears Ptolemy's name is the *Tetrabiblos* (Four Books), a discourse on astrology and natural philosophy.

Ptolemy's *Almagest* is the oldest complete discussion of astronomy surviving from the ancient world, though older discussions are known from the Babylonian astronomers. Ptolemy claimed to have used notes from older astronomers going back more than 800 years before him in *Almagest,* yet these sources have largely disappeared with time. *Almagest* contains star tables, a star catalog, and a description of the 48 ancient constellations. Ptolemy used a geocentric model for the solar system and universe, which would be later challenged by Polish astronomer Nicolaus Copernicus, who advocated a heliocentric model. Ptolemy used a system of nested spheres to describe his geocentric model of the universe and calculated that the Sun was located 1,210 Earth radii from the Earth, while so-called fixed stars were located at 20,000 radii from the Earth, which was located at the center of the universe. *Almagest* contained a group of tables that contained the data to predict the locations of the Sun, Moon, planets, and eclipses, and these became popular. Ptolemy used them to produce a star calendar and almanac that was favored among astrologers. The popularity of *Almagest* meant that it was translated into Arabic and Latin from Greek, hence was preserved well and survives to this day.

The second main scientific work of Ptolemy was his *Geographia* (geography), a compilation of the known geographical features of the Roman Empire at the time. This included Europe and the Mediterranean area, North Africa, Arabia, Persia, and southern Asia. Many of the details of the maps shown by Ptolemy came from the earlier Greek geographer Marinos of Tyre, as well as maps from the Persians. *Geographia* contains several sections. The first includes data and methods and defines a coordinate system that included latitude measured from the equator. Ptolemy measured latitude in terms of the length of the longest day of the year at that location, which increases from 12 hours to 24 hours between the equator and the Arctic (and Antarctic) circles. For longitude Ptolemy used degrees, plotted from a meridian he placed at the westernmost known lands (Cape Verde Islands), which he called the Fortunata Islands. Ptolemy's maps of the east show China extending southward well into the Pacific Ocean, suggesting the Greek geographers of the time knew less about Asia than about the Mediterranean. Ptolemy worked on map projections, which are ways to project the three-dimensional globe onto a two-dimensional flat map. Ptolemy's maps appear quite distorted compared to modern maps, partly because of the primitive nature of the knowledge of the land, and partly because Ptolemy used an estimate for the size of the Earth that is much smaller than the real value.

The *Tetrabiblos,* Ptolemy's treatise on astrology, was very popular for predicting horoscopes through celestial positions and was widely translated into Arabic and Latin. The *Tetrabiblos* was written in general terms that common people could understand. In it, Ptolemy cautioned readers to use astrology as a compilation of astronomical data and not to be overly interpretive of the numerological significance of the astronomical data as was popular at the time. Most historians suggest that Ptolemy compiled the

Tetrabiblos from earlier sources and that his main contribution was to organize it in a rational and systematic way.

Ptolemy also wrote on music, publishing *Harmonics* about the mathematics of music, suggesting that music should be based on mathematic ratios, and he demonstrated how music could be translated into mathematics. He called this technique Pythagorean tuning, after Pythagoras, who first described the relation between music and math. Ptolemy wrote about light and optics, although his works in these fields are not well preserved.

See also ASTRONOMY; CONSTELLATION; COPERNICUS, NICOLAUS; HIPPARCHUS; SOLAR SYSTEM.

FURTHER READING

Berggren, J. Lennart, and Alexander Jones. *Ptolemy's Geography: An Annotated Translation of the Theoretical Chapters.* Princeton, N.J.: Princeton University Press, 2000.

Stevenson, Edward Luther, trans. and ed. *Claudius Ptolemy: The Geography.* 1932. Reprint: Mineola, N.Y.: Dover Publications, 1991.

pulsar   Pulsars are an unusual class of highly magnetized neutron stars that rotate quickly and emit a beam of radiation in the form of radio waves. The origin of the apparent pulsation of the beam is from the rotation of the star, and the difference between the magnetic pole of the neutron star and the rotation axis. Like the Earth, the magnetic axis of the neutron star can be offset from the rotational axis. Pulsars emit a strong beam of radiation whose direction is controlled by the orientation of the magnetic field. When the magnetic pole and the rotational pole are not coincident, the radiation beam swings wildly in space, defining a cone whose aperture is defined by the angular distance between the rotational and magnetic poles. This beam can be detected only if the beam of radiation is pointed at the observer (Earth) at one point along this cone of rotation, so there are many more pulsars in the universe than those observed from Earth. Pulsars have very regular rotation periods due to their large mass, so the observed pulses of radiation can be used as very precise clocks, with accuracies as good as atomic clocks.

The first pulsar was discovered in 1967 by Irish and British astronomers Jocelyn Bell Burnell and Antony Hewish, but the origin of the radiation was a puzzle, and some scientists even speculated that the pulses might be signals from extraterrestrial life and advanced civilizations. In 1968 Austrian astronomer Thomas Gold and his colleague Franco Pacini sug-

gested that pulsars were rotating neutron stars. In 1974 Antony Hewish became the first astronomer to be awarded the Nobel Prize in physics, although his student Jocelyn Bell, who made the discovery while working under Hewish, was not awarded the prize. Pulsars remain a poorly understood system in the universe but those that are known fall into four distinct classes based on the source of the energy that powers the emitted radiation.

- Rotation-powered pulsars are those where the loss of rotational energy of the stars powers the radiation.
- Accretion-powered pulsars are those where the release of gravitational potential energy of infalling matter produces the energy and emits X-ray radiation.
- Magnetars are those where the decay of a tremendously strong magnetic field powers the radiation.
- Gamma ray pulsars are poorly known, but emit only gamma ray radiation.

Understanding of pulsar evolution is rapidly evolving, and some theories now suggest connections between the different types of pulsars. X-ray pulsars may be old rotation-powered pulsars that lost most of their energy and became visible again after their binary companion stars expanded and began transferring matter to the neutron star. Pulsars show an evolutionary trend, with young pulsars being fast and energetic, and old ones being slow and weak. The fastest pulsar has a period of 1.4 milliseconds, with the upper limit on rotational speed being determined by the maximum speed the neutron star can rotate at without causing the neutron-degenerate matter in its core to break up.

Observations of pulsars have been applied to many other fields in astronomy and physics. Because of the large masses and strong radiation, they have been used to test, and confirm, gravitational radiation as predicted by Einstein's theory of general relativity and aided in the first detection of an extrasolar planetary system.

See also ASTRONOMY; STELLAR EVOLUTION.

FURTHER READING

Animation of a Pulsar. Available online. URL: http://www.einstein-online.info/de/images/einsteiger/pulsar.gif. Accessed January 16, 2010.

Chaisson, Eric, and Steve McMillan. *Astronomy Today.* 6th ed. Upper Saddle River, N.J.: Addison-Wesley, 2007.

Comins, Neil F. *Discovering the Universe.* 8th ed. New York: W. H. Freeman, 2008.

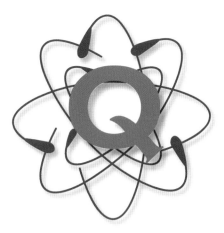

quasars Quasars are extremely powerful and distant galactic cores that are strong emitters of electromagnetic energy across the whole range of the spectrum, including X-rays, radio waves and visible light, and are associated with large redshifts, meaning that they are receding rapidly from the Earth and are located at great distances. Most quasars emit their energy almost equally across the electromagnetic spectrum, with a peak in the ultraviolet to optical bands. The name quasar is an abbreviation for QUASistellar radio source. The radiation from quasars comes from small regions, typically 10–10,000 Schwarzchild radii for the mass, which has led most astronomers to conclude that quasars represent supermassive black holes at the cores of galaxies. Being that quasars are so distant and are still observable, they must be some of the largest producers of radiation of any astronomical phenomena. Some of the more luminous quasars are more luminous than entire galaxies, emitting more radiation than a trillion stars like the Sun. The largest redshift known for any quasar is 6.43, corresponding to a distance of 28 billion light-years. This distance is seemingly impossible given the currently accepted age of the universe of 13.7 billion years, but is explained by some subtleties in the distance definitions for large distances in cosmology and general relativity. In any case, quasars are the most distant and powerful object known in the universe. Since the energy from the quasars travels to Earth at the speed of light, we know also that quasars are the oldest known objects in the universe. In many cases the radiation now being detected on Earth was emitted from the quasar shortly after the big bang.

Some quasars change rapidly in luminosity much like pulsars, particularly in the optical and X-ray bands. This observation suggests that quasars must be small objects since they cannot change faster than it takes for light to travel from one end of the object to the other. For such small objects to emit such high amounts of radiation they must use an energy source that is more efficient than the nuclear fusion that occurs in stars. The only known process that produces the required energy for sustained periods of time is the release of gravitational energy during the falling and accretion of matter to a massive star or black hole.

Quasars may be thought of as a special class of active galaxies. They are powered by release of gravitational energy during accretion of material into super massive black holes in the centers of distant galaxies. As the matter falls into the black hole it orbits the massive core, forming an accretion disk, from which much of the radiation is emitted.

There are more than 100,000 known quasars, each with a redshift between 0.06 and 6.4, which is equated with distances of 780 million to 28 billion light-years, meaning that they are among the most distant, brightest objects in the universe. Since distance is equated with time, the radiation from quasars provides a picture of what the earliest universe looked like soon after the big bang. The brightest known quasar is in the Virgo constellation and has a luminosity about 2 trillion times that of the Sun, or 100 times more than the whole Milky Way Galaxy. It would appear as bright as the Sun if it were located 33 light-years from Earth.

The fact that quasars all have large redshifts and are located at great distances from Earth means that they were more common in the early history of the universe soon after the big bang. This can be understood by considering that it takes the light billions of light years to travel from the quasars to Earth, and there are 100,000 known quasars of this old age. No quasars are known to be younger than 780 million light-years, and the further back in space/time telescopes can probe, the more abundant quasars are relative to other objects in the universe.

See also ASTRONOMY; ASTROPHYSICS; BINARY STAR SYSTEMS; BLACK HOLES; CONSTELLATION; COSMOLOGY; EINSTEIN, ALBERT; GALAXIES; HUBBLE, EDWIN; ORIGIN AND EVOLUTION OF THE UNIVERSE; PLANETARY NEBULA.

FURTHER READING

Chaisson, Eric, and Steve McMillan. *Astronomy Today.* 6th ed. Upper Saddle River, N.J.: Addison-Wesley, 2007.

Comins, Neil F. *Discovering the Universe.* 8th ed. New York: W. H. Freeman, 2008.

Melia, Fulvio. *The Edge of Infinity. Supermassive Black Holes in the Universe.* Cambridge: Cambridge University Press. 2003.

Quaternary   The last 1.8 million years of Earth history are known as the Quaternary period, which is divided into the older Pleistocene and the younger Holocene. Jules Desnoyers was the first to recognize that the rocks and unconsolidated deposits formed during this period were different from older deposits. Their characteristic boulder clays and other units deposited by glaciers in Europe reflected globally cool climates for the first part of the Quaternary, since glacier deposits were found in many parts of both the Northern and Southern Hemispheres. Global climate zones condensed near the equator, ice sheets covered about one-third of the continental surfaces, and desert regions converted to moist grasslands. Grasses, plants, and mammals experienced a rapid expansion.

Another major important discovery from this period was the recognition of the first human fossils, which became the basis for dividing the Quaternary into the older Pleistocene and the younger Holocene, in which human fossils appear abundantly about 10,000 years ago. Older primate fossils including early hominids are found in the older record going back several million years, and the record of human habitation in the Western Hemisphere extends back to approximately 13,000 or 14,000 years ago. Genetic evidence suggests that humans are all descendants of a single female ancestor that lived somewhere in East Africa about 100,000 to 300,000 years ago, with the first hominids appearing about 4 million years ago.

See also NEOGENE; PLEISTOCENE; TERTIARY.

FURTHER READING

Charlesworth, J. Kaye. *The Quaternary Era.* London: Arnold, 1957.

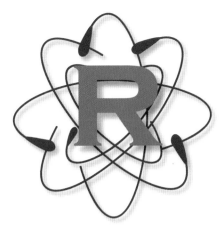

radiation Radiation refers to any process in which radiation is emitted as particles or waves (such as heat, light, alpha particles, or beta particles) from one body, travels through a medium, and is absorbed by another body. Heat transfer by infrared rays is also known as radiation or radiative heat transfer. Infrared radiation travels at the speed of light, can travel through a vacuum, gets reflected and refracted, and does not affect the medium that it passes through. Ionizing radiation occurs in processes associated with nuclear weapons and reactors and the decay of radioactive substances, whereas electromagnetic radiation refers to energy transferred through the whole range of wavelengths and frequencies, as described by the electromagnetic spectrum.

IONIZING AND NONIONIZING RADIATION

Ionizing radiation includes any radiation process in which individual quanta of energy are capable of ionizing atoms or molecules within the material that absorbs the radiation. Ionizing radiation is produced by the natural radioactive decay of rocks and radioactive materials, by cosmic rays, nuclear fission, nuclear fusion, and by similar processes that occur in nuclear weapons, nuclear reactors, X-ray equipment, and high-energy physics experiments. Ionizing radiation can cause chemical changes to the material and can damage biological tissues as well as rock and structural materials.

Particle radiation is a type of radiation that is sometimes ionizing, and sometimes not. When fast-moving subatomic particles such as electrons, protons, and neutrons carry enough energy, they produce an ionizing effect on the material through which they pass, but if their energy is too low they do not.

Nonionizing radiation is any type of radiation that does not carry enough energy per quantum to ionize atoms or molecules. In most cases nonionizing radiation consists of the lower-energy parts of the electromagnetic spectrum, including radio waves, microwaves, terahertz radiation, infrared light, and visible light. These lower-energy forms of electromagnetic radiation do not ionize or damage tissue but can excite electrons in the tissue to a higher energy state.

ELECTROMAGNETIC RADIATION

The electromagnetic spectrum categorizes types of radiation according to wavelength, with the shortest wavelengths being cosmic rays, and in increasing wavelength, gamma rays, X-rays, ultraviolet rays, visible rays, infrared rays, microwave rays, radio waves, and television waves. The environment contains a low level of constant background radiation, mostly from the radioactive decay of minerals and radioactive gases such as radon and thoron. Some background radiation, known as cosmic radiation, comes from space. The Sun emits solar radiation consisting of visible light, ultraviolet radiation, and infrared waves spanning the entire spectrum of electromagnetic wavelengths from radio waves to X-rays. The Sun also emits high-energy particles such as electrons, especially from solar flares. X-rays, because of their extremely short wavelength, are able to penetrate soft tissue and some sands and soils and reflect off internal denser material such as bones or rocks. This property has made X-rays useful for diagnostic medicine and geologic mapping of subsurface materials. Short-wavelength radiation is measured in nanometers (nm), where 1 nm equals 3.937×10^{-8} inches). Short-wavelength, high-frequency electromagnetic waves from 3.93×10^{-7} to 1.57×10^{-5} inch (10–400 nm) are known as ultraviolet radiation, which is powerful and useful to people for many applications but harmful in strong doses. Visible radiation includes all that humans see with their eyes, including the wide range of colors of the rainbow. Higher energy forms of electromagnetic radiation such as infrared, microwave, X-rays, and gamma rays can ionize materials and damage tissue.

CHARACTERISTICS OF ELECTROMAGNETIC RADIATION

Radiation from different wavelengths in the electromagnetic spectrum has very different characteristics and uses.

Radio waves have wavelengths ranging from one millimeter to hundreds of meters and frequencies of about 3 Hz to 300 GHz. They are commonly used by people to transmit data for television, mobile phones, wireless internet connections, and many other applications. The technology to encode radio waves with data is complex but involves changing the amplitude and frequency and phase relations of waves within a specific frequency band.

Microwave radiation has wavelengths ranging from one millimeter to one meter and frequencies between 0.3 GHz and 300 GHz. It includes super high frequency (SHF) and extremely high frequency classes. Microwaves are absorbed by molecules with dipolar covalent bonds, a property that is used to heat material uniformly and rapidly in microwave ovens. Microwave radiation is also used for some communication applications.

Terahertz radiation has wavelengths between the far infrared and microwaves and frequencies between 300 GHz and 3 terahertz. Radiation in this region can be used for imaging and communications and in electronic warfare to disable electronic equipment.

Infrared radiation has wavelengths between visible light and terahertz radiation and frequencies of 300 GHz (1 mm) to 400 THz (750 nm). Far-infrared radiation (300 GHz to 30 THz) is absorbed by the rotation of many gas molecules, by the molecular motions in liquids, and by phonons (a quantized mode of vibration of a crystal lattice) in solid phases. Most of the far-infrared radiation that enters the Earth's atmosphere is absorbed except for a few wavelength ranges (called windows) where some energy can penetrate. Mid-infrared radiation has frequencies from 30–120 THz and includes thermal radiation from blackbodies (i.e., bodies that absorb all energy at all wavelengths when they are cold). Near-infrared radiation is similar to visible light and has frequencies from 120 to 400 THz.

Higher frequency radiation (between 400 and 790 THz) with wavelengths between 400 and 700

nanometers is detectable by the human eye. Known as visible light, this form of radiation is also the range of most of the radiation emitted from the Sun and stars. When objects reflect or emit light in the visible range, the human eye and brain are able to process data from these wavelengths into an optical image of the object. The details of how the human brain perceives radiation from these wavelengths and processes it into an image are not completely understood. Molecular biologists, neuroscientists, psychologists, and biophysicists are actively studying these processes.

Ultraviolet radiation has wavelengths shorter than visible light and longer than X-rays, falling between 400 and 10 nm, and has energies between 3 and 124 electron volts. Ultraviolet radiation is emitted by the Sun and is a highly energetic ionizing radiation that can induce chemical reactions, may cause some substances to glow or fluoresce, and can cause sunburn on human skin. The ultraviolet radiation from the Sun is poisonous to most living organisms but is absorbed by the atmospheric ozone layer, preventing significant damage to life on Earth. If the ozone layer is depleted, ultraviolet radiation will cause significant damage to life on the surface of the Earth. During the early history of the Earth, no ozone layer existed, so the surface was constantly drenched in ultraviolet radiation. This may have prevented life from inhabiting the surface until a few billion years after the formation of the planet, when the ozone layer developed.

X-rays have wavelengths from 10 to 0.01 nanometers with frequencies between 30 petahertz and 30 exahertz (30×10^{15} Hz to 30×10^{18} Hz) and energies between 120 eV to 120 keV. X-rays can see through some objects (like flesh) but not others (like bones) and can be used to produce images for diagnostic radiography and crystallography. In the cosmos X-rays are emitted by neutron stars, some nebulae, and the accretion disks around black holes.

Gamma rays are the most energetic photons; they have no lower limit to their wavelength. Their frequency is greater than 10^{19} Hz, their energies are more than 100 keV, and their wavelengths are less than 10 picometers. Gamma rays are highly energetic and ionizing so they can cause serious damage to human tissue and represent a serious health hazard.

See also ARCHEAN; ASTRONOMY; COSMIC MICROWAVE BACKGROUND RADIATION; ELECTROMAGNETIC SPECTRUM; ENERGY IN THE EARTH SYSTEM; LIFE'S ORIGINS AND EARLY EVOLUTION; OZONE HOLE; PULSAR; QUASAR; RADIOACTIVE DECAY; RADIO GALAXIES; REMOTE SENSING.

FURTHER READING

Bossavit, A. *Computational Electromagnetism.* Boston: Academic, 1998.

LeRoy, Claude. *Principles of Radiation Interaction in Matter and Detection.* Hackensack, N.J.: World Scientific Publication Incorporated, 2009.

Saslow, W. M. *Electricity, Magnetism, and Light.* Boston, London, New York: Academic Press, 2002.

Sutcliffe, Jill. *Natural Background Radiation.* London: Imperial College Press, 2009.

radioactive decay The nuclei of unstable radioactive elements spontaneously break down to become more stable, emitting radiation as alpha particles, beta particles, or gamma rays. When these particles and rays are emitted by radioactive decay, they move through matter and knock electrons out of surrounding atoms, ionizing these atoms. Alpha decay of an atomic nucleus releases heavy, slow-moving alpha particles, the most ionizing form of radioactive radiation, consisting of two neutrons and two positively charged protons. Beta decay of a nucleus converts a neutron into a proton, emitting a high-speed electron and an electron-antineutrino, increasing the atomic number by one and leaving the mass number the same. The emitted beta particles are high-speed electrons that are moderately ionizing and penetrate deeper than alpha particles. They can travel about 10 feet (several meters) in the air and are easily deflected by electromagnetic fields. Gamma rays carry no charge and are weakly ionizing; they consist of a very high frequency type of electromagnetic radiation emitted by the nuclei of radioactive elements during decay, typically as part of alpha or beta decay. Gamma rays may also form from the interaction of high-energy electrons with matter. Gamma rays are deeply penetrating, are not deflected by electromagnetic fields, and are used to kill bacteria in food or to treat malignant tumors. Space-based observatories have detected cosmic gamma radiation coming from distant pulsars, quasars, and radio galaxies, but this cosmic gamma radiation cannot penetrate the Earth's atmosphere.

When radioactive elements or radioactive isotopes of stable isotopes decay to more stable elements, the atomic mass number of the element is changed, transmuting the parent element into a different element known as a daughter isotope, emitting atomic radiation in the process. For each radioactive element or isotope, decay occurs at a constant rate known as the half-life, determined by the time taken for half of any mass of that isotope to decay from the parent isotope to the daughter isotope. Radioactive decay is an exponential process, with half of the original starting material decaying in the first step, half of the remaining material (25 percent of the original material) decaying after the second step, half of the remaining material (12.5 percent of the original material) decaying after the third step, and so on.

Radioactive decay may occur in one step, or more commonly, in a series of steps known as a decay series. In some decay series the intermediate steps may be moderately or very short-lived, and the daughter isotope may be more or less radioactive than the parent isotope. There is a very wide range in half-lives for different radioactive isotopes, ranging from 4.4×10^{-22} seconds for lithium 5, through 4.551×10^9 years for uranium 238, to 1.5×10^{24} years for tellurium 128. The final product of all decay schemes is a stable element.

RADON
Radon is a poisonous gas released during radioactive decay of the uranium decay series. Radon is a heavy gas, and it presents a serious indoor hazard in every part of the country because it accumulates in poorly ventilated basements and well-insulated homes that are built on specific types of soil or bedrock rich in uranium minerals. Radon causes lung cancer, and since it is an odorless, colorless gas, its presence can go unnoticed in homes for years. However, the hazard of radon is easily mitigated, and homes can be made safe once the hazard is identified.

Uranium is a radioactive mineral that spontaneously decays to lighter daughter elements by losing high-energy particles at a predictable rate. Uranium decays to radium through a long series of steps with a cumulative half-life of 4.4 billion years. During these steps, intermediate daughter products are produced, and high-energy particles including alpha particles, consisting of two protons and two neutrons, are released, producing heat. The daughter mineral radium is itself radioactive, and it decays with a half-life of 1,620 years by losing an alpha particle, forming the heavy radon gas. Since radon is a gas, it escapes out of the minerals and ground, and makes its way to the atmosphere where it is dispersed, unless it gets trapped in people's homes, where it can cause damage by inhalation. Radon is a radioactive gas, and it decays with a half-life of 3.8 days, producing daughter products of polonium, bismuth, and lead. If this decay occurs while the gas is in someone's lungs, then the solid daughter products become lodged in the lungs, where they can cause damage. Most of the health risks from radon are associated with the daughter product polonium, which is easily lodged in lung tissue. Polonium is radioactive, and its decay and emission of high-energy particles in the lungs can damage lung tissue and eventually cause lung cancer.

Different geographic regions and specific places within those regions display huge variations in the concentration of radon. The concentration of radon gas also varies in the soil, home, and atmosphere. This variation is related to the concentration and type of radioactive elements present at a location. Radioactivity is measured in a unit known as a pico-curie (pCi), which is approximately equal to the amount of radiation produced by the decay of two atoms per minute.

Soils have gases trapped between the individual grains that make up the soil, and these soil gases have typical radon levels of 20 pCi per liter to 100,000 pCi per liter, with most soils in the United States falling in the range of 200–2,000 pCi/L. Radon can also be dissolved in groundwater, with typical levels falling between 100 and 2 million pCi/L. Outdoor air typically has 0.1–20 pCi/L, and radon in the air inside people's homes ranges from 1–3,000 pCi/L, with 0.2 pCi/L being typical.

Different parts of the country and world exhibit large natural variation in radon levels. One of the main variables controlling the concentration of radon at any site is the initial concentration of the parent element uranium in the underlying bedrock and soil. If the underlying materials have high concentrations of uranium, homes built in the area are likely to have higher concentrations of radon. Most natural geologic materials contain a small amount of uranium, typically about 1–3 parts per million (ppm). The concentration of uranium in soils derived from a rock is typically about the same as in the original source rock; however, some rock (and soil) types have much higher initial concentrations of uranium, ranging up to and above 100 ppm. Granites, some types of volcanic rocks (especially rhyolites), phosphate-bearing sedimentary rocks, and the metamorphosed equivalents of all of these rocks have the highest uranium contents.

As the uranium in the soil gradually decays, it leaves behind its daughter product, radium, in concentrations proportional to the initial concentration of uranium. The radium then decays by forcefully ejecting an alpha particle from its nucleus. This ejection is an important step in the formation of radon, since every action has a reaction. In this case the reaction is the recoil of the nucleus of the newly formed radon. Most radon remains trapped in minerals once it forms. However, if the decay of radium happens near the surface of a mineral, and if the recoil of the new nucleus of radon is away from the center of the grain, the radon gas can escape the bondage of the mineral. Once free, it can move into the intergranular space between minerals, soil, or cracks in the bedrock or become absorbed in groundwater between the mineral grains. Less than half (10–50 percent) of the radon produced by decay of radium actually escapes the host mineral. Most remains trapped inside, where it eventually decays leaving the solid daughter products behind as impurities in the mineral.

Once the radon is free in the open or water-filled pore spaces of the soil or bedrock it can move rather quickly. The exact rate of movement is critical to whether or not the radon enters homes, because radon does not stay around for very long with a half-life of only 3.8 days. The rate at which radon moves through a typical soil depends on the amount of pore space in the soil (or rock), the degree of connectedness between these pore spaces, and the exact geometry and size of the openings. Radon moves slowly through less permeable materials such as clay, but quickly through porous and permeable soils such as sand and gravel, and quickly through fractured material, whether it is bedrock, clay, or concrete.

The large variation in the concentration of radon from place to place partly results from the influence of the geometry of the pore spaces in a soil or bedrock underlying a home, on the rates of movement, and also on the initial concentration of uranium in the bedrock. Homes built on dry permeable soils can accumulate radon quickly because it can migrate through the soil quickly. Conversely, homes built on impermeable soils and bedrock are unlikely to concentrate radon beyond its natural background levels.

GEOCHRONOLOGY AND THE AGE OF THE EARTH

Why do geologists say that the Earth is 4.6 billion years old? For many hundreds of years, most people in European, Western, and other cultures believed the Earth to be about 6,000 years old, based on interpretations of passages in the Torah and Old Testament. However, based on the principles of uniformitarianism outlined by James Hutton and Charles Lyell, geologists in the late 1700s and 1800s began to understand the immense amount of time required to form the geologic units and structures on the planet and argued for a much greater antiquity of the planet. When Charles Darwin advanced his ideas about evolution of species, he added his voice to those calling for tens to hundreds of millions of years required to explain the natural history of the planet and its biota. In 1846, the physicist Lord Kelvin joined the argument, but he advocated an even more ancient Earth. He noted that the temperature increased with depth, and he assumed that this heat was acquired during the initial accretion and formation of the planet and has been escaping slowly ever since. Using heat flow equations Kelvin calculated that the Earth must be 20–30 million years old. However, Kelvin assumed that there were no new inputs of heat to the planet since it formed, and he did not know about radioactivity and heat produced by radioactive decay. In 1896, Madame Curie, working in the labs of Henri Becquerel in France, exposed film to uranium in a light-tight container and found that the film became exposed by a kind of radiation that was invisible to the eye. Soon, many elements were found to have isotopes, or nucleii of the same element with different amounts of neutrons in the nucleus. Some isotopes are unstable and decay from one state to another, releasing radioactivity. Radioactive decay occurs at a very specific and fixed average rate that is characteristic of any given isotope. In 1903, Pierre Curie and Albert Laborde recognized that radioactive decay releases heat, a discovery that was immediately used by geologists to reconcile geologic evidence of uniformitarianism with Lord Kelvin's calculated age of the Earth.

In 1905, Ernest Rutherford suggested that the constant rate of decay of radioactive isotopes could be used to date minerals and rocks. Because radioactivity happens at a statistically regular rate for each isotope, it can be used to date rocks. For each isotope an average rate of decay is defined by the time that it takes half of the sample to decay from its parent to daughter product, a time known as the half-life of the isotope. Thus, to date a rock we need to know the ratio of the parent to daughter isotopes, and simply multiply by the decay rate of the parent. Half-life is best thought of as the time it takes for half of any size sample to decay, since radioactive decay is a non-linear exponential process.

The rate of decay of each isotope determines which isotopic systems can be used to date rocks of certain ages. Also, the isotopes must occur naturally in the type of rock being dated, and the daughter products must be present only from decay of the parent isotope. Some of the most accurate geochronologic clocks are made by comparing the ratios of daughter products from two different decay schemes, since both daughters are present only as a result of decay from their parents, and their ratios provide special highly sensitive clocks.

Isotopes and their decay products provide the most powerful way to determine the age of the Earth. Most elements formed during thermonuclear reactions in pre-solar system stars that experienced supernovae explosions. The main constraints we have on the age of the Earth are that it must be younger than 6–7 billion years, because it still contains elements such as K-40, with a half-life of 1.25 billion years. If the Earth were any older, all of the parent product would have decayed. Isotopic ages represent the time that that particular element-isotope system got incorporated in a mineral structure. Since isotopes have been decaying since they were incorporated, the oldest age from an Earth rock gives a minimum age of the Earth. So far, the oldest known rock is the 4.03 billion-year-old Acasta gneiss of the Slave Province in northwest Canada, and the oldest mineral is a 4.2 billion-year-old zircon from Western Australia. From these data, we can infer that the Earth is between 4.2 and 6 billion years old.

The crust on the Moon is 4.2–4.5 billion years old, and the Earth, Moon, and meteorites all formed when the solar system formed. The U-Pb isotopic system is one of the most useful for determining the age of the Earth, although many other systems give identical results. Some meteorites contain lead, but no uranium or thorium parents. Since the proportions of the various lead isotopes have remained fixed since they formed, their relative proportions can be used to measure the primordial lead ratios in the early Earth. Then, by looking at the ratios of the four lead isotopes in rocks on Earth from various ages, we can extrapolate back to when they had the same primordial lead ratio. These types of estimates give an age of 4.6–4.7 billion years for the Earth, and 4.3–4.6 billion years for meteorites. So, the best estimate for the age of the Earth is 4.6 billion years, a teenager in the universe.

See also GEOCHRONOLOGY; HUTTON, JAMES.

FURTHER READING

Dalrymple, G. Brent. *The Age of the Earth*. Stanford, Calif.: Stanford University Press, 1994.

Dickin, Alan P. *Radiogenic Isotope Geology*. Cambridge: Cambridge University Press, 2005.

Faure, Gunter. *Principles of Isotope Geology*. New York: John Wiley & Sons, 1986.

radio galaxies Radio galaxies are types of active galaxies that emit large amounts of electromagnetic radiation as radio waves. They are related to other types of active galaxies including quasars, pulsars, and blazars, all of which emit the radiation through the synchrotron process. Most of the host galaxies for radio galaxies are large elliptical types, and the included radio galaxies are characterized by twin jets of material that beam strong radio-wave radiation out two opposed sides of the galaxy. Radio galaxies are used by cosmologists for calculating distances between Earth and astronomical bodies in the universe and studied for understanding processes with these unusual objects and their effects on the surrounding interstellar medium.

Emissions from radio galaxies are characterized by a smooth and polarized field, showing a broad band of distribution, so astronomers infer that the emissions are generated by the synchrotron process. Synchrotron radiation is formed by acceleration of charged particles traveling close to the speed of light in a magnetic field, so it is governed by relativistic effects described by quantum mechanics instead of the more familiar Newtonian mechanics. The material in the radio galaxies is inferred to be a plasma, or partially ionized gas in which some proportion of the electrons are free rather than bound to atoms or molecules. The charged particles that are accelerated and that emit the synchrotron radiation may be electrons, protons, or positrons. Positrons are the antiparticles, or antimatter counterparts to electrons, containing an electric charge of +1 and having the same mass as an electron. When a positron collides with an electron they annihilate each other, emitting two gamma-ray photons.

Radio galaxies show a wide range of structures when mapped using the distribution of radio-wave emissions. The most-common structure consists of lobes, which are double, generally symmetrical ellipsoidal structures extending far beyond the central nucleus of the radio galaxy. If the lobes are very elongated they are called plumes, and some radio galaxies show very thin beams known as jets that emit the radiation in narrow channels like flashlight beams at radio wavelengths. These lobes, plumes, and jets are powered by beams of high-energy particles accelerated in the nucleus of the galaxy, and they are focused by some process to form the lobe and jet structures. Radio galaxies are classified into two main types, FRI and FRII types. FRI types have the brightest emissions toward the center of the galaxy, whereas FRII types are the brightest near their edges. In general FRI types have low luminosity whereas FRII types have a high luminosity. This classification, based on morphology, also relates to the modes of energy transportation in the radio source. FRI objects transport the radiation through bright jets in the center of the galaxy, whereas FRII objects have faint jets but bright hotspots at the ends of the lobes. Energy transportation in FRII objects must be efficient to transfer the energy to the ends of the lobes, but in FRI objects the energy transportation radiates more of their energy away as it is moved along the jets.

The lobes of the largest radio galaxies extend to megaparsec scales, suggesting that it took them tens to hundreds of millions of years to grow. It is thought that the radio sources start small and grow larger gradually with time. Lobes extend from the galactic core and are fed by the jets from the cores, and as the pressure from the jets increases it causes the lobes to progressively expand. The greater the pressure exerted from the jets, the faster the lobes can expand into the interstellar medium.

Radio galaxies are commonly used by astronomers to measure large distances in the universe. Since they are so bright they can be observed from very large distances. The redshift is then measured to determine the distance to the object. Radio galaxies are being developed into a standard ruler to measure distances in the universe.

See also ASTRONOMY; ASTROPHYSICS; BLACK HOLES; COSMIC RAYS; ELECTROMAGNETIC SPECTRUM; GALAXIES; HUBBLE, EDWIN; INTERSTELLAR MEDIUM; PULSAR; QUASAR; REMOTE SENSING; TELESCOPES; UNIVERSE.

FURTHER READING

Atlas of DRAGNs. Available online. URL: http://www. jb.man.ac.uk/atlas/. Modified November 5, 2000.

Chaisson, Eric, and Steve McMillan. *Astronomy Today.* 6th ed. Upper Saddle River, N.J.: Addison-Wesley, 2007.

Comins, Neil F. *Discovering the Universe.* 8th ed. New York: W. H. Freeman, 2008.

Encyclopedia of Astronomy and Astrophysics. CRC Press, Taylor and Francis Group Publishers. Available online. URL: http://eaa.crcpress.com/. Accessed October 24, 2008.

ScienceDaily: Astrophysics News. ScienceDaily LLC. Available online. URL: http://www.sciencedaily.com/news/ space_time/astrophysics/. Accessed October 24, 2008.

Snow, Theodore P. *Essentials of the Dynamic Universe: An Introduction to Astronomy.* 4th ed. St. Paul, Minn.: West Publishing Company, 1991.

remote sensing Remote sensing is the acquisition of information about an object by recording devices that are not in physical contact with the object. Different types include airborne or spaceborne techniques and sensors that measure different properties of earth materials, ground-based sensors that measure properties of distant objects, and techniques that penetrate the ground to map subsurface properties. Common use of the term *remote sensing* refers to the airborne and space-based observation systems, with ground-based systems more commonly referred to as geophysical techniques.

Remote sensing grew out of airplane-based photogeologic reconnaissance studies, designed to give geologists a vertically downward-looking regional view of an area of interest, providing information and a perspective not readily appreciated from the ground. Many geological mapping programs include the use of stereo aerial photographs, produced by taking downward-looking photographs at regular intervals along a flight path from an aircraft, with every area on the ground covered by at least two frames. The resolution of typical aerial photographs is such that objects less than 3.2 feet (1 m) across can be easily identified. The camera and lens geometry is set so that the photographs can be viewed with a stereoscope, where each eye looks at one of the overlapping images, producing a visual display of greatly exaggerated topography. This view can be used to pick out details and variations in topography, geology, and surface characteristics that greatly aid geologic mapping. Geologic structures, rock dips, general rock types, and the distribution of these features can be mapped from aerial photographs.

Modern techniques of remote sensing employ a greater range of the electromagnetic spectrum than aerial photographs. Photographs are limited to a narrow range of the electromagnetic spectrum between the visible and infrared wavelengths that are reflected off the land's surface from the Sun's rays. Since the 1960s a wide range of sensors that can detect and measure different parts of the electromagnetic spectrum have been developed, along with a range of different optical-mechanical and digital measuring and recording devices used for measuring the reflected spectrum. In addition the establishment of many satellite-based systems has provided stable observation platforms and continuous or repeated coverage of most parts of the globe. One technique uses a mirror that rapidly sweeps back and forth across an area, measuring the radiation reflected in different wavelengths. Another technique uses a line-scanning technique, where thousands of detectors are arranged to electronically measure the reflected strength of radiation from different wavelengths in equally divided time intervals as the scanner sweeps across the surface, producing a digital image consisting of thousands of lines of small picture elements (pixels) representing each of the measured intervals. The strength of the signal for each pixel is converted to a digital number (dn) for ease of data storage and manipulation to produce a variety of different digital image products. Information from the reflected spectrum is picked up by the sensors. The digital data encodes this information, and digital image processing converts the strength of the signal from different bands into the strength of the mixture of red, green, and blue, with the mixture producing a colored image of the region. Different bands may be assigned different colors such as red, green, and blue to produce a colored image. Different electromagnetic bands may even be numerically or digitally combined or ratioed to highlight different geologic features.

Optical and infrared imagery are now widely used for regional geologic studies, with common satellite platforms including the United States–based Landsat systems, the French SPOT satellite, and more recently some multispectral sensors including Advanced Spaceborne Thermal Emission and Reflection Radiometer (ASTER) and Advanced Very High Resolution Radiometer (AVHRR) data. Optical and infrared imagery can detect differences in rock and mineral types because the reflection is sensitive to molecular interactions with solar radiation, highlighting differences between Al-OH bonds, C-O bonds, and Mg-OH bonds, effectively discriminating between different minerals such as micas, Mg-silicates, quartz, and carbonates. Bands greater than 2.4 microns are sensitive to the temperature of the surface instead of the reflected light, and studies of surface temperature have proven useful for identifying rock types, moisture content, water and hydrocarbon seeps, and caves.

Microwave remote sensing (of wavelengths less than 0.04 inches, or 1 mm) uses artificial illumination of the surface since natural emissions are too low to be useful. Satellite and aircraft-based radar systems are used to shoot energy of specific wavelength and orientation to the surface, which reflects it back to the detector. Radar remote sensing is very complex, depending on the geometry and wavelength of the system and on the nature of the surface. The strength of the received signal is dependent on features such as surface inclination, steepness, orientation, roughness, composition, and water content. Nonetheless, radar remote sensing has proven to be immensely useful for both military and scientific purposes, producing images of topography, surface roughness, and structural features such as faults, foliations, and other features that are highlighted by radar reflecting off sharp edges. Under certain circumstances, radar penetrates the surface of some geological materials (such as dry sand) and can produce images of what lies beneath the surface, including buried geologic structures, pipelines, and areas of soil moisture.

TYPES OF SATELLITE IMAGERY

Satellite imagery forms one of the basic tools for remote sensing. The types of satellite images available to the geologist, environmental scientist, and others are expanding rapidly, and only the most common in use are discussed here.

The Earth Resources Technology Satellite (ERTS-1), the first unmanned digital imaging satellite, was launched on July 23, 1972. Four other satellites from the same series, later named Landsat, were launched at intervals of a few years. The Landsat spacecraft carried a Multi-Spectral Scanner (MSS), a Return Beam Vidicon (RBV), and later, Thematic Mapper (TM) imaging systems.

Landsat Multi-Spectral Scanners produce images representing four different bands of the electromagnetic spectrum. The four bands are designated band 4 for the green spectral region (0.5 to 0.6 microns); band 5 for the red spectral region (0.6 to 0.7 microns); band 6 for the near-infrared region (0.7 to 0.8 microns); and band 7 for another near-infrared region (0.8 to 1.1 microns).

Radiation reflectance data from the four scanner channels are converted first into electrical signals, then into digital form for transmission to receiving stations on Earth. The recorded digital data are reformatted into what we know as computer compatible tapes (CCT) and/or converted at special processing laboratories to black-and-white images. These images are recorded on four black-and-white films from which photographic prints are made in the usual manner.

The black-and-white images of each band provide different sorts of information because each of the four bands records a different range of radiation. For example, the green band (band 4) most clearly shows underwater features because light with wavelengths in the green region of the visible spectrum is able to penetrate shallow water, and is therefore useful in coastal studies. The two near-infrared bands, which measure the reflectance of the Sun's rays outside the sensitivity of the human eye (visible range), are useful in the study of vegetation cover.

When these black-and-white bands are combined, false-color images are produced. For example, in the most popular combination of bands 4, 5, and 7, the red color is assigned to the near-infrared band number 7 (and green and blue to bands 4 and 5 respectively). Vegetation appears red because plant tissue is one of the most highly reflective materials in the infrared portion of the spectrum, and thus, the healthier the vegetation, the redder the color of the image. Because water absorbs nearly all infrared rays, clear water appears black on band 7. Therefore, one cannot use this band to study features beneath water even in the very shallow coastal zones, but it is useful in delineating the contact between water bodies and land areas.

The geologic mapping community originally was most interested in flying the RBV, since it offered better geometric accuracy and ground resolution (130 feet; 40 m) than was available from the MSS (260 feet/80 m resolution) with which the RBV shared space on Landsats 1, 2, and 3. The RBV system contained three cameras that operated in different spectral bands: blue-green, green-yellow, and red-infrared. Each camera contained an optical lens, a shutter, the RBV sensor, a thermoelectric cooler, deflection and focus coils, erase lamps, and the sensor electronics. The three RBV cameras were aligned in the spacecraft to view the same 70-square-mile- (185-km^2-) ground scene as the MSS of Landsat. Although the RBV is not in operation today, images are available and can be utilized in mapping.

The TM is a sensor that was carried first on Landsat 4 and 5 with seven spectral bands covering the visible, near-infrared, and thermal infrared regions of the spectrum. With a ground resolution of 100 feet (30 m), the TM was designed to satisfy more demanding performance parameters, using experience gained from the operation of the MSS.

The seven spectral bands were selected for their band passes and radiometric resolutions. For example, band 1 of the TM coincides with the maximum transmissivity of water and demonstrates coastal water-mapping capabilities superior to those of the MSS; it also has beneficial features for the differentiation of coniferous and deciduous vegetation. Bands 2–4 cover the spectral region that is most significant for the characterization of vegetation. Band 5 readings allow estimation of vegetation and soil moisture, and thermal mapping in band 6 allows estimation of plant transpiration rates. Band 7 is primarily motivated by geological applications, including the identification of rocks altered by percolating fluids during mineralization. The band profiles, which are narrower than those of the MSS, are specified with stringent tolerances, including steep slopes in spectral response and minimal out-of-band sensitivity.

Geologic studies commonly use TM band combinations of 7 (2.08–2.35 μm), 4 (0.76–0.90 μm), and 2 (0.50–0.60 μm), due to the ability of this combination to discriminate features of interest, such as soil moisture anomalies, lithological variations, and to some extent, mineralogical composition of rocks and sediments. Band 7 is typically assigned to the red channel, band 4 to green, and band 2 to blue. This procedure results in a color composite image; the color of any given pixel represents a combination of brightness values of the three bands. With the full dynamic range of the sensors, there are 16.77×10^6 possible colors. By convention, this false-color combination is referred to as TM 742 (RGB). In addition to the TM 742 band combination, geologists sometimes use the thermal band (TM band 6; 10.4–12.5 μm) because it contains useful information potentially relevant to hydrogeology.

The French Système pour l'Observation de la Terre (SPOT) obtains data from a series of satellites in a sun-synchronous 500-mile- (830-km-) high orbit, with an inclination of 98.7°. The Centre Nationale d'Etudes Spaciales (CNES) designed the SPOT system, and the French industry in association with partners in Belgium and Sweden built it. Like the American Landsat, SPOT consists of remote-sensing satellites and ground receiving stations. The imaging is accomplished by two High-Resolution Visible (HRV) instruments that operate in either a panchromatic (black-and-white) mode for observation over a broad spectrum, or a multispectral (color) mode for sensing in narrow spectral bands. The ground resolutions are 33 and 66 feet (10 and 20 m) respectively. For viewing directly beneath the spacecraft, the two instruments can be pointed to cover adjacent areas. By pointing a mirror that directs ground radiation to the sensors, observation of any region within 280 miles (450 km) from the nadir is possible, thus

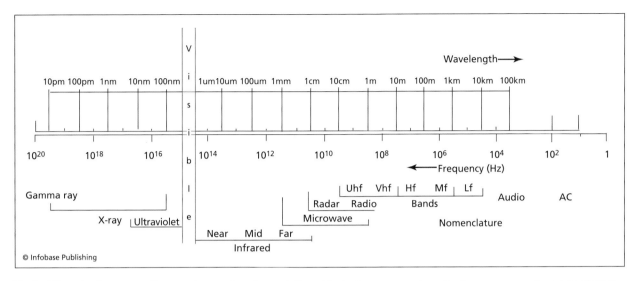

Part of the electromagnetic spectrum, showing relationship between wavelength, frequency, and nomenclature for electromagnetic radiation with different characteristics

allowing the acquisition of stereo photographs for three-dimensional viewing and imaging of scenes as frequently as every four days.

Radar is an active form of remote sensing, where the system provides a source of electromagnetic energy to illuminate the terrain. The energy returned from the terrain is detected by the same system and is recorded as a digital signal that is converted into images. Radar systems can be operated independently of light conditions and can penetrate cloud cover. A special characteristic of radar is the ability to illuminate the terrain from an optimum position to enhance features of interest.

Airborne radar imaging has been extensively used to reveal land surface features. However, until recently it has not been suitable for use on satellites because: (1) power requirements were excessive; and (2) for real-aperture systems, the azimuth resolution at the long slant ranges of spacecraft would be too poor for imaging purposes. The development of new power systems and radar techniques has overcome the first problem and synthetic-aperture radar systems have remedied the second.

The first flight of NASA's Shuttle Imaging Radar (SIR-A) in November of 1981 acquired images of a variety of features including faults, folds, outcrops, and dunes. Among the revealed features are the sand-buried channels of ancient river and stream courses in the Western Desert of Egypt. The second flight, SIR-B, had a short life; however, the more advanced and higher resolution SIR-C was flown in April 1994 (and was again utilized in August 1994). The SIR-C system measures both horizontal and vertical polarizations simultaneously at two wavelengths: L-band (23.5 cm) and C-band (5.8 cm).

This provides dual frequency and dual polarization data, with a swath width between 18 and 42 miles (30 and 70 km), yielding precise data with large ground coverage.

Different combinations of polarizations are used to produce images showing much more detail about surface geometric structure and subsurface discontinuities than a single-polarization-mode image. Similarly, different wavelengths are used to produce images showing different roughness levels since radar brightness is most strongly influenced by objects comparable in size to the radar wavelength; hence, the shorter wavelength C-band increases the perceived roughness.

Interpretation of a radar image is not intuitive. The mechanics of imaging and the measured characteristics of the target are significantly different for microwave wavelengths than the more familiar optical wavelengths. Hence, possible geometric and electromagnetic interactions of the radar waves with anticipated surface types have to be assessed prior to their examination. In decreasing order of effect, these qualities are surface slope, incidence angle, surface roughness, and the dielectric constant of the surface material.

Radar is uniquely able to map the geology at the surface and, in the dry desert environments, up to a maximum 30 feet (10 m) below the surface. Radar images are most useful for mapping structural and morphological features, especially fractures and drainage patterns, as well as the texture of rock types, in addition to revealing sand-covered paleochannels. The information contained in the radar images complements that in the TM images and eliminates the limitations of Landsat when only sporadic measurements can be made; radar sensors have the ability to "see" at night and through thick cloud cover since they are active rather than passive sensors.

RADARSAT is an Earth observation satellite developed by Canada, designed to support both research on environmental change and research on resource development. It was launched in 1995 on a Delta II rocket with an expected life span of five years. RADARSAT operates with an advanced radar sensor called Synthetic Aperture Radar (SAR). The synthetic aperture increases the effective resolution of the imaged area by means of an antenna design in which the spatial resolution of a large antenna is synthesized by multiple sampling from a small antenna. RADARSAT's SAR-based technology provides its own microwave illumination, thus can operate day or night, regardless of weather conditions. Thus, resulting images are not affected by the presence of clouds, fog, smoke, or darkness. This provides significant advantages in viewing under conditions that preclude observation by optical satellites. Using a single frequency, 2-inch (5-cm) horizontally polarized C band, the RADARSAT SAR can shape and steer its radar beam to image swaths between 20 and 300 miles (35 km to 500 km), with resolutions from 33 feet to 330 feet (10 m to 100 m), respectively. Incidence angles can range from less than 20° to more than 50°.

The Space Shuttle orbiters have the capability of reaching various altitudes, which allows the selection of the required photographic coverage. A camera that was specifically designed for mapping the Earth from space using stereo photographs was first flown in October 1984 on the Space Shuttle Challenger Mission 41-G. It used an advanced, specifically designed system to obtain mapping-quality photographs from Earth orbit. This system consisted of the Large Format Camera (LFC) and the supporting Attitude Reference System (ARS). The LFC derives its name from the size of its individual frames, which are 26 inches (66 cm) in length and 9 inches (23 cm) in width. The 992-pound (450-kg) camera has a 12-inch (305-mm) f/6 lens with a 40° × 74° field of view. The film, which is three-fourths of a mile (1,200 m) in length, is driven by a forward motion compensation mechanism as it is exposed on a vacuum plate, which keeps it perfectly flat. The spectral range of the LFC is 400 to 900 nanometers, and its photo-optical ground resolution ranges from 33–66 feet (10 to 20 m) from an altitude of 135 miles (225 km) in the 34,200 square mile (57,000 km²) area that is covered by each photograph.

The ARS is composed of two cameras with normal axes that take 35-millimeter photographs of star fields at the same instant as the LFC takes a photograph of the Earth's surface. The precisely known positions of the stars allow the calculation of the exact orientation of the Shuttle orbiter, and particularly of the LFC in the Shuttle cargo bay. This accurate orientation data, together with the LFC characteristics, allows the location of each frame with an accuracy of less than half a mile (1 km) and the making of topographic maps of photographed areas at scales of up to 1:50,000.

SYNTHETIC APERTURE RADAR

Spaceborne remotely sensed imagery has been routinely used as a reconnaissance tool by geologists since the initial launch of the Landsat series of satellites in 1972. More recently, spaceborne sensors such as Thematic Mapper (TM), Seasat Synthetic Aperture Radar (SAR), Shuttle Imaging Radar (SIR-A and SIR-B), and Système pour l'Observation de la Terre (SPOT) have scanned the Earth's surface with other portions of the electromagnetic spectrum in order to sense different features, particularly surface roughness and relief, and to improve spatial resolution. While TM and SPOT images have proven spectacularly effective at differentiating between various rock types, synthetic aperture radar (SAR) is particularly useful at delineating topographically expressed structures. Spaceborne SAR systems also play a major role in exploration of other bodies in the solar system.

Synthetic aperture radar (SAR) is an active sensor where energy is sent from a satellite (or airplane) to the surface at specific intervals in the ultrahigh frequency range of radar. The radar band refers to the specific wavelength sent by the source, and may typically include X-band (4 cm), K-band (2 cm), P-band (1 meter), L-band (23.5 cm), C-band, or others. SAR allows the user to acquire detailed images at any time of day or night and also in inclement weather. A complicated system, SAR basically works by first obtaining a two-dimensional image and then fine tuning that image with computers and sensors to create a decisively more accurate image. SAR provides detailed resolutions of a particular area for governments, military applications, and scientists, but is expensive to others who may wish to use it. The advancement of technology, however, is making it possible and economical in other applications.

The effectiveness of orbital SAR for geologic, particularly structural, studies depends primarily on three factors: (1) roughness contrasts; (2) local incidence angle variations (i.e., topography); and (3) look azimuth relative to topographic trends. Atmospheric or soil moisture can attenuate the strength of the radar signal, as can the types of atomic bonds in the minerals present in surface materials to some degree. Bodies of water are generally smoother than land, and appear as dark, radar smooth terrain. Structure is delineated on land by variations in local incidence angle, with surface roughness controlling the precise backscatter dependence. Different SAR satellites have different radar incidence or look angles, and some, such as RADARSAT, are adjustable and specifiable by the user. The 20° look angle chosen for Seasat was intended to maximize the definition of sea conditions, but had the incidental benefit of producing stronger sensitivity to terrain than would larger angles. Look azimuth has been shown to be an extremely important factor for low relief terrain of uniform roughness, with topography within about 20° of the normal to look azimuth being strongly highlighted.

FURTHER READING

Campbell, James B. *Introduction to Remote Sensing.* 4th ed. New York: Guilford Press, 2007.

Canty, Morton J. *Image Analysis, Classification and Change Detection in Remote Sensing.* New York: Taylor & Francis, 2007.

Drury, Steven A. *Image Interpretation in Geology.* London: Chapman and Hall, 1993.

Jensen, John R. *Remote Sensing of the Environment: An Earth Resource Perspective.* 2nd ed. Upper Saddle River, N.J.: Prentice Hall, 2007.

Lillesand, T. M., R. W. Kiefer, and J. W. Chipma. *Remote Sensing and Image Interpretation.* 6th ed. New York: John Wiley & Sons, 2007.

Sabins, Floyd F. *Remote Sensing, Principles and Interpretation.* New York: W. H. Freeman, 1997.

river system Stream and river valleys have served as preferred sites for human habitation for millions of years because they provide routes of easy access through rugged mountainous terrain and water for drinking, watering animals, and irrigation. Most of the world's large river valleys are located in structural or tectonic depressions such as rifts, including the Nile, Amazon, Mississippi, Hudson, Niger, Limpopo, Rhine, Indus, Ganges, Yenisei, Yangtze, Amur, and Lena. The soils in river valleys are also some of the most fertile that can be found, as they are replenished by yearly or less frequent floods. The ancient Egyptians, whose entire culture developed in the Nile River valley and revolved around the flooding cycles of the river, appreciated this characteristic of river systems. Rivers now provide easy and relatively cheap transportation on barges, and the river valleys are preferred routes for roads and railways as they are relatively flat and easier to build in than over mountains. Many streams and rivers have also become polluted as industry has dumped billions of gallons of chemical waste into our nation's waterways.

Stream and rivers are dynamic environments—their banks are prone to erosion, and the rivers periodically flood over their banks. During floods, rivers typically cover their floodplains with several or more feet of water, dropping layers of silt and mud. Ancient civilizations relied on this normal part of a river's cycle for replenishing and fertilizing their fields. Now that many floodplains are industrialized or populated by residential neighborhoods, the floods are no longer welcome and natural floods are regarded as disasters. On average, floods kill a couple of hundred people each year in the United States. Dikes and levees have been built around many rivers in attempts to keep the floodwaters out of towns. This exacerbates the flooding problem because it confines the river to a narrow channel, and the waters rise more quickly and cannot seep into the ground of the floodplain.

Streams are important geologic agents critical for other Earth systems. They carry most of the water from the land to the sea, they transport billions of tons of sediment to the beaches and oceans, and they erode and reshape the land's surface, forming deep valleys and floodplains and passing through mountains.

Streams and rivers are dynamic systems that constantly change their patterns, the amount of water (discharge) they carry, and the sediment transported by the system. Rivers can transport orders of magnitude more water and sediment during spring floods compared to low-flow times of winter or drought. Since rivers are dynamic systems, and the amount of water flowing through the channel changes, the channel responds by changing its size and shape to accommodate the extra flow. Five factors control how a river behaves: (1) width and depth of channel, measured in feet (m), (2) gradient, measured in feet per mile (m/km), (3) average velocity, measured in feet per second (m/sec), (4) discharge, measured in cubic feet per second (m^3/sec), and (5) load, measured as tons per cubic yard (metric tons/m^3). These factors continually interplay to determine the behavior of the river system. As one factor, such as discharge, changes, so do the others, expressed as:

$$Q = w \times d \times v$$

where Q represents discharge, w represents channel width, d represents channel depth, and v represents the velocity of the water in the channel.

All factors vary across stream, so they are expressed as averages. If one term changes then all or one of the others must change too. For example, with increased discharge, the river erodes and widens, and deepens its channel. The river may also respond by increasing the number of bends, known as meanders, and their curvature (measured as sinuosity), effectively creating more space for the water to flow

Kuskokwim River, just up from Bethel, Alaska, showing distributary channels, meanders, oxbow lakes, point bars, sand bars, and cut banks *(Paul Andrew Lawrence/Alamy)*

in and occupy by adding length to the river. The meanders can develop quickly during floods because the increased stream velocity adds more energy to the river system, and this can rapidly erode the cut banks enhancing the meanders.

The amount of sediment load available to the river is also independent of the river's discharge, so different types of river channels develop in response to different amounts of sediment load availability. If the sediment load is low, rivers tend to have simple channels, whereas braided stream and river channels develop where the sediment load is greater than the stream's capacity to carry that load. If a large amount of sediment is dumped into a river, it will respond by straightening, thus increasing the gradient and velocity and increasing its ability to remove the added sediment.

When rivers enter lakes or reservoirs along their path to the sea, the velocity of the water suddenly decreases. This causes the sediment load of the stream or river to be dropped as a delta on the lake bottom, and the river attempts to fill the entire lake with sediment in this manner. The river is effectively attempting to regain its gradient by filling the lake, then eroding the dam or ridge that created the lake in the first place. When the water of the river flows over the dam, it does so without its sediment load and therefore has greater power to erode the dam more effectively.

Rivers carry a variety of materials as they make their way to the sea. These materials range from minute dissolved particles and pollutants to giant boulders moved only during the most massive floods. The bed load consists of the coarse particles that move along or close to the bottom of the river bed. Particles move more slowly than the stream, by rolling or sliding. Saltation is the movement of a particle by short intermittent jumps caused by the current lifting the particles. Bed load typically constitutes between 5 and 50 percent of the total load carried by the river, with a greater proportion carried during high-discharge floods. The suspended load consists of the fine particles suspended in the river that make many rivers muddy. The particles of silt and clay move at the same velocity as the river. The suspended load generally accounts for 50–90 percent of the total load carried by the river. The dissolved load of a river consists of dissolved chemicals, such as bicarbonate, calcium, sulfate, chloride, sodium, magnesium, and potassium. The dissolved load tends to be high in rivers fed by groundwater. Rivers also carry pollutants, such as fertilizers and pesticides from agriculture and industrial chemicals, as dissolved load.

The range of sizes and amounts of material that a river can transport varies widely. The competence of a stream refers to the size of particles a river can transport under a given set of hydraulic conditions, measured in diameter of largest bed load. A river's capacity is the potential load it can carry, measured in the amount (volume) of sediment passing a given point in a set amount of time. The amount of material carried by rivers depends on a number of factors. Climate studies show erosion rates are greatest in climates between a true desert and grasslands. Topography affects river load, as rugged topography contributes more detritus, and some rocks are more erodable. Human activity, such as farming, deforestation, and urbanization, all strongly affect erosion rates and river transport. Deforestation and farming greatly increase erosion rates and supply more sediment to rivers, increasing their loads. Urbanization has complex effects, including decreased infiltration and decreased times between rainfall events and floods.

FLOODPLAINS

Floodplains are generally flat or low-lying areas that are adjacent and run parallel to river channels and are covered by water during flood stages of the river. The floodplain of a river is built by alluvium carried by the river and deposited in overbank environments, forming layers of silt, clay, and sand. Many narrow elongate channels filled by sands and gravels typically cut overbank deposits, marking places where the river formerly flowed and meandered away from during the course of river evolution. Sandy or gravelly levee deposits formed during flood stages of the river typically separate active and buried channels from floodplain deposits. These form because the velocity of floodwater decreases rapidly as it moves out of the channel, causing the current to drop heavy coarse-grained material near the river, forming a levee. Floodplains are also found around some lake basins that experience flood stages.

The increasing development and construction over floodplains creates potential and real hazards during floods. Floodplains are characterized by fertile soils and make excellent farmlands, which are nourished by yearly, decadal, and centurial floods, whereas buildings, towns, and cities have a much more difficult time dealing with periodic flooding.

FLUVIAL SYSTEMS AND CHANNEL TYPES

The term *fluvial* means "of the river," referring to different and diverse aspects of rivers. The term may be used to refer to the environment around rivers and streams, or more commonly, to refer to sediments deposited by a stream or river system.

Sediments deposited by rivers tend to become finer-grained and more rounded with increasing transport distance from the eroded source terrain, typically an uplifted mountain range. The sediments also tend to become more enriched in the stable chemical components, such as quartz and micas, and depleted in chemically vulnerable particles, such as feldspars.

Stream channels are rarely straight, and the velocity of flow changes in different places. Friction makes the flow slower on the bottom and sides of the channel, and the bends in the river make the zone of fastest flow swing from side to side. The character of channels changes in different settings because of differences in slope, discharge, and load. Straight channels are very rare, and those that do occur share many properties with curving streams. The thalweg is a line connecting the deepest parts of the channel. In straight segments, the thalweg typically meanders from side to side of the stream. In places where the thalweg is on one side of the channel, a bar may form on the other side. A bar (for example, a sand bar) is a deposit of alluvium in a stream

Most streams move through a series of bends known as meanders. Meanders are always migrating across the floodplain by the process of the deposition of the point bar deposits and the erosion of the bank on the opposite side of the stream with the fastest flow. The erosion typically occurs through slumping of the stream bank. Meanders typically migrate back and forth, and also down valley at a slow rate. If the downstream portion of a meander encounters a slowly erodable rock, the upstream part may catch up and cut off the meander. This forms an oxbow lake, which is an elongate and curved lake formed from the former stream channel.

A braided stream consists of two or more adjacent but interconnected channels separated by bars or an island. They form in many different settings, including mountain valleys, broad lowland valleys, and in front of glaciers. Braided streams tend to form where there are large variations in the volume of water flowing in the stream and a large amount of sediment is available to be transported during times of high flow. The channels typically branch, separate and reunite, forming a pattern similar to a complex braid. Braided streams have constantly shifting channels, which move as the bars are eroded and redeposited (during large fluctuations in discharge). Most braided streams have highly variable discharge in different seasons, and they carry more load than meandering streams. Braided streams form where the stream load exceeds the stream's capacity to carry the load.

The two or more adjacent but interconnected channels separated by bars or islands on braided streams constantly shift, moving as the bars are eroded and redeposited during large fluctuations in discharge. The highly variable discharge of braided streams enables them to carry more load than meandering streams.

River and stream channel deposits tend to be composed of sands and gravelly sands that exhibit large-scale three-dimensional ripples, shown in cross section as cross-bedding. These cross-bedded sands are common around the inner bends of channels and mark the former positions of point bars. They are often interbedded with planar-bedded sands marking flood stage deposits and gravelly sands deposited during higher flood stages. The tops of channel deposits are characterized by finer-grained sands with small scale ripples and mud drapes forming flaser-bedding, interbedded with muds, and grading up into overbank floodplain deposits. This upward-fining sequence is typical of fluvial deposits, especially those of meandering streams. In contrast, braided stream deposits show less order and are characteristically dominated by bed load material such as gravel and sand. They include imbricated gravels, gravels deposited in shallow scours, horizontally bedded sands, and gravels deposited in bars.

Fluvial channel deposits form a variety of geometric patterns on a more regional scale. Shoestring sands form a branched pattern of river channels enclosed in overbank shales and muds, formed by meandering and anastomosing river channels. Sheets and wedges of fluvial sediments form in front of uplifted mountain chains in foreland and rift basins, and they may pass basinward into deltaic or shallow marine sediments and mountainward into alluvial fan deposits. The type of tectonic setting for a basin can be deduced by changes or migration of different fluvial facies with stratigraphic height. Fluvial sediments are widely exploited for hydrocarbon deposits, and also are known for placer deposits of gold and other valuable minerals.

See also FLOOD; FLUVIAL.

FURTHER READING

Gordon, N. D., T. A. McMahon, and B. L. Finlayson. *Stream Hydrology: An Introduction for Ecologists.* New York: John Wiley & Sons, 1992.

Leopold, L. B. *A View of the River.* Cambridge, Mass.: Harvard University Press, 1994.

Maddock, Thomas, Jr. "A Primer on Floodplain Dynamics." *Journal of Soil and Water Conservation* 31 (1976): 44–47.

Noble, C. C. "The Mississippi River Flood of 1973." In *Geomorphology and Engineering,* edited by D. R. Coates, 79–98. London: Allen and Unwin, 1980.

Ritter, D. F., R. C. Kochel, and J. R. Miller. *Process Geomorphology.* 3rd ed. Dubuque, Iowa: W. C. Brown, 1995.

Rosgen, D. *Applied River Morphology.* Pasoga Springs, Colo.: Wildland Hydrology, 1966.

Schumm, S. A. *The Fluvial System.* New York: Wiley-Interscience, 1977.

United States Geological Survey. Water Resources. Available online. URL: http://water.usgs.gov/. Accessed December 10, 2007.

Russian geology The geology of Russia is dominated by the Archean Russian (East European) craton in the west, the Siberian craton in the north, and the large Altaid orogenic belt in the south. The Siberian craton is divided into the Aldan and Anabar shields. The Ural Mountains separate the Russian craton from the Siberian craton and the Altaids, and the Verkhoyansk-Kolyma block occupies northeasternmost Russia. Large parts of Siberia are covered by the Siberian flood basalts that formed during massive eruptions at the Permian-Triassic boundary and probably initiated the largest mass extinction known in Earth history. Siberia also stores vast amounts of the global carbon supply in the organic soils of the tundra and taiga.

RUSSIAN (EAST EUROPEAN) CRATON

The Russian or East European craton is well exposed in the Baltic states and in the Ukrainian shield, but it is mostly buried beneath late Precambrian to Phanerozoic cover in the Russian craton. The amalgamated East European craton, which formed the core of the Baltica block in the Proterozoic supercontinents of Rodinia and Gondwana, consisted of the Fennoscandian block (Baltic shield) in the northwest, the Volgo-Uralia block in the east, and the Sarmatia block in the south. The Baltic shield has a diverse Archean and Proterozoic crustal history including

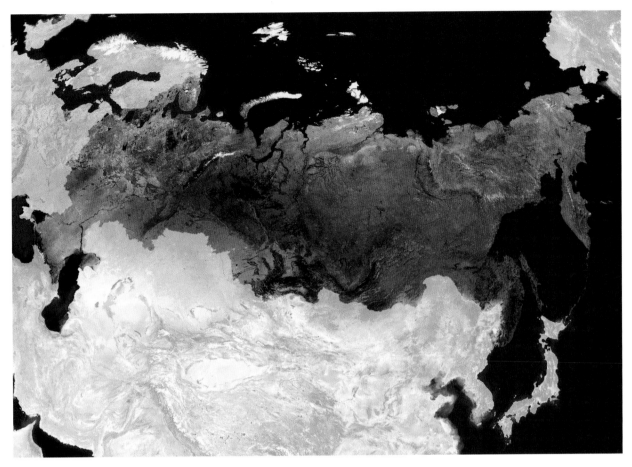

Satellite image of Russia *(M-Sat Ltd/Photo Researchers, Inc.)*

several convergent margin accretionary events, while the core of Sarmatia appears to be older than the Baltic shield. Most of the Volgo-Uralia block is buried beneath thick younger cover, but deep drill holes have revealed Archean rocks at depth. Most of the East European craton is covered by a thick sequence of middle to late Proterozoic sedimentary cover that is 1.8 miles (3 km) thick, whereas most of the Baltic shield is exposed down to the Archean and late Proterozoic basement.

The Sarmatia block consists of the Ukrainian shield and the Voronezh uplift. The Ukrainian shield in the southern part of the East European craton consists of 3.8–3.2 billion-year-old rocks, exposed along the big bend of the Dneiper River. These rocks include five main granitoid-greenstone rich blocks, each separated by structurally complex belts containing banded-iron formations and other metasedimentary rocks. The Voronezh uplift north of the Ukrainian shield contains similar rocks and is separated from the Ukrainian shield by a younger rift, the Dneiper-Donets aulacogen.

The Ural Mountains, where the craton collided with the Siberian craton in the Late Paleozoic, mark the eastern margin of the East European craton. Phanerozoic sediments largely shed from the Alpine orogen bury the southern margin of the craton. The southwestern boundary of the East European craton is marked by the Trans-European suture zone, separating the craton from the Alpine and Variscan belts of western Europe. The Early Paleozoic Caledonian orogen truncates the northwestern margin of the craton.

SIBERIAN CRATON (ALDAN AND ANABAR SHIELDS)

The Siberian craton has Archean crust exposed in the Aldan and Anabar shields and in some of the Phanerozoic fold belts that surround the craton. Much of the craton is covered by thick platform sediments, and these are also underlain by Archean crust.

The Aldan shield contains a gneissic and migmatitic basement deformed into large oval gneiss domes. Granulite facies gneisses, known as the Aldan Supergroup, have yielded ages of 3.4 to 3.2 billion years old. Interspersed greenstone and schist belts known as the Subgan or Olondo Group have been dated to be about 3.0 billion years old. Both of these rock suites are cut by granitoid plutons that fall into two

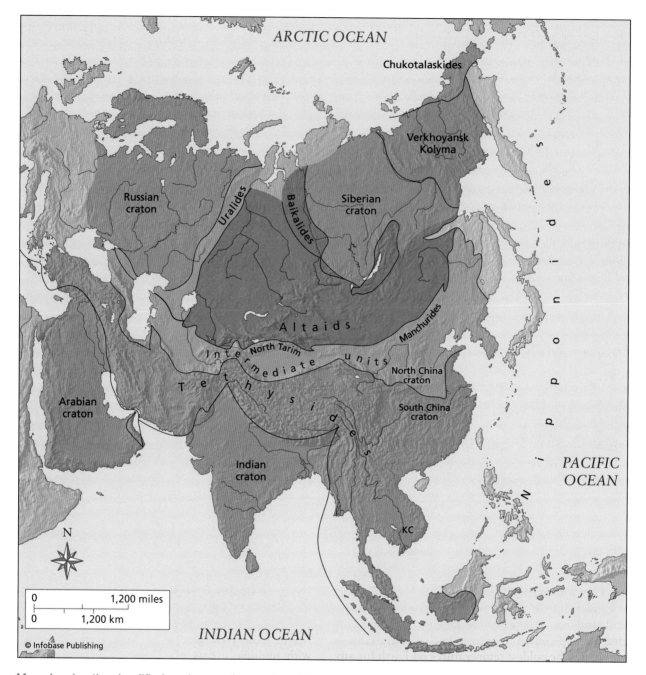

Map showing the simplified geology and tectonics of Russia and surrounding areas, including the Russian (East European) craton, Siberian (Angara) craton, Urals, Altaids, and the Verkhoyansk Kolyma region.

age groups, 3.1–2.9 billion years and 2.6–2.5 billion years. The gneisses form domes that separate the belts of greenstones and schist, which are typically complexly deformed and show evidence of polyphase folding.

The Aldan Supergroup consists mostly of granulite facies gneiss with interleaved slivers of quartzites, marble, and calc-silicate rocks, reflecting that a deformed and highly metamorphosed shallow-water sedimentary type sequence is deformed along with the gneiss. The Aldan Supergroup can be divided into three main belts: the Aldan-Timpton block contains a lower assemblage of thick quartzite overlain by pyroxene-amphibole gneiss; this is unconformably overlain by hypersthene-tonalite gneiss, marble and cal-silicate rocks, with quartzite and mica-gneiss, and an upper sequence consisting of interlayered calcsilicates and graphitic gneiss.

The Subgan Group includes greenstone belts made predominantly of mafic to felsic metavolcanic rocks along with minor amounts of quartzites, banded-iron formation, and schist metamorphosed to greenschist to amphibolite facies. There are approximately 30 greenstone belts in the Aldan shield, most 15–100 miles (30–150 km) long, and 1.5–2 miles (2.5–4 km) wide. The structural geology of the greenstone belts is difficult to decipher, but it is clear that they are cut by many shear zones and have been through polyphase deformation. Their relationships with older gneisses are not well established.

The Aldan shield is cut by abundant granitoid plutons, including a 3.4–3.2 billion-year-old tonalite-charnockite series and a 2.6–2.5 billion-year-old group of biotite-rich granites and pegmatites.

The Anabar shield forms a plateau sitting 1,500 to 2,800 feet (500–900 m) above sea level. Geologically, the shield is divided into Archean granulites and anorthosites of the Anabar Complex and the early Proterozoic (1.9 billion years old) Lamyuka Complex. Most interpretations of the Lamyuka Complex suggest that it represents amphibolite-facies reworking of the Anabar Complex along 6–20 mile- (10–30 km-) wide and 120 mile- (200 km-) long belts of deep shearing and basement reactivation.

The Anabar Complex consists of high grade granulitic gneiss, schist, and migmatite that are strongly folded about north-northwest axes and deformed into broad domes. The gneisses have yielded ages of about 3.2 billion years old and show strong metamorphism at 2.7 billion years ago. Granodiorite plutons cut the older rocks and are elongate parallel to regional foliations and structural trends.

VERKHOYANSK-KOLYMA BLOCK
The Verkhoyansk Ranges of northeastern Siberia stretch about 600 miles (1,000 km) from the Lena to Aldan Rivers, rising to 8,150 feet (2,480 m). This mountain belt is located just to the west of the Eurasian-North American plate boundary and is known to host significant deposits of coal, silver, lead, and zinc. These remote ranges, which are covered in deep snow for most of the year, are famous for having the coldest temperatures on Earth for any inhabited region.

ALTAIDS
The Altaid orogenic belt stretches across southern Russia and several former Soviet Republics including Kazakhstan, Uzbekistan, Mongolia, and parts of China. This huge, poorly known mountain system forms about half of northern Asia and is bounded by the Siberian (Angara) and Russian cratons in the north and by the Alpine-Himalaya mountains, Tarim, and North China blocks in the south. The Urals bound the Altaids in the west, and the Baikal Mountains bound them in the east. There are two main schools of thought about the main tectonic events that led to the formation of the Altaid Moun-

tains. The first suggests that a large number of different island arcs and small continents collided with each other in an oceanic setting and were then later accreted to continental land masses like the Siberian craton. The second group of models, championed by the Turkish geologist A. M. Celail Sengor, suggests that the Altaids are made of a collage of subduction-accretion materials that were added to a relatively few magmatic arcs, which then accreted to the Siberian and East European continents and were later disrupted and repeated by fault imbrication during large-scale strike-slip motions.

The Altaids grew around the Siberian craton from 600 to 144 million years ago. Consisting largely of continental crust that had already formed before the mountain ranges existed, they were brought together in convergent margin and collisional processes. Subduction-accretion complexes and magmatic arcs formed during the evolution of the Altaids, and blocks of continental crust of older ages were all stretched and thinned during the evolution of the mountain belt.

According to the model of Sengor, the Altaids are divided into 44 distinct tectonic units. Each of these different units may be composed of parts of cratons, magmatic arcs, passive margin fragments, and subduction/accretion complexes, now arranged into long curvilinear belts bounded by large strike-slip faults. Despite the complexity, three main rock assemblages of different rock types dominate the Altaids. These include

- turbidites and their low-grade metamorphic equivalents in flysch terranes
- pelagic sediments including chert, limestone, and shale, typically associated with the flysch
- mafic-ultramafic rock groups forming incomplete ophiolite suites

All of these rock groups are complexly deformed with multiple events of folding and faulting, and some belts are so deformed that they form tectonic mélanges. These three main rock assemblages are interspersed with terranes dominated by older, gneissic rocks, some with Mesozoic ages and others that are presumed to be Precambrian in age.

Tectonic Evolution of the Altaids
The tectonic evolution of the Altaids began in the Late Precambrian (Vendian) as a continental margin arc that had formed on the Baikalide-Uralide basement fringing the Angara (Siberian) and Russian (East European) cratons. This arc separated from the craton in the Early Cambrian, forming a long island arc known as the Kipchak arc. The arc was attached to continental crust on the Angara craton but lay

offshore the East European craton. As the Russian and Angara cratons rifted and began rotating toward each other, the free end of the Kipchak arc collided with the Russian craton in an arc/continent collision event, followed by major strike-slip faulting. The arc was sliced into many pieces that then were stacked and repeated along the continental margins to form the Kazakhstan microcontinent. The tectonic collage was then caught between the colliding Russian and Angara cratons in the Late Carboniferous, causing the whole Altaid tectonic assemblage to be further flattened. By the Middle Jurassic all motion between the cratons had ceased.

South of the Angara craton (present reference frame), the Mongolian and Far East sections of the Altaids developed mostly within the realms of the Tuva-Mongolian basement, between Late Precambrian and Late Jurassic times. The evolution of this part of the Altaids was terminated by the final docking of a group of rocks called the "intermediate units" in Asia, including the Tarim basement and the North China craton. These blocks show a very complex history because Tethyan Ocean margin processes were affecting their southern margins, while their northern edges had not yet collided with the Altaids in the Jurassic. For some time they had subduction zones dipping beneath both their northern and southern margins, from the Altaids and from the Tethys Ocean.

URAL MOUNTAINS

The boundary between Europe and Asia is typically taken to be the Ural Mountains, a particularly straight mountain range that stretches 1,500 miles (2,400 km) from the Arctic tundra to the deserts north of the Caspian Sea. Naroda (6,212 feet; 1,894 m) and Telpos-Iz (meaning "nest of winds," 5,304 feet; 1,617 m) are the highest peaks, found in the barren rocky and tundra-covered northern parts of the range. Southern parts of the mountain range rise to 5,377 feet (1,639 m) at Yaman-Tau, in the Mugodzhar Hills. The southern parts of the range are densely forested, whereas the northern parts are barren and covered by tundra or bare rock.

The Ural River flows out of the southern Urals into the Caspian Sea, and the western side of the range is drained by the Kama and Belaya Rivers, tributaries that also feed into the Caspian Sea, providing more than 75 percent of the water that flows into this shallow, closed basin. The eastern side of the range is drained by the Ob-Irtysh drainage system that flows into the Ob Gulf on the Kara Sea.

The Urals are extremely rich in mineral resources, including iron ore in the south and large deposits of coal, copper, manganese, gold, aluminum, and potash. Ophiolitic rocks in the south are also rich in chromite and platinum, plus deposits of bauxite, zinc, lead, silver, and tungsten are mined. Basins on the western side of the Urals produce large amounts of oil, and regions to the south in the Caspian are yielding many new discoveries. The Urals are also very rich in rare minerals and gems, yielding many excellent samples of emeralds, beryl, and topaz.

The Urals form part of the Ural-Okhotsk mobile belt, a Late Proterozoic to Mesozoic orogen that bordered the Paleoasian Ocean. The Ural Mountains section of this orogen saw a history that began with Early Paleozoic, probably Cambrian rifting of Baikalian basement, and Late Ordovician spreading to form a back-arc or oceanic basin that was active until the mid-Carboniferous. Oceanic arcs grew in this basin, but by the Middle Devonian began colliding with the East European continent, forming flysch basins. The Kazakhstan microcontinent collided with the Laurussian continent in the Permian, forming a series of foredeep basins on the Russian and Pechora platforms. These foredeeps are filled with molasse and economically important Middle to Late Permian coal deposits as well as potassium salts.

The Urals show a tectonic zonation from the Permian flysch basins on the East European craton to the Permian molasse basins on the western slopes of the Urals, then into belts of allochthonous carbonate platform rocks derived from the East European craton and thrust to the west over the Permian foredeeps. These rocks are all involved in westward-vergent fold-thrust belt structures, including duplex structures, indicating westward tectonic transport in the Permian. The axial zone of the Urals includes a chain of anticlinoria bringing up Riphean rocks, whose eastern contact is known as the Main Uralian fault. This major fault zone brings oceanic and island arc rocks in large nappe and klippe structures, placing them over the passive margin sequence.

The eastern slope of the Urals consists of a number of Ordovician to Carboniferous oceanic and island arc synformal nappes, imbricated with slices of the Precambrian crystalline basement. It is uncertain if these Precambrian gneisses are part of the East European craton, part of the accreted Kazakhstan microcontinent, or an exotic terrane. The eastern slopes of the Urals are intruded by many Devonian-Permian granites.

SIBERIAN TRAPS

A large part of the central Siberian plateau northwest of Lake Baikal is covered by a thick series of mafic volcanic flows. They are more than half a mile thick (1 km) over an area of 210,000 square miles (543,900 km^2) but have been significantly eroded from an estimated volume of 1,240,000 cubic miles (5,168,545 km^3). This extraordinary sequence of lavas was erupted over a remarkably short period of less than 1 million years, 250 million years ago, at the Permian-Triassic boundary. Within the resolution of measurements, their age is coincident in time with the major Permian-Triassic extinction, implying a causal link. The Permian-Triassic boundary at 250 million years ago marks the greatest extinction in Earth history, when 90 percent of marine species and 70 percent of terrestrial vertebrates became extinct. It has been postulated that the rapid volcanism and degassing released enough sulfur dioxide to cause a rapid global cooling, inducing a short ice age with an associated rapid fall of sea level. Soon after the ice age took hold, the effects of the carbon dioxide took over and the atmosphere heated, resulting in a global warming. The rapidly fluctuating climate postulated to have been caused by the volcanic gases is thought to have killed off many organisms, which were simply unable to cope with the wildly fluctuating climate extremes.

SIBERIAN TAIGA FOREST AND GLOBAL CARBON SINK

The northern third of Asia, stretching from the Ural Mountains in the west to the Pacific coast into the east, is known as Siberia. The southern border of Siberia is generally taken to be the Kazakh steppes in the southwest, the Altai and Sayan Mountains in the south, and the Mongolian steppes in the southeast. This region occupies approximately 3,000,000 square miles (7,500,000 km^2). The western third of Siberia is occupied by the Siberian lowland, stretching from the Urals to the Yenisei River. This low marshy area is drained by the Ob River and its tributaries, and it hosts agriculture, industry, and most of Siberia's human population in the wooded steppe. Eastern Siberia stretches from the Yenisei River to a chain of mountains including the Yablonovy, Stanovoy, Verkhoyansk, Kolyma, and Cherskogo Ranges. The eastern half of Siberia is an upland plateau, drained by the Vitim and Aldan Rivers. The Lena runs along the eastern margin of the region, and Lake Baikal, the world's deepest lake, is located in the southeast. Northeasternmost Siberia hosts a smaller plain on the arctic coast between the Lena and Kolyma Rivers, in the Republic of Yakutia (Sakha).

Siberia shows a strong zonation of vegetation, including a zone of tundra that extends inland about 200 miles (300 km) from the coast, followed by the taiga forest, a mixed forest belt, and the southern steppes. Siberia's taiga forest accounts for about 20 percent of the world's total forested land, covering about two-thirds of the region. This region accounts for about half of the world's evergreen forest and buffers global warming by acting as a large sink for carbon that otherwise could be released into the

atmosphere as carbon dioxide, a greenhouse gas. The forest and the rich soils derived from the decay of dead trees represents a very significant sink for global carbon. Much of the taiga forest is currently being logged at an alarming rate of loss of 12 million hectares per year. Much of this is being done by clear-cutting, where 90 percent of the timber is harvested, leading to increased erosion of the soil and runoff into streams. The effects of deforestation could be dramatic for global climate. With so much carbon stored in the taiga forest, both in the trees and in the peat and soils, any logging or development that releases this carbon to the atmosphere will increase global carbon dioxide levels, contributing to global warming. Acid rain and other pollution largely emitted from the coal, nickel, aluminum, and lead smelting plants in the west is causing additional loss of forest. Additionally, large tracts of forest are being torn up to explore for and extract oil, natural gas, iron ore, and diamonds.

See also ACCRETIONARY WEDGE; ARCHEAN; ASIAN GEOLOGY; CLIMATE CHANGE; CONVERGENT PLATE MARGIN PROCESSES; CRATON; DEFORMATION OF ROCKS; DIVERGENT PLATE MARGIN PROCESSES; EUROPEAN GEOLOGY; FLYSCH; GRANITE, GRANITE BATHOLITH; GREENSTONE BELTS; IGNEOUS ROCKS; LARGE IGNEOUS PROVINCES, FLOOD BASALT; MÉLANGE; PHANEROZOIC; PLATE TECTONICS; PROTEROZOIC; STRUCTURAL GEOLOGY; TRANSFORM PLATE MARGIN PROCESSES.

FURTHER READING

Berzin, R., O. Oncken, J. H. Knapp, A. Perez-Estaun, T. Hismatulin, N. Yunusov, and A. Lipilin. "Orogenic Evolution of the Ural Mountains: Results from an Integrated Seismic Experiment." *Science* 274 (1996): 220–221.

Bogdanova, Svetlana V., R. Gorbatschev, and R. G. Garetsky. "The East European Craton." In *Encyclopedia of Geology*, vol. 5, edited by R. C. Selley, L. R. Cocks, and I. R. Plimer, 34–49. Amsterdam; London: Elsevier Academic, 2005.

Condie, Kent C., and Robert Sloan. *Origin and Evolution of Earth: Principles of Historical Geology.* Upper Saddle River, N.J.: Prentice Hall, 1997.

Goodwin, Alan M. *Precambrian Geology*. London: Academic Press, 1991.

Kusky, Timothy M. *Precambrian Ophiolites and Related Rocks.* Amsterdam: Elsevier, 2004.

Sengor, A. M. C., and B. A. Natal'in. "Tectonics of the Altaids: An Example of a Turkic Type Orogen." In *Earth Structure,* 2nd ed., edited by B. A. van der Pluijm and S. Marshak, 535–546. New York: W. W. Norton, 2004.

Windley, Brian F. *The Evolving Continents.* 3rd ed. Chichester, England: John Wiley & Sons, 1995.

Saturn   Saturn is the sixth planet, residing between Jupiter and Uranus, orbiting at 9.54 astronomical units (888 million miles, or 1,430 million kilometers) from the Sun, twice the distance from the center of the solar system as Jupiter, and having an orbital period of 29.5 Earth years. The mass of Saturn is 95 times that of Earth, yet it rotates at more than twice the rate of Earth. The average density of this giant gaseous planet is only 0.7 grams/cm^3, less than water. The planet has a molecular hydrogen interior with a radius of 37,282 miles (60,000 km), a metallic hydrogen core with a radius of 18,641 miles (30,000 km), and a rocky/icy inner core with a radius of 9,320 miles (15,000 km).

The most striking features of Saturn are its many rings and moons, with the rings circling the planet along its equatorial plane and their appearance from Earth changing with the seasons because of the different tilt of the planet as it orbits the Sun. The rings are more than 124,275 miles (200,000 km) in diameter but are less than 30 feet (10 m) thick. They are composed of numerous small particles, most of which are ice between less than an inch (a few millimeters) and about 50 feet (a few tens of meters) in diameter. The breaks in the rings result from gravitational dynamics between the planet and its many moons.

Saturn has a yellowish-tan color produced largely by gaseous methane and ammonia, but the atmosphere consists of 92.4 percent molecular hydrogen, 7.4 percent helium, 0.2 percent methane, and 0.02 percent ammonia. These gases are stratified into three main layers, including a 62–124-mile- (100–200-km-) thick outer layer of ammonia, a 31–62-mile- (50–100-km-) thick layer of ammonium hydrosulfide ice, and a deeper 31–62-mile- (50–100-km-) thick layer of water ice. The atmosphere of Saturn is somewhat colder and thicker than that of Jupiter. Atmospheric winds on Saturn reach a maximum eastward-flowing velocity of 930 miles per hour (1,500 km/hr) at the equator and diminish with a few belts of high velocity toward the poles. Like Jupiter, Saturn has atmospheric bands related to these velocity variations, as well as turbulent storms that show as spots, and a few westward-flowing bands.

Many moons circle Saturn, including the large rocky Titan, possessing a thick nitrogen-argon-rich atmosphere that contains hydrocarbons including methane, similar to the basic building blocks of life on Earth. Other large to midsize moons include, in increasing distance from the planet, Mimas, Enceladus, Tethys, Dione, Rhea, and Iapetus. About a dozen other moons of significant size also circle the planet.

See also EARTH; JUPITER; MARS; MERCURY; NEPTUNE; SOLAR SYSTEM; URANUS; VENUS.

FURTHER READING

Chaisson, Eric, and Steve McMillan. *Astronomy Today.* 6th ed. Upper Saddle River, N.J.: Addison-Wesley, 2007.

Comins, Neil F. *Discovering the Universe.* 8th ed. New York: W. H. Freeman, 2008.

National Aeronautic and Space Administration. "Solar System Exploration page. Saturn." Available online. URL: http://solarsystem.nasa.gov/planets/profile.cfm?Object=Saturn. Last updated June 25, 2008.

Snow, Theodore P. *Essentials of the Dynamic Universe: An Introduction to Astronomy.* 4th ed. St. Paul, Minn.: West Publishing Company, 1991.

Mosaic of Saturn taken by *Cassini,* October 2004: The mosaic is made from 126 images processed to show the planet in natural color. Saturn's equator is tilted relative to its orbit by 27 degrees, very similar to the 23-degree tilt of the Earth. As Saturn moves along its orbit, first one hemisphere, then the other is tilted toward the Sun. This cyclical change causes seasons on Saturn, just as the changing orientation of Earth's tilt causes seasons on our planet. Saturn's rings are incredibly thin, with a thickness of only about 30 feet (10 m). The rings are made of dusty water ice, in the form of boulder-sized and smaller chunks that gently collide with each other as they orbit around Saturn. Saturn's gravitational field constantly disrupts these ice chunks, keeping them spread out and preventing them from combining to form a moon. The rings have a slight pale reddish color due to the presence of organic material mixed with the water ice. Saturn is about 75,000 miles (120,000 km) across and is flattened at the poles because of its very rapid rotation. A day is only 10 hours long on Saturn. Strong winds account for the horizontal bands in the atmosphere of this giant gas planet. The delicate color variations in the clouds are due to smog in the upper atmosphere, produced when ultraviolet radiation from the Sun shines on methane gas. Deeper in the atmosphere, the visible clouds and gases merge gradually into hotter and denser gases. *(NASA Jet Propulsion Laboratory)*

sea-level rise Global sea levels are currently rising as a result of the melting of the Greenland and Antarctica ice sheets and thermal expansion of the world's ocean waters due to global warming. Earth is presently in an interglacial stage of an ice age. Sea levels have risen nearly 400 feet (122 m) since the last glacial maximum 20,000 years ago and about 6 inches (15 cm) in the past 100 years. The rate of sea-level rise seems to be accelerating, and may presently be as much as an inch (2.5 cm) every eight to 10 years. If all of the ice on both the Antarctic and Greenland ice sheets were to melt, global sea levels would rise by 230 feet (70 m), inundating most of the world's major cities and submerging large parts of the continents under shallow seas. The coastal regions of the world are densely populated and are experiencing rapid population growth. Approximately 100 million people presently live within three feet (1 m) above the present-day sea level. If sea level were to rise rapidly and significantly the world would experience an economic and social disaster of a magnitude not yet experienced by civilization. Many areas would become permanently flooded or subject to inundation by storms, beach erosion would be accelerated, and water tables would rise.

Sea level has risen and fallen by hundreds of feet many times in the billions of years represented by Earth history, and it is presently slowly rising at about one foot (0.3 m) per century, a rate that may be accelerating from the effects of global warming. The causes of sea-level rise and fall are complex, operate on vastly different time scales, and may affect local regions or the entire planet. These include growth and melting of glaciers, changes in the volume of the mid-ocean ridges, thermal expansion of water from global warming, and other complex interactions of the distribution of the continental landmass in mountains and plains during periods of faulting, mountain-building, and basin-forming activity. Deciphering the amount of sea-level rise depends critically on correct identification and separation of the local effects of the rising and sinking of the land, known as relative sea-level changes, from global changes in sea level, that are referred to as eustatic events.

Sometimes individual large earthquakes may displace the land surface vertically, resulting in subsidence or uplift of the land relative to the sea. One of the largest and best-documented cases of earthquake-induced subsidence resulted from the March 27, 1964, magnitude 9.2 earthquake in southern Alaska. This earthquake tilted a huge approximately 125,000-square-mile (200,000 square km) area of the Earth's crust. Significant changes in ground level were recorded along the coast for more than 600 miles (1,000 km), including uplifts of up to 36 feet (11 m), subsidence of up to 6.5 feet (2 m), and lateral shifts of 30 or more feet (several to tens of meters). Much of the area that subsided was along Cook Inlet, north to Anchorage and Valdez, and south to Kodiak Island. Towns that were built around docks prior to the earthquake were suddenly located below the high tide mark, and entire towns had to move to higher ground. Forests that subsided found their root systems suddenly inundated by salt water, leading to the death of the forests. Populated areas located at previously safe distances from the high-tide (and storm) line became prone to flooding and storm surges and had to be relocated.

The fastest changes in sea level are caused by instantaneous geologic catastrophes such as meteorite impacts into the ocean, but luckily these types of events do not happen often. Seasonal changes can blow or move water to greater heights on one side of a basin, move it to lower heights on another side, and cause water to expand and contract with changes in temperature, causing small changes in sea level. Climate changes can cause glaciers and ice caps to melt and reform, causing sea levels to rise and fall on time scales of hundreds to thousands of years. Longer-term climate variations related to variations in the orbit of the Earth around the Sun can lengthen the time scale of sea-level changes related to climate change to hundreds of thousands of years for individual cycles. Plate tectonics also influences sea-level changes, but on much longer time scales than climate variations. If the processes of seafloor spreading and submarine volcanism become accelerated in any geologic time period, the volume of material that makes up the mid-ocean ridge system will be larger, and this extra volume of elevated seafloor will displace an equal amount of seawater and raise sea levels. This process typically operates with time variations on the order of tens of millions of years. An even longer-term variation in sea levels is caused by the motions and collisions of continents in supercontinent cycles. When continents collide, large amounts of continen-

tal material are uplifted above sea level, effectively taking this material out of the oceans, making the ocean basins bigger, and lowering sea levels. When continents rift apart, the opposite happens—more material is added to the ocean basins, and sea levels rise on the continents. These slow tectonic variations can change sea levels on time scales of tens to hundreds of millions of years.

Rising sea levels cause the shoreline to move landward, whereas a fall in sea level causes the shoreline to move oceanward. With the present sea-level rise, coastal cliffs are eroding, barrier islands are migrating (or being submerged if they were heavily protected from erosion), beaches are moving landward, and the sea is flooding estuaries. At some point in the not too distant future, low-lying coastal cities will be flooded under several feet of water, and eventually the water could be hundreds of feet deep. Cities including New Orleans, New York, Washington, Houston, London, Shanghai, Tokyo, and Cairo will be inundated; the world's nations need to begin planning how to handle this inevitable geologic hazard and encroachment of the sea.

About 70 percent of the world's sandy beaches are being eroded. The reasons for this erosion include rising sea levels, increased storminess, a decrease in sediment transport to beaches from the damming of rivers, and perhaps shifts in global climate belts. Construction of sea walls to reduce erosion of coastal cliffs also causes a decreased supply of sand to replenish the beach, so also increases beach retreat. Pumping of groundwater from coastal aquifers also results in coastal erosion, because pumping causes the surface to subside, leading to a relative sea-level rise.

When sea level rises, beaches try to maintain their equilibrium profile and move each beach element landward. A sea-level rise of 1 inch (2.5 cm) is generally equated with a landward shift of beach elements of more than four feet (>1 m). Most sandy beaches worldwide are retreating landward at rates of 20 inches to 3 feet per year (0.5–1 m), consistent with sea-level rise of an inch (2.5 cm) every 10 years.

CAUSES OF CHANGING SEA LEVELS

The average position of the median sea level may appear to rise or fall with respect to the land surface to an observer on a shoreline, and this is called relative sea-level rise or fall. However, it is difficult for the observer on the local shoreline to know if the height of the water is changing, or if the height of the continent is rising or falling. In many places plate tectonics causes areas of the crust to rise slowly out of the sea or to sink gradually downward below sea level, while the water level is actually staying at the same height. The weight of glaciers or sedimentary deposits can also cause local shorelines to sink, or to rise if the weight is removed. Therefore geologists need a way to differentiate between local changes in relative sea level and true global sea-level changes. This is a difficult problem and is best done by obtaining accurate ages on the times of sea-level rise and fall and correlating these changes with other places around the world. This has been done through many years of study, and now there is a fairly well-established curve of global sea-level heights going back in geological time. Local or apparent changes in sea level are called apparent sea level, whereas global changes in the height of sea level are called eustatic sea-level changes.

Measuring global eustatic sea-level rise and fall is difficult because many factors influence the relative height of the sea along any coastline. These vertical motions of continents are called epeirogenic movements, and they may be related to plate tectonics, to rebound from being buried by glaciers, or to changes in the amount of heat added to the base of the continent by mantle convection. Continents may rise or sink vertically, causing apparent sea-level change, but these sea-level changes are relatively slow compared to changes induced by global warming and glacial melting. Slow, long-term sea-level changes can also be induced by changes in the amount of seafloor volcanism associated with seafloor spreading. At some times during Earth history seafloor spreading was particularly vigorous, and the increased volume of volcanoes and the mid-ocean ridge system caused global sea levels to rise.

INFLUENCE OF SHORT-TERM CLIMATE CHANGES ON SEA LEVEL

Minor changes in sea level of up to about a foot (30 cm) happen in many places in yearly seasonal cycles. Many of these are caused by changes in the wind patterns, as the Sun alternately heats different belts of the ocean, and the winds blow water from one side of the ocean to the other. When water is heated in the summer months it also expands slightly, accounting for sea-level changes of an inch (2.5 cm) or so. Thermal expansion associated with global warming may raise sea levels about 12 inches (30 cm) by the year 2050 and 20 inches (50 cm) by 2100. Seasonal development of regional high- and low-pressure systems that characterize some areas also change sea levels on short time scales. High pressure areas, such as the Bermuda high that often develops over the central Atlantic in the summer, lower local sea levels because the high atmospheric pressure (weight) pushes sea levels lower than in other times.

Steady winds and currents can amass water against a particular coastline, causing a local and temporary sea-level rise. Such a phenomenon is associated with the El Niño-Southern Oscillation

(ENSO), causing sea levels to rise by 4–8 inches (10–20 cm) in the Australia-Asia region. When the warm water moves east in an ENSO event, sea levels may rise may 4–20 inches (10–50 cm) across much of the North and South American coastlines. The irregular El Niño event results from changes in atmospheric heating that then cause a warm current to move from the western Pacific to the eastern Pacific. This can raise sea levels off the coast of Peru (and sometimes as far as California) by up to 2 feet (60 cm), enough to cause enhanced erosion, landslides, and considerable damage to the coastal environment in South America. Other atmospheric phenomena can also change sea level by up to several feet (up to a few meters) locally, on short time scales. Changes in atmospheric pressure, salinity of seawaters, coastal upwelling, onshore winds, and storm surges all cause short-term fluctuations along segments of coastline. Global or local warming of waters can cause them to expand slightly, causing a local sea-level rise. The extraction and use of groundwater and its subsequent release into the sea might be causing sea-level rise of about 0.78 inch per year (1.3 mm/yr). Seasonal changes in river discharge can temporarily change sea levels along some coastlines, especially where winter cooling locks up large amounts of snow that melt in the spring.

Attempts to estimate eustatic sea-level changes must be able to average out the numerous local and tectonic effects to arrive at a globally meaningful estimate of sea-level change. Most coastlines seem to be dominated by local fluctuations that are larger in magnitude than any global sea-level rise. Recently, satellite radar technology has been employed to measure sea surface height precisely and to document annual changes in sea level. Radar altimetry is able to map sea-surface elevations to the subinch scale and to do this globally, providing an unprecedented level of understanding of sea-surface topography. Satellite techniques support the concept that global sea levels are rising at about 0.01 inch per year (0.025 cm/yr).

INFLUENCE OF LONG-TERM CLIMATE EFFECTS ON SEA LEVEL

Many changes in the Earth's climate that control relative sea level are caused by variations in the amount of incoming solar energy, which in turn are caused by systematic changes in the way the Earth orbits the Sun. These systematic changes in the amount of incoming solar radiation caused by variations in Earth's orbital parameters are known as Milankovitch cycles, after the Serbian mathematician Milutin Milankovitch, who first clearly described these cycles. These changes can affect many Earth systems, causing glaciations, global warming, dramatic sea-level

Tourists and residents wading in flooded St. Mark's Square in Venice, Italy, October 31, 2004 *(AP Images)*

changes, and changes in the patterns of climate and sedimentation.

Astronomical effects that influence the amount of incoming solar radiation include minor variations in the path of the Earth in its orbit around the Sun and the inclination or tilt of its axis, causing variations in the amount of solar energy reaching the top of the atmosphere. These variations are thought to be responsible for the advance and retreat of the Northern and Southern Hemisphere ice sheets in the past few million years and the associated huge sea-level changes. In the past 2 million years alone, the Earth has seen the ice sheets advance and retreat approximately 20 times. The climate record as deduced from ice-core records from Greenland and isotopic tracer studies from deep ocean, lake, and cave sediments suggests that the ice builds up gradually over periods of about 100,000 years, then retreats rapidly over a period of decades to a few thousand years. These patterns result from the cumulative effects of different astronomical phenomena.

Several movements are involved in changing the amount of incoming solar radiation. The Earth rotates around the Sun following an elliptical orbit. The shape of this elliptical orbit, called its eccentricity, changes cyclically with time over a period of 100,000 years, alternately bringing the Earth closer to and farther from the Sun in summer and winter. This 100,000-year cycle is about the same as the general pattern of glaciers advancing and retreating every 100,000 years in the past 2 million years, suggesting that this is the main cause of variations within the present-day ice age.

The Earth's axis is presently tilting by 23.5°N/S away from the orbital plane, and the tilt varies between 21.5°N/S and 24.5°N/S. The tilt changes by plus or minus 1.5°N/S from a tilt of 23°N/S every 41,000 years. When the tilt is greater, there is greater seasonal variation in temperature.

Wobble of the rotation axis describes a motion much like a top rapidly spinning and rotating with a wobbling motion, such that the direction of tilt toward or away from the Sun changes, even though the tilt amount stays the same. This wobbling phenomenon is known as precession of the equinoxes, and it has the effect of placing different hemispheres closest to the Sun in different seasons. Presently the precession of the equinoxes is such that the Earth is closest to the Sun during the Northern Hemisphere winter. This precession changes with a double cycle, with periodicities of 23,000 years and 19,000 years.

Because each of these astronomical factors acts on a different time scale, their effects are combined in a more complex cycle. These factors interact in a complicated way, known as Milankovitch cycles. Using the power of understanding these cycles, it is possible to make predictions of where the Earth's climate is heading, whether into a warming or cooling period, and whether sea levels will rise or fall, or if some regions may experience desertification, glaciation, floods, or droughts.

Present data shows that temperatures were about 3–5°F (2–3°C) cooler at the height of the glacial advances 12,000 years ago than they are today and that temperatures may warm an additional 3–4°C by the year 2100. If this warming occurs as predicted, then large amounts of the glacial ice on Antarctica and Greenland will melt, raising sea levels dramatically. Many scientists predict sea levels will rise at least a foot (0.3 m) by 2100, others predict more. The sea-level rise will likely continue past the year 2100, with at least 16 feet (5 m) over the next few centuries. When this happens, most of the world's large port cities will be partly to largely underwater and world civilizations will have needed to find ways to move huge populations to higher ground. There is a current debate about how much humans are contributing to global warming and the consequent sea-level rise. Most data suggest that human-induced warming is about or slightly less than 2°F (1°C) over the past 100 years, but that warming is superimposed on the longer-term cycles described above. What is not known is how these long-term natural cycles may change. Warming may continue, or the natural cycles may reverse, or other sudden catastrophic cooling events may occur, such as a volcanic eruption on the scale of Tambora in Indonesia in 1815 that lowered global temperatures by about 2°F (1°C).

SEA-LEVEL CHANGES CAUSED BY CHANGES IN WATER/ICE VOLUME

Global sea levels are currently rising, in part from thermal expansion of the seawater in a warmer climate, and partly as a result of the melting of the Greenland and Antarctic ice sheets. The Greenland and Antarctic ice sheets have some significant differences that cause them to respond differently to changes in air and water temperatures. The Antarctic ice sheet is about 10 times as large as the Greenland ice sheet, and since it sits on the South Pole, Antarctica dominates its own climate. The surrounding ocean is cold even during summer, and much of Ant-

arctica is a cold desert with low precipitation rates and high evaporation potential. Most meltwater in Antarctica seeps into underlying snow and simply refreezes, with little running off into the sea. Antarctica hosts several large ice shelves fed by glaciers moving at rates of up to 1,000 feet (305 m) per year. Most ice loss in Antarctica is accomplished through calving and basal melting of the ice shelves, at rates of 10–15 inches (25–38 cm) per year.

In contrast, Greenland's climate is influenced by warm North Atlantic currents and by its proximity to other landmasses. Climate data measured from ice cores taken from the top of the Greenland ice cap show that temperatures have varied significantly in cycles of years to decades. Greenland also experiences significant summer melting, abundant snowfall, has few ice shelves, and its glaciers move quickly at rates of up to miles per year. These fast-moving glaciers are able to drain a large amount of ice from Greenland in relatively short amounts of time.

The Greenland ice sheet is thinning rapidly along its edges, losing an average of 15–20 feet (4.5–6 m) in the past decade. In addition, tidewater glaciers and the small ice shelves in Greenland are melting an order of magnitude faster than the Antarctic ice sheets, with rates of melting between 25–65 feet (7–20 m) per year, a rate that is apparently increasing. About half of the ice lost from Greenland is through surface melting that runs off into the sea. The other half of ice loss is through calving of outlet glaciers and melting along the tidewater glaciers and ice shelf bases.

These differences between the Greenland and Antarctic ice sheets lead them to play different roles in global sea level rise. Greenland contributes more to the rapid short-term fluctuations in sea level, responding to short-term changes in climate. In contrast, most of the world's water available for raising sea level is locked up in the slowly changing Antarctic ice sheet. Antarctica contributes more to the gradual, long-term sea-level rise.

PLATE TECTONICS, SUPERCONTINENT CYCLES, AND SEA LEVEL

Movement of the tectonic plates on Earth causes the semi-regular grouping of the planet's landmasses into a single or several large continents that remain stable for a long period of time, then disperse, and eventually come back together as new amalgamated landmasses with a different distribution. This cycle is known as the supercontinent cycle. At several times in Earth history, the continents have joined together to form one large supercontinent, with the last supercontinent Pangaea (meaning all land) breaking up approximately 160 million years ago. This process of supercontinent formation and dispersal and reamalgamation seems to be grossly cyclic, perhaps reflecting mantle convection patterns, but also influencing sea level, climate, and biological evolution.

The basic idea of the supercontinent cycle is that continents drift about on the surface until they all collide, stay together, and come to rest relative to the mantle. The continents are only half as efficient at conducting heat as oceans, so after the continents are joined together, heat accumulates at their base, causing doming and breakup of the continent. For small continents, heat can flow sideways and not heat up the base of the plate, but for large continents the lateral distance is too great for the heat to be transported sideways. The heat rising from within the Earth therefore breaks up the supercontinent after a heating period of several tens or hundreds of millions of years. The heat then flows away and is transferred to the ocean and atmosphere system, and continents move away until they come back together forming a new supercontinent.

The supercontinent cycle has many effects that greatly affect other Earth systems. First, the breakup of continents causes sudden bursts of heat release, associated with periods of increased, intense magmatism. It also explains some of the large-scale sea-level changes, episodes of rapid and widespread orogeny, episodes of glaciation, and many of the changes in life on Earth.

Sea level has changed by hundreds of meters above and below current levels at many times in Earth history. In fact, sea level is constantly changing in response to a number of different variables, many of them related to plate tectonics. The diversity of fauna on the globe is closely related to sea levels, with greater diversity during sea-level high stands, and lower diversity during sea-level lows. For instance, sea level was 1,970 feet (600 m) higher than now during the Ordovician Period, and the sea level high stand was associated with a biotic explosion. Sea levels reached a low stand at the end of the Permian Period, and this low was associated with a great mass extinction. Sea levels were high again in the Cretaceous.

SEA-LEVEL CHANGES RELATED TO CHANGES IN MID-OCEAN RIDGE VOLUME

Sea levels may change at different rates and amounts in response to changes in several other Earth systems. Local tectonic effects may mimic sea-level changes through regional subsidence or uplift, and these effects must be taken into account and filtered out when trying to deduce ancient, global (eustatic) sea-level changes. The global volume of the mid-ocean ridges can change dramatically, either by increasing the total length of ridges, or changing the rate of seafloor spreading. The total length of ridges typi-

cally increases during continental breakup, since continents are being rifted apart and some continental rifts can evolve into mid-ocean ridges. Additionally, if seafloor spreading rates are increased, the amount of young, topographically elevated ridges is increased relative to the slower, older topographically lower ridges that occupy a smaller volume. If the volume of the ridges increases by either mechanism, then a volume of water equal to the increased ridge volume is displaced and sea level rises, inundating the continents. Changes in ridge volume are able to change sea levels positively or negatively by about 985 feet (300 m) from present values, at rates of about 0.4 inch (1 cm) every 1,000 years.

SEA-LEVEL CHANGES RELATED TO CHANGES IN CONTINENTAL AREA

Continent-continent collisions can lower sea levels by reducing the area of the continents. When continents collide, mountains and plateaus are uplifted, and the amount of material that is taken from below sea level to higher elevations no longer displaces seawater, causing sea levels to drop. The ongoing India-Asia collision has caused sea levels to drop by 33 feet (10 m).

Other things, such as mid-plate volcanism, can also change sea levels. The Hawaiian Islands are hot-spot style mid-plate volcanoes that have been erupted onto the seafloor, displacing an amount of water equal to their volume. Although this effect is not large at present, at some periods in Earth history there were many more hot spots (such as in the Cretaceous Period), and the effect may have been larger.

The effects of the supercontinent cycle on sea level may be summarized as follows: Continent assembly favors regression, whereas continental fragmentation and dispersal favors transgression.

SUMMARY

Sea level is rising presently at a rate of one foot (0.3 m) per century, although this rate seems to be accelerating. This rising sea level will obviously change the coastline dramatically—a one-foot (0.3-m) rise in sea level along a gentle coastal plain can be equated with a 1,000-foot (300-m) landward migration of the shoreline. The world will look significantly different when sea levels rise significantly. Many of the world's low-lying cities like New York, New Orleans, London, Cairo, Tokyo, and most other cities in the world may look like Venice in a hundred or several hundred years. The world's rich farmlands on coastal plains, like the East Coast of the United States, northern Europe, Bangladesh and much of China will be covered by shallow seas. If sea levels rise more significantly, as they have in the past, then vast parts of the interior plains of North America will be covered by inland seas, and much of the world's climate and vegetation zones will be shifted to different latitudes.

Governments must begin to plan for how to deal with rising sea levels, yet very little has been done so far. It is time that groups of scientists and government planners begin to meet to first understand the magnitude of the problem, then to study and recommend which tactics to initiate to mitigate the effects of rising sea levels.

See also El Niño and the Southern Oscillation (ENSO); glacier, glacial systems; plate tectonics.

FURTHER READING

Botkin, D., and E. Keller. *Environmental Science*. Hoboken, N.J.: John Wiley & Sons, 2003.

Burkett, Virginia R., D. B. Zikoski, and D. A. Hart. "Sea-Level Rise and Subsidence: Implications for Flooding in New Orleans, Louisiana." In *U.S. Geological Survey Subsidence Interest Group Conference, Proceedings for the Technical Meeting*, 63–70. Reston, Va.: U.S. Geological Survey, 2003.

Davis, R., and D. Fitzgerald. *Beaches and Coasts*. Malden, Mass.: Blackwell Publishing, 2004.

Douglas, Bruce C., Michael S. Kearney, and Stephen P. Leatherman. *Sea Level Rise: History and Consequences*. San Diego, Calif.: Academic Press, 2000.

Intergovernmental Panel on Climate Change. Available online. URL: http://www.ipcc.ch/index.htm. Accessed January 30, 2008.

Intergovernmental Panel on Climate Change. *Climate Change 2007: The Physical Science Basis. Contributions of Working Group I to the Fourth Assessment Report of the Intergovernmental Panel on Climate Change*, edited by S. Solomon, D. Qin, M. Manning, Z. Chen, M. Marquis, K. B. Averyt, M. Tignor, and H. L. Miller. Cambridge: Cambridge University Press, 2007.

Kusky, T. M. *Climate Change: Shifting Deserts, Glaciers, and Climate Belts*. New York: Facts On File, 2008.

———. *The Coast: Hazardous Interactions within the Coastal Zone*. New York: Facts On File, 2008.

Schneider, D. "The Rising Seas." *Scientific American* (March 1997): 112–117.

seawater   The oceans cover more than 70 percent or the Earth's surface and extend to an average depth of a couple of miles (several kilometers). As part of the hydrologic cycle, each year approximately 1.27×10^{16} cubic feet (3.6×10^{14} m^3) of water evaporate from the oceans with about 90 percent of this returning to the oceans as rainfall. The remaining 10 percent falls as precipitation on the continents, where it forms freshwater lakes and streams and seeps into the groundwater system for temporary storage before eventually returning to the sea. During its passage over and in the land, the water erodes huge quantities of rock, soil, and sediment, and dissolves chemical elements such as salts from the continents, carrying these and other sediments as dissolved, suspended, and bed load to the oceans. Water transports more than 50 million tons of continental material into the oceans each year. Most of the suspended and bed load material is deposited as sedimentary layers near passive margins, but the dissolved salts and ions derived from the continents play a major role in determining seawater chemistry, as listed in the table "Composition of Typical Seawater." The most abundant dissolved salts are chloride and sodium, which together with sulfate, magnesium, calcium, potassium, bicarbonate, bromide, borate, strontium, and fluoride form more than 99.99 percent of the total material dissolved in seawater.

In addition to the elements listed in the table "Composition of Typical Seawater," a number of additional minor and trace elements dissolved in seawater are important for the life cycle of many organisms. For instance, nitrogen, phosphorus, silicon, zinc, iron, and copper play important roles in the growth of tests and other parts of some marine organisms. Seawater also contains dissolved gases, including nitrogen, oxygen, and carbon dioxide. The amount of oxygen dissolved in the surface layers of

COMPOSITION OF TYPICAL SEAWATER

Name	Symbol	Concentration in Parts per Thousand	Percentage of Dissolved Material
Chloride	Cl^-	18.980	55.05
Sodium	Na^+	10.556	30.61
Sulfate	SO_4^{2-}	2.649	7.68
Magnesium	Mg^{2+}	1.272	3.69
Calcium	Ca^{2+}	0.400	1.16
Potassium	K^+	0.380	1.10
Bicarbonate	HCO_3^-	0.140	0.41
Bromide	Br	0.065	0.19
Borate	$H_3BO_3^-$	0.026	0.07
Strontium	Sr^{2+}	0.008	0.03
Fluoride	F^-	0.001	0.00
Total		34.477	99.99

seawater is about 34 percent of the total dissolved gases, significantly higher than the 21 percent of the total atmospheric gases. Marine organisms generate this oxygen through photosynthesis. Some of it exchanges with the atmosphere across the air-water interface, and some sinks and is used by deep aerobic organisms. The amount of carbon dioxide dissolved in seawater is about 50 times greater than its concentration in the atmosphere. CO_2 plays an important role in buffering the acidity and alkalinity of seawater where, through a series of chemical reactions, it keeps the pH of seawater between 7.5 and 8.5. Marine organisms make carbonate shells out of the dissolved CO_2, and some is incorporated into marine sediments where it is effectively isolated from the atmosphere. The total amount of CO_2 stored in the ocean is very large, and as a greenhouse gas, if it were to be released to the atmosphere, it would have a profound effect on global climate.

The salinity and temperature of seawater are important in controlling mixing between surface and deep water and in determining ocean currents. Temperature is controlled largely by latitude, whereas river input, evaporation from restricted basins, and other factors determine the total dissolved salt concentration. Density differences caused by temperature and salinity variations induce ocean currents and thermohaline circulation, distributing heat and nutrients around the globe.

See also GEOCHEMICAL CYCLES; HYDROSPHERE; OCEAN BASIN; OCEAN CURRENTS; OCEANOGRAPHY; THERMOHALINE CIRCULATION.

FURTHER READING

Allaby, Alisa, and Michael Allaby. *A Dictionary of Earth Sciences*. 2nd ed. Oxford: Oxford University Press, 1999.

Erickson, Jon. *Marine Geology: Exploring the New Frontiers of the Ocean*. Rev. ed. New York: Facts On File, 2003.

Sedgwick, Adam (1785–1873) British *Geologist* One of the founders of geology as a science, Adam Sedgwick was born on March 22, 1785, in Yorkshire, England. He was the third of seven children of the Anglican vicar of the town of Dent, and he spent a happy childhood with many hours exploring the countryside collecting fossils and rocks. Sedgwick attended the Sedberg School in Yorkshire, then he was admitted to Trinity College at Cambridge University on a special scholarship, obtaining his bachelor of arts in natural sciences in 1808. The college made him a fellow in 1810, when he was charged with supervising six students, which he noted seriously held him back in his own studies.

In 1818 Sedgwick was appointed the Woodwardian professor of geology at Cambridge, which had been endowed by natural historian John Woodward in the early 1700s. Until this time, Sedgwick had not had any formal studies in geology, and is credited with saying, "Hitherto I have never turned a stone; henceforth I will leave no stone unturned," upon his appointment to the post. While in this post Sedgwick taught himself geology and paleontology and paid great attention to expanding the geological collections of Cambridge University, while gaining experience by doing field work throughout the British Isles. Sedgwick became a very popular lecturer and went against tradition of the times by allowing women to attend his courses. In 1829 he was elected president of the Geological Society of London, and in 1845 he became a vice-master of Trinity College. Sedgwick's health began faltering in the 1850s, and he stopped giving lectures due to his health in 1871.

Adam Sedgwick was exploring the field of geology in England at a time when the science was in its infancy. He met and worked with gentleman geologist Roderick Murchison, and the two jointly presented their research on some fossiliferous rocks of Devonshire, England. The distinctive fossil assemblage in these rocks led Sedgwick and Murchison to propose a new division of geologic time for these rocks—the Devonian period. In the early 1830s the two began working on the folded and faulted rocks of Wales. Murchison worked on the fossil assemblages and determined that they appeared more primitive (containing fewer fish) than the rocks of Devonshire, so he assigned these rocks to an older period, naming it the Silurian after the Silures, a Celtic tribe who lived in the Welsh borderlands in Roman times. Sedgwick then suggested that even older rocks existed in central Wales, and he named these the Cambrian, after Cambria, the Latin name for Wales. Sedgwick and Murchison then presented their descriptions of the rocks and stratigraphic divisions of England and Wales in a famous paper called "On the Silurian and Cambrian Systems, Exhibiting the Order in Which the Older Sedimentary Strata Succeed Each Other in England and Wales." The paper became famous as it offered the first division of lower Paleozoic time. During these studies Sedgwick became the first geologist to clearly distinguish between the structures of jointing, slaty cleavage, and stratification.

There was some overlap between the upper part of the Cambrian as proposed by Sedgwick and the lower part of the Silurian as proposed by Murchison. This led to a major dispute between Sedgwick and Murchison, with both claiming they were correct. At stake was the honor of being the first person to describe the rocks that seemingly contained the earliest record of life on Earth since, at that time, the oldest fossils known were Cambrian. Murchison claimed that Sedgwick's Cambrian rocks were not sufficiently different from his Silurian rocks to warrant a further division of geologic time. The debate was resolved in 1879 when British geologist Charles Lapworth proposed a new division of geologic time between the Cambrian and Silurian, which he called the Ordovician after a Celtic tribe in Wales. The Ordovician included both the disputed Upper Cambrian and Lower Silurian strata.

During some of his field work in Wales, Sedgwick took a student from Cambridge along as a field assistant—the young Charles Darwin. Darwin was in Sedgwick's geology lecture course and wanted more experience so took the employment from Sedgwick. This experience proved invaluable, as during Darwin's famous voyage on the HMS *Beagle* (1831–36) Darwin sent many rock samples and descriptions of South America back to Sedgwick, who read and helped interpret the work. Sedgwick also highly recommended Darwin's work to the Geological Society of London, improving the career and reputation of his former student and then colleague. However, later Sedgwick did not approve of Darwin's theory of evolution, writing in a letter to Darwin after reading his *On the Origin of Species* that "other parts I read with absolute sorrow; because I think them utterly false and grievously mischievous—You have deserted—after a start in that tram-road of all solid physical truth—the true method of induction."

Sedgwick was a geologic catastrophist, believing most Earth history events could be explained by a series of major catastrophes, much as described by the French geologist Georges Cuvier (1769–1832), and opposed to the gradualistic models of Sir Charles Lyell. Sedgwick's main opposition to Darwin's model for evolution was its apparent lack of any involvement of a divine being or creation. Although Sedgwick believed in the great lengths of geological time, he thought that there was a god in the evolution of life and the Earth, arguing with Darwin that "there is a moral or metaphysical part of nature as well as a physical."

See also CAMBRIAN; CARBONIFEROUS; DARWIN, CHARLES; LIFE'S ORIGINS AND EARLY EVOLUTION; LYELL, SIR CHARLES; ORDOVICIAN; SORBY, HENRY CLIFTON.

FURTHER READING

Clark, J. W., and T. M. Hughes. *The Life and Letters of the Reverend Adam Sedgwick.* 2 vols. Cambridge: Cambridge University Press, 1890.

sedimentary rock, sedimentation   Sedimentary rocks are rocks that have consolidated from accumulations of loose sediment produced by physical, chemical, or biological processes. Common physical processes involved in the formation of sediments include the breaking, transportation of fragments, and accumulation of older rocks; chemical processes include the precipitation of minerals by chemical processes or evaporation of water; common biological processes include the accumulation of organic remains.

Soils and other products of weathering of rocks are continuously being removed from their sources and deposited elsewhere as sediments. This process can be observed as gravel in streambeds, on alluvial fans, and in wind-blown deposits. When these sediments are cemented together, commonly by minerals deposited from water percolating through the ground, they become sedimentary rocks. Other types of sedimentary rocks are purely chemical in origin, and were formed by the precipitation of minerals from an aqueous solution.

Clastic sediments (also detritus) are the accumulated particles of broken rocks, some with the remains of dead organisms. The word clastic is from the Greek word *klastos*, meaning broken. Most clastic particles have undergone various amounts of chemical change, and some have a continuous gradation in size from huge boulders to submicroscopic particles. Size is the main basis for classifying clastic sediments and sedimentary rocks. The textures of the sedimentary rocks or individual sedimentary particles act as additional criteria for the classification of sedimentary rocks.

Clastic sediments can be transported by wind, water, ice, or gravity, and each method of transport leaves specific clues as to how it was transported and deposited. For instance, deposits from sediments transported by gravity in a landslide will consist of a poorly sorted mixture including everything that was in the path, whereas sediment transported by wind will have a very uniform grain size and typically forms large dunes. Clastic sediments are deposited when the transporting agent can no longer carry them. For instance, if the wind stops, the dust and sand will fall out, whereas sediments transported by streams are deposited when the river velocity slows down. This happens either where the stream enters a lake or the ocean or when a flood stage lowers and the stream returns to a normal velocity and clears up. Geologists can look at old rocks and tell how fast the water was flowing during deposition and can also use clues such as the types of fossils or the arrangement of the individual particles to decipher the ancient environment.

Chemical sediment is sediment formed when minerals precipitate from solution. They may result from biochemical activities of plants and animals that live in the water, or they may form from inorganic reactions in the water, induced by things such as hot springs or simply the evaporation of seawater. This produces a variety of salts, including ordinary table salt. Chemical sedimentary rocks are classified according to their main chemical component, with common types including limestone (made of predominantly calcite), dolostone (consisting of more than 50 percent dolomite), rock salt (composed of NaCl), and chert (whose major component is SiO_2).

Evaporite sediments include salts precipitated from aqueous solutions, typically associated with the evaporation of desert lake basins known as playas, or the evaporation of ocean waters trapped in restricted marine basins associated with tectonic movements and sea level changes. They are also associated with sabkha environments along some coastlines such as along the southern side of the Persian (Arabian) Gulf, where seawater is drawn inland by capillary action and evaporates, leaving salt deposits on the surface.

Evaporites are typically associated with continental breakup and the initial stages of the formation of ocean basins. For instance, the opening of the south Atlantic Ocean about 110 million years ago is associated with the formation of up to 3,280 feet (1,000 m) of salts north of the Walvus-Rio Grande Ridge. This ridge probably acted as a barrier that episodically (during short sea-level rises) let seawater spill into the opening Atlantic Ocean, where it evaporated in the narrow rift basin. A column of ocean water about 18.5 miles (30 km) thick would be necessary to form the salt deposits in the south Atlantic, suggesting that water spilled over the ridge many times during the opening of the basin. The evaporate-forming stage in the opening of the Atlantic probably lasted about 3 million years, perhaps involving as many as 350 individual spills of seawater into the restricted basin. Salts that form during the opening of ocean basins are economically important because when they get buried under thick piles of passive continental margin sediments, the salts typically become mobilized and intrude overlying sediments as salt diapirs, forming salt domes and other oil traps exploited by the petroleum industry.

Salts can also form during ocean closure, with examples known from the Messinian (Late Miocene) of the Mediterranean region. In this case thick deposits of salt with concentric compositional zones reflect progressive evaporation of shrinking basins, when water spilled out of the Black Sea and Atlantic into a restricted Mediterranean basin. So-called closing salts are also known from the Hercenian orogen north of the Caspian Sea and in the European Permian Zechstein basin in the foreland of the collision.

As seawater evaporates, a progressive sequence of different salts forms from the concentrated brines.

Typically, anhydrite ($CaSO_4$) is followed by halite (NaCl), which forms the bulk of the salt deposits. A variety of other salts can form depending on the environment, composition of the water being evaporated, when new water is added to the brine solution, and whether or not it partly dissolves existing salts.

Most sedimentary rocks display a variety of internal and surface markings known as sedimentary structures that can be used to interpret the conditions of formation. Stratification results from a layered arrangement of particles in a sediment or sedimentary rock that accumulated at the surface of the Earth. The layers are visible and distinct from adjacent layers because of differences (such as size, shape, or composition) in the particles between successive layers and because of differences in the way the particles are arranged between different layers. Bedding is the layered arrangement of strata in a body of rock. Parallel strata are sedimentary layers in which individual layers lie parallel to one another. The presence of parallel strata usually means that the sediments were deposited underwater, such as in lakes or in the deep sea. Some sediments with parallel layers have a regular alternation between two or more types of layers, indicating a cycle in the depositional environment. These can be daily, yearly, or some other rhythm influenced by solar cycles. One unusual type of layered rock is a varve, which is a lake sediment that forms a repeating cycle of coarse-grained sediments with spring tides, and fine clay with winter conditions, when the suspended sediments gradually settle out of the water column. Cross strata are layers that are inclined with respect to larger layers in which they occur. Most cross-laminated deposits are sandy or coarser, and they form as ripples that move along the surface. The direction of inclination of the cross strata is the direction that the water formerly flowed.

Sorting is a sedimentary characteristic that refers to the distribution of grain sizes within a sediment or sedimentary rock. Sediments deposited by wind are typically well sorted, but those deposited by water may show a range of sorting. A bed is called uniform if its layers contain grains with the same size throughout. A gradual transition from coarse- to fine-grained, or fine- to coarse-grained, is known as a graded bed. Graded beds typically reflect a change in current velocity during deposition. Nonsorted layers represent a mixture of different grain sizes, without any apparent order. These are common in rock falls, avalanche deposits, landslides, and from some glaciers. *Rounding* is a textural term that describes the relative shape or roundness of grains. When sediments first break off from their source area, they tend to be angular and reflect the shape of joints or internal

Cross-bedded Navajo sandstone from the Jurassic period in Zion National Park, Utah *(François Gohier/ Photo Researchers, Inc.)*

GEOLOGY; OCEAN BASIN; PETROLEUM GEOLOGY; PETTIJOHN, FRANCIS JOHN; SEQUENCE STRATIGRAPHY; STRATIGRAPHY, STRATIFICATION, CYCLOTHEM.

FURTHER READING

Allen, P. A., and J. R. Allen. *Basin Analysis, Principles and Applications.* Oxford: Blackwell Scientific Publications, 1990.

Bouma, Arnold H. *Sedimentology of Some Flysch Deposits: A Graphic Approach to Facies Interpretation.* Amsterdam, Elsevier, 1962.

Cefrey, Holly. *Sedimentary Rocks.* New York: Rosen Publishing Group, 2003.

Pettijohn, Francis J. *Sedimentary Rocks.* London: Harper, 1957.

Prothero, Donald, and Robert Dott. *Evolution of the Earth.* 6th ed. New York: McGraw-Hill, 2002.

Stanley, Steven M. *Earth and Life through Time.* New York: Freeman, 1986.

seismology The study of the propagation of seismic waves through the Earth, including analysis of earthquake sources, mechanisms, and the determination of the structure of the Earth through variations in the properties of seismic waves is called seismology.

Determination of the structure of the deep parts of the Earth can be achieved only by remote geophysical methods such as seismology. Seismographs are stationed all over the world, and studying the propagation of seismic waves from natural and artificial source earthquake and seismic events allows for the calculation of changes in the properties of the Earth in different places. If the Earth had a uniform composition, seismic wave velocity would increase smoothly with depth, because increased density is equated with higher seismic velocities. However, by plotting the observed arrival time of seismic waves, seismologists have found that the velocity does not increase steadily with depth but that several dramatic changes occur at discrete boundaries and in transition zones deep within the Earth.

One can calculate the positions and changes across these zones by noting several different properties of seismic waves. Some are reflected off interfaces, just like light is reflected off surfaces, and other waves are refracted, changing the velocity and path of the rays. These reflection and refraction events happen at specific sites in the Earth, and the positions of the boundaries are calculated using wave velocities. The core-mantle boundary at 1,802 miles (2,900 km) depth in the Earth strongly influences both P and S waves. It refracts P-waves, causing a P-wave shadow and, because liquids cannot transmit S waves, none get through, causing a huge S-wave shadow. These

mineral forms. With progressive transportation by wind or water, abrasion tends to smooth the grains and make them rounded. The greater the transport distance, in general, the greater the rounding.

Surface features on sedimentary layers also yield clues about the depositional environment. Like ripple marks or footprints on the beach, many features preserved on the surface of strata offer clues about the origin of sedimentary rocks and the environments in which they formed. Ripple marks show the direction of ancient currents, whereas tool marks record places where an object was dragged by a current across a surface. Turbulent eddies in a current produce grooves in the underlying sediment called flute marks, by scouring out small pockets on the paleosurface. Mud cracks reveal that the surface was wet, then desiccated by subaerial exposure. Other types of surface marks include footprints and animal tracks in shallow water environments, and raindrop impressions in subaerial settings.

Fossils are remains of animals and plants preserved in the rock that can also reveal clues about past environments. For instance, deep marine fossils are not found in lake environments, and dinosaur footprints are not found in deep marine environments.

See also ARABIAN GEOLOGY; BASIN, SEDIMENTARY BASIN; CONTINENTAL MARGIN; HISTORICAL

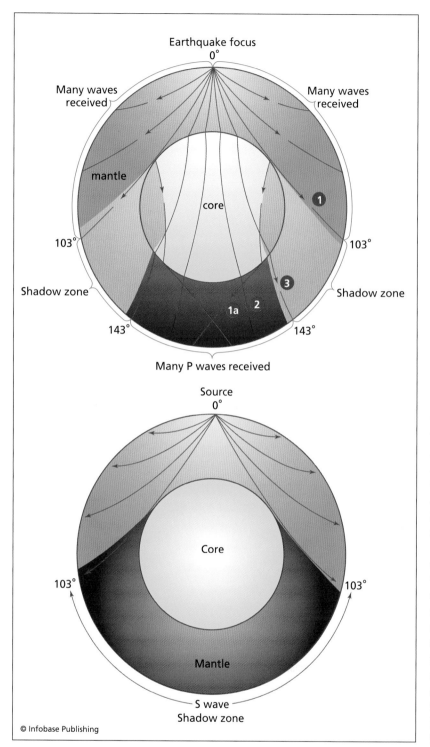

Cross sections of the Earth showing shadow zones that develop in bands around the planet due to refraction of P and S waves at internal boundaries

about 62 miles (100 km), where the velocity drops slightly between 62 and 124 miles (100–200 km) depth, in a region known as the low velocity zone. The reason for this drop in velocity is thought to be small amounts of partial melt in the rock, corresponding to the asthenosphere, the weak sphere on which the plates move, lubricated by partial melts.

Another seismic discontinuity exists at 248.5 miles (400 km) depth, where velocity increases sharply due to a rearrangement of the atoms within olivine in a polymorphic transition into spinel structure, corresponding to an approximate 10 percent increase in density.

A major seismic discontinuity at 416 miles (670 km) could be either another polymorphic transition or a compositional change, the topic of many current investigations. Some models suggest that this boundary separates two fundamentally different types of mantle, circulating in different convection cells, whereas other models suggest that there is more interaction between rocks above and below this discontinuity.

The core-mantle boundary is one of the most fundamental on the planet, with a huge density contrast from 5.5 g/cm^3 above, to 10–11 g/cm^3 below, a contrast greater than that between rocks and air on the surface of the Earth. The outer core is made dominantly of molten iron. An additional discontinuity occurs inside the core at the boundary between the liquid outer core and the solid, iron-nickel inner core.

The properties of seismic waves can also be used to understand the structure of the Earth's crust. Andrija Mohorovičić (a Yugoslavian seismologist) noticed slow and fast arrivals from nearby earthquake source events. He proposed that some seismic waves traveled through the crust, some along the surface, and that some were reflected off a deep seismic discontinuity between seismically slow and fast material at about 18.6

contrasting properties of P and S waves can be used to accurately map the position of the core-mantle boundary.

Variations in the propagation of seismic waves illustrate several other main properties of the deep Earth. Velocity gradually increases with depth to

PREDICTING FUTURE EARTHQUAKES IN THE WESTERN UNITED STATES

The Earth is a dynamic planet composed of different internal layers that are in constant motion, driven by a vast heat engine deep in the planet's interior. The cool surface layer is broken into dozens of rigid tectonic plates that move around on the surface at rates of up to a few inches (~5 cm) per year, driven by forces from the internal heat and motion in the partly molten layers within the planet. Most destructive earthquakes are associated with motions of continents and ocean floor rocks that are part of these rigid tectonic plates riding on moving parts of the Earth's interior. Plate tectonics is a model that describes the process related to the slow motions of more than a dozen of these rigid plates of solid rock riding around on the surface of the Earth. The plates ride on a deeper layer of partially molten material that is found at depths starting at 60–200 miles (100–320 km) beneath the surface of the continents, and 1–100 (1–160 km) miles beneath the oceans. The motions of these plates involve grinding, sticking, and sliding where the different plates are in contact and moving in different directions, causing earthquakes when sudden sliding motions occur along faults. These earthquakes release tremendous amounts of energy, raising mountains and, unfortunately, sometimes causing enormous destruction.

For many years there has been much speculation about the effects of the anticipated magnitude 7–9 earthquake, or the "big one," in southern California. This speculation and fear is not unfounded. The unrepentant forces of plate tectonics continue to slide the Pacific plate northward relative to North America along the San Andreas Fault, and the stick-slip type of behavior that characterizes segments of this fault system generates many earthquakes. Several segments of the fault have seismic gaps, where the tectonic stresses may be particularly built up. One of these gaps released its stress during the Loma Prieta earthquake of 1989, but other gaps remain, including the large area generally east of Los Angeles. Some models predict that the seismic energy may be released in this area by a series of moderate earthquakes (magnitude 7), but other, more sinister predictions also remain plausible. The last major rupture along the southern California segment was in 1680, and the event before that occurred between 1330–1480. Studies of prehistoric earthquakes in this region show that major, catastrophic events recur roughly every 350 years. In this scenario, the next major event could be expected sometime around the year 2030, with a large margin of error, and the magnitude could reach or exceed 8 on the Richter scale. Such an event would be truly devastating to urban areas in southern California, and would result in tremendous loss of life and property. Residents should prepare, having emergency plans and supplies ready, and understand the different risks in their specific locations.

What Can the Pacific Northwest Expect?

Only 15 years ago many people in the Seattle area thought they were far enough away from active faults to not have to worry about earthquakes. Scientists thought the Cascadia subduction zone beneath the Pacific Northwest was aseismic since it had not produced any earthquakes in at least a couple of hundred years. Then, on February 28, 2001, a magnitude 6.8 earthquake lasting 45 seconds and injuring 250 people sent Seattle a rude wake-up call. Since then, scientists and residents have come to realize that the Cascadia subduction zone forms a potential threat for earthquakes, even for great earthquakes (magnitudes greater than 8) similar in magnitude to the December 26, 2004, Indonesian earthquake. In addition to the active seismicity, this realization comes in part from studies that have identified paleoearth-

quakes and tsunami deposits from local earthquakes.

The U.S. Geological Survey has identified three main potential sources for Pacific Northwest earthquakes, including deep ruptures that originate along the subduction zone thrust and have in the past generated magnitude 6.5–7.1 earthquakes in this area. The second potential source is within the upper plate over the subduction zone, where events greater than magnitude 7 have occurred in the past (1872, 1918, and 1946). The third source is potentially the most destructive, including the shallow segment of the subduction zone where ruptures can be large and include significant surface movements, generating tsunami. The most recent giant (~ magnitude 9) shallow subduction zone quake in the area was in 1700, where the earthquake triggered a huge tsunami that damaged parts of Japan and left deposits around the Pacific Northwest area. It is estimated that events of this size occur every 400–600 years in the Cascadia subduction zone, and the clock is ticking.

A magnitude 9 earthquake in the Cascadia subduction zone would be devastating to the Seattle-Tacoma-Aberdeen-Bellingham area, including at least several minutes of shaking (for comparison, the Sumatra earthquake had up to 10 minutes of shaking), with peak ground accelerations exceeding half the force of gravity. Like San Francisco's infamous Nimitz Freeway that collapsed in the 1989 Loma Prieta earthquake, Seattle has double-decker freeways built on tidal flat deposits and uncompacted landfill-deposits that are prone to severe shaking and liquefaction during earthquakes, potentially leading to highway collapse. Faults have recently been discovered running through downtown Seattle, suggesting that the region may be susceptible to strong earthquakes. The Seattle fault runs through Puget Sound, then close to the Kingdome in downtown, and appears

(continues)

(continued)

to have accommodated about 20 feet (6 m) of movement of a block measuring 10 miles (16 km) long by four miles (6.4 km) wide about 1,000 years ago. Shallow faults like this one have the potential to generate devastating earthquakes since their energy is released near the surface, generating large amounts of ground shaking. Other geologic and natural features in the Seattle area all indicate that a major earthquake occurred here, such as landslide scars that mar many hillsides, some of which dammed valleys and formed lakes. Forests now observed on the bottoms of lakes died between 800 and 1,400 years ago, perhaps from drowning associated with earthquake-induced changes in ground level. Sand deposits from tsunamis that inundated

bayhead marshes have been tentatively correlated with a sudden uplift of parts of Puget Sound by more than 20 feet (6 m) sometime between 500 and 1,700 years ago during a catastrophic earthquake, and other areas record sudden subsidence of about 5 feet (1.5 m) at 1,000 years ago.

Even more devastating would be an earthquake along the Olympia subduction zone, with potential magnitudes exceeding 9. In addition to tremendous ground shaking, huge tsunamis could be generated that would quickly sweep along the coastline. Sand layers in coastal swamps attest to a succession of tsunamis in the region as recently as 300 years ago, though research is only beginning on these to understand their frequency and whether they were generated from local or distant earth-

quakes. Some coastal forests with Sitka spruce are dead, having been killed 300 years ago when the land dropped six to eight feet (1.8–2.5 m), putting the roots of the trees into saltwater, killing them. The potential for great (magnitude greater than 9) earthquakes in the Cascadia subduction zone is clear, but the cities of Seattle, Portland, and regions in Washington, British Columbia, Oregon, and northern California are not adequately prepared. Building codes are not as strict as in earthquake-prone southern California, yet the potential for exceptionally large earthquakes is greater in the Pacific Northwest.

FURTHER READING

Kusky, T. M. *Earthquakes: Plate Tectonics and Earthquake Hazards.* New York: Facts On File, 2008.

---

miles (30 km) depth. Geologists now recognize this boundary to be the base of the crust and call it the Mohorovičić (or Moho) boundary and use its seismically determined position to measure the thickness of the crust, typically between 6.2–43.5 miles (10–70 km).

SEISMOGRAPH

Seismographs are sensitive instruments that can detect, amplify, and record ground vibrations, especially earthquakes, producing a seismogram. Numerous seismographs have been installed in the ground throughout the world and form a seismograph network, monitoring earthquakes, explosions, and other ground-shaking events.

The first very crude seismograph was constructed in 1890. While the seismograph could tell that an earthquake was occurring, it was unable to actually record the earthquake. Modern-era seismographs display movements of the Earth by means of an ink-filled stylus on a continuously turning roll of graph paper. When the ground shakes, the needle wiggles and leaves a characteristic zigzag line on the paper. In early models, the ink-filled stylus recorded real movement between the ground and the stylus, and was not very accurate. In recent models, the ink-filled stylus records motions that are detected through modern, ultra-sensitive seismographs, and converted into an electronic signal that is used to move the stylus and make the trace on the paper on the moving drum.

Seismographs are built using a few simple principles of physics. To measure the vibrations of the Earth during an earthquake, the point of reference must be separate from the ground and free from shaking. To this end, engineers have designed an instrument known as an inertial seismograph. These make use of the principle of inertia, which is the resistance of a large mass to sudden movement. When a heavy weight is hung from a string or thin spring, the string can be shaken and the heavy weight will remain stationary. Using an inertial seismograph, the ink-filled stylus is attached to the heavy weight, and remains stationary during an earthquake. The continuously turning graph paper is attached to the ground, and moves back and forth during the quake, resulting in the zigzag trace of the record of the earthquake motion on the graph paper.

Seismographs are used in groups, each recording a different type of motion of the ground. Some seismographs are set up as pendulums and some others as springs, to measure ground motion in many directions. Engineers have made seismographs that can record motions as small as one hundred-millionth of an inch, about equivalent to being able to detect the ground motion caused by a car driving by several blocks away. The ground motions recorded by seismographs are very distinctive, and geologists who study them have methods of distinguishing between earthquakes produced along faults and earthquake swarms associated with magma moving into volcanoes, and even between explosions from different types of construction, accidents, and nuclear blasts. It is even possible to infer the size and other characteristics of different nuclear and other explosions with detailed analysis of the seismic signal from a specific event. Interpreting seismograph traces has therefore become an important aspect of nuclear test ban treaty verification.

In the late 19th century, seismologist and engineer E. Wiechert introduced a seismograph with a large, damped pendulum used as the sensor, with the damping reducing the magnitude of the pendulum's oscillations. This early seismograph recorded horizontal motions using a photographic recording device. Wiechert soon introduced a new seismograph with a mechanical recording device, with an inverted pendulum that could vibrate in all horizontal directions. The pendulum was supported by springs that helped stabilize the oscillations and furthered the productivity of the seismograph. Wiechert's assistant, named Schluter, introduced a vertical recording device. He moved the mass horizontally away from the axis of rotation and maintained it there with a vertical spring. In doing so he was able to record vertical displacement, which helped record many of the complex movements associated with earthquakes.

In the 20th century, seismographs that recorded movements using a pen on a rotating paper-covered drum were introduced, with alternative devices including those that recorded movements using a light spot on photographic film. More sophisticated seismographs can record movements in three directions (up-down, north-south, and east-west), and electronic recording of relative motions is now commonplace.

See also CONVECTION AND THE EARTH'S MANTLE; EARTHQUAKES; MANTLE; PLATE TECTONICS.

FURTHER READING

Keary, P., Keith Klepeis, and Frederick J. Vine. *Global Tectonics*. Oxford: Blackwell Publishers, 2008.

Shearer, Peter M. *Introduction to Seismology*. Cambridge: Cambridge University Press, 2009.

Sheriff, Robert E. *Encyclopedic Dictionary of Applied Geophysics*, 4th ed. Tulsa, Okla.: Society of Exploration Geophysicists, 2002.

Stein, S. *Introduction to Seismology*. Oxford: Blackwell Publishing, 2000.

Turcotte, Donald L., and Gerald Schubert. *Geodynamics*. 2nd ed. Cambridge: Cambridge University Press, 2002.

sequence stratigraphy Sequence stratigraphy is the study of the large-scale three-dimensional arrangement of sedimentary strata and the factors that influence the geometry of these sedimentary packages. Sequences are defined as groups of strata that are bounded above and below by identifiable surfaces that are at least partly unconformities. Many sequence boundaries show up well in seismic reflection profiles, enabling their identification in deeply buried rock packages. Sequence stratigraphy differs from classical stratigraphy in that it groups together different sedimentary facies and depositional environments that were deposited in the same time interval, whereas classical stratigraphy would separate these units into different formations. By analyzing the three-dimensional shape of time-equivalent packages, the depositional geometry and factors that influenced the deposition are more easily identified. Some of the major factors that control the shape of depositional sequences include global sea-level changes, local tectonic or thermal subsidence or uplift, sediment supply, and differential biologic responses to subsidence in different climate conditions. For instance, carbonate reefs may be expected to keep pace with subsidence in tropical climates, but to be absent in temperate or polar climates. Sedimentologists and tectonicists use the techniques of sequence stratigraphy in the petroleum industry to understand regional controls on sedimentation and to correlate sequences of similar age worldwide.

See also PASSIVE MARGIN; SEDIMENTARY ROCK, SEDIMENTATION; STRATIGRAPHY, STRATIFICATION, CYCLOTHEM.

Silurian The Silurian refers to the third period of Paleozoic time ranging from 443 Ma to 415 Ma, falling between the Ordovician and Devonian Periods, and the corresponding system of rocks. From base to top it is divided into the Llandoverian and Wenlockian Ages or Series (comprising the Early Silurian) and the Ludlovian and Pridolian Ages or Series (comprising the Late Silurian). The period is named after a Celtic tribe called the Silures, who inhabited a region of Wales where rocks of the Silurian system are well exposed. The Silurian is also known as the age of fishes.

Rocks of the Silurian system are well exposed on most continents, with carbonates and evaporites covering parts of the Midwest of North America, the Russian platform, and China. Silurian clastic sequences form thick orogenic wedges in eastern and western North America, central Asia, western Europe, China, and Australia. Much of Gondwana was together in the Southern Hemisphere, and included the present-day landmasses of South America, Africa, Arabia, India, Antarctica, Australia, and a fragmented China. North America, Baltica, Kazakhstan, and Siberia formed separate landmasses

Fossilized crinoid, or sea lily, in a mudstone deposit, from the Silurian period *(Kaj R. Svensson/Photo Researchers, Inc.)*

in equatorial and northern latitudes. Much of Gondwana was bordered by convergent margins, and subduction was active beneath the Cordillera of North America. Baltica and Laurentia had collided during early stages of the Acadian-Caledonian orogeny, following an arc-accretion event in the Middle to Late Ordovician, known as the Taconic orogeny in eastern North America.

Land plants first appeared in the Early Silurian and were abundant by the middle of the period. Scorpionlike eurypterids and arthropods inhabited freshwater environments and may have scurried across the land. In the marine realm, trilobites, brachiopods, cephalopods, gastropods, bryozoans, crinoids, corals, and echinoderms inhabited shallow waters. Stromatoporoids and rugose and tabulate corals built conspicuous reefs, while jawed fish fed on plankton and nekton.

See also PLEOZOIC; PHANEROZOIC.

FURTHER READING

Kious, Jacquelyne, and Robert I. Tilling. U.S. Geological Survey. "This Dynamic Earth: The Story of Plate Tectonics." Available online. URL: http://pubs.usgs.gov/gip/dynamic/dynamic.html. Last modified March 27, 2007.

Prothero, Donald, and Robert Dott. *Evolution of the Earth.* 6th ed. New York: McGraw-Hill, 2002.

Stanley, Steven M. *Earth and Life through Time.* New York: Freeman, 1986.

Smith, William (1769–1839) English *Geologist*

William Smith was a self-taught surveyor who recognized the regular succession of strata across England and proposed that lithologically similar rock beds could be distinguished by the groups of characteristic fossils embedded within. Using this information, he created the world's first large-scale geologic map of an entire country, showing more than 20 different units, topography, description of the stratigraphy, and structural cross sections. During the same time, Smith produced his works on "Strata identified by organic remains," in which he illustrated the fossils in the rocks through a series of wood engravings. In 1831, the Geological Society of London awarded William Smith the first Wollaston Medal, its highest honor. Since his death, William Smith has become known as the "Father of English Geology."

EARLY YEARS

The eldest of four children, William Smith was born on March 23, 1769, to John and Ann Smith of Churchill, Oxfordshire, England. A blacksmith and a mechanic, John died when William was only eight years old. (Ann remarried a few years later.) As a child he was attracted to the pound stones that he found on Oxfordshire fields. These were round, dome-shaped stones that weighed approximately one pound and were used as a standard weight measure by dairymaids. Sometimes they had an interesting pattern shaped like a five-point star. It turns out that these were fossilized remains of sea urchins. William also collected pundibs, spherical-shaped rocks the size of acorns that actually were the remains of terebratulids, a type of brachiopod. As a suitable substitute for marbles, to young William they were merely play toys, though hindsight shows they were an unrecognized symbol of his future achievements.

CAREER AS A SURVEYOR

William continued studying on his own after leaving the village school, yet had to ask to borrow from his inheritance in order to purchase books. One such book was *The Art of Measuring* by Daniel Fenning. One day when he was 18, he met a man named Edward Webb who was visiting Oxfordshire. He was a professional surveyor, someone who determined boundaries of areas and measured land elevations using geometry and trigonometry. Webb needed an apprentice to help him divide up some farming fields and hired William as an assistant.

Smith learned about the soil and rocks of Oxfordshire and the methods of surveying quickly. Within a few months he could skillfully use a pantograph, a theodolite, dividers, and a great steel chain, all tools for geographic surveying. By the summer of 1788, Smith was doing his own work. He traveled with Webb and eventually moved in with Webb's family 10 miles away at Stow-on-the-Wold. As he traveled, he kept diaries of his observations, especially geologic findings.

In 1791 Smith traveled to Somerset to do a valuation survey in the village of Stowey, near Bath. He ended up staying there and working for Webb for a few years, getting to know the terrain and making contacts with the residents. One influential woman he met was Lady Elizabeth Jones, whose land he originally came out to survey. He rented a farmhouse from her called Rugborne, on the eastern side of High Littleton. This estate later became known as the birthplace of geology.

Somerset was a coal mining community, and Lady Jones was the director of the High Littleton Coal Company. Under her employment, Smith surveyed, planned, and drained land. In 1792 his main work site was the Mearns Pit. The first time Smith went into that mine, he was puzzled by what he observed. As he descended into the mine, he passed grass, gravel, and topsoil, and then a more solid rock layer consisting of limestone, marlstone, and shales. Below this rocky layer was an abrupt transition. The color suddenly changed from reddish green to grayish brown, and there was a surprising, deep, downward slope. This layer was warped and all broken up, in sharp contrast to the nicely laid out horizontal sheets of varying thicknesses in the upper layers. Something very different had controlled the arrangement of layers below the flat coal in the upper part of the mine, and William Smith was determined to figure out what caused such differences.

Approximately 300 million years ago, the European and African tectonic plates collided with each other, forming Pangaea. an enormous former supercontinent composed of all the existing continents. The resulting contraction and folding persisted for millions of years, forming many geological structures such as numerous folds and faults throughout Britain, including the folds and faults that William Smith observed in the coal mine. This process was called the Variscan Orogeny, and it left a complexly deformed mess of pre-Permian rocks. It also made coal mining difficult in north Somerset since rocks containing useful coal were deposited during the Upper Carboniferous period, 290–320 million years ago.

During his time in the mines, Smith made other astute observations. For example, he found the same pattern in mine after mine. From top to bottom, this pattern was sandstone, siltstone, mudstone, nonmarine bands, marine bands, coal, seat earth, and then back to sandstone, in a cyclical pattern. Particular seams of coal were always located in the same relative position. He also noticed that all sedimentary rocks laid down at the same time were similar and that the same fossil types appeared in the same stratigraphical order. Smith began to view geology as more of a science, his interest grew, and he formulated many scientific questions that he next began to test. Smith inquired if the patterns he observed in the coal mine could be applied to other rocks which lay below the ground but miles away, and what if any relationships the rocks underground had to rocks above the ground such as in mountains. He started to use his observations to predict where certain rock types could be found in other places.

Meanwhile, the miners were fretful. Across the Avon River, the Welsh were building a canal to help transport their coal. Somerset could not compete for coal sales if they could not move their coal efficiently, so they decided to build a canal as well. A surveyor was needed to determine the best route for what was to be called the Somerset Coal Canal. The canal eventually connected Limpley Stoke, at a junction with a larger canal, to Camerton, where the coal was located.

A Scotsman named John Rennie initially signed up to make the survey, but he was too busy. Lady Jones suggested Smith as an apprentice. In 1793 Smith started inspecting the structure of the land to choose a route for the canal that would be easy to dig and retain water. This was a wonderful opportunity for him to collect geologic information from the rock exposed by the digging. He eventually recommended that two parallel canals be built, a northerly Dunkerton Line and a southerly Radstock Line. This allowed him to examine the geology of even more land area. He noticed a uniform dip to the rocks between Dunkerton and Midford. This taught him that strata did not always exist as horizontal lines. He also observed that there was a distinctive sequence to the rock layers and was anxious to learn if his ideas and observations applied to the entire nation.

In early 1794, Smith traveled to London as a witness before Parliament in order to obtain authorization to build the canal. This was simply a bureaucratic process necessary before commencing with the project. He had a lot of spare time while in London, and he spent it at libraries and bookstores trying to learn if anyone else had published anything similar to the ideas that were forming in his mind. He was unsuccessful but still worried that someone else might be developing similar ideas to his own.

THE BIRTH OF STRATIGRAPHY

Smith took a carriage trip with two other members of the canal committee later that year. The purpose of the 900-mile (1,448-km) trip over England and Wales was to see how others were building canals. While on this excursion, he continually jumped off the stagecoach to take samples of rock and fossils and took frequent notes of geologic observations. He commented to his companions that he could tell from the landscape just what type of rock lay underneath. He demonstrated this skill to the older men, who were no doubt amused by his enthusiasm.

Smith was becoming quite skilled at identifying different strata, but some strata looked very similar. If rock layers were deposited under the same conditions, even if during different time periods, they can look alike. Likewise, rock layers that were deposited at the same time and were of similar composition can look different due to physical disturbances such as volcanic activity or sweeping currents. In one instance, Smith had observed separate outcrops of limestone that looked very similar but were separated by great distances, representing long periods of time. In addition, his knowledge of dip and strike had convinced him they were in fact different strata from different time periods. Dip is the angle by which a rock layer deviates from the horizontal plane, and strike is the direction 90 degrees to the dip. So, how could one distinguish these rock layers? After years of careful examination of the terrain across England, Smith formulated his principle of fossil succession, which he memorialized by writing in his journal of 1796.

The principle of fossil succession states that the sequence of fossils in rock strata is so regular, that fossils can be used to identify the rocks in which they are embedded. Fossils can be used to establish the time sequence by which the rocks were laid. This was a new concept for geologists in the early 1800s, but today it is a basic principle of stratigraphy. Around the same time, Georges Cuvier and Alexandre Brongniart also were recognizing the utility of fossils in geologic chronologies.

In 1798 Smith purchased a home near Bath called the Tucking Mill. Bath was uniquely suited for geological study since the Middle Jurassic rock outcrop was blatantly exposed. Many strata were apparent, including outcrops of rocks spanning several time periods. While living there, he created what is technically considered the first true geological map, a circular map with a five-mile (8-km) radius and Bath at its center. Smith noted the locations of different fossil types, and used dip and strike information to estimate locations of various strata. Significantly, he used color to specifically depict the locations of oolite (yellow), lias (dirty blue) and red marls (brick red).

Smith was suddenly fired from the Somerset Canal Company in 1799 for an unknown reason. He worked as an independent mineral surveyor and drainage engineer for the next two decades. His expertise was in constant demand, and he made decent money, but unfortunately, most of his earnings were spent on a project that occupied him until 1815, the construction of a large-scale geologic map.

Smith was elected to the Bath Agricultural Society in 1796. This association led to many connections that had later influences on Smith's life and accomplishments. Two encouraging members, Reverend Benjamin Richardson of Farleigh and Reverend Joseph Townsend of Pewsey, were also fossil collectors. One night in 1799, the clerics pulled out some paper and wrote out, as Smith dictated, a list of strata. This was in the form of a table that included information not only on the succession of 23 strata from chalk to coal, but their thicknesses, lithologic characteristics, and distinguishing types of embedded fossils. They made three copies of the table that night, one for each of them, with the understanding that the men would recopy and disseminate this information to whoever was interested. Years later, Smith heard that copies of this table were being distributed on different continents.

In 1801 Richardson suggested that Smith write a prospectus, outlining his intent to publish a work on the natural order of strata in England and Wales. He obtained the sponsorship of Sir Joseph Banks, the president of the Royal Society of London, for this project. Years passed, however, and no progress was made. Smith was too busy working, trying to make enough money to pay his mortgages.

In 1805 Smith had leased a large house in London. Being of common birth and never formally educated, he felt it was important to keep up appearances of success in order to obtain respectable employment and sponsorship; however, he still owned his Tucking Mill estate. In addition, he had made a bad investment a few years earlier. He took out a second mortgage on Tucking Mill to invest in a quarry for the excavation of oolitic limestone, a popular building material. Unfortunately, times were bad, and in the early 1800s people stopped building altogether. Furthermore, the quality of the stone was inferior. Smith also rented an office in Bath, serving as his base for a brief partnership he had in a firm, Smith and Cruse, Land Surveyors. At the expense of his long-awaited mapping project, he furiously toiled away just trying to make ends meet.

In the midst of this, Smith got married. His 17-year-old bride, Mary Ann, was uneducated, often physically ill, and eventually became mentally deranged, adding to Smith's troubles. In 1807 he became the guardian of his orphaned nephew, John

Phillips. John later became Smith's assistant and then a notable geologist, but at the time he was another financial burden.

THE WORLD'S FIRST GEOLOGIC MAP

John Cary, a highly regarded English cartographer, agreed to publish Smith's map in 1812. A topographical map was engraved, on top of which Smith added his geological information. The actual construction of the map was quite a task itself, involving 16 engraved plates and three years to accomplish. The data collection and assimilation of the information took Smith 14 years to complete. The completed map, *A Delineation of the Strata of England and Wales, with Part of Scotland,* was published in 1815 with a 50-page textual explanation.

The map was slightly larger than eight by six feet (2.4 by 1.8 m) with the scale being five miles (8 km) to one inch (2.54 cm). One striking feature was the color. Smith used a variety of shades to depict certain types of rock: gray for Tertiary outcrops, blue green for chalk, brown for coral rag and carstone, yellow for oolites, blue for lias, and red for red ground. Not only was the use of color original, but he colored the base of each rock formation darker than its top. Thus, if one stood back, an immediate pattern was apparent. This piece of work became a classic in cartography. Modern geological maps use the same principles and even the same color scheme that Smith used almost 200 years ago.

Four hundred copies were made, but they sold poorly, partly due to George Bellas Greenough, one of the original founders of the Geological Society of London, a small, elite club of rich intellectuals. Though Smith was hurt by the lack of inclusion in their society, in 1808 he had invited them to view his impressive fossil collection, which was carefully organized by chronological succession and beautifully displayed on a series of sloping shelves meant to represent the sequence of strata. The visit of the Geological Society was a disappointing one. Smith received neither the praise he deserved nor the invitation to join the Society that he so desperately craved. Little did he know that the president of the society was not only impressed but also jealous. Greenough was about to embark on a mission of scientific pilfering that would be personally and professionally catastrophic to Smith.

Greenough announced the intention of the Geological Society to publish a geologic map of England, similar to Smith's. Potential buyers of Smith's map decided to wait until Greenough's map was published rather than buy Smith's map. After all, Smith's did not have the backing of the Geological Society, and Greenough's promised to be cheaper. When the map did come out in 1819, it did not fare much better than Smith's map, nor did it contain any new information. Later Greenough was forced to admit that he stole much of Smith's work in constructing his map. In a ridiculous effort to make amends, Greenough apologized and presented a copy of his map to Smith.

Financial troubles intensified, and Smith was forced to sell his extensive fossil collection to the British Museum (the present-day Natural History Museum of London). In hopes of earning a little money, Smith published *Strata Identified by Organized Fossils* (1816) and *Stratigraphical System of Organized Fossils Part I* (1817). The latter was a catalog of the collection now owned by the museum. Neither sold well, and he continued creating and publishing geologic maps of several counties in England (1819–24). He also released *A Geological Section from London to Snowdon* (1817), showing the relative thicknesses and arrangements of rocks. In 1819, unfortunately, his financial difficulties became too great. He was sent to debtor's prison for 11 weeks, during which time he lost his home and his few remaining personal belongings. He would have lost his papers and maps too, but an anonymous friend purchased and returned them to Smith. After his release, he gathered his sickly wife and his nephew and moved away from London, where he had been so horribly treated, into obscurity, where he remained for the next 12 years.

They traveled to Yorkshire, where Smith enjoyed lecturing on geology, but he had to give this up due to poor health. He settled in Scarborough from 1824 to 1828 and continued to study geology. He also designed a museum and helped with the town water supply.

RESPECTED AT LAST

Sir John Vanden Bempde Johnstone hired Smith as his land steward in Hackness in 1828. Johnstone was a member of the Geological Society and a fossil collector himself. He was aware of Smith's accomplishments, and with the assistance of a friend, he championed for an annuity to be purchased for the aging geologist.

In 1831 Smith was awarded the first Wollaston Medal by the Geological Society of London in recognition of his research into the mineral structure of the Earth. In return, Smith presented the Society with his original table of 23 strata (1799), his colored geologic map of Bath and the surrounding area (1799), and an original rough sketch of his geologic masterpiece (1801). He received his gold medal the next year followed by a government pension. Trinity College in Dublin awarded him an honorary doctorate degree in 1835.

Smith's last job was serving as part of a committee selected by the government to choose the new building material for the British House of Parliament, as the old building had burned down in 1834. The committee selected a magnesium limestone from a quarry in Derbyshire. The supply ran short, and a quick substitution had to be found. The substitute stone turned out to be unsuitable. One wonders if Smith might have recognized this and corrected the error before it was too late, if he had lived longer. Within 10 years, the exterior of the buildings deteriorated.

On the way to a British Association meeting in Birmingham, Smith stopped to visit a friend in Northampton. He caught a cold that turned fatal. The father of English geology died on August 28, 1839. He was buried nearby in Saint Peter's Church.

Smith freely shared his knowledge of England's geology. His geologic maps were practically applied to the fields of mining, agriculture, road building, water draining, and canal building. His 1815 map of England and Wales is considered a milestone in geological cartography. Though Smith's major accomplishments went unnoticed by the scientific community initially, Smith's contributions to geography and biostratigraphy were just beginning at be recognized at the time of his death. In 1865 the Geological Society added Smith's name to Greenough's map, rightfully acknowledging his intellectual contribution. The Geological Society and the Oxford Museum display busts of Smith, and signposts and plaques adorn his former residences. Since 1977 the Society has awarded the William Smith Medal for contributions to applied and economic aspects of geology. The man who revealed his vision of the underworld has finally received the recognition he deserves.

See also EUROPEAN GEOLOGY; STRATIGRAPHY, STRATIFICATION, CYCLOTHEM.

FURTHER READING

Phillips, John. *Memoirs of William Smith*. 1844. Reprint, with additional material by Hugh Torrens, Bath, U.K.: Bath Royal Literary and Scientific Society, 2003.
Winchester, Simon. *The Map That Changed the World*. Rockland, Mass.: Wheeler, 2001.

soils Soils include all the unconsolidated material resting above bedrock and serve as the natural medium for plant growth. Differences in soil profile and type result from differences in climate, the type of the original rock source, the types of vegetation and organisms, topography, and time. Normal weathering produces a characteristic soil profile, marked by a succession of distinctive horizons in a soil from the surface downward. The A horizon is closest to the surface, and usually has a gray or black color because

Peat layer over bedrock in a cliff near Quito, Ecuador *(Dr. Morley Read/Photo Researchers, Inc.)*

of high concentrations of humus (decomposed plant and animal tissues). The A horizon typically loses some substances through downward leaching. The B horizon is commonly brown or reddish, enriched in clay produced in the same place that the rock it was weathered from was located, and transported downward from the A horizon. The C horizon of a typical soil consists of slightly weathered parent material. Young soils regularly lack a B horizon, and the B horizon grows in thickness with increasing age.

Some unique soils form under unusual climate conditions. Polar climates are typically cold and dry, and the soils produced in polar regions are typically well drained and lack an A horizon, sometimes underlying layers of frost-heaved stones. In wetter polar climates, tundra may overlie permafrost, which prevents the downward draining of water. These soils are saturated in water and rich in organic matter. These polar soils play a crucial role in maintaining the global environment. They contain an abundance of organic material, effectively isolating it from the atmosphere and locking up much of the planet's carbon. Cutting down of northern forests as is occurring in Siberia may affect the global carbon dioxide budget by releasing much of this organic material as carbon dioxide to the atmosphere, possibly contributing to climate change and global warming.

Dry climates limit the leaching of unstable minerals, such as carbonate, from the A horizon. Leaching may also be impeded by evaporation of groundwater. Extensive evaporation of groundwater over prolonged times leads to the formation of caliche crusts. These are hard, generally white carbonate minerals and salts that were dissolved in the groundwater but precipitated when the groundwater moved up through the surface and evaporated, leaving the initially dissolved minerals behind.

In warm, wet climates, most elements (except for aluminum and iron) are leached from the soil profile, forming laterite and bauxite. Laterites, which are typically deep red in color, are found in many tropical regions. Some of these soils are so hard that they are used for bricks.

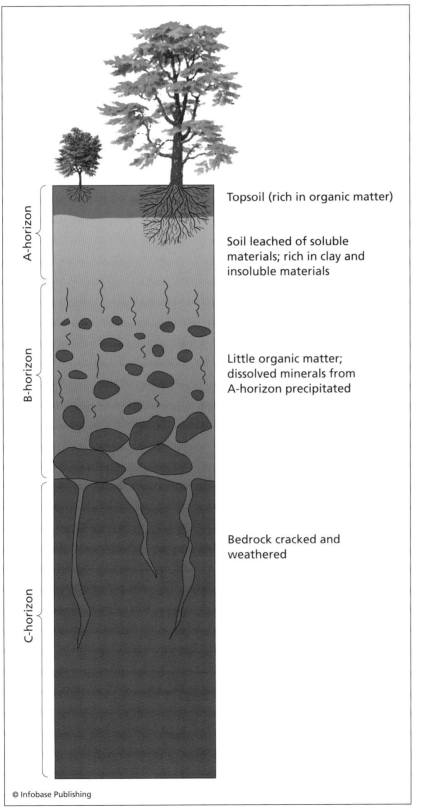

Topsoil (rich in organic matter)

Soil leached of soluble materials; rich in clay and insoluble materials

Little organic matter; dissolved minerals from A-horizon precipitated

Bedrock cracked and weathered

A-horizon

B-horizon

C-horizon

© Infobase Publishing

Typical soil profile, showing organic zone, and the A, B, and C horizons

Soils form at various rates in different climates and other conditions, ranging from about 50 years in moderate temperatures and wet climates, to about 10,000 to 100,000 years for a good soil profile to develop in dry climates, such as the desert southwest of the United States. Some mature soils, such as those in the tropics, have been forming for several million years. Deforestation causes soil erosion, which cannot be repaired easily. In many places, such as parts of Madagascar, South America, and Indonesia, deforestation has led to accelerated rates of soil erosion, removing thick soils that have been forming for millions of years. These soils supported a rich diversity of life, and it is unlikely that the soils will ever be restored in these regions.

See also MASS WASTING; WEATHERING.

FURTHER READING

Birkland, P. W. *Soils and Geomorphology.* New York: Oxford University Press, 1984.

solar system The Earth's solar system represents the remnants of a solar nebula that formed in one of the spiral arms of the Milky Way galaxy. After the condensation of the nebula, the solar system consisted of eight major planets, the moons of these planets, and many smaller bodies in interplanetary space. From the Sun outward, these planetary bodies include Mercury, Venus, Earth, Mars, the asteroids, Jupiter, Saturn, Uranus, and Neptune. The physical properties of these bodies are listed in the table "Physical Properties of Objects in the Solar System." Until 2006 Pluto was regarded as a planet, but in 2006 a team of astronomers voted that Pluto did not meet the criteria of being a planet in that its orbit was too erratic and its size too small, and demoted Pluto to the status equivalent of a captured asteroid. Most asteroids are concentrated in a broad band called the asteroid belt located between the orbits of Mars and Jupiter. Although none of the asteroids are larger than the Earth's moon, they are considered by many to be "minor planets," since they are orbiting the Sun.

The asteroids are small rocky, metallic bodies, most of which orbit in the asteroid belt located between Mars and Jupiter, although some have different erratic orbits. Others are located in different belts further from the Sun. Comets include icy bodies and rocky bodies, and are thought by many astronomers to be material left over from the formation of the solar system that was not incorporated into any planetary bodies. Thus, comets may have clues about the early composition of the solar nebula.

The planets and asteroids orbit the Sun counterclockwise when viewed from above the Earth's North Pole, and most have roughly circular orbits that are confined to a relatively flat plane called the ecliptic plane. The spacing between the different orbits increases with increasing distance from the Sun. The inner four planets (Mercury, Venus, Earth, and Mars), referred to as the terrestrial planets, have densities and properties that are roughly similar to Earth and are generally rocky in character. In contrast the outer planets (Jupiter, Saturn, Uranus, and Neptune), known as the Jovian planets, have much lower densities and are mostly gaseous or liquid in form. The Jovian planets are much more massive than the terrestrial planets, rotate more rapidly, have stronger magnetic fields than the terrestrial planets, and have systems of rings that circle the planets.

FORMATION OF SOLAR SYSTEM

The solar system began to form from a spinning solar nebula about 5 billion years ago, 9 billion years after the universe started expanding from nothing in the big bang some 14 billion years before the present. This solar nebula consisted of a mass of gas, dust, and fragments that began spinning faster as gravitational forces caused the material to collapse on itself. Temperatures ranged from extremely hot in inner parts of the solar nebula to cold in the outer reaches. Planets began accreting by accumulating more and more dust and small fragments to form the bigger

PHYSICAL PROPERTIES OF OBJECTS IN THE SOLAR SYSTEM

Object	Orbital Distance (AU)	Mass (earths)	Diameter (Earths)	Rotational Period (Days) (- Means Retrograde)	Orbital Period (Years)	Density (Earths)	Surface Gravity (Earths)	Moons
Sun	0.0	332,000	109.2	25.8	. . .	1.42	28	. . .
Mercury	0.39	0.06	0.38	59	0.24	0.98	0.38	0
Venus	0.72	0.81	0.95	-243	0.62	0.95	0.90	0
Earth	1.0	1.00	1.00	1.00	1.0	1.00	1.00	1
Mars	1.5	0.11	0.53	1.03	1.9	0.71	0.38	2
Ceres asteroid	2.8	0.0002	0.07	0.38	4.7	0.38	0.03	0
Jupiter	5.2	317.8	11.2	0.42	11.9	0.24	2.34	63
Saturn	9.5	95.2	9.5	0.44	29.5	0.12	1.16	60
Uranus	19.2	14.5	4.0	-0.69	83.7	0.23	1.15	27
Neptune	30.1	17.2	3.9	0.72	163.7	0.30	1.19	13
Pluto (dwarf planet)	39.5	0.002	0.18	-6.40	248.0	0.37	0.04	3
Eris (dwarf planet discovered June 2007)	67.7	0.002	0.18	~8	557	?	?	1

planetesimals that began rotating around the large mass accumulating as the Sun at the center of the disk. These early planetesimals grew into protoplanets, still experiencing many impacts with large asteroids and comets by 4.56 billion years ago. As the main planets formed, they differentiated into core-mantle-crust systems, and in the late bombardment period from about 4.5–3.5 billion years ago, these planets suffered many impacts with large asteroids and comets. Several large, differentiated bodies in what is now the asteroid belt were destroyed by large impacts, forming billions of fragments that now form the bulk of meteorites that hit the Earth. The inner planets are made dominantly of silicate minerals and are called the rocky or terrestrial planets, whereas the outer planets are mainly gaseous, often called the Jovian planets after the largest body, Jupiter. Comets come from further out in the solar system, most from beyond the orbit of Neptune in a region called the Oort Cloud.

See also ASTEROID; ASTRONOMY; COMET; EARTH; GALILEI, GALILEO; JUPITER; MARS; MERCURY; NEPTUNE; ORIGIN AND EVOLUTION OF THE EARTH AND SOLAR SYSTEM; PLUTO; SATURN; URANUS; VENUS.

FURTHER READING

Chaisson, Eric, and Steve McMillan. *Astronomy Today.* 6th ed. Upper Saddle River, N.J.: Addison-Wesley, 2007.

Cloud, Preston. *Oasis in Space.* New York: W.W. Norton, 1988.

Comins, Neil F. *Discovering the Universe.* 8th ed. New York: W. H. Freeman, 2008.

Condie, Kent C., and Robert E. Sloan. *Origin and Evolution of Earth, Principles of Historical Geology.* Upper Saddle River, N.J.: Prentice Hall, 1997.

Snow, Theodore P. *Essentials of the Dynamic Universe: An Introduction to Astronomy.* St. Paul, Minn.: West, 1984.

Sorby, Henry Clifton (1826–1908) British *Geologist, Biologist, Microscopist, Metallurgist*

Henry Sorby was a well-known British scientist whose most influential scientific work was done from 1849–64. His work was based on the application of the microscope to geology and metallurgy. In these two fields, his work included simple quantitative observation and the building and meticulous use of new experimental equipment and interpretation based on the application of elementary physicochemical principles to complex natural phenomena. His goal was to "apply experimental physics to the study of rocks." Sorby's most famous achievement was the development of the basic techniques of petrography by using the polarizing microscope to study the structure of thin rock sections.

Henry Sorby was born in Woodbourne near Sheffield in Yorkshire on May 10, 1826, and died March 9, 1908. He attended the Sheffield Collegiate School, where he developed a keen interest in natural sciences, especially geography, and began to study some of the excavated valleys of Yorkshire. In these studies he became interested in the older geological periods, sedimentary layers, and the formation of structures during the deformation of these rocks. He began working on sedimentary rocks, and by 1851 he was involved in a debate on the origin of slaty cleavage. His paper "On the Origin of Slaty Cleavage" in 1853 showed that cleavage was a result of the reorientation of particles of mica accompanying the deformation flow of the deposit under anisotropic pressure. Basically Sorby demonstrated that when a rock is flattened by deformation, the flat mica grains tend to rotate so that most of them are close to parallel, forming the planar structure called slaty cleavage in the rocks. This work was followed by the publication in 1858 of an important memoir in the *Quarterly Journal of the Geological Society of London*, which Sorby titled "On the Microscopical Structure of Crystals." He went on to study organisms in limestone and discussed the significance of microorganisms in chalk. Sorby then moved from slate to schist and metamorphic rocks in general. His paper on liquid inclusions in crystals, both natural and artificial, was very important since later study of these inclusions yielded information about the pressure and temperature conditions during the deformation and metamorphism of these and other rocks. His use of the microscope helped him to find abundant smaller inclusions within the microcrystals of many metamorphic rocks.

Henry Sorby was a member of the Royal Microscopic Society and was awarded the Wollaston Medal by the Geological Society of London in 1869, the highest award given by that society. In 1882 he was elected president of Firth College in Sheffield. The International Association of Sedimentologists and the Yorkshire Geological Society both have Sorby Medals, named after Henry, and award these to individuals with outstanding achievements in geology.

See also SEDIMENTARY ROCK, SEDIMENTATION; STRUCTURAL GEOLOGY.

FURTHER READING

Sorby, Henry C. "On the Application of Quantitative Methods to the Study of the Structure and History of Rocks." *Quarterly Journal of the Geological Society* 64 (1908): 171–233.

———. "On the Origin of Slaty Cleavage." *Edinburgh New Philosophical Journal* 55 (1853): 137–148.

———. "On the Theory of the Origin of Slaty Cleavage." *Philosophical Magazine* 12 (1856): 127–129.

South American geology South America has a diverse and long geological history. The Andean Mountain chain on the western side of the continent has been active for a couple of hundred million years since before the breakup of Gondwana, with most activity in the Mesozoic-Cenozoic-Tertiary recent times, and it is still one of the world's most active continental margin arcs. These ranges are drained by the Amazon and Paraná River systems, which cut through other ranges including the Pampean, Austral, and Tandilia ranges. In the north, the Caribbean oceanic plateau, formed by the Caribbean mountain system and the Andean Cordillera in the northwest, is in an oblique collision with the South American continent. The southern boundary of South America is marked by the Magellan mountain system and the Falkland Plateau to the southeast.

The core of the South American continent consists of the Precambrian Guiana and Brazilian shields and the Río de la Plata craton. The Guiana shield is bordered, in subsurface strata, on the north and northeast by gently folded Paleozoic strata, and by metamorphic belts in the south and northwest. The Brazilian shield is bordered in the north along the Amazon basin by gently folded

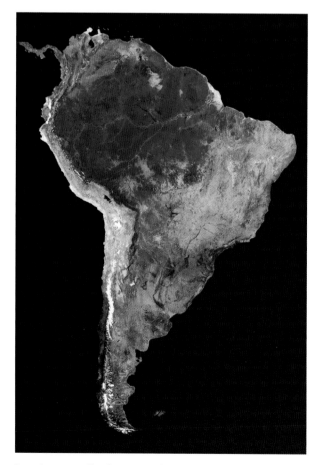

Landsat satellite image of South America
(*Bill Howe/Alamy*)

Paleozoic strata, and by Paleozoic and Mesozoic strata in the east.

ANDES

The Andean mountain chain is a 5,000-mile- (8,000-km-) long belt of deformed igneous, metamorphic, and sedimentary rocks in western South America, running generally parallel to the coast, between the Caribbean coast of Venezuela in the north and Tierra del Fuego in the south. Some sections are characterized by active volcanoes, others by their absence. Most of the Andes are located above the subducting Nazca plate except for the southern Andes, which are located above the subducting Antarctic plate. The mountains merge with ranges in Central America and the West Indies in the north, and with ranges in the Falklands and Antarctica in the south. Many snow-covered peaks rise higher than 22,000 feet (6,000 m), making the Andes the second largest mountain belt in the world, after the Himalayan chain. The highest range in the Andes is the Aconcagua on the central and northern Argentine-Chile border. The high cold Atacama Desert is located in the northern Chile sub-Andean range, and the high Altiplano Plateau is situated along the great bend in the Andes in Bolivia and Peru.

The southern part of South America consists of a series of different terranes added to the margin of the Gondwanan supercontinent in the late Proterozoic and early Paleozoic. Subduction and the accretion of oceanic terranes continued through the Paleozoic, forming an accretionary wedge 155 miles (250 km) wide. The Andes formed as a continental margin volcanic arc system on the older accreted terranes, above a complex system of subducting plates from the Pacific Ocean. They are geologically young, having been uplifted mainly in the Cretaceous and Tertiary, with active volcanism, uplift, and earthquakes. The specific styles of volcanism, plutonism, earthquakes, and uplift are found to be segmented strongly into seven main zones in the Andes and related to the nature of the subducting part of the Nazca plate, including its dip and age. Where the subducting slab dips more than 30 degrees beneath South America, the Andes have active volcanism on the surface. In contrast, regions above places where the subduction zone is sub-horizontal do not have active volcanoes, but instead have contraction and sedimentary basin formation.

Although the history of the Andes extends back into the Paleozoic with the accretion of different terranes, especially in the southern parts of the mountain belt, most of the current-day Andean ranges were formed in Mesozoic to recent times. The dominant processes in their formation are related to their location on the leading edge of the convergent plate boundary between South America and the Nazca plate.

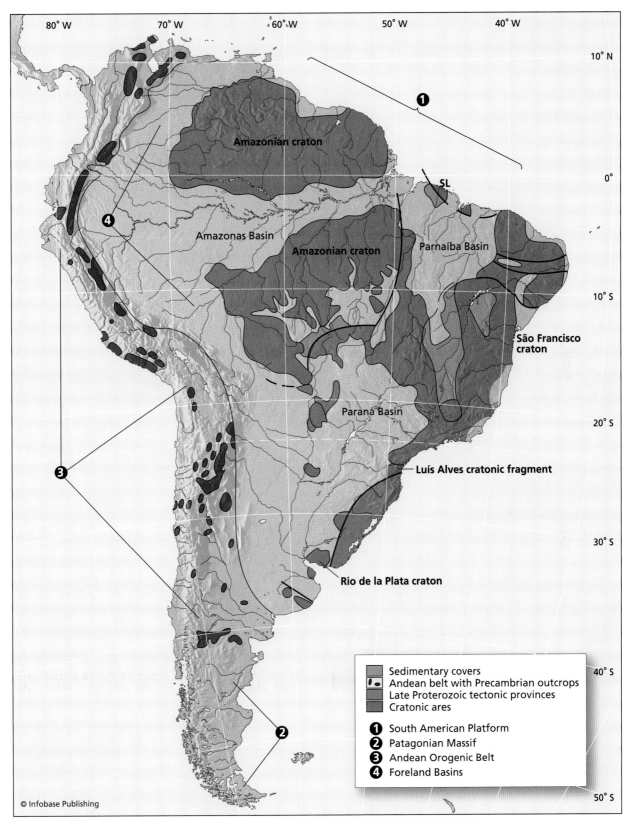

Simple geological map of South America showing the location of the Archean cratons, including the Amazonian craton (AM), São Francisco craton (SF), Rio de la Plata craton (RP), Sao Luis cratonic fragment (SL), Luis Alves cratonic fragment (LA). Late Proterozoic orogens and basins surrounding these cratons. The Andean belt forms the western side of South America and includes some Precambrian outcrops (shown in purple). Younger basins including the Amazona, Paraná, and Parnaiba contain thick sequences of Phanerozoic sediments and are drained by large rivers including the Amazon, Río Negro, and Paraná.

Altiplano

The Altiplano is a large uplifted plateau in the Bolivian and Peruvian Andes of South America. The plateau has an area of about 65,536 square miles (170,000 km^2) and an average elevation of 12,000 feet (3,660 m) above sea level. The Altiplano is a sedimentary basin caught between the mountain ranges of the Cordillera Oriental on the east and the Cordillera Occidental on the west. Lake Titicaca, the largest high-altitude lake in the world, is located at the northern end of the Altiplano, a dry region with sparse vegetation and scattered salt flats. Villagers grow potatoes and grains, and a variety of minerals are extracted from the plateau and surrounding mountain ranges.

The Atacama Desert is an elevated arid region located in northern Chile, extending over 384 square miles (1,000 km^2) south from the border with Peru. The desert is located 2,000 feet (600 m) above sea level and is characterized by numerous dry salt basins (playas), flanked on the east by the Andes and on the west by the Pacific coastal range. The Atacama is one of the driest places on Earth, with no rain ever recorded in many places, and practically no vegetation in the region. Nitrate and copper are mined extensively in the region.

The Atacama is first known to have been crossed by the Spanish conquistador Diego de Almagro in 1537, but it was ignored until the middle 19th century, when mining of nitrates in the desert began. However, after World War I, synthetic nitrates were developed and the region has been experiencing economic decline ever since, as it is too expensive to mine the natural nitrates from the desert.

AMAZON RIVER

The Amazon is the world's second-longest river, stretching 3,900 miles (6,275 km) from the foothills of the Andes to the Atlantic Ocean. The Amazon begins where the Ucayali and Marañon tributaries merge, and it drains into the Atlantic near the city of Belém. The Amazon carries the most water and has the largest discharge of any river in the world, averaging 150 feet (45 m) deep. Its drainage basin amounts to about 35 percent of South America, covering 2,500,000 square miles (6,475,000 km^2). The Amazon lowlands in Brazil include the largest tropical rainforest in the world. In this region, the Amazon is a muddy, silt-rich river with many channels that wind around numerous islands in a complex maze. The delta region of the Amazon is marked by numerous fluvial islands and distributaries, as the muddy waters of the river get dispersed by strong currents and waves into the Atlantic. A strong tidal bore, up to 12 feet (3.7 m) high, runs up to 500 miles (800 km) upstream.

Waters of the Amazon River flow through a sediment-filled rift basin, between the Precambrian crystalline basements of the Brazil and Guiana shields. The area hosts economic deposits of gold, manganese, and other metals in the highlands, and detrital gold in lower elevations. Much of the region's economy relies on the lumber industry, with timber, rubber, vegetable oils, Brazil nuts, and medicinal plants sold worldwide.

Spanish commander Vincent Pinzon in 1500 was probably the first European to explore the lower part of the river basin, followed by the Spanish explorer Francisco de Orellana in 1540–41. Orellana's tales of tall strong female warriors gave the river its name, borrowing from Greek mythology. Further exploration by Pedro Teixeira, Charles Darwin, and Louis Agassiz led to greater understanding of the river's course, peoples, and environment settlements did not appear until steamship service began in the middle 1800s.

THE MAGELLAN RANGES, CAPE HORN, AND THE FALKLAND PLATEAU

The southern tip of South America has consistently horrid weather with high winds, rain and ice storms, and large sea waves. Southernmost South America is a large island archipelago known as Tierra del Fuego, separated from the mainland by the Strait of Magellan, and the southern tip of which is known as Cape Horn. The Drake Passage separates Cape Horn from the northern tip of the Antarctic Peninsula. Tierra del Fuego and the Strait of Magellan were discovered by Magellan in 1520 and settled by Europeans, Argentineans, and Chileans after the discovery of gold in the 1880s. These peoples brought diseases that spread to and killed off all of the indigenous people of the islands.

The Falkland Plateau is a shallow-water shelf extending 1,200 miles (2,000 km) eastward from Tierra del Fuego on the southern tip of South America, past South Georgia Island. The plateau includes the Falkland Islands 300 miles (480 km) east of the coast of South America and is bounded on the south by the Scotia Ridge and on the north by the Agulhas-Falkland fracture zone. The Falkland Islands include two main islands (East and West Falkland) and about 200 small islands and are administered by the British but also claimed by Argentina, with the capital at Stanley. The islands are stark rocky outposts, plagued by severe cold rains and wind, but have abundant seals and whales in surrounding waters. The highest elevation is 2,315 feet (705 m) on Mount Adam. Thick peat deposits support a sheep-farming community among the dominantly Scottish and Welsh population.

The Falkland Plateau formed as a remnant of the southern tip of Africa that remained attached to

South America during the breakup of Gondwana and the movement of South America away from Africa. The Agulhas-Falkland fracture zone extends to the tip of Africa and represents the transform along which divergence of the two continents occurred. Numerous Mesozoic rift basins on the plateau are the site of intensive oil exploration. The geology of the Falklands was first described by Charles Darwin from his expedition with the HMS *Beagle* in 1833 and reported in 1846, and Johan G. Andersson completed later pioneering studies.

Precambrian granite, schist, and gneiss are found on the southwest part of West Falkland Island, probably correlated with the Nama of South Africa. The Precambrian basement is overlain by a generally flat to gently tilted Paleozoic sequence including 1.7 miles (3,000 m) of Devonian quartzite, sandstone, and shale. A Permo-Triassic sequence 2.2 miles (3,500 m) thick unconformably overlies the Paleozoic sequence and includes tillites and varves indicating glacial influence. These rocks are cut by Triassic-Jurassic dioritic to diabasic dikes and sills related to the Karoo and Parana flood basalts. Diamictites and long lobes of gravel interpreted as mudflows deposited in a periglacial environment overlie quaternary interglacial deposits.

The Falklands are folded into a series of northwest-southeast trending folds that intensify to the south and swing to east-west on the east of the plateau.

PRECAMBRIAN CRATONS AND SHIELDS

The core of South America is made of its Precambrian cratons, including the Guiana and Brazilian (also called the Amazonian) shields and the Rio de la Plata craton. These cratonic blocks also formed central regions of several past supercontinents including Gondwana and Rodinia.

The Guiana shield has rocks as old as 3.4 billion years, and other major groups of rocks formed at 1.5 and 0.9 billion years ago. Highland regions of the Guiana shield, called the Guiana Highlands, are characterized by a series of beautiful table-top mountains called *tequis*. These regions host some of the world's most spectacular waterfalls such as Angel Falls, Kaieteur Falls, and Cuquenan Falls. The main part of the Archean section of the Guiana shield is largely confined to the northern part of the shield in Venezuela, where it shows major tectonic and magmatic events at 2.7–2.5 billion years ago in the Aroan-Jaquie event, from 2.2–1.8 billion years ago in the Transamazonian orogeny, at 1.75–1.5 billion years ago in Uruacuan event, then the Parguazan event at 1.5–1.4 billion years ago, the Orinocan-Nickerie event from 1.3–1.2 billion years ago, and 1.0–0.65 billion years ago in the Brazilian orogeny. The older Archean rocks include granulite facies, mafic volcanic rocks, iron-rich cherts, and gneisses

cut by granitic intrusives. Most Archean areas are separated from the late Archean and Paleoproterozoic sections by major shear zones. Late Archean rocks include amphibolite-facies belts in both the Guiana and Brazilian shields, with both mafic and ultramafic magmatic rocks in greenstone belts, and with tectonic boundaries with other units.

The Uruacuan event at 1.75–1.55 Ga affected the southern border of the Guiana shield and the north central border of the Brazilian shield and is associated with eruption and deposition of pyroclastic and continental clastic rocks. This was followed by intrusion of granitic plutons from 1.5 to 1.4 billion years ago. Both shields appear to have stabilized around this time, with the intrusion of post-tectonic rapakivi granitoids from 1.75–1.55 billion years ago. Brasiliano events from 1.0–0.9, and from 0.75–0.65 billion years ago later affected the edges of the shields.

A thick platformal sedimentary cover developed over these Precambrian rocks in the Paleozoic, which includes Silurian-Devonian marine sequences, Carboniferous marine and continental deposits, and Permian to Triassic continental rocks in the Parana-Chaco-Amazon region. The Triassic saw active rifting associated with the opening of the Atlantic, with the activation of many faults on the shields. The faulting was associated with tholeiitic magma eruption in the Triassic followed by alkaline ultramafic intrusions in the Jurassic to Eocene in Brazil and Paraguay and is associated with carbonatites and kimberlites.

The Rio de la Plata craton forms a small Archean cratonic block in southeastern South America, mostly in Uruguay. This craton has a core of ancient greenstone belts and gneisses in the Quadrilatero Ferrifero area, with a long complex deformation history. Dikes that cut the craton at 1.73 billion years ago may be related to a mantle plume event.

See also ARCHEAN; CONTINENTAL CRUST; CONVERGENT PLATE MARGIN PROCESSES; CRATON; GREENSTONE BELTS; OROGENY; PALEOZOIC; PLATE TECTONICS; RIVER SYSTEM; STRUCTURAL GEOLOGY.

FURTHER READING

Bahlburg, H., and F. Herve. "Geodynamic Evolution and Tecnostratigraphic Terranes of Northwestern Argentina and Northern Chile. *Geological Society of America Bulletin* 109 (1997): 869–884.

Condie, Kent C., and Robert Sloan. *Origin and Evolution of Earth, Principles of Historical Geology.* Upper Saddle River, N.J.: Prentice Hall, 1997.

Darwin, Charles. "Geology of the Falkland Islands." *Quarterly Journal of the Geological Society* 2 (1846): 267–274.

Ericksen George, E., T. C. Pinochet, and J. A. Reinemund, eds. *Geology of the Andes and Its Relation to Hydrocarbon and Mineral Resources.* Houston, Tex.: United

States, Circum-Pacific Council for Energy and Mineral Resources, 1990.

Goodwin, Alan M. *Precambrian Geology.* London: Academic Press, 1991.

———. *Principles of Precambrian Geology.* London: Academic Press, 1996.

Kusky, Timothy M. *Precambrian Ophiolites and Related Rocks.* Amsterdam: Elsevier, 2004.

Ramos, V. A. "The Birth of Southern South America." *American Scientist* 77 (1989): 444–450.

Windley, Brian F. *The Evolving Continents.* 3rd ed. Chichester, England: John Wiley & Sons, 1995.

star formation   Stars form by the condensation or gravitational contraction of particles in interstellar gas and dust found in giant molecular clouds. As these cold clouds contract, they heat up and eventually become hot enough to stimulate the process of nuclear fusion, at which point a star is born. Although this process sounds simple, the process of star formation involves many uncertainties and unknown triggers. Why do some interstellar clouds collapse to form stars and why do others not collapse? What processes occur in interstellar molecular clouds as they collapse, and what leads to different types of star formation? Some of the answers to these questions relate to variations in the initial molecular clouds, and some relate to events that happen during cloud collapse. One of the main factors is the relative strength of two forces, those of gravity and temperature. Gravitational attraction between atoms in a molecular cloud or any other object is proportional to the masses involved, and since atoms have very low mass the gravitational attraction between

atoms can be very small. Temperature of a gas is a measure of the average speed of atoms or molecules in the gas, with higher temperatures equated with higher speeds of the particles. Many processes in the formation of stars relate to the relative strengths of the two effects, and the ability of the force of gravity to overcome the effects of heat. The balance between these forces is also what stops stars from collapsing, since the outward pressure from the heated, fast-moving gases exactly equals the inward pull of gravity in these stars.

Rotation or spin of molecular clouds also acts to oppose the inward pull of gravity. Rotation of molecular clouds causes them to develop a bulge around their center. As the cloud contracts the law of conservation of angular momentum states that it must spin faster, and material in its outer reaches may spin off into outer space. In order for material not to fly off into space an opposing force must be applied to pull the particles inward, and this force is gravity. If the mass of the spinning molecular cloud is great enough to exert a gravitational pull that is greater than the outward force from the spin of the cloud, it will condense further to form a spinning disk. The cloud will eventually flatten out into a pancake shape with a bulge in its center and the bulge will become a star. If there is less mass, then the outward force from the spin may be great enough to cause the cloud simply to disperse with the particles flying off into space.

Most interstellar clouds are also characterized by magnetic fields, which can influence whether or not the cloud condenses into a solar disk. As molecular clouds contract and heat up, the number of collisions between particles increases dramatically, causing the

Birth of stars in a molecular cloud. The shroud of dust shown here in constellation Cygnus hides an exceptionally bright source of radio emission called DR21. Visible light traces reveal no evidence of what is happening in this region because of heavy dust obscuration. NASA's *Spitzer Space Telescope* took this image and is able to peek behind the veil of dust at one of the most massive embryonic stars in the Milky Way galaxy. This star is more than 100,000 times as bright as the Sun. The image also shows a powerful outflow of hot gas emanating from the star and bursting through the molecular cloud. *(NASA)*

gas to become ionized. Magnetic fields are generated, and these fields cause particles to move along the magnetic field lines but prevent the particles from moving against the magnetic field, causing the shape of the condensing molecular cloud to become distorted. Since the formation of the charged particles in the densest part of the collapsing cloud is tied to the formation of the magnetic field, the two processes are linked and the magnetic field intensifies and is pulled in toward the center of the condensing cloud. The interactions between these forces are complex and may explain many of the variations in the shapes and other characteristics of condensing molecular clouds and young star systems.

Composition can also affect the way a star evolves and matures. Stars that evolve from interstellar molecular clouds with higher concentrations of heavy elements tend to be cooler and a little less luminous than stars with more light elements.

STAGES TO MAIN SEQUENCE EVOLUTION

Stars that have roughly the same mass as the Earth's Sun follow a similar evolutionary path, much of which lies along a general trend on a diagram of solar luminosity versus surface temperature (this diagram is called a Hertzsprung-Russell [H-R] diagram). The H-R diagram shows how a star's size changes as it evolves with respect to luminosity and temperature over time, with most stars the size of the Sun following a curved S-shaped trajectory across the diagram, called the main sequence of stellar evolution. For about 90 percent of a star's evolutionary history it will burn hydrogen quietly without changing much and will remain close to stationary on the H-R diagram. However, a star's early and late stages of evolution can be quite different from its average or steady-state main-sequence characteristics. The evolutionary stages and track of a star are shown on the H-R diagram through its history of luminosity and surface temperature. Stars with the mass of the Sun go through several stages, from interstellar cloud, to collapsing interstellar cloud fragment, to protostar, to main sequence star.

Interstellar Cloud

Stars that are close to a solar mass start their evolution as an ordinary, dense, and cold interstellar cloud that may be tens of parsecs across (1 parsec equals 3.3 light years), having temperatures of about 10 K, and densities of around 10^9 particles per cubic meter. The amount of mass in these types of interstellar clouds may be thousands or millions of times a solar mass at this stage. Typical giant molecular clouds that collapse to form solar-type stars are 6,000,000 solar masses and 100 light years across. Something must happen to this interstellar cloud to make it

collapse and break into smaller pieces, but the exact mechanisms that trigger such a collapse are poorly known. There may be external triggers, such as pressure waves from nearby events (such as supernovas), gravitational collapse of smaller gas pockets in the interstellar cloud, or interactions of the magnetic fields and ionized particles. Once collapse begins the molecular cloud tends to fragment into tens to thousands of smaller and smaller clumps of matter, controlled by the gravitational instabilities within the nebula. Each of these clumps can then eventually develop into a star, with individual interstellar clouds thus capable of producing tens to thousands of new stars. Collapse from a stable interstellar molecular cloud to a group of collapsing cloud fragments may take only a couple of million years. Examples of various stages of collapse of interstellar molecular clouds are abundant in the universe and include spectacular glowing clouds such as the emission nebula M20 that shows bright red dust clouds, dark regions, and young glowing protostars.

Collapsing Interstellar Cloud Fragment

As the huge interstellar cloud collapses into many fragments, it is useful to consider the processes inside one of the individual cloud fragments as it continues to develop into a star. Most of these fragments are about one to two solar masses but can be about 100 times the size of the Earth's present solar system. The temperature of the cloud is about the same as when it started to condense, but it would have an increased density of about 10^{12} particles per cubic meter in the center of the cloud fragment. The temperature has not changed because the density of the cloud is still too low to capture photons emitted from the gas, and the energy of the photons escapes to space instead of being absorbed by the cloud. However the very center of the cloud may experience significant warming by this stage, perhaps to 100 K, as the gas is denser there and can absorb more of the radiation produced in the gas. As the cloud continues to shrink, it becomes denser so that eventually the cloud begins to trap the radiation across large regions, and the temperature of the whole cloud increases. This causes an increase in the internal pressure (equated with temperature and speed of particle movement), which grows strong enough to overcome the force of gravity that was pulling the cloud together. At this stage, the contraction of the cloud stops and the fragmentation of the original cloud stops. The Orion nebula, in the constellation Orion, provides beautiful examples of cloud fragments that are lit by the absorption of radiation produced in the cloud.

The time from initial contraction to the end of fragmentation of the interstellar cloud may take only

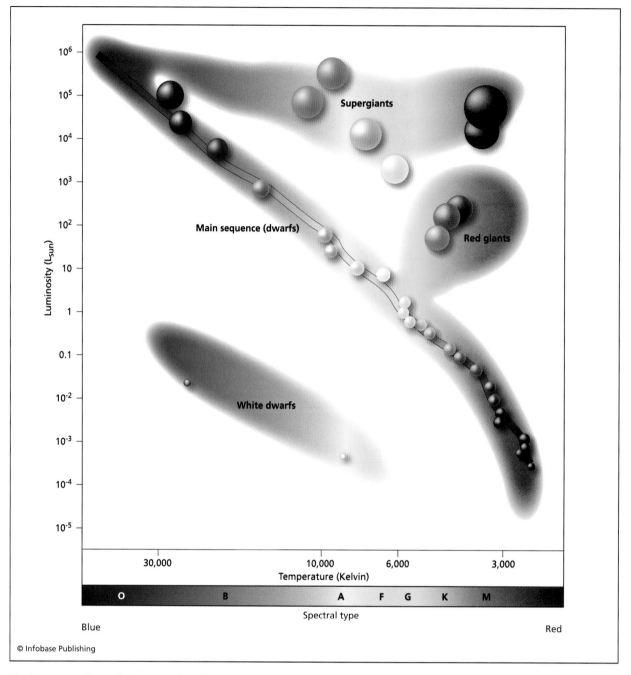

Hertzsprung-Russell diagram showing stellar evolution. The diagram plots luminosity on the vertical axis and surface temperature or spectral class on the horizontal axis.

a few tens of thousands of years, and by this stage the size of the cloud fragment is roughly the same as Earth's present solar system, but the central temperature of the cloud has now reached about 10,000 K, while the peripheral temperatures at the edge of the cloud are still close to their starting temperatures. Since the edges of the cloud are cooler than the denser interior, they are also thinner, and the cloud takes on the shape of a thick ball in the center surrounded by a flattened and outward thinning disk.

The central density in this stage may be 10^{18} particles per cubic meter.

The center of the collapsed fragment is now roughly spherical, dense, and hot and begins to resemble an embryonic protostar, as it continues to grow in mass as gravity attracts more material into its core, although the size of the internal embryonic protostar continues to shrink since the force of gravity in the core remains greater than the internal pressure generated by the gas temperature. At this stage

Infrared and visible light composite of Orion Nebula, 1,500 light-years from Earth, taken by *Spitzer* and *Hubble Space Telescopes*, November 7, 2006. The image shows a region of star birth, including four massive stars at the center of the cloud, occupying the region that resembles a yellow smudge near the center of the image. Swirls of green reveal hydrogen and sulfur gas that is heated and ionized by the intense ultraviolet radiation from the massive young stars at the center of the cloud. The wisps of red and orange are carbon-rich organic molecules in the cloud. The orange-yellow dots scattered throughout the cloud are infant stars embedded in a cocoon of dust and gas. The ridges and cavities in the cloud were formed by winds emanating from the four super massive stars at the center of the cloud. *(NASA Jet Propulsion Laboratory)*

the embryonic protostar develops a protosphere, a surface below which the material is opaque to the radiation it emits.

Protostar

The protostar in the center of the collapsed disk continues to shrink and grow in density, its internal temperature increases, and the surface temperature on its protosphere continues to rise, generating higher pressures. About 100,000 years after the initial fragment formed from the interstellar cloud, the center of the protostar reaches about 100,000,000 K, and free electrons and protons swirl around at hundreds of miles (km) per second, but temperatures remain below the critical value (10^7 K) necessary to start nuclear fusion to burn hydrogen into helium. The protosphere has a temperature of a few thousand K, and the radius of the protosphere places it at about the distance of Mercury from the Sun.

At this stage the protostar can be plotted on an H-R diagram, where it would have a radius of 100 or more times that of the present Sun, a temperature of about half of the Sun's present temperature, and a luminosity that is several thousand times that of the Sun. The luminosity is so high because of the large size of the protostar, even though the temperature is lower than the present temperature of the Sun. The energy source for the luminosity and elevated temperatures in this protostar is from the release of gravitational potential energy during collapse of the interstellar cloud.

Pressures build up inside the protostar at this stage from the elevated temperatures, but the gravitational force is still stronger than the thermal pressures, so contraction continues albeit at a slower rate. Heat diffuses from the core of the protostar to the surface, where it is radiated into space, limiting the rise in temperature but allowing contraction to continue. If this did not happen then the temperatures would rise in the star and it would not contract enough to reach densities at which nuclear fusion would begin, and it would not form into a true star, but would remain a dim protostar.

The protostar continues to move down on the evolutionary H-R diagram toward higher temperature and lower luminosity as the surface area shrinks. Internal densities and temperatures increase while the surface temperature remains about the same,

Horsehead Nebula in the Orion molecular cloud complex. This image was produced from three images obtained with multimode FORS2 instrument at the second VLT Unit Telescope. *(European Southern Observatory)*

but the surface can be intensely active and be associated with intense solar winds such as those that characterize T-Tauri stars. A possible example of a protostar in this stage of evolution in the Orion molecular cloud is known as the Kleinmann-Low Nebula, which emits strong infrared electromagnetic radiation at about 1,000 times the solar luminosity. Some protostars are surrounded by dense dark dust clouds that absorb most of the ultraviolet radiation emitted by the protostars. This dust then reemits the radiation at infrared wavelengths, where it appears as bright objects. Since the source of the radiation is cloaked in a blanket of dust, these types of structures have become known as cocoon nebula.

By the time the protostar has shrunk to about 10 times the size of the Sun, the surface temperature is about 4,000 K and its luminosity is now about 10 times that of the Sun. However, the central temperature has risen to about 5,000,000 K, which is enough to completely ionize all the gas in the core but not high enough to start nuclear fusion. The high pressures cause the gravitational contraction to slow, with the rate of continued contraction controlled by the rate at which the heat can be transported to the surface and radiated away from the protosphere. Strong presolar winds at this stage blow hydrogen and carbon monoxide molecules away from the protostar at velocities of 60 miles per second (100 km/sec). These winds encounter less resistance in a direction that is perpendicular to the plane of the disk formed from the flattened interstellar cloud fragment, since there is less dust in these directions. In this stage, therefore, some protostar nebulae exhibit a strong bipolar flow structure, in which strong winds blow jets of matter out in the two directions perpendicular to the plane of the disk. This strange-looking structure eventually decays, though, as the strong winds blow the dust cloud away in all directions.

Approximately 10 million years after becoming a protostar, when the temperature reaches about 10,000,000 K and the radius about 6,200 miles (10,000 km), nuclear fusion begins in the center of the protostar. At this stage, a true star is born and is typically a little larger than our present Sun, the surface temperature is about 4,500 K, and the luminosity is about two-thirds that of the present Sun.

Main Sequence Star

The newly formed star continues to contract for about another 30 million years after nuclear fusion has begun; the central density increases to 10^{32} particles per cubic meter, the central temperatures rise to 15,000,000 K, and the surface heats to about 6,000 K. At this stage the thermal pressures balance the gravitational contraction forces, and the rate that nuclear energy is generated in the core equals that at which it is radiated at the surface. At this stage the star has reached the main sequence, where it can burn stably for about another 10 billion years without significantly changing.

VARIATIONS IN STELLAR FORMATION WITH SOLAR MASS

The sequence of star formation described above is based on the evolutionary history of a one-solar-mass star from the interstellar cloud to main sequence stages. Stars with larger sizes form from initial interstellar molecular cloud fragments that condense into larger fragments, and smaller stars form from smaller fragments. Each of the stages for larger or smaller stars may be similar to the stages described above, but the magnitudes for size, density, temperature, and time for reaching different stages can vary significantly for stars of different mass. Larger mass embryonic stars generally have higher luminosities, densities, and surface temperatures at different stages compared to the lower-mass objects, and they move along the evolutionary paths much faster than smaller-mass objects. The most massive stars can progress from the interstellar cloud stage to being a main sequence star in only a million years, compared to 50 million years for stars with only one solar mass. At the other end of the spectrum, stars with lower masses are cooler, smaller, and can take much longer to evolve into main sequence stars, even a billion years or more in cases.

The zero-age main sequence for a star is the time at which the stellar properties become stable and the star enters a steady period of burning or fusion. This is the time the star is effectively born or joins the main sequence. Stars do not evolve along the main sequence trend on the H-R diagram; rather the main sequence is just the point at which most stars stop evolving for an extended period of time and show a stable relationship between luminosity, temperature, and star mass. Higher mass stars plot on the upper left part of the H-R diagram; lower mass stars plot in the lower right.

Some cloud fragments are too small to ever produce a star but end up producing other gaseous bodies in the universe. The giant gaseous planets of Jupiter and Saturn are examples of parts of condensed interstellar clouds that formed from parts of the cloud that were too small to produce a star. These planets collapsed from the interstellar cloud like the Sun but were too small to continue contracting to produce a star or to start nuclear fusion. Other interstellar cloud fragments that are too small to produce stars become isolated in space as fragments of unburned cool matter. These objects may be abundant in interstellar space but are difficult to detect since they are small and cold. They are called brown dwarfs and may account for much of the unknown mass in the universe.

EMISSION NEBULAE

The sections above focused on the formation of stars from the collapse of interstellar dust clouds but did not focus on the effects these processes have on the surrounding intergalactic medium.

Star formation as described above does not usually result in the formation of just one star, but rather a group or cluster of stars with similar characteristics and ages in a region that represents the original collapsed interstellar cloud. The more massive the original collapsed dust cloud, the more new stars that form. Dispersed within this region is a variable amount of unused gas and dust that did not make it into the newly formed stars or associated planetary systems. This gas is commonly ionized by high-energy photons and glows in different colors as an emission nebula.

Observations of newly formed star regions and star clusters show that low-mass stars form more commonly than high-mass stars, but at present astronomers do not understand the reasons that control the types and spacing of stars, nor why in some cases the process is very efficient and uses much of the original dust and gas from the collapsed dust cloud, whereas in other cases the process is not efficient.

If very massive stars form they have such strong winds and radiation that they typically blow away all the gas from the surrounding nebula, revealing the brightly glowing new star clusters that formed from the collapse of the nebula. Such star clusters have also been detected in collapsed nebulae without such large stars in them, but dust that absorbs the ultraviolet radiation surrounds them, so they can be detected only with infrared observations. Analysis using data from infrared observatories of the collapsed dust clouds in the Orion Nebula clearly show many areas where there are star clusters hidden optically by the emission nebula but shining brightly and identifiable in infrared images. Astronomical observations show that in these regions, for every large star that forms there may be tens to hundreds of smaller stars in the same region. Typical emission nebula may be 10 parsecs (33 light years) across and contain roughly 1,000 new stars formed during the collapse of the interplanetary dust cloud. Other star clusters seem to form in a more dispersed way, spanning greater distances but having an order of magnitude fewer (about 100) stars; these are referred to as star associations.

Over time stars from these clusters and associations interact. Gravitational forces between the stars tend to eject the lighter stars out of the clusters, and the clusters eventually degenerate into individual stars that are not associated with large clusters. Most of the young clusters observed in emission nebulae last a few hundred million years before they are dispersed, but some of the larger clusters may last a few billion years.

See also ASTRONOMY; ASTROPHYSICS; CONSTELLATION; COSMOLOGY; DWARFS (STARS); GALAXIES; INTERSTELLAR MEDIUM; STAR FORMATION; STELLAR EVOLUTION.

FURTHER READING

Chaisson, Eric, and Steve McMillan. *Astronomy Today*. 6th ed. Upper Saddle River, N.J.: Addison-Wesley, 2007.

Comins, Neil F. *Discovering the Universe*. 8th ed. New York: W. H. Freeman, 2008.

Encyclopedia of Astronomy and Astrophysics. CRC/ Taylor and Francis Press. Available online. URL: http://eaa.crcpress.com/. Accessed October 24, 2008.

ScienceDaily: Astrophysics News. ScienceDaily LLC. Available online. URL: http://www.sciencedaily.com/news/space_time/astrophysics/. Accessed October 24, 2008.

Snow, Theodore P. *Essentials of the Dynamic Universe: An Introduction to Astronomy*. 4th ed. St. Paul, Minn.: West Publishing Company, 1991.

stellar evolution During their lifetime, stars undergo a predictable sequence of changes that are a function of their mass. The lifetimes of individual stars range from a few millions of years for the most massive stars to trillions of years for less massive stars. Some stars have lifetimes longer than the age of the universe. To understand stellar evolution astronomers and astrophysicists study large numbers of stars at different stages in their evolutionary history and integrate these observations with computer models that simulate stellar evolution and structure. Stars change dramatically during their formation stages until they reach the main sequence, when they burn uniformly for about 90 percent of their history. As the nuclear hydrogen fuel runs out near the end of a star's history, it may once again undergo dramatic changes as it leaves the main sequence. The path a star takes near the end of its life depends primarily on its mass and somewhat less on its interactions with other nearby stars. The end states of stars can be extremely spectacular and strange, or they can just fade away from visible detection.

THE MAIN SEQUENCE

After stars form over a period of a few to a few tens of millions of years, they reach a steady state of hydrogen burning that can last for 10 billion years or more before changing significantly again. Stars on the main sequence plot on a fuzzy line on a diagram

Hubble image of bright blue newly formed stars that are blowing a cavity in the center of star-forming region N90 *(NASA)*

depicting the relationship between solar luminosity and surface temperature (called a Hertzsprung-Russell or H-R diagram [see page 698]), indicating a balance between a star's temperature, luminosity, and mass during the hydrogen-burning stage. Once a star uses up its hydrogen fuel, these relationships change, the star is out of equilibrium, and it leaves the main sequence, plotting in different places on the H-R diagram, depending on its mass. The time that a star leaves the main sequence represents the beginning of its end, and its lifetime after departure from the main sequence will be relatively short.

Low-mass stars burn so slowly that most still exist, as they have not yet exhausted their hydrogen fuel supply. In contrast high-mass stars burn more quickly; some of the largest stars stayed on the main sequence for only a few tens of millions of years. Most of the very massive stars that were created in the history of the universe have left the main sequence, gone through their death throes, and moved to the end states of existence. Stars of intermediate mass are experiencing or will experience intermediate fates. In general low-mass stars have a gentle end of their existence as bright objects, whereas high-mass stars generally have an explosive ending. The exact boundary between high- and low-mass is fuzzy but ranges somewhere between five and ten solar masses.

While a star is burning hydrogen on the main sequence, it maintains a balance or equilibrium between gravitational forces that tend to draw all the atoms together to the center of the star and gas

pressure from the hydrogen burning inside the star pressing outward against gravity. This equilibrium is maintained for about 90 percent of a star's lifetime, and as long as it is maintained nothing spectacular will happen to the star; it will simply continue to burn hydrogen in fusion reactions, converting it into helium. The nuclear fusion process is a multistep process in which four hydrogen nuclei (protons) combine to produce one helium 4 nucleus, emitting gamma ray radiation and two neutrinos. Over this time period the star surface may occasionally erupt, forming giant solar flares or sunspots, and is constantly emitting large amounts of photons and other particles from the nuclear reactions inside. A star typically grows in luminosity during its history on the main sequence. The Earth's Sun, for instance, is now about 30 percent more luminous than when the Earth, Sun, and solar system formed nearly 5 billion years ago.

FUEL SHORTAGE STAGE

After about 10 billion years, a star the size of the Earth's Sun will start to run out of hydrogen fuel in its central core, in a gradual process that has a catastrophic ending. The core of the star initially is composed of mostly hydrogen and about 10 percent helium, but the amount of helium gradually increases over billions of years until at about 10 billion years the core consists of all helium. Outer parts of the star will still contain large amounts of hydrogen at this stage, but the critical, high-temperature, high-pressure core has at this stage depleted its hydrogen fuel supply. The reason some hydrogen remains in the outer parts of the star is that the temperatures are lower in those regions, and the hydrogen burning proceeded at a lower rate than in the core. At this stage, the inner core stops burning hydrogen, and the main area of fusion reactions moves higher (farther from the star center) in the star interior. With time the inner core of non-burning helium grows. This leads to an imbalance between the forces of gravity and gas pressure in the inner core, because the lack of hydrogen burning in the inner core decreases the gas pressures in that region, while the gravity remains the same. Eventually, the force imbalance grows to the point that something serious must happen.

CORE CONTRACTION AND HYDROGEN SHELL BURNING

As the helium in the core builds up it cannot burn, since it requires much higher temperatures to fuse than does hydrogen. The result is that when the hydrogen is used up, the core contracts, since the gas pressure without high temperatures from nuclear fusion is not sufficient to counteract the force of gravity. The shrinkage of the core then releases gravitational potential energy that creates heat, raising the temperature of the core. As the core grows increasingly hot in this way (but not hot enough to burn helium), it raises the temperatures of the outer parts of the star where there is still abundant hydrogen, causing this hydrogen to burn much faster than before. The hydrogen burns especially fast around the non-burning helium core, forming a hot-burning shell within the interior of the star. This is called the hydrogen shell–burning stage. The burning of the hydrogen in this shell generates more energy faster than when the star entered onto the main sequence, and the star begins to significantly increase in luminosity.

As the hydrogen shell burning continues, the star begins to leave the main sequence and is no longer in equilibrium. The helium core is shrinking and becoming hotter, nearing temperatures sufficient for helium burning. The hydrogen shell burns faster, increasing the gas pressures and temperatures in the outer parts of the core and causing the outer parts of the star to expand in response, which lowers temperatures on the surface.

The overall consequences of this phase of star evolution, then, are that the core shrinks and heats up, while the outer surface expands dramatically, becomes more luminous, and cools by about 2,000 K, over a period of about 100 million years. The luminosity of the star continues to rise as the temperature falls and the radius of the star increases to about 70 times that of the initial, one-solar-radius star. At this stage the one-solar-mass main sequence star has become a red giant.

RED GIANTS

Red giants are huge stars, roughly 70 times as large as the Earth's Sun, yet with a helium core only a few times larger than the Earth, or about 1/10,000 the size of the star. With the lack of nuclear fusion in the core, the gas pressures allow gravitational forces to compress the helium to high densities, such that about 25 percent of the star's mass exists within the core, and the density rises to about 1,000,000,000 kg per cubic meter. Strong stellar winds in this stage blow away a significant part of the star's mass, typically 20–30 percent of the original mass.

HELIUM FUSION STAGE

The core of the red giant continues to contract while the outer layers of the star continue to expand for a couple hundred million years after the star leaves the main sequence. This would lead to the destruction of the star with the core collapsing and expelling the outer shells of the planet to space, but before that happens the temperatures in the core rise above the 100,000,000 K required to start burning helium.

When this temperature is reached helium fusion begins, and the dense hot core starts burning the helium, causing it to fuse to form carbon and releasing tremendous energy and heat in the process.

By the time the helium burning begins in the core of the red giant star, pressures have risen so high that a new state of matter, called electron degenerate matter, exists in the star's center. Pressures have squeezed the free electrons so close together that they cannot be physically compressed further and have degenerated to a low energy quantum level in which they cannot accommodate the addition of more heat. The result of the onset of helium fusion in this electron degenerate matter is therefore unstable. The gas pressure in the electron degenerate matter cannot rise because the electrons cannot be pushed farther apart; instead, the new energy added to the system by nuclear fusion causes it to undergo a rapid acceleration of the helium burning, causing a massive explosive reaction called the helium flash. This uncontrolled helium burning lasts only a few hours, but releases enough energy to expand the core, lowering its density out of the range of the electron degenerate matter quantum state to a condition where gas pressures in the core can build up to equalize the inward pull of gravity, and the core stabilizes.

After the helium flash, the red giant changes its evolutionary path on the H-R diagram. It stops expanding and becoming more luminous, reversing its direction to become less luminous with a higher surface temperature, and will then reside on a different branch of the H-R diagram known as the horizontal branch, where stars plot for the time period that they burn helium stably in their cores. The position the star plots in this horizontal field depends on how much of the star's mass remains after the strong solar winds from the red giant stage blow away large parts of the mass. Stars in the helium-burning stage lie on a horizontal line because they are all characterized by a similar luminosity, although stars with higher masses have lower surface temperatures in this stage.

Helium burning proceeds very rapidly in the core of the star after the helium flash, producing carbon that builds up as a carbon ash shell in the core. The high temperatures in the core help the helium to burn very quickly toward depletion, using up the helium fuel within a few tens of millions of years after the helium flash. The carbon in the core builds up and shrinks, releasing gravitational potential energy and heating up as it shrinks, and the helium fuel is progressively used up. This higher temperature causes more rapid burning of helium in a shell around the core, which is then progressively surrounded by a hydrogen-burning shell and a non-burning shell beneath the star's surface. The high temperatures in this stage cause the non-burning shell to expand, once again becoming a huge red giant, this time burning helium in its core. The helium burning proceeds much more rapidly than the hydrogen burning that occurred when the star first became a red giant, and the star moves up a different path on the H-R diagram called the asymptotic giant branch. The much higher luminosity and radius in this stage classify the star as a red supergiant. The carbon core continues to shrink, while the helium- and hydrogen-burning shells move progressively outward and reach higher temperatures and luminosities. This intense, furious burning marks the beginning of the end for the star; its death is near.

STAR DEATH: PLANETARY NEBULAE

For stars with masses approximately equal to that of the Earth's Sun, the period of expansion of the shell and core shrinkage will continue, but the gravitational contraction will not release enough energy to reach the 600 million K needed to start nuclear fusion of carbon. Densities in the core continue to increase until about 10,000,000,000 kg per cubic meter, when the carbon ash in the core attains an electron degenerate state and can be compressed no further. At this stage temperatures stop rising, since there is no internal nuclear fusion and gravitation contraction has stopped, so no more potential energy is released. Thus, when the helium fuel is all burned, the fires go out.

When all the helium is used from the core and just a dense carbon ash remains, the core of the star is essentially dead. The inner core continues to accumulate the spent carbon while outer core layers form shells that burn helium and hydrogen at an intense pace and the outermost part of the star continues to expand. The intense burning in the helium shell is very unstable in this stage, and stars typically experience a series of helium shell flashes from the fluctuating temperatures at high pressures. These helium flashes send intense radiation waves to the star's surface, causing it to violently pulsate. Expansion of the outer layers of the star causes the temperature of the outer layers to drop, which has the effect of allowing electrons that were dissociated by the earlier higher temperatures to recombine with atoms, releasing additional photons in the process. The energy from these photons causes the outer envelope of the star to move progressively further from the core of the star, causing the radius to oscillate in progressively larger variations in distance from the core. This process causes the outer layers of the star to be ejected to space at tens of miles (km) per second.

The wild oscillations of the star and ejection of material from the outer layers forms a new highly unusual looking stellar structure called a planetary nebula. This nebula has a small, dense core made of the spent carbon ash with a shell of material still burning helium into carbon. A span of relatively empty space about the size of Earth's solar system surrounds the core, succeeded outward by a glowing ring consisting of the ejected outer layers of the former giant star. The Milky Way Galaxy contains more than 1,000 known planetary nebulae, which last only for a very brief period of stellar evolution, typically a few tens of thousands of years as compared to the billions of years of the star's previous life history. The planetary nebula stage marks the death, or the final active stage of the star, so the thousand or so planetary nebula known from the Milky Way represent the stars currently (when the light was emitted) going through their death throes. When stars die, however, they do not go away, they simply disappear from sight. As the dead stars fade from view they go through two more stages, the white dwarf and the black dwarf stages.

WHITE DWARF STAGE

Over a relatively short period of time, tens of thousands of years, the expanding shell of material on the outer parts of the planetary nebula disperses into interstellar space and fades away as a visible nebula. The atoms of hydrogen, helium, and a small amount of carbon are in this way added to the interstellar medium, having spent the past few billions of years in the interior of a star.

The carbon core of the dead star, however, continues to go through a few more stages of evolution or degeneration. After ejection of the outer layers of the former red giant star, the remnant core, which is about the size of the Earth, glows white-hot, giving the name white dwarf. The initial surface temperature of the white dwarf is about 50,000 K, but the heat is only heat stored from its formerly active processes of nuclear fusion and gravitational collapse; no new heat is generated in the white dwarf. White dwarfs can be extremely dense, however; for example, Sirius B is about a million times denser than Earth. With time this core of a formerly great red giant will continue to cool, fade, and become a virtually invisible black dwarf.

BLACK DWARF STAGE

As the white dwarf continues to cool, its surface becomes less and less luminous until it becomes a cold, dark, carbon-rich, burned-out sphere floating in space. Even as it cools, the black dwarf no longer contracts. This is because its atoms are so dense that they are in the electron degenerate state and cannot be squeezed together any further. Therefore most solar-sized stars end up as dense, cold, dark, Earth-sized objects, perpetually cooling in space. Such is the fate of most solar-sized stars—to cool to near 0 K and float invisibly in space, waiting for some chance encounter with another star, black hole, or object that could possibly give its atoms a new life.

EVOLUTION OF HIGH-MASS STARS

Stars that have much higher masses than the Earth's Sun evolve faster than the typical scenario for a one-solar-mass stellar evolution outlined in this entry. The evolutionary speed is particularly dramatic for the nuclear fusion stage, since these stars burn their fuel supply at a much higher rate than solar-size stars. For instance, whereas a Sun-type star will typically spend about 10 billion years in the hydrogen-burning stage on the main sequence, a five-solar-mass star will spend only a few hundred million years in this stage.

After the hydrogen-burning stage is complete, high-mass stars also leave the main sequence and burn helium, albeit at a higher rate than their lower-mass cousins. However, since high-mass stars enter the helium-burning stage with lower internal densities than their lower-mass equivalents, electron degeneracy has not been attained in the core, and the star can burn helium smoothly and without the helium flashes that characterized the low- or normal-mass stars. Another difference is that since the core of a high-mass star can continue to contract after burning helium, it can attain temperatures high enough to start burning carbon (600 million K), and the high-mass star does not therefore enter a white dwarf stage. Instead, it continues to burn heavier and heavier elements, fusing hydrogen to helium, helium to carbon, carbon to oxygen, oxygen to neon, neon to magnesium, magnesium to silicon, and silicon to iron. As the star burns different fuels, its radius expands and contracts and its surface temperature goes up and down multiple times, instead of the simpler history of low-mass stars. The fusion reactions proceed at faster and faster rates as the star burns heavier elements, as if some catastrophic event is approaching. The fate of high-mass stars is an explosive ending, as the accelerating reactions and burning lead to one of the universe's most spectacular events, that of a supernova, or stellar explosion.

See also ASTRONOMY; ASTROPHYSICS; BINARY STAR SYSTEMS; CONSTELLATION; COSMOLOGY; DWARFS (STARS); GALAXIES; INTERSTELLAR MEDIUM; PLANETARY NEBULA; STAR FORMATION; SUPERNOVA.

FURTHER READING

Chaisson, Eric, and Steve McMillan. *Astronomy Today.* 6th ed. Upper Saddle River, N.J.: Addison-Wesley, 2007.
Comins, Neil F. *Discovering the Universe.* 8th ed. New York: W. H. Freeman, 2008.
Encyclopedia of Astronomy and Astrophysics. CRC / Taylor and Francis Group Publishers. Available online. URL: http://eaa.crcpress.com/. Accessed October 24, 2008.
ScienceDaily: Astrophysics News. ScienceDaily LLC. Available online. URL: http://www.sciencedaily.com/news/space_time/astrophysics/. Accessed October 24, 2008.
Snow, Theodore P. *Essentials of the Dynamic Universe: An Introduction to Astronomy.* 4th ed. St. Paul, Minn.: West Publishing Company, 1991.

Steno, Nicolaus (1638–1686) Danish *Anatomist, Bishop, Geologist, Paleontologist* Nicolaus Steno is probably the first scientist to clearly show that fossils are organic remains of formerly living organisms. He studied anatomy at Copenhagen and Leiden, and Florence. While dissecting a shark, he noted the similarity between the teeth of the modern shark and fossil shark teeth in local strata. After this revelation, Steno traveled around Tuscany collecting as many fossils as he could, becoming obsessed with understanding the origin of fossils. He produced a major work on the origin of fossils, the *Prodromus*, in 1669, that led him to ponder differences between his observations and the history of the world as described in his version of the Bible. Steno proposed the law of stratal superposition, clearly outlining that younger rocks are deposited over older rocks, and he recognized a sequence of changing fossil forms in the stratigraphic record. He also described metalliferous mineral deposits, recognizing crosscutting veins, and he described volcanic mountain building, erosion, and faulting. Nicolaus Steno is considered the father of geology for formulating three major principles in a single geological work: the law of superposition, the principle of original horizontality, and the principle of lateral continuity.

EARLY YEARS

Niels Stensen was born on January 1, 1638, to Sten Pedersen and Anne Nielsdatter in Copenhagen, Denmark. Sten was a skilled goldsmith, and one of his regular customers was the Danish king. Niels suffered from an unknown illness from the age of three to six. As he recovered, his father died, leaving the family without a source of income. Anne remarried quickly, but her new husband died the following year. She did marry again, but Niels's childhood was quite unstable, foreshadowing the rest of his life. In addition, the mid-1600s was a rough period for Denmark. The Thirty Years' War had ravaged Europe since its inception in 1618. Between 1654 and 1655, the plague stole the lives of one-third of the Danish population.

In the midst of all this, Niels received an education. He attended the Lutheran academy Vor Frue

Skole, which lost half its student population to the plague. As was common for educated people at the time, his name was Latinized to Nicolai Stenosis, which has since been altered to Nicolaus Steno. Ole Borch taught Steno Latin at Vor Frue, but Borch was interested in many subjects and was an admired physician as well. He is credited with pointing Steno toward science. He performed many scientific demonstrations that impressed Steno. The two men developed a friendship based on a common love of natural and experimental philosophy.

Steno entered the University of Copenhagen in 1656 to study medicine. It was an unfortunate time to attend college, as the country went to war with Sweden. Food and fuel were in short supply on campus. Many professors and fellow students had joined the war effort, leaving few students behind who essentially had to teach themselves. This was not a problem for Steno, who read voraciously. During this time, he kept a journal titled *Chaos*. Much of what is known about Steno's studies, inner struggles, personal characteristics, and reflections on literature were recorded in this journal.

THE SEASHELL QUESTION

Thomas Bartholin, an anatomy professor from the University of Copenhagen, was famous for his discovery of the vessels that carry lymph throughout the body. Lymph is a transparent, yellowish fluid that plays an important role in the immune system and in the transportation of certain materials throughout the body. Bartholin not only conveyed an appreciation for anatomy to Steno but introduced the famous seashell question to him.

In mountainous regions, objects which resembled seashells and other marine life-forms were found embedded in the rock of the mountains. Though their shape resembled that of marine life, their composition was of a different material, more similar to hardened rock than brittle shells. Did they grow naturally out of the Earth itself? Or might they be remains of past marine life? One thing on which Catholics and Protestants did agree was the creation of the Earth and all life by an omnipotent God, but land and water were separated on the third day of creation, whereas fowl and water life were created on the fifth day. So, if the fossils were the remains of past marine life, how did they get to be embedded in dry land? (At the time the word *fossil* referred to anything that came from the Earth.) One possible explanation was the great flood described in Genesis. However, given the short period the entire Earth was covered by water, there was not enough time for slow clams to travel the distance to the remote locations where the fossils were sometimes found. Besides, they were made of a different material. The arguments went back and

forth. This paradox bothered some more than others. Steno listened to Bartholin's debate with intrigue and recorded several notes about fossils in his *Chaos* journal; nonetheless, he continued with his medical studies. He was particularly interested in anatomy.

Steno originally had wanted to study mathematics, but medicine offered better prospects for a career. Anatomy seemed very clear and logical to Steno. Perhaps it appeased his mathematical yearnings. After three years at Copenhagen, Steno left for the Netherlands in possession of a letter of introduction from Bartholin. He stopped by Amsterdam and was hosted by a physician friend of Bartholin, Gerhard Bläes. In Amsterdam, Bläes gave Steno private anatomy lessons.

DISCOVERY OF A SALIVARY DUCT

During his studies Steno was examining the arteries and veins surrounding the jaws on a butchered sheep head and inserted his skinny metal probe through a small tunnel and heard a clinking noise from hitting teeth. After close examination, he realized he had discovered a previously unrecognized duct leading from the parotid gland to the oral cavity. The parotid glands supply saliva to the mouth. He pointed this out to his teacher. Bläes immediately cast off Steno's finding as a blunder. He thought Steno must have clumsily probed through the side of the cheek accidentally, but Steno had faith in his own dissecting skills. When he demonstrated to Bläes that he did not puncture the cheek, Bläes then retorted that it must be a deformity.

After staying in Amsterdam for three months, Steno made his way on to Leiden where he enrolled at the University of Leiden in 1660. He repeated his dissection to his new professors who excitedly accepted the duct as a new and real discovery. They presented his findings. Word got back to Amsterdam, and Bläes angrily responded by corresponding that it was his own discovery and that Steno had stolen credit. Bläes rushed to publish his account of the newly discovered duct. This drove Steno to work even harder to prove his skills as an anatomist. He continued his dissection studies and, in 1662, published *Anatomical Observations on Glands*, in which he described not only the parotids but all of the glands in the head. He built up a highly regarded reputation and was able to reveal the inaccuracies of Bläes's phony report by providing a description that only an extremely skilled anatomist could give. The duct leading from the parotid to the oral cavity is referred to today as Stensen's duct, *ductus stenosis*.

When his stepfather died in 1663, Steno briefly returned to Copenhagen. In 1664 he published the results of his years of research at Leiden, *On Muscles and Glands*. That same year Steno was awarded a doctorate of medicine from the University of Leiden in absentia. He was hopeful of obtaining a position at the University of Copenhagen but was rejected. Thus he set out again, landing in Paris for a year.

QUESTIONING DESCARTES

Steno believed the best way to learn was to study the objects of interest directly. For example, to understand poetry one should not simply read one scholar's interpretations of a verse, but he should read the verse himself. If one was curious about botany, she should observe plants, not only rely on pictures in books. Though Steno was well read, he did not believe books were the utmost authority. Proof was necessary for progress. The 17th-century French philosopher and mathematician René Descartes had popularized the method of systematically doubting everything at first. Direct observation or other reliable proof was necessary in order to attain absolute certainty. The young Dane subscribed to this new philosophy.

By 1665 Steno was engrossed in the anatomy of the brain. Though much had been written about the structure and function of the brain, Steno confessed at a public lecture in Paris to knowing nothing about the brain. After shocking the audience by making such a bold declaration, the prominent anatomist proceeded to explain his philosophy on learning. He said he was starting from scratch. Rather than relying on the many varied, written descriptions from centuries of so-called anatomists who had mangled brains and followed only prescribed methods of dissection, he planned to explore it carefully on his own. He would accept only that which he directly observed. His presentation, *Discourse on the Anatomy of the Brain,* is remembered for his scientific philosophy as well as the content itself.

Steno had been initially introduced to Cartesian philosophy by Ole Borch. While Steno subscribed to Descartes's method of doubting first, he was very bothered by Descartes's lack of practicing what he preached. For example, the function of the heart had mystified physicians for centuries. The ancient Greek philosopher Aristotle believed the heart was responsible for a person's emotions and intelligence. In the second century Galen said the heart was the body's source of heat and that it was the seat of the soul. It prepared "vital spirits" that were transported around the body in the blood, giving life to the organism. Descartes said the heart was a furnace. When the blood passed through it, it became heated and expanded, and as a result, the blood rushed into the arteries. Back in Leiden, Steno had been intrigued by this and decided to investigate. He purchased an ox heart from the local butcher. He cooked the organ and carefully peeled away the outer protective layer,

noticing fibers similar to those present in muscle tissue. The fibers were arranged in a manner such that contraction would force the blood out through the arteries. Having a rabbit on hand, he dissected its muscles to compare to the ox's heart. Not surprisingly, he found them to be essentially the same. He concluded that the heart was simply a muscle that worked to pump blood around the body. Sometimes things are as simple as they seem. Yet this instance troubled him, and he began losing faith in Cartesian philosophy. He brought these doubts with him from Amsterdam to Paris.

In *On Man,* Descartes claimed that the human body was simply a machine and that the pineal gland coordinated movements of the body in accordance with the feelings of the soul. He thought the pineal gland caused the actual moving by pulling strings like a puppeteer. It seemed that Descartes based an awful lot of assumptions either on some shoddy dissections or on the inaccurate records of others. In his Parisian brain dissections, Steno found the pineal gland to be completely stationary. There was no way it could perform the functions Descartes claimed. Cartesian anatomy seemed to be based on faulty deductive reasoning and conjecture rather than experimentation, something that bothered Steno extremely.

How could one accept Descartes's rational proof of the existence of God when Descartes could not even verify basic anatomical facts? From *Chaos,* it is known that Steno was always deeply religious, but this sequence of events caused him much spiritual anxiety. Nevertheless, he stood by the mechanical approach to science, even if Descartes had not. He became frustrated with Paris, where people did not want to hear him question the venerated Descartes. So he picked up and moved again.

STUDIES ON MUSCULAR CONTRACTION

In 1666 he arrived in Florence, Italy, by way of the Alps, where he was reminded of the seashell question. He was awed by the massive mountains and delighted to observe the rock strata firsthand. He was also pleased to find a group of similar-minded philosophers who thrived on experimental science. One of these men was Francesco Redi, the grand duke's physician. Redi had disproved spontaneous generation. People had thought that flies came to life from dung or rotting meat, but Redi showed that if the meat was covered with netting, then no flies appeared. Preexisting flies needed access to the organic matter to lay the eggs that developed into maggots. Redi was also a member of the Accademia del Cimento, a group devoted to experimental science. This group was sponsored by the grand duke of Tuscany, Ferdinando II de'Medici, and his brother Prince Leopoldo. The intelligent Medici brothers were not only formal philanthropists, but they actively participated in the experiments and the discussions of the Cimento. They generously provided materials for experiments and welcomed Steno into their association. The grand duke gave him an appointment as a physician at the Hospital of Santa Maria Nuova, leaving him plenty of time to pursue independent research.

Steno had been working on a new line of research involving muscle contraction. Anatomists believed that muscles moved because something pushed on them, yet muscles seemed to contract on their own. How did this happen? It certainly was not the pineal gland. One hypothesis was that fluids rushed in, causing the muscle to swell. With support from other members of the Cimento, Steno pursued this problem. Geometrically, he showed that when a muscle contracted, it neither grew nor shrank. The overall volume was maintained though the shape of the muscle fibers changed by contracting. These results were published in *Elements of Muscular Knowledge* in 1667.

GLOSSOPETRAE

While waiting for these results to be published in the fall of 1666, an enormous great white shark that weighed about 2,800 pounds (1,270 kg) was captured and killed on a beach off Livorno. Ferdinando asked Steno to dissect its head, which was brought to Florence. Before a large audience, Steno carefully dissected away the skin and soft tissues and examined the nerves and the tiny brain. The excitement at the scene must have been incredible. The beast's teeth were almost three inches (7.6 cm) long, and each jaw had 13 rows, but the shape of the teeth was what drew Steno's attention. They resembled a type of fossil about which he had first learned from Bartholin and had since viewed for himself. Glossopetrae, sometimes called tongue stones, were a type of hard, blackish, serrated, triangular stone. They were believed to have magical powers and were used to treat everything from speech impediments to poisoning. No one knew exactly where they came from. Some thought they were hardened woodpecker tongues. Others thought they fell from the heavens. They seemed abundant after heavy rain, so maybe they were jagged edges from lightning bolts. One explanation was biblical. The apostle Paul had been bitten by a poisonous snake on the island of Malta but was unaffected by the bite. The Maltese people thought Paul cursed the viper, making its venom harmless, and that nature honored this miracle by growing the glossopetrae in the shape of vipers' teeth. Steno thought they looked suspiciously similar to shark teeth.

He questioned all the stories he had heard and became obsessed with determining the nature of these glossopetrae. Steno was not the first to compare the tongue stones to shark teeth, but if they were shark teeth, how did they get onto dry land? And why was their composition different from live shark teeth? These questions reminded him of the seashell paradox. Actually, the marine bodies found in the Earth were often very near the glossopetrae. But the popular belief was that they came from the Earth itself. The Earth contained lapidifying juices (mineral-laden solutions) and unexplained "plastic forces" that gave form where none existed previously. People claimed that they could almost see rocks and other inanimate objects grow and multiply on pathways and in fields. The fact that some of the fossils resembled current marine creatures was just a trick of nature.

Steno began his examination. He compared glossopetrae side by side with shark teeth and found they were the same. Most of the arguments supporting the spontaneous growth theory were easily dismissed through logical reasoning. He described anatomical evidence supporting his belief. He explained that the chemical composition could change during fossilization though the shape would be preserved. He very cautiously worded his report to the grand duke, timidly stating that one would not be so far from the truth in saying the glossopetrae might be fossilized shark teeth.

Mentally he was engaged deep in the Earth following his preliminary studies and wanted to continue studying natural history. Just as the structure of body parts divulged their function, he believed the structure of the Earth also had something to say. Now he wondered not only how the seashell got on the mountaintop but also how did the mountain itself get there? Rather than rely on past speculation or biblical stories, he figured that studying the Earth was a good place to start finding certain truths about its own history. He was still deeply religious but began applying his critical thinking to his own religious beliefs. Whereas the Lutherans believed the Bible should be accepted literally, the Catholics were more lenient. After spending time observing the nature surrounding him, Steno had a hard time accepting that creation took place over only several days. What he observed from the Earth, its strata, the mountains, and the fossils told a different story.

The fact that the shapes of some fossils were true to the original forms of some current marine creatures told him that the shells must have been laid in a soft muddy layer that molded around the shell. Other subsequent events led to the hardening of the sediment while preserving the original shape. A solid became enclosed in another solid. He carried this further. Noting the ordered horizontal layering of the Earth's strata, he said that the layers must have resulted from sediment settling at the bottom of a liquid. The liquid evenly spread out across the surface, and the minerals and particles contained within the liquid fell heaviest first over the surface of the preceding layer. Each layer also spread out continuously over the Earth's surface, except where other large solid structures impeded the flow of the liquid. Furthermore, the layer underneath must have solidified before the upper layer hardened. This process occurred repeatedly, with each layer containing within it pieces of history from that geological age. Steno believed this took much longer than the 6,000 years that the Earth was believed to be in existence based on a biblical chronology. Scientists have since determined the Earth to be about 4.6 billion years old. Steno published some of his findings along with his shark head dissection report in 1667. The report was added as an addendum to his muscle paper, which was still at the printer from the previous fall.

A FORERUNNER

Impressed, the grand duke granted him a salary, and Steno became a full member of the Cimento. He was able to explore fully his new geological interests with all expenses paid. He traveled around Tuscany collecting fossils, climbing mountains, and examining strata. He also continued dabbling in anatomy and arrived at a startling conclusion during this time. It was obvious that females of many species of animals laid eggs, but Steno proved that females that gave birth to live organisms also produced eggs in their ovaries. This was important because people generally believed that the female only acted as an incubator for the seed that the man placed inside of her. However, most of Steno's time was spent deciphering the history of the Earth as told by its anatomy, that is, its distinct and unique geological formations.

Steno was thoroughly enjoying his scientific freedom and making remarkable progress. Unfortunately, in the fall of 1667 three events had a profound effect on the direction his life would soon take. First, Leopoldo was elected a cardinal and would no longer be able to manage the Cimento. Second, Steno received a letter from the king of Denmark requesting he return home and offering him a decent salary. Lastly, Steno converted to Catholicism. This would not please the Danish king, as Catholicism was banned in Denmark. Thus Steno wrote a letter to the king in reply explaining the new circumstance.

Previously, he would have been thrilled at being given such a position, but Denmark could not provide the rich geological material for his research that Italy could. He felt his new studies were coming to an end. He hastily continued observing shells and strata and crystals. By 1668 he started writing up his interpretations, *De solido intra solidum naturaliter contento dissertationis prodromus* (Forerun-

ner to a dissertation on solids naturally enclosed in solids). This work is commonly referred to simply as *Prodromus*, Steno's magnum opus that earned him the title "founder of geology." The full dissertation never materialized, but *Prodromus* was packed full of clearly explained rational ideas, setting the framework for the new science of geology.

The principles outlined in *Prodromus* truly set the stage for all future geological studies. In attempting to address the central question of how a solid becomes enclosed in another solid, Steno revealed three chief geological principles. The law of superposition stated that the layers of strata in the Earth's crust represent a relative geological chronology. Each layer of sedimentary rock was older than the one above it and younger than the one below it. When a layer formed, all the matter on top of it was still fluid, and the layer below it was already hardened. The principle of original horizontality maintained that the layers of sediment were leveled after being deposited by water and, to a lesser effect, wind. If strata lie positioned at an angle, it was the result of a calamitous event that disrupted the crustal arrangement after the deposition. For example, a volcanic eruption or strong water current could cause a disturbance in an ordered stratigraphic sequence. The principle of lateral extension stated that strata were encompassed on all sides, or completely extended around the spherical Earth. Though these principles might seem quite simple today, Steno was the first to clearly state them. Steno also noticed that in crystal formation, the angles between the faces of the crystals were constant regardless of the shape or size of the crystal. This is referred to as the law of angle constancy.

In those days, books could not be published until they were approved by censors. The first *Prodromus* censor, Vincenzo Viviani, gave it a favorable recommendation. The second one, Redi, was more conservative and withheld it for several months. In the meantime, Steno received a reply from the king of Denmark granting him a position as royal anatomist in Copenhagen. A professorship was out of the question since he was now a Catholic. Steno left the arrangements for publication of *Prodromus* to Viviani and returned to Copenhagen for two years. They were not happy years. His research and his spirit suffered, prompting him to request a leave from the king so he could return to Florence, which he did in 1674.

FINAL YEARS

The new grand duke hired Steno to tutor his 11-year-old son. Then in 1675 Steno was ordained and took a vow of poverty. The pope appointed him a bishop in 1677. The last years of his life were fairly dismal.

In 1684 he wrote the pope, pleading for a release from his obligations. He wanted to return to Florence, where his days had been happiest. He was officially granted his request, but right before leaving he was asked to make a detour to help strengthen a new church in Schwerin. Unfortunately, the priest whom Steno was to be helping became ill, and Steno's brief delay in northern Germany was extended two years. Years of overwork, sleep deprivation, fasting, and not taking care of his physical needs took their toll, and Steno himself fell gravely ill. He died at age 48, on November 25, 1686, in Schwerin. His only belongings were a few worn-out garments.

Steno was not buried until 11 months later. His body was shipped from Hamburg to Florence in a crate that supposedly contained books. Seamen would have been hesitant to transport the crate if they had known its true contents. Steno's remains were buried in the crypt of the Medici in the San Lorenzo church. Steno's scientific career was brief, but his *Prodromus* is as relevant today as it was when it was published more than 300 years ago, and his ideas are presented in all current basic geology textbooks. In honor of the 300-year anniversary of the publication of *Prodromus*, in 1969 the Danish Geological Society started awarding a Steno Medal for outstanding achievements in the field of geology.

Steno's greatness was in clearly and formally stating that which seems simple and obvious, though it stirred up controversy at the time. People did not want to believe the heart, the seat of the soul, was simply a muscle. And what was fantastic about the organic material of sharks' teeth being replaced with mineral matter? Most upsetting, however, was Steno's insistence that scientists look to the Earth itself rather than supernatural revelation to learn about its creation. It took a frail Danish priest to open up the field of geology by reading a story that had been waiting to be told for billions of years.

See also EVOLUTION; FOSSIL; GONDWANA, GONDWANALAND.

FURTHER READING

Cloud, Preston. *Adventures in Earth History; being a volume of significant writings from original sources, on cosmology, geology, climatology, oceanography, organic evolution, and related topics of interest to students of earth history, from the time of Nicolaus Steno to the present.* San Francisco, Calif.: W. H. Freeman, 1970.

Cutler, Alan. *The Seashell on the Mountaintop: A Story of Science, Sainthood and the Humble Genius Who Discovered a New History of the Earth.* New York: Dutton, 2003.

Steno, Nicolaus. *De Solido Intra Solidum Naturaliter Contento Dissertationis Prodromus.* Florence, 1669.

———. *Discours sur l'anatomie du cerveau.* Paris, 1669.

———. *Steno: Geological Papers.* Edited by Gustav Scherz. Translated by Alex J. Pollock. Odense, 1969.

Stille, Wilhelm Hans (1876–1966) German *Tectonic Geologist* Wilhelm Hans Stille began his studies in stratigraphy and tectonics near his home in Hannover, Germany, and continued to do his fieldwork in this area for many years. His exploration work in Colombia introduced him to the continent of South America, where he continued to do most of his research later on. Although he did not do much research abroad, he helped other students who worked on the Mediterranean region. His constant reading helped him to become a leader in developing the field of global tectonics by synthesizing the geologic history of many regions. He is best known for proposing a system of about 50 orogenic phases in the Phanerozoic eon, many of which still bear the names he proposed. Stille used a concept known as geosynclinal theory that was later replaced by the plate tectonic paradigm. The basic idea of the discredited geosynclinal theory is that sediments accumulate in basins that cause the crust to sag downward, and when they become several miles deep the heat from deep in the Earth causes the sediments to partially melt, generating belts of magmas that rise to the surface forming magmatic arcs. Although the geosyncline concept is no longer used in geology or tectonics, the work of Stille on correlating tectonic events across Europe and the world is noteworthy. The idea of seafloor spreading was not proposed until a year after Stille died, so he never had the opportunity to place his models and observations into a plate tectonic framework. Stille was instrumental in establishing the global correlation of many geologic events, such as major deformation and magmatic episodes in mountain belts, and this was later important in establishing supercontinent cycles. In this concept, many orogenic events are globally correlated because they occur when most of the planet's landmasses aggregate into one large continent, forming widespread global orogenic events. When the supercontinent breaks up, there is widespread rifting and magmatism around the margins of the fragments breaking off, and plate tectonic drift will separate these fragments so that they are later widely dispersed. Stille's work enabled later geologists to propose many of the events in different parts of the supercontinent cycle.

Stille started out as a chemistry student, but his interest shifted to geology under the influence of German geologist Adolf von Koenen of Göttingen. Stille worked for the Prussian geological survey and then taught in Hannover. In 1912 he became a professor of geology and the director of the Royal Saxon Geological Survey at Leipzig. He was later named professor at the University of Berlin in 1932 and developed a reputation of being an outstanding teacher and philosopher of global tectonics.

Stille was known as a leader in German geology, an outstanding investigator and collator of the history of global tectonic events, and a great teacher. He directed attention to the explanation of the relationships among large crustal features, and his studies of the eugeosynclinal belts led to the interest in their magmatic history. Stille received many honorary doctorates and was elected an honorary member in numerous academics of science, geological societies, and other scientific organizations. He became the honorary president of the German Geological Society, which awarded him the Leopold von Buch Medal and later had the Hans Stille Medal instituted in his honor.

See also BASIN, SEDIMENTARY BASIN; CONTINENTAL DRIFT; DEFORMATION OF ROCKS; EUROPEAN GEOLOGY; PASSIVE MARGIN; PHANEROZOIC; PLATE TECTONICS; SUPERCONTINENT CYCLES.

stratigraphy, stratification, cyclothem
Stratigraphy is the study of rock strata or layers, especially with concern for their succession, age relationships, lithologic composition, geometry, distribution, correlation, fossil content, and other aspects of the strata. The main aim of stratigraphy is to understand and interpret the rock record in terms of paleoenvironments, mode of origin of the rocks, and the causes of similarities and differences between different stratigraphic units. Because sedimentary rocks are laid down one on top of another, examination of successively lower layers in a thick pile of sedimentary rocks, such as those in the Grand Canyon, represents a backward progression in time. The time difference between rocks at the top of the Grand Canyon and those at the base is nearly 2 billion years. Thus, by looking at the different layers, we can reconstruct the past conditions on the planet at this particular place.

These relationships are expressed in several laws of stratigraphy. The first, known as the law of original horizontality, states that water-laid sediments form horizontal strata, parallel to the Earth's surface. So sedimentary rocks that are inclined have been deformed. The second law, the principle of stratigraphic superposition, states that the order in which the strata were deposited is from bottom to top, assuming that the strata have not since been overturned.

The principle of strata superposition permits definition of the relative ages of two different sedimentary units. Simply put, the older rocks are below the younger ones—this is useful for correlating geologic strata from well-exposed to poorly exposed areas, for once the relative age of a unit is known, then it is possible to know which rocks are above and below it. Where rocks are folded or tectonically deformed, some may be upside down, and knowledge of the sequence of units is important for determining whether the rocks in a particular area are right-side up or upside down. One way to tell if rocks are upside down or not is to use the geometry of sedimentary features that formed when the rock was a sediment. For instance, if the original rock showed graded bedding, from coarse-grained at the bottom to fine-grained at the top, and now the rock has fine material at the base and course at the top, it may be upside down. Sand ripples or cross laminations on the bottoms of the beds instead of the tops would also suggest that the strata are upside down.

Although one can determine the relative ages of strata by their position with respect to one another, the absolute ages cannot be determined in this manner, nor can the intervals of time between the different units. One reason for this is that deposition is not continuous; the stratigraphic record might contain breaks or discontinuities, represented by unconformities.

STRATIGRAPHIC CLASSIFICATION
Because rocks laid down in succession each record environmental conditions on the Earth when they were deposited, experienced geologists can read the record in the stratigraphic pile like a book recording the history of time. Places like the Grand Canyon are especially spectacular because they record billions of years of history.

Classical stratigraphy is based on the correlation of distinct rock stratigraphic units, or unconformity surfaces, that are internally homogeneous and occur over large geographic areas. The basic unit of rock stratigraphy is the formation, defined as a group of strata which constitutes a distinctive recognizable unit for geologic mapping purposes. Thus, a formation must be thick enough to show up on a map, must be laterally extensive, and must be distinguishable from surrounding strata. Formations are named according to a code (the Code of North American Stratigraphic Nomenclature), using a prominent local geographic feature. Formations are divided into members and beds, according to local differences or regionally distinctive horizons. Formations are sometimes combined together with other formations into groups.

A more recent advance in stratigraphy is time stratigraphy, that is, the delineation of stratigraphic units by time. Units divided in this manner have lower and upper boundaries that are the same age everywhere, but may look very different and comprise different rock types. Fossils known to occur only during a certain period or correlating unconformities (erosional surfaces) that have approximately the same age in different places help identify time-stratigraphic units. The primary unit of time stratigraphy is the system, which is an interval so great that it can be recognized over the entire planet. Most systems represent time periods of at least tens of millions of years. Larger groups of systems are called erathems or eras for short. Time units smaller than the system are series and stage, terms typically used for describing correlations on a single continent or within a geographic province.

TIMELINES AND DIACHRONOUS BOUNDARIES

In many sedimentary systems, such as the continental shelf, slope, and rise, different types of sediments are deposited in different places at the same time. We can draw time lines through these sequences to represent all the sediments deposited at a given time or to represent the old sediment/water interference at a given time. In these types of systems, the transition from one rock type to another, such as from the sandy delta front to the marsh facies, will be diachronous in time (it will have different ages in different places).

CORRELATION OF ROCKS

If a geologist has studied a stratigraphic unit or system in one location and figured out conditions on the Earth at that point when the rock was deposited, this information can be related to the rest of the planet or simply to nearby areas. In order to accomplish this task, the geologist first needs to determine the relative ages of strata in a column, then estimate the absolute ages relative to a fixed time scale. One can determine correlations between stratigraphic units locally using various physical criteria, such as continuous exposure where a formation is recognizable over large areas. Typically, a group of characteristics for each formation distinguishes it from other formations. These include gross lithology or rock type, mineral content, grain size, grain shape, color, or distinctive sedimentary structures such as cross-laminations. Occasionally, key beds with characteristics so distinctive that they are easily recognized are used for correlating rock sections.

Most sedimentary rocks lie buried beneath the surface layer on the Earth, and geologists and oil companies interested in correlating different rock units have to rely on data taken from tiny drill holes. The oil companies in particular have developed many clever ways of correlating rocks with distinctive (oil rich) properties. One common method is to use well logs, where the electrical and physical properties of the rocks on the side of the drill hole are measured, and distinctive patterns between different wells are correlated. This helps the oil companies in relocating specific horizons that may be petroleum-rich.

Index fossils are those that have a wide geographic distribution and occur commonly but have a restricted time interval in which they formed. Because the best index fossils should be found in many environments, most are floating organisms that can travel quickly around the planet. If the index fossil is found at a certain stratigraphic level, often its age is well known, and it can be correlated with other rocks of the same age.

See also BASIN, SEDIMENTARY BASIN; MILANKOVITCH CYCLES; SEDIMENTARY ROCK, SEDIMENTATION; SEQUENCE STRATIGRAPHY.

FURTHER READING

Allen, P. A., and J. R. Allen. *Basin Analysis, Principles and Applications.* Oxford: Blackwell Scientific Publications, 1990.

Bouma, Arnold H. *Sedimentology of Some Flysch Deposits: A Graphic Approach to Facies Interpretation.* Amsterdam: Elsevier, 1962.

Goldhammer, Robert K., Paul A. Dunn, and Lawrence A. Hardie. "High-Frequency Glacial-Eustatic Sea Level Oscillations with Milankovitch Characteristics Recorded in Middle Triassic Platform Carbonates in Northern Italy." *American Journal of Science* 287 (1987): 853–892.

Grotzinger, John P. "Upward Shallowing Platform Cycles: A Response to 2.2 Billion Years of Low-Amplitude, High-Frequency (Milankovitch Band) Sea Level Oscillations." *Paleoceanography* 1 (1986): 403–416.

Hayes, James D., John Imbrie, and Nicholas J. Shakelton. "Variations in the Earth's Orbit: Pacemaker of the Ice Ages." *Science* 194 (1976): 2,212–2,232.

Imbrie, John. "Astronomical Theory of the Pleistocene Ice Ages: A Brief Historical Review." *Icarus* 50 (1982): 408–422.

Prothero, Donald, and Robert Dott. *Evolution of the Earth.* 6th ed. New York: McGraw-Hill, 2002.

Stanley, Steven M. *Earth and Life Through Time.* New York: Freeman, 1986.

structural geology Structural geology is the study of the deformation of the Earth's crust or lithosphere. The surface of the Earth is actively deforming, as demonstrated by evidence such as earthquakes and active volcanism and from rocks at the surface of the Earth that have been uplifted from great depths. The rates of processes (or time scales) of structural geology are very slow compared to ordinary events. For instance, the San Andreas fault moves only about an inch (a couple of centimeters) a year and is considered relatively fast for a geological process. Even this process is discontinuous near the surface, with major earthquakes happening every 50–150 years. At great depths the movement between the plates may be accommodated by more continuous flowing types of deformation, instead of the stick-slip type of behavior that occurs near the surface. Mountain ranges such as the Alps, Himalayas, or those in the American West are uplifted at rates of a fraction of an inch (a few millimeters) a year, with heights of a mile or two (several kilometers) being reached in a few million years. These types of processes have been happening for billions of years, and structural geology attempts to understand the current activity and this past history of the Earth's crust.

Structural geology and tectonics are both concerned with reconstructing the motions of the outer layers of the Earth. The terms have similar roots—structure comes from the Latin *struere*, meaning to build, whereas tectonics comes from the Greek *tektos*, meaning builder.

Rigid body rotations are one type of motion of the surface of the Earth in which a unit of rock is transported from one place to another without a change in size and shape. These types of motions fall under the scope of tectonics. In contrast, deformations are motions involving a change in the shape and size of a unit of rock, something that falls under the realm of structural geology.

When motions occur at faults or when mountains are uplifted, rocks break at shallow levels of the crust and flow like soft plastic at deeper levels of the crust. These processes occur at all scales, ranging from the scale of plates, continents, and regional maps to what is observable only using electron microscopes.

Structural geology and tectonics have changed dramatically since the 1960s. Before 1960, structural geology was a purely descriptive science, and since then has become an increasingly quantitative discipline, especially applying principles of continuum mechanics, with increasing use of laboratory experiments and the microscope to understand the mechanisms of deformation.

Tectonics has also undergone a recent revolution (since the understanding of plate tectonics in the 1960s) that provided a framework for understanding the large-scale deformation of the crust and upper mantle. Both structural geology and tectonics have made extensive use of new tools since the 1960s, including geophysical data (e.g., seismic lines), paleomagnetism, electron microscopes, petrology, and geochemistry.

Most studies in structural geology rely on field observations of deformed rocks at the Earth's surface and proceed either downscale to microscopic observations or upscale to regional observations. None of these observations alone provides a complete view of structural and tectonic processes, so structural geologists must integrate observations at all scales and use the results of laboratory experiments and mathematical calculations to interpret observations better.

To work out the structural or tectonic history of an area, the geologist will usually proceed in a logical order. First, the geologist systematically observes and records structures (folds, fractures, contacts) in the rock, usually in the field. This consists of determining the geometry of the structures, including their geographical location, orientation, and characteristics. Additionally, the structural geologist is concerned with determining the number of times a rock has been deformed and which structures belong to which deformation episode.

The term *attitude* means the orientation of a plane or line in space. Attitude is measured using two angles—one measured from geographic north and the other from a horizontal plane. The attitude of a plane is represented by a strike and a dip, whereas the attitude of a line is represented by a trend and plunge. Strike is the horizontal angle, measured relative to geographic north of the horizontal line in a given planar structure. The horizontal line is referred to as the strike line, and is the intersection of a horizontal plane with the planar structure. One can easily measure strike in the field with a compass, by holding the compass against the plane and keeping the compass horizontal. Dip is the slope of the plane defined by the dip angle and the dip direction, which must be specified. It is the acute angle between a horizontal plane and the planar structure, measured in a vertical plane perpendicular to the strike line.

To understand the processes that occurred in the Earth, structural geologists must also examine the kinematics of formation of the structures; that is, the motions that occurred in producing them. This will lead to a better understanding of the mechanics of formation, including the forces that were applied, how they were applied, and how the rocks reacted to the forces to form the structures.

To improve understanding of these aspects of structural geology, geologists make conceptual models of how the structures form and test predictions of these models against observations. Kinematic models describe a specific history of motion that could have carried the system from one configuration to another (typically from an undeformed to a deformed state). Such models are not concerned with why or how motion occurred or the physical properties of the system (plate tectonics is a kinematic model).

Mechanical models are based on continuum mechanics (conservation of mass, momentum, angular momentum, and energy) and an understanding of how rocks respond to applied forces (based on laboratory experiments). With mechanical models geologists can calculate the theoretical deformation of a body subjected to a given set of physical conditions of forces, temperatures, and pressures (an example of this is the driving forces of tectonics based on convection in the mantle). Mechanical models represent a deeper level of analysis than kinematic models, constrained by geometry, physical conditions of deformation, and the mechanical properties of rocks.

One must remember, however, that models are only models, and they only approximate the true Earth. Models are built through observations and allow one to make predictions that can be tested to draw conclusions concerning the model's relevance to the real Earth. New observations can support or refute a model. If new observations contradict predictions, models must be modified or abandoned.

STRUCTURAL GEOLOGY AND THE INTERIOR OF THE EARTH

Structures at the surface of the Earth reflect processes occurring at deeper levels. We know that the Earth is divided into three concentric shells—the core, mantle, and crust. The core is a very dense iron-nickel alloy, comprising the solid inner core and the liquid outer core. The mantle is composed of lower-density, solid magnesium-iron silicates, and is actively convecting, transporting heat from the interior of the Earth to the surface. This heat transfer is the main driving mechanism of plate tectonics. The crust is the thin, low-density rock material making up the outer shell of the Earth.

Temperature increases with depth in the Earth at a gradient of about 54°F per half a mile (30°C/km) in the crust and upper mantle, and with a much smaller gradient deeper within Earth. The heat of the Earth comes from several different sources, including residual heat trapped from initial accretion, radioactive decay, latent heat of crystallization of the outer core, and dissipation of tidal energy of the Sun-Earth-Moon system.

Heat flows out of the interior of the Earth toward the surface through convection cells in the outer core and mantle. The top of the mantle and the crust form a relatively cold and rigid boundary layer called the lithosphere, which is about 61 miles (100 km) thick. Heat escapes through the lithosphere largely by conduction, transport of heat in igneous melts, and in convection cells of water through mid-ocean ridges.

Structural geologists predominantly study only the outer 12–18 miles (20–30 km) of the lithosphere. This puts into perspective how much structural geology infers a great deal about the interior of the Earth by examining only its skin.

CHARACTERISTICS OF THE CRUST

Earth's crust is divisible broadly into continental crust of granitic composition and oceanic crust of basaltic composition. Continents comprise 29.22 percent of the surface, whereas 34.7 percent of Earth's surface is underlain by continental crust (continental crust under submerged continental shelves accounts for the difference). The continents are in turn divided into orogens, made of linear belts of concentrated deformation, and cratons, the stable, typically older interiors of the continents.

Hypsometric diagrams reflect the strongly bimodal distribution of surface elevation. Continental freeboard, the difference in elevation between the continents and ocean floor, results from differences in thickness and density between continental and oceanic crust, tectonic activity, erosion, sea level, and strength of continental rocks.

CONTROLS OF DEFORMATION

Deformation of the lithosphere is controlled by the strength of rocks, which in turn depends mostly on temperature and pressure. Strength increases with pressure and decreases exponentially with increasing temperature. Because temperature and pressure both increase downward, a cross section through the crust or lithosphere will have different zones where the effects of either pressure or temperature dominate.

In the upper layers of the crust, effects of pressure dominate, and rocks that are subject to high stress will fail with brittle processes such as fracturing. Rocks grow stronger with depth with increasing pressure until about nine miles (15 km) depth. At this point, the effects of temperature become increasingly more important; the rocks get weaker as they get hotter, and the rocks deform by different mechanisms including flowing ductile deformation.

Other important properties that determine how the lithosphere deforms are composition (e.g., quartz versus olivine in crust, mantle, continents, and oceans) and strain rate, a measure of how fast a rock is deformed. The greatest variations in strain rate occur along plate boundaries, and most structures develop as a consequence of plate interactions along plate boundaries.

STRUCTURAL GEOLOGY AND PLATE TECTONICS

The surface of the Earth is divided into 12 major and several minor plates that are in motion with respect to each other. Plate tectonics describes these relative motions, which are, to a first approximation, rigid body rotations. However, deformation of the plates does occur (primarily in belts tens to hundreds of kilometers in width along the plate boundaries), and in a few places, extends into the plate interiors. Structural geology deals with these deformations,

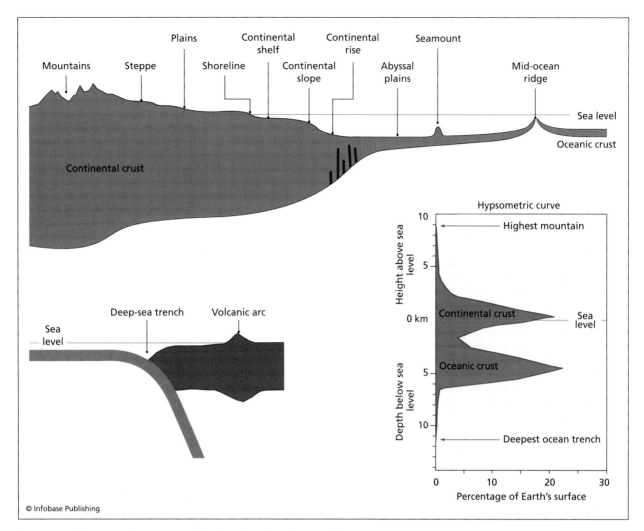

Hypsometric curve showing the distribution of land with different elevations on the planet. Note the bimodal distribution, reflecting two fundamentally different types of crust (oceanic and continental) that have different isostatic compensation levels. Cross sections show a typical continental margin–ocean transition and ocean trench–island arc boundary.

which in turn give clues to the types of plate boundary motions that have occurred and to the tectonic causes of the deformation.

Plate boundaries may be divergent, convergent, or conservative/transform. The most direct evidence for plate tectonics comes from oceanic crust, which has magnetic anomalies or stripes recording plate motions. However, the seafloor magnetic record goes only as far back as 180 million years, the age of the oldest in-place oceanic crust. Any evidence for plate tectonics in the preceding 96 percent of Earth history must come from studying the continents (structural geology).

Highly deformed continental rocks are concentrated in long linear belts called orogens, comparable to those associated with modern plate boundaries. This observation suggests that these belts represent former plate boundaries. The structural geologist examines these orogens, determines the geometry, kinematics, and mechanics of these zones, and makes models for the types of plate boundaries that created them. The types of structures that develop during deformation depend on the orientation and intensity of applied forces, the physical conditions (temperature and pressure) of deformation, and the mechanical properties of rocks.

The most important forces acting on the lithosphere that drive plate tectonics and cause the deformation of rocks are the gravitational "ridge push" down the flanks of oceanic ridges, gravitational "trench pull" of subducting lithosphere caused by its greater density than surrounding asthenosphere, and the resistance of trenches and mountain belts.

At low temperature and pressure and high intensity of applied forces, rocks undergo brittle deformation, forming fractures and faults. At high temperature

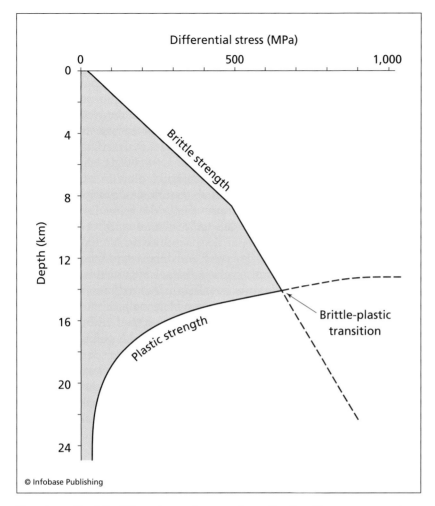

The strength of the lithosphere changes dramatically with pressure and temperature, based on the mechanical properties of the minerals in the rocks at different depths. There are several weak and strong zones at different depths that vary in depth in different parts of the world based on the different rock types in different places.

and pressure and low intensity of applied forces, rocks undergo ductile deformation by flow, coherent changes in shape, folding, stretching, thinning, and many other mechanisms.

Different styles of deformation characterize different types of plate boundaries. For instance, at mid-ocean ridges new material is added to the crust, and relative divergent motion of the plates creates systems of extensional normal faults and ductile thinning at depth. At convergent plate boundaries, one plate is typically subducted beneath another, and material is scraped off the down-going plate in a system of thrust faults and folds. Along transform plate boundaries, systems of strike-slip faults merge downward with zones of ductile deformation with horizontal relative displacements. All types of plate boundaries have small-scale structures in common, so it is necessary to carefully examine regional patterns before making inferences about the nature of ancient plate boundaries.

STRESS

Stress is the total force exerted by all the atoms on one side of an arbitrary plane upon the atoms immediately on the other side of the plane. Stress is what causes rocks to deform, so understanding the concepts of stress is essential to structural geology. Body forces are those that act from a distance (e.g., gravity) and are proportional to the amount of material present. Surface forces are those that act across surfaces of contact between parts of bodies, including all possible internal surfaces. There are two kinds of surface forces, including normal (compressive and tensile) that act perpendicular to the surface and shear (clockwise and anticlockwise) that act parallel to the surface. The state of stress equals the force divided by the area across which it acts.

For any applied force, it is possible to find a choice of coordinate axes such that all shear stresses are equal to zero, and only three perpendicular principal stresses have nonzero values. The principal stresses, commonly abbreviated $\sigma1$, $\sigma2$, and $\sigma3$, are parallel to the semimajor axes of an ellipsoid called the stress ellipsoid, parallel to the coordinate axes chosen such that they are the only nonzero stresses.

The deviatoric stress, or the difference between the principal stresses, is most important for forming structures in rocks, because it drives the deformation. However, the mean stress, given by

$$\frac{(\sigma1 + \sigma2 + \sigma3)}{3},$$

is important for determining which deformation mechanisms operate and the strength of materials.

Stress has dimensions of force per unit area. In the SI system, stress is reported in Pascals (Pa), which is equal to one Newton per meter squared (N/m^2).

$$1\,Pa = \frac{N}{M^2},$$

The sign convention that geologists use considers compressive stresses to have a positive sign.

STRAIN

Strain is a measure of the relationship between size and shape of a body before and after deformation. Strain is one component of a *deformation,* a term that includes a description of the collective displacement of all points in a body. Deformation consists of three components: rigid body rotation, a rigid body translation, and a distortion known as strain. Strain is typically the only visible component of deformation, manifest as distorted objects, layers, or geometric constructs.

There are many measures of strain: changes in lengths of lines, changes in angles between lines, changes in shapes of objects, and changes in volume or area. The change in the length of lines can be quantified using several different strain measures.

$$\text{Extension } (\varepsilon) = \frac{(L' - L)}{L}$$

where L is the original length of line, and L' is the final length of line;

$$\text{Stretch } (S) = \frac{L'}{L} = (1 - \varepsilon).$$

The quadratic elongation

$$\lambda = \left(\frac{L'}{L}\right)^2 = (1 - \varepsilon)^2,$$

whereas the natural or logarithmic strain is expressed as: (ε)

$$\varepsilon = \log e\left(\frac{L'}{L}\right) = \log e(1 - \varepsilon).$$

The change of angles is typically measured using the angular shear (ψ angular shear), which is the change in the angle between two lines that were initially perpendicular. More commonly, structural geologists measure angular strain using the tangent of the angular shear, known as the shear strain:

$$\gamma = \tan \psi$$

Volumetric strain is a measure of the change in volume of an object, layer, or region. Dilation (δ) measures the change in volume:

$$(\Delta) = \frac{(V' - V)}{V},$$

whereas the volume ratio

$$\frac{V'}{V}$$

measures the ratio of the volume after and before deformation.

Strains may be homogeneous or heterogeneous. Heterogeneous strains are extremely difficult to analyze, so the structural geologist interested in determining strain typically focuses on homogeneous domains with the heterogeneous strain field. In contrast, the geologist interested in tectonic problems involving large-scale translation and rotation often finds it necessary to focus on zones of discontinuity in the homogeneous strain field, as these are often sites of faults and high strain zones along which mountain belts and orogens have been transported. For homogeneous strains, the following five general principles hold true: straight lines remain straight and flat planes remain flat; parallel lines remain parallel and are extended or contracted by the same amount; perpendicular lines do not remain perpendicular unless they are oriented parallel to the principal strain axes; circular markers are deformed into ellipses; finally, there is one special initial ellipsoid that becomes a sphere when deformed. When these conditions are met, the strain field is homogeneous, and strain analysis of deformed objects indicates the strain of the whole body.

Structural geologists often find it important to measure the strain in deformed rocks in order to reconstruct the history of mountain belts, to determine the amount of displacement across a fault or shear zone, or to accurately delineate the distribution of an ore body—this process is called strain analysis. To measure strain in deformed rocks, the geologist searches for features that had initial shapes that are known and can be quantified, such as spheres (circles), linear objects, or objects like fossils that had initial angles between lines that are known. In most cases, geologists cannot directly see the three-dimensional shape of deformed objects in rocks. Strain analysis proceeds by measuring the two-dimensional shapes of the objects on several different planes at angles to each other. The deformed shapes are graphically or algebraically fitted together to get the three-dimensional shape of the deformed object and ultimately the three-dimensional shape and orienta-

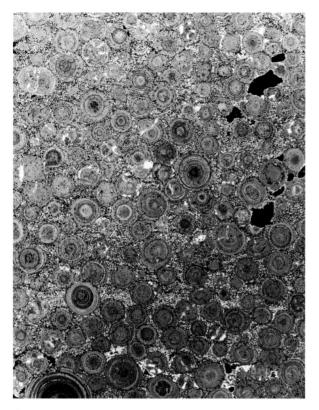

Ooids in limestone showing a small amount of deformation. Sample is 1.3 inches (3.3 cm) tall. *(Dirk Wiersma/Photo Researchers, Inc.)*

tion of the strain ellipsoid. The strain ellipsoid has major, intermediate, and minor axes of X, Y, and Z, parallel to the principal axes of strain.

Structural geologists interested in determining the strain of a body search for appropriate objects to measure the strain. Initially spherical objects prove to be among the most suitable for estimating strain. Any homogeneous deformation transforms an initial sphere into an ellipsoid whose principal axes are parallel to the principal strains, and whose lengths are proportional to the principal stretches S1, S2, and S3. Using elliptical markers that were originally circular, one can immediately tell the orientation of the principal strains on that surface and their relative magnitudes. However, the true values of the strains are not immediately apparent, because the original volumes are not typically known. Strain markers in rocks that serve as particularly good recorders of strain and approximate initially circular or spherical shapes include conglomerate clasts, ooids, reduction spots in slates, certain fossils, and accretionary lapilli.

Angular strain is often measured using the change in angles of bilaterally symmetric fossils, or igneous dikes cutting across shear zones. Many fossils, such as trilobites and clams, are bilaterally symmetric, so if line of symmetry can be found, similar points on opposite sides of the plane of symmetry can be joined, and the change in the angles from the initially right angles in the undeformed fossils can be constructed. When the fossils are deformed, the right angles are also deformed, and the same relationships derived above for change in angles can be used to determine the angular strain of the sample.

Strain represents the change from the initial to the final configuration of a body, but it tells us very little about the path the body took to get to the final shape, known as the deformation path. The strain represents the combination of all events that occurred, but they are by no means unique. Fortunately, rocks have a memory, and there are many small-scale structures and textures in the rocks that tell us much about where the rock has been, or what its deformation path was. One of the most important attributes of the strain path to determine is whether the principal strain axes were parallel between each successive strain increment or not. A coaxial deformation is one in which the principal axes of strain are parallel with each successive increment. A noncoaxial deformation is one in which the principal axes of strain rotate with respect to the material during deformation.

Two special geometric cases of strain history are pure shear and simple shear. A pure shear is a coaxial strain (with no change in volume). Simple shear is analogous to sliding a deck of cards over itself and is a two-dimensional noncoaxial rotational strain, with constant volume and no flattening perpendicular to the plane of slip.

In simple shear, the principal axes rotate in a regular manner. The principal strain axes start out at 45° to the shear plane, and strain S1 rotates into parallelism with it at very high (infinite) strains. The

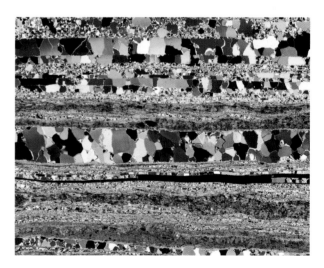

Microscopic view in polarized light of mylonite. Mica rich layers are dark, while quartz layers are colorful, with polygonal shapes of individual grains. *(Dirk Wiersma/Photo Researchers, Inc.)*

principal axes remain perpendicular, but some other lines will be lengthening with each increment, and others will be shortening. There are some orientations that experience shortening first, and then lengthening. This leads to some complicated structures in rocks deformed by simple shear; for instance, folds produced by the shortening, and then extensional structures, such as faults or pull-apart structures known as boudens, superimposed on the early contractional structures.

Natural strains in rocks deform initially spherical objects into ellipsoids with elongate (prolate) or flattened (oblate) ellipsoids. All natural strains may be represented graphically on a graph known as a Flynn Diagram, which plots a = (X/Y) versus b = (Y/Z). The number

$$k = \frac{(a-1)}{(b-1)} \ .$$

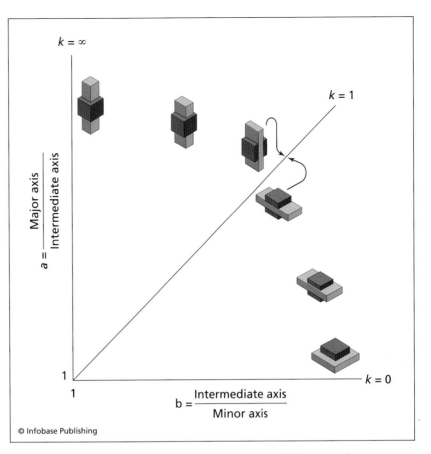

Flynn diagram showing fields of prolate, oblate, and plane strain

For k= 0, strain ellipsoids are uniaxial oblate ellipsoids or pancakes. For 0 > k > 1, deformation is a flattening deformation, forming an oblate ellipsoid. For k = 1 the deformation is plane strain if the volume has remained constant. All simple shear deformations lie on this line. For 1 > k > infinity, the strain ellipsoids are uniaxial prolate ellipsoids, or cigar shapes.

See also CONVERGENT PLATE MARGIN PROCESSES; DEFORMATION OF ROCKS; DIVERGENT PLATE MARGIN PROCESSES; FRACTURE; MÉLANGE; METAMORPHISM AND METAMORPHIC ROCKS; PLATE TECTONICS; TRANSFORM PLATE MARGIN PROCESSES.

FURTHER READING

Hatcher, Robert D. *Structural Geology, Principles, Concepts, and Problems.* 2nd ed. Englewood Cliffs, N.J.: Prentice Hall, 1995.

Kious, Jacquelyne, and Robert I. Tilling. U.S. Geological Survey. This Dynamic Earth: The Story of Plate Tectonics. Available online. URL: http://pubs.usgs.gov/gip/dynamic/dynamic.html. Last modified March 27, 2007.

Twiss, Robert J., and Eldredge M. Moores. *Structural Geology.* New York: Macmillan Press, 1992.

van der Pluijm, Ben A., and Stephen Marshak. *Earth Structure, An Introduction to Structural Geology and Tectonics.* Boston: WCB-McGraw-Hill, 1997.

subduction, subduction zone Subduction zones are long, narrow belts where an oceanic lithospheric plate descends beneath another lithospheric plate and enters the mantle in the processes of subduction. Two basic types of subduction zones exist, the first of which being where oceanic lithosphere of one plate descends beneath another oceanic plate, such as in the Philippines and Marianas of the southwest Pacific. The second type of subduction zone forms where an oceanic plate descends beneath a continental upper plate, such as in the Andes of South America. Deep-sea trenches typically mark the place on the surface where the subducting plate bends to enter the mantle, and oceanic or continental margin arc systems form above subduction zones a couple of hundred miles (a few hundred kilometers) from the trench. As the oceanic plate enters the trench it must bend, forming a flexural bulge up to a few thousand feet (a couple of hundred meters) high, typically about 100 miles (161 km) before the oceanic plate enters the trench. A series of down-to-the-trench normal faults marks the outer trench slope on the downgoing plate, in most cases. Trenches may be partly or nearly entirely filled with sediments, many of which become offscraped and attached to the accretionary prism on the overriding plate. The inner trench slope on the overriding plate typically is marked by these folded and complexly faulted and offscraped sediments, and distinctive disrupted complexes known as mélanges may form in this environment.

In ocean-ocean subduction systems the arc develops about 100–150 miles (150–200 km) from the trench. Immature or young oceanic island arcs are dominated by basaltic volcanism and may be mostly underwater, whereas mature systems have more intermediate volcanics and have more of the volcanic edifice protruding above sea level. A forearc basin, filled by sediments derived from the arc and uplifted parts of the accretionary prism, typically occupies the area between the arc and the accretionary prism. Many island arcs have back-arc basins developed on the opposite side of the arc, separating the arc from an older rifted arc or a continent.

Ocean-continent subduction systems are broadly similar to ocean-ocean systems, but the magmas must rise through continental crust so are chemically contaminated by this crust, becoming more silicic and enriched in certain sialic elements. Basalts, andesites, dacites, and even rhyolites are common in continental margin arc systems. Ocean-continent subduction systems tend to also have concentrated deformation, including deep thrust faults, fold/thrust belts on the back-arc side of the arc, and significant crustal thickening. Other continental margin arcs experience extension and may see rifting events that open back-arc basins that extend into marginal seas, or close. Extensive magmatic underplating also aids crustal thickening in continental margin subduction systems.

Oceanic plates may be thought of as conductively cooling upper boundary layers of the Earth's convection cells, and in this context subduction zones are the descending limbs of the mantle convection cells. Once subduction is initiated, the sinking of the dense downgoing slabs provides most of the driving forces needed to move the lithospheric plates and force seafloor spreading at divergent boundaries where the mantle cells are upwelling.

The amount of material cycled from the lithosphere back into the mantle of the Earth in subduction zones is enormous, making subduction zones the planet's largest chemical recycling systems. Many of the sedimentary layers and some of the upper oceanic crust are scraped off the downgoing slabs and added to accretionary prisms on the front of the overlying arc systems. Hydrated minerals and sediments release much of their trapped seawater in the upper couple hundred miles (few hundred kilometers) of the descent into the deep Earth, adding water to the overlying mantle wedge and triggering melting that supplies the overlying arcs with magma. The material

that is not released or offscraped and underplated in the upper couple hundred miles (few hundred kilometers) of subduction forms a dense slab that may go through several phase transitions and either flatten out at the 416-mile (670-km) mantle discontinuity or descend all the way to the core-mantle boundary. The slab material then rests and is heated at the core-mantle boundary for about a billion years, after which it may form a mantle plume that rises through the mantle to the surface. In this way, an overall material balance is maintained in subduction zone-mantle convection-plume systems.

Most continental crust has been created in subduction zone-arc systems of various ages stretching back to the Early Archean.

See also ACCRETIONARY WEDGE; ANDES MOUNTAINS; CONVERGENT PLATE MARGIN PROCESSES; MANTLE PLUMES; MÉLANGE; OCEAN BASIN; OPHIOLITES; PLATE TECTONICS.

FURTHER READING

Kious, Jacquelyne, and Robert I. Tilling. U.S. Geological Survey. This Dynamic Earth: The Story of Plate Tectonics. Available online. URL: http://pubs.usgs.gov/gip/dynamic/dynamic.html. Last modified March 27, 2007.

Kusky, Timothy M., and Ali Polat. "Growth of Granite-Greenstone Terranes at Convergent Margins and Stabilization of Archean Cratons." In *Tectonics of Continental Interiors,* edited by Stephen Marshak and Ben van der Pluijm, *Tectonophysics* 305 (1999): 43–73.

Moores, Eldridge M., and Robert Twiss. *Tectonics.* New York: W. H. Freeman, 1995.

Skinner, Brian, and B. J. Porter. *The Dynamic Earth: An Introduction to Physical Geology,* 5th ed. New York: John Wiley & Sons, 2004.

Stern, Robert J. "Subduction Zones." *Reviews of Geophysics* 40 (2002): 3.1–3.38.

subsidence Natural geologic subsidence is the sinking of land relative to sea level or some other uniform surface. Subsidence may be a gradual, barely perceptible process, or it may occur as a catastrophic collapse of the surface. Subsidence occurs naturally along some coastlines and in areas where groundwater has dissolved cave systems in rocks such as limestone. It may occur on a regional scale, affecting an entire coastline or it may be local in scale, such as when a sinkhole suddenly opens and collapses in the middle of a neighborhood. Other subsidence events reflect the interaction of humans with the environment and include ground surface subsidence as a result of mining excavations, groundwater and petroleum extraction, and several other processes.

Compaction is a related phenomenon, where the pore spaces of a material are gradually reduced, condensing the material and causing the surface to subside. Subsidence and compaction do not typically result in death or even injury, but they do cost Americans alone tens of millions of dollars per year. The main hazard of subsidence and compaction is damage to property.

Coastal subsidence, which is equated with a local sea level rise, can also result in more sinister long-term effects. Many coastal cities are experiencing slow subsidence so that surfaces once above sea level sink to many feet below sea level over hundreds of years. This phenomenon results in putting cities including Venice, New Orleans, and many others below sea level. In the case of New Orleans, the subsidence has caused the surrounding wetlands to have sunk below sea level, placing the city—now partly below sea level—much closer to the coast than when it was built. Subsidence has therefore contributed greatly to the increased damage to the city from recent hurricanes, including Katrina and Rita in 2005, and continues to place the city at ever-increasing risk.

Subsidence and compaction of the land (and relative rise of sea level) directly affect millions of people. Residents of New Orleans live below sea level and are constantly struggling with the consequences of living on a slowly subsiding delta. Coastal residents in the Netherlands have constructed massive dike systems to try to keep the North Sea out of their slowly subsiding land. The city of Venice, Italy, has dealt with subsidence in a uniquely charming way, drawing tourists from around the world. Millions of people live below the high-tide level in Tokyo. The coastline of Texas along the Gulf of Mexico is slowly subsiding, placing residents of Baytown and other Houston suburbs close to sea level and in danger of hurricane-induced storm surges and other more frequent flooding events. In Florida, sinkholes have episodically opened up swallowing homes and businesses, particularly during times of drought.

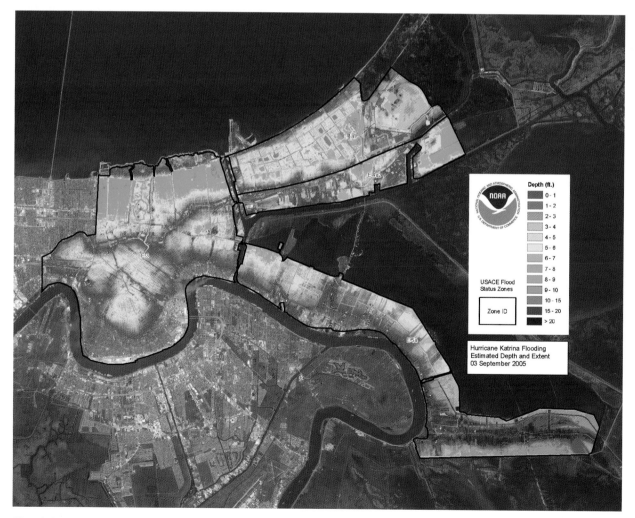

Flood depth estimation map of New Orleans, September 3, 2005 *(NOAA)*

The driving force of subsidence is gravity, with the style and amount of subsidence controlled by the physical properties of the soil, regolith, and bedrock underlying the area that is subsiding. Subsidence does not require a transporting medium, but it is aided by other processes such as groundwater dissolution that can remove mineral material and carry it away in solution, creating underground caverns that are prone to collapse.

Natural subsidence has many causes, all of which may operate in the coastal environment. Dissolution of limestone by underground streams and water systems is one of the most common, creating large open spaces that collapse under the influence of gravity. Groundwater dissolution results in the formation of sinkholes, large, generally circular depressions caused by collapse of the surface into underground open spaces.

Earthquakes may raise or lower the land suddenly, as in the case of the 1964 Alaskan earthquake where tens of thousands of square miles suddenly sank or rose 35 feet (11.5 m), causing massive disruption to coastal communities and ecosystems. Earthquake-induced ground shaking can also cause liquefaction and compaction of unconsolidated surface sediments, also leading to subsidence. Regional lowering of the land surface by liquefaction and compaction was widespread in the magnitude 6.9 Kobe, Japan, earthquake of 1995.

Volcanic activity can cause subsidence, as when underground magma chambers empty out during an eruption. In this case, subsidence is often the lesser of many hazards that local residents need to fear. Subsidence may also occur on lava flows, when lava empties out of tubes or underground chambers. The eruption of Krakatua in Indonesia in 1883 was associated with rapid collapse of the coastal caldera, and the sea rushed into the exposed magma chamber, generating a huge tsunami that killed 36,000 people in nearby coastal villages.

Some natural subsidence on the regional scale is associated with continental scale tectonic processes. The weight of sediments deposited along continental shelves can cause the entire continental margin to sink, causing coastal subsidence and a landward migration of the shoreline. Tectonic processes associated with extension, continental rifting, strike-slip faulting, and even collision can cause local or regional subsidence, sometimes at rates of several inches (7–10 cm) per year.

TYPES OF SURFACE SUBSIDENCE AND COLLAPSE

Some subsidence occurs because of processes that happen at depths of thousands of feet beneath the surface, and is referred to as deep subsidence. Other subsidence is caused by shallow near-surface processes and is known as shallow subsidence. Tectonic subsidence is a result of the movement of the plates on a lithospheric scale, whereas human-induced subsidence refers to cases where the activities of people, such as extraction of fluids from depth, have resulted in lowering of the land surface.

Compaction-related subsidence may be defined as the slow sinking of the ground surface because of reduced pore space, lowered pore pressure, and other processes that cause the regolith to become more condensed and occupy a smaller volume. Most subsidence and compaction mechanisms are slow and result in gradual sinking of the land surface, whereas sometimes the process may occur catastrophically, and is known as collapse.

HUMAN-INDUCED SUBSIDENCE

Several types of human activity can result in the formation of sinkholes or cause other surface subsidence phenomena. Withdrawals of fluids from underground aquifers, depletion of the source of replenishment to these aquifers, and collapse of underground mines can all cause surface subsidence. In addition, vibrations from drilling, construction, or blasting can trigger collapse events, and the extra load of buildings over unknown deep collapse structures can cause them to propagate to the surface, forming a sinkhole. These processes reflect geologic hazards caused by humans' interaction with the natural geologic environment.

Groundwater Extraction

The extraction of groundwater, oil, gas, or other fluids from underground reservoirs can cause significant subsidence of the land's surface. In some cases the removal of underground water is natural. During times of severe drought, soil moisture may decrease dramatically and drought-resistant plants with deep root systems can draw water from great depths, reaching a hundred feet or more (many tens of meters) in some cases. In most cases, however, subsidence caused by deep fluid extraction is caused by human activity.

This deep subsidence mechanism operates because the fluids that are extracted served to help support the weight of the overlying regolith. The weight of the overlying material places the fluids under significant pressure, known as hydrostatic pressure, that keeps the pressure between individual grains in the regolith at a minimum. This in turns helps prevent the grains from becoming closely packed or compacted. If the fluids are removed, the pressure between individual grains increases and the grains become more closely packed and compacted, occupying less space than before the fluid was extracted. This can cause the surface to subside. A small amount of this subsidence may be temporary, or recoverable, but generally once surface subsidence related to fluid extraction occurs, it is non-recoverable. When this process occurs on a regional scale, the effect can be subsidence of a relatively large area. Subsidence associated with underground fluid extraction is usually gradual but still costs millions of dollars in damage every year in the United States.

The amount of surface subsidence is related to the amount of fluid withdrawn from the ground and also to the compressibility of the layer that the fluid has been removed from. If water is removed from cracks in a solid igneous, metamorphic, or sedimentary rock, then the strength of the rock around the cracks will be great enough to support the overlying material and no surface subsidence is likely to occur. In contrast, if fluids are removed from a compressible layer such as sand, shale, or clay, then significant surface subsidence may result from fluid extraction. Clay and shale have a greater porosity and compressibility than sand, so extraction of water from clay-rich sediments results in greater subsidence than the same amount of fluid withdrawn from a sandy layer.

One of the most common causes of fluid extraction–related subsidence is the over-pumping of groundwater from aquifers. If many wells are pumping water from the same aquifer the cones of depression surrounding each well begin to merge, lowering the regional groundwater level. Lowering of the groundwater table can lead to gradual, irreversible subsidence.

Surface subsidence associated with groundwater extraction is a serious problem in many parts of the southwestern United States and in coastal cities such as New Orleans. Many cities such as Tucson, Phoenix, Los Angeles, Salt Lake City, Las Vegas, and San Diego, rely heavily on groundwater pumped from compressible layers in underground aquifers.

The San Joaquin Valley of California offers a dramatic example of the effects of groundwater extraction. Extraction of groundwater for irrigation over a period of 50 years has resulted in nearly 30 feet (9 m) of surface subsidence. Parts of the Tucson Basin in Arizona are presently subsiding at an accelerating rate, and many investigators fear that the increasing rate of subsidence reflects a transition from temporary recoverable subsidence to a permanent compaction of the water-bearing layers at depth.

The world's most-famous subsiding city is Venice, Italy. Venice is sinking at a rate of about one foot per century. The city has subsided more than 10 feet since it was founded near sea level. Much of the city is below sea level or just above sea level and prone to floods from storm surges and astronomical high tides in the Adriatic Sea. These *acqua altas* (meaning high water in Italian) flood streets as far as the famous Piazza San Marco. Venice has been subsiding for a combination of reasons, including compaction of the coastal mud that the city was built on. One of the main causes of the sinking of Venice has been groundwater extraction. Nearly 20,000 groundwater wells pumped water from compressible sediment beneath the city, with the result being the city sank into the empty space created by the withdrawal of water. The Italian government has now built an aqueduct system to bring drinking water to residents, and has closed most of the 20,000 wells. This action has slowed the subsidence of the city, but it is still sinking, and this action may be too little too late to spare Venice from the future effects of storm surges and astronomical high tides.

Mexico City is also plagued with subsidence problems caused by groundwater extraction. Mexico City is built on a several-thousand-foot-thick sequence of sedimentary and volcanic rocks, including a large dried lake bed on the surface. Most of the groundwater is extracted from the upper 200 feet (60 m) of these sediments. Parts of Mexico City have subsided dramatically, whereas others have not. The northeast part of the city has subsided about 20 feet (6 m). Many of the subsidence patterns in Mexico City can be related to the underlying geology. In places like the northeast part of the city that are underlain by loose compressible sediments, the subsidence has been large. In other places underlain by volcanic rocks, the subsidence has been minor.

The extraction of oil, natural gas, and other fluids from the Earth also may result in surface subsidence. In the United States, subsidence related to petroleum extraction is a large problem in Texas, Louisiana, and parts of California. One of the worst cases of oil field subsidence is that of Long Beach, California, where the ground surface has subsided 30 feet (9 m) in response to extraction of underground oil. There are approximately 2,000 oil wells in Long Beach, pumping oil from beneath the city. Much of Long Beach's coastal area subsided below sea level, forcing the city to construct a series of dikes to keep the water out. When the subsidence problem was recognized and understood, the city began a program of reinjecting water into the oil field to replace the extracted fluids and to prevent further subsidence. This reinjection program was initiated in 1958, and since then the subsidence has stopped, but the land surface can not be pumped up again to its former levels.

Pumping of oil from an oil field west of Marina del Rey along the Newport-Inglewood fault resulted in subsidence beneath the Baldwin Hills Dam and Reservoir, leading to the dam's catastrophic failure on December 14, 1963. Oil extraction from the Inglewood oil field resulted in subsidence-related slip on a fault beneath the dam and reservoir, which was enough to initiate a crack in the dam foundation. The crack was quickly expanded by pressure from the water in the reservoir, which led to the dam's catastrophic failure at 3:38 P.M. on December 14, 1963. Sixty-five million gallons of water were suddenly released, destroying dozens of homes, killing five people, and causing $12 million in damage.

TECTONIC SUBSIDENCE

Plate tectonics is associated with subsidence of many types and scales, particularly on or near plate boundaries. Plate tectonics is associated with the large-scale vertical motions that uplift entire mountain ranges, drop basins to lower elevations, and form elongate depressions in the Earth's surface known as rifts that can be thousands of feet (km) deep. Plate tectonics also causes the broad flat coastal plains and passive margins to slowly subside relative to sea level, causing the sea to encroach slowly onto the continents. More local scale folding and faulting can cause areas of the land surface to rise or sink, although at rates that rarely exceed half an inch (1 cm) per year.

Extensional or divergent plate boundaries are naturally associated with subsidence, since these boundaries are places where the crust is being pulled apart, thinning, and sinking relative to sea level. Places where the continental crust has ruptured and is extending are known as continental rifts. In the United States, the Rio Grande rift in New Mexico represents a place where the crust has begun to rupture, and it is subsiding relative to surrounding mountain ranges. In this area, the actual subsidence does not present much of a hazard, since the land is not near the sea, and a large region is subsiding. The net effect is that the valley floor is slightly lower in elevation every year than it was the year before. The rifting and subsidence is sometimes associated with faulting when the basin floor suddenly drops, and the earthquakes are associated with their own sets of hazards. Rifting in the Rio Grande is also associated with the rise of a large body of magma beneath Socorro, and if this magma body has an eruption it is likely to be catastrophic.

The world's most extensive continental rift province is found in East Africa. An elongate subsiding rift depression extends from Ethiopia and Somalia in the north, south through Kenya, Uganda, Rwanda, Burundi, and Tanzania, then swings back toward the coast through Malawi and Mozambique. The East African rift system contains the oldest hominid fossils, and is also host to areas of rapid land surface subsidence. Earthquakes are common, as are volcanic eruptions such as the catastrophic eruption of Nyiragongo in Congo in January of 2002. Lava flows from Nyiragongo covered large parts of the town of Goma, forcing residents to flee to neighboring Rwanda.

Subsidence in the East African rift system has formed a series of very deep elongate lakes, including Lakes Edward, Albert, Kivu, Malawi, and Tanganyika. These lakes sit on narrow basin floors, bounded on their east and west sides by steep rift escarpments. The shoulders of the rifts slope away from the center of the rift, so sediments carried by streams do not enter the rift, but are carried away from it. This allows the rift lakes to become very deep without being filled by sediments. It also means that additional subsidence can cause parts of the rift floor to subside well below sea level, such as Lake Abe in the Awash depression in the Afar rift. This lake and several other areas near Djibouti rest hundreds of feet below sea level. These lakes, by virtue of being so deep, become stratified with respect to dissolved oxygen, methane, and other gases. Methane is locally extracted from these lakes for fuel, although periodic overturning of the lake's water can lead to hazardous release of gases.

When continental rifts continue to extend and subside, they eventually extend far enough that a young narrow ocean forms in the middle of the rift. An example of where a rift has evolved into such a young ocean is the Red Sea in the Middle East. The borders of the Red Sea are marked by large faults that down drop blocks of crust toward the center of the sea, and the blocks rotate and subside dramatically in this process. Most areas on the margin of the Red Sea are not heavily developed, but some areas, such as Sharm Al Sheikh on the southern tip of the Sinai Peninsula, have large resorts along the coast. These areas are prone to rapid subsidence by faulting and pose significant risks to the development in this and similar areas.

Transform plate boundaries, where one plate slides past another, can also be sites of hazardous subsidence. The strike-slip faults that comprise transform plate boundaries are rarely perfectly straight. Places where the faults bend may be sites of uplift of mountains, or rapid subsidence of narrow elongate basins. The orientation of the bend in the fault system determines whether the bend is associated with contraction and the formation of mountains or extension, subsidence, and the formation of the elongate basins known as pull-apart basins. Pull-apart basins typically subside quickly, have steep escarpments marked by active faults on at least two sides, and may have volcanic activity. Some of the topographically lowest places on Earth are in pull-apart basins, including the Salton Sea in California and the Dead Sea along the border between Israel and Jordan. The hazards in pull-apart basins are very much like those in continental rifts. An example of a transform boundary with coastal subsidence and uplift problems is found in southern California along the San Andreas fault. Many areas south of Los Angeles are characterized by faults that down-drop the coast because the faults have an extensional component along them. Further north, near Ventura and Santa Barbara, the San Andreas fault bends so that there is compression across the fault, and many areas along this segment of the coast are experiencing tectonic uplift instead of subsidence.

Convergent plate boundaries are known for tectonic uplift, although they may also be associated with regional subsidence. When a mountain range is pushed along a fault on top of a plate boundary, the underlying plate may subside rapidly. In most situations erosion of the overriding mountain range sheds enormous amounts of loose sediment onto the under-riding plate, so the land surface does not actually subside, although any particular marker surface will be buried and subside rapidly.

On March 27, 1964, southern Alaska was hit by a massive magnitude 9.2 earthquake that serves as an example of the vertical motions of coastal areas associated with a convergent margin. Ground displacements above the area that slipped were remarkable—much of the Prince William Sound and Kenai Peninsula area moved horizontally almost 65 feet (20 m), and moved upward by more than 35 feet (11.5 m). Other areas more landward of the uplifted zone were down dropped by several to 10 feet. Overall, almost 125,000 square miles (200,000 km²) of land saw significant movements upward, downward, and laterally during this huge earthquake.

COMPACTION-RELATED SUBSIDENCE ON DELTAS AND PASSIVE MARGINS

Subsidence related to compaction and removal of water from sediments deposited on continental margin deltas, in lake beds, and in other wetlands poses a serious problem to residents trying to cope with the hazards of life at sea level in coastal environments. Deltas are especially prone to subsidence because the sediments that are deposited on deltas are very water-rich, and the weight of overlying new sediments compacts existing material, forcing the water out of pore spaces. Deltas are also constructed along continental shelves that are prone to regional-scale tectonic subsidence and are subject to additional subsidence forced by the weight of the sedimentary burden deposited on the entire margin. Continental margin deltas are rarely more than a few feet above sea level, so are prone to the effects of tides, storm surges, river floods, and other coastal disasters. Any decrease in the sediment supply to keep the land at sea level has serious ramifications, subjecting the area to subsidence below sea level.

Some of the world's thickest sedimentary deposits are formed in deltas on the continental shelves, and these are of considerable economic importance because they also host the world's largest petroleum reserves. The continental shelves are divided into many different sedimentary environments. Many of the sediments transported by rivers are deposited in estuaries, which are semi-enclosed bodies of water near the coast in which fresh water and seawater mix. Near-shore sediments deposited in estuaries include thick layers of mud, sand, and silt. Many estuaries are slowly subsiding, and they get filled with thick sedimentary deposits. Deltas are formed where streams and rivers meet the ocean and drop their loads because of the reduced flow velocity. Deltas are complex sedimentary systems, with coarse stream channels, fine-grained inter-channel sediments, and a gradation seaward to deepwater deposits of silt and mud.

All of the sediments deposited in the coastal environments tend to be water rich when deposited, and thus subject to water loss and compaction. Subsidence poses the greatest hazard on deltas, since these sediments tend to be thickest of all deposited on continental shelves. They are typically fine-grained mud and shale that suffer the greatest water loss and compaction. Unfortunately, deltas are also the sites of some of the world's largest cities, since they offer great river ports. New Orleans, Shanghai, and many other major cities have been built on delta deposits and have subsided 10 or more feet (several m) since they were first built. Many other cities built on these very compactable shelf sediments are also experiencing dangerous amounts of subsidence, as shown in the table "Subsidence Statistics for the 10 Worst-Case Coastal Cities" on page 726. The response to this subsidence will be costly. Some urban and government planners estimate that protecting the populace from sea level rise on subsiding coasts will be the costliest endeavor ever undertaken by humans.

The fate of these and other coastal cities that are plagued with natural and human-induced subsidence in a time of global sea level rise is subaqueous. The natural subsidence in these cities is accelerated by human activities. First of all, construction of tall heavy buildings on loose, compactable water-rich sediments forces water out of the pore spaces of the sediment underlying each building, causing that building to subside. The weight of cities has a cumulative effect, and big cities built on deltas and other compactable sediment cause a regional flow of water out of underlying sediments, leading to subsidence of the city as a whole.

New Orleans has one of the worst subsidence problems of coastal cities in the United States. Its rate and total amount of subsidence are not the highest,

SUBSIDENCE STATISTICS FOR THE 10 WORST-CASE COASTAL CITIES

City/State or Country	Maximum Subsidence Feet (m)	Area Affected square miles (km²)	Tectonic Environment
Los Angeles (Long Beach), California	29.5 (9.0)	20 (50)	Oilfield subsidence
Tokyo, Japan	14.8 (4.5)	1,170 (3,000)	Delta
San Jose, California	12.8 (3.9)	312 (800)	Delta
Osaka, Japan	9.8 (3.0)	195 (500)	Delta
Houston, Texas	9.0 (2.7)	4,720 (12,100)	Oilfield and coastal marsh
Shanghai, China	8.6 (2.63)	47 (121)	Delta
Niigata, Japan	8.2 (2.5)	3,237 (8,300)	Delta
Nagoya, Japan	7.8 (2.37)	507 (1,300)	Delta
New Orleans, Louisiana	6.6 (2.0)	68 (175)	Delta
Taipei, Taiwan	6.2 (1.9)	51 (130)	

but since nearly half of the city is at or below sea level, any additional subsidence will put the city dangerously far below sea level. Already, the Mississippi River is higher than downtown streets, and ships float by at the second story level of buildings. Dikes keep the river at bay and usually keep storm surges from inundating the city. However, the catastrophes of Hurricanes Katrina and Rita in 2005, of Hurricane Camille in 1969, and many before this, show that the levees cannot be trusted to hold. Additional subsidence will make these measures unpractical, and lead to greater disasters than Hurricane Katrina. New Orleans, Houston, and other coastal cities have been accelerating their own sinking by withdrawing groundwater and oil from compactable sediments beneath the cities. They are literally pulling the ground out from under their own feet.

The combined effects of natural and human-induced subsidence with global sea level rise have resulted in increased urban flooding of many cities, and greater destruction during storms. Storm barriers have been built in some cases, but this is only the beginning. Thousands of miles of barriers will need to be built to protect these cities unless billions of people are willing to relocate to inland areas, an unlikely prospect.

Something must be done to reduce the risks from coastal subsidence. First, a more intelligent regulation of groundwater extraction from coastal aquifers, and oil from coastal regions, must be enforced. If oil is pumped out of an oil reservoir then water should be pumped back in to prevent subsidence. Sea level is rising, partly from natural astronomical effects and partly from human-induced changes to the atmosphere. It is not too early to start planning for sea level rises of a few feet (about 2 m). Sea walls should be designed and tested before given on massive scales. Consideration to moving many operations inland to higher ground should be given.

See also BASIN, SEDIMENTARY BASIN; DELTAS; GROUNDWATER; KARST; PASSIVE MARGIN; PLATE TECTONICS.

FURTHER READING

Beck, B. F. *Engineering and Environmental Implications of Sinkholes and Karst.* Rotterdam: Balkema, 1989.

Dolan, Robert, and H. Grant Goodell. "Sinking Cities." *American Scientist* 74, no. 1 (1986): 38–47.

Holzer, Thomas L., ed. *Man-Induced Land Subsidence.* Reviews in Engineering Geology VI. Boulder, Colo.: Geological Society of America, 1984.

Whittaker, Barry N. *Subsidence: Occurrence, Prediction, and Control.* Amsterdam: Elsevier, 1989.

Sun The Sun is an average star that sits very close to Earth, a mere eight light minutes away, and 300,000 times closer than the next closest star, Alpha Centauri, which is located 4.3 light-years distant. It is a glowing ball of gas held together by its own gravity and powered by nuclear fusion in its core. Because of its proximity and fairly average characteristics,

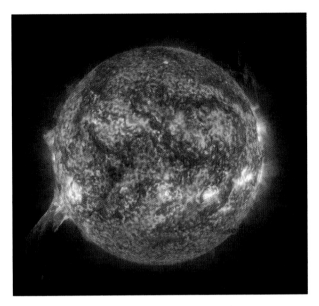

Solar flare image taken by NASA's SOHO satellite, July 1, 2002 *(AP Images)*

the Sun has been extensively studied and forms the basis for many of the concepts and models about other stars in the universe. The Sun is also the only source of light and the main source of heat for life on Earth.

PHYSICAL PROPERTIES OF THE SUN

The Sun contains no solid material, but the apparent surface visible as a glowing solar disk is known as the photosphere. Above the photosphere is the Sun's lower atmosphere, called the chromosphere, and extending far past that is the corona, an outer atmosphere (visible during eclipses) that gradually merges into the solar wind and consists of particles that flow through the whole solar system. Extending below the photosphere toward the deep interior of the Sun are three more main zones. The region just below the photosphere is called the convection zone, where material is in constant motion. The solar interior has two main parts. The center of the Sun consists of the solar core, about 124,000 miles (200,000 km) in diameter, where nuclear fusion reactions burning hydrogen into helium power the entire Sun, generating light and heat for the whole solar system. Between the solar core and extending to the base of the convection zone is the radiation zone, a second interior zone, about 186,000 miles (300,000 km) thick, where the main heat transfer process is radiative.

The Sun has a diameter of 870,000 miles (1.4 million km) to the top of the photosphere, about 100 times the diameter of the Earth. The mass of the Sun is about 300,000 times that of Earth, approximately 4.3×10^{30} pounds (2.0×10^{30} kg). The table "Sun Reference Data" lists these and other basic properties of the Sun. The Sun rotates at different velocities at the poles and the equator, spinning faster at the equator than the poles in a manner similar to the gaseous Jovian planets. The temperature at the top of the photosphere is estimated to be 9,932°F (5,500°C).

One of the most important properties of the Sun is that it radiates energy into space, providing the energy to drive surface processes and provide conditions necessary for life on Earth. There are several ways to measure how much energy the Sun radiates. One way is to measure how much energy is received

SUN REFERENCE DATA

DIAMETER	870,000 miles (1.4 million km)	AGE	4.5 billion years
MASS	4.3×10^{30} pounds (2.0×10^{30} kg)	DISTANCE FROM EARTH	93 million miles (150 million km)
DENSITY	1.41 (relative to water being 1)	DISTANCE TO NEXT NEAREST STAR	4.3 light-years (Alpha Centauri)
SOLAR WIND SPEED	1,860,000 miles/hr (3 million km/hr)	LUMINOSITY	390 billion billion megawatts
SOLAR CYCLE	8–11 years	TEMPERATURE AT SURFACE	9,932°F (5,500°C)
TEMPERATURE AT CORE	22.5 million°F (14 million°C)	TEMPERATURE OF SUNSPOTS	7,232°F (4,000°C)
ROTATION PERIOD AT EQUATOR	27 Earth days	ROTATION PERIOD AT POLES	31 Earth days

in a specific area in a specific time. The solar constant is defined as how much solar energy reaches the Earth per unit time; its value is about 1,400 watts per 1.2 square yards (1 m^2). Solar luminosity is a measure of the total energy emitted by the Sun. It is defined as the total energy reaching an imaginary sphere with a radius of 1 astronomical unit (A.U., the distance from the Sun to the Earth). The Sun has a luminosity of 4×10^{26} watts. This is a huge amount of energy, equivalent to the energy released by 100 billion one-megaton nuclear bombs every second for the past several billion years.

THE INTERIOR OF THE SUN

Direct observation of the deep interior of the Sun is impossible, so most knowledge about the interior is based on computer modeling of remotely sensed surface phenomena. Observations have been grouped together to form what is known as the standard solar model.

One of the techniques of studying the solar interior is called helioseismology. This science studies complex patterns of pressure waves that alternately cause the surface of the photosphere to move up and down. These pressure waves then bounce off the undersurface of the photosphere and move through the interior. As the pressure waves interact they produce complex patterns on the surface that can be studied using techniques similar to earthquake seismology to yield information on the deep interior of the Sun. Much of the knowledge about the deep interior, including the density and temperature distributions, is based on analyses of these complex patterns.

Numerical models of the Sun predict that the zone beneath the photosphere is strongly convecting, as occurs in places where a cooler liquid overlies a warmer liquid. Convection evens out the temperature in the region where it occurs and is in agreement with the temperature distribution model based on helioseismology. The motion of gases transfers heat within this zone, flowing upward as warm currents and back down as cool currents; the gases then pick up more heat and flow upward once again. These currents form complex characteristic cell patterns that have many different scales and depths of overturning cells. These cells are tens of thousands of miles (km) in diameter 124,000 miles (200,000 km) below the surface, but only about 600 miles (1,000 km) across in the upper 600 miles (1,000 km) of the convection zone to the surface of the photosphere. The tops of these smaller convection cells form the visible surface of the Sun. Observations of these convection cells shows that they form a granular-looking surface, with many continent-sized light and dark patches that appear and disappear every 5–10

minutes. Spectroscopic observation of these granules shows that the bright patches are hot and moving upward, and the dark patches are colder and moving downward. The granules correspond to the tops of the convection cells, bringing the heat from the deep interior to the surface. A larger scale of granulation, forming supergranules, is superimposed on top of these granules.

Heat transfer beneath the convection zone occurs by radiation, a fundamentally different process. The gas in the deep interior above the core is completely ionized, and the high temperatures mean that the particles are moving quickly and encountering many collisions with each other. Nearly all atoms have been stripped of their electrons in the ionized medium, and photons produced by nuclear reactions in the core cannot be trapped by the atoms without electrons. Therefore, energy transfer in this zone is by the very efficient process of radiation. However, near the outer edge of the radiation zone in the interior, temperatures become progressively lower outward, and more and more atoms retain their electrons. This means that photons produced in the core begin to be absorbed in the outer part of the radiation zone. At a distance of about 124,000 miles (200,000 km) below the photosphere, the gas is almost totally opaque to photons, and by this distance, all of the photons produced by nuclear reactions in the core have been absorbed. The energy from these photons is then transferred as heat to the base of the convection zone, where it is transferred to the surface by convection. The photosphere, or surface of the Sun, is only about 300 miles (500 km) thick, so appears very sharp when viewed with telescopes.

SOLAR ATMOSPHERE

The lower part of the solar atmosphere, resting directly above the photosphere, is called the chromosphere. The chromosphere emits very little light compared to the photosphere, so is visible only during total solar eclipses, as a bright and somewhat irregular diffuse band around the Sun. The chromosphere has relatively few gas particles (hydrogen), so emits few photons. However, the chromosphere is a dynamic environment. Every few minutes the convection cells on the surface of the Sun emit spicules, storms of matter that shoot upward several thousand miles (km) into the upper atmosphere of the Sun at velocities of about 60 miles (100 km) per second. These spicules form mostly around the edges of the supergranules and may result from magnetic field interactions along the edges of the supergranules.

The corona is a diffuse zone outside the chromosphere, visible only during eclipses that block the photosphere as well as the chromosphere. Gases including hydrogen and iron in the corona are highly ionized (stripped of electrons) by the very high temperatures, which are higher than in the photosphere below. The temperature at the top of the photosphere is 6,000 K (10,340°F; 5,727°C), sinking to about 4,500 K (7,640°F; 4,227°C) in the chromosphere, rising through a transition zone out to 6,000 miles (10,000 km) into the corona where temperatures exceed 1,000,000 K (1,799,540°F; 999,727°C) out to distances well past 12,000 miles (20,000 km). The intense heating of the corona is probably caused by magnetic disturbances from the photosphere.

Fast-moving particles, including protons and electrons, and electromagnetic radiation are constantly moving away from the Sun at high velocity, forming the solar wind. The radiation travels at the speed of light but the physical particles move more slowly, at approximately 310 miles (500 km) per second, reaching the Earth a few days after they leave the Sun. The solar wind originates in the corona. About 6,000,000 miles (10,000,000 km) above the photosphere the temperatures are so high that the speed of the particles exceeds the escape velocity of the Sun's gravity, and material flows outward or evaporates in all directions into space. The material is constantly replaced from below, shedding about a million tons of matter each second. Remarkably, only 0.1 percent of the solar mass has been lost by this mechanism in the past 4.5 billion years. Most of the charged particles in the solar wind escape through areas of especially low density, called coronal holes. With less matter to interact with, the charged particles stream into space along magnetic field lines, reaching far into the solar system.

SUNSPOTS, FLARES, AND THE SOLAR CYCLE

The Sun produces a steady stream of electromagnetic radiation from the photosphere, essentially unchanging with time. Superimposed on this steady, quiet process are several dynamic, active, or changing events and cycles that show the Sun also has some unpredictable and explosive behavioral traits. These features are not significant in terms of total solar energy output but do influence the electromagnetic radiation received on Earth. They include sunspots, solar flares, magnetic storms, the solar cycle, and changes in the solar corona.

The surface of the Sun is covered by a number of dark spots, typically about the size of Earth (6,000 miles or 10,000 km across), that appear, disappear, and change in size and shape over periods of a day to 100 days. In detail they show a change in color from darkest in the center (called the umbra) to an intermediate color around the edges (penumbra) to merge with the photosphere on their edges. The colors correspond to changes in temperature, with the darkest regions representing a drop in temperature from 6,000 K (10,340°F; 5,727°C) on the photosphere, to 5,500 K (9,440°F; 5,227°C) in the penumbra, to 4,500 (7,640°F; 4,227°C) in the umbra.

Sunspots are intricately related to the magnetic field of the Sun. The magnetic field strength is measured to be about 1,000 times stronger in the sunspots than elsewhere on the photosphere. The strong magnetic field in these regions can block the normal convective flow from rising in these spots, explaining why they are cooler than their surroundings.

Sunspots typically occur in pairs, and the magnetic field has opposite polarity (positive or negative) in adjacent spots, with magnetic field lines looping out of one spot high into the solar atmosphere and into the other, forming a high-reaching arc in between. Another interesting observation is that at any time, all of the loops between sunspots have the same configuration in each hemisphere with respect to the rotation of the Sun. If the field lines loop from west to east in one pair, then all the other pairs in that hemisphere orient in the same direction, and all the pairs in the opposite hemisphere orient in the opposite direction. This phenomenon results from the differential rotation of the Sun. The different speeds of rotation between the poles and the equator cause the N-S magnetic axes to be distorted and sometimes reoriented to E-W, and the convection cells then distort these lines further forming the loops. These loops are occasionally so attenuated that they become shaped like narrow tubes which distort the convection cells and become manifested as sunspots.

The numbers and distributions of sunspots change in a regular fashion, a phenomenon known as the sunspot cycle. More than a century of observation shows that the number of spots peaks at about 100 per year every 11 years, and then decreases to about zero in between peaks. There is some variation in the cycle from 7–15 years, but it is a well-established cycle. Sunspots do not move in position, but during the course of a cycle, older spots at high latitudes are gradually replaced by more spots in more equatorial regions as the cycle reaches a climax every 11 years, with most spots appearing 15–20° from the equator. The solar minimum is when the spots are fewest in number, and the solar maximum is when most spots are observed, especially in equatorial regions. As the solar maximum grades into the next minimum, the highest latitude spots tend to disappear first, leaving the last few spots near the equator. As the next maximum approaches, the new spots will appear in high latitudes.

The sunspot cycle is complicated further by the fact that the entire magnetic field of the Sun reverses polarity every 11 years as part of a full 22-year cycle, so that the leading spots in both hemispheres switch from positive to negative, and negative to positive with each successive 11-year sunspot cycle. This cyclicity allows for a fairly regular interaction of the magnetic field with the convection in the outer layers of the Sun, in a manner similar to the generation of the Earth's geodynamo and magnetic field. The present 11-year cycle has not always been so regular, for instance, from the mid-1600s to the early 1700s, there were very few sunspots. This reflects the complex dynamics of the Sun's magnetic field and interaction with the convection system.

Although sunspots are dark, cool, and relatively inactive areas, they are sometimes surrounded by active regions that emit huge quantities of energetic particles into the surrounding corona. These active regions also follow the solar cycle and, like the sunspots, are most abundant and active during the solar maximum.

Solar prominences are giant loops or sheets of glowing ionized gas that erupt from the photosphere and move through the lower corona under the influence of the magnetic field, following the loops between two sunspots. They may be caused by magnetic instabilities near the sunspots where the magnetic field lines are extremely concentrated and unstable. At times their height reaches about half the solar diameter. Some prominences last for days or weeks in a fairly stable configuration, whereas others surge and disappear in a matter of hours.

Active regions also release highly energetic flares that erupt and move around on the surface near the active regions, releasing huge amounts of energy, especially in the X-ray and ultraviolet wavelengths. Temperatures in the cores of flares can exceed 10,000,000 K (17,999,540°F; 9,999,726°C), and the amount of energy released in these flares in a few minutes can exceed that from the larger prominences over a period of weeks. The particles released in the flares are so energetic that they escape the Sun's gravity and are blasted into space where they disrupt communications and other electronics.

The sunspot cycle also affects the solar corona, which is smooth and uniform during solar minimums, and irregular, much larger, and contains many beams or streamers moving away from the Sun at solar maximums. The corona may be heated mostly by flares and prominences; during periods of more activity the corona becomes more active as well, and the solar wind increases.

SOLAR CORE

The energy from the Sun comes from nuclear reactions in its core, generating a luminosity of 4×10^{26} W, or 2×10^4 W/kg. The Sun has been producing approximately this vast amount of energy for the past 4.5 billion years. For the entire lifetime of the Sun, the total amount of energy generated has been 3×10^{13} joule/kg. The generation of this energy has been remarkably steady and is expected to continue for another 5 billion years, through the process of nuclear fusion.

Fusion works by combining two atomic nuclei to form one and releases energy according to the law of conservation of mass and energy. During fusion reactions, mass is lost but is converted to energy according to Einstein's equation

$$E = mc^2$$

where E = energy, m = mass, and c = the speed of light.

During nuclear fusion the total amount of mass and energy is conserved, but the mass is gradually converted to energy, which is the energy that has been emitted from the Sun for the past 4.5 billion years. Nuclear energy is generated by fusion along a reaction series called the proton-proton chain. Positively charged atomic nuclei naturally repel each other, but if they collide at very high speeds (such as generated by high temperatures in the core of the Sun) then they can overcome the repulsive forces between the nuclei and then be influenced by a different force called the strong nuclear force. This force, which acts at distances smaller than 10^{-15} m, can bind the two positively charged nuclei together, releasing energy in the process. Temperatures in excess of 10^7 K are needed to generate the speeds that cause this nuclear fusion.

When the two nuclei of hydrogen (protons) interact in the fusion reaction, they produce a new proton, a neutron, a positron, and a neutrino. The positron has the same properties as an electron except that it has a positive charge, and it is classified as an antiparticle of an electron, or as antimatter. When the fusion reaction occurs and the positron is formed, it is immediately released into a sea of free electrons in the solar core. When matter and antimatter particles meet they violently annihilate each other and release energy in the form of gamma rays.

The other particle released in the fusion reaction is a neutrino, a particle with no charge and a mass so low that it is approximately equal to about 1/10,000 that of an electron. They move at nearly the speed of light and are nearly (but not quite) impossible to detect since they can penetrate anything, even a wall of lead several light-years thick.

The proton and neutron produced in the fusion reaction merge to form a deuteron, which is the nucleus of deuterium or heavy hydrogen. It is called "heavy" since it has an extra neutron relative to the most common form of hydrogen, which lacks a neutron. Nuclei that have the same number of protons but different numbers of neutrons are known as isotopes of the same element, and deuterium is an isotope of hydrogen. The isotopic number is written as a prefix to the element symbol in standard notation. Thus, the hydrogen fusion reaction that powers the Sun can be written as

$$^1H + {}^1H \rightarrow {}^2H + positron + neutrino$$

This reaction is only the first in the proton-proton chain that powers the Sun, followed quickly by the formation of helium (He) by the interaction of deuterium with an isotope of helium according to the following equation:

$$^2H + {}^1H \rightarrow {}^3He + energy\ (gamma\ ray\ photons)$$

Next, helium 4 is produced by the fusion of two helium 3 isotopes according to the following equation:

$$^3He + {}^3He \rightarrow {}^4He + {}^1H + {}^1H$$
$$+ energy\ (gamma\ ray\ photons)$$

The net effect of the proton-proton reaction chain is that four hydrogen nuclei are fused into one helium 4 isotope plus two neutrinos, releasing a large amount of energy in the form of gamma rays. As the gamma ray photons move through the Sun they slowly lose energy as it is absorbed by ions and electrons, getting converted to heat by the convecting layer and then is emitted at the photosphere in the form of visible light that is observed from Earth. The helium remains in the core of the Sun, and the neutrinos escape at close to the speed of light.

Calculations of the mass to energy conversion show that about 600 million tons of hydrogen have been fused into helium in the core of the Sun every second for the past several billion years.

See also ASTRONOMY; ASTROPHYSICS; AURORA, AURORA BOREALIS, AURORA AUSTRALIS; COSMIC RAYS; GREENHOUSE EFFECT; ORIGIN AND EVOLUTION OF THE EARTH AND SOLAR SYSTEM; SOLAR SYSTEM; STAR FORMATION; STELLAR EVOLUTION; SUN HALOS, SUNDOGS, AND SUN PILLARS.

FURTHER READING

Chaisson, Eric, and Steve McMillan. *Astronomy Today.* 6th ed. Upper Saddle River, N.J.: Addison-Wesley, 2007.
Comins, Neil F. *Discovering the Universe.* 8th ed. New York: W. H. Freeman, 2008.
NASA. Imagine the Universe! The Sun. Available online. URL: http://imagine.gsfc.nasa.gov/docs/science/know_l1/sun.html. Modified August 22, 2008.
NASA. Worldbook, Sun. Available online. URL: http://www.nasa.gov/worldbook/sun_worldbook.html. Modified November 29, 2007.
ScienceDaily: Astrophysics News. ScienceDaily LLC. Available online. URL: http://www.sciencedaily.com/news/space_time/astrophysics/. Accessed October 24, 2008.
Snow, Theodore P. *Essentials of the Dynamic Universe: An Introduction to Astronomy.* 4th ed. St. Paul, Minn.: West Publishing Company, 1991.

sun halos, sundogs, and sun pillars A number of unusual phenomena are related to the interaction of the Sun's rays with ice crystals in the upper atmosphere. A ring of light that circles and extends outward from the Sun, or the Moon, is known as a halo. The halo forms when ice crystals in high-level cirriform clouds refract the Sun's or Moon's rays. Most sun halos form at an angle of 22° from the Sun because randomly oriented small ice particles refract the light at this angle. Occasionally a 46° halo is visible, formed when subhorizontally oriented columnar ice crystals refract the light at this higher angle. Most sun halos are simply bright bands of light but some exhibit rainbowlike zones of color. These form when the light is dispersed by the ice crystals and light of different wavelengths (colors) is refracted by different amounts depending on its speed

Atmospheric crystals sometimes cause hexagonal or platy ice crystals to fall slowly through the atmosphere, and this vertical motion causes the crystals to become uniformly oriented with their long dimensions in a horizontal direction. This orientation prevents light that is refracted through the ice crystals from forming a halo, but when the Sun approaches the horizon it causes two bright spots or colored bright spots to appear on either side of the sun. These spots are commonly called sundogs, or parhelia.

Sun pillars are a similar phenomenon but are formed by light that is reflected off the ice crystals instead of refracted through them. In this case, usually at sunset or sunrise, the Sun's rays reflect off the subhorizontally oriented ice crystals and form a long column of light extending downward from the Sun.

Rainbows are a somewhat related phenomena. Rainbows are translucent concentric arcs of colored bands that are visible in the air under certain conditions when rain or mist is present in the air and the Sun is at the observer's back. Rainbows form where sunlight enters the rain or water drops in the air, and a small portion of this light is reflected off the back of the raindrops and directed back to the observer. When the sunlight enters the rain drops, it is bent and slows, and as in a prism, violet light is refracted the most and red the least. The amount of light that is reflected off the back of each raindrop is small compared to the amount that enters each drop, and only the rays that hit the back of the drop at angles greater than the critical angle are reflected. Since the sun's rays are refracted and split by color when they enter the water drops, each color hits the back of the raindrop at a slightly different angle, and the reflected light emerges from the raindrop at different angles for each color. Red light emerges at 42° from the incoming beam, whereas violet light emerges at 40°. An observer sees only one color from each drop, but with millions of drops in the sky an observer is able to see a range of colors formed from different raindrops with light reflected at slightly different angles to the observer. Rainbows appear to move as an observer moves, since each ray of light is entering the observer's eyes from a single raindrop, and as the observer moves, light from different drops enters the observer's eyes.

See also ATMOSPHERE; SUN.

FURTHER READING

Ahrens, C. Donald. *Meteorology Today.* 7th ed. Pacific Grove, Calif.: Brooks/Cole, 2002.

Ashworth, William, and Charles E. Little. *Encyclopedia of Environmental Studies.* New ed. New York: Facts On File, 2001.

supercontinent cycles Supercontinent cycles are semiregular groupings of the planet's landmasses into single or large continents that remain stable for a period of time, then disperse, and eventually come back together as new amalgamated landmasses with a different distribution. At several times in Earth history, the continents have joined together forming one large supercontinent, with the last supercontinent, Pangaea (meaning all land), breaking up approximately 160 million years ago. This process of supercontinent formation, dispersal, and reamalgamation seems to be grossly cyclic, perhaps reflecting mantle convection patterns, but also influencing climate and biological evolution. Early investigators noted global "peaks" in age distributions of igneous and metamorphic rocks and suggested that these represent global orogenic or mountain building episodes, related to supercontinent amalgamation.

The basic idea of the supercontinent cycle is that continents drift about on the surface until they all collide, stay together, and come to rest relative to the mantle in a place where the gravitational potential surface (geoid) has a global low. The continents are only one-half as efficient at conducting heat as oceans, so after the continents are joined together, heat accumulates at their base, causing doming and breakup of the continent. For small continents, heat can flow sideways and not heat up the base of the

plate, but for large continents the lateral distance is too great for the heat to be transported sideways. The heat rising from within the Earth therefore breaks up the supercontinent after a heating period of several tens or hundreds of millions of years. The heat then disperses and is transferred to the ocean/atmosphere system, and continents move apart until they come back together to form a new supercontinent.

The supercontinent cycle greatly affects other Earth systems. The breakup of continents causes sudden bursts of heat release, associated with periods of increased, intense magmatism. It also explains some of the large-scale sea level changes, episodes of rapid and widespread orogeny, episodes of glaciation, and many of the changes in life on Earth.

Compilations of Precambrian isotopic ages of metamorphism and tectonic activity suggest that the Earth experiences a periodicity of global orogeny of 400 million years. Peaks have been noted at time periods including 3.5, 3.1, 2.9, 2.6, 2.1, 1.8, 1.6, and 1.1 billion years ago, as well as at 650 and 250 million years ago. One hundred million years after these periods of convergent tectonism and metamorphism, rifting is common and widespread. Geologist A. H. Sutton (1963) proposed the term *chelogenic cycle,* in which continents assemble and desegregate in antipodal supercontinents.

RELATIONSHIP OF SUPERCONTINENTS, LOWER MANTLE CONVECTION, AND THE GEOID

Some models for the formation and dispersal of supercontinents suggest a link between mantle convection, heat flow, and the supercontinent cycle. Stationary supercontinents insulate the mantle, causing it to heat up, because the cooling effects of subduction and seafloor spreading are absent. As the mantle then heats up, convective upwelling is initiated, causing dynamic and isostatic uplift of the continent, injection of melts into the continental crust, and extensive crustal melting. These crustal melts are widespread in the interiors of some reconstructed supercontinents, such as the Proterozoic anorogenic granites in interior North America, which were situated in the center of the supercontinent of Rodinia when they formed between 1 billion and 800 million years ago.

After intrusion of the anorogenic magmas, the lithosphere is weakened and can be more easily driven apart by divergent flow in the asthenosphere. Thermal effects in the lower mantle lag behind surface motions. So, the present Atlantic geoid high and associated hot spots represent a "memory" of heating beneath Pangaea. Likewise, the circum Pangaea subduction zones may have memory in a global ring of geoid lows.

Other models for relationships between supercontinents and mantle convection suggest that super-

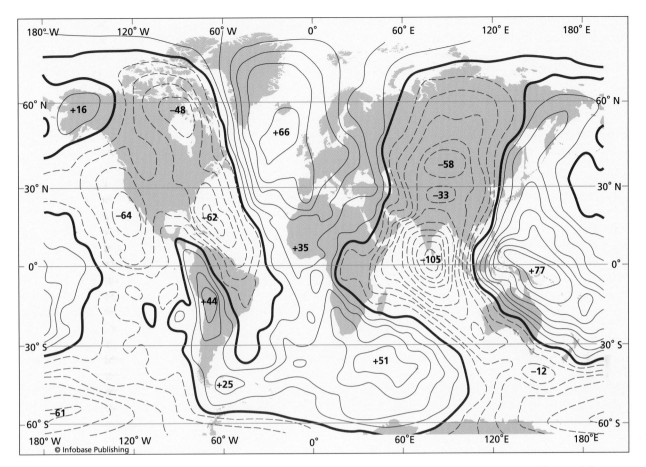

The geoid is an imaginary surface that would be sea level if it extended through the continents. The geoid surface is perpendicular to the gravity plumb lines at every location.

continents result from mantle convection patterns. Continental fragments may be swept toward convective downwellings, where they reaggregate as supercontinents.

PLATE TECTONICS, SUPERCONTINENTS, AND LIFE

Plate tectonic motions, especially the supercontinent cycle, profoundly affect the distribution and evolution of life on Earth. Plate tectonic activity such as rifting, continental collision, and drifting continents affects the distribution of life-forms, the formation and destruction of ecological niches, and radiation and extinction blooms. Plate tectonic effects also can induce sea level changes, initiate periods of global glaciation, change the global climate from hothouse to icehouse conditions, and affect seawater salinity and nutrient supply. All of these consequences of plate tectonics profoundly influence life on Earth.

Changes in latitude brought on by continental drift bring land areas into latitudes with better or worse climate conditions. This has different consequences for different organisms, depending on their temperature tolerance, as well as food availability in their environment. Biological diversity generally increases toward the equator, so, in general, as continents drift poleward more organisms tend to go extinct, and as they drift equatorward, diversification increases.

Tectonics and supercontinent dispersal break apart and separate faunal provinces, which then evolve separately. Continental collisions and supercontinent amalgamation build barriers to migration but eventually bring isolated fauna together. One of the biggest mass extinctions (at the end of the Permian) occurred with the formation of a supercontinent (Pangaea), sea level regression, evaporite formation, and global warming. At the boundary between the Permian and Triassic Periods and between the Paleozoic and Mesozoic Periods (250 million years ago), 70–90 percent of all species became extinct. Casualties included the rugose corals, trilobites, many types of brachiopods, and marine organisms including many foraminifer species.

The Siberian flood basalts cover a large area of the central Siberian Plateau northwest of Lake Baikal. They are more than one-half mile thick over an area of 210,000 square miles (547,000 km^2) but have significantly eroded from an estimated volume of 1,240,000 cubic miles (3,3133,000 km³). They were erupted over a period of less than 1 million years, 250 million years ago at the end of the Permian at the Permian-Triassic boundary. They are remarkably coincident in time with the major Permian-Triassic extinction, implying a causal link. The Permian-Triassic boundary at 250 million years ago marks the greatest extinction in Earth history, where 90 percent of marine species and 70 percent of terrestrial vertebrates became extinct. The rapid volcanism and degassing could have released enough sulfur dioxide to cause a rapid global cooling, inducing a short ice age with associated rapid fall of sea level. Soon after the ice age took hold, the effects of the carbon dioxide that was also emitted by the volcanism took over and the atmosphere heated, resulting in a global warming. The rapidly fluctuating climate postulated to have been caused by the volcanic gases is thought to have killed off many organisms, which were simply unable to cope with the wild climate changes.

Continental breakup can physically isolate species that cannot swim or fly between the diverging continents. Physical isolation (via tectonics) produces adaptive radiation—continental dispersal thus increases biotic diversity. Mammals had an explosive radiation (in 10–20 million years) in the Paleocene-Eocene, right after breakup of Pangaea.

SEA LEVEL CHANGES, SUPERCONTINENTS, AND LIFE

Sea level has changed by about a thousand feet (hundreds of meters) above and below current levels many times in Earth history. In fact, sea level constantly changes in response to a number of different variables, many of them related to plate tectonics. The diversity of fauna on the globe closely relates to sea levels, with greater diversity during sea level high stands, and lower diversity during sea level lows. For instance, sea level was 1,970 feet (600 m) higher than now during the Ordovician, and the sea level high stand was associated with a biotic explosion. Sea levels reached a low stand at the end of the Permian, and this low was associated with a great mass extinction. Sea levels rose again in the Cretaceous.

Sea levels change at different rates and amounts in response to changes in several other Earth systems. Local tectonic effects may mimic sea level changes through regional subsidence or uplift, and these effects must be taken into account and filtered out when trying to deduce ancient, global (eustatic) sea level changes. The global volume of the mid-ocean ridges can change dramatically, either by increasing the total length of ridges or changing the rate of seafloor spreading. The total length of ridges typically increases during continental breakup, since conti-

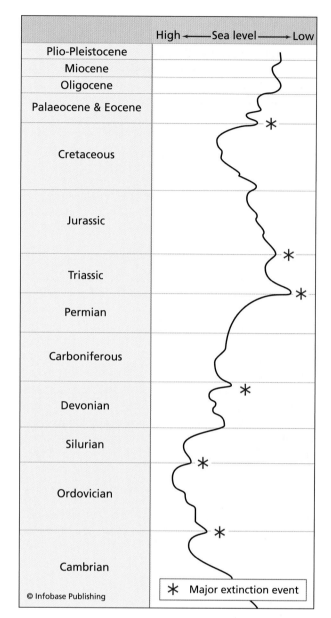

Sea level has risen and fallen dramatically through Earth history. Global sea level rise and fall must be separated from local subsidence and uplift events along individual coastlines, by correlating events between different continents. Global eustatic sea level curves (Vail curves) also show the height of the world's oceans after local effects have been removed.

nents are being rifted apart and some continental rifts can evolve into mid-ocean ridges. Additionally, if seafloor spreading rates are increased, the amount of young, topographically elevated ridges is increased relative to the slower, older, topographically lower ridges that occupy a smaller volume. If the volume of the ridges increases by either mechanism, then a volume of water equal to the increased ridge volume is displaced and sea level rises, inundating the continents. Changes in ridge volume are able to change sea levels positively or negatively by about 985 feet (300 m) from present values, at rates of about 0.4 inch (1 cm) every 1,000 years.

Continent-continent collisions, such as those associated with supercontinent formation, can lower sea levels by reducing the area of the continents. When continents collide, mountains and plateaus are uplifted, and the amount of material that is taken from below sea level to higher elevations no longer displaces seawater, causing sea levels to drop. The contemporaneous India-Asia collision has caused sea levels to drop by 33 feet (10 m).

Other factors such as mid-plate volcanism can also change sea levels. The Hawaiian Islands are hot-spot style, mid-plate volcanoes that have been erupted onto the seafloor, displacing an amount of water equal to their volume. Although this effect is not large at present, at some periods in Earth history many more hot spots existed (such as in the Cretaceous), and the effect would have been larger.

The effects of the supercontinent cycle on sea level may be summarized as follows: Continent assembly favors regression, whereas continental fragmentation and dispersal favors transgression. Regressions followed formation of the supercontinents of Rodinia and Pangaea, whereas transgressions followed the fragmentation of Rodinia and the Jurassic-Cretaceous breakup of Pangaea.

EFFECTS OF TRANSGRESSIONS AND REGRESSIONS

During sea level transgressions, continental shelves are covered by water, and available habitats are enlarged, increasing the diversity of fauna. Transgressions are generally associated with greater diversification of species. Regressions cause extinctions through loss of environments, both shallow marine and beach. There is a close association between Phanerozoic extinctions and sea level low stands. Salinity fluctuations also affect diversity—the formation of evaporites (during supercontinent dispersal) causes reduction in oceanic salinity. For instance, Permian-Triassic rifts formed during the breakup of Pangaea had lots of evaporites (up to 4.4 miles, or 7 km thick), which lowered the salinity of oceans.

Supercontinents affect the supply of nutrients to the oceans, and thus, the ability of life to proliferate. Large supercontinents cause increased seasonality, and thus lead to an increase in the nutrient supply through overturning of the ocean waters. During breakup and dispersal, smaller continents have less seasonality, yielding decreased vertical mixing, leaving fewer nutrients in shelf waters. Seafloor spreading also increases the nutrient supply to the ocean; the more active the seafloor spreading system, the more interaction there is between ocean waters and

crustal minerals that get dissolved to form nutrients in the seawater.

See also GREENHOUSE EFFECT; ICE AGES; PLATE TECTONICS.

FURTHER READING

de Wit, Maarten J., Margaret Jeffry, Hugh Bergh, and Louis Nicolaysen. *Geological Map of Sectors of Gondwana Reconstructed to Their Disposition at ~150 Ma.* Tulsa, Okla.: American Association of Petroleum Geologists, Map Scale 1:10,000,000, 1988.

Hoffman, Paul F. "Did the Breakout of Laurentia Turn Gondwana Inside-out?" *Science* 252 (1991): 1,409–1,412.

Kusky, Timothy M., Mohamed Abdelsalam, Robert Tucker, and Robert Stern, eds., *Evolution of The East African and Related Orogens, and the Assembly of Gondwana.* Precambrian Research, 2003.

Murphy, J. Brendan, and Richard Damian Nance. "Mountain Belts and the Supercontinent Cycle." *Scientific American* 266, no. 4 (1992): 84–91.

Rogers, J. J. W., and M. Santosh. *Continents and Supercontinents.* Oxford: Oxford University Press, 2004.

supernova Supernovas are extremely luminous stellar explosions associated with large bursts of radiation that can exceed the brightness of the entire galaxy it is associated with for a period of weeks or months. The energy released in a supernova event can exceed the amount of energy produced by the Earth's Sun over its entire life span. In these stellar explosions material is expelled from the core star at speeds of about one-10th the speed of light and is associated with a shock wave that moves through the interstellar medium, sweeping up an expanding shell of gas and dust that is known as a supernova remnant. A nova is a general name for a type of star that vastly increases in brightness (by up to 10,000 times) over very short periods of time, typically weeks. The word *nova* comes from the Latin for "new" and stems from early astronomers who thought that nova were new stars appearing in the sky, since the parent stars were too faint to be observed from the Earth before powerful telescopes were invented. The term *nova* is used for any star that increases rapidly in brightness by ejecting some of its material in the form of a cloud, whereas supernova are associated with stellar explosions, and are much more, even millions of times as luminous and energetic as nova.

Several different types of supernovas have been identified. The most common type is produced when a massive star ages and stops generating energy from nuclear fusion because it has spent all of its fuel. The massive star can undergo sudden gravitational collapse to form a neutron star or black hole, releasing

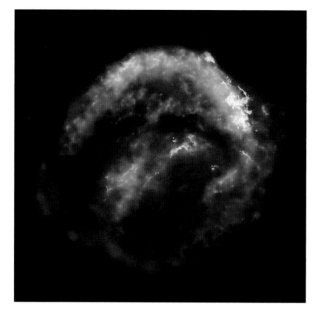

Kepler's supernova taken from combined data from *Hubble Telescope, Spitzer Telescope,* and *Chandra X-Ray Observatory (NASA Marshall Space Flight Center)*

huge amounts of gravitational potential energy that heats up and then expels the star's outer layers. A second type of supernova forms when a white dwarf star accumulates enough material from a stellar companion, typically in a binary star system, to raise the core temperature enough to start fusion of carbon, which forms a runaway nuclear fusion reaction, causing the star to explode. This may happen after the white dwarf has gone through hundreds of smaller nova events, each time exploding away an outer shell of gas, but leaving some behind that eventually contributes to the final, death-blow supernova explosion of the entire star.

A supernova event occurs about once every 50 years in the Milky Way Galaxy and other galaxies of similar size. The expanding shock waves from supernovas play an important role in synthesizing heavier elements and in triggering the formation of new stars.

DEATH OF HIGH-MASS STARS

High-mass stars containing more than eight solar masses have a very different evolution from low mass stars and have an explosive death that is responsible for forming many of the heavier elements in the universe.

Evolved high-mass stars have a core with concentric zones of progressively heavier fuel, with the burnt "ash" of one layer forming the fuel for the next deeper layer in the star. At the edge of the core, hydrogen burns to fuse helium, then helium burns to fuse into carbon, which fuses into oxygen. The oxy-

gen goes through nuclear fusion to form neon, which then forms magnesium, then silicon, and then iron nuclei ash in the core of the massive star. As each fuel is used up, the core of the star contracts, heats up, then starts to burn the fuel of the ash of the previous episode of burning. Each successive burning stage is hotter, proceeds faster than the one before, and lasts for a much shorter time. A typical massive star that is about 20 solar masses may burn hydrogen for 10 million years, helium for 1 million years, carbon 1,000 years, oxygen for 1 year, silicon for a week, and iron for less than one day. After this the core of the massive star collapses, and one of the universe's most spectacular events unfolds.

CORE COLLAPSE IN GIANT STARS

When iron is produced in the core of the massive star, the internal processes suddenly change. The iron is so dense, consisting of 26 protons and 30 neutrons, that when it burns no energy can be extracted by combining this to make heavier elements. The result is that when significant quantities of iron accumulate in the core the internal fires suddenly are extinguished, and the equilibrium of the star is destroyed. The temperature in the core is several billion K, but without continued nuclear fusion the force of gravity at this stage begins to overwhelm the outward gas pressure and the star begins a fateful collapse and implosion.

As the star begins to collapse the core temperature shoots up to about 10 billion K, and the photons take on a high energy state, splitting the iron into lighter nuclei, quickly breaking down the elements in a process called photodisintegration so that only protons and neutrons remain. In less than one second all of the heavy elements produced by nuclear fusion in the entire core of the star have broken down into simple protons and neutrons, undoing 10 million years of nuclear reactions. Photodisintegration requires a huge amount of energy from the high-temperature core and transferring this energy from the core to break down these elements cools the core, which reduces the outward fluid pressure that is resisting the inward pull of gravity. The sudden loss of pressure in the core causes the collapse of the core of the star to accelerate rapidly.

As the core continues to shrink, the pressures on the mixture of free electrons, protons, neutrons, and photons, and the rapidly rising pressures crush the protons and electrons together to form neutrons and neutrinos in a process called neutronization. Neutrinos are such high-energy particles that they rarely interact with matter, even matter as dense as that in the core of the collapsing giant star. Most of the neutrinos therefore pass right through the core and escape to space carrying large amounts of energy with them. At this stage the core finds itself in a state where the electrons and neutrinos have escaped, so that the neutrons in the core are coming into contact with each other, at a super high density of 10^{15} kg/m³. At this high density the neutrons influence and oppose each other with a very strong force, generating strong pressure called the neutron degeneracy pressure, which acts against the inward pull of gravity. By the time this force is able to oppose the ongoing rapid gravitational collapse of the star, however, the core has gone beyond its point of equilibrium and has reached densities between 10^{17} to 10^{18} kg/m³. These forces interact; the rapid gravitational collapse hits the super-dense neutron mass in the core with the strong neutron degeneracy pressure, and the inward collapse bounces backward to produce a rapid expansion. The total time elapsed from the start of the collapse of the core to the beginning of the outward expansion is less than one second. The outward expansion is associated with a tremendously powerful shock wave that moves outward through the star, blasting all of the star's outer layers into space in one of the most powerful events known in the universe, a supernova. These outer layers contain many heavy elements around the core, as well as light elements near the surface, and as the star explodes these are all blown into interstellar space in a bright flash that may last only a few days at its peak. Supernovas are among the brightest events known in the universe, being about a million times brighter than novas, and often being about the same brightness—for a few days—as a galaxy with a trillion stars. The supernova may have an initial flash that is more than a billion times brighter than the Earth's Sun and produces more energy over the time from its initial flash until it completely fades away a few months later than the Sun produces in its entire history.

DIFFERENT TYPES OF SUPERNOVAS

Enough supernovas have been observed to characterize some differences between them. Some supernovas have very little hydrogen associated with them (called Type-I supernovas), whereas others are hydrogen-rich (Type-II supernovas) and are associated with the star collapse or implosion described above. These two types of supernovas that have observationally different luminosity vs. time curves also have fundamentally different origins.

A Type-I supernova, also known as a carbon-detonation supernova, is produced from a white dwarf star that has gone through a number of nova events after accreting hydrogen and helium from a companion main sequence star and then burning the hydrogen to form helium. In some white dwarf binary star systems, each nova does not expel all of the accreted helium from around the core of the white dwarf, and eventually this builds up to a critical mass, at which

point the star becomes unstable because the gravitational force exceeds the electron degeneracy force in the core, causing it to collapse and explode for a final time in a supernova. The mass at which a white dwarf binary system becomes unstable has been calculated to be about 1.4 solar masses, by Indian astronomer (and Nobel Laureate) Subramanyan Chandrasekhar, and is known as the Chandrasekhar mass. These supernovas are hydrogen-poor because there is very little hydrogen in the system when it explodes.

During the collapse of such a white dwarf to form a Type-I supernova, the star heats up to the point at which carbon begins to fuse into heavier elements almost everywhere throughout the star at the same time, causing the star to explode in a massive carbon-detonation explosion that is comparable in violence to the Type-II supernova formed by the implosion of very massive stars. Type-I or carbon-detonation supernovas may also be caused by two white dwarf stars in a binary system that collide to form a massive unstable star that explodes in a supernova.

SUPERNOVA REMNANTS

Supernovas are observed only about every hundred years from Earth, but many supernova remnants are still observable long after their peak of luminosity and radiance. The most famous of these is the Crab Nebula, now a dim nebula sitting about 5,940 light-years (1,800 parsecs) from the Earth and having a visible angular diameter about one-fifth that of the Moon. The Crab Nebula is so interesting because in 1054 its initial explosion was recorded by Chinese, Native American, and Middle Eastern astronomers, who recorded its brightness to be greater than Venus and rivaling the Moon. The explosion was so bright that it was visible in broad daylight for about one month, and the material is still moving outward from the central region at a couple of thousand miles per second (several thousand km/sec). Another historically famous supernova is Tycho's supernova, named after Danish nobleman and astronomer Tycho Brahe (1546–1601). It caused a sensation throughout the world during the Renaissance, causing many people to abandon existing ideas that the universe was constant and nonchanging.

SUPERNOVAS AND THE FORMATION OF THE HEAVY ELEMENTS

Supernovas are fundamentally important for life and the state of the universe, since nearly all of the elements heavier than carbon are formed in massive stars, and the elements heavier than bismuth 209 are all formed in supernova explosions. Only the elements hydrogen and helium are primordial in the universe, meaning that they have been in existence since the earliest moments of the universe. All of the other elements have been produced by nucleosynthesis, or the combination of large atomic nuclei from smaller ones, in stars and in more energetic events such as supernovas.

Heavy elements are created by successive nuclear fusion reactions, beginning with the fusion of two hydrogen atoms to form helium; then helium can fuse to carbon in some star cores. The temperature in very massive stars can be high enough to fuse carbon into magnesium, but it is very rare to synthesize any elements that require the fusion of two nuclei larger than carbon because the nuclear forces between the protons become prohibitively large with larger atomic nuclei. Production of heavier elements typically happens by a different process—the capture of a helium atom by a larger atomic nucleus—to produce heavier elements. In this way, a carbon 12 nucleus can collide with a helium 4 nucleus to produce oxygen 16, and oxygen 16 can collide with helium 4 to produce neon 20. The process of helium capture is thought to have produced many of the heavier elements in the universe, because a plot of the abundance of the elements shows that elements with nuclear masses of 4 units (helium), 12 units (carbon), 16 units (oxygen),

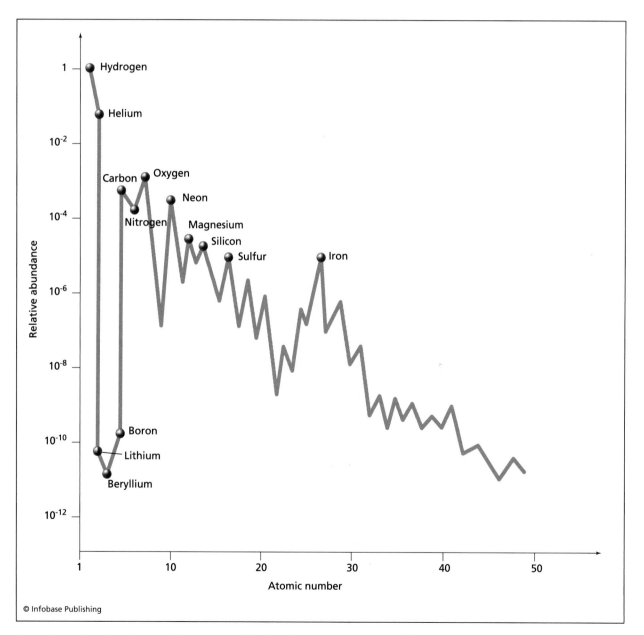

Plot of cosmic abundance of elements and their isotopes expressed relative to the abundance of hydrogen. The horizontal axis shows atomic number. Note how many of the common elements are located on peaks, with other elements being tens to hundreds of times less abundant. The elements on the peaks (e.g., iron) were produced in stellar nucleosynthesis.

20 units (neon), 24 units (magnesium), and 28 units (silicon) stand out as prominent peaks. The process continues in large-mass stars with sulfur 32, argon 36, calcium 40, titanium 44, chromium 48, iron 52, and nickel 56. However, nickel 56 is unstable and quickly decays into cobalt 56, then iron 56, which is a very stable nucleus (the most stable of all nuclei), consisting of 26 protons and 30 neutrons. This process therefore inevitably leads to the buildup of stable iron in the core of massive stars. Many other nuclear reactions occur in large evolved stars, but the plot of the relative abundance of the elements shows that helium capture is one of the most important in synthesizing the heavier elements.

A different process is needed to make elements heavier than iron. This process, called the slow, or s-process by astronomers, involves the capture and absorption of neutrons by other nuclei. Neutron capture occurs in the cores of large evolved stars, where the nuclei of iron atoms capture some of the neutrons produced as by-products of the nuclear reactions going on in the core. The process of adding neutrons to an atomic nucleus changes the isotope of that element to a heavier isotope, but it is still the same element. At some point, however, there are so many neutrons that the isotope decays radioactively to produce a nucleus of a new element. For instance, iron 56 will add neutrons and become iron 59, which will decay to cobalt 59. Cobalt 59 will then add neutrons to become cobalt 60, which decays to nickel 60, and the process goes on and on, producing successively heavier elements. It typically takes an atomic nucleus about a year to capture a neutron by this process so each unstable nucleus decays to the more stable form before the next neutron is added. The s-process is responsible for the synthesis of most heavy elements on Earth and in the solar system and universe, including the atoms in common things such as gold in jewelry, lead in batteries, and nearly all of the other heavy metals and elements.

Elements heavier than bismuth 209 can not be produced by the s-process since any nuclei heavier than bismuth 209 produced by neutron capture are unstable and immediately decay back to bismuth 209. Another mechanism, called the rapid or r-process is the only one known that can synthesize the heaviest elements such as thorium 232 and uranium 238, and this process occurs only in supernova explosions.

The violence of the first 15 minutes of a supernova explosion creates huge numbers of free neutrons so that the neutron capture rate of nuclei is so high that even unstable nuclei capture new neutrons before they can decay to more stable forms. The rapid bombardment of these nuclei with many neutrons in the first 15 minutes of the supernova creates all of the elements heavier than bismuth 209, explaining why these elements are so rare in the universe. This explains why the abundance of the heaviest elements (heavier than iron) is a billion times lower than the abundance of hydrogen and helium.

The early or primordial universe contained only hydrogen and helium, and all of the heavier elements were created in nucleosynthesis reactions inside stars or in supernova explosions. This model is supported by the observation that older globular clusters have more hydrogen and helium in them, and younger clusters are enriched in the heavier elements, having concentrated the remnants of novas and supernovas over time. Stars form when interstellar clouds are compressed by shock waves; then the stars evolve. Solar-sized stars evolve along the main sequence and end up as white dwarfs, and more massive stars end their lives in spectacular supernova explosions. Both processes spew heavy elements into interstellar space, where they may be captured in new interstellar clouds, and compressed into new stars by shock waves from supernova and other events.

See also ASTRONOMY; ASTROPHYSICS; CONSTELLATION; DWARFS (STARS); NOVA; ORIGIN AND EVOLUTION OF THE UNIVERSE.

FURTHER READING

Chaisson, Eric, and Steve McMillan. *Astronomy Today.* 6th ed. Upper Saddle River, N.J.: Addison-Wesley, 2007.

Comins, Neil F. *Discovering the Universe.* 8th ed. New York: W. H. Freeman, 2008.

Dibon-Smith, Richard. The Constellations Web Page. Available online. URL: http://www.dibonsmith.com/index.htm. Last update November 8, 2007.

Prialnik, Dina. "Novae." In *Encyclopedia of Astronomy and Astrophysics,* edited by Paul Murdin, 1,846–1,856. London: Institute of Physics Publishing Ltd and Nature Publishing Group, 2001.

Snow, Theodore P. *Essentials of the Dynamic Universe: An Introduction to Astronomy.* 4th ed. St. Paul, Minn.: West Publishing Company, 1991.

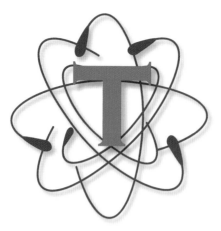

telescopes The word telescope comes from the Greek *tele* (far) and *skopein* (to look or see), meaning far-seeing. The Greek mathematician Giovanni Demisiani coined the word in 1611 for a refracting instrument designed by Galileo Galilei, who modified an instrument built a few years earlier in 1608 in the Netherlands by two spectacle makers, Hans Lippershey and Zacharias Janssen. In 1616 the Italian Jesuit astronomer and physicist Niccolo Zucchi invented the first reflecting telescope, which Isaac Newton improved in 1668. Now the term *telescope* is used to describe a wide range of scientific instruments that observe remote objects by collecting electromagnetic radiation from them and enhancing this radiation by different processes in different types of telescopes. In the 20th century a wide range of types of telescopes were designed and constructed to collect and enhance radiation from a wide variety of wavelengths in the spectrum.

Many different types of telescopes exist, with the most common being optical telescopes. Optical telescopes are widely used in astronomy, and similar technology is also used in many other practical instruments such as in spotting telescopes, binoculars and monoculars, camera lenses, and theodolites for surveying instruments. Optical telescopes collect and focus light from the visible part of the electromagnetic spectrum, whereas other types of telescopes work in the infrared and ultraviolet wavelengths. These telescopes increase the angular size and apparent brightness of distant objects by using a series of curved optical elements (lenses or mirrors) to gather the light and focus it at a focal point where it is enhanced from the original strength. The different types of optical telescopes include the following:

- refracting telescopes, which use lenses to enhance the light and form an image
- reflecting telescopes, which use mirrors to form the image
- catadioptric telescopes, which use a combination of lenses and mirrors to form the image

Radio telescopes collect electromagnetic radiation from distant objects using directional radio antennas with a parabolic shape, and these are often arranged in groups. They are designed using a conductive wire mesh with openings that are smaller than the wavelength being observed. When these large antennae are arranged in groups they can collect data with a wavelength that is similar in size to the separation between the antenna dishes. One such array of radio telescopes is the Very Large Array located in Socorro, New Mexico. The individual telescopes in this array can be moved so that they have different separations; in this way they can be used to collect data from a wide variety of wavelengths. This process is known as aperture synthesis. Distant radio telescopes can be linked in this process to study very long wavelengths, a process known as Very Long Baseline Interferometry (VLBI). The largest array size exceeds the diameter of the Earth. The VLBI Space Observation Program satellite uses a space-based system established by Japan in 2005. Radio telescopes can also be used to collect and study microwave radiation, such as signals from distant and faint quasars.

X-ray and gamma-ray telescopes collect radiation of these wavelengths that can pass through most metal and glass. Since the Earth's atmosphere is opaque to X-rays and gamma-rays, these telescopes must be based in space or from high-flying balloons. Most of these telescopes use a system of ring-shaped glancing mirrors that reflect the rays only a few degrees and do not completely focus the radiation. Instead the signal is interpreted using a system called coded aperture masks, where the patterns of shadows on the altered images can be interpreted to form an image.

See also ASTRONOMY; BRAHE, TYCHO; COSMIC MICROWAVE BACKGROUND RADIATION; ELECTRO-MAGNETIC SPECTRUM; GALILEI, GALILEO; KEPLER, JOHANNES; QUASAR; RADIO GALAXIES; REMOTE SENSING.

FURTHER READING

Chaisson, Eric, and Steve McMillan. *Astronomy Today.* 6th ed. Upper Saddle River, N.J.: Addison-Wesley, 2007.

Comins, Neil F., *Discovering the Universe.* 8th ed. New York: W. H. Freeman, 2008.

Hewitt, Adelaide, ed. *Optical and Infrared Telescopes for the 1990s.* Proceedings. Tucson: Kitt Peak National Observatory, 1980.

Snow, Theodore P. *Essentials of the Dynamic Universe: An Introduction to Astronomy.* 4th ed. St. Paul, Minn.: West Publishing Company, 1991.

Tertiary The Tertiary is first period of the Cenozoic era, extending from the end of the Cretaceous of the Mesozoic at 66 million years ago until the beginning of the Quaternary 1.6 million years ago. The Tertiary is divided into two periods, the older Paleogene (66–23.8 Ma) and the younger Neogene (23.8–1.8 Ma), and further divided into five epochs, the Paleocene (66–54.8 Ma), Eocene (54.8–33.7 Ma), Oligocene (38.7–23.8 Ma), Miocene (23.8–5.3 Ma), and Pliocene (5.3–1.6 Ma). The term *Tertiary* was first coined by the Italian geologist Giovanni Arduino in 1758 and later adopted by Charles Lyell in 1833 for his post-Mesozoic sequences in western Europe. The term *Tertiary* is being gradually replaced by the terms *Paleogene* and *Neogene* periods.

The Tertiary is informally known as the "age of mammals" for its remarkable diverse group of mammals, including marsupial and placental forms that appeared abruptly after the extinction of the dinosaurs. The mammals radiated rapidly in the Tertiary while climates and seawater became cooler. The continents moved close to their present positions by the end of the Tertiary, with major events including the uplift of the Himalayan-Alpine mountain chain.

Pangaea continued to break apart through the early Tertiary, while the African and Indian plates began colliding with Eurasia, forming the Alpine-

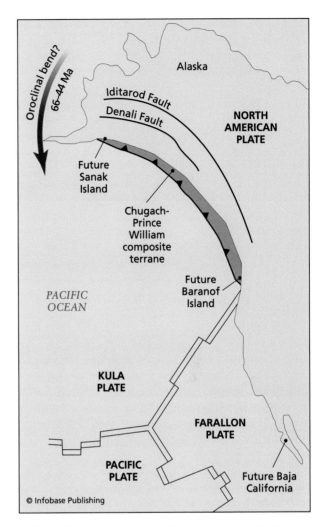

Schematic plate reconstruction of western North America and the NE Pacific Ocean for the Tertiary, showing the Kula-Farallon spreading center interacting with the convergent margin of North America and meeting the Kula-Pacific and Farallon-Pacific spreading centers far offshore in the paleo–Pacific Ocean

Himalayan mountain chain. Parts of the Cordilleran mountain chain experienced considerable amounts of strike-slip translation of accreted terranes, with some models suggesting thousands of kilometers of displacement of individual terranes. The Cordillera of western North America experienced an unusual geologic event with the subduction of at least one oceanic spreading ridge beneath the convergent margin. The boundaries between three plates moved rapidly along the convergent margin from about 60 million years ago in the north to about 35 million years ago in the south, initiating a series of geological consequences including anomalous magmatism, metamorphism, and deformation. New subduction zones were initiated in the southwest Pacific (Southeast Asia) and in the Scotia arc in the south Atlantic. The Hawaiian-Emperor sea mount chain formed as a hot-spot track, with the oldest preserved record starting about 70 million years ago and a major change in the direction of motion of the Pacific plate recorded by a bend in the track near Midway Island formed 43 million years ago.

The San Andreas fault system was initiated about 30 million years ago as the East Pacific rise was subducted beneath western North America, and the relative motions between the Pacific plate and the North American plate became parallel to the margin. Around 3.5 million years ago the Panama arc grew, connecting North and South America, dramatically changing the circulation patterns of the world's oceans and influencing global climate. The East African rift system began opening about 5–2 million years ago, forming the sheltered environments that hosted the first known Homo sapiens.

Climate records show a general cooling of ocean waters and the atmosphere from the earliest Tertiary through the Paleocene, with warming then cooling in the Eocene. The oceans apparently became stratified with cold bottom waters and warmer surface waters in the Eocene, with further cooling reflecting southern glaciations in the Oligocene. Late Oligocene through Early Miocene records indicate a period of warming, followed by additional cooling in the mid-Miocene with the expansion of the Antarctic ice sheet that continued through the end of the Miocene. Pliocene climates began fluctuating wildly from warm to cold, perhaps as a precursor to the Pleistocene ice ages and interglacial periods. The Late Pliocene climates and change into the Pleistocene ice ages were strongly influenced by the growth of the Panama arc and the closing of the ocean circulation routes between the Pacific and Atlantic oceans. The Panama isthmus blocked warm Caribbean waters from moving west into the Pacific Ocean, but forced these waters into the Gulf Stream that brings warm water northward into the Arctic Ocean basin. Warm waters here cause increased evaporation and precipitation, leading to rapid growth of the northern glaciers.

Nearly all of the mammals present on the Earth today appeared in the Cenozoic, and most in the Tertiary, with the exception of a primitive group known as the pantotheres, which arose in the Middle Cretaceous. The pantotheres evolved into the first marsupial, the opossum, which in turn branched into the first placental mammals that spread over much of the northern continents, India, and Africa by the Late Cretaceous. Pantotheres and earlier mammals laid eggs, whereas marsupial offspring emerge from an eggshell-like structure in the uterus early in their development but then develop further in an external pouch. In contrast, placental mammals evolve more fully inside the uterus and emerge stronger with a higher likelihood of surviving infancy. It is believed that this evolutionary advantage led to the dominance of placental mammals and the extinction of the pantotheres.

Mammalian evolution in the Tertiary was strongly influenced by continental distributions. Some continents like Africa, Madagascar, India, and Australia were largely isolated. Connections or landbridges between some of these and other continents, such as the Bering landbridge between Alaska and Siberia, allowed communication of taxa between continents. With the land distribution patterns, certain families and orders evolved on one continent and others on other continents. Rhinoceroses, pigs, cattle, sheep, antelope, deer, cats, and related families evolved primarily in Asia, whereas horses, dogs, and camels evolved chiefly in North America, with some families reaching Europe. Horses have been used as a model of evolution with progressive changes in the size of the animals, as well as the complexity of their teeth and feet.

Marine faunas included gastropods, echinoids, and pelecypods along with bryozoans, mollusks, and sand dollars in shallow water. Coiled nautiloids floated in open waters, whereas sea mammals including whales, sea cows, seals, and sea lions inhabited coastal waters. The Eocene-Oligocene boundary is marked by minor extinctions, and the end of the Pliocene saw major marine extinctions caused by changes in oceanic circulation with massive amounts of cold waters pouring in from the Arctic and from meltwater from growing glaciers.

See also CENOZOIC; HISTORICAL GEOLOGY; NEOGENE; PLATE TECTONICS.

FURTHER READING

Bradley, Dwight C., Timothy M. Kusky, Peter Haeussler, David C. Rowley, Richard Goldfarb, and Steve Nelson. "Geologic Signature of Early Ridge Subduction in the Accretionary Wedge, Forearc Basin, and Magmatic Arc of South-Central Alaska." In *Geology of a Transpressional Orogen Developed During a Ridge-Trench Interaction along the North Pacific Margin,* Special Paper, edited by Virginia B. Sisson, Sarah Roeske, and Terry L. Pavlis. Denver: Geological Society of America, 2003.

Pomeral, C. *The Cenozoic Era.* New York: John Wiley & Sons, 1982.

Savage, R. J. G., and M. R. Long. *Mammal Evolution: An Illustrated Guide.* New York: Facts On File, 1986.

thermodynamics Thermodynamics is the study of the transformation of heat into and from other forms of energy, particularly mechanical, chemical, and electrical energy. The science is concerned with energy conversions into heat and the relations of this conversion to variables including pressure, temperature, and volume. The name comes from the Greek *therme,* meaning heat, and *dynamis,* meaning power. Thermodynamics forms the basis of many principles of chemistry, physics, and earth sciences. The core of the science is based on statistical predictions of the collective motion of atoms and molecules based on their microscopic behavior. In this sense *heat* means energy in transit, and *dynamics* refers to movement, so thermodynamics can also be thought of as the study of the movement of energy. To study the movement of heat and energy between different objects, it is important to define systems and surroundings. For thermodynamics a system is defined as a group of particles whose average motion defines its properties, which are related to each other by equations of state (thermodynamic equations that describe the state of matter under a given set of physical conditions such as temperature, pressure, volume, or internal energy). Thermodynamics uses these equations to describe how systems respond to changes in their surroundings.

The study of thermodynamics rose from the study of steam engines and efforts to find ways to make them more efficient. The first law of thermodynamics states that energy can be neither created nor destroyed and that heat and mechanical work are mutually convertible. This is why moving engines get hot: The mechanical energy is transformed into heat energy. The second law of thermodynamics states that it is impossible for an unaided self-acting machine to transfer heat from a low-temperature body to a higher-temperature body. As an example, an ice cube can not make a cup of coffee warmer. Fundamental to the second law of thermodynamics is the quantity entropy (abbreviated as *S*), which is a measure of the unavailability of a system's energy to do work and is basically a measure of the randomness of the molecules in the system. The third law of thermodynamics states that it is impossible to reduce any system to absolute zero temperature (0°K, -273°C, or -459°F).

Energy is the capacity to do work, and it can exist in many different forms. Potential energy is energy of position, such as when an elevated body exhibits gravitational potential in that it can move to a lower elevation under the influence of gravity. Kinetic energy is the energy of motion and can be measured as the mean speed of the constituent molecules of a body. Einstein's theory of relativity showed that mass too can be converted to energy, as

$$E=mc^2$$

where *E* = energy, *m* = mass, and *c* = the speed of light. This remarkable relationship forms the basis of atomic power, and many mysteries of the universe.

Heat is a form of kinetic energy that manifests itself as motion of the constituent atoms of a substance. According to the laws of thermodynamics, heat may be transferred only from high-temperature bodies to lower-temperature bodies, and it does so by convection, conduction, or radiation. The specific heat of a substance is the ratio of the quantity of heat required to raise the temperature of a unit mass of the substance through a given range of temperature to the heat required to raise the temperature of an equal mass of water through the same range.

Conduction is the flow of heat through a material without the movement of any part of the material. The heat is transferred as kinetic energy of the vibrating molecules, which is passed from one molecule or atom to another. Convection is the transfer of heat through a fluid (liquid, gas, or slow-moving solid such as the Earth's mantle) by moving currents. Radiation is a heat transfer by infrared rays. All materials radiate heat, but hotter objects emit more heat energy than cold objects. Infrared radiation can pass through a vacuum and operates at the speed of light. Radiative heat can be reflected and refracted across boundaries, but it does not affect the medium through which it passes.

See also ATMOSPHERE; BLACK HOLES; CLOUDS; CONVECTION AND THE EARTH'S MANTLE; ENERGY IN THE EARTH SYSTEM; GEOCHEMISTRY; GEOPHYSICS; GRANITE, GRANITE BATHOLITH; HOT SPOT; MANTLE PLUMES; RADIOACTIVE DECAY; THUNDERSTORMS, TORNADOES.

FURTHER READING

Cengel, Yunus A., and Michael A. Boles. *Thermodynamics: An Engineering Approach.* New York: McGraw-Hill, 2005.

Dunning-Davies, Jeremy. *Concise Thermodynamics: Principles and Applications.* Chichester, U.K.: Horwood Publishing, 1997.

Van Ness, H. C. *Understanding Thermodynamics.* New York: Dover Publications, 1969.

thermohaline circulation Thermohaline circulation refers to the vertical mixing of seawater driven by density differences caused by variations in temperature and salinity. Variations in formation and circulation of ocean water driven by thermohaline circulation may cause some of the thousands-of-years to decadal scale variations in climate. Cold water forms in the Arctic and Weddell Seas. This cold salty water is denser than other water in the ocean, so it sinks to the bottom and gets ponded behind seafloor topographic ridges, periodically spilling over into other parts of the oceans. The formation and redistribution of North Atlantic cold bottom water accounts for about 30 percent of the solar energy budget input to the Arctic Ocean every year. Eventually, this cold bottom water works its way to the Indian and Pacific Oceans where it upwells, gets heated, and returns to the North Atlantic. Variations in temperature and salinity that drive thermohaline circulation are found in waters that occupy different ocean basins and in those found at different levels in the water column. When the density of water at one level is greater than or equal to that below that level, the water column becomes unstable and the denser water sinks, displacing the deeper, less-dense waters below. When the dense water reaches the level at which it is stable it spreads out laterally and forms a thin sheet, forming intricately stratified ocean waters. Thermohaline circulation is the main mechanism responsible for the movement of water out of cold polar regions, and it exerts a strong influence on global climate. The upward movement of water in other regions balances the sinking of dense cold water, and these upwelling regions typically bring deep water, rich in nutrients, to the surface. Thus, regions of intense biological activity are often associated with upwelling regions.

The coldest water on the planet is formed in the polar regions, with large quantities of cold water originating off the coast of Greenland, and in the Weddell Sea of Antarctica. The planet's saltiest ocean water is found in the Atlantic Ocean, and this is moved northward by the Gulf Stream. As this water moves near Greenland it is cooled and then sinks to flow as a deep cold current along the bottom of the western North Atlantic. The cold water of the Weddell Sea is the densest on the planet, where surface waters are cooled to -35.4°F (-1.9°C), then sink to form a cold current that moves around Antarctica. Some of this deep cold water moves northward into all three major ocean basins, mixing with other waters and warming slightly. Most of these deep ocean currents move at a few to 10 centimeters per second.

Presently, the age of bottom water in the equatorial Pacific is 1,600 years, and in the Atlantic it is 350 years. Glacial stages in the North Atlantic have been correlated with the presence of older cold bottom waters, approximately twice the age of the water today. This suggests that the thermohaline

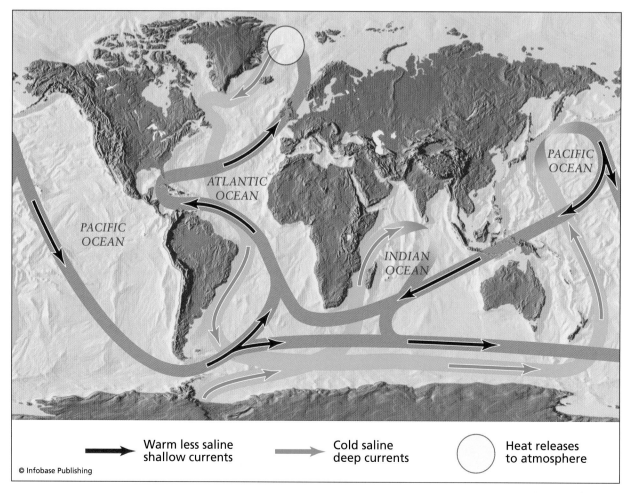

PACIFIC OCEAN

ATLANTIC OCEAN

PACIFIC OCEAN

INDIAN OCEAN

PACIFIC OCEAN

→ Warm less saline shallow currents
→ Cold saline deep currents
○ Heat releases to atmosphere

© Infobase Publishing

Map of the world's oceans showing main warm and cold currents driven by thermohaline circulation

circulation system was only half as effective at recycling water during recent glacial stages, with less cold bottom water being produced during the glacial periods. These changes in production of cold bottom water may in turn be driven by changes in the North American ice sheet, perhaps itself driven by 23,000-year orbital (Milankovitch) cycles. Scientists suggest that a growth in the ice sheet would cause the polar front to shift southward, decreasing the inflow of cold saline surface water into the system required for efficient thermohaline circulation. Several periods of glaciation in the past 14,500 years (known as the Dryas) are thought to have been caused by sudden, even catastrophic injections of glacial meltwater into the North Atlantic, which would decrease the salinity and hence density of the surface water. This in turn would prohibit the surface water from sinking to the deep ocean, inducing another glacial interval.

Shorter-term decadal variations in climate in the past million years are indicated by so-called Heinrich events, defined as specific intervals in the sedimentary record showing ice-rafted debris in the North Atlantic. These periods of exceptionally large iceberg discharges reflect decadal-scale sea-surface and atmospheric cooling and are related to thickening of the North American ice sheet followed by ice stream surges associated with the discharge of the icebergs. These events flood the surface waters with low-salinity freshwater, leading to a decrease in flux to the cold bottom waters, and hence a short-period global cooling.

Changes in the thermohaline circulation rigor have also been related to other global climate changes. Droughts in the Sahel and elsewhere are correlated with periods of ineffective or reduced thermohaline circulation, because this reduces the amount of water drawn into the North Atlantic, in turn cooling surface waters and reducing the amount of evaporation. Reduced thermohaline circulation also reduces the amount of water that upwells in the equatorial regions, in turn decreasing the amount of moisture transferred to the atmosphere, reducing precipitation at high latitudes.

Atmospheric levels of greenhouse gases such as carbon dioxide (CO_2) and atmospheric temperatures show a correlation to variations in the thermohaline circulation patterns and production of cold bottom waters. CO_2 is dissolved in warm surface water and transported to cold surface water, which acts as a sink for the CO_2. During times of decreased flow from cold, high-latitude surface water to the deep ocean reservoir, CO_2 can build up in the cold polar waters, removing it from the atmosphere and decreasing global temperatures. In contrast, when the thermohaline circulation is vigorous, cold oxygen-rich surface waters downwell, and dissolve buried CO_2 and even carbonates, releasing this CO_2 to the atmosphere and increasing global temperatures.

The present-day ice sheet in Antarctica grew in the Middle Miocene, related to active thermohaline circulation that caused prolific upwelling of warm water that put more moisture in the atmosphere, falling as snow on the cold southern continent. The growth of the southern ice sheet increased the global atmospheric temperature gradients, which in turn increased the desertification of mid-latitude continental regions. The increased temperature gradient also induced stronger oceanic circulation, including upwelling and removal of CO_2 from the atmosphere, lowering global temperatures, and bringing on late Neogene glaciations.

Ocean bottom topography exerts a strong influence on dense bottom currents. Ridges deflect currents from one part of a basin to another and may restrict access to other regions, whereas trenches and deeps may focus flow from one region to another.

See also CLIMATE; CLIMATE CHANGE; OCEAN BASIN; OCEAN CURRENTS.

FURTHER READING

Ashworth, William, and Charles E. Little. *Encyclopedia of Environmental Studies.* New ed. New York: Facts On File, 2001.

Botkin, D., and E. Keller. *Environmental Science.* Hoboken, N.J.: John Wiley & Sons, 2003.

Intergovernmental Panel on Climate Change. Available online. URL: http://www.ipcc.ch/index.htm. Accessed January 30, 2008.

Intergovernmental Panel on Climate Change. *Climate Change 2007: The Physical Science Basis. Contributions of Working Group I to the Fourth Assessment Report of the Intergovernmental Panel on Climate Change,* edited by S. Solomon, D. Qin, M. Manning, Z. Chen, M. Marquis, K. B. Averyt, M. Tignor, and H. L. Miller, 996. Cambridge, U.K.: Cambridge University Press, 2007.

Windows to the Universe, Thermohaline Ocean Circulation home page. University Corporation for Atmospheric Research. Available online. URL: http://www.windows.ucar.edu/tour/link=/earth/Water/thermohaline_ocean_circulation.html. Accessed October 10, 2008.

thunderstorms, tornadoes Any storm that contains lightning and thunder may be called a thunderstorm. However, the term normally implies a gusty heavy rainfall event with numerous lightning strikes and thunder, emanating from a cumulonimbus cloud or cluster or line of cumulonimbus clouds. There is a large range in the severity of thunderstorms from minor to severe, with some causing extreme damage through high winds, lightning, tornadoes, and flooding rains.

Thunderstorms are convective systems that form in unstable rising warm and humid air currents. The air may start rising as part of a converging air system, along a frontal system, as a result of surface topography, or from unequal surface heating. The warmer the rising air is than the surrounding air, the greater the buoyancy forces acting on the rising air. Scattered thunderstorms that typically form in summer months are referred to as ordinary thunderstorms, and these typically are short-lived, produce only minor to moderate rainfall, and do not have severe winds. However, severe thunderstorms associated with fronts or combinations of unstable conditions may have heavy rain, hail, strong winds or tornadoes, and drenching or flooding rains.

Ordinary thunderstorms are most likely to form in regions where surface winds converge, causing parcels of air to rise, and where there is not significant wind shear or change in the wind speed and direction with height. These storms evolve through several stages, beginning with the cumulus or growth stage, where the warm air rises and condenses into cumulus clouds. As the water vapor condenses it releases a large amount of latent heat that keeps the cloud warmer than the air surrounding it and causes it to continue to rise and build as long as it is fed from air below. Simple cumulus clouds may quickly grow into towering cumulus congestus clouds in this way. As the cloud builds above the freezing level in the atmosphere, the particles in the cloud get larger and heavier and eventually are too large to be kept entrained in the air currents, and they fall as precipitation. As this precipitation is falling, drier air from around the storm is drawn into the cloud, but as the rain falls through this dry air it may evaporate, cooling the air. This cool air is then denser than the surrounding air and it may fall as a sudden downdraft, in some cases enhanced by air pulled downward by the falling rain.

The development of downdrafts marks the passage of the thunderstorm into the mature stage, in which the upward and downward movement of air constitutes a convective cell. In this stage the top of the storm typically bulges outward in stable levels of the stratosphere, often around 40,000 feet (12,192 m), forming the anvil shape characteristic of mature thunderstorms. Heavy rain, hail, lightning, and strong, turbulent winds may come out of the base of the storms, which can be several miles in diameter. Cold downwelling air often expands out of the cloud base, forming a gust front along its leading edge, forcing warm air up into the storm. Most mature storm cells begin to dissipate after half an hour or so, as the gust front expands away from the storm and can no longer enhance the updrafts that feed the storm. These storms may quickly turn into gentle rains, and then evaporate, but the moisture may be quickly incorporated into new, actively forming thunderstorm cells.

Severe thunderstorms are more intense than ordinary storms, producing large hail, wind gusts of greater than 50 knots (57.5 m/hr, or 92.5 km/hr), more lightning, and heavy rain. Like ordinary thunderstorms, severe storms form in areas of upwelling unstable moist warm air, but severe storms tend to develop in regions where there is also strong wind shear. The high level winds have the effect of causing the rain that falls out of the storm to fall away from the region of upwelling air so that it does not have the effect of weakening the upwelling. In this way the cell becomes much longer lived and grows stronger and taller than ordinary thunderstorms, often reaching heights of 60,000 feet (18,288 m). Hail may be entrained for long times in the strong air currents and even thrown out of the cloud system at height, falling several kilometers from the base of the cloud. Downdrafts from severe storms are marked by bulbous mammatus clouds.

Supercell thunderstorms form where strong wind shear aloft is such that the cold downwelling air does not cut off the upwelling air, and a giant rotating storm with balanced updrafts and downdrafts may be maintained for hours. These storms may produce severe tornadoes, strong downbursts of wind, large (grapefruit-sized) hail, very heavy rains, and strong winds exceeding 90 knots (103.5 m/hr, or 167 km/hr).

Unusual winds are associated with some thunderstorms, especially severe storms. Gust fronts may be quite strong with winds exceeding 60 miles per hour (97 km/hr), followed by cold gusty and shifty winds. Gust fronts may be marked by lines of dust kicked up by the strong winds, or ominous-looking shelf clouds formed by warm moist air rising above the cold descending air of the gust front. In severe cases, gust fronts may force so much air upward that they generate new multi-celled thunderstorms with their own gust fronts that merge, forming an intense gust front called an outflow boundary. Intense downdrafts beneath some thunderstorms spread laterally outward at speeds sometimes exceeding 90 miles per hour (145 km/hr) when they hit the ground and are termed downbursts, microbursts, or macrobursts depending on their size. Some clusters of thunderstorms produce another type of unusual wind called a straight line wind, or derecho. These winds may exceed 90 miles per hour (145 km/hr), and extend for tens or even hundreds of miles.

Thunderstorms often form either in groups called mesoscale convective systems or as lines of storms called squall lines. Squall lines typically form along

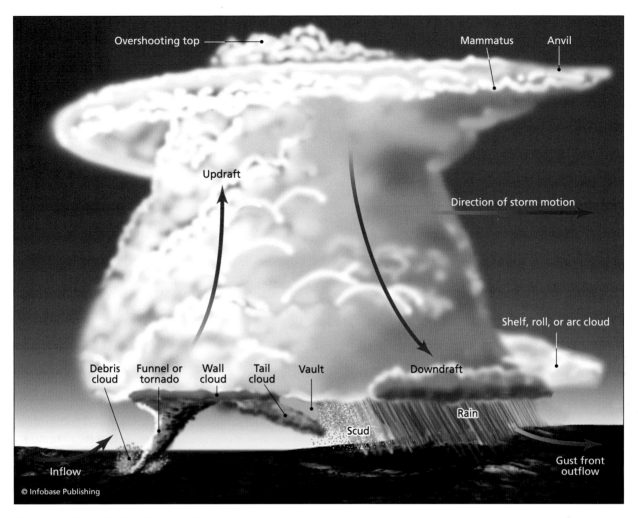

Cross section of typical thunderstorm

or within a zone up to a couple of hundred miles in front of the cold front where warm air is compressed and forced upward. Squall lines may form lines of thunderstorms hundreds or even a thousand miles long, and many of the storms along the line may be severe with associated heavy rain, winds, hail, and tornadoes. Mesoscale convective complexes form when many individual thunderstorm cells across a region start to act together, forming an exceedingly large convective system that may cover more than 50,000 square miles (130,000 km²). These systems move slowly and may be associated with many hours of flooding rains, hail, tornadoes, and wind.

Cumulonimbus clouds typically become electrically charged during the development of thunderstorms, although the processes that lead to the unequal charge distribution are not well known. About 20 percent of the lightning generated in thunderstorms strikes the ground, with most passing from cloud to cloud. Lightning is an electrical discharge that heats the surrounding air to 54,000°F (30,000°C), causing the air to expand explosively,

creating the sound waves heard as thunder. As the air expands along different parts of the lightning stroke, the sound is generated from several different places, causing the thunder to have a rolling or echoing sound, enhanced by the sound waves bouncing off hills, buildings, and the ground. Cloud-to-ground lightning forms when negative electrical charges build up in the base of the cloud, causing positive charges to build in the ground. When the electrical potential gradient reaches 3 million volts per meter along several tens of meters, electrons rush to the cloud base and form a series of stepped leaders that reach toward the ground. At this stage, a strong current of positive charge moves up, typically along an elevated object, from the ground to the descending leader. As the two columns meet huge numbers of electrons rush to the ground, and a several-centimeter wide column of positively charged ions shoots up along the lightning stroke, all within one ten-thousandth of a second. The process then may be repeated several or even dozens of times along the same path, all within a fraction of a second.

TORNADOES

Tornadoes are a rapidly circulating column of air with a central zone of intense low pressure that reaches the ground. Most tornadoes extend from the bottom of severe thunderstorms or supercells as funnel-shaped clouds that kick up massive amounts of dust and debris as they rip across the surface. The exact shape of tornadoes is quite variable, from thin rope-like funnels, to classic cylindrical shapes, to powerful and massive columns that have almost the same diameter on the ground as at the base of the cloud. Many tornadoes evolve from an immature upward swirling mass of dust to progressively larger funnels that may shrink and tilt as their strength diminishes. Funnel clouds are essentially tornadoes that have not reached the ground. More tornadoes occur in the United States than anywhere else in the world, with most of these occurring in a region known as "tornado alley," extending from Texas through Oklahoma, Nebraska, Kansas, Iowa, Missouri, and Arkansas.

Most tornadoes rotate counterclockwise (as viewed from above) and have diameters from a few hundred feet (100 m) to a mile or more (2 km). Wind speeds in tornadoes range from about 40 miles per hour to more than 300 miles per hour (65–480 km/hr) and may move forward at a few miles per hour to more than 70 miles per hour (2–115 km/hr). Most tornadoes last for only a few minutes, but some last longer, with some reports of massive storms lasting for hours and leaving trails of destruction hundreds of miles long. Some supercell thunderstorms produce families or outbreaks of tornadoes, with half a dozen or more individual funnel clouds produced over the course of a couple of hours from a single storm.

The strength of winds and potential damage of tornadoes is measured by the Fujita scale, proposed by the tornado expert Dr. Theodore Fujita. The

Tornado crossing road near Manchester, South Dakota, June 24, 2003 *(Mike Berger/Photo Researchers, Inc.)*

scale measures the rotational speed (not the forward speed) and classifies the tornadoes into F0–F5 categories.

Many tornadoes form in a region of the midwestern states known as tornado alley. This most commonly occurs in the springtime when cold air

FUJITA TORNADO SCALE

Scale	Category	Wind Speed	Damage Potential
F0	weak	40–72 mph	Minor; broken tree branches, damaged signs
F1	weak	73–112 mph	Moderate; broken windows, trees snapped
F2	strong	113–157 mph	Considerable; large trees uprooted, mobile homes tipped, weak structures destroyed
F3	strong	158–206 mph	Severe; trees leveled, walls torn from buildings, cars flipped
F4	violent	207–260 mph	Devastating; frame homes destroyed
F5	violent	261–318 mph	Incredible; strong structures damaged, cars thrown hundreds of yards

from the north overruns warm moist air from the Gulf of Mexico. Tornadic supercell thunderstorms form in front of the cold front as the warm moist air is forced upward in front of the cold air in this region. Supercell thunderstorms that have large rotating updrafts also spawn many tornadoes. Spinning roll clouds and vortexes may form as these storms roll across the plains, and if these horizontally spinning clouds are sucked into the storm by an updraft, the circulation may be rotated to form a tornadic condition that may evolve into a tornado. Before the supercell spawns a tornado, rotating clouds may be visible, and then a wall cloud may descend from the rotating vortex. Funnel clouds are often hidden behind the wall cloud, so these types of clouds should be eyed with caution.

Some tornadoes have formed from smaller and even nonsevere thunderstorms, from squall lines, and even smaller cumulus clouds. These types of tornadoes are usually short-lived, and less severe (F0–F1) than the supercell tornadoes. Waterspouts are related phenomena and include tornadoes that have migrated over bodies of water; they may also form in fair weather over warm shallow coastal waters. These weak (F0) funnel clouds form in updrafts, usually when cumulus clouds are beginning to form above the coastal region. Their formation is aided by converging surface air, such as when sea breezes and other systems meet.

See also ATMOSPHERE; CLOUDS; ENERGY IN THE EARTH SYSTEM; HURRICANES; METEOROLOGY; PRECIPITATION; THERMODYNAMICS.

FURTHER READING
Ahrens, C. Donald. *Meteorology Today: An Introduction to Weather, Climate, and the Environment.* 7th ed. Pacific Grove, Calif.: Thomson Brooks/Cole, 2003.

National Weather Service, National Oceanic and Atmospheric Administration, home page. Available online. URL: http://www.nws.noaa.gov/. Last modified September 15, 2008. Data updated continuously.

Schaefer, Vincent, and John Day. *A Field Guide to the Atmosphere: The Peterson Field Guide Series.* Boston: Houghton Mifflin, 1981.

transform plate margin processes Processes that occur where two plates are sliding past each other along a transform plate boundary, either in the oceans or on the continents, are known as transform plate margin processes. Famous examples of transform plate boundaries on land include the San Andreas fault in California, the Dead Sea Transform in the Middle East, the East Anatolian transform in Turkey, and the Alpine fault in New Zealand. Transform boundaries in the oceans are numerous, including the many transform faults that separate segments of the mid-ocean ridge system. Some of the larger transform faults in the oceans include the Romanche in the Atlantic, the Cayman fault zone on the northern edge of the Caribbean plate, and the Eltanin, Galápagos, Pioneer, and Mendocino fault zones in the Pacific Ocean.

Three main types of transform faults are those that connect segments of divergent boundaries (ridge-ridge transforms), offsets in convergent boundaries, and those that transform the motion between convergent and divergent boundaries. Ridge-ridge transforms connect spreading centers and develop with this geometry because it minimizes the ridge segment lengths and therefore minimizes the dynamic resistance to spreading. Ideal transforms have purely strike-slip motions and maintain a constant distance from the pole of rotation for the plate.

Transform segments in subduction boundaries are largely inherited configurations formed in an earlier tectonic regime. In collisional boundaries the inability of either plate to be subducted yields a long-lived boundary instability, often formed to compensate the relative motion of minor plates in complex collisional zones, such as that between Africa and Eurasia.

The evolution of the San Andreas-Fairweather fault system best represents the development of a divergent-convergent boundary. When North America overrode the East Pacific rise, the relative velocity structure was such that a transform resulted, with a migrating triple junction that lengthened the transform boundary.

TRANSFORM BOUNDARIES IN THE CONTINENTS
Transform boundaries on the continents include the San Andreas fault in California, the North Anatolian fault in Turkey, the Alpine fault in New Zealand, and, by some definitions, the Altyn Tagh and Red River faults in Asia. Transform faults in continents show strike-slip offsets during earthquakes and are high angle faults with dips greater than 70°. They never occur as a single fault, but rather as a set of subparallel faults. The faults are typically subparallel because they form along theoretical slip lines (along small circles about the pole of rotation), but the structural grain of the rocks interferes with this prediction. The differences between theoretical and actual fault orientations leads to the formation of segments that have pure strike-slip motions and segments with compressional and extensional components of motion.

Extensional segments of transform boundaries form at left steps in left-slipping (left lateral) faults and at right steps in right-slipping (right lateral) faults. Movement along fault segments with exten-

San Andreas Fault in Carrizo Plain, California, marking the transform plate boundary between North America and Pacific plates *(USGS)*

sion of the extra volume of crust compressed into the bend in the fault. Examples of compressional (or restraining) bends include the Transverse Ranges along the San Andreas fault and Mount McKinley along the Denali fault in Alaska. Many of the faults that form along compressional bends have low-angle dips toward the main strike-slip fault but progressively steeper dips toward the center of the main fault. This forms a distinctive geometry known as a flower or palm tree structure, with a vertical strike-slip fault in the center and branches of mixed thrust/strike-slip faults branching off the main fault.

In a few places along compressional bends, two thrust-faulted mountain ranges may converge, forming a rapidly subsiding basin between the faults. These basins are known as ramp valleys. Many ramp valleys started as pull-apart basins, and became ramp valleys when the fault geometries changed.

A distinctive suite of structures that form in predictable orientations characterizes transform plate margins. Compressional bends form at high angles to the principal compressive stress and at about 30–45° from the main strike-slip zone. These are often associated with flower structures, containing a strike-slip fault at depth, and folds and thrusts near the surface. Dilational bends often initiate with their long axes perpendicular to the compressional bends, but large

sional bends generates gaps where deep basins known as pull-apart basins form. The planet presently has about 60 active pull-apart basins, including locations like the Salton trough along the San Andreas Fault and the Dead Sea along the Dead Sea transform. Pull-apart basins tend to form with an initially sigmoid form, but as movement on the fault continues, the basin becomes very elongate parallel to the bounding faults. In some cases the basin may extend so much that oceanic crust is generated in the center of the pull-apart, such as along the Cayman trough in the Caribbean. Pull-apart basins have stratigraphic and sedimentologic characteristics similar to rifts, including rapid lateral facies variations, basin-marginal fanglomerate and conglomerate deposits, interior lake basins, and local bimodal volcanic rocks. They are typically deformed soon after they form, however, with folds and faults typical of strike-slip regime deformation.

Compressional bends form at right bends in left lateral faults, and left bends in right lateral faults. These areas are characterized by mountain ranges and thrust-faulted terrain that uplift and aid ero-

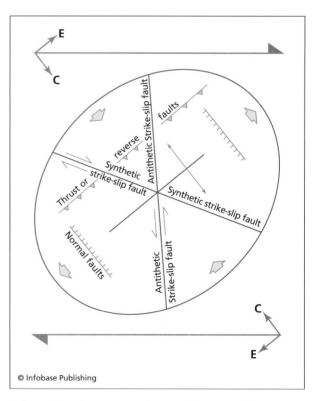

Orientation of structures in transform margins, including strike-slip faults, normal faults, and thrust faults

amounts of extension may lead to the long axis being parallel to the main fault zone. Folds, often arranged in en echelon or a stepped manner, typically form at about 45° from the main fault zone, with the fold axes developed perpendicular to the main compressive stress. The sense of obliquity of many of these structures can be used to infer the sense of shear along the main transform faults.

Strike-slip faults along transform margins often develop from a series of en echelon fractures that initially develop in the rock. As the strain builds up, the fractures are cut by new sets of fractures known as Riedel fractures, in new orientations. Eventually, after several sets of oblique fractures have cut the rock, the main strike-slip fault finds the weakest part of the newly fractured rock to propagate through, forming the main fault.

TRANSFORM BOUNDARIES IN THE OCEANS

Transform plate boundaries in the oceans include the system of ridge-ridge transform faults that are an integral part of the mid-ocean ridge system. Magma upwells along the ridge segments, cools and crystallizes, becoming part of one of the diverging plates. The two plates then slide past each other along the transform fault between the two ridge segments, until the plate on one side of the transform meets the ridge on the other side of the transform. At this point, the transform fault is typically intruded by mid-ocean ridge magma, and the apparent extension of the transform, known as a fracture zone, juxtaposes two segments of the same plate that move together horizontally. Fracture zones are not extensions of the transform faults, and they are no longer considered plate boundaries. After the ridge/transform intersection is passed, the fracture zone juxtaposes two segments of the same plate. There is typically some vertical motion along this segment of the fracture zone, since the two segments of the plate have different ages, and subside at different rates.

The transform and ridge segments preserve an orthogonal relationship in almost all cases, because this geometry creates a least work configuration, creating the shortest length of ridge possible on the spherical Earth.

Transform faults generate very complex geological relationships. They juxtapose rocks from very different crustal and even mantle horizons, show complex structures, exhibit intense alteration by high-temperature metamorphism, and have numerous igneous intrusions. Rock types along oceanic transforms typically include suites of serpentinite, gabbro, pillow lavas, lherzolites, harzburgites, amphibolite-tectonites, and even mafic granulites.

Transform faults record a very complex history of motion between the two oceanic plates. The relative motion includes dip-slip (vertical) motions due to subsidence related to the cooling of the oceanic crust. A component of dip-slip motion occurs all along the transform, except at one critical point, known as the

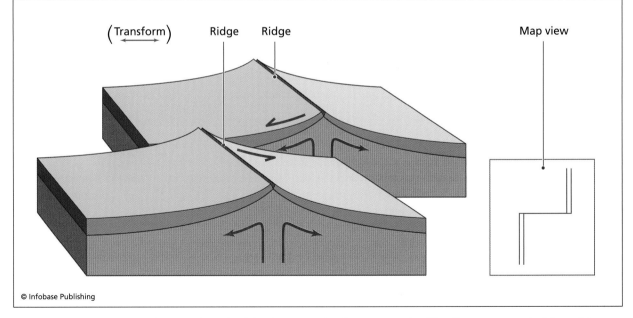

© Infobase Publishing

Three-dimensional view of a transform fault in the ocean basin, apparently offsetting a segment of the mid-ocean ridge. The sense of motion on the transform is opposite the apparent offset. Note that the lateral motion between the two segments of the oceanic crust ceases once the opposite ridge segment is passed. At this point, magmas from the ridge intrude the transform, and the contact becomes an igneous contact.

crossover point, where the transform juxtaposes oceanic lithosphere of the same age formed at the two different ridge segments. This dip-slip motion occurs along with the dominant strike-slip motion, recording the sliding of one plate past the other.

Fracture zones are also called nontransform extension regions. The motion along the fracture zone is purely dip-slip, due to the different ages of the crust with different subsidence rates on either side of the fracture zone. The amount of differential subsidence decreases with increasing distance from the ridge, and the amount of dip-slip motion decreases to near zero after about 60 million years. Subsidence decreases according to the square root of age.

Transform faults in the ocean may juxtapose crust with vastly different ages, thickness, temperature, and elevation. These contrasts often lead to the development of a deep topographic hole on the ridge axis at the intersection of the ridge and transform. The cooling effects of the older plate against the ridge of the opposing plate influences the axial rift topography all along the whole ridge segment, with the highest topographic point on the ridge being halfway between two transform segments. Near transform zones, magma will not reach its level of hydrostatic equilibrium because of the cooling effects of the older cold plate adjacent to it. Therefore, the types and amounts of magma erupted along the ridge are influenced by the location of the transforms.

Transform faults are neither typically vertical planes nor are they always straight lines connecting two ridge segments. The fault planes typically curve toward the younger plate with depth, since they tend to seek the shortest distance through the lithosphere to the region of melt. This is a least energy configuration, and it is easier to slide a plate along a vertically short transform than along an unnecessarily thick fault. This vertical curvature of the fault causes a slight change in the position and orientation of the fault on the surface, causing it to bend toward each ridge segment. These relationships cause the depth of earthquakes to decrease away from the crossover point, due to the different depth of transform fault penetration. Motion on these curved faults also influences the shape and depth of the transform-ridge intersection, enhancing the topographic depression and in many causing the ridge to curve slightly into the direction of the transform. Faults and igneous dikes also curve away from the strike of the ridge, toward the direction of the transform in the intersection regions.

Many of the features of ridge-transform intersections are observable in some ophiolite complexes (on-land fragments of ancient oceanic lithosphere), including the Arakapas transform in Troodos ophiolite in Cyprus and the Coastal complex in the Bay of Islands ophiolite in Newfoundland.

EXAMPLES OF EARTHQUAKE DISASTERS ALONG TRANSFORM PLATE BOUNDARIES

Transform plate margins sometimes also have large earthquakes, typically up to magnitude 8. These events can be quite devastating but are not as big as the magnitude 9+ events that may occasionally strike convergent margins. However, the distinction between a magnitude 9 and a magnitude 8 earthquake will not matter to people who are in collapsing buildings and cities devastated by these massive destructive events. Some plate boundaries have characteristics of both convergent margins and transform margins, such as where a plate is being subducted obliquely and part of the overriding plate moves sideways along the plate margin. These margins, such as southeastern Alaska or northern Sumatra–Andaman Islands, may have both subduction zone earthquakes (along the Benioff or subduction zone) and transform margin earthquakes (with the hypocenter located along the transform fault in the overriding plate). Some of the most famous of all earthquakes have occurred along transform plate boundaries. The following is a set of examples that describes some of the more significant earthquakes to strike transform margins in history.

San Francisco, 1906 (magnitude 7.8)

Perhaps the most infamous earthquake of all time is the magnitude 7.8 temblor that shook San Francisco at 5:12 A.M. on April 18, 1906, virtually destroying the city, crushing 315 people to death and killing 700 people throughout the region. Many of the unreinforced masonry buildings that were common in San Francisco immediately collapsed, but most steel and wooden frame structures remained upright. Ground shaking and destruction were most intense where structures were built on areas filled in with gravel and sand and least intense where the buildings were anchored in solid bedrock. Most of the destruction from the earthquake came not from ground shaking, but from the intense firestorm that followed. Gas lines and water lines were ruptured, and fires started near the waterfront that worked their way into the city. Other fires were inadvertently started by people cooking in residential neighborhoods and by people dynamiting buildings trying to avoid collapses and stop the spread of the huge fire. It has even been reported that some fires were started by individuals in attempts to collect insurance money on their slightly damaged homes. In all, 490 city blocks were burned.

The problems did not cease after the earthquake and fires but continued to worsen as a result of the poor sanitary and health conditions that followed as a consequence of the poor infrastructure. Hundreds of cases of bubonic plague claimed lives, and dysentery and other diseases combined to bring the death total to as high as 5,000.

One of the lessons that could have been learned from this earthquake was not appreciated until many years later—that is, that loose unconsolidated fill tends to shake more than solid bedrock during earthquakes. San Franciscans noted that some of the areas that shook the most and suffered from the most destruction were built on unconsolidated fill. After the 1906 earthquake, much of the rubble was bulldozed into San Francisco Bay, and later construction on this fill became the Marina district, which saw some of the worst damage during the 1989 Loma Prieta earthquake.

Loma Prieta, 1989 (magnitude 7.1)

The San Francisco area and smaller cities to the south, especially Santa Cruz, were hit by a moderate-sized earthquake (magnitude 7.1) at 5:04 P.M. on Tuesday, October 17, 1989, during live broadcast of the World Series baseball game. Sixty-seven people died, 3,757 people were injured, and 12,000 left homeless. Tens of millions of people watched on television as the earthquake struck just before the beginning of game three, and the news coverage that followed was unprecedented in the history of earthquakes.

The earthquake was caused by a rupture along a 26-mile- (42-km-) long segment of the San Andreas fault near Loma Prieta peak in the Santa Cruz Mountains south of San Francisco. The segment of the fault that had ruptured was the southern part of the same segment that ruptured in the 1906 earthquake, but this rupture occurred at greater depths and involved some vertical motion as well as horizontal motion. The actual rupturing lasted only 11 seconds, during which time the western (Pacific) plate slid almost six feet (1.9 m) to the northwest, and parts of the Santa Cruz Mountains were uplifted by up to four feet (1.3 m). The rupture propagated at 1.24 miles per second (2 km/sec) and was a relatively short-duration earthquake for one of this magnitude. Had it been much longer, the damage would have been much more extensive. As it was, the damage totals amounted to more than 6 billion dollars.

The actual fault plane did not rupture the surface, although many cracks appeared and slumps formed along steep slopes. The Loma Prieta earthquake had been predicted by seismologists, because the segment of the fault that slipped had a noticeable paucity of seismic events since the 1906 earthquake and was identified as a seismic gap with a high potential for slipping and causing a significant earthquake. The magnitude 7.1 event and the numerous aftershocks filled in this seismic gap, and the potential for large earthquakes along this segment of the San Andreas fault is now significantly lower. There are, however, other seismic gaps along the San Andreas fault in heavily populated areas that should be monitored closely.

Turkey, 1999 (magnitude 7.8)

On August 17, 1999, a devastating earthquake measuring 7.4 on the Richter scale hit heavily populated areas in northwestern Turkey at 3:02 A.M. local time. The epicenter of the earthquake was near the industrial city of Izmit about 60 miles (100 km) east of Istanbul, near the western segment of the notorious North Anatolian strike-slip fault. The earthquake formed a surface rupture more than 75 miles (120 km) long, along which offsets were measured between four and 15 feet (1.5–5 m). This was the deadliest and most destructive earthquake in the region in more than 60 years, causing more than 30,000 deaths and the largest property losses in Turkey's recorded history. The World Bank estimated that direct losses from the earthquake were approximately 6.5 billion dollars, with perhaps another 20 billion dollars in economic impact from secondary and related losses.

The losses from this moderate-sized earthquake were so high because

- The region in which it occurred is home to approximately 25 percent of Turkey's population.
- The region hosts much of the country's industrial activity.
- In a recent construction boom, building codes were ignored.
- Large numbers of high-rise apartment buildings were constructed with substandard materials including extra-coarse and sandy cement.
- too few reinforcement bars in concrete structures
- general lack of support structures

Haiti, 2010 (magnitude 7.0)

On Tuesday, January 10, 2010, at 4:53 P.M., Port-au-Prince, the capital city of Haiti, was hit by a magnitude 7.0 earthquake that essentially leveled the city, killing an estimated 200,000 people and severely affecting another 3 million. As of January 31, 2010, about 150,000 bodies had been recovered, but of necessity many were dumped in mass graves outside the capital, so exact estimates of the number of people killed may never be known. Cities and towns outside the capital, such as Jacmel, Léogone, and Miragoune, were also severely damaged, with many buildings collapsing. At the time of this writing, the death toll from these outlying areas was not known, and aid was only beginning to reach the places three weeks after the quake. The quake was the largest in the region in the past 200 years and was felt throughout Haiti and the Dominican Republic, southeastern Cuba, eastern Jamaica, parts of Puerto Rico, the Bahamas, and as far as Florida and

Venezuela. Nearly every building in Port-au-Prince was crushed or severely damaged, including the presidential palace and the grand Hotel Montana, where many hundreds of foreign tourists, UN workers, and World Bank employees perished, and the UN headquarters, where more were killed. As of January 30, 2010, there were 54 aftershocks with magnitudes greater than 4.5, many causing additional damage and spreading fear among the population. The U.S. Geological Survey estimates that aftershocks will continue for months to years but diminish in intensity and frequency over time.

The earthquake occurred along the transform plate boundary between North America and the Caribbean plates, where the Caribbean plate is moving about 0.8 inches (2 cm) per year toward the east with respect to North America. Motion between North America and the Caribbean plates is divided between two major east-west trending strike-slip fault systems. The Septentrional fault system cuts across northern Haiti, whereas the Enriquillo-Plantain Garden fault system cuts through southern Haiti and the Port-au-Prince area. The earthquake was located on the Enriquillo-Plantain Garden fault system, at a depth of 8.1 miles (13 km). The shallow depth of this earthquake, coupled with the very lax building standards in Haiti (the poorest country in the Western Hemisphere), explain why the damage and death toll were so exceptionally high from this earthquake compared to larger earthquakes such as the magnitude 7.9 Wenchuan China earthquake of May 12, 2008, that killed an estimated 87,652 people.

The infrastructure of the port, airport, and roads leading into Port-au-Prince was all severely damaged by the earthquake, so it was initially very difficult to get rescue workers, aid, and relief supplies to the city. Some of the first rescuers to reach Port-au-Prince were from the Israeli Defense Forces, who set up a field hospital and performed search-and-rescue operations; the U.S. military, who sent paratroopers to maintain law and order and perform rescue operations; and the Dutch who sent rescue personnel. Many other agencies were able to get rescue teams and aid to the people of the city a few days later when the airport was secured by the U.S. military, and remarkably, survivors were still being pulled from the rubble 15 days after the main shock. After about two weeks the United States was able to get a hospital ship anchored offshore, where many thousands of injured people were treated.

The city prison in Port-au-Prince collapsed during the earthquake, and about 1,000 dangerous criminals are believed to have escaped. In the week after the quake, some desperate survivors started looting collapsed shops for food, water, and valuables, and some of the escaped prisoners are believed to have been behind some of the looting and crime.

SUMMARY

Transform boundaries develop where one plate simply slides past the other along a transform fault, with the most famous example being the San Andreas fault along the western North America transform boundary. Historical accounts of many earthquakes along transform boundaries have shown that the amount of death and destruction is closely related to the population density and quality of buildings in an area. If a huge earthquake hits an unpopulated area, it is of little consequence. However, even moderate-sized earthquakes have killed tens and even hundreds of thousands of people in areas with homes made of unreinforced concrete, piles of stones, or loose earth. Southern California is the most vulnerable area in the United States for future large earthquakes along a transform boundary in a densely populated area, and residents in that area need to be diligent in application of strict building codes, in development of earthquake warning systems, and in preparation of sophisticated emergency response plans.

Southern California is most likely to suffer a major earthquake in the next 50 years, although it is currently impossible to tell exactly when this earthquake might strike. Government disaster planners in potentially affected regions are devising emergency response plans, and building codes have been improved to make the people and infrastructure in these areas less prone to injury and damage. Continued studies and monitoring can continue to improve the science of earthquake prediction, perhaps one day saving thousands of lives.

See also CONVERGENT PLATE MARGIN PROCESSES; DIVERGENT PLATE MARGIN PROCESSES; OPHIOLITES; PLATE TECTONICS; STRUCTURAL GEOLOGY.

FURTHER READING

Kious, W. Jacquelyne, and Robert I. Tilling. *This Dynamic Earth: The Story of Plate Tectonics.* Online edition. URL: http://pubs.usgs.gov/gip/dynamic/dynamic.html. Last modified March 27, 2007.

Skinner, B. J., and B. J. Porter. *The Dynamic Earth: An Introduction to Physical Geology,* 5th ed. New York: John Wiley & Sons, 2004.

Wallace, R. E., ed. "The San Andreas Fault System." Professional Paper 1515. Calif.: U.S. Geological Survey, 1990.

tsunami, generation mechanisms A tsunami is a long-wavelength seismic sea wave generated by the sudden displacement of the seafloor. The name is of Japanese origin, meaning *harbor wave.* Tsunamis are also commonly called tidal waves, although this is improper because they have nothing to do with tides. Every few years, these giant sea waves rise unexpectedly out of the ocean and sweep over coastal communities, killing hundreds or thousands of people and causing millions of dollars in damage. Major tsunamis hit coastal areas in 1946, 1960, 1964, 1992, 1993, and 1998 in coastal Pacific regions, and in 2004 the Indian Ocean was swept by a tsunami that killed 283,000 people. In 1998 a catastrophic 50-foot- (15.2-m-) high wave unexpectedly struck Papua New Guinea, killing more than 2,000 people and leaving more than 10,000 homeless.

Tsunamis may be generated by any event that suddenly displaces the seafloor, which in turn causes the seawater to move suddenly to compensate for the displacement. Most tsunamis are caused by earthquakes on the seafloor or are induced by volcanic eruptions that suddenly boil or displace large amounts of water. Giant submarine landslides have initiated other tsunami, and it is even possible that gases dissolved on the seafloor may suddenly be released, forming a huge bubble that erupts upward to the surface, generating a tsunami. The most catastrophic tsunami in the geological record may have been thousands of feet (hundreds of meters) tall, generated when asteroids or meteorites impacted with Earth in the oceans, displacing huge amounts of water in a geological instant.

PLATE TECTONICS AND TSUNAMI GENERATION

The movement of the tectonic plates causes earthquakes. Nearly all of the convergent plate boundaries on the planet are located in the oceans, because the bending of oceanic plates into subduction zones causes the surface of the crust to be pulled down to several miles (several km) depth. Most of the largest earthquakes occur along convergent plate boundaries when large amounts of crust move at one time. The sudden movement of the seafloor must move huge volumes of water, and this displacement of water is what causes many tsunamis.

Earthquake-Induced Tsunamis

Earthquakes that strike offshore or near the coast have generated most of the world's tsunamis. In general, the larger the earthquake, the larger the potential tsunami, but this is not always the case. Some large earthquakes produce large tsunami, whereas others do not. Earthquakes that have large amounts of vertical displacement of the seafloor result in larger tsunamis than earthquakes that have predominantly horizontal movements of the seafloor. This difference is approximately a factor of 10, probably because earthquakes with vertical displacements are much more effective at pushing large volumes of water upward or downward, generating tsunamis. Another factor that influences

the size of tsunami generated by an earthquake is the speed at which the seafloor breaks during the earthquake—slower ruptures tend to produce larger tsunamis. In general, earthquakes with a magnitude of 6.5 or greater, with a shallow focus or place of rupture, are required to generate a tsunami.

Tsunami earthquakes are a special category of earthquakes that generate tsunamis unusually large for the earthquake's magnitude. Tsunami earthquakes are generated by large displacements that occur along faults near the seafloor. Most are generated on steeply dipping seafloor surface penetrating faults that have vertical displacements along them during the earthquake, displacing the maximum amount of water. These types of earthquakes also frequently cause large submarine (underwater) landslides or slumps, which also generate tsunamis. In contrast to tsunamis generated by vertical slip on vertical faults, which cause a small region to experience a large uplift, other tsunamis are generated by movement on very shallowly dipping faults. These are capable of causing large regions to experience minor uplift, displacing large volumes of water and generating a tsunami. Some of the largest tsunamis may have been generated by earthquake-induced slumps along convergent tectonic plate boundaries. For example, in 1896 an earthquake-induced submarine slump generated a huge 75-foot (23-m) tsunami in Sanriku, Japan, killing 26,000 people in the wave. Another famous tsunami generated by a slump from an earthquake is the 1946 wave that hit Hilo, Hawaii. This tsunami was 50 feet (15 m) high, killed 150 people, and caused about $25 million in damage to Hilo and surrounding areas. The amazing thing about this tsunami is that it was generated by an earthquake-induced slump off Unimak Island in the Aleutian Chain of Alaska 4.5 hours earlier. This tsunami traveled at 500 miles per hour (800 km/hour) across the Pacific, hitting Hawaii without warning.

Another potent type of tsunami-generating earthquake occurs along subduction zones. Sometimes, when certain kinds of earthquakes strike in this environment, the entire forearc region above the subducting plate may snap upwards by up to a few tens of feet, displacing a huge amount of water. The devastating 2004 Indian Ocean tsunami was generated by motion of about a 600-mile- (1,000-km-) long segment of the forearc of the Sumatra arc and subduction zone. The tsunami generated during the 1964 magnitude 9.2 Alaskan earthquake also formed a tsunami of this sort, and it caused numerous deaths and extensive destruction in places as far away as California.

More rarely horizontal movements along vertical strike-slip faults may generate tsunamis. Sideways motion along strike-slip faults rarely generates tsunamis because the sideways motion on these faults

Diagram showing types of faults that generate large tsunamis, including (1) thrust faults in forearc regions, (2) vertical displacement of the seafloor, and (3) slumping of large blocks into the water

does not cause the water surface to be directly disturbed.

Volcanic Eruption–induced Tsunamis

Some of the largest recorded tsunamis have been generated by volcanic eruptions. These may be associated with the collapse of volcanic slopes, debris and ash flows that displace large amounts of water, or submarine eruptions that explosively displace water above the volcano. Approximately 20 percent of volcanic-induced tsunamis form when volcanic ash or pyroclastic flows hit the ocean, displacing large amounts of water, and 20 percent form from earthquakes associated with the eruption. About 15 percent result from eruptions beneath the water, and 7 percent result from collapse of the volcano and landslides into the sea. The remaining causes are not known. The most famous volcanic eruption–induced tsunamis include the series of huge waves generated by the eruption of Krakatau in 1883, which reached run-up heights of 130 feet (36 m) and killed 36,500 people. The number of people that perished in the eruption of Santorini in 1600 B.C.E. is not known, but the toll must have been huge. The waves reached 800 feet (240 m) in height on islands close to the volcanic vent of Santorini. Flood deposits have been found 300 feet (90 m) above sea level in parts of the Mediterranean Sea and extend as far as 200 miles (320 km) southward up the Nile River. Several geologists suggest that these were formed from a tsunami generated by the eruption of Santorini. The floods from this eruption may also, according to

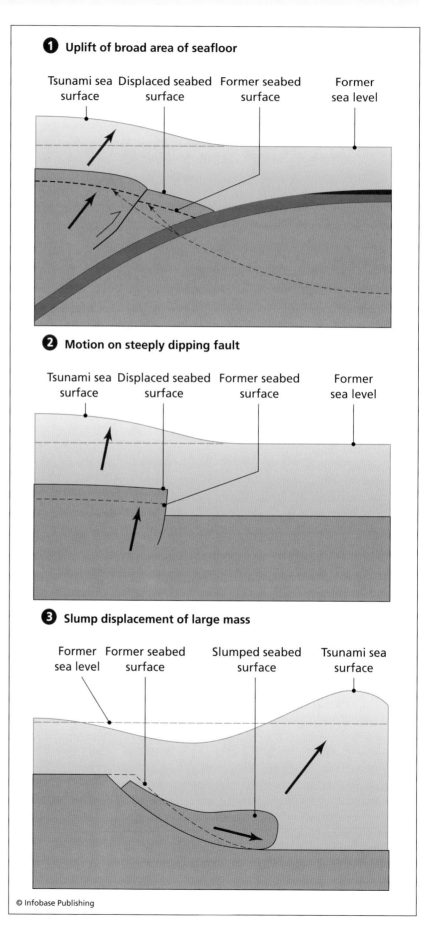

1 Uplift of broad area of seafloor

Tsunami sea surface   Displaced seabed surface   Former seabed surface   Former sea level

2 Motion on steeply dipping fault

Tsunami sea surface   Displaced seabed surface   Former seabed surface   Former sea level

3 Slump displacement of large mass

Former sea level   Former seabed surface   Slumped seabed surface   Tsunami sea surface

© Infobase Publishing

some scientists, account for some historical legends such as the great biblical flood, the parting of the Red Sea during the exodus of the Israelites from Egypt, and the destruction of the Minoan civilization of the island of Crete.

The specific mechanisms by which volcanic eruptions can form tsunami are diverse but number fewer than a dozen. Perhaps the most common is from earthquakes associated with the volcanism. Many volcanic eruptions are accompanied by swarms of moderate-sized earthquakes, and some of these may be large enough to trigger tsunamis. This is especially true in cases of volcanoes built at convergent margins and that are partway under sea level, such as Mount Vesuvius in Italy. The 79 C.E. eruption of Vesuvius was associated with many earthquake-induced tsunamis, some being triggered before the main eruption, and others during and after the eruption.

When pyroclastic flows or nuées ardentes surge down the slopes of volcanoes they may eventually reach the ocean, where they displace water and spread laterally. Some pyroclastic flows are less dense than water and produce layers of ash and pumice that ride over the ocean surface, whereas others are denser than seawater and can suddenly displace large volumes of water, producing a tsunami. In these cases, the larger the flow, the larger the resulting tsunami. Some pyroclastic flows are dense, and continue to flow in a surgelike manner on the seafloor, pushing a wall of water ahead of the flow, thus generating a tsunami. The August 26, 1883, eruption of Krakatau in Indonesia generated an 33-foot- (10-m-) high tsunami by such an undersea surge of a pyroclastic flow, whereas the March 5, 1871, eruption of Ruang volcano, also in Indonesia, produced an 82-foot- (25-m-) high tsunami by this mechanism.

Submarine eruptions in shallow water can generate tsunamis by displacing water when the volcano erupts. When the water is deeper than about 1,650 feet (500 m), the weight of the water is so great that it suppresses the formation of surface waves. This is fortunate, since there are many volcanic eruptions along the 8,900-foot- (2,700-m-) deep mid-ocean ridges, and if each of these generated a tsunami, the coastal zone on most continents would be constantly plagued with tsunamis. When submarine eruptions expose the magma chamber of a volcano to seawater, larger tsunamis result from the sudden steam explosion as the cold seawater vaporizes immediately after touching the extremely hot magma. An explosion of this sort generated the deadly 1883 tsunami from the eruption of Krakatau in Indonesia, when a steam explosion formed a 130-foot (40-m) high tsunami that killed tens of thousands of people.

One of the most catastrophic ways to bring large volumes of seawater into sudden contact with a magma chamber is through the formation of a caldera, where the top of the volcanic complex suddenly collapses into the magma chamber, forming a large depression on the surface. If the volcano is located at or near sea level, ocean waters can suddenly rush into the depression, where they will encounter the hot magma and form giant steam eruptions. Many volcanoes around the Pacific Rim have formed caldera structures, and many appear to have generated tsunamis. The most famous is the tsunami associated with the formation of the caldera on Krakatau in 1883, in which the tsunami devastated the Sunda straits, killing thousands of people. Another example of a tsunami generated by the seawater rushing into a collapsing caldera is from Mount Ritter, in Papua New Guinea. On March 13, 1888, a 1.5-mile- (2.5-km-) wide caldera collapsed beneath the volcano, and the sea rushed in, generating a 50-foot- (15-m-) high tsunami that swept local shores. The actual amount of seawater displaced in tsunamis formed by caldera collapse is small, typically many orders of magnitude smaller than the amount displaced during large earthquakes. Therefore tsunamis associated with caldera collapse tend not to travel very far, but to decay in height quickly, according to the inverse of the square root of the distance from the source.

Tsunamis can be generated from volcanoes during collapse of the slopes and formation of landslides and debris avalanches that move into the ocean. Most tsunamis generated by the collapse of slopes of volcanoes are small, localized, and very directional, in that they propagate directly away from the landslide or debris avalanche and fade rapidly in other directions. Many volcanoes around the Pacific Rim have steep slopes near the sea and pose hazards for tsunamis generated by slope collapse. Some of these can be locally quite powerful, as shown by the example of Mount Unzen in Japan. On May 21, 1792, a large debris avalanche that roared off the slope of the volcano traveled four miles (6.5 km) and hit the seawater of the Ariake Sea. This displaced enough water to generate a tsunami with run-up heights of up to 170 feet (55 m) along a 50-mile- (77-km-) long section of coast on the Shimabara Peninsula, killing 14,524 people, destroying 6,000 homes, and sinking 1,650 ships. Other volcanic islands, such as Hawaii and the Canary Islands, pose a different kind of landslide threat, in which large sections of the volcano collapse, forming giant undersea landslides with associated tsunamis.

A final way that volcanoes can form tsunamis is through lateral blasts or sideways eruptions from the volcano. The most famous lateral blast from a volcano is the well-documented 1980 eruption of Mount St. Helens, but this was far from the ocean and did not generate a tsunami. Some volcanoes have a tendency to erupt sideways, one of these being Taal volcano in the Philippines, in which at least five lateral blasts have occurred in the past 250 years, each generating a deadly tsunami.

Mass-wasting processes on volcanic slopes have generated a number of other tsunamis, but these can generally be classified as being caused by landslides, or as one of the mechanisms from movement of lava discussed in this section. Mudflows and lahars have generated many local tsunamis; lava flows, if large and fast enough, can also generate local tsunamis. Most of these minor mechanisms generate only small, localized, and strongly directional tsunamis.

Landslide-induced Tsunamis

Landslides that displace large amounts of water generate many tsunamis. These may be from rock falls and other debris that falls off cliffs into the water, such as the huge avalanche that triggered a 200-foot- (60-m-) high tsunami in Lituya Bay, Alaska. Submarine landslides tend to be larger than avalanches that originate above the water line, and they have generated some of the largest tsunamis on records. Many submarine landslides are earthquake-induced, whereas others are triggered by storm events and by increases in pressure on the sediments on the continental shelf induced by rises in sea level. A deeper water column above the sediments on shelf or slope environment can significantly increase the pressure in the pores of these sediments, causing them to become unstable and slide downslope. After the last glacial retreat 6,000–10,000 years ago, sea levels have risen by 320–425 feet (100–130 m), which has greatly increased the pore pressure on continental slope sediments around the world. This increase in pressure is thought to have initiated many submarine landslides, including the large Storegga slides from 7,950 years ago off the coast of Norway.

Tsunamis are suspected of being landslide-induced when the earthquake is not large enough to produce the observed size of the associated tsunami. Many areas beneath the sea are characterized by steep slopes, including areas along most continental margins, around islands, and along convergent plate boundaries. Sediments near deep-sea trenches are often saturated in water and close to point of failure, where the slope gives out and collapses, causing the pile of sediments to slide suddenly down to deeper water depths. When an earthquake strikes these areas, large parts of the submarine slopes may give out simultaneously, displacing water and generating a tsunami. The 1964 magnitude 9.2 earthquake in Alaska generated more than 20 tsunamis, and these were responsible for most of the damage and deaths from this earthquake.

Some steep submarine slopes that are not characterized by earthquakes may also be capable of generating huge tsunamis. Recent studies along the east coast of North America, off the coast of Atlantic City, New Jersey, have revealed significant tsunami hazards. The submarine geology off the coast of eastern North America consists of a pile of unconsolidated sediments several thousands of feet (hundreds of meters) thick on the continental slope. These sediments are so porous and saturated with water that the entire slope is on the verge of collapsing under its own weight. A storm or minor earthquake may be enough to trigger a giant submarine landslide in this area, possibly generating a tsunami that could sweep across the beaches of Long Island, New Jersey, Delaware, and much of the east coast of the United States.

Storms are capable of generating submarine landslides even if the storm waves do not reach and disrupt the seafloor. Large storms are associated with storm surges that form a mound of water in front of the storm that may sometimes reach 20–32 feet (6–10 m) in height. As the storm surge moves onto the continental shelf it is often preceded by a drop in sea level caused by a drop in air pressure, so the storm surge may be associated with large pressure changes on the seafloor and in the pores of unconsolidated sediments. A famous example of a storm surge–induced tsunami is the catastrophic event that occurred in Tokyo, Japan, on September 1, 1923. On this day, a powerful typhoon swept across Tokyo and was followed that evening by a huge submarine landslide and earthquake that generated a 36-foot- (11-m-) tall tsunami that killed 143,000 people. Surveys of the seabed after the tsunami revealed that large sections had slid further into the sea, deepening the bay in many places by 300–650 feet (100–200 m), and locally by as much as 1,300 feet (400 m). Similar storm-induced submarine slides are known from many continental slopes and delta environments, including the Mississippi delta in the Gulf of Mexico and the coast of Central America.

Submarine slides are part of a larger group of processes that can move material downslope on the seafloor and that includes other related processes such as slumps, debris flows, grain flows, and turbidity currents. Submarine slumps are a type of sliding slope failure in which a downward and outward rotational movement of the slope occurs along a concave up-slip surface or fault. This produces either a singular or a series of rotated blocks, each with the original seafloor surface tilted in the same direction. Slumps can rapidly move large amounts of material short distances and are capable of generating tsunamis. Debris flows involve the downslope movement of unconsolidated sediment and water, most of which is coarser than sand. Some debris flows begin as slumps but then continue to flow downslope as debris flows. They typically fan out and come to rest when they emerge out of submarine canyons onto flat abyssal plains on the deep seafloor. Rates of movement in debris flows vary from several feet per year to several hundred miles per hour. Debris flows are commonly shaped like a tongue with numerous ridges and depressions. Large debris flows can suddenly move large volumes of sediment, so are also capable of generating tsunamis. Turbidity currents are sudden movements of water-saturated sediments that move downhill under the force of gravity. Typically, a water-saturated sediment on a shelf or shallow water setting is disturbed by a storm, earthquake, or some other mechanism that triggers the sliding of the sediment downslope. The sediment-laden water mixture then moves rapidly downslope as a density current and may travel tens or even hundreds of miles at tens of miles (km) per hour until the slope decreases, and the velocity of current decreases. As the velocity of the current decreases, the ability of the current to hold coarse material in suspension decreases. The current drops first its coarsest load, then progressively finer material as the current decreases further. Turbidity currents do not usually generate tsunamis, but many are associated with slumps and debris flows that may generate tsunamis. However, some turbidity flows are so massive that they may form tsunamis.

Volcanic hot-spot islands in the middle of some oceans have a long record of producing tsunamis from submarine landslides. These islands include Hawaii in the Pacific Ocean, the Cape Verde Islands in the North Atlantic, and Réunion in the Indian Ocean. The shape of many of these islands bears the telltale starfish shape with cuspate scars indicating the locations of old curved landslide surfaces. On average, a significant tsunami is generated somewhere in the world every 100 years by a collapse and submarine landslide from a mid-ocean volcanic island. These islands are volcanically active. Lava flows move across the surface and then cool and crystallize quickly as the lava enters the water. This causes the islands to grow upward as very steep-sided columns, whose sides are prone to massive collapse and submarine sliding. Many volcanic islands are built up with a series of volcanic growth periods followed by massive submarine landslides, effectively widening the island as it grows. However, island growth by deposition of a series of volcanic flows over older landslide scars causes the island to be unstable—the old landslide scars are prone to slip later since they are weak surfaces, and the added stress of the new material piled on top of them makes them unstable. Other processes may also contribute to making these surfaces and the

parts of the island above them unstable. For instance, on the Hawaiian Islands, volcanic dikes have intruded along some old landslide scars, which can reduce the strength across the old surfaces by large amounts. Some parts of Hawaii are moving away from the main parts of the island by up to 0.5–4 inches (1–10 cm) per year by the intrusion of volcanic dikes along old slip surfaces. Also, many landslide surfaces are characterized by accumulations of weathered material and blocks of rubble that, under the additional weight of new volcanic flows, can help to reduce the friction on the old slip surfaces, aiding the generation of new landslides. Therefore, as the islands grow, they are prone to additional large submarine slides that may generate tsunamis.

The Cape Verde Islands, located in the eastern Atlantic off the coast of west Africa, were constructed from hot-spot style volcanism (i.e., not associated with the mid-ocean ridge or island arcs). The islands have very steep western slopes, and these cliffs are unstable. If new magma enters the volcanic islands, it may heat the groundwater in the fractures in the rock, creating enough pressure to induce the giant cliffs to collapse. Any landslides generated from such an anticipated collapse would have the potential to generate giant tsunamis, several thousand feet (several hundred meters) high, that could sweep the shores of the Atlantic. The wave height will diminish with distance from the Cape Verde Islands, but the effects on the shores of eastern North America, the Caribbean, and eastern South America are expected to be devastating, if this event ever occurs. The waves will probably also wrap around the Cape Verde Islands, hitting the United Kingdom and the west coast of Africa with smaller, but still damaging waves.

The characteristics of tsunamis generated by landslides depend on the amount of material that moves downslope, the depth that the material moves from and to, and the speed at which the slide moves. Tsunamis generated by submarine landslides are usually quite different from those generated by other processes such as displacements of the seafloor caused by earthquakes. Submarine slides move material in one direction, and the resulting tsunami tends also to be more focused in slide-induced events than from other triggering mechanisms. Therefore, tsunamis produced by submarine slides are characterized by a wave located above the slide that moves offshore parallel to the direction the slide has moved. A complementary wave is produced that moves in the opposite direction, upslope and toward shore.

Tsunamis resulting from submarine slides also have wave shapes that are different from waves generated by earthquake-induced displacement of the seafloor. Slide-induced tsunamis typically have a first wave with a small crest, followed by a deep trough that may be three times deeper than the height of the first wave. The next wave will have the height of the trough, but may change into several waves with time. The large height difference between the first wave and the succeeding trough often leads tsunamis generated by landslides to have greater run-up heights than tsunamis generated by other mechanisms. Slide-induced tsunamis also differ from other tsunamis in that they start with slow velocities as the submarine slide forms, then must accelerate as they take form. Therefore, the arrival time of a slide-induced tsunami at shore locations is typically later than expected. The wavelength of tsunamis from submarine slides is typically 0.5–6 miles (1–10 km), and the periods range from one to five minutes. The period may increase as the area of the slide increases and the slope of the seafloor decreases. Submarine landslides rarely move faster than 160 feet per second (50 m/sec), whereas the resulting tsunamis quickly accelerate to 325–650 feet per second (100–200 m/sec). The characteristics of slide-induced tsunamis are therefore quite different from tsunamis produced by other mechanisms.

Natural Gas and Gas Hydrate Eruption–induced Tsunamis

The continental shelves and slopes around most continents are the sites of deposition of very thick piles of sediments. River deltas such as the Mississippi delta may add even more sediments to these environments, in some cases forming piles of sediment that are 10 miles (16 km) thick, deposited over many millions of years. Natural gas is produced in submarine sediments by the anaerobic decay of organic matter that becomes buried with the sediments. The gas produced by the decay of these organic particles may escape or become trapped within the sediments. When the gas gets trapped between the pore spaces of the sediments, it causes the pressure to build up within the seafloor sediments; this process is called underconsolidation. These pressures may become quite large and even be greater than the pressure exerted by weight of the overlying water. The pressure from the weight of the water is called hydrostatic pressure. When the gas pressure within a layer or larger section of sediments on the continental shelves and slopes becomes greater than the hydrostatic pressure, the sediments on the slope are on the verge of failure, such that any small disturbance could cause a massive submarine landslide, in turn inducing a tsunami.

Decaying organic matter on the seafloor releases large volumes of gas, such as methane. Under some circumstances, including in cold water at deep depths, these gases may coagulate, forming gels called gas hydrates. It has recently been recognized that these gas hydrates occasionally spontaneously release their trapped gases in giant bubbles that rapidly erupt to the surface. Such catastrophic degassing of gas hydrates poses a significant tsunami threat to regions not previously thought to have a significant threat from this type of tsunami, such as along the eastern and gulf coasts of the United States.

Other Tsunamis

The largest tsunamis in the geological record are generated by the impact of giant asteroids with the earth. These types of events do not happen very often and none are known from historical records, but when they do occur they are cataclysmic. Geologists are beginning to recognize deposits of impact-generated tsunamis and now estimate that they may reach several thousand feet (1 km) in height. One such tsunami was generated about 66 million years ago by an impact that struck the shoreline of the Yucatán Peninsula, producing the Chicxulub impact structure. This impact produced a huge crater and sent a 3,000-foot- (1-km-) high tsunami around the Atlantic, devastating the Caribbean and the U.S. gulf coast. Subsequent fires and atmospheric dust that blocked the Sun for several years killed off many of the planet's species, including the dinosaurs. Even relatively small meteorites that hit the ocean have the potential to generate significant tsunamis. A meteorite only 1,000 feet (300 m) in diameter would produce a tsunami seven feet (2 m) tall that could strongly affect coastal regions for 600 miles (1,000 km) around the impact site. Statistical analysis predicts about a 1 percent chance of impact-related tsunami events happening once every 50 years.

Weather-related phenomena may also rarely generate tsunamis. In some special situations, large variations in atmospheric pressure, especially at temperate latitudes, can generate long-wavelength waves (tsunamis) that resonate, or become larger, in bays and estuaries. Although these types of tsunamis are not generated by displacement of the seafloor, they do have all the waveform characteristics of other tsunamis and are therefore classified as such.

PHYSICS OF TSUNAMI MOVEMENT

Witnessing the approach of a large tsunami to shore can be one of the most awe-inspiring and deadly sights ever witnessed by many residents of low-lying coastal areas. Tsunamis are very different from normal storm- or wind-generated waves, in that they have exceptionally long wavelengths, the distance between successive crests. Whereas most storm waves will rise, break, and dissipate most of their energy in the surf zone, tsunamis are characterized by rapid rise of sea level that leads the wave to break at the shoreline, then to keep on rising or running up into the coastal zone for extended periods of time. This is related to the long wavelength of tsunamis. Most waves have distances between the crests of several hundreds of feet (tens of m) and rise and fall relatively quickly as each wave crest passes. Tsunamis, however, may have distances between each wave crest of a hundred or even many hundreds of miles (km), so it may take half an hour, an hour, or more for the wave to stop its incessant and destructive rise into coastal areas and retreat into the sea, before the next crest of the tsunami train crashes into shore. Tsunamis are like normal wind-generated waves in that they have a series of wave crests separated by troughs. Therefore, like normal waves, after the first, typically hour-long wave sweeps through a coastal region and retreats, the tsunami event is not over. A series of wave crests follows, sometimes with the second or third crest being the largest, sweeping into the coastal areas, each crest causing its own destruction. Many of the deaths reported from tsunami disasters are associated with the fact that many people do not understand this basic physical principal about tsunamis, and they rush to the coastal area after the first destructive wave to rescue injured people and become victims of the second or third crest's incursion onto the land.

Movement in Open Ocean

Tsunamis are waves with exceptionally large distances between individual crests, and they move like other waves across the ocean. Waves are described using the terms related to the regular geometrically repeating pattern of the waves. Waves in a series are called a wave train, with regularly repeating crests and troughs. Wavelength is the distance between crests, wave-height is the vertical distance from the crest to the bottom of the trough, and the amplitude is one-half of the wave height. The period of the wave is the time between the passage of two successive crests. Most tsunamis have wave periods of 1.6 to 33 minutes. Most ocean waves have wavelengths of 300 feet (100 m) or less. Tsunamis are exceptional in that they have wavelengths that can exceed 120 miles (200 km). The particle motion in deepwater waves follows roughly circular paths, where particles move approximately in a circle, and return back to their starting position after the wave passes. The amount of circular motion decreases gradually with depth, until a depth that equals one-half of the wavelength. At this depth all motion associated with the passage of the wave stops, and the water beneath this point experiences no effect from the passage of the wave above. This depth is known as the wave base. The movement of deepwater waves is therefore associated with the transfer of energy, but not the transfer of water from place to place.

The particle motion of individual molecules of water during the passage of tsunamis is elliptical, similar to the circular motion of other deepwater waves. The motion of any particles during the passage of the waves follows elliptical paths, first forward, down, then up and back to near the starting point during the passage of individual wave crests. The passage of deep-water tsunamis is therefore associated with the movement and transportation of individual particles. The reason that particle paths in tsunamis are elliptical, and in other deepwater waves the motion is circular, is that tsunamis have such long wavelengths that the ocean's depth of 2–3 miles (3–5 km) is less than the wave base. Tsunamis therefore travel as shallow-water waves across the open ocean where the water depth is less than wave base, and they feel some frictional effects from the ocean bottom. This friction distorts the preferred circular particle paths into elliptical paths that are observed in tsunamis. The end result is that tsunamis are associated with motion of the entire water column during passage, whereas wind-generated waves have motion only down to the wave base, at a distance equal to half the wavelength.

Since the wavelength of tsunamis is typically about 120 miles (200 km), movement associated with the passage of the waves could be felt to a depth of 60 miles (100 km), much greater than the depth of the oceans. Tsunamis therefore are felt at much greater depths than ordinary waves, and the normally still, deep-ocean environment will experience sudden elliptical or back and forth motions, plus pressure differences, during the passage of tsunamis. These effects may be used with deep-ocean-bottom tsunami detectors to help warn coastal communities when tsunamis are approaching.

Many tsunamis are different from regular waves in that they may have highly irregular wave-train patterns. Some tsunamis have a high initial peak, followed by successively smaller wave crests, whereas other tsunamis have the highest crest located several crests behind the initial crest. The reasons for these differences are complex, but most are related to the nature of the triggering mechanism that formed the wave train. A splash or meteorite impact may create an initial large crest, whereas an undersea explosion may cause an initially small crest, followed by a larger one related to the interaction of the waves that fill the hole in the water column related to the explosion. Many variables contribute to the initial shape of the wave train, and each wave train needs to be examined separately to understand what caused its shape.

Different mechanisms of tsunami generation may form several sets of wave trains with different wavelengths. Longer-wavelength wave trains travel at higher speeds than shorter-wavelength wave trains, so the farther the tsunami travels from the source, the more spread out the waves of different wavelengths will become. This phenomenon is known as wave dispersion. The effect of dispersion is such that locations near the source will see complex waves where the short- and long-wavelength sets are superimposed, and these waves combine to make taller or shorter waves of each set as they crash into shore. Locations more distant from the tsunami source will first experience the fast-traveling, long-wavelength waves, and then later be hit by the shorter-wavelength (and -period) waves. The time difference between these different sets becomes greater with increasing distance from the source.

When tsunamis are traveling across deep-ocean water, their amplitudes are typically less than three feet (1 m), even though the wavelength may be more than 100 miles (160 km). A passenger on a ship would probably not notice even the largest of tsunamis if the ship was in the deep ocean.

Since tsunamis are essentially shallow-water waves when they travel across the open ocean, they will experience different amount of friction by different water depths and topographic features, such as submerged mountains, on the seafloor. The shallower the feature, the greater the friction, and the more it will slow the passage of the tsunami wave-front above that feature. Tsunamis are like other wave features—when they encounter objects that slow their travel, they bend around that object, and are said to be refracted. Refraction of tsunamis is quite common in the Pacific and Indian Oceans. In the Pacific, this phenomenon is typically seen when waves generated along the Pacific subduction zones travel to the center of the ocean and get refracted around the islands of the Hawaiian chain. The bending of tsunami wave-fronts around objects such as Hawaii can have two main effects on the wave energy—it can focus the energy in some locations where the wave fronts that bend around the object from either side merge and add together, or it can spread apart this energy and disperse it so that it is less intense. In the Pacific Ocean, the island of Hawaii tends to focus the energy from earthquake-generated tsunamis that form along the western coasts of North and South America on the island of Japan. This is one reason that Japan has endured so many tsunami events in history. Japan must accommodate tsunamis generated locally by earthquakes in its own vicinity, and the shape of the seafloor in the Pacific focuses the energy from distant earthquake-generated tsunamis onto this island nation. Tsunamis that are amplified in this way from distant earthquakes are known as teleseismic tsunamis. An example of the opposite effect, the defocusing of seismic energy is commonly afforded by the small island of Tahiti in French Polynesia. The seafloor topography of the Pacific commonly causes incoming tsunamis to be defocused or dispersed as the waves approach this island.

Encounter with Shallow Water

Waves with long wavelengths travel faster than waves with short wavelengths. Since the longer the wavelength the faster the wave in deep open water, tsunamis travel extremely fast across the ocean. Normal ocean waves travel at less than 55 miles per hour (90 km/hr), whereas many tsunamis travel at 375 to 600 miles per hour (800 to 900 km/hr), faster than most commercial airliners. The wave speeds slow down as the tsunamis encounter shallow water, typically in the range of 60–180 miles per hour (100–300 km/hr) across the continental shelves, and about 22 miles per hour (36 km/hr) at the shore. This slowing of the wave speed as it begins to encounter shallow water causes the waves at the back of the train to move faster than those in the front. When this occurs the wave must become taller and narrower to accommodate the waves moving into the same space from behind; thus as the tsunami moves from deep water into shallow waters, it becomes taller (larger amplitude), has a shorter distance between crests (shorter wavelength), and moves slower (velocity). In some cases many of the crests will merge and the troughs will disappear during this process, producing huge solitary waves, whose height from base to top is entirely above sea level.

When waves encounter shallow water, the friction of the seafloor along the base of the wave becomes greater than when the waves were traveling in deep water, causing them to slow down dramatically, and the waves effectively pile up on themselves as successive waves move into shore. This causes the wave height or amplitude to increase dramatically, sometimes 10 to 150 feet (3–45 m) above the normal stillwater line for tsunamis.

One of the main effects of the friction at the base of the tsunami as it enters shallow water is that the wave fronts tend to be strongly refracted, or bent, so that they approach land at less than 10° no matter what the original angle of approach to the shore was. This refraction occurs because the part of the wave that encounters shallow water first will be slowed down by the increased friction, whereas the other part of the wave still in deepwater will continue to move faster, until it catches up with the rest of the wave by being in the same water depth, then moves at the same rate. This effect bends tsunamis, like other waves, so that they hit most shoreline areas nearly head-on. Seafloor topography very close to the shore can modify this refraction, and either focus the energy into specific locations, or disperse it across the shoreline.

The friction at the base of the wave dissipates or takes some of the energy away from the tsunami. In most settings the amount of dissipation of energy by friction is minor (less than 3 percent), but in some cases where the continental shelf areas are very wide and narrow, the dissipation may be significant enough to reduce the tsunami threat to the region dramatically. Such is the case for the part of the northeast seas of China (Yellow Sea, Bohai, and related areas) where the water depth is quite shallow for many hundreds of miles, making the northeast coast of China much less susceptible to tsunamis than the southeastern coast, where the shelf area is deeper and narrower. Much of the east coast of the United States has a moderately wide and shallow shelf that is able to dissipate about 20 percent of the energy of most tsunamis by friction along the wave base. Areas that have narrow and steep continental shelves offshore are prone to the most severe tsunamis, since these areas do not have the ability to reduce the wave energy by friction.

Tsunami waves exhibit a phenomenon called diffraction when they enter bays through a narrow entrance. Diffraction occurs when energy moves or is leaked sideways along a wave crest, enabling the wave to grow along the wave front to fill the available area inside a wide bay that the wave has entered through a narrow opening. The wave front must enter through the narrow passage to the ocean, but then spreads across the bay as a longer wave. The process of dispersion moves energy from the initially high wave crest sideways, and in doing so takes energy away from the central area, decreasing the height of the wave. Thus, the dispersion of energy during the wave's entrance and spread into the harbor is a good thing, reducing the threat to areas inside bays with narrow entrances. An example of a bay with a shape that would disperse tsunami energy is San Francisco Bay. If a tsunami were to pass under the Golden Gate Bridge it would crash through the narrows there, but then spread, losing height and ferocity, as it moved into the Bay Area.

Amplification, the opposite effect of dispersion, occurs in some bays where the opening of the bay or estuary is wide, and the bay narrows progressively inland. Tsunamis that enter such treacherous waters will find their wave crests being amplified or increased in height, transferring energy along the wave crest as they are forced to become shorter lengthwise along the crests of the waves. There are many examples of tsunami disasters that occurred because the shape of the bay amplified the tsunami that would otherwise have been minor. A famous example of this effect is the 1964 tsunami that hit Crescent City, California, from the magnitude 9.2 earthquake in Alaska. Most areas along the California coast experienced a relatively minor tsunami (less than two feet, or half a meter in height), but the shape of the bay and seafloor at Crescent City amplified the wave until it consisted of a series of five tsunami crests. The fifth

was a 21-foot- (6.3-m-) high crest that swept into downtown, washing away much of the waterfront district and killing 11 people.

When tsunamis strike the coastal environment, the first effect is sometimes a significant retreat or drawdown of the water level, whereas in other cases the water just starts to rise quickly. Since tsunamis have long wavelengths, it typically takes several minutes for the water to rise to its full height. Also, since there is no trough right behind the crest of the wave, on account of the very long wavelength of tsunamis, the water does not recede for a considerable time after the initial crest rises onto land. The rate of rise of the water in a tsunami depends in part on the shape of the seafloor and coastline. If the seafloor rises slowly, the tsunami may crest slowly, giving people time to outrun the rising water. In other cases, especially where the seafloor rises steeply, or the shape of the bay causes the wave to be amplified, tsunamis may come crashing in huge walls of water with breaking waves that pummel the coast with a thundering roar and wreaking utmost destruction.

Because tsunamis are waves, they travel in successive crests and troughs. Many deaths in tsunami events are related to people going to the shoreline to investigate the effects of the first wave, or to rescue those injured or killed in the initial crest, only to be drowned or swept away in a succeeding crest. Tsunamis have long wavelengths, so successive waves have a long lag time between individual crests. The period of a wave is the time between the passage of individual crests, and for tsunamis the period can be an hour or more. Thus, a tsunami may devastate a shoreline area and retreat, and then another crest may strike an hour later, then another, and another in sequence.

The specific shape of any shoreline has large effects on the tsunami height and the way it approaches the shoreline. The study of the effects of local coastal features on the tsunami is called morphodynamics. As the water from one wave crest retreats, it must move back to sea, and interact with the next incoming wave. Some of this water moves quickly sideways along the coast, setting up a new independent set of waves that oscillates up and down in amplitude along the coast, typically with a wavelength that is double that of the original tsunami. These secondary waves are called edge waves, and may be nearly as large as the original tsunami. When the following tsunami crests approach the shoreline, they may interact with a positive crest and produce a wave that is larger than the initial tsunami, or they may interact with a negative trough, and produce a smaller wave. These edge waves and local morphodynamics explain much of the variability of the height of tsunamis along some shorelines. In some locations the tsunami crest may be 30 feet

(10 m) high, while in other nearby areas it may only be six feet (2 m) high.

The name tsunami means *harbor wave* in Japanese, and the term describes another physical phenomenon of waves called resonance. When waves enter harbors or bays, they have a characteristic period that in many cases matches a natural harmonic frequency of that particular harbor. This means that many tsunamis enter a bay, and bounce back and forth across the harbor with the exact period that causes the wave to dramatically increase in height. The effect is similar to slowly moving a glass of water back and forth, and gradually increasing the speed until suddenly the waves in the glass start to become amplified and then leap out of the glass. This happens when the period of the wave equals that of the natural frequency of the glass. Many tsunamis that enter bays will resonate, or oscillate back and forth in the bay for 24 hours or more, causing disruption of activities for an extended period.

Tsunami Run-up

Run-up is the height of a tsunami above sea level at the farthest point it reaches on the shore. This height may be considerably different from the height of the wave where it first hits the shore, and is commonly twice that of the height of the wave at the shore. Run-up heights of 30 feet (10 m) are fairly common for tsunamis, while heights of 150–300 feet (45–90 m) are rare, and heights greater than this in the range of 300–1,700 feet (90–525 m) are very rare, but have been observed in the past hundred years. Many things influence the run-up of tsunamis, including the size of the wave, the shape of the shoreline, the pro-

file of the water depth, diffraction, formation of edge waves that move along the coast, and other irregularities particular to individual areas. Some bays and other places along some shorelines may amplify the effects of waves that come in from a certain direction, making run-ups higher than average. These areas are called wave traps, and in many cases the incoming waves form a moving crest of breaking water, called a bore, that smashes into coastal areas with great force. Tsunami magnitudes are commonly reported using the maximum run-up height along a particular coastline.

When tsunamis approach the shore, the wave fronts pile up and the wave changes form from a sinusoidal wave to a solitary wave, with the entire wave form above sea level. These types of waves maintain their forms, and since the kinetic energy in the wave is evenly distributed throughout the wave, the waves lose very little energy as they approach the shore. Steep coastlines may experience larger run-ups, since they have the least amount of energy dissipation. The shape and angles of cliffs along the beach can also amplify tsunami heights, in some cases tripling the height of the wave at the shore. Embayments that become narrower inland may amplify waves, and in some bays it may take two or even several tsunami crests for the amplification process to reach a maximum. Refraction effects can increase tsunami run-up around promontories where narrow strips of land jut out into the sea.

In some cases, tsunamis are refracted around the shores of islands, or both sides of bays, producing large edge waves that move parallel to shore. These edge waves must merge on the back sides of islands,

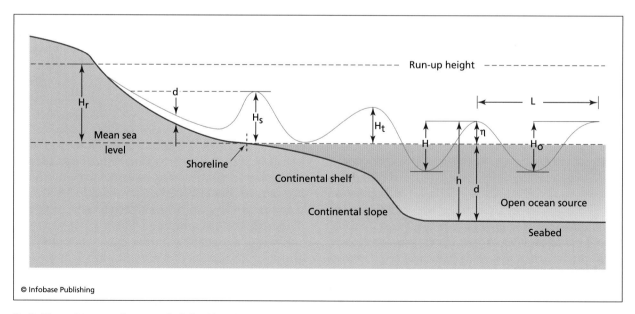

Definition of tsunami run-up height. Note how this is higher than the wave height at the coast.

or in the ends of embayments, and in these locations particularly large run-ups have been recorded. Movement of edge waves around islands accounts for many of the large run-ups on the leeward sides of the Hawaiian Islands from the April 1, 1946, tsunami in Hilo, Hawaii, where many bays on the back sides of the islands experience run-ups almost as high as those facing the initial wave front. The 1992 Flores Island tsunami also saw many villages on the leeward sides (facing away from the wave front) of islands washed away by large tsunamis. These waves also formed by the waves being refracted around the islands, forming edge waves that combined to cause unusually large run-ups in specific locations. More than 2,000 people died when the 16–23-foot (5–7-m) waves washed into these villages on the back sides of the islands. Similarly, in a 1993 tsunami in Japan, the island of Okushiri focused the energy of a tsunami on the town of Hamatsumae lying behind the island. On July 12, a 100-foot (30-m) wave grew behind the island and washed into the town, killing 330 people.

Force of Tsunami Impact and Backwash

When tsunamis crash into coastal areas they are typically moving at about 22 miles per hour (35 km/hr). The speed as the wave moves inland changes dramatically, decreasing to a few miles per hour (several km/hr) over short distances, depending on the slope of the beach or shore environment and how much resistance the wave encounters from obstacles on shore. The force associated with a debris-laden wall of water 50–70 miles (80–120 km) wide moving inland at that speed is tremendous. As tsunamis impact the shoreline and move inland they rapidly pick up debris and move this with the wave front, and these objects smash into whatever is in the path of the water, destroying almost anything in the way. The force of the tsunami can be appreciated by considering the impact of a series of rocks thrown from a moving car, or a train hitting a building at 22 miles per hour (35 km/hr). After the first impact the force of the wave does not stop but keeps on pounding into the coast until the crest passes; then the water continues to move inland and remain high for 30, 40, 50 minutes or more before retreating back to sea. The force of the tsunami backwash can be just as strong, and in some cases stronger than the initial impact. Some waves take five minutes or more to move inland, and less than two minutes to wash back out to sea, so the outgoing velocity may be greater than the initial surge. The outgoing waves often take the loose debris from the destruction of the incoming wave with them, placing projectiles in the water for the next crest to launch when it moves inland.

Many tsunamis have been observed to form a thin wedge of turbulent water that shoots out in front of the wave crest with tremendous speed and force. These turbulent wedges of foamy debris-laden water do a lot of damage to buildings and vegetation just before the wave crest hits, and may be associated with much higher velocities of projectiles than found in the main wave. These wedges may form by the weight of the wave compressing air trapped in front of the wave and shooting this air-water-debris mixture out as the wave moves inland.

In many cases the area of land that is flooded by a tsunami is roughly equal to the area found beneath the wave crest when it is close to shore. Larger tsunamis flood larger areas. The amount of flooding is greatest for flat open areas such as mudflats, pastures, etc., where the wave can move uninterrupted inland. The amount of inland penetration decreases for areas that have forests, buildings, or other obstacles that slow the wave down. A moderate tsunami, about 30 feet (10 m) high, might penetrate a little less than a mile (1.6 km) inland in flat but developed coastal areas, half a mile (less than a kilometer) in a developed downtown city environment, and perhaps four miles (6 km) on an undeveloped open coast. Dense coastal forests are able to significantly decrease the amount of inland penetration, taking much of the energy of the wave away as it must snap and move the trees to move further inland. Large tsunamis, greater than 160 feet (50 m) tall, can move inland 5–7 miles (9–12 km), while great tsunamis can theoretically reach tens of miles (km) inland.

REDUCING THE THREAT FROM TSUNAMIS

As more and more people move to the coastline in the United States and worldwide, ways must be found to reduce the threat from tsunamis. Tsunamis will inevitably form in the world's oceans and strike many shorelines in the years to come, but they do not have to kill a quarter million people like the Indian Ocean tsunami of 2004. Zones of high tsunami risk need to be mapped, such as places that tend to amplify the waves leading to higher run-ups, and the public needs to be educated about the signs of an impending tsunami. It is important for the public to know not only how to recognize a tsunami, but what to do when one may be approaching. Different types of tsunami warning systems can be built, and new methods of building along coastlines to reduce tsunami damage should be encouraged.

Monitoring Tsunami Threats

Many countries around the Pacific cooperate in monitoring the generation and movement of tsunamis. The seismic sea wave warning system was established and became operational after the great 1946 tsunami that devastated Hilo, Hawaii, parts of Japan, and many other circum-Pacific coastlines. The seismic sea wave warning system and other tsunami warning systems generally operate by monitoring seismograms to detect potentially seismogeneic earthquakes, then monitor tide gauges to determine if a tsunami has been generated. Warnings are issued if a tsunami is detected, and special attention is paid to areas that have greater potential for being inundated by the waves.

It takes several different specialists to be able to warn the public of impending tsunami danger and reduce the threat from tsunamis. First, seismologists are needed to monitor and quickly interpret the earthquakes and determine which ones are potentially dangerous for tsunami generation. Second, oceanographers are needed to predict the travel characteristics of the tsunami. Coastal geomorphologists must interpret the shape of coastlines and submarine topography to determine which areas may be the most prone to being hit by tsunamis, and geologists are needed to search for any possible ancient tsunami deposits to see what the history of tsunami run-up is along specific coastlines. Finally, engineers are needed to try to modify coastlines to reduce the risk from tsunamis. Features such as sea walls and breakwaters can be built, and buildings can be sited in places that are outside of reasonable tsunami striking distance. Loss control engineers typically work with insurance underwriters to identify areas and buildings that are particularly prone to tsunami-related flooding.

Several detailed reports have described areas that are particularly prone to repeated tsunami hazards. The United States Army Engineer Waterways Experiment Station has produced several of these reports useful for city planners, the Federal Insurance Administration, and state and local governments.

The National Oceanic and Atmospheric Administration (NOAA) has been operating tsunami gauges in the deep ocean since 1986. These instruments must be placed on the deep seafloor (typically up to 1,500–2,000 feet [1,000 m] depth) and recovered and redeployed each year. Cables send the recordings from these instruments back to shore. The information derived from these tsunami gauges is used for tsunami warning systems and also used for planning coastal development, since the pressure changes associated with tsunami can be accurately recorded over long periods of time, and the history of tsunami heights in given areas assessed before coastal zones are further developed.

The NOAA runs Pacific Tsunami Warning Center in Honolulu. The United States Geological Survey also has been actively engaged in mapping tsunami hazard areas and establishing ancient tsunami run-up heights on coastlines prone to tsunamis, to help in

predicting future behavior in individual areas. The results from these mapping programs are routinely presented to local government planning boards, to help in protecting people in coastal areas and assigning risks to development in areas prone to tsunamis.

Predicting Tsunamis

Great progress has been made in predicting tsunamis, both in the long term and in the short term, following tsunami earthquakes. Much of the long-term progress reflects recognition of the association of tsunamis with plate tectonic boundaries, particularly convergent margins. Certain areas along these convergent margins are susceptible to tsunami-generating earthquakes, either because of the types of earthquakes that characterize that region, or because thick deposits of loose unconsolidated sediments characteristically slide into trenches in other areas. Progress in short-term prediction of tsunamis stems from the recognition of the specific types of seismic wave signatures that are associated with tsunami-generating earthquakes. Seismologists are in many cases able to immediately recognize certain earthquakes as potentially tsunami-generating and issue an immediate warning for possibly affected areas.

Tsunamis are generated mainly along convergent tectonic zones, mostly in subduction zones. The motion of the tsunami-generating faults in these areas is typically at right angles to the trench axis. After many years of study, geologists have documented a relationship between the direction of the motion of the fault block and the direction toward which most of the tsunami energy (expressed as wave height) is directed. Most earthquakes along subduction zones move at right angles to the trench, and the tsunamis are also preferentially directed at right angles away from the trench. This relationship causes certain areas around the Pacific to be hit by more tsunamis than others, because there is a preferential orientation of trenches around the Pacific. Most tsunamis are generated in southern Alaska (the tsunami capital of the world), are directed toward Hawaii, and glance the west coast of the lower 48 states, whereas earthquakes in South America direct most of their energy at Hawaii and Japan.

Tsunami Hazard Zones and Risk Mapping

The USGS and other Civil Defense agencies have mapped many areas that are particularly prone to tsunamis. Recent tsunamis, historical records, and deposits of ancient tsunamis identify some of these areas. Many coastal communities, especially those in Hawaii, have posted coastal areas with tsunami warning systems, showing maps of specific areas prone to tsunami inundation. Tsunami warning signals are in place and residents are told what to do and where to go if the alarms are sounded. Residents and visitors in these areas must understand what to do in the event of a tsunami warning, and where the escape routes may be. If there is a local earthquake, landslide, or undersea volcanic eruption, there may be only minutes before a tsunami hits, so it is essential not to waste time if the tsunami warning sirens are sounded.

Tsunami Warning Systems

Tsunami warning systems have been developed that are capable of saving many lives by alerting residents of coastal areas that a tsunami is approaching their location. These systems are most effective for areas located more than 500 miles (750 km), or one hour away from the source region of the tsunami, but may also prove effective at saving lives in closer areas. The tsunami warning system operating in the Pacific Ocean basin integrates data from several different sources, and involves several different government agencies. The National Oceanic and Atmospheric Administration operates the Pacific Tsunami Warning Center in Honolulu, which includes many seismic stations that record earthquakes, and quickly sorts out those earthquakes that are likely to be tsunamogenic based on the earthquake's characteristics. A series of tidal gauges placed around the Pacific monitors the passage of any tsunamis past their locations, and if these stations detect a tsunami, warnings are quickly issued for local and regional areas likely to be affected. Analyzing all of this information takes time, however, so this Pacific-wide system is most effective for areas located far from the earthquake source.

Tsunami warning systems designed for shorter-term, more local, warnings are also in place in many communities, including Japan, Alaska, Hawaii, and many other Pacific islands. These warnings are based mainly on quickly estimating the magnitude of nearby earthquakes, and the ability of public authorities to rapidly issue the warning so that the population has time to respond. For local earthquakes, the time between the shock event and the tsunami hitting the shoreline may be only a few minutes. Anyone in a coastal area that feels a strong earthquake should take that as a natural warning that a tsunami may be imminent and leave low-lying coastal areas. This is especially important considering that approximately 99 percent of all tsunami-related fatalities have historically occurred within 150 miles (250 km) of the tsunami's origin, or within 30 minutes of when the tsunami was generated.

Knowing When a Tsunami Is Imminent

Anybody who is near the sea or in an area prone to tsunamis (as indicated by warning signs in places like Hawaii) needs to pay particular attention to some of the subtle and not so subtle warning signs that a tsunami may be imminent. First, there may be warning sirens in areas that are equipped with a tsunami warning system. If the sirens are sounded, it is imperative to move to high ground immediately. People in more remote locations may need to pay attention to the natural warning signs. Anyone on the shore who feels an earthquake should run for higher ground. A tsunami may hit within minutes, an hour or two, or not at all, but it is better to be safe than sorry. Tsunamis travel in groups with periods between crests that can be an hour or more, so there are likely to be several crests over a period of many hours. Many people have died when they returned to the beach to investigate the damage after the first crest passes. If the tsunami-generating earthquake occurred far away, there may not be any detectable ground motion before a tsunami hits, and residents of remote areas may not have any warning of the impending tsunami, except for the thunderous crash of waves right before it hits the beachface. In other cases, the water may suddenly recede to unprecedented levels right before it quickly rises up again in the tsunami crest. In either case, any one enjoying the beachfront needs to remain aware of the dangers. In general, the heads of bays receive the highest run-ups, and the sides and mouths record lower run-up heights. but this may vary considerably depending on the submarine topography and other factors.

Public Education and Knowing How to Respond

Communities situated in tsunami-prone areas need to educate their populace on how to respond in a tsunami emergency, especially how to follow directions of emergency officials and to stay away from the waterfront. People in low-lying areas should have planned routes to quickly seek high ground, such as by climbing a hill, or entering into a tall building designed to withstand a tsunami. Most important, people should know not to return to the seafront until they are instructed by authorities that it is safe to do so, because there may be more waves coming, spaced an hour or more apart.

Many United States and foreign agencies have completed exhaustive studies of ways to mitigate or reduce the hazards of tsunamis. For instance, the Japanese Disaster Control Research Center has worked with the Ministry of Construction to evaluate the effects of tsunamis on their coastal road network, to plan better for inundation by the next tsunami. They considered historical records of types of damage to the road systems (wash-outs, flooding, blockage by debris from destroyed homes and cars, etc.) and detailed surveys of the region to devise a plan of alternate roads to use during tsunami emergencies. They have devised a mechanism of communicating the immediate danger and alternate plans to motorists. Their system includes plans for the installation of multiple wireless electronic bulletin boards at key locations, warning motorists to steer away from hazardous areas. Similar studies have been undertaken by other agencies that deal with the coastal area, such as the Fisheries Agency. They have suggested building a series of levees, emergency gates, and cut-off facilities to maintain the fresh water supply to residents.

SUMMARY

Tsunamis are long-wavelength deep sea waves formed by the sudden displacement of large volumes of seawater. When these waves encounter shallow water, they may form huge breaking waves with walls of water tens or a hundred feet (tens of m) tall that slam ashore. Every few years some of these giant waves rise unexpectedly out of the ocean and sweep over coastal communities, killing thousands of people and causing millions of dollars worth of damage. Triggering mechanisms for tsunamis include earthquake-related displacements of the seafloor, submarine slumps and landslides that displace seawater, submarine volcanism, explosive release of methane gas from deep-ocean sediments, and asteroid impacts.

Tsunamis have wavelengths of 120 miles (200 km) or greater, periods of 1.6–33 minutes, and travel at speeds of 375–600 miles per hour (800–900 km/hr), compared to 55 miles per hour (90 km/hr) for normal wind-blown ocean waves. The effective wave base for tsunamis is therefore deeper than the ocean basins. Tsunamis feel friction from the deep ocean basins and are effectively refracted and reflected around the world's oceans. When tsunamis encounter shallow water they slow their forward velocity, and the waves behind the waves in the front move faster, pile up behind the first waves, and increase the amplitude of the waves crashing into the beach. Run-up is the height of the tsunami above sea level at the farthest point it reaches on shore. This height may be considerably different from the height of the wave where it first hits the shore, and is commonly twice that of the height of the wave at the shore. When tsunamis crash into coastal areas they are commonly moving at 22 miles per hour (35 km/hr). The force associated with the tsunami hitting the shoreline is tremendous, as it consists at this point of a debris-laden wall of water, 50–70 miles (80–120 km) wide, moving steadily inland like an unstoppable derailed locomotive. Since tsunamis are long-wavelength waves, they continue to move inland and remain high for 30–50 minutes before rapidly retreating with a force that may exceed the force of initial incursion. Tsunamis travel in wave trains, so this process may repeat itself six or seven times over the course of many hours, with the second or third wave often being the tallest.

The threat to coastal communities from tsunamis can be reduced. First, the historical record of tsunamis in any area should be determined through geological mapping, and areas at greatest risk need to be identified. Residents and visitors to these areas should know escape routes, and what to do in the event of a tsunami emergency. Most ocean basins now have seismic sea wave warning systems installed, allowing scientists to monitor triggering mechanisms (such as earthquakes) and sea-bottom pressure and motion detectors that tell of passing tsunamis. Warnings can be issued to coastal communities when it is determined that a tsunami may be approaching. Public education programs should teach the public about the hazards of tsunamis, since many of the deaths from tsunamis have been preventable. For instance, many people have perished because they have returned to coastal areas after the first wave crest has passed, not realizing that the second or third crest may be the largest, and many crests hit hours after the first wave crashes ashore. Areas in the United States most at risk include the Pacific Northwest, where a potentially tsunamogenic forearc zone is located close to densely populated coastal areas. Any large earthquake in this area could form a tsunami that crashes in coastal communities within minutes of the earthquake, barely giving residents enough time to evacuate to higher ground. Most other coastal areas of the United States also have some risk of tsunamis, so all coastal residents need to be aware of the risks and how to respond in a tsunami emergency.

See also BEACHES AND SHORELINES; EARTHQUAKES; GEOLOGICAL HAZARDS; PLATE TECTONICS; TSUNAMIS, HISTORICAL ACCOUNTS.

FURTHER READING

Bernard, E. N., ed. *Tsunami Hazard: A Practical Guide for Tsunami Hazard Reduction.* Dordrecht, The Netherlands: Kluwer Academic Publishers, 1991.

Booth, J. S., D. W. O'Leary, P. Popencoe, and W. W. Danforth. "U.S. Atlantic Continental Slope Landslides: Their Distribution, General Attributes, and Implications." *United States Geological Survey Bulletin* 2002. (1993): 14–22.

Bryant, E. *Tsunami: The Underrated Hazard.* Cambridge: Cambridge University Press, 2001.

Dawson, A. G., and S. Shi. "Tsunami Deposits." *Pure and Applied Geophysics* 157 (2000): 493–511.

Driscoll, N. W., J. K. Weissel, and J. A. Goff. "Potential for Large-Scale Submarine Slope Failure and Tsunami Generation along the United States' Mid-Atlantic Coast." *Geology* 28 (2000): 407–410.

Dvorak, J., and T. Peek. "Swept Away." *Earth* 2, no. 4 (1993): 52–59.

Kusky, T. M. *Tsunamis: Giant Waves from the Sea.* New York: Facts On File, 2008.

Latter, J. H. "Tsunami of Volcanic Origin: Summary of Causes, with Particular Reference to Krakatau, 1883." *Journal of Volcanology* 44 (1981): 467–490.

Los Alamos National Laboratory, Tsunami Society. Available online. URL: http://library.lanl.gov/tsunami/. Accessed January 26, 2009.

McCoy, F., and G. Heiken. "Tsunami Generated by the Late Bronze Age Eruption of Thera (Santorini), Greece." *Pure and Applied Geophysics* 157 (2000): 1,227–1,256.

Minoura, K., F. Inamura, T. Nakamura, A. Papadopoulos, T. Takahashi, and A. Yalciner. "Discovery of Minoan Tsunami Deposits." *Geology* 28 (2000): 59–62.

National Oceanographic and Atmospheric Association, tsunami research program. Available online. URL: http://www.pmel.noaa.gov/tsunami/. Accessed January 26, 2009.

National Tsunami Hazard Mitigation Program home page. Available online. URL: http://nthmp.tsunami.gov/. Accessed October 10, 2008.

Revkin, A. C. "Tidal Waves Called Threat to East Coast." *The New York Times* (2000): A18.

Satake, K. "Tsunamis." *Encyclopedia of Earth System Science* 4 (1992): 389–397.

Steinbrugge, K. V. *Earthquakes, Volcanoes, and Tsunamis: An Anatomy of Hazards.* New York: Skandia America Group, 1982.

Tsuchiya, Y., and N. Shuto, eds. *Tsunami: Progress in Prediction Disaster Prevention and Warning.* Boston: Kluwer Academic Publishers, 1995.

Tsunamis.com. Available online. URL: http://www.tsunamis.com/tsunami-pictures.html. Accessed January 26, 2009.

United States Geological Survey. "Surviving a Tsunami— Lesson from Chile, Hawaii, and Japan." United States Geological Survey Circular 1187, 1987. Available online. URL: http://pubs.usgs.gov/circ/c1187/. Accessed September 8, 2009.

tsunamis, historical accounts Tsunamis have taken hundreds of thousands of lives in the past few hundred years, and some of the larger tsunamis have caused millions to billions of dollars in damage. The table "Historical Tsunamis" lists the more significant tsunamis that are well documented in recorded history.

EARTHQUAKE-INDUCED TSUNAMIS

The most common type of tsunamis are those generated by earthquakes that displace the seafloor. The majority of these are initiated at convergent margins such as along trench-forearc accretionary wedges, where large displacements of the seafloor are relatively common. Most convergent margins are located around the edges of the Pacific Ocean, so it follows that most tsunamis strike the margins of the Pacific Ocean. Many others are generated along the convergent margins in the Indonesian region, including the devastating December 26, 2004, Indian Ocean tsunami.

December 26, 2004, Indian Ocean Tsunami

One of the deadliest natural disasters in history unfolded on December 26, 2004, when a great undersea earthquake with a magnitude of 9.0 triggered a tsunami that devastated many coastal areas of the Indian Ocean, killing an estimated 283,000 people. The Sumatra-Andaman earthquake had an epicenter located 100 miles (160 km) off the west coast of Sumatra, and struck at 7:58 A.M. local time. This was a very unusual earthquake, in that the rupture (and quake) lasted between eight and 10 minutes, one of the longest times ever recorded for an earthquake. The hypocenter, or point of first energy release, was located 19 miles (30 km) below sea level, and the rupture length of the fault extended to a remarkable 750 miles (1,200 km) along the coast of Sumatra. Shaking from the earthquake was felt as far away as India, Thailand, Singapore, and the Maldives. The energy released by this earthquake was so great that it set the whole planet into a set of slow oscillations where all locations on the planet were vibrating back and forth by 8–12 inches (20–30 cm) initially, with a force roughly equivalent to the attraction between the Earth and Moon. The surface waves from the earthquake also traveled around the planet, producing a vibration of at least one-third of an inch (1 cm) everywhere on the planet. These vibrations gradually diminished in intensity over a period of a week, until they became so small that they were difficult to measure. The amount of energy released by this earthquake alone was roughly one-eighth of the energy released by all earthquakes on the planet in the past 100 years.

The tectonic setting of the Sumatra-Andaman earthquake was in the forearc of the active convergent margin between the Indian-Australian plate and the Burma plate of Eurasia. Oceanic crust of the Indian-Australian plate is being subducted beneath Sumatra and Indonesia at about two inches (6 cm) per year, forming a complex of volcanoes and active fault zones. The December 26, 2004, Sumatra-Andaman earthquake had characteristics that were extremely favorable for producing a large tsunami. A huge section of the forearc, 750 miles (1,200 km) long, was pushed upward and sideways by 50 feet (15 m), displacing a vast amount of water and sending it across the Indian Ocean as a giant tsunami. This displacement took place in two stages. First a 250-mile- (400-km-) long by 60-mile- (100-km-)

HISTORICAL TSUNAMIS

Location	Year	Number of Deaths
Indian Ocean	2004	> 283,000 dead across region
Santorini	1500 B.C.E.	devastation of Mediterranean
Eastern Atlantic	November 1, 1755	60,000 dead in Lisbon, estimated 100,000 total
Messina, Italy	1908	70,000 dead
Taiwan	May 22, 1782	50,000 dead
Krakatau	August 27, 1883	36,500 dead
Nankaido, Japan	October 28, 1707	30,000 dead
Sanriku, Japan	June 15, 1896	27,122 dead
Nankaido, Japan	September 20, 1498	26,000 dead
Arica, Chile, Peru	August 13, 1868	25,674 dead
Sagami Bay, Japan	May 27, 1293	23,024 dead
Guatemala	February 4, 1976	22,778 dead
Lima, Peru	October 29, 1746	18,000 dead
Bali, Indonesia	January 21, 1917	15,000 dead
Unzen, Japan	May 21, 1792	14,524 dead
Ryukyu, Japan	April 24, 1771	13,486 dead
Bali, Indonesia	November 22, 1815	10,253 dead
Guangzhou, China	May 1765	10,000 dead
Moro Bay, Philippines	August 16, 1976	8,000 dead
Honshu, Japan	March 2, 1933	3,000 dead
Indonesia	December 12, 1992	2,000 dead
Chile	May 22, 1960	2,231 known dead/missing, estimated 10,000 dead
Aleutians	April 1, 1946	150 in Hawaii, $25 million damage
Alaska	March 28, 1964	119 dead in California, $104 million damage
Nicaragua	September 2, 1992	170 dead
Indonesia	December 2, 1992	137 dead

wide rupture formed, ripping the rocks of the seafloor at a rate of 1.7 miles per second (2.8 km/sec), or in other words, at 6,300 miles per hour (10,000 km/hr). This rupture not only occurred rapidly, but represents the biggest rupture known to have ever been created by a single earthquake. After a break of less than two minutes, the rupture continued to propagate northward from the Aceh area at a slower rate (1.3 miles per second, or 2.1 km/sec) for another 500 miles (800 km) toward the Andaman and Nicobar Islands. Displacements of the seafloor changed the capacity of the Indian Ocean basin to hold water, slightly raising global sea levels by about 0.03 inch (0.01 cm).

Submarine sonar surveys of the seafloor by the British Navy vessel H.M.S. *Scott* revealed that several huge, fault-related submarine ridges collapsed during the earthquake, creating submarine landslides, some as large as 7 miles (10 km) across. The amount that these contributed to the formation of the tsunami is not known, but certainly less than the huge displacement of the entire seafloor of the forearc.

The vertical component of motion of the seafloor during the earthquake is estimated to have displaced about seven cubic miles (30 km³) of seawater, producing a tsunami that radiated outward from the entire 750-mile- (1,200-km-) long rupture area, eventually reaching most of the world's oceans including the Pacific, Atlantic, and even the Arctic Ocean. The tsunami continued to travel around the Earth for days, with very small amplitudes.

Thousands of aftershocks followed the main earthquake for days and months after the main event, gradually decreasing in strength and frequency. The largest of these was a magnitude 8.7 event that occurred on March 28, 2005, in virtually the same location along the same fault, with events of up to magnitude 6.7 continuing for more than four months after the main earthquake. The distribution of aftershocks greater than magnitude 4 outlines an area beneath the forearc of the Sumatra-Andaman arc that moved or slipped as a result of the earthquake. The slipped area is roughly the size of the state of California.

The amount of energy that was released in the December 26, 2004, Sumatra earthquake is staggering. Estimates vary between energy equivalents of 250–800 megatons of TNT, or the amount of energy consumed within the entire United States over 3–11 days. This energy caused some interesting effects. First, the change in the shape of the Earth caused by the displacement changed the length of the day by a minute amount (2.58 microseconds). However, this effect is already worn off since the tidal friction of the moon increases the length of the day by about 15 microseconds per year. The change in mass distribution also changed the amount the Earth wobbles about its rotational axis by about an inch (2.5 cm), but since the natural wobble is about 50 feet (15 m), this is not a large amount and will be evened out by future earthquakes.

In addition to triggering one of the worst tsunami disasters of history, the December 26, 2004, Sumatra earthquake awakened the dormant volcano of Mount Talang, which erupted in Aceh Province in April 2005. This is one of the rare cases where energy from an earthquake can convincingly be shown to have initiated other geologic activity.

The tsunami from the December 26 earthquake was unprecedented in the amount of observation that was possible from satellites. The satellite data were not analyzed until after the event, and the satellites were therefore not used to help provide a warning to areas about to be hit by the tsunami, but the

People flee as tsunami crashes at Koh Roya, part of Thailand's territory in the Andaman Islands, December 26, 2004. *(John Russell/AFP/Getty Images)*

results show that it is possible to develop a satellite-based tsunami warning system. Radar satellites showed that while in deep water, in the middle of the Indian Ocean, the tsunami had a maximum height of only 2 feet (60 cm) and that the wave rose to enormous height when it moved into shallow water. For instance, the wave was more than 80 feet (24 m) high as it approached much of Aceh Province in Indonesia and rose to 100 feet (30 m) or more in some places as it moved inland. In some places the wave moved inland by more than 1.25 miles (2 km).

Although the energy of the tsunami was much less than that of the earthquake, it still had a remarkably high energy equivalent of about five megatons of TNT. For comparison, this is more than double the amount of energy released by all the bombs and explosions (including the atomic bombs) in all of World War II.

Since the fault that produced the December 26 earthquake and tsunami was oriented nearly north-south, and the motion of the upper plate was up and to the west, most of the energy of the tsunami was focused into an east-west direction. Therefore the biggest tsunami waves moved westward from the 750-mile- (1,200-km-) long fault rupture, while smaller waves moved out in all directions. Many areas along the coast of northern Indonesia were hit quickly, less than 15 minutes from the initial earthquake, while it took the waves 90 minutes to two hours to reach the southern end of India and Sri Lanka. The tsunami hit parts of Somalia several hours later, and swept down the African coast until it struck South Africa about 16 hours after the quake with a crest five feet (1.5 m) high. Tidal stations in Antarctica recorded a wave three feet (1 m) high, with oscillations lasting for a couple of days after the first wave. The energy from the wave next moved into the Pacific and Atlantic Oceans. Some unusual focusing of the wave energy may have occurred, perhaps by the mid-ocean–ridge system, since some of the waves that hit the west coast of Mexico were 8.5 feet (2.6 m) high.

The Indian Ocean tsunami was particularly tragic because there was no tsunami warning system in place in the Indian Ocean, and most victims were taken by total surprise when the tsunami struck their areas. Even though the wave took hours to move around the ocean, there was not even a simple telephone communication network set up to alert residents of one coastal community that others had just perished in a tsunami, and the tsunami was moving their way. Such a simple warning system could have saved tens of thousands of lives. Since the tragedy of December 26, 2004, countries of the Indian Ocean have worked with the United Nations and other countries to establish a tsunami warning system that includes not only the Indian Ocean but also the Atlantic and Caribbean.

In most places the tsunami struck as a series of waves after an initial retreat of the sea, followed by a large crest moving ashore. There were about 30 minutes between each wave crest that rose into coastal areas. In most places the third wave was the largest, although many smaller tsunami crests continued to strike throughout the day. In a few locations people recognized early warning signs of the tsunami and successfully evacuated to safety. The most famous case is that of 10-year-old British citizen Tilly Smith, who paid attention to her studies about tsunamis in school and knew that when the waters on the beach rapidly retreat it is a sign of an impending tsunami. Tilly frantically warned her parents, who led an evacuation of the beach, saving many lives. Likewise, a Scottish science teacher named John Chroston recognized similar warning signs in a bay north of Phuket beach on Indonesia, and took a busload of tourists away from the beach front, saving lives. Some native communities also recognized early warning signs and retreated to safety. For instance, islanders on Simeulue, near the epicenter, fled away from the coast after they felt the initial earthquake, undoubtedly saving many lives. In some of the more puzzling responses to precursory phenomena, it is reported that elephants and other animals on some Indian Ocean islands of the Maldives chain south of India fled inland before the tsunami struck. Elephants have very astute hearing and may have felt the ground shaking from the approaching tsunami, so ran in fear into the dense forest.

Many countries in southeast Asia have developing economies and often sacrifice environmental concerns to advance economic development. To this end, some of these countries including Indonesia have been promoting the development of the shipping industry and the growth of shrimp farming in coastal regions. Many coral reefs have been blown up and coastal mangrove forests destroyed to let these industries grow faster. In other places coastal sand dunes have been removed to enhance growth of the coastal region. All of these natural ecosystems are fragile coastal systems that not only preserve habitats for a diverse set of species but also serve as a powerful shield from the force of incoming tsunamis. In areas where the reefs and forests were preserved, for instance the Surin Island chain off Thailand, the force of the tsunami was broken by these natural barriers. In other places, where these barriers have been removed, the waves crashed ashore with much greater force, killing those who would seemingly benefit by the destruction of their natural protective barrier. Many governments began to realize the value of these natural barriers after the tsunami and may make efforts to reverse, stop, or slow their destruction.

1755 Eastern Atlantic Tsunami

Tsunamis do not regularly strike Atlantic regions, with only about 10 percent of all tsunamis occurring in the Atlantic Ocean. A few tsunamis have been associated with earthquakes in the Caribbean region, such as in 1867 in the Virgin Islands, 1918 in Puerto Rico, and on June 6, 1692, when 3,000 people were killed by a tsunami that leveled Port Royal, Jamaica. The most destructive historical tsunami to hit the Atlantic region struck on November 1, 1755. Lisbon, Portugal, was the worst hit because it was near the epicenter of the earthquake that initiated the tsunami. At least three large waves, each ranging in height from 14 to 40 feet (4–12 m), struck Lisbon in quick succession, killing at least 60,000 people in Lisbon alone. England was hit by waves 6–10 feet (2–3 m) in height, and the tsunami even affected the Caribbean region, hitting Antigua with 12-foot (4 m) waves, and waves more than 20 and 15 feet (6 and 5 m) respectively in height swept Saba and St. Martin.

The Lisbon-Eastern Atlantic tsunami was generated by a large earthquake whose epicenter was located about 60 miles (100 km) southwest of Lisbon, probably on the boundary between the European and Azores-Gibraltar plates. The earthquake had an estimated magnitude of 9.0, and the shaking lasted for 10 minutes. During this time, three exceptionally large jolts occurred, causing massive destruction in Lisbon and the Moroccan towns of Fez and Mequinez. Most of Europe and Scandinavia reported seiche waves from lakes and inland water bodies.

The earthquake caused extensive damage in the city of Lisbon, toppling many buildings and causing widespread and uncontrolled fires in the city. In fear, many residents ran to the city docks in the harbor and on the Tagus River. As the buildings continued to collapse and the fire raged through the city, driving much of the population of 275,000 people to the waterfront, the worst part of this disaster was about to unfold. About 40–60 minutes after the massive earthquake, residents of Lisbon watched from the docks as the water rapidly drained out of the harbor, as if someone had pulled the plug in the bottom of a bathtub. A few minutes later a massive wall of water 50 feet (15 m) high swept up the harbor and over the docks, and then swept more than 10 miles (16 km) upriver. The tsunami was associated with a powerful backwash that dragged tens of thousands of people, bodies, and debris into the harbor. Two more giant waves rushed into the city an hour apart, and killed more of the terrified residents trapped between raging tsunami waves and a city crumbling under the forces of fire and earthquake aftershocks. About 60,000 people, nearly a quarter of the city's population, perished in the tsunami.

The November 1, 1755, earthquake caused extensive damage across the eastern Atlantic Ocean, with tsunamis sweeping the coasts of North Africa, Portugal, Spain, France, and the British Isles, and even islands in the Caribbean were affected. The tsunami moved inland by 1.5 miles (2.5 km) across low-lying areas of Portugal, and run-up heights around Portugal locally reached 65–100 feet (20–30 m) above sea level. Southern Portugal was the worst hit, where medieval fortresses and towns were destroyed or suffered heavy damage. The waves washed over the ancient walled city of Lagos, whose walls are anchored 36 feet (15 m) above sea level. The walls reduced the force of the waves, but the city was flooded, and the water had to drain out of the narrow city gates. The initial wave was followed by at least 18 secondary crests in southern Portugal, each adding damage upon the effects of the previous wave.

Western Europe was strongly hit by the tsunami as it spread northward. The south coast of England saw massive waves that tore up the coastal muds and sandbars, and the waves swept the shores of the Bay of Biscay. Boats in the North Sea were ripped from their moorings. The English Channel was swept by waves 10–13 feet (3–4 m) high at high tide, followed by even higher waves that oscillated every 10–20 minutes in the channel over the next five hours. The Azores, located offshore from Portugal, were hit by 50-foot (15-m) waves that raced across the Atlantic to the eastern seaboard of North America. Caribbean islands were hit by tsunamis with run-up heights in the range of 20–25 feet (6–7.5 m), with the worst hit areas reported to be St. Martin and Saba. Elsewhere in the Caribbean, 10–15-foot (3–4-m) waves were reported to have oscillated every five minutes for about three hours, affecting many harbor and coastal areas.

The 1992 Flores, Indonesia, Tsunami
One of the more deadly tsunamis in recent history hit the island of Flores, located in Indonesia several hundred miles from the coast of northern Australia near the popular resort island of Bali. The tsunami hit on December 12, 1992, and was triggered by a magnitude 7.9 earthquake, with the earthquake faulting event lasting for a long 70 seconds. The tsunami had run-up heights of 15 to 90 (4–27 m) feet along the northeastern part of Flores Island, where more than 2,080 people were killed and at least another 2,000 injured. Large amounts of sediment slumped underwater on the north side of the island during the earthquake, generating the unusually large and destructive tsunami.

Flores is an island located above the subduction zone where the Indian-Australian plate is being pushed beneath the Eurasian plate, along a convergent boundary. This subduction zone produces some of the largest earthquakes and tsunamis in the world, including the December 24, 2004, massive Indian Ocean tsunami. The earthquake that generated the Flores tsunami had its epicenter near the coast, and produced subsidence of 1.5–3.5 feet (0.5–1.0 m). Unlike most other tsunami-generating earthquakes in this part of the world, this earthquake was located on the back-arc side of the island, away from the trench and forearc. The type of fault that forms on the back side of the arc is called a back-thrust, which is what slipped producing the Flores tsunami.

There was very little warning time for the residents of Flores Island, because the epicenter of the earthquake that generated the tsunami was only 30 miles (50 km) off the northern coast of the island. The first waves hit less than five minutes after the initial earthquake shock, with five or six individual waves being recorded by residents in different places. Many places recorded three large waves, the first one preceded by a rapid withdrawal of water from the coast, then the wave arrived as a wall of water. In most locations the second wave was the largest, with run-up heights typically between 6–16 feet (2–5 m). However, run-up heights in the village of

TSUNAMI NIGHTMARE

One of the worst natural disasters of the 21st century unfolded on December 26, 2004, following a magnitude 9.0 earthquake off the coast of northern Sumatra in the Indian Ocean. The earthquake was the largest since the 1964 magnitude 9.2 event in southern Alaska and released more energy than all the earthquakes on the planet in the last 25 years combined. During this catastrophic earthquake, a segment of the seafloor the size of the state of California, lying above the Sumatra subduction zone trench, suddenly moved upward and seaward by more than 30 feet (9 m). The sudden displacement of this volume of undersea floor displaced a huge amount of water and generated the most destructive tsunami known in recorded history.

Within minutes of the initial earthquake a mountain of water more than 100 feet (30 m) high was ravaging northern Sumatra, sweeping into coastal villages and resort communities with a fury that crushed all in its path, removing buildings and vegetation and in many cases eroding shoreline areas down to bedrock, leaving no traces of the previous inhabitants or structures. Similar scenes of destruction and devastation rapidly moved up the coast of nearby Indonesia, where residents and tourists were enjoying a holiday weekend. Firsthand accounts and numerous videos made of the catastrophe reveal similar scenes of horror, where unsuspecting tourists and residents were enjoying themselves in beachfront playgrounds, resorts, and villages and reacted as large breaking waves appeared off the coast. Many moved toward the shore to watch the high surf with interest, then ran in panic as the sea rapidly rose beyond expectations, and walls of water engulfed entire beachfronts, rose above hotel lobbies, and washed through towns with the force of Niagara Falls. In some cases the sea retreated to unprecedented low levels before the waves struck, causing many people to move to the shore to investigate the phenomenon; in other cases, the sea waves simply came crashing inland without warning. Buildings, vehicles, trees, boats, and other debris were washed along with the ocean waters, forming projectiles that smashed at speeds of up to 30 miles per hour (50 km/hr) into other structures, leveling all in their paths, and killing more than a quarter million people.

The displaced water formed a deepwater tsunami that moved at speeds of 500 miles per hour (805 km/hr) across the Indian Ocean, smashing within an hour into Sri Lanka and southern India, wiping away entire fishing communities and causing additional widespread destruction of the shore environment. Ancient Indian legends speak of villages that have disappeared into the sea, stories that many locals now understand as relating times of previous tsunamis, long since forgotten by modern residents. South of India lie many small islands including the Maldives, Chagos, and Seychelles, many of which have maximum elevations of only a few to a few tens of feet (1–10 m) above normal sea level. As the tsunami approached these islands, many wildlife species and primitive tribal residents fled to the deep forest, perhaps sensing the danger as the sea retreated and the ground trembled with the approaching wall of water. As the tsunami heights were higher than many of the maximum elevations of some of these islands, the forest was able to protect and save many lives in places where the tsunami caused sea levels to rise with less force than in places where the shoreline geometry caused large breaking waves to crash ashore.

Several hours later the tsunami reached the shores of Africa and Madagascar, and though its height was diminished to less than 10 feet (3 m) with distance from the source, several hundred people were killed by the waves and high water. Kenya and Somalia were hit severely, with harbors experiencing rapid and unpredictable rises and falls in sea level and many boats and people washed to sea. Villages in coastal eastern Madagascar, recently devastated by tropical cyclones, were hit by large waves, washing homes and people into the sea, and forming new coastal shoreline patterns.

The tsunami traveled around the world with progressively decreasing height with time, as measured by satellites and ocean bottom pressure sensors more than 24 hours later in the North Atlantic and Pacific Oceans. Overall, more than 283,000 people perished in the December 26th Indian Ocean tsunami, though many might have been saved if a tsunami warning system had been in place in the Indian Ocean. Tsunami warning systems have been developed that are capable of saving many lives by alerting residents of coastal areas that a tsunami is approaching their location. These systems are most effective for areas located more than 500 miles (805 km), or one hour away, from the source region of the tsunami but may be effective at saving lives in closer areas. The tsunami warning system operating in the Pacific Ocean basin integrates data from several different sources and involves several different government agencies. The National Oceanic and Atmospheric Administration operates the Pacific Tsunami Warning Center in Honolulu. It includes many seismic stations that record earthquakes and quickly sorts out those earthquakes that are likely to be tsunamogenic based on the earthquake's characteristics. A series of tidal gauges placed around the Pacific monitors the passage of any tsunamis past their locations, and if these stations detect a tsunami, warnings are quickly issued for local and regional areas likely to be affected. Analyzing all of this information takes time, however, so this Pacific-wide system is most effective for areas located far from the earthquake source.

Tsunami warning systems designed for shorter-term, more local, warnings are also in place in many communities, including Japan, Alaska, Hawaii, and many other Pacific islands. These warnings are based mainly on quickly estimating the magnitude of nearby earthquakes and the ability of public authorities to rapidly issue the warning so that the population has time to respond. For local earthquakes, the time between the shock event and the tsunami hitting the shoreline may be only a few minutes. Anyone in a coastal area that feels a strong earthquake should take that as a natural warning that a tsunami may be imminent, and should quickly leave low-lying coastal areas. This is especially important considering that approximately 99 percent of all tsunami-related fatalities have historically occurred within 150 miles (250 km) of the tsunami's origin, or within 30 minutes of when the tsunami was generated.

The magnitude 9.0 Sumatra earthquake that caused the Indian Ocean tsunami was detected by American scientists who tried to warn countries in soon-to-be-affected regions that a tsunami might be approaching. However, despite efforts by some scientists over the past few years, no systematic warning system was in place in the Indian Ocean. Initial cost estimates for a crude system were about 20 million dollars, deemed too expensive by poor nations who needed the funds for more obviously pressing humanitarian causes. When the earthquake struck on a Sunday, scientists who tried calling and e-mailing countries and communities surrounding the Indian Ocean to warn them of the impending disaster typically found no one in the office and no systematic list of phone numbers of emergency response personnel. Having a simple phone-pyramid list could have potentially saved tens

of thousands of lives. Indian Ocean communities are now establishing a regional tsunami warning system.

The areas in the United States at greatest risk for the largest tsunami are along the Pacific coast, including Hawaii, Alaska, Washington, Oregon, and California. Most of the future tsunamis in these regions will be generated in subduction zones in Alaska, along the western and southwestern Pacific, and most frighteningly, along the Cascadia subduction zone in northern California, Oregon, and Washington. This region has experienced catastrophic tsunamis in the past, with geologists recently recognizing a huge wave that devastated the coast about 300 years ago. The reason this area has the present greatest risk in the United States for the largest loss of life and destruction in a tsunami is that it is heavily populated (unlike Alaska), and coastal areas lie very close to a potentially tsunami-generating subduction zone. Tsunamis travel faster than regular wind-generated waves, at close to 500 miles per hour (800 km/hr), so if the Cascadia subduction zone were to generate a tsunami, coastal areas in this region would have very little time to respond. If distant subduction zones generate a tsunami, the Pacific tsunami warning system could effectively warn coastal areas hours in advance of any crashing waves. However, a large earthquake in the Cascadia subduction zone would immediately wreak havoc on the land by passage of the seismic waves, then minutes to an hour later, potentially send huge waves into coastal Washington, Oregon, and California. There would be little time to react. It is these regions that need to invest most in more sophisticated warning systems, with coastal defenses, warning sirens, publicized and posted evacuation plans, and education of the public about how to behave (run and stay on high ground) in a tsunami emergency. Other coastal areas should initiate ocean basin–wide warning systems, install warning sirens, and post information on tsunami warnings and evacuation plans. Finally, the nation's

general public should be better educated about how to recognize and react to tsunamis and other natural geologic hazards.

What are the lessons to be learned from the tragic Indian Ocean tsunami? People who are near the sea or in an area prone to tsunamis (as indicated by warning signs in places like Hawaii), need to pay particular attention to some of the subtle and not so subtle warning signs that a tsunami may be imminent. First, there may be warning sirens in an area that is equipped with a tsunami warning system. If the sirens sound an alert, do not waste time—run to high ground immediately. People in more remote locations may need to pay attention to the natural warning signs. Anyone who feels an earthquake while on the coast should run for higher ground. There may only be minutes before a tsunami hits, or maybe an hour or two, or never, but it is better to be safe than sorry. It is important to remember that tsunamis travel in groups, with periods between crests that can be an hour or more, so do not go back to the beach to investigate the damage after the first crest passes. If the tsunami-generating earthquake occurred far away, and there is no ground motion in the area, there may not be any warning of the impending tsunami, except for the thunderous crash of waves as it rises into the coastal area. In other cases, the water may suddenly recede to unprecedented levels right before it quickly rises up again in the tsunami crest. In either case, anyone enjoying the beach needs to remain aware of the dangers. Campers should pick a sheltered spot where the waves might be refracted and not run up so far. In general, the heads of bays receive the highest run-ups, and the sides and mouths record lower run-up heights. But this may vary considerably, depending on the submarine topography and other factors.

FURTHER READING

Kusky, T. M. *Tsunamis: Giant Waves from the Sea.* New York: Facts On File, 2008.

Riang-Kroko were amplified to an astounding 86 feet (26.2 m), explaining why 137 of the 406 residents of the village were killed. Another hard-hit area was Babi Island, located three miles (5 km) offshore from Flores. The tsunami approached the island from the northeast and refracted around the island, where the edge waves met on the southwest corner and combined to produce a larger wave with run-up heights of 24 feet (7.2 m). The first, direct wave from the tsunami had a run-up velocity of 3.2 feet per second (1 m/sec), whereas the wave produced by the combined edge waves had a faster run-up velocity of up to 10 feet per second (3 m/sec). The island had 1,093 inhabitants, and 263 were washed away and drowned by the tsunami.

The residents of Flores were unaware that they lived in a tsunami hazard area, and they did not connect the ground shaking with possible sea hazards. Therefore, no warnings were issued and nobody fled the coastal areas after the quake. The villagers had built their homes and villages right along the coastline not far above the high-tide line, further compounding the threat. Many of the homes were made of bricks, but even these relatively strong structures were washed away by the powerful tsunami. In some cases, the concrete foundations of the basement moved many tens of feet inland as coherent blocks as the tsunami rushed across the area, with its associated strong currents. In this way, the tsunami made a less-than-subtle suggestion about where the safe building zone should begin. When the tsunami receded, 2,080 people were dead, 28,118 homes were washed away, and thousands of other structures were leveled. In many cases only white wave-washed beaches remained where there were once villages with populations of hundreds of people. Houses, furniture, clothing, animals and human remains were scattered through the forests behind the villages.

The Flores tsunami also caused severe coastal erosion, including cliff collapse and the scouring of coral reef complexes. Forests, brush, and grasses were removed, leaving vegetated hill tops and barren areas along the lower slopes of the coastal hills. Thick deposits of loose sediment were both moved inland and also redeposited as sheets of sediment in deeper water by the tsunami backwash, causing an overall sudden erosion of the land. Some large coral reef boulders, up to 4 feet (1 m) in diameter were moved inland many hundreds of feet (100 m) from the shoreline.

1992 Nicaragua Tsunami

The west coast of Nicaragua and Central America is a convergent margin where oceanic crust of the Cocos plate (a small plate attached to the Pacific Ocean plate) is being subducted beneath the western edge of the Caribbean plate. A volcanic arc with active volcanoes, earthquakes, and steep mountains has formed above this convergent margin subduction zone. The area is prone to large earthquakes in the forearc and to explosive volcanic eruptions and also suffers from many hurricanes, landslides, and other natural disasters.

On September 1, 1992, a relatively small (magnitude 7) earthquake centered 30 miles (50 km) off the coast of Managua, Nicaragua, generated a huge tsunami that swept across 200 miles (320 km) of coastline, killing 170 people and making 14,500 more homeless. This tsunami was generated by a tsunamogenic earthquake that ruptured near the surface, with a shallow focus of only 28 miles (45 km). The earthquake struck in an area where thick sediments have been subducted and are now plastered along the interface between the downgoing oceanic plate and the overriding continental plate of Central America. It has been suggested that these sediments lubricated the fault plane and caused the earthquake to rupture slowly and last a particularly long time, generating an unusually large tsunami for an earthquake of this size. After the fault plane initially slipped at depth, the rupture propagated outward and to the surface at a speed of 0.5–1 mile per second (1–1.5 km/sec), with the movement lasting for two minutes. The slow speed of the movement of the ground in this earthquake made the motion barely perceptible to people on the surface, but it was the perfect speed for effectively moving large volumes of water as the seafloor was displaced. Slow earthquakes like this one, now known to be particularly dangerous for generating tsunamis, are called tsunamogenic earthquakes. The height of the Nicaraguan tsunami was about 10 times greater than predicted for an earthquake of this magnitude.

The tsunami that was generated by the slow Nicaraguan earthquake of 1992 was up to 40 feet (12 m) high at nearby beachfronts, and it swept homes, vehicles, and unsuspecting people to sea within 40 minutes of the earthquake. The tsunami had a relatively slow run-up velocity, so that many people were able to outrun the wave, but elderly, sick, sleeping, and sedentary people and young children could not escape. The beachfront was heavily damaged, and 170 people were drowned by the tsunami. After the initial disaster, the hazards were not over, because the tsunami had ripped through water storage and sewage treatment plants, and the resulting contamination caused an outbreak of cholera that took even more lives in the following weeks.

1964 Alaska Earthquake-related Tsunami

Southern Alaska is hit by a significant tsunami every 10 to 30 years. The earthquakes are generated along the convergent plate boundary where oceanic crust of the Pacific plate is being pushed back into the mantle beneath Alaska. Continental crust of the North American plate in Alaska is moving over the Pacific plate, along a huge fault zone known as the Aleutian-Alaska megathrust zone, which is part of the subduction zone between the Pacific and North American plates. This fault dips at only 20° north, so motion on the fault causes large displacements of the seafloor, and hence generates large tsunamis. This fault zone has generated large Pacific Ocean tsunamis in 1878, 1946, 1957, and 1964, with many smaller events in between these major events. During the magnitude 9.2 March 27, 1964, earthquake in Alaska, approximately 35,000 square miles (90,650 km²) of seafloor and adjacent land were suddenly thrust upward by up to 35 feet (11.5 m), as part of more regional movements of the land during this event. About 83,000 square miles (215,000 km²) of seafloor was displaced significantly upward, while other areas of southern Alaska moved downward. This mass movement generated a huge tsunami as well as many related seiche-like waves in surrounding bays. The tsunami generated from this earthquake caused widespread destruction in Alaska, especially in Seward, Valdez, and Whittier. These towns all experienced large earthquake-induced submarine landslides, some of which tore away parts of the towns' waterfronts. The submarine landslides also generated large tsunamis that cascaded over these towns barely after the ground stopped shaking from the earthquake.

The earthquake that struck southern Alaska at 5:36 P.M. on Friday, March 27, 1964, was one of the largest earthquakes ever recorded, second in the amount of energy released only to the 1960 Chile earthquake, and followed closely by the 2004 Sumatra magnitude 9.0 earthquake. The epicenter was located in northern Prince William Sound, and the focus, or point of initial rupture, was located at a remarkably shallow 14 miles (23 km) beneath the surface. The energy released during the Valdez earthquake was more than the world's largest nuclear explosion, and greater than the Earth's total average annual release of seismic energy; yet, remarkably, only 131 people died during this event. Damage is estimated at 240 million dollars (1964 dollars), a surprisingly small figure for an earthquake this size. During the initial shock and several other shocks that followed in the next one to two minutes, a 600-mile- (1,000-km-) long by 250-mile- (400-km-) wide slab of subducting oceanic crust slipped further beneath the North American crust of southern Alaska. Ground displacements above the area that slipped were tremendous—much of the Prince William Sound and Kenai Peninsula area moved horizontally almost 65 feet (20 m) and moved upward

by more than 35 feet (11.5 m). Other areas more landward of the uplifted zone were down dropped by several to 10 feet (3 m). Overall, almost 200,000 square miles (520,000 km²) of land saw significant movements upward, downward, and laterally during this huge earthquake. These movements suddenly displaced an estimated volume of water of 6,000 cubic miles (25,000 km³).

The ground shook in most places for three to four minutes during the March 27, 1964, earthquake, but lasted for as much as seven minutes in a few places such as Anchorage and Valdez, where unconsolidated sediment and fill amplified and prolonged the shaking. The shaking caused widespread destruction in southern Alaska and damage as far away as southern California and induced noticeable effects across the planet. Entire neighborhoods and towns slipped into the sea during this earthquake, and ground breaks, landslides, and slumps were reported across the entire region. The Hanning Bay fault on Montague Island, near the epicenter, broke through the surface, forming a spectacular fault scarp with a displacement of more than 15 feet (3 m), uplifting beach terraces and mussel beds above the high-water mark, many parts of which rapidly eroded to a more stable configuration. Urban areas such as Anchorage suffered numerous landslides and slumps, with tremendous damage done by translational slumps where huge blocks of soils and rocks slid on curved faults downslope, in many cases toward the sea. Houses ended up in neighbors' backyards, and some homes were split in two by ground breaks. A neighborhood in Anchorage known as Turnagain Heights suffered extensive damage when huge sections of the underlying ground slid toward the sea on a weak layer in the bedrock known as the Bootlegger Shale, which lost cohesion during the earthquake shaking.

The transportation system in Alaska was severely disrupted by the earthquake. All major highways and most secondary roads suffered damage to varying degrees—186 of 830 miles (300 of 1,340 km) of roads were damaged, and 83 miles (125 km) of roadway needed replacement. Seventy-five percent of all bridges collapsed, became unusable, or suffered severe damage. Many railroad tracks were severed or bent by movement on faults, sliding and slumping into streams, and other ground motions. In Seward, Valdez, Kodiak, and other coastal communities, a series of three to 10 tsunami waves tore trains from their tracks, throwing them explosively onto higher ground. The shipping industry was devastated, which was especially difficult as Alaskans use shipping for more than 90 percent of their transportation needs, and the main industry in the state is fishing. Submarine slides, tsunamis, tectonic uplift and subsidence, and earthquake-induced fires totally destroyed all port facilities in southern Alaska except those in Anchorage. Huge portions of the waterfront facilities at Seward and Valdez slid under the sea during a series of submarine landslides, resulting in the loss of the harbor facilities and necessitating the eventual moving of the cities to higher, more stable ground. Being thrown to higher ground by tsunamis destroyed hundreds of boats, although no large vessels were lost. Uplift in many shipping channels formed new hazards and obstacles that had to be mapped to avoid grounding and puncturing hulls. Downed lines disrupted communication systems, and initial communications with remote communities were taken over by small, independently powered radio operators. Water, sewer, and petroleum storage tanks and gas lines were broken, exploded, and generally disrupted by slumping, landslides, and ground movements. Residents were forced to obtain water and fuel trucked in to areas for many months while supply lines were restored. Groundwater levels generally dropped, in some cases below well levels, further compounding the problems of access to freshwater.

Most damage from the 1964 Alaskan earthquake was associated with the numerous tsunamis and seiche waves generated during this event. A major Pacific-wide tsunami was generated by the large movements of the seafloor and continental shelf, whereas other tsunamis were generated by displacements of the seafloor near Montague Island in Prince William Sound, by landslides, and by natural resonance in the many bays and fjords of southern Alaska.

The town of Seward on the Kenai Peninsula experienced strong ground shaking for about four minutes during the earthquake, and this initiated a series of particularly large submarine landslides and seichelike waves that removed much of the waterfront docks, train-yards, and streets. During the earthquake, a large section of the waterfront at Seward, several hundred feet (100 m) wide and more than half a mile long (1 km), slid into the ocean along a curved fault surface. The movement of such significant quantities of material underwater generated a tsunami with a 30-foot (9-m) run-up that washed over Seward about 20 minutes after the earthquake. The earthquake and landslides tore apart many seafront oil storage facilities, and the oil caught fire and exploded, sending flames 200 feet (60 m) into the air. The oil on the waves caught fire, and rolled inland on the 30–40-foot- (9–12 m-) high tsunami crests. These flaming waves crashed into a train loaded with oil, causing each of 40 successive tankers to explode one after the other as the train was torn from the tracks. The waves moved inland, carrying boats, train cars, houses, and other debris, much of it in flames. The mixture moved overland at 50–60 miles per hour (80–100 km/hr), raced up the airport runway, and

blocked the narrow valley marking the exit from the town into the mountains. The tsunami had several large crests, with many reports of the third being the largest. Twelve people died in Seward, and the town was declared a total loss after the earthquake and ensuing tsunami. Seward was then moved to a new location a few miles away, on ground thought to be more stable from submarine landslides. Steep mountains mark the southwest side of the current town.

As the tsunami reached into the many bays and fjords in southern Alaska, it caused the water in many of these bays to oscillate back and forth in series of standing waves that caused widespread and repeated destruction. In Whittier, wave run-ups were reported to be up to 107 feet (32 m), and the town suffered submarine landslides, which destroyed oil and train facilities. Submarine landslides, seichelike waves, 23-foot- (7-m-) high tsunamis, and exploding oil tanks similarly affected Valdez. Near Valdez, the tsunami broke large trees, leaving only stumps more than 100 feet (30 meters) above high-tide mark. The tsunami deposited driftwood, sand, and other debris up to 170 feet (67 m) above sea level near Valdez. Like Seward, Valdez was built on a glacial outwash delta at the head of the fjord, and large sections of the delta collapsed and slid under the sea during the earthquake. A section of the town 600 feet (180 m) wide and nearly a mile (1.3 km) long slipped beneath the waters of the fjord during the earthquake, followed within minutes by a 30-foot- (9-m-) high tsunami that swept into town, killing 32 people. The natural resonance of the bay at Valdez caused the water to oscillate for hours, such that five to six hours after the initial tsunami, another series of waves grew, and continued to wash into the town with crests moving through the town near midnight on March 27; then another tsunami crest hit at 1:45 the following afternoon, moving through downtown as a tidal bore.

The town of Kodiak, 500 miles (800 km) from the epicenter, also experienced extensive damage, and 18 people were killed. The land at Kodiak subsided 5.6 feet (1.7 m) during the earthquake, and then the town was hit by a series of at least 10 tsunami crests up to 20 feet (6 m) high that destroyed the port and dock facilities as well as more than 200 other buildings.

Within 25 minutes of the earthquake, the huge displacements of water had organized into a deep-water tsunami that was moving southward into the open Pacific Ocean. This wave had a period of more than an hour, and in many places the first wave to hit was characterized by a slow rise of water into coastal areas, but the second wave was more powerful and associated with a steep breaking wave. When the tsunami reached the state of Washington four hours after the earthquake, it washed up in most places as a five-foot- (1.5 m-) tall wall of water, and was decreased in amplitude to two feet by the time it got to Astoria, Oregon. Local bays and other effects caused some variations, and the maximum wave heights in all of Washington, Oregon, and California generally were in the range of 14–15 feet (4.3–4.5 m). However, when the wave got to Crescent City, California, the shapes of the seafloor and bay were able to focus the energy from this wave into a series of five tsunami crests. The fourth was a 21-foot- (6.3-m-) high crest that swept into downtown, washing away much of the waterfront district and killing 11 people, after they had returned to assess the damage from the earlier wave crests. This fourth wave was preceded by a large withdrawal of water from the bay, such that boats were resting on the seafloor. When the large crest of the fourth wave hit the city, it washed inland through 30 city blocks and destroyed waterfront piers and docks. Later analysis of the Crescent City area has shown that the bay has been hit by at least 13 significant historical tsunamis, and the shape of the seafloor serves to amplify waves that enter the bay.

Other locations along the California coast were luckier than Crescent City but still suffered destruction of piers, boats, and other waterfront facilities. One potential disaster was averted, perhaps fortuitously, in San Francisco. About 10,000 people had rushed to the waterfront to watch the tsunami as it was supposed to pass the city. These people did not know the dangers, but the shape of the seafloor and bays in this region did not amplify the waves as occurred in Crescent City, so no lives were lost. The tsunami continued to move around the Pacific, being recorded as 7.5-foot (2.3-m) waves in Hawaii and smaller crests across Japan, South America, and Antarctica.

1946 Hilo, Hawaii, Tsunami
On April 1, 1946, an earthquake generated near the Unimak Islands in Alaska devastated vast regions of the Pacific Ocean. This tsunami was one of the largest and most widespread tsunamis to spread across the Pacific Ocean this century.

A lighthouse at Scotch Cap in the Aleutians was the first to feel the effect of the tsunami. At about 1:30 P.M., the crew of five at the lighthouse recorded feeling an earthquake lasting 30–40 seconds, but no serious damage occurred. A second quake was felt nearly half an hour later, also with no damage. Fifty minutes after the earthquake, the crew of a nearby ship recorded a "terrific roaring of the sea, followed by huge seas." They reported a wave (tsunami) that rose over the top of the Scotch Cap lighthouse and over the cliffs behind the station, totally destroying the lighthouse and coast guard station. The lighthouse was built of steel-reinforced concrete and sat on a bluff 46 feet (14 m) above sea level. The wave is estimated to have been 90–100 feet (27–30 m) high. A rescue crew sent to Scotch Cap lighthouse five hours after the disaster reported that the station was gone, and debris (including human organs) was strewn all over the place. There were no survivors.

Many hours later, the tsunami had traveled halfway across the Pacific and was encroaching on Hawaii. Residents of Hilo, Hawaii, first noticed that Hilo Bay drained, and springs of water sprouted from the dry seafloor that was littered with dying fish. As the residents wondered at the cause of the water's suddenly draining from their bay, a series of huge waves came crashing in from the ocean and quickly moved into the downtown district. Buildings were ripped from their foundations and thrown into adjacent structures, bridges were pushed hundreds of feet (hundreds of meters) upstream from their crossings, and boats and railroad cars were tossed about like toys. After the tsunami receded, Hilo was devastated, with a third of the city destroyed and 96 people dead. Outside Hilo, entire villages disappeared and were washed into the sea along with their residents.

1896 Kamaishi (Sanriku) Tsunami
On June 15, 1896, 27,000 people died in a huge tsunami that swept over the seaport of Kamaishi, Japan. A local earthquake caused mild shaking of the port city, which was not unusual in this tectonically active area. However, 20 minutes later the bay began to recede, then 45 minutes after the earthquake, the port city was inundated with a 90-foot- (27-m-) high wall of water that came in with a tremendous roar. The town was almost completely obliterated, and 27,000 people, mostly women and children, perished in a few short moments. Kamaishi was a fishing port, and when the tsunami struck, the fishing fleet was at sea and did not notice the tsunami, since it had a small amplitude in the deep ocean. When the fishing fleet returned the next morning, they sailed through many miles of debris, thousands of bodies, and reached their homes only to find a few smoldering fires among a totally devastated community.

1960 Chile Earthquake and Tsunami
The great magnitude 9.5 Chilean earthquake of May 22, 1960, generated a huge tsunami that killed more than 1,000 people near the earthquake epicenter and almost 1,000 more as the wave propagated across the Pacific Ocean. This tsunami was generated along the convergent boundary between the small Nazca oceanic plate in the Pacific Ocean and southern South America. This part of the "Ring of Fire" convergent boundary generates more tsunamogenic earthquakes than anywhere else on the planet, unleashing a large tsunami about every 30 years. Damage from the

1960 earthquake led directly to the establishment of the modern Pacific tsunami warning system.

Saturday, May 21, began as a normal day in Chile, a morning soon interrupted by a series of about 50 significant earthquakes that shook the continent beginning at 6:02 A.M. The first tremor destroyed much of the area around Concepción, with a sequence of aftershocks that continued until 3:11 P.M. the following day. Then, on May 22, there were two massive earthquakes with magnitudes of 8.9 and 9.5, with their hypocenters located a mere 20 miles (33 km) below the surface. A section of the seafloor and coast nearly 200 miles (300 km) long experienced sudden uplift of 3.3 feet (1 m), and subsidence of the land of 5 feet (1.6 m) occurred across an area of 5,000 square miles (13,000 km²), extending 18 miles (29 km) inland.

Since the area was (and still is) prone to tsunamis generated from earthquakes, local coastal fishermen knew that such a large earthquake would likely be followed by a giant tsunami, so they rapidly took their families and ran their boats to the open ocean. This knowledge of how to respond to tsunamis undoubtedly saved many lives initially, since 10 to 15 minutes after the large quake, a 16-foot- (5-m-) high tsunami rolled into many shoreline areas, causing destruction of dock areas and coastal villages. However, after this wave washed back to sea, many fishermen returned, or stayed close to shore. This was a mistake, since 50 minutes after the first wave retreated, another, larger crest struck, this time as a 26-foot- (8-m-) tall wall of water that crashed into the shore at a remarkable speed of 125 miles per hour (200 km/hr). Many of the deaths in Chile were reportedly related to the fact that after the first tsunami crest passed, many Chileans assumed the danger was over and returned to the shoreline. The second crest had a run-up of 26 feet (7.8 m) and was followed by a third crest with a height of 36 feet (11 m) that moved inland at about half the speed of the second wave, but was so massive that it inflicted considerable damage. For the next several hours the coast was pounded by a series of waves that virtually destroyed most of the development along the coastline between Concepción and Isla Chiloe. Run-up heights were variable along the coast in the southern part of Chile, with most between 28 and 82 feet (8.5 and 25 m). The number of people that were killed in Chile by this tsunami is unknown because of poor documentation, but most estimates place it between 5,000 and 10,000 people.

The massive earthquakes of May 22 sent a series of tsunami crests racing across the Pacific Ocean at 415–460 miles per hour (670–740 km/hr) and around the world for the next 24 or more hours. The tsunami train had wavelengths of 300–500 miles

(500–800 km), periods of 40–80 minutes between passing crests, and was only a little more than a foot high (40 cm) in the open ocean. The waves swept up the coast of South America, then the western United States, where run-up heights were small, typically less than 4 feet (1.2 m), but locally up to 12 feet. The tsunami was accurately predicted to hit Hawaii 14.8 hours after the earthquake, and it arrived within a minute of the predicted time. This should have saved lives, but 61 people were killed by this tsunami in Hawaii, including many who heard the warnings but rushed to the coast to watch the waves strike.

The response of the population of Hilo to tsunami warnings in 1960 is a lesson in the need to understand the hazards of tsunamis to save lives. As the tsunami approached Hawaii, the wave crests refracted around to the north side of the islands and hit the city of Hilo particularly hard, because of the shape of the shoreline features in Hilo Bay. Even though residents were warned of the danger, less than one-third of the population evacuated the coastal area when the warnings were issued, and about half remained after the first crest hit Hilo. Like many tsunamis, the first crests to invade Hilo were not the largest, and sightseers were surprised, and many killed, when the third crest pounded the downtown harbor and business district with a 20-foot (6-m) wall of water that carried 20-ton projectiles of pieces of the city wharf and waterfront buildings. The city suffered an inundation of 5 city blocks as the run-up reached 35 feet (10.7 m). Amazingly, about 15 percent of the city population stayed even through the largest waves. More than 540 homes and businesses were destroyed with tens of millions of dollars of damage (24 million 1960 dollars), and 61 residents of Hilo were dead.

Islands in the Western Pacific were widely affected, with the height of the waves largely determined by factors such as the shape of the shoreline, slope of the seafloor, and orientation of the beach with respect to the source in Peru. Pitcairn Island was the strongest hit, with run-up heights of more than 40 feet (12.2 m) reported. The effects of wave refraction around some Pacific islands focused some of the tsunami energy on Japan, where the run-up heights exceeded 20 feet (6.4 m). The combined effects of this refraction of wave energy from the distant earthquake, and natural resonance effects of some harbors that further amplified the waves, made the 1960 earthquake an unexpectedly large and devastating disaster in Japan. Approximately 22 hours after the earthquake the coast of Japan began to feel its destruction. The resonance effects caused the largest waves to grow hours after the initial wave crests hit Japan, whose east coast saw an average run-up of 9 feet (2.7 m). The resonance effects and seiching in the harbors caused the greatest damage, with 5,000 homes destroyed in Hokkaido and Honshu, 191 people killed and another 854 injured, and 50,000 left homeless. Property damage in Japan was estimated to exceed 400 million (1960) dollars.

VOLCANIC ERUPTION-INDUCED TSUNAMI

Some of history's most devastating tsunamis have been generated by volcanic processes, either during eruptions and landslides from the slopes of volcanoes or during collapse of the volcanic edifice into a caldera complex. Two of the most severe tsunamis from volcanic processes were from the 1815 eruption of Tambora, and from the 1883 eruption of Krakatau, both in Indonesia.

Tambora, Indonesia, 1815

The largest volcanic eruption ever recorded is that of the Indonesian island arc volcano Tambora in 1815. This eruption initially killed an estimated 92,000 people, largely from the associated tsunami. The eruption sent so much particulate matter into the atmosphere that it influenced the climate of the planet, cooling the surface and changing patterns of rainfall globally. The year after the eruption is known as "the year without a summer" in reference to the global cooling caused by the eruption, although people at the time did not know the reason for the cooling. In cooler climates, the year without a summer saw snow throughout the summer and crops were not able to grow. In response, great masses of farmers moved from New England in the United States to the Midwest and Central Plains, seeking a better climate for growing crops.

Tambora is located in the Indonesian region, a chain of thousands of islands that stretch from Southeast Asia to Australia. The tectonic origins of these islands are complex and varied, but many of the islands along the southwest part of the chain are volcanic in origin, formed above the Sumatra-Sunda trench system. This trench marks the edge of subduction of the Indian-Australia plate beneath the Philippine-Eurasian plates, which formed a chain of convergent margin island arc volcanoes above the subduction zone. Tambora is one of these volcanoes, located on the island of Sumbawa, east of Java. Tambora is unusual among the volcanoes of the Indonesian chain, as it is located further from the trench (210 miles; 340 km) and further above the subduction zone (110 miles; 175 km) than other volcanoes in the chain. This is related to the fact that Tambora is located at the junction of subducting continental crust from the Australian plate and subducting oceanic crust from the Indian plate. A major fault cutting across the convergent boundary is related to this transition, and the magmas that feed Tambora seem to have risen along fractures along this fault.

Tambora has a history of volcanic eruptions extending back at least 50,000 years. The age difference between successive volcanic layers is large, and there appear to have been as much as 5,000 years between individual large eruptions. This comparatively large time interval may be related to Tambora's unusual tectonic setting far from the trench along a fault zone whose origin is related to differences between the types of material being subducted on either side of the fault.

In 1812 Tambora started reawakening with a series of earthquakes plus small steam and ash eruptions. People of the region did not pay much attention to these warnings, having not remembered the ancient eruptions of 5,000 years past. On April 5, 1815, Tambora erupted with an explosion that was heard 800 miles (1,300 km) away in Jakarta. Ash probably reached more than 15 miles (25 km) in the atmosphere but this was only the beginning of what was to be one of history's greatest eruptions. Five days after the initial blast a series of huge explosions rocked the island, sending ash and pumice 25 miles (40 km) into the atmosphere and sending hot pyroclastic flows (nuée ardentes) tumbling down the flanks of the volcano and into the sea, generating tsunamis. When the hot flows entered the cold water, steam eruptions sent additional material into the atmosphere, creating a scene of massive explosive volcanism and wreaking havoc on the surrounding land and marine ecosystems. More than 36 cubic miles (150 km^3) were erupted during these explosions from Tambora, more than 100 times the volume of the Mount St. Helens eruption of 1980.

Ash and other volcanic particles such as pumice from the April eruptions of Tambora coved huge areas that stretched many hundreds of miles across Indonesia. Towns located within a few tens of miles experienced strong, hurricane-force winds that carried rock fragments and ash, burying much in their path and causing widespread death and destruction. The ash caused a darkness like night that lasted for days even in locations 40 miles (65 km) from the eruption center, so dense was the ash. Roofs collapsed from the weight of the ash, and 15-foot- (4.5-m-) tall tsunamis were formed when the pyroclastic flows entered the sea. These tsunamis swept far inland in low-lying areas, killing and sweeping away many people and livestock. A solid layer of ash, lumber, and bodies formed on the sea extending several miles west from the island of Sumbawa, and pieces of this floating mass drifted off across the Java sea. Although it is difficult to estimate, at least 92,000 people were killed in this eruption. Crops were incinerated or poisoned and irrigation systems destroyed, resulting in additional famine and disease after the eruption ceased, killing tens of thousands of people that survived the initial eruption, and forcing hundreds of thousands of people to migrate to neighboring islands.

The year of 1816 is known as the year without a summer, caused by the atmospheric cooling from the sulfur dioxide released from Tambora. Snow fell in many areas across Europe and in some places was colored yellow and red from the volcanic particles in the atmosphere. Crops failed, people suffered, and social and economic unrest resulted from the poor weather, and the Napoleonic wars soon erupted. Famine swept Europe hitting France especially hard, with food and antitax riots erupting in many places. The number of deaths from the famine in Europe is estimated at another 100,000 people.

Krakatau, Indonesia, 1883

Indonesia has seen catastrophic volcanic eruptions and associated tsunamis other than from Tambora. The island nation of Indonesia has more volcanoes than any other country in the world, with more than 130 known active volcanoes. These volcanoes have been responsible for about one-third of all the deaths attributed to volcanic eruptions and associated tsunamis in the world. Indonesia stretches for more than 3,000 miles (5,000 km) between Southeast Asia and Australia, is characterized by very fertile soils and a warm climate, and is one of the most densely populated places on Earth. The main islands in Indonesia include, from Northwest to Southeast, Sumatra, Java, Kalimantan (formerly Borneo), Sulawesi (formerly Celebes), and the Sunda Islands. The country averages one volcanic eruption per month and because of the dense population, Indonesia suffers from approximately one-third of the world's fatalities from volcanic eruptions and associated phenomena such as tsunamis.

One of the most spectacular and devastating eruptions of all time was that of 1883 from Krakatau, an uninhabited island in the Sunda Strait off the coast of the islands of Java and Sumatra. This eruption generated a sonic blast that was heard thousands of miles away, spewed enormous quantities of ash into the atmosphere, and initiated a huge tsunami that killed roughly 40,000 people and wiped out more than 160 towns. The main eruption lasted for three days, and the huge amounts of ash ejected into the atmosphere circled the globe, remaining in the atmosphere for more than three years, forming spectacular sunsets, and affecting global climate. Locally, the ash covered nearby islands, killing crops, natural jungle vegetation, and wildlife, but most natural species returned within a few years.

Legends in the Indonesian islands discuss several huge eruptions from the Sunda Strait area, and geological investigations confirm many deposits and calderas from ancient events, and recent work has revealed the presence of ancient tsunami deposits around the straits. Prior to the 1883 eruption, Krakatau consisted of several different islands including Perbuwatan in the north, Danan, and Rakata in the south. The 1883 eruption emptied a large underground magma chamber, resulting in the formation of a large caldera complex. During the 1883 eruption, Perbuwatan, Danan, and half of Rakata collapsed into the caldera and sank below sea level. Since then a resurgent dome has grown out of the caldera, emerging above sea level as a new island in 1927. The new island is named Anak Krakatau (child of Krakatau), growing to repeat the cycle of cataclysmic eruptions in the Sunda Strait.

Prior to the 1883 eruption, the Sunda Strait was densely populated with many small villages built from bamboo and palm-thatched roofs and other local materials. Krakatau is located in the middle of the strait, with many starfish-shaped arms of the strait extending into the islands of Sumatra and Java. Many villages, such as Telok Betong, lay at the ends of these progressively narrowing bays, pointed directly at Krakatau. These villages were popular stops for trading ships from the Indian Ocean to stop and obtain supplies before heading through the Sunda Strait to the East Indies. The group of islands centered on Krakatau in the center of the strait was a familiar landmark for these sailors.

Although not widely appreciated as such at the time, the first signs that Krakatau was not a dormant volcano but was about to become very active appeared in 1860 and 1861 with small eruptions, then a series of earthquakes between 1877 and 1880. On May 20, 1883, Krakatau entered a violent eruption phase, witnessed by ships sailing through the Sunda Strait. The initial eruption sent a seven-mile-(11-km-) high plume above the strait, with the eruptions heard 100 miles (160 km) away in Jakarta. As the eruption expanded, ash covered villages in a 40-mile (60 km) radius. For several months the volcano continued to sporadically erupt covering the straits and surrounding villages with ash and pumice, while the earthquakes continued.

On August 26, the style of the eruptions took a severe turn for the worse. A series of extremely explosive eruptions sent an ash column 15 miles (25 km) into the atmosphere, sending many pyroclastic flows and nuées ardentes spilling down the island's slopes and into the sea. Tsunamis associated with the flows and earthquakes sent waves into the coastal areas surrounding the Sunda Strait, destroying or damaging many villages on Sumatra and Java. Ships passing through the straits were covered with ash, while others were washed ashore and shipwrecked by the many and increasingly large tsunamis.

On August 27, Krakatau put on its final show, exploding with a massive eruption that pulverized the island and sent a large towering Plinian eruption column 25 miles (40 km) into the atmosphere. The blasts from the eruption were heard as far away as Australia, the Philippines, and Sri Lanka. Atmospheric pressure waves broke windows on surrounding islands and traveled around the world as many as seven times, reaching the antipode (area on the exactly opposite side of the Earth from the eruption) at Bogotá, Columbia, 19 hours after the eruption. The amount of lava and debris erupted is estimated at 18–20 cubic miles (75–80 km^3), making this one of the largest eruptions known in the past several centuries. Many sections of the volcano collapsed into the sea, forming steep-walled escarpments cutting through the volcanic core, some of which are preserved to this day. These massive landslides were related to the collapse of the caldera beneath Krakatau and contributed to huge tsunamis that ravaged the shores of the Sunda Strait, with average heights of 50 feet (15 m), but reaching up to 140 feet (40 m) where the V-shaped bays amplified wave height. Many of the small villages were swept away with no trace, boats were swept miles inland or ripped from their moorings, and thousands of residents in isolated villages in the Sunda Strait perished.

The tsunami was so powerful that many trees were ripped from the soil, leaving only shattered stumps remaining as vestiges of the previous forest. In some places the forest was uprooted to elevations of 130 feet (40 m) above sea level. Bodies were strewn around the shores of the Sunda Strait and formed horrible scenes of death and destruction that survivors were not equipped to clean up. The population was decimated, food supplies and farm land were destroyed, and entire villages and roads were wiped off the islands or buried in deep layers of mud. Survivors were in a state of shock and despair after the disaster and soon had to deal with more loss when disease and famine took additional lives. Soon a state of anarchy took over as rural people and farmers from the mountains descended to the coastal region and engaged in ganglike tribal looting and robbery, creating a state of chaos. Within a few months, however, troops sent by the colonial Dutch government regained control and began the rebuilding the region. Nevertheless, many of the coastal crop lands had their soil horizons removed and were not arable for many decades to come. Coastal reefs that served as fishing grounds were also destroyed, so without fishing or farming resources, many of the surviving residents moved from the coast into interior regions.

Although it is uncertain how many people died in the volcanic eruption and associated tsunamis, the Dutch colonial government estimated in 1883 that 36,417 people died, most of them (perhaps 90 percent) from the tsunamis. Several thousand people were also killed by extremely powerful nuées ardentes, or glowing clouds of hot ash, that raced across the Sunda Strait on cushions of hot air and steam. These clouds burned and suffocated all who were unfortunate enough to be in their direct paths.

Tsunamis from the eruption spread out across the Indian Ocean, caused destruction across many of the coastal regions of the entire Indian Ocean, and spread around the world. Although documentation of this Indonesian tsunami is not nearly as good as that from the 2004 tsunami, many reports of the tsunami generated from Krakatau document this event. Residents of coastal India reported the sea suddenly receding to unprecedented levels, stranding fish that were quickly picked up by residents, many of whom were then washed away by large waves. The waves spread into the Atlantic Ocean and were detected in France, and a seven-foot- (2-m-) high tsunami beached fishing vessels in Auckland, New Zealand.

Weeks after the eruption, huge floating piles of debris and bodies were still floating in the Sunda Strait, Java Strait, and Indian Ocean, providing grim reminders of the disaster to sailors in the area. Some areas were so densely packed with debris that sailors reported they appeared to be solid ground, and people were able to walk across the surface. Fields of pumice from Krakatau reportedly washed up on the shores of Africa a year after the eruption, some even mixed with human skeletal remains. Other pumice rafts carried live plant seeds and species to distant shores, introducing exotic species across oceans that normally acted as barriers to plant migration.

On western Java, one of the most densely populated regions in the world, destruction on the Ujong Kulon Peninsula was so intense that the peninsula was designated a national park, as a reminder of the power and continued potential for destruction from Krakatau. Such designation of hazardous coastal and other areas of potential destruction as national parks and monuments is good practice for decreasing the severity of future natural eruptions and processes.

Krakatau began rebuilding new cinder cones that emerged from beneath the waves in 1927 through 1929, when the new island, named Anak Krakatau (child of Krakatau) went into a rapid growth phase. Several cinder cones have now risen to heights approaching 600 feet (190 m) above sea level. The cinder cones will undoubtedly continue to grow until Krakatau's next catastrophic caldera collapse eruption.

LANDSLIDE-INDUCED TSUNAMIS

Many tsunamis are initiated by landslides that displace large amounts of water. These include rockfalls and avalanches, such as the huge avalanche that triggered a 200-foot- (60-m-) high tsunami in Lituya Bay, Alaska. Submarine landslides tend to be larger than avalanches that originate above the water line, and they have generated some of the largest tsunamis on records. Many submarine landslides are earthquake-induced, whereas others are triggered by storm events and by increases in pressure on the sediments on the continental shelf induced by rises in sea level.

The 1958 Lituya Bay, Alaska, Tsunami

One of the largest-known landslide-induced tsunamis struck Lituya Bay of southeastern Alaska on July 9, 1958. Lituya Bay is located about 150 miles (240 km) southeast of Juneau and is a steep-sided, seven-mile- (11-km-) long, glacially carved fjord with T-shaped arms at the head of the bay where the Lituya and Crillon Glaciers flow down to the sea. The glaciers are rapidly retreating, a rock spit known as La Chausse Spit blocks the entrance to the bay, and a large island, Cenotaph Island, rests in the center of the fjord. Forest-covered mountains rise 6,000 feet (1,800 m) out of the water. As the fjord is a glacially carved valley, it has a rounded floor under the sea, with a depth of only 720 feet (220 m). The rocks surrounding the bay are part of the Pacific plate, and the Fairweather fault, the boundary between the Pacific and North American Plates, lies just inboard of the bay.

At 10:15 P.M. on July 9, 1958, a magnitude 7.9–8.3 earthquake struck the region along the Fairweather fault 13 miles (20.8 km) Southeast of the bay. The ground surface was displaced by up to 3.5 feet (1.1 m) vertically and 20 feet (6.3 m) horizontally in Crillon and Lituya Bays. In some locations, ground accelerations in a horizontal direction exceeded twice the force of gravity and approached the force of gravity in the vertical direction. The earthquake sent a huge mass of rock plunging into the water below, released from about 3,000 feet (900 m) up the cliffs near the head of the bay next to Lituya Glacier. This material landed in the water at the head of Lituya Bay and created a huge semicircular crater 800 feet (250 m) across, circling the rockfall. The force of the impact of the rockfall was so great that it tore off the outer 1,300 feet (400 m) of Lituya Glacier and threw it high into the air (an observer at sea reported seeing the glacier rise above the surrounding ridges). The wave generated by this massive collapse was enormous. The first wave (really a splash) soared up to 1,720 feet (524 m) on the opposite side of the bay, removing trees and soil with the force of the wave and the backwash. This splash reached heights that were three times deeper than the water in the bay. It washed over Cenotaph Island in the middle of the bay, destroying a government research station and killing two geologists stationed there who happened to be investigating the possibilities of tsunami hazards in the bay. A 100–170-foot- (30–51-m-) high tsunami was generated that moved at 96–130 miles per hour (155–210 km/hr) out toward the mouth of the bay, erasing shoreline features along its path and shooting a fishing troller out of the bay into open water.

The size of the tsunami produced by this landslide was exceptional, being eight times higher than the next highest known landslide- or rockfall-induced tsunami from a Norwegian fjord. Most landslides that fall into the water do not produce large tsunamis because only about 4 percent of the energy from the rockfall is transferred to the water to form waves in these events. Therefore, some geologists have suggested that the Lituya Bay tsunami may have had help from an additional source, such as a huge surge of water from an ice-dammed lake on Lituya glacier that may have been suddenly released during the earthquake, and the rockfall may have landed on this huge surge of water. However, even if this speculative release of water occurred, it still would not be enough to create such a large wave. A better understanding of this wave generation phenomenon is needed and awaits further study.

Lituya Bay and others like it have experienced numerous tsunamis as shown by distinctive scour marks and debris deposits found in the bay. Studies have shown that tsunamis in the bay in 1853 and 1936 produced run-up heights of 400 and 500 feet (120 and 150 m) above sea level. This phenomenon was also well known to the native Tlingit, who had legends of spirits who lived in the bay who would send huge waves out to punish those who angered them.

Giant Landslide-induced Tsunamis in Hawaii

Many volcanic islands, such as those of the Hawaiian chain in the Pacific, Reunion in the Indian Ocean, and the Canary Islands and Tristan da Cunha in the Atlantic, are built through a combination of volcanic flows adding material to a small area in the center of the island. Frequent submarine landslides cause the islands to collapse, spreading the rocks from these flows across a wide area. Undersea mapping of the Hawaiian chain using sonar systems that can produce detailed views of the seafloor have shown that the island chain is completely surrounded by a series of debris fans and aprons from undersea landslides, covering a much larger area than the islands themselves. Submarine mapping efforts have discovered more than 70 giant landslide deposits along the 1,360-mile- (2,200-km-) long segment of the island chain from Hawaii to Midway Island. The age of the islands and flows increases from Hawaii to Midway, and studies have suggested that the average recurrence time between giant submarine slides along the Hawaiian chain is 350,000 years. The youngest giant slides on Hawaii, the Alika slides, are estimated to be several hundred thousand years old, suggesting that parts of the Hawaiian chain could be close to being ready to produce another giant slide.

Many of the landslides on the Hawaiian Islands are 100–200 miles (150–300 km) long, but seem to be somewhat shorter for the older volcanoes to the west. The largest ones have displaced volumes of material up to 1,200 cubic miles (5,000 km³). Many started on the top of the volcano near where different rift zones meet at the flank of the volcano. Giant slides that start near the topographic highs of the islands carve out a semicircular amphitheater on the island, and repeated slides from different directions can carve out the island into the shape of a star. The starlike shape of many volcanic islands is therefore the result of repeated volcanism-landslide cycles, acting together to build a high and wide volcanic edifice.

Giant landslides on volcanic islands may be initiated by many causes. Most seem to be triggered by earthquakes or by the collapse of slopes of volcanoes that are inflated by magma and ready to erupt. However, in other cases the volcanoes have built up such steep and unstable slopes that relatively minor events have triggered the release of giant landslides. These events have included the stresses from storm surges, internal waves in the ocean at depth, or pronounced rainfall events.

When the slopes of the volcanoes collapse, they typically produce several different types of submarine slides simultaneously. The submarine slides may begin at the surface or underwater as slumps, where large volumes of material move outward and downward on curved fault surfaces, some of which extend to about six miles (10 km) depth. These slumps can carve out huge sections of the island, but usually move too slowly to produce tsunamis. However, this is not always the case, and the Hawaiian Islands are famous for having earthquakes generated by fast-moving slumps that in turn do produce tsunamis. For instance, in 1868 a magnitude 7.5 earthquake was generated by a slump that produced a 66-foot- (20-m-) high tsunami, killing 81 people. In 1975 a 3.5 magnitude quake occurred on Kilauea when a 37-mile- (60-km-) long section of the flank of the volcano slumped 25 feet (8 m) laterally and 12 feet downward (3.5 m), forming a 47-foot- (14.3-m-) high tsunami that killed 16 people on the shoreline. Similar slumps and earthquakes are frequent occurrences on the Hawaiian Islands, but only some produce tsunamis.

The most dangerous submarine slides are the giant and fast moving debris avalanches. These chaotic flows can start as slumps, then break into incoherent masses of moving debris that flow downslope at hundreds of miles (km) per hour. The large volumes and high speeds of these flows make them very potent tsunami generators. The largest known submarine debris flow around the Hawaiian Islands is the 200,000-year-old Nuuanu debris avalanche on the northern side of Oahu. This flow deposit is 150 miles (230 km) long, covers an area of 14,300 square miles (23,000 km²) and is more than a mile (2 km) thick at its source, making it one of the largest debris avalanches known on Earth. The sheer volume of the material in this flow, which probably moved at hundreds of miles (km) per hour, would have sent huge tsunamis moving around the Pacific. Models suggest that this debris avalanche would have caused tsunami run-ups of more than 65 feet (20 m) along the west coast of the United States.

Some of the younger submarine slide deposits around Hawaii that are much smaller than the Nuuanu slide have produced wave run-ups of up to 1,000 feet (305 m) on nearby islands, showing how locally devastating these slide-generated tsunamis may become. On the southwest side of the main island of Hawaii, two moderate-size slides, the 900-square-mile Alika 1 slide and the 650-square-mile Alika 2 slide (2,300 and 1,700 km², respectively) released about 150 cubic miles (600 km³) of rock from Mauna Loa, excavating steep-sided amphitheaters on the island and sending the debris shooting downslope to the Pacific seafloor. Nearby islands have uncharacteristically high beach deposits that are a couple of hundred thousand years old and are probably tsunami deposits related to these slides. For instance, on the islands of Oahu, Molokai, and Maui, tsunami-related beach deposits are found at elevations of 213–260 feet (65–80 m) above sea level. On Lanai, boulder ridges form dunelike features that were deposited at more than 1,000 feet (326 m) above sea level by a catastrophic tsunami from this event. A wave with a run-up height of 1,000 feet (305 m) would need to be at least 100 feet (30 m) tall when it crashed on the coast. Areas closer to the coast on Lanai and Kahoolawe were stripped of their cover and soil to heights of 300 feet (100 m), dumping this material near the shore where it was redeposited in tsunami beds from later waves in this series of crests. This wave was so powerful when it struck the little island of Lanai that it not only removed the soil cover, but cracked the bedrock to 30 feet (10 m) depth, removing huge pieces of fractured bedrock, and filled the fractures with tsunami debris.

Landslide-induced Tsunamis of the Canary Islands, Atlantic Ocean

The Canary Islands form a hot-spot chain of small rugged volcanic islands off the northwest coast of Africa. They constitute two provinces of Spain, Santa Cruz de Tenerife and Las Palmas. The highest point on the islands is Mt. Teide on Tenerife at 12,162 feet (3,709 m) above sea level, although the volcanoes actually rise more than 10,000–13,000 feet (3,000–4,000 m) from the seafloor before the rise above sea level. The Canary Islands are hot-spot type shield volcanoes, very similar to Hawaii, yet they have very steep and rugged topography with amphitheater-shaped cliffs rising out of the sea, forming steep-sided mountain horns where different amphitheaters intersect. The star shape of several of the islands, together with the amphitheater shapes of many of the bays, suggests that these islands, like the Hawaiian Islands, have been built by a combination of volcanic eruptions and construction of tall volcanic edifices that have in turn collapsed and spread through the action of large landslides that move the material further out to sea.

Mapping of the seafloor around the Canary Islands has revealed that they are surrounded by many debris avalanche deposits, giant landslide deposits, debris aprons, and far-reaching turbidites derived from the repeated collapse of the volcanic islands. A series of seven large debris flow and turbidite deposits all less than 650,000 years old are located on the west sides of the islands. Some of these include 35,000 cubic feet (1,000 m³) of material that suddenly collapsed from the islands and was deposited at sea. On the Canary Islands, it is possible to trace landslide scars offshore into debris avalanche deposits that in turn grade into the far-traveled turbidite deposits. For instance, nearly 200 cubic miles (800 km³) of material derived from a five-mile- (8-km-) long, 3,000-foot- (900-m-) high headwall scarp on El Hierro Island include huge boulders near shore, some up to three-quarters of a mile (1.2 km) across, grading in deeper water into a debris avalanche deposit that is up to 250 feet (75 m) thick. This debris flow covers about 580 square miles (1,500 km²), then grades oceanward into a turbidite flow that extends 370 miles (600 km) to the northwest. This deposit is estimated to be 13,000–17,000 years old and undoubtedly initiated a tsunami. Tsunami deposits have been identified on some of the Canary Islands, most lying at heights up to 300 feet (90 m) above sea level (and one that is 650 feet, or 200 m high), but so far specific slides have not been correlated with specific historical tsunamis. Older deposits, such as a 133,000-year-old slide also from El Hierro, is thought to have generated a tsunami that devastated the Bahamas during the last interglacial period.

Tsunamis from Submarine Slides on Continental Margins of the Grand Banks of Newfoundland and Storegga, Norway

Submarine landslides on continental margins with steep slopes are known to have produced a number of destructive tsunamis. The causes of these slides are not always clear. Some seem to have been produced by small earthquakes, others by storms, some by rapid loading of thick and weak sedimentary layers by new sediments, and some may have been caused by the buildup and release of gas in the sedimentary sections. Several slides on passive continental margins and slopes have occurred soon after sea levels rose with the retreat of the glaciers, suggesting that the increased load of the additional water on oversteepened margins may play a role in slope failure and tsunami generation.

On November 18, 1929, a magnitude 7.2 earthquake initiated a large submarine slide from the southern edge of the Grand Banks of Newfoundland. The earthquake's epicenter was located 170 miles (280 km) south of Newfoundland, and the rupture occurred 12 miles (20 km) beneath the surface. The slide involved about 50 cubic miles (200 km³) of material that slumped eastward, then mixed with more water and transformed into a giant turbidity current that traveled 600 miles (1,000 km) eastward at speeds of 35–60 miles per hour (60–100 km/h), as determined by the times at which a series of 12 submarine telegraph cables were snapped.

The submarine slide associated with the 1929 earthquake generated the most catastrophic tsunami known in Canadian history. The tsunami had run-up heights of 40 feet (13 m) along the south coast of Newfoundland, where the waves killed 27 people. The waves penetrated about half a mile (1 km) inland, and caused a total of about 400,000 (1929) dollars damage, including the damage to the submarine cables. The waves propagated down the east coast of North America, killing one person in Nova Scotia, and were observed as far as Charleston, South Carolina, and in Bermuda. The waves also crossed the Atlantic and were recorded in the Azores and along the coast of Portugal.

The Grand Banks tsunami is a very important event for understanding the present-day hazards of tsunamis along the Atlantic seaboard and in the Gulf of Mexico. The failure of the slope from an earthquake in a plate interior shows that significant risks exist anywhere there may be oversteepened thick piles of loosely consolidated sediments. Residents of coastal areas need to understand the threat from tsunamis and build structures accordingly.

The eastern coast of Norway has been the site of several large submarine landslides that have sent tsunamis raging across the Norwegian and North Seas, and into the open Atlantic. The largest of these are the Storegga slides, where masses of sediment slid from the shelf to deepwater at rates of 160 feet per second (50 m/sec), or 109 mph (175 km/hr) depositing material 250–1,500 feet (80–450 m) thick at the base of the slope and forming turbidite layers up to 65 feet (20 m) thick that traveled 300 miles (500 km) into the Norwegian Sea. The first well-documented slide occurred about 30,000 years ago, when about 930 cubic miles (3,880 km³) of sediment suddenly slipped down the steep continental slope. At this time, sea levels were low because much of the world's ocean water was being used to make the continental glaciers that covered much of North America, Europe, and Asia. Therefore, the tsunami resulting from this slide did not affect the present-day coast line. However, two younger slides at Storegga, occurring between 8,000 and 6,000 years ago, struck during higher sea levels and left a strong imprint on the modern coastline. The later two slides were much smaller than the first, involving a total of 400 cubic miles (1,700 km³) of sediment. The second slide has been dated to have happened 7,950 +/- 190 years ago and had a height in the open ocean at its source of 25–40 feet (8–12 m). The waves crashed into Iceland, Greenland, and Scotland within a couple of hours. Greenland and Iceland saw the maximum run-up heights of 30–50 feet (10–15 m), and the waves refracted into the North Sea, causing variable run-ups of 10–65 feet (3–20 m) feet on the north coast of Scotland. The wide range in run-up heights is due to the variable topography of the coast of Scotland and some uncertainty in the models to calculate run-up heights. These tsunamis scoured the coastline of Scotland and deposited tsunami sands and gravels 30–60 feet (10–20 m) above sea level in many places.

SUMMARY

One of the deadliest natural disasters in history unfolded on December 26, 2004, when a great undersea earthquake with a magnitude between 9.0–9.3 triggered a tsunami that devastated many coastal areas of the Indian Ocean, killing an estimated 283,000 people. The tsunami devastated large parts of coastal Indonesia such as Banda Aceh, then swept across islands in the Indian Ocean to strike Sri Lanka, India, and east Africa. The tsunami propagated into all of the world's oceans where it was locally amplified by local coastal effects, but generally did little damage outside the Indian Ocean. The Indian Ocean was not equipped with any tsunami warning system, so the wave successively surprised coastal residents and tourists visiting many beach resorts. If there had been even a simple warning system in place, then tens of thousands of lives may have been saved. Nations of the Indian Ocean have

since invested in a tsunami warning system in order to prepare for future disasters.

Examination of historical tsunami events reveals that most are caused by large undersea earthquakes, since they displace vast amounts of water and can form waves that travel great distances before they dissipate. Undersea volcanic eruptions or collapse of calderas along the coast have also formed historical deadly tsunamis, with the most famous examples being the eruptions of Krakatau and Tambora in Indonesia. Volcanic-eruption-induced tsunamis can form large waves, but since they displace less water than earthquake-induced tsunamis, the waves dissipate faster than those from earthquakes. Giant landslides around volcanic islands and along continental margins have also generated large tsunamis in the historical and geological records.

See also BEACHES AND SHORELINES; EARTHQUAKES; GEOLOGICAL HAZARDS; PLATE TECTONICS; TSUNAMIS, GENERATION MECHANISMS.

FURTHER READING

Booth, J. S., D. W. O'Leary, P. Popencoe, and W. W. Danforth. "U.S. Atlantic Continental Slope Landslides: Their Distribution, General Attributes, and Implications." *United States Geological Survey Bulletin* 2002, 1993: 14–22.

Driscoll, N. W., J. K. Weissel, and J. A. Goff. "Potential for Large-Scale Submarine Slope Failure and Tsunami Generation along the United States' Mid-Atlantic Coast." *Geology* 28 (2000): 407–410.

Dvorak, J., and T. Peek. "Swept Away." *Earth* 2, no. 4 (1993): 52–59.

Kusky, T. M. *Tsunamis: Giant Waves from the Sea.* New York: Facts On File, 2008.

Latter, J. H. "Tsunami of Volcanic Origin: Summary of Causes, with Particular Reference to Krakatau, 1883." *Journal of Volcanology* 44 (1981): 467–490.

McCoy, F., and G. Heiken. "Tsunami Generated by the Late Bronze Age Eruption of Thera (Santorini), Greece." *Pure and Applied Geophysics* 157 (2000): 1,227–1,256.

Minoura, K., F. Inamura, T. Nakamura, A. Papadopoulos, T. Takahashi, and A. Yalciner. "Discovery of Minoan Tsunami Deposits." *Geology* 28 (2000): 59–62.

Revkin, A. C. "Tidal Waves Called Threat to East Coast." *The New York Times* (2000): A18.

Tsunamis.com. Available online. URL: http://www.tsunamis.com/tsunami-pictures.html Accessed January 26, 2009.

unconformities An unconformity is a substantial break or gap in a stratigraphic sequence that marks the absence of part of the rock record. These breaks may result from tectonic activity with uplift and erosion of the land, sea level changes, climate changes, or simple hiatuses in deposition. Unconformities normally imply that part of the stratigraphic sequence has been eroded but may also indicate that part of the sequence was not ever deposited in that location.

Angular unconformity, looking north from Moran Point, at Grand Canyon National Park, Arizona. Colorado River at bottom *(USGS)*

There are several different types of unconformities. Angular unconformities are angular discordances between older and younger rocks. Angular unconformities form in places where older layers were deformed and partly eroded before the younger layers were deposited. Disconformities represent a significant erosion interval between parallel strata. They are typically recognized by their irregular surfaces, missing strata, or large breaks between dated strata. Nonconformities are surfaces where strata overlie igneous or metamorphic rocks. Unconformities are significant in that they record an unusual event, such as tectonism, erosion, sea level change, or climate change.

Unconformities are typically overlain by a progradational marine sequence, starting with shallow water sandstone, conglomerate or quartzite, and succeeded by progressively deeper water deposits such as sandstone, shale, and limestone. Unconformities are often used by stratigraphers and other geologists to separate different packages of rocks deposited during different tectonic, climatic, or time systems.

See also HISTORICAL GEOLOGY; PLATE TECTONICS; STRATIGRAPHY, STRATIFICATION, CYCLOTHEM.

FURTHER READING

Prothero, Donald, and Robert Dott. *Evolution of the Earth*. 6th ed. New York: McGraw-Hill, 2002.

universe The universe includes all matter and energy everywhere, including the Earth, solar system, galaxies, interstellar matter and space, regarded as a whole. The large-scale structure of the universe refers to mapping and characterization of the observable distributions of matter, light, and space, on scales of billions of light-years. Sky surveys from Earth and space-based observation platforms using a wide range of the electromagnetic spectrum have been used to determine what is presently known about the large scale structure of the universe and the types of matter and energy contained within the cosmos. Astronomers have been able to determine that there is a hierarchal organization to the universe, with matter organized into progressively larger structures from atoms, to solar systems, to galaxies, to clusters and superclusters, to filaments, then a continued structure known as the End of Greatness.

The small-scale structure of matter and energy in the universe includes many scales of observation, ranging from subatomic particles, through stellar and solar system scales, and through the galactic scales. The large-scale structure begins at the galactic scale. Galaxies are gravitationally-bound assemblages of stars, dust, gas, radiation, and dark matter.

Most contain vast numbers of star systems and are located at enormous distances from the Milky Way Galaxy, such that the light reaching Earth from these galaxies was generated billions of years ago. Galaxies are classified into four basic types including spiral galaxies, barred spiral galaxies, elliptical galaxies, and irregular galaxies.

Galaxies are organized into clusters and superclusters held together by their mutual gravitational attraction. These groups or clusters are separated by voids characterized by relatively empty space that is apparently devoid of luminous matter. The Milky Way Galaxy is part of the Local Group, which also includes the large Andromeda Spiral Galaxy, the Large and Small Magellanic Clouds, and about 20 smaller galaxies. Galaxy clusters are of several different types. Regular clusters are spherical with a dense central core and are classified based on how many galaxies reside within 1.5 megaparsecs of the cluster center. Most regular clusters have a radius between one and 10 megaparsecs and masses of 10^{15} solar masses. Irregular clusters generally have slightly lower mass ($\sim 10^{12}$–10^{14} solar masses) than regular clusters and have no well-defined center. An example of an irregular cluster is the Virgo Cluster.

Galaxies and galaxy clusters are further grouped into even larger structures. Superclusters typically consist of groups or chains of clusters with masses of about 10^{16} solar masses. The Milky Way Galaxy is part of one supercluster, centered on the Virgo Cluster, and has a size of about 15 megaparsecs. In contrast, the largest known superclusters, like that associated with the Coma Cluster, are about 100 megaparsecs across. Recent advances in astronomers' ability to map the distribution of matter in space reveal that about 90 percent of all galaxies are located within a network of superclusters that permeates the known universe.

The motions of galaxies show patterns on different scales of observation. The motion of individual galaxies within clusters of galaxies appears random, but the clusters show some very ordered patterns to their motions at some of the largest scales of observation in the universe. Every known galaxy shows a redshifted spectrum, meaning they are all moving away from the Earth, in all directions. Individual galaxies that are not in clusters are moving away, as are the groups of galaxies in clusters, even though they have some random motions within the clusters. The farther away the galaxy or cluster is from Earth, the greater the redshift, and the faster it is receding. The same is true for other observational pairs—the farther apart any two galaxies or clusters are, the greater the redshift between them, showing that the entire universe is expanding like the surface of an inflating balloon.

Before 1989 astronomers thought that superclusters were the largest-scale structures in the universe. However, in 1989 a survey of redshifts by American astronomers Margaret Geller (Smithsonian Astrophysical Observatory) and John Huchra (Harvard-Smithsonian Center for Astrophysics) discovered a larger-scale structure dubbed "The Great Wall," consisting of an organized sheet of galaxies that is more than 500 million light-years long and 200 million light-years wide, yet only 15 million light-years thick. In 2003 another similar but larger sheet of galaxies, the "Sloan Great Wall," was discovered by Princeton University astronomers J. Richard Gott III and Mario Juric. This wall is 1.37 billion light-years long.

Sometimes there can be organization to emptiness. Much of the large-scale structure of the universe looks like a bubble-bath with sheets and walls of galaxies surrounding voids inside the bubbles. One of the largest known voids is the Capricornus Void, with a diameter larger than 230 million light-years. There appears to also be a hierarchal organization to voids that is only recently being observed and understood. In 2007 a possible supervoid was mapped in the constellation Eridanus by the Wilkinson Microwave Anisotropy Probe (WMAP), corresponding to the WMAP Cold Spot, a region that is cold in microwave wavelengths. The Eridanus Supervoid is about 500 million to 1 billion light-years across, making it one of the largest known structures in the universe.

Mapping the cosmic microwave background radiation, distribution of matter, energy, and light shows an overall bubble-like distribution of voids throughout the universe, separated by many sheets and filaments (string-like clusters between bubble-like voids) of galaxies, with superclusters appearing as very dense nodes, in some ways corresponding to places where three of the bubble-like voids would meet.

The "End of Greatness" proposes an even larger-scale structure to the universe at observational scales of about 300 million light-years (100 Mpc). At this scale, the lumpiness in the distribution of superclusters, the walls and filaments, and distribution of voids and supervoids all become homogenous (equal density of mass) and isotropic (similar in every direction). This largest-scale organization of the universe is consistent with models of the big bang and the inexorable expansion of the universe.

See also ASTRONOMY; ASTROPHYSICS; CONSTELLATION; COSMIC MICROWAVE BACKGROUND RADIATION; COSMOLOGY; DARK MATTER; ELECTROMAGNETIC SPECTRUM; GALAXIES; GALAXY CLUSTERS; HUBBLE, EDWIN; INTERSTELLAR MEDIUM; ORIGIN AND EVOLUTION OF THE UNIVERSE.

FURTHER READING

Chaisson, Eric, and Steve McMillan. *Astronomy Today.* 6th ed. Upper Saddle River, N.J.: Addison-Wesley, 2007.

Comins, Neil F. *Discovering the Universe.* 8th ed. New York: W. H. Freeman, 2008.

National Aeronautics and Space Administration. "Universe 101, Our Universe, Big Bang Theory. Cosmology: The Study of the Universe" Web page. Available online. URL: http://map.gsfc.nasa.gov/universe/. Last updated May 8, 2008.

Snow, Theodore P. *Essentials of the Dynamic Universe: An Introduction to Astronomy.* 4th ed. St. Paul, Minn.: West Publishing Company, 1991.

The 2dF Galaxy Redshift Survey. A Map of the Universe. Available online. URL: http://www2.aao.gov.au/~TDFgg/. Accessed January 13, 2009.

Uranus   The seventh planet from the Sun, Uranus is a giant gaseous sphere with a mass 15 times that of the Earth and a diameter four times as large as Earth's (51,100 km). The equatorial plane is circled by a system of rings, some associated with the smaller of the 15 known moons circling the planet. Uranus orbits the Sun at a distance of 19.2 astronomical units (Earth-Sun distance) with a period of 84 Earth years and has a retrograde rotation of 0.69 Earth days. Its density is only 1.2 grams per cubic centimeter, compared to the Earth's average density of 5.5 grams per cubic centimeter. The density is higher, however, than Jupiter's or Saturn's, suggesting that Uranus has a proportionally larger rocky core than either one of these giant gaseous planets.

Most of the planets in the solar system have their rotational axis roughly perpendicular to the plane of the ecliptic, the plane within which the planets approximately orbit the Sun. However, the rotational axis of Uranus is one of the most unusual in the solar system as it lies roughly within the plane of the ecliptic, as if it is tipped over on its side. The cause of this unusual orientation of the planet is not known but some astronomers have speculated that it may be a result of a large impact early in the planet's history. As a consequence of this unusual orientation, as Uranus orbits the Sun, it goes through seasons where the north and south poles are alternately pointing directly at the Sun, and periods in between (spring and autumn) when the poles are aligned in between these extremes. The poles experience very long summers and winters because of the long orbital period of Uranus and are alternately plunged into icy cold darkness for 42 years, then exposed to the distant Sun for 42 years. With the rapid rotation rate and unusual orientation of the planet, an observer on the pole experiencing the change from winter to spring

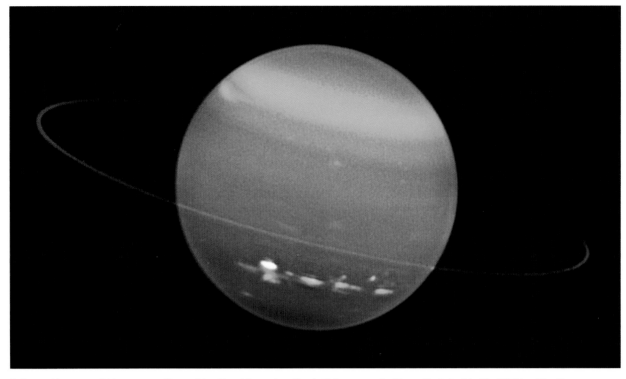

Infrared image of Uranus gathered by the 10-meter Keck Telescope in Hawaii, July 11–12, 2004—Northern Hemisphere is left of rings *(California Association for Research in Astronomy/Photo Researchers, Inc.)*

would first observe the distant Sun rising above the horizon and tracing out part of a circular path, then sinking below the horizon. Eventually the Sun would finally emerge totally and trace out complete circle paths every 17 hours. The position of the Sun would progressively change over the next 42 years until it sank below the horizon for the following 42 years.

The atmosphere of Uranus is roughly similar to Jupiter's and Saturn's, consisting mostly of molecular hydrogen (84 percent), helium (14 percent), and methane (2 percent). Ammonia seems to be largely absent from the atmosphere of Uranus, part of a trend that has ammonia decreasing in abundance outward in the outer solar system, with less at lower temperatures. The reason is that ammonia freezes into ammonia ice crystals at -335°F (70 K), and Jupiter's and Saturn's upper atmospheres are warmer than this, whereas Uranus's is -355°F (58 K). Any ammonia therefore would have crystallized and fallen to the surface. The atmosphere of Uranus appears a blue-green color because of the amount of methane, but so far relatively few weather systems have been detected on the planet. Enhancement of imagery however has revealed that the atmosphere is characterized by winds that are blowing around the planet at 125 to 310 miles per hour (200–500 km/hr) in the same sense as the planet's rotation, with detectable channeling of the winds into bands. The winds are responsible for transporting heat

from the warm to the cold hemisphere during the long winter months.

The magnetic field of Uranus is surprisingly strong, about 100 times as strong as the Earth's. However, since the rocky core of the planet is so far below the cloud level the strength of the magnetic field at the cloud tops on Uranus is actually similar to that on the surface of the Earth. Like the rotational axis, the orientation of the magnetic poles and field on Uranus are highly unusual. The magnetic axis is tilted at about 60° from the spin axis, and is not centered on the core of the planet, but is displaced about one-third of the planetary radius from the planet's center. The origin of these unusual magnetic field properties is not well understood, but may be related to a slurry of electrically conducting ammonia clouds near the planet's rocky surface. Whatever the cause, a similar unusual field exists on nearby Neptune.

Uranus has 15 known large moons with diameters of more than 25 miles (40 km), orbiting between 31,070 miles (50,000 km) from the planet (for the smallest moon) to 362,260 miles (583,000 km) for the largest moon. From largest to smallest, these moons include Oberon, Titania, Umbriel, Ariel, Miranda, Puck, Belinda, Rosalind, Portia, Juliet, Desdemona, Cressida, Bianca, Ophelia, and Cordelia. The 10 smallest moons all orbit inside the orbit of Miranda and are associated with the ring system around Uranus, and all of the moons rotate in the planet's equatorial plane, not the ecliptic of the solar system. Most of these moons are relatively dark, heavily cratered, and geologically inactive bodies, with the exception of Miranda, which shows a series of ridges, valleys, and different morphological terrains. One of the most unusual is a series of oval wrinkled or faulted terrains of uncertain origin, but perhaps related to subsurface magmatism, impacts, or volcanism.

The ring system around Uranus was only discovered recently, in 1977, when the planet passed in front of a bright star and the rings were observed by astronomers studying the planet's atmosphere. There are nine known rings between 27,340 and 31,690 miles (44,000–51,000 km) from the planet's center, and each of these rings appears to be made of many much smaller rings. The rings are generally dark and narrow with widths of up to six miles (10 km), with wide spaces between the main rings ranging from 125–620 miles (200–1,000 km). The rings are only a few tens of meters thick. Most of the particles that make up the rings are dust- to boulder-sized, dark-colored, and trapped in place by the gravitational forces between Uranus and its many moons.

See also EARTH; JUPITER; MARS; MERCURY; NEPTUNE; SATURN; SOLAR SYSTEM; VENUS.

FURTHER READING

Chaisson, Eric, and Steve McMillan. *Astronomy Today.* 6th ed. Upper Saddle River, N.J.: Addison-Wesley, 2007.

Comins, Neil F. *Discovering the Universe.* 8th ed. New York: W. H. Freeman, 2008.

National Aeronautic and Space Administration. Solar System Exploration page. "Uranus." Available online. URL: http://solarsystem.nasa.gov/planets/profile.cfm?Object=Uranus. Last updated June 25, 2008.

Snow, Theodore P. *Essentials of the Dynamic Universe: An Introduction to Astronomy.* 4th ed. St. Paul, Minn.: West Publishing Company, 1991.

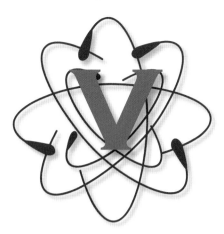

Venus The second planet from the Sun, Venus is the planet in our solar system that most closely resembles Earth, with a planetary radius of 3,761 miles (6,053 km), or 95 percent of the Earth's radius. Venus orbits the Sun in a nearly circular path at 0.72 astronomical units (Earth-Sun distance) and has a mass equal to 81 percent of Earth's and density of 5.2 grams per cubic centimeter, very similar to Earth's 5.5 grams per cubic centimeter. The orbital period (year) of Venus is 0.62 Earth years, but it has a retrograde rotation about its axis of 243 days, with its north pole turned essentially upside down so that its equatorial pole is inclined at 177.4° from the orbital plane. One result of this tilt and the slow retrograde rotation is that the two effects combine to make each day on Venus take the equivalent of 117 Earth days. Another effect is that the slow rotation has not set up a geodynamo current in the planet's core, so Venus has no detectable magnetic field. Thus, it lacks a magnetosphere to protect it from the solar wind, so it is constantly bombarded by high-energy particles from the Sun. These particles lead to constant ionization of the upper levels of the atmosphere. Venus is usually one of the brightest objects in the sky (excepting the Sun and Moon), and is usually visible just before sunrise or just after sunset, since its orbit is close to the Sun.

The atmosphere of Venus is very dense, is composed mostly (96.5 percent) of carbon dioxide, and is nearly opaque to visible radiation, so most observations of the planet's surface are based on radar reflectivity. Spacecraft and Earth-based observations of Venus show that the atmospheric and cloud patterns on the planet are more visible in ultraviolet (UV) wavelengths, since some of the outer clouds seem to be made of mostly sulfuric acid, which absorbs UV radiation, whereas other clouds reflect this wavelength, producing a highly contrasted image. These observations show that the atmosphere contains many large clouds moving around 250 miles per hour (400 km/hr) that rotate around the planet on average once every four days. The atmospheric patterns on Venus resemble the jet stream systems on Earth. Aside from carbon dioxide, the atmosphere contains nitrogen plus minor or trace amounts of water vapor, carbon monoxide, sulfur dioxide, and argon.

Although many basic physical Venusian properties are similar to Earth's, the atmosphere on Venus is about 90 times more massive and extends to much greater heights than Earth's atmosphere. The mass of the Venusian atmosphere causes pressures to be exceedingly high at the surface, a value of 90 bars, compared to Earth's one bar. The troposphere, or region in which the weather occurs, extends to approximately 62 miles (100 km) above the surface. The upper layers of the Venusian atmosphere, from about 75–45 miles (120–70 km) height, are composed of sulfuric acid cloud layers, underlain by a mixing zone that is underlain by a layer of sulfuric acid haze from 30–20 miles (50–30 km). Below about 18 miles (30 km) the air is clear.

The carbon dioxide–rich and water-poor nature of the Venusian atmosphere has several important consequences for the planet. First of all, these gases are greenhouse gases that trap solar infrared radiation, an effect that has raised the surface temperature to an astounding 750 K (900°F, or 475°C, and compared to 273 K for Earth). Second, surface processes are much different on Venus than Earth because of the elevated temperatures and pressures. There is no running water, rock behaves mechanically

Computer simulated view of northern hemisphere of Venus, May 26, 1993 *(NASA Jet Propulsion Laboratory)*

dissimilarly under high temperatures and pressures, and heat flow from the interior is buffered with a drastically different surface temperature.

Earth and Venus had essentially the same amounts of gaseous carbon dioxide, nitrogen, and water in their atmospheres soon after the planets formed. However, since Venus is closer to the Sun (it is 72 percent of the distance from the Sun that Earth is) it receives about two times as much solar radiation as the Earth, which is enough to prevent the oceans from condensing from vapor. Without oceans, carbon dioxide did not dissolve in seawater or combine with other ions to form carbonates, so the CO_2 and water stayed in the atmosphere. Because the water vapor was lighter than the CO_2, it rose to high atmospheric levels and was dissociated into hydrogen (H) and oxygen (O) ions; the hydrogen escaped to space, whereas the oxygen combined with other ions. Thus, oceans never formed on Venus, and the water that could have formed them dissociated and is lost in space. Earth is only slightly further from the Sun, but conditions here are exactly balanced to allow water to condense and the atmosphere to remain near the equilibrium (triple) point of solid, liquid, and vapor water. These conditions allowed life to develop on Earth, and life further modified the atmosphere-ocean system to maintain its ability to support further life. Venus never had a chance.

With Venus's thick atmosphere, the surface must be mapped with cloud-penetrating radar from Earth and spacecraft. The surface shows many remarkable features, including a division into a bimodal crustal

elevation distribution reminiscent of Earth's continents and oceans. Most of the planet is topographically low, including about 27 percent flat volcanic lowlands and about 65 percent relatively flat plains, probably basaltic in composition, surrounded by volcanic flows. The plains are punctuated by thousands of volcanic structures, including volcanoes and elongate narrow flows, including one that stretches 4,225 miles (6,800 km) across the surface. Some of the volcanoes are huge, with more than 1,500 having diameters of more than 13 miles (20 km), and one (Sapas Mons) more than 250 miles (400 km) across and almost one mile (1.5 km) high. About 8 percent of the planet consists of highlands made of elevated plateaus and mountain ranges. The largest continent-like elevated landmasses include the Australian-sized Ishtar Terra in the southern hemisphere, and Africa-sized Aphrodite Terra in equatorial regions. Ishtar Terra has interior plains rimmed by what appear to be folded mountain chains and Venus's tallest mountain, Maxwell Mons, reaching 7 miles (11 km) in height above surrounding plains. Aphrodite Terra also has large areas of linear folded mountain ranges, many lava flows, and some fissures that probably formed from lava upwelling from crustal magma chambers, then collapsing back into the chamber instead of erupting.

The surface of Venus preserves numerous impact structures and unusual circular to oval structures and craters that are most likely volcanic in origin. Some of the most unusual-appearing are a series of rounded pancake-like bulges that overlap each other on a small northern hemisphere elevated terrane named Alpha Regio. These domes are about 15.5 miles (25 km) across and probably represent lava domes that filled, then had the magma withdrawn from them, forming a flat, cracked lava skin on the surface. There are many basaltic shield volcanoes scattered about the surface and some huge volcanic structures known as coronae. These are hundreds of kilometers across and are characterized by a series of circular fractures reflecting a broad upwarped dome, probably formed as a result of a plume from below. Many volcanoes dot the surface in and around coronae, and lava flows emanate and flow outward from some of them. Impact craters are known from many regions on Venus, but their abundance is much less than expected for a planet that has had no changes to its surface since formation. No impacts less than two miles (3 km) across are known since small meteorites burn up in the thick atmosphere before hitting the surface. The paucity of other larger impacts reflects the fact that the surface has been reworked and plated by basalt in the recent history of the planet, as confirmed by the abundant volcanoes, lava flows, and the atmospheric composition of the planet. It is likely that volcanism is still active on Venus.

Many of the surface features on Venus indicate some crustal movements. For instance, the folded mountain ranges show dramatic evidence of crustal shortening, and there are many regions of parallel fractures. Despite these features, there have not been any features found that are indicative of plate tectonic types of processes operating. Most of the structures could be produced by crustal downsagging or convergence between rising convective plumes, in a manner similar to that postulated for the early Earth before plate tectonics was recognized.

See also EARTH; JUPITER; MARS; MERCURY; NEPTUNE; SATURN; SOLAR SYSTEM; URANUS.

FURTHER READING

Chaisson, Eric, and Steve McMillan. *Astronomy Today.* 6th ed. Upper Saddle River, N.J.: Addison-Wesley, 2007.

Comins, Neil F. *Discovering the Universe.* 8th ed. New York: W. H. Freeman, 2008.

National Aeronautic and Space Administration. Solar System Exploration Page. Venus. Available online. URL: http://solarsystem.nasa.gov/planets/profile.cfm?Object=Venus. Last updated May 7, 2008.

Snow, Theodore P. *Essentials of the Dynamic Universe: An Introduction to Astronomy.* 4th ed. St. Paul, Minn.: West Publishing Company, 1991.

volcano A volcano is a mountain or other constructive landform built by a singular volcanic eruption or a sequence of eruptions of molten lava and pyroclastic material from a volcanic vent. Volcanoes have many forms, ranging from simple vents in the Earth's surface, through elongate fissures that erupt magma, to tall mountains with volcanic vents near their peaks. Volcanic landforms and landscapes are as varied as the volcanic rocks and eruptions that produce them. Volcanic eruptions provide one of the most spectacular of all natural phenomena, yet they also rank among the most dangerous of geological hazards. More than 500 million people worldwide live near active volcanoes and need to understand the risk associated with volcanic eruptions and how to respond in the event of volcanic emergencies. Eruptions may send blocks of rock, ash, and gas tens of thousands of feet into the atmosphere in beautiful eruption plumes, yet individual eruptions have also killed tens of thousands of people. The hazards associated with volcanic eruptions are not limited to the immediate threat from the flowing lava and ash, but include longer-term atmospheric and climate effects and changes to land-use patterns and the livelihood of human populations.

People have been awed by the power and fury of volcanoes for thousands of years, as evidenced by biblical passages referring to eruptions, and more recently by the destruction of Pompeii and Herculaneum by the eruption of Italy's Mount Vesuvius in the year 79 C.E. Sixteen thousand people died in Pompeii alone, buried by a fast-moving hot incandescent ash flow known as a nuée ardente. This famous eruption buried Pompeii in thick ash that quickly solidified and preserved the city and its inhabitants remarkably well. In the 16th century Pompeii was rediscovered and has since then been the focus of archeological investigations. Mount Vesuvius is still active, looming over the present-day city of Naples, Italy. Most residents of Naples rarely think about the threat looming over their city. There are many apparently dormant volcanoes similar to Vesuvius around the world, and people who live near these volcanoes need to understand the potential threats and hazards posed by these sleeping giants to know how to react in the event of a major eruption from the volcanoes.

VOLCANOES AND PLATE TECTONICS

Most volcanic eruptions on the planet are associated with the boundaries of tectonic plates. Extensional or divergent plate boundaries where plates are being pulled apart, such as along the mid-ocean ridges, have the greatest volume of magma erupted each year for any volcanic province on the Earth. However, these eruptions are not generally hazardous, especially since most of these eruptions occur many miles (km) below sea level. A few exceptions to this rule are noted where the mid-ocean ridges rise above sea level, such as in Iceland. There, volcanoes including the famous Hekla volcano have caused significant damage. Hekla has even erupted beneath a glacier, causing instant melting and generating fast-moving catastrophic floods called Jökulhlaups. Despite these hazards, Icelanders have learned to benefit from living in a volcanically active area, tapping a large amount of heat in geothermal power-generating systems.

Larger volcanoes are associated with extensional plate boundaries located in continents. For instance, the East African rift system is where the African continent is being ripped apart, and it hosts some spectacular volcanic cones, including Kilimanjaro, Ol Doinyo Lengai, Nyiragongo, and many others.

The most hazardous volcanoes on the planet are associated with convergent plate boundaries, where one plate is sliding beneath another in a subduction zone. The famous "Ring of Fire" rims the Pacific Ocean and refers to the ring of abundant volcanoes located above subduction zones around the Pacific Ocean. The Ring of Fire extends through the western Americas, Alaska, Kamchatka, Japan, Southeast Asia, and Indonesia. Volcanoes in this

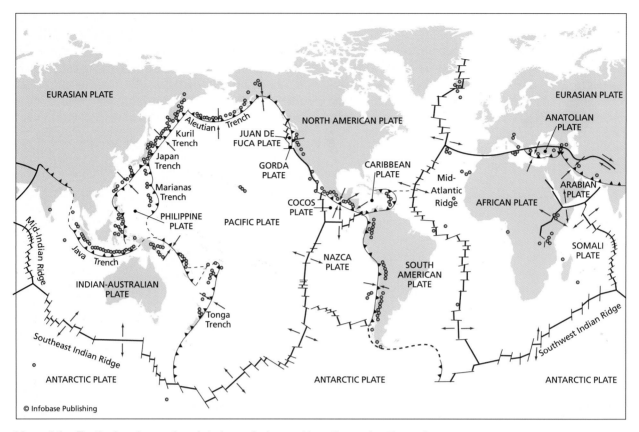

Map of the Earth showing major plate boundaries and locations of active volcanoes

belt, including Mount Saint Helens in Washington State, typically have violent explosive eruptions that have killed many people and altered the landscape over wide regions.

An unusual style of volcanic activity is not associated with plate boundaries, but forms large broad volcanic shields in the interior of plates. These volcanoes typically have spectacular but not extremely explosive volcanic eruptions. The most famous of these "intraplate" volcanoes is Mauna Loa on Hawaii. Hazards associated with these volcanoes include lava flows and traffic jams from tourists trying to see the flows.

TYPES OF VOLCANIC ERUPTIONS AND LANDFORMS

The style of volcanic eruptions varies tremendously, both between volcanoes and from a single volcano during the course of an eruptive phase. This variety is related to the different types of magma produced by the different mechanisms described above. Geologists have found it useful to classify volcanic eruptions based on how explosive the eruption was, on which materials erupted, and by the type of landform produced by the volcanic eruption.

Tephra is material that comes out of a volcano during an eruption, and it may be thrown through the air or transported over the land as part of a hot moving flow. Tephra includes both new magma from the volcano and older broken rock fragments caught during the eruption. It includes ash and pyroclasts, rocks ejected by the volcano. Large pyroclasts are called volcanic bombs; smaller fragments are lapilli, and the smallest grade into ash.

While the most famous volcanic eruptions produce huge explosions, many eruptions are relatively quiet and nonexplosive. Nonexplosive eruptions have magma types that have low amounts of dissolved gases, and they tend to be basaltic in composition. Basalt flows easily and for long distances and tends not to have difficulty flowing out of volcanic necks. Nonexplosive eruptions may still be spectacular, as any visitor to Hawaii lucky enough to witness the fury of Pele, the Hawaiian goddess of the volcano, can testify. Mauna Loa, Kilauea, and other nonexplosive volcanoes produce a variety of eruption styles, including fast-moving flows and liquid rivers of lava, lava fountains that spew fingers of lava trailing streamers of light hundreds of feet into the air, and thick sticky lava flows that gradually creep downhill. The Hawaiians devised clever names for these flows, including *aas* for blocky rubble flows because walking across these flows in bare feet makes one exclaim "ah, ah" in pain. Pahoehoes are ropy-textured flows, after the Hawaiian term for rope.

Explosive volcanic eruptions are among the most dramatic of natural events on Earth. With little warning, long-dormant volcanoes can explode with the force of hundreds of atomic bombs, pulverizing whole mountains and sending the existing material together with millions of tons of ash into the stratosphere. Explosive volcanic eruptions tend to be associated with volcanoes that produce andesitic or rhyolitic magma and have high contents of dissolved gases. These are mostly associated with convergent plate boundaries. Volcanoes that erupt magma with high contents of dissolved gases often produce a distinctive type of volcanic rock known as pumice, which is full of bubble holes, in some cases making the rock light enough to float on water.

When the most explosive volcanoes erupt they produce huge eruption columns known as Plinian columns (named after Pliny the Elder, the Roman statesman, naturalist, and naval officer who died in 79 C.E. while struggling through thick ash while trying to rescue friends during the eruption of Mount Vesuvius). These eruption columns can reach 28

Lava fountain about 1,000 feet (304.8 m) high from Kilauea eruption, Hawaii, April 4, 1983 *(J. D. Griggs/ USGS)*

miles (45 km) in height, and they spew hot turbulent mixtures of ash, gas, and tephra into the atmosphere where winds may disperse them around the planet. Large ashfalls and tephra deposits may be spread across thousands of square miles (km²). These explosive volcanoes also produce one of the scariest and most dangerous clouds on the planet. Nuées ardentes are hot glowing clouds of dense gas and ash that may reach temperatures of nearly 1,830°F (1,000°C), rush down volcanic flanks at 450 miles per hour (700 km/h), and travel more than 60 miles (100 km) from the volcanic vent. Nuées ardentes have been the nemesis of many a volcanologist and curious observer, as well as thousands upon thousands of unsuspecting or trusting villagers. Nuées ardentes are but one type of pyroclastic flow, which include a variety of mixtures of volcanic blocks, ash, gas, and lapilli that produce volcanic rocks called ignimbrites.

Most volcanic eruptions emanate from the central vents at the top of volcanic cones. However, many other flank eruptions have been recorded, where eruptions blast out of fissures on the side of the volcano. Occasionally volcanoes blow out their sides, forming a lateral blast like the one that initiated the 1980 eruption of Mount Saint Helens in the state of Washington. This blast was so forceful that it began at the speed of sound, killing everything in the initial blast zone.

Volcanic landforms and landscapes are wonderful, dreadful, beautiful, and barren. They are as varied as the volcanic rocks and eruptions that produce them. Shield volcanoes include the largest and broadest mountains on the planet (Mauna Loa is more than 100 times as large as Mount Everest). These have slopes of only a few degrees, produced by basaltic lavas that flow long distances before cooling and solidifying. Stratovolcanoes, in contrast, are the familiar steep-sided cones like Mount Fuji, made of stickier lavas such as andesites and rhyolites, and they may have slopes of 30°. Other volcanic constructs include cinder or tephra cones, including the San Francisco Peaks in Arizona, which are loose piles of cinder and tephra. Calderas, like Crater Lake in Oregon, are huge circular depressions, often many miles (km) in diameter, that are produced when deep magma chambers under a volcano empty out (during an eruption). Such eruptions form huge empty spaces below the surface, and the overlying land collapses inward producing a topographic depression known as a caldera. Yellowstone Valley occupies one of the largest calderas in the United States. Many geysers, hot springs, and fumaroles in the valley are related to groundwater circulating to depths, being heated by shallow magma, and mixing with volcanic gases that escape through minor cracks in the crust of the Earth.

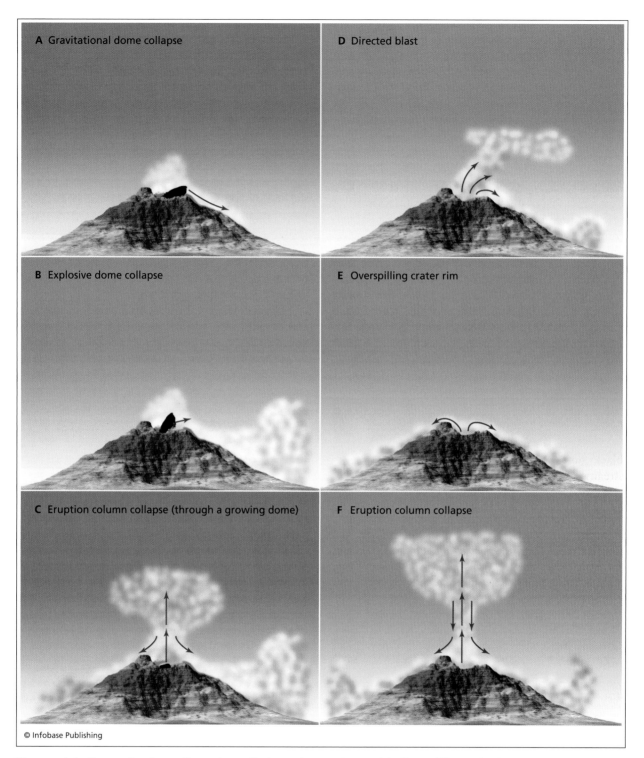

A Gravitational dome collapse

D Directed blast

B Explosive dome collapse

E Overspilling crater rim

C Eruption column collapse (through a growing dome)

F Eruption column collapse

© Infobase Publishing

Diagram labeling parts of eruption column that may form catastrophic flows. (A) gravitational dome collapse; (B) explosive dome collapse; (C) eruption column collapse blasted through a growing dome; (D) directed blast, with Mount St. Helens as a recent example; (E) eruption overspilling crater rim; (F) eruption column collapse. If continuous this produces large volume deposits, and if discontinuous from discrete explosions, the deposits are smaller.

VOLCANISM IN RELATION TO PLATE TECTONIC SETTING

The types of volcanism and associated volcanic hazards differ in various tectonic settings because each tectonic setting produces a different type of magma, through the processes described above. Mid-ocean ridges and intraplate hot-spot types of volcanoes typically produce nonexplosive eruptions, whereas convergent tectonic margin volcanoes may produce tremendously explosive and destructive eruptions. Much of the variability in the eruption style may be related to the different types of magma produced in these different settings and also to the amount of dissolved gases, or volatiles, in these magmas. Magmas with large amounts of volatiles tend to be highly explosive, whereas magmas with low contents of dissolved volatiles tend to be nonexplosive. The difference is very much like shaking two bottles, one containing soda and one containing water. The soda contains a high concentration of dissolved volatiles (carbon dioxide) and explodes when opened. In contrast, the water has a low concentration of dissolved volatiles and does not explode when opened.

Eruptions from mid-ocean ridges are mainly basaltic flows, with low amounts of dissolved gases. These eruptions are relatively quiet, with basaltic magma flowing in underwater tubes and breaking off in bulbous shapes called pillow lavas. The eruption style in these underwater volcanoes is analogous to toothpaste being squeezed out of a tube. Eruptions from mid-ocean ridges may be observed in the few rare places where the ridges emerge above sea level, such as Iceland. Eruptions there include lava fountaining, where basaltic cinders are thrown a few hundred feet in the air and accumulate as cones of black glassy fragments, and they also include long stream-like flows of basalt.

Hot-spot volcanism tends to be much like that at mid-ocean ridges, particularly where the hot spots are located in the middle of oceanic plates. The Hawaiian Islands are the most famous hot-spot type of volcano in the world, with the active volcanoes on the island of Hawaii known as Kilauea and Mauna Loa. Mauna Loa is a huge shield volcano, characterized by a very gentle slope of a few degrees from the base to the top. This gentle slope is produced by lava flows that have a very low viscosity (meaning they flow easily) and can flow and thin out over large distances before they solidify. Magmas with high viscosity would be much stickier and would solidify in short distances, producing volcanoes with steep slopes. Measured from its base on the Pacific Ocean seafloor to its summit, Mauna Loa is the tallest mountain in the world, a fact attributed to the large distances that its low-viscosity lavas flow and to the large volume of magma produced by this hot-spot volcano.

Volcanoes associated with convergent plate boundaries produce by far the most violent and destructive eruptions. Recent convergent margin eruptions include Mount Saint Helens and two volcanoes in the Philippines, Mount Pinatubo and Mayon volcano. The magmas from these volcanoes tend to be much more viscous, are higher in silica content, and have the highest concentration of dissolved gasses. Many of the dissolved gasses and volatiles, such as water, are released from the subducting oceanic plate as high mantle temperatures heat it up as it slides beneath the convergent margin volcanoes.

VOLCANIC HAZARDS

Volcanic eruptions have been responsible for the deaths of hundreds of thousands of people, and they directly affect large portions of the world's population, land-use patterns, and climate. The worst volcanic disasters have killed tens of thousands of people, whereas others may kill only a few thousand, hundreds, or even none. The following sections discuss specific phenomena associated with volcanic eruptions that have been responsible for the greatest loss of life in individual eruptions. Understanding the hazards associated with volcanoes is important for reducing losses from future eruptions, especially considering that millions of people live close to active volcanoes. Some of the hazards are obvious, such as being overrun by lava flows, being buried by layers of ash, or being hit by hot glowing avalanche clouds known as nuées ardentes. Other hazards are less obvious, such as poisonous gases that can seep out of volcanoes, suffocating people nearby, and changes to global climate as a consequence of large volcanic eruptions. The table "Examples of Volcanic Disasters" lists examples of some of the worst volcanic disasters of the last 200 years.

Hazards of Lava Flows

In some types of volcanic eruptions lava may bubble up or effuse from volcanic vents and cracks and flow like thick water across the land surface. During other eruptions lava oozes out more slowly, producing different types of flows with different hazards. Variations in magma composition, temperature, dissolved gas content, surface slope, and other factors lead to the formation of three major types of lava flows: aa, pahoehoe, and block lava. Aas are characterized by rough surfaces of spiny and angular fragments, whereas pahoehoe have smooth ropelike or billowing surfaces. Block lavas have larger fragments than aa flows and are typically formed by stickier, more silicic (quartz rich) lavas than aa and pahoehoe flows. Some flows are transitional between these main types or may change from one type to another as surface slopes and flow rates change. It is common to see pahoehoe flows change into aa flows with increasing distance from the volcanic source.

Lava flows are most common around volcanoes that are characterized by eruptions of basalt with low contents of dissolved gasses. About 90 percent of all lava flows worldwide are made of magma with basaltic composition, followed by andesitic (8 percent) and rhyolitic (2 percent). Places with abundant basaltic flows include Hawaii, Iceland, and other exposures of oceanic islands and mid-oceanic

EXAMPLES OF VOLCANIC DISASTERS

Volcano, Location	Year	Deaths
Tambora, Indonesia	1815	92,000
Krakatau, Indonesia	1883	36,500
Mount Pelée, Martinique	1902	29,000
Nevada del Ruiz, Colombia	1985	24,000
Santa Maria, Guatemala	1902	6,000
Galunggung, Indonesia	1822	5,500
Awu, Indonesia	1826	3,000
Lamington, Papua New Guinea	1951	2,950
Agung, Indonesia	1963	1,900
El Chichon, Mexico	1982	1,700

ridges, all characterized by nonexplosive eruptions. Virtually the entire islands of the Hawaiian chain are made of a series of lava flows piled high, one on top of the other. Most other volcanic islands are similar, including the Canary Islands in the Atlantic, Reunion in the Indian Ocean, and the Galapagos in the Pacific. In January of 2002, massive lava flows were erupted from Nyiragongo volcano in Congo, devastating the town of Goma and forcing 300,000 people to flee from their homes as the lava advanced through the town.

Lava flows generally follow topography, flowing from the volcanic vents downslope in valleys, much as streams or water from a flood would travel. Some lava flows may move as fast as water, up to almost 40 miles per hour (65 km/h), on steep slopes, but most lava flows move considerably slower. More typical rates of movement are about 10 feet (several m) per hour, to 10 feet (3 m) per day for slower flows. These rates of movement of lava allow most people to move out of danger to higher ground, but lava flows are responsible for significant amounts of property damage in places like Hawaii. Lava flows also are known to bury roads, farmlands, and other low-lying areas. It must be kept in mind, however, that the entire Hawaiian island chain was built by lava flows, and the real estate that is being damaged would not even be there if it were not for the lava flows. In general, pahoehoe flows flow the fastest, aa are intermediate, and blocky flows are the slowest.

Basaltic lava is extremely hot (typically about 1,900–2,100°F, or 1,000–1,150°C) when it flows across the surface, so when it encounters buildings, trees, and other flammable objects, they typically burst into flame and are destroyed. More silicic lavas may be slightly cooler, typically in the range of 1,550–1,920°F (850–1,050°C). Most lavas will become semisolid and stop flowing at temperatures of around 1,380°F (750°C). Lavas typically cool quickly at first, until a crust or hard skin forms on the flow, then they cool much more slowly. One of the greatest hazards of lava flows is caused by this property of cooling. A lava flow may appear hard, cool, and safe to walk on, yet just below the thin surface may lay a thick layer of molten lava at temperatures of about 1,380°F (750°C). Many people have mistakenly thought it was safe to walk across a recent crusty lava flow, only to plunge through the crust to a fiery death. Thick flows may take years to crystallize and cool, and residents of some volcanic areas have learned to use the heat from flows for heating water and piping it to nearby towns.

It is significantly easier to avoid hazardous lava flows than some other volcanic hazards since in most cases it is possible to simply walk away from the hot moving lava. It is generally unwise to build or buy homes in low-lying areas adjacent to volcanoes, as these are the preferential sites for future lava flows to fill. Lava flows have been successfully diverted in a few examples in Hawaii, Iceland, and elsewhere in the world. One of the better ways is to build large barriers of rock and soil to divert the flow from its natural course to a place where it will not damage property. More creative methods have also had limited success: Lava flows have been bombed in Hawaii and Sicily, and spraying large amounts of water on active flows in Hawaii and Iceland has chilled these flows enough to stop their advance into harbors and populated areas.

Hazards of Pyroclastic Flows

While some volcanoes spew massive amounts of lava in relatively nonthreatening flows, other volcanoes are extremely explosive and send huge eruption clouds tens of thousands of feet into the air. Violent pyroclastic flows present one of the most severe hazards associated with volcanism. Unlike slow-moving lava flows, pyroclastic flows may move at hundreds of miles (km) per hour by riding on a cushion of air, burying entire villages or cities before anyone has a chance to escape. There are several varieties of pyroclastic flows and related volcanic emissions. Nuées ardentes are particularly hazardous varieties of pyroclastic flows, with temperatures that may exceed 1,470°F (800°C), and down slope velocities measured in many hundreds of miles per hour. One of the most devastating pyroclastic flows of modern time was generated from a convergent margin arc volcano in the Lesser Antilles arc on the eastern edge of the Caribbean, between North and South America. On an otherwise quiet day in 1902, the city of St. Pierre on the beautiful Caribbean island of Martinique was quickly buried by a nuée ardente from Mount Pelée that killed more than 29,000 people. Pyroclastic surges are mixtures of gas and volcanic tephra that move sideways in a turbulent mixture that may flow over topography and fill in low-lying areas. A third type of pyroclastic flow is a lateral blast, where an explosion removes large sections of a volcano and blows the material out sideways, typically with disastrous results. The May 1980 eruption of Mount Saint Helens was a laterally directed blast.

One of the most famous volcanic eruptions of all time was a pyroclastic flow. In 79 c.e., Mount Vesuvius of Italy erupted and buried the towns of Pompeii and Herculaneum under pyroclastic ash clouds, permanently encapsulating these cities and their inhabitants in volcanic ash. Later discovery and excavation of parts of these cities by archaeologists led to the world's better understanding of the horrors of being caught in a pyroclastic flow. People were found buried in ash, preserved in various poses of fear, suffocation, and running and in attempts to shelter their children from the suffocating cloud of gas and ash. Crater Lake in Oregon, the result of a pyroclastic eruption 760,000 years ago, provides a thought-provoking example of the volume of volcanic ash that may be produced. The eruption that caused this caldera to collapse covered an area of 5,000 square miles (13,000 km²) with more than six inches (15 cm) of ash, and blanketed most of present-day Washington, Oregon, Idaho, and large parts of Montana, Nevada, British Colombia and Alberta with ash. Larger eruptions have occurred: for instance, a thick rock unit known as the Bishop tuff (a term for hardened ash) covers a large part of the southwestern United States. If any of the active volcanoes in the western United States were to produce such an eruption, it would be devastating for the economy and huge numbers of lives would be lost.

Volcanoes spew a wide variety of sizes and shapes of hardened magma and wall rock fragments during eruptions. *Tephra* is a term that is used to describe all airborne products of an eruption except for gases. Volcanic ash and material erupted during pyroclastic flows may therefore be called tephra, as are other hard objects such as volcanic bombs, blocks, and lapilli. These terms describe material that is ejected out of the volcanic vent on ballistic trajectories, potentially harming people and structures in its path or intended landing area. Volcanic bombs are clots of magma, more than 2.5 inches (6 cm) in diameter, ejected from a volcanic vent. Bombs have various shapes and sizes, with their shape being determined by magma composition, gas content, and other factors. Volcanic blocks are pieces of the wall rock plucked from the walls of the vent and sent shooting through the air by the force of the eruption. These blocks tend to retain angular and blocky shapes. Lapilli are magma clots from ash-size to 2.5 inches (up to 6 cm) in diameter ejected from the volcano and are of many shapes and internal structures as well. Many lapilli form by rain drops moving through ash clouds, causing concentric layers of ash to build around the drop and fall to the ground.

During eruptions projectiles may be thrown thousands of feet into the air and land thousands of feet from the volcanic vent. Some large volcanic bombs and blocks have been found at distances of up to several miles from their sources. Most projectiles leave the volcanic vent at speeds of 300–2,000 feet per second (100–600 m/s), and land on a near vertical trajectory, often forming impact craters where they land. The volume of projectiles landing from a volcanic vent increases toward the vent, with steep-sided volcanoes often covered by fields of projectiles that land and then fragment and roll downhill.

During eruptions large tephra clouds may explode upward and outward from the volcanic vent, expanding as the cloud incorporates more air as it rises. As it expands it becomes less dense, and the heavy particles in the clouds fall out, covering the area below the cloud with blankets of tephra. The thickest tephra blankets are deposited closest to the volcano with the fallout pattern determined by the direction of the winds that carried the tephra cloud. Tephra blankets from major eruptions may cover hundreds of thousands of square miles with volcanic ash and tephra particles. Some of this may be so thick as to collapse buildings, suffocate plants, animals, and people and sometimes may be hot enough to burn objects on impact. Tephra falls are associated with intense darkness for many hours or even days, with many historical accounts relating difficulty in seeing objects only inches from one's face. Some types of tephra are associated with high concentrations of poisonous sulfur compounds, leading to long-term crop and animal disease. Rains and wind quickly wash some tephra falls away. Other types of tephra blankets tend to form hard crusts or thick layers that solidify and become a new land surface in the affected areas.

Tephra falls tend to be so hazardous for many reasons, most relating to the size and physical characteristics of the ejecta. Eruption columns are typically 1.5–6 miles high (3–10 km) but may reach heights of more than 35 miles (55 km), covering huge areas downwind with thick layers of ash and tephra. Many falls occur rapidly, with downwind transport velocities of 5–60 miles per hour (10–100 km/hr). Ash from these clouds may be semitoxic, creating hazards to breath and for vegetation. Additionally, ash is electrically conductive and magnetic when wet, causing problems for electronics and other infrastructure.

Hazards of Mudflows, Floods, Debris Flows, and Avalanches

When pyroclastic flows and nuées ardentes move into large rivers, they quickly cool and mix with water, becoming fast-moving mudflows known as lahars. Lahars may also result from the extremely rapid melting of icecaps on volcanoes. A type of lahar in which ash, blocks of rock, trees, and other material is chaotically mixed together is known as a debris flow. Some lahars originate directly from a pyroclastic flow moving out of a volcano and into a river, whereas other lahars are secondary and form after the main eruption. These secondary, or rain-lahars, form when rain soaks volcanic ash that typically covers a region after an eruption, causing the ash to be mobilized and flow downhill into streams and rivers. It is estimated that it takes only about 30 percent water to mobilize an ash flow into a lahar. The hazards from lahars may continue for months or years after an ash fall on a volcano, as long as the ash remains in place and rains may come to remobilize the ash into flowing slurries. There is a gradual transition between pyroclastic flows, lahars, mud-laden river flows, and floods, with a progressively increasing water content and decreasing amount of suspended material. This transition generally takes place gradually as the flow moves away from the volcano. In other cases earthquakes or the collapse of crater lakes may initiate lahars, such as occurred on Mount Pinatubo during its massive 1991 eruption. Jökulhlaups, massive floods produced by volcanic eruptions beneath glaciers, are common in Iceland where volcanoes of the mid-Atlantic ridge rise to the surface but are locally covered by glaciers. Other Jökulhlaups are known to have originated beneath the snow-covered peaks of the high Andes Mountains in South America.

Lahars and mudflows were responsible for much of the burial of buildings and deaths from the trapping of automobiles during the 1980 eruption of Mount Saint Helens. Big river valleys were also filled by lahars and mudflows during the 1991 eruption of Mount Pinatubo in the Philippines, resulting in extensive property damage and loss of life. One of the greatest volcanic disasters of the 20th century resulted from the generation of a huge mudflow when the icecap on the volcano Nevado del Ruiz in Colombia catastrophically melted during the 1985 eruption. The water mixed with the volcanic ash on the slopes, forming a wet slurry that was able to flow rapidly under the force of gravity. When the mudflow moved downhill it buried the towns along the rivers leading to the volcano, killing 23,000 people and causing more than $200 million of damage to property.

Pyroclastic flows may leave thick unstable masses of unconsolidated volcanic ash that can be easily remobilized long after an eruption during heavy rains or earthquakes. These thick piles of ash may remain for many years after an eruption and not be remobilized until a hurricane or other heavy rain storm (itself a catastrophe) fluidizes the ash, initiating destructive mudflows.

The hazard potential from lahars depends on a wide range of their properties. The thicker and faster-moving lahars are the most dangerous and move the greatest distance from the volcanic source. In places where volcanoes have crater lakes, the hazard potential from lahars and floods is directly proportional to the amount of water that is stored in the lake and may be catastrophically released. Thick piles of ash and tephra may be mobilized into lahars under different conditions, depending on the size of volcanic particles, thickness of the deposits, the angle of the slope upon which they are deposited, and the amount of rain or snow that infiltrates the deposit. Communities located downhill (or down river valley) from volcanic slopes that are covered with ash need to make hazard assessments of the potential for remobilization of ash into lahars to protect citizens and property in low-lying areas.

Slope failures resulting from collapse of oversteepened volcanic constructs on volcanoes sometimes produce debris avalanches and flows that can travel large distances in very short intervals of time. Globally there are more large debris avalanche catastrophes than there are caldera collapse events. Numerous successive debris avalanche deposits, suggesting that collapse of the slopes may be one of the most important processes that modify the slopes of volcanoes, flank many volcanoes. These deposits typically consist of unsorted masses of angular to subangular fragments with hummocky surfaces. Variations in the internal structure of the deposits depend on many factors, such as the amount of air and water present in the mixture as it collapses. Water-rich debris avalanche deposits are transitional in nature with lahars. Some slope failures are associated with very dangerous lateral blasts of volcanoes, such as the 1980 eruption of Mount Saint Helens.

Hazards of Poisonous Gases

One of the lesser-known hazards of active volcanoes stems from their emission of gases. These gases normally escape through geysers, fumaroles, and fractures in the rock. In some instances, however, volcanoes emit poisonous gases, including carbon monoxide, carbon dioxide, and sulfurous gases.

These may also mix with water to produce acidic pools of hydrochloric, hydrofluoric, and sulfuric acid. Because of this, it is generally not advisable to swim in strange-colored or unusual-smelling ponds on active volcanoes.

Some of the more devastating emissions of poisonous gases such as carbon dioxide from volcanoes occurred in 1984 and 1986 in Africa. In the larger of the emissions in 1986, approximately 1,700 people and thousands of cattle were killed when a huge cloud of invisible and odorless carbon dioxide bubbled out of Lakes Nyos and Monoun in volcanic craters in Cameroon, quickly suffocating the people and animals downwind from the vent lakes. More than 3.5 billion cubic feet (100 million m^3) of gas emissions escaped without warning and spread out over the area in less than two hours, highlighting the dangers of living on and near active volcanoes. Lakes similar to Lakes Nyos and Manoun are found in many other active volcanic areas, including heavily populated parts of Japan, Zaire, and Indonesia.

Steps have recently been taken to reduce the hazards of additional gas emissions from these lakes. In 2001 a team of scientists from Cameroon, France, and the United States installed the first of a series of degassing pipes into the depths of Lake Nyos, in an attempt to release the gases from depths of the lake gradually, before they erupt catastrophically. The first pipe extends to 672 feet (205 m) deep in the lake and causes a pillar of gas-rich water to squirt up the pipe and form a fountain on the surface, slowly releasing the gas from depth. The scientific team estimates that they need five additional pipes to keep the gas at a safe level, which will cost an additional 2 million dollars.

Hazards of Volcanic-induced Earthquakes and Tsunamis

Minor earthquakes generally accompany volcanic eruptions. The earthquakes are generated by magma forcing its way upward through cracks and fissures into the volcano from the magma chamber at depth. Gas explosions in the magma conduits under the volcano generate other earthquakes. The collapse of large blocks of rock into calderas or the shifting of mass in the volcano may also initiate earthquakes. Some of these earthquakes happen with a regular frequency, or time between individual shocks, and are known as harmonic tremors. Harmonic tremors have been noted immediately before many volcanic eruptions. These earthquakes have therefore become one of the more reliable methods of predicting exactly when an eruption is imminent, as geologists can trace the movement of the magma by very detailed seismic monitoring. Harmonic tremor earthquakes typically form a continuous, low frequency rhythmic ground

shaking that is distinct from the more isolated shocks associated with movement on faults. Some swarms of harmonic tremors precede an eruption by as little as a few hours or a day, and other cases are reported where the harmonic tremors have gone on for a year before the eruption. Many volcanoes are located in tectonically active areas, and these regions also experience earthquakes that are related to plate movements instead of magma movement and impending volcanic eruptions. There is an ongoing debate in the scientific community about whether or not some specific historical earthquakes and plate movements may have triggered magma migration and volcanic eruptions. For instance, the 1990 eruption of Mount Pinatubo in the Philippines was preceded by a large earthquake, but there is no direct evidence that the earthquake caused the eruption. It may, however, have opened cracks that allowed magma to rise to the surface, helping the eruption proceed.

Tsunamis, or giant seismic sea waves, may be generated by volcanic eruptions, particularly if the eruptions occur underwater. These giant waves may inundate coastlines with little warning, and tsunamis account for the greatest death toll for all volcanic hazards. For instance, in 1883 more than 36,000 people in Indonesia were killed by a tsunami generated by the eruption of Krakatau volcano. Most volcanic-induced tsunamis are produced by collapse of the upper part of the volcanic center into a caldera, displacing large amounts of water, forming a tsunami. Other volcanic tsunamis are induced by giant landslides, by volcanic-induced earthquakes, by submarine explosions, and by pyroclastic material or lahars hitting sea water. Volcanic-induced atmospheric shock waves may also initiate some tsunamis. The tsunami then moves radially away from the source at speeds of about 500 miles per hour (800 km/hr). The tsunami may have heights of a few feet at most in the open ocean and have wavelengths (distance between crests of successive waves) of a hundred or more miles (~200 km). When the tsunamis encounter and run up onto shorelines, the shape of the sea floor, bays, promontories, and the coastline helps determine the height of the wave (of course the initial height and distance also determine the height). Some bays that get progressively narrower tend to amplify the height of tsunamis, causing more destruction at their ends than their mouths.

When tsunamis strike the coastal environment, the first effect is sometimes a significant retreat or drawdown of the water level, whereas in other cases the water just starts to rise quickly. Since tsunamis have long wavelengths, it typically takes several minutes for the water to rise to its full height. Being that there is no trough right behind the crest of the wave, on account of the very long wavelength of the tsunami, the water does not recede for a considerable time after the initial crest rises onto land. The rate of rise of the water in a tsunami depends in part on the shape of the sea floor and coastline. If the sea floor rises slowly, the tsunami may crest slowly, giving people time to outrun the rising water. In other cases, especially where the sea floor rises steeply, or the shape of the bay causes the wave to be amplified, tsunamis may come crashing in huge walls of water with breaking waves that pummel the coast with a thundering roar and wreaking utmost destruction.

Because tsunamis are waves, they travel in successive crests and troughs. Many deaths in tsunami events are related to people going to the shoreline to investigate the effects of the first wave, or to rescue those injured or killed in the initial crest, only to be drowned or swept away in a succeeding crest. Tsunamis have long wavelengths, so successive waves have a long lag time between individual crests. The period of a wave is the time between the passage of individual crests, and for tsunamis the period can be an hour or more. A tsunami may therefore devastate a shoreline area and retreat, and then another crest may strike an hour later, then another, and another, in sequence.

Some of the largest recorded tsunamis have been generated by volcanic eruptions. The most famous volcanic eruption–induced tsunamis include the series of huge waves generated by the eruption of Krakatau in 1883, which reached run-up heights of more than 200 feet (60 m) and killed 36,500 people. The number of people that perished in the eruption of Santorini in 1650 B.C.E. is not known, but the toll must have been huge. The waves reached 800 feet (240 m) in height on islands close to the volcanic vent of Santorini. Flood deposits have been found 300 feet (90 m) above sea level in parts of the Mediterranean Sea and extend as far as 200 miles (320 km) southward up the Nile River. Several geologists suggest that these were formed from a tsunami generated by the eruption of Santorini. The floods from this eruption may also, according to some scientists, account for the biblical parting of the Red Sea during the exodus of the Israelites from Egypt and the destruction of the Minoan civilization of the island of Crete.

Hazards of Atmospheric Sound and Shock Waves

Large volcanic eruptions are associated with rapid expansion of gases during explosive phases, and these have been known to produce some of the loudest sounds and atmospheric pressure waves known on Earth. For instance, explosions from the 1883 eruption of Krakatau were heard almost 3,000 miles (4,700 km) away, in places as diverse as the Indian Ocean islands of Diego Garcia and Rodriguez, in south India, in southeast Asia across Myan-

mar, Thailand, and Vietnam, in the Philippines, and across western and central Australia. Other volcanic eruptions that produced blasts heard for hundreds to thousands of miles include the 1835 eruption of Cosiguina in Nicaragua, the eruption of Mount Pelée in 1902, and Katmai (Alaska) in 1912. Although interesting, sounds from eruptions are relatively harmless. Explosions that produce sound waves may also be associated with much more powerful atmospheric shock waves that can be destructive.

Atmospheric shock waves are produced by the pressure changes caused by the sudden explosive release of rapidly expanding steam and gases that may sometimes exceed the speed of sound. When the gas eruptions proceed at supersonic velocities (1,000–2,700 feet [305–823 m] per second depending on the temperature and density of the gas), the shock waves may be associated with huge, expanding flashing arcs of light that pulsate out of the volcano. Firsthand accounts of these flashing arcs are rare, but observations of the 1906 eruption of Vesuvius told of flashes ranging from several times a second to every few seconds. Deafening explosive sounds followed these flashes.

Large atmospheric shock waves may be powerful enough to damage or knock down buildings and may travel completely around the world. The eruptions of Krakatau in 1883, Pelée in 1902, Asama (Japan) in 1783 and 1973, and others were recorded at atmospheric weather stations around the globe. Some of these shock waves destroyed or damaged buildings across hundreds of square miles surrounding the volcano.

Many volcanic eruptions are associated with spectacular lightning storms in the ash clouds, or in the expanding gas clouds. As this material is ejected from the volcano, many particles may rub together, creating electrical discharges seen as lightning. Intense lightning storms may typically extend for 5–10 miles (8–16 km) from the eruption, posing threats to people brave enough to remain close to the eruption, as well as to communication systems. A phenomenon known as St. Elmo's fire is sometimes associated with volcanic eruptions. It refers to a glowing blue or green electrical discharge that emanates from tall objects that are near an intense electrical charge and has been observed on ships near eruptions, including Krakatau in 1883, Vesuvius in 1906, and even during the 1980 eruption of Mount Saint Helens.

Hazards from Changes in Climate

Some of the larger, more explosive volcanic eruptions spew vast amounts of ash and finer particles called aerosols into the atmosphere and stratosphere, and it may take years for these particles to settle back down to Earth. They get distributed about the planet by high-level winds, and they have the effect of blocking out some of the Sun's rays, which lowers global temperatures. This happens because particles and aerosol gases in the upper atmosphere tend to scatter sunlight back to space, lowering the amount of incoming solar energy. In contrast, particles that get injected only into the lower atmosphere absorb sunlight and contribute to greenhouse warming. A side effect is that the extra particles in the atmosphere also produce more spectacular sunsets and sunrises, as does extra pollution in the atmosphere. These effects were readily observed after the 1991 eruption of Mount Pinatubo, which spewed more than 172 billion cubic feet (5 billion m³) of ash and aerosols into the atmosphere, causing global cooling for two years after the eruption. Even more spectacularly, the 1815 eruption of Tambora in Indonesia caused three days of total darkness for approximately 300 miles (500 km) from the volcano, and it initiated the famous "year without a summer" in Europe, because the ash from this eruption lowered global temperatures by more than a degree. Even these amounts of gases and small airborne particles are dwarfed by the amount of material placed into the atmosphere during some of Earth's largest eruptions, known as flood basalts. No flood basalts have been formed on Earth for several tens of millions of years, which is a good thing, since their eruption may be associated with severe changes in climate. For instance, 66 million years ago a huge flood basalt field was erupted over parts of what is now western India, the Seychelles Islands, and Madagascar (these places were closer together then, but since separated by plate tectonics). At this time in Earth history the climate changed drastically, and it is thought that these severe climate changes contributed to massive global extinction of most living things on Earth at the time, including the dinosaurs. The atmospheric changes stressed the global environment to such an extent that when another catastrophic event, a meteorite impact, occurred, the additional environmental stresses caused by the impact were too great for most life-forms to handle, and they died and became extinct.

Aside from global cooling or warming associated with major volcanic eruptions, volcanic gases and ash can have some severe effects on regional environmental conditions. Gases and aerosols (fine particles suspended in gas) may be acidic and may be carried far from the source as gases, aerosols, or salts, or adsorbed on the ash and tephra particles. Some of these compounds, including sulfur and chlorine, are acidic and may mix with water and other cations to form sulfuric and sulfurous acids, as well as hydrochloric acid, hydrofluoric acid, carbonic acid, and ammonia. Ash falls and rains that move through ash clouds and deposits may spread harmful and acidic fluids that may have harmful effects on vegetation, crops, and water supplies. Other volcanic eruptions have been associated with more toxic gases, such as concentrated fluorine, which has occasionally posed a volcanic hazard in Iceland.

Other Hazards and Long-Term Effects of Volcanic Eruptions

Dispersed ash may leave thin layers on agricultural fields, which may be beneficial or detrimental, depending on the composition of the ash. Much of the richest farmland on the planet is on volcanic ash layers near volcanoes. Some ash basically fertilizes soil, whereas other ash is toxic to livestock. Volcanic ash consists of tiny but jagged and rough particles that may conduct electricity when wet. These properties make ash also pose severe hazards to electronics and machinery that may last many months past an eruption. In this way ash has been known to disrupt power generation and telecommunications. The particles are abrasive and may cause serious heart and lung ailments if inhaled.

Ash clouds have some unexpected and long-term consequences to the planet and its inhabitants. Airplane pilots have sometimes mistakenly flown into ash clouds, thinking they are normal clouds, which has caused engines to fail. For instance, KLM Flight 867 with 231 people aboard flew through the ash cloud produced during the 1989 eruption of Mount Redoubt in Alaska, causing engine failure. The plane suddenly dropped two miles (3.2 km) in altitude before the pilots were able to restart the engines, narrowly averting disaster. This event led the United States Geological Survey to formulate a series of warning codes for eruptions in Alaska and their level of danger to aircraft.

After volcanic eruptions, large populations of people may be displaced from their homes and livelihoods for extended or permanent time periods. In many cases these people are placed into temporary refugee camps, which all too often become permanent shanty villages, riddled with disease, poverty, and famine. Many of the casualties from volcanic eruptions result from these long-term effects, and not the initial eruption. More needs to be done to insure that populations displaced by volcanic disasters are relocated into safe settings.

PREDICTING AND MONITORING VOLCANIC ERUPTIONS

One of the best ways to understand what to anticipate from an active volcano is to study its history. One can examine historical records to learn about geologically recent eruptions. Geological mapping and analysis can reveal what types of material the volcano has spewed forth in the more distant past.

A geologist who studies volcanic deposits can tell through examination of these deposits whether the volcano is characterized by explosive or nonexplosive eruptions, whether it has nuée ardentes or mudflows, and how frequently the volcano has erupted over long intervals of time. This type of information is crucial for estimating what the risks are for any individual volcano. Programs of risk assessment and volcanic risk–mapping need to be done around all of the nearly 600 active volcanoes on continents of the world. These risk assessments will determine which areas are prone to ash falls and which areas have been repeatedly hit by mudflows. They will help residents determine if there are any areas characterized by periodic emissions of poisonous gas. Approximately 60 eruptions occur globally every year, so these data would prove immediately useful when eruptions appear imminent.

Nearly a quarter million people have died in volcanic eruptions in the past 400 years, with a couple of dozen volcanic eruptions killing more than a thousand people each. Eruptions in remote areas have little consequence except for global climate change. In contrast, eruptions in populated areas can cause billions of dollars in damage and result in entire towns and cities being relocated. The eruption of Mount Saint Helens in 1980 caused 1 billion dollars worth of damage, and this was not even a very catastrophic eruption in terms of the volume of material emitted. The Mount Saint Helens eruption spewed about a cubic mile (4 km³) of debris into the atmosphere, whereas 10 years later the eruption of Mount Pinatubo in the Philippines sent 5–6 cubic miles (20–25 km³) of material skyward. The table "Catastrophic Volcanic Eruptions of the Last 200 Years" lists the deadliest volcanic eruptions of the last 200 years.

The United States Geological Survey is the main organization responsible for monitoring volcanoes and eruptions in the United States and assumes this responsibility for many other places around the world. In cases of severe eruptions or eruptions that threaten populated areas, other agencies such as the Federal Emergency Management Agency (FEMA) will join the U.S. Geological Survey and help in disseminating information and evacuating the population.

CATASTROPHIC VOLCANIC ERUPTIONS OF THE LAST 200 YEARS

Location	Year	Deaths
Tambora, Indonesia	1815	92,000
Krakatau, Indonesia	1883	36,500
Mount Pelée, Martinique	1902	32,000
Nevada del Ruiz, Colombia	1985	24,000
Unzen, Japan	1792	14,300
Laki, Iceland	1783	9,350
Kelut, Indonesia	1919	5,110
Santa Maria, Guatemala	1902	6,000
Galunggung, Indonesia	1822	5,500
Vesuvius, Italy	1631	3,500
Vesuvius, Italy	79	3,360
Awu, Indonesia	1826	3,000
Papandayan, Indonesia	1772	2,957
Lamington, Papua New Guinea	1951	2,950
El Chichon, Mexico	1982	2,000
Agung, Indonesia	1963	1,900
Sofriere, St. Vincent	1902	1,680
Oshima, Japan	1741	1,475
Asama, Japan	1783	1,377
Taal, Philippines	1911	1,335
Mayon, Philippines	1814	1,200
Agung, Indonesia	1963	1,184
Cotopaxi, Ecuador	1877	1,000
Pinatubo, Philippines	1991	800

Precursors to Eruptions

Volcanic eruptions are sometimes preceded by a number of precursory phenomena, or warnings that an eruption may be imminent. Many of these involve subtle changes in the shape or other physical characteristics of the volcano. Many volcanoes develop bulges, swells, or domes on their flanks when magma rises in the volcano before an eruption. These shape changes can be measured using sensitive devices called tilt meters, which measure tilting of the ground surface, or devices that precisely measure distances between points, such as geodolites and laser measuring devices. Bulges were measured on the flanks of Mount Saint Helens before the 1980 eruption.

Eruptions may also be preceded by other more subtle precursory events, such as increase in the temperature or heat flow from the volcano, measurable both on the surface and in crater lakes, hot springs, fumaroles, and hot springs on the volcano. There may also be detectable changes in the composition of gases emitted by the volcano, such as increases in the hydrochloric acid and sulfur dioxide gases before an eruption.

One of the most reliable precursors to an eruption is the initiation of the harmonic seismic tremors that reflect the movement of magma into the volcano. These tremors typically begin days or weeks before an eruption, and steadily change their characteristics, enabling successively more accurate predictions of how imminent the eruption is before it actually happens. Careful analysis of precursor phenomena including the harmonic tremors, change in the shape of the volcano, and emission of gases has enabled accurate prediction of volcanic eruptions, including Mount Saint Helens and Mount Pinatubo. These predictions saved innumerable lives.

Volcano Monitoring

Signs that a volcano may be about to erupt may be observed only if volcanoes are carefully and routinely monitored. Volcanic monitoring is aimed at detecting the precursory phenomena described above and tracking the movement of magma beneath volcanoes. In the United States, the United States Geological Survey is in charge of comprehensive volcano moni-

toring programs in the Pacific Northwest, Alaska, and Hawaii.

One of the most accurate methods of determining the position and movement of magma in volcanoes is using seismology, or the study of the passage of seismic waves through the volcano. These can be natural seismic waves generated by earthquakes beneath the volcano or seismic energy released by geologists who set off explosions and monitor how the energy propagates through the volcano. Certain types of seismic waves travel through fluids like magma (compressional- or P-waves), whereas other types of seismic waves do not (shear- or S-waves). The position of the magma beneath a volcano can be determined by detonating an explosion on one side of the volcano, and having seismic receivers placed around the volcano to determine the position of a "shadow zone" where P-waves are received, but S-waves are not. The body of magma that creates the shadow zone can be mapped out in three dimensions by using data from the numerous seismic receiver stations. Repeated experiments over time can track the movement of the magma.

Other precursory phenomena are also monitored to track their changes with time, which can further refine estimates of impending eruptions. Changes in

Arching lava fountain on Kilauea, Hawaii, February 25, 1983 *(J. D. Griggs/USGS)*

the temperature of the surface can be monitored by thermal infrared satellite imagery, and other changes, such as shifts in the composition of emitted gases, are monitored. Other promising precursors may be found in changes to physical properties, such as the electrical and magnetic field around volcanoes prior to eruptions.

Changes in the geochemical nature of gases and fluids coming out of volcanic vents and fumaroles can be used as indicators of activity beneath volcanoes. These changes depend largely on the changing rates of magma degassing beneath the volcano, and interactions of the magma with the groundwater system. Monitoring of gases is largely designed to look for rapid changes or nonequilibrium conditions in hydrochloric and sulfurous acids, carbonic acids, oxygen, nitrogen, and hydrogen sulfide. Convergent margin andesitic types of explosive volcanoes show the greatest variation in composition of gases prior to eruption, since magma in these volcanoes is ultimately derived from fluids carried to depth by the subducting oceanic lithospheric slabs.

The details of geophysical volcano monitoring are complex and have undergone a rapid explosion in sophistication in recent years. One of the more common techniques in use now involves the use of a dense array or group of very sensitive seismographs called broadband seismometers that can detect a variety of different kinds of earthquakes. Broadband seismometers can detect seismic waves with frequencies of 0.1–100 seconds, a great improvement over earlier short-period seismometers that detected only frequencies between 0.1–1 second. Swarms of small earthquakes, known as harmonic tremors, are sometimes associated with the movement of magma upward or laterally beneath a volcano, and they characteristically increase in number before an eruption. These are different from tectonic earthquakes that generally follow a pattern of main shock–aftershocks. By analysis of the seismic data from the array of seismographs geologists are able to build a three-dimensional image of the area beneath the volcano, much like a tomographic image or a CAT scan, and thereby monitor the distribution and movement of magma beneath the volcano. When the magma gets closer to the surface, an eruption is more likely to occur. Movement of magma is also sometimes associated with explosion-type earthquakes, easily differentiated from earthquakes associated with movement on faults.

Many explosive volcanic eruptions are preceded by swelling, bulging, or other deformation of the ground surface on the volcano, so one method to predict eruptions involves measuring and monitoring this bulging. Ground deformation is commonly measured using a variety of devices. Some instruments precisely measure shifts in the level surface, others measure tilting, and still others make electronic distance measurements. These types of measurements have recently increased in accuracy with the advent of the use of precise Global Positioning System instruments that allow measurements of latitude, longitude, and elevation to be made that are accurate to less than a half-inch (1 cm), even in very remote locations.

Some observations have been made of phenomena that precede some eruptions, even though their cause is not clearly understood. Electrical and magnetic fields have been observed to show changes at many volcanoes, especially those with basaltic magma that has a high concentration of magnetic minerals. These changes may be related to movement of magma (and the magnetic minerals), changes in heating, movement of gases, or other causes. Recent studies have linked small changes in the microgravity fields around active volcanoes, especially explosive andesitic volcanoes, to movement of magma beneath the cones.

Satellite images are now commonly used to map volcanic deposits and features and to monitor eruptions. There is now a wide range in types of features satellites can measure and monitor, including a large range of the visible and other parts of the electromagnetic spectrum. Changes in the volcanic surface, growth of domes, and opening and closure of fissures on the volcano can be observed from satellites. Some satellites use radar technology that is able to see through clouds and some ash, and thus are particularly helpful for monitoring volcanoes in remote areas, in bad weather, at night, and during eruptions. A technique called radar interferometry can measure ground deformation at the sub-inch (cm) scale, showing bulges and swelling related to buildup of magma beneath the volcano. Some satellites can measure and monitor the temperature of the surface, and others can watch eruption plumes, ash clouds, and other atmospheric effects on a global scale.

Together all these techniques have given seismologists and geologists tools they need to make more accurate predictions of when an eruption may occur, saving lives and property. When many of the techniques are integrated into one monitoring program, then scientists are better able to predict when the next eruption may occur. Several volcano monitoring programs in the United States use many different types of observations to provide for the safety of citizens. These include the Alaskan Volcano Observatory, the Cascades Volcano Observatory, and the Hawaiian Volcano Observatory.

See also CONVERGENT PLATE MARGIN PROCESSES; ENERGY IN THE EARTH SYSTEM; GEOCHEMICAL CYCLES; MAGMA; PLATE TECTONICS; TSUNAMI, GENERATION MECHANISMS.

FURTHER READING

Blong, Russel J. *Volcanic Hazards: A Sourcebook on the Effects of Eruptions.* New York: Academic Press, 1984.

Fisher, R. V. *Out of the Crater: Chronicles of a Volcanologist.* Princeton, N.J.: Princeton University Press, 2000.

Fisher, R. V., G. Heiken, and J. B. Hulen. *Volcanoes: Crucibles of Change.* Princeton, N.J.: Princeton University Press, 1998.

Oregon Space Grant Consortium. Volcanoworld. Available online. URL: http://volcano.oregonstate.edu/. Accessed October 10, 2008.

Scarpa, Roberto, and Robert I. Tilling. *Monitoring and Mitigation of Volcano Hazards.* New York: Springer, 1996.

Simkin, T., and R. S. Fiske. *Krakatau 1883: The Volcanic Eruption and Its Effects.* Washington, D.C.: Smithsonian Institution Press, 1993.

U.S. Geological Survey. Volcano Hazards Program home page. Available online. URL: http://volcanoes.usgs.gov/. Last modified September 11, 2008. Data updated daily or more frequently.

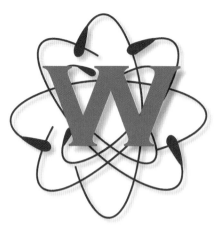

weathering Weathering is the process of mechanical and chemical alteration of rock marked by the interaction of the lithosphere, atmosphere, hydrosphere, and biosphere. The resistance to weathering varies with climate, composition, texture, and how much a rock is exposed to the elements of weather. Weathering processes occur at the lithosphere/atmosphere interface. This is actually a zone that extends down into the ground to the depth that air and water can penetrate—in some regions this is a few feet (meters), in others it is a mile (kilometer) or more. In this zone, the rocks make up a porous network, with air and water migrating through cracks, fractures, and pore space. The effects of weathering can often be seen in outcrops on the sides of roads, where they cut through the zone of alteration into underlying bedrock. These roadcuts and weathered outcroppings of rock show some similar properties. The upper zone near the surface is made of soil or regolith in which the texture of the fresh rock is not apparent, a middle zone in which the rock is altered but retains some of its organized appearance, and a lower zone, of fresh unaltered bedrock.

PROCESSES OF WEATHERING
There are three main types of weathering. Chemical weathering, the decomposition of rocks through the alteration of individual mineral grains, is a common process in the soil profile. Mechanical weathering is the disintegration of rocks, generally by abrasion. Mechanical weathering is common in the talus slopes at the bottom of mountains, along beaches, and along river bottoms. Biological weathering involves the breaking down of rocks and minerals by biological agents. Some organisms attack rocks for nutritional purposes; for instance, chitons bore holes through limestone along the seashore, extracting their nutrients from the rock.

Generally, mechanical and chemical weathering are the most important, and they work hand-in-hand to break down rocks into the regolith. The combination of chemical, mechanical, and biological weathering produces soils, or a weathering profile.

Mechanical Weathering
There are several different types of mechanical weathering which may act separately or together to break down rocks. The most common process of mechanical weathering is abrasion, where movement of rock particles in streams, along beaches, in deserts, or along the bases of slopes causes fragments to knock into each other. These collisions cause small pieces of each rock particle to break off, gradually rounding the particles and making them smaller, and creating more surface area for processes of chemical weathering to act upon.

Some rocks develop joints, or parallel sets of fractures, from differential cooling, the pressures exerted by overlying rocks, or tectonic forces. Joints are fractures along which no observable movement has occurred. Joints promote weathering in two ways: they are planes of weakness across which the rock can break easily, and they act as passageways for fluids to percolate along, promoting chemical weathering.

Crystal growth may aid mechanical weathering. When water percolates through joints or fractures, it can precipitate minerals such as salts, which grow larger and exert large pressures on the rock along the joint planes. If the blocks of rock are close enough to a free surface such as a cliff, large pieces of rock may be forced off in a rockfall, initiated by the gradual growth of small crystals along joints.

Sheets of granite carved by Tuolumne River and ancient glaciers in Yosemite National Park, California *(Joseph H. Bailey/National Geographic/Getty Images)*

When water freezes to form ice, its volume increases by 9 percent. Water is constantly seeping into the open spaces provided by joints in rocks. When water filling the space in a joint freezes, it exerts large pressures on the surrounding rock. These forces are very effective agents of mechanical weathering, especially in areas with freeze-thaw cycles. They are responsible for most rock debris on talus slopes of mountains and cracks in concrete in areas with cold climates.

Heat may also aid mechanical weathering, especially in desert regions where the daily temperature range may be extreme. Rapid heating and cooling of rocks sometimes exerts enough pressure on the rocks to shatter them to pieces, thus breaking large rocks into smaller fragments.

Plants and animals may also aid mechanical weathering. Plants grow in cracks and push rocks apart. This process may be accelerated if plants such as trees become uprooted, or blown over by wind, exposing more of the underlying rock to erosion. Burrowing animals, worms, and other organisms bring an enormous amount of chemically weathered soil to the surface, and continually turn the soils over and over, greatly assisting the weathering process.

Chemical Weathering

Minerals that form in igneous and metamorphic rocks at high temperatures and pressures may be unstable at temperatures and pressures at the Earth's surface, so they react with the water and atmosphere to produce new minerals. This process is known as chemical weathering. The most effective chemical agents are weakly acidic solutions in water. Therefore, chemical weathering is most effective in hot and wet climates.

Rainwater mixes with CO_2 from the atmosphere and from decaying organic matter, including smog, to produce carbonic acid according to the following reaction:

$$H_2O + CO_2 \rightarrow H_2CO_3$$

Water + carbon dioxide → carbonic acid

Carbonic acid ionizes to produce the hydrogen ion (H^+), which readily combines with rock forming minerals to produce alteration products. These alteration products may then rest in place and become soils, or be eroded and accumulate somewhere else.

Hydrolysis is a process that occurs when the hydrogen ion from carbonic acid combines with potassium feldspar to produce kaolinite, a clay mineral, according to the following reaction:

$$2KAlSi_3O_8 + 2H_2CO_3 + H_2O \rightarrow$$
$$Al_2Si_2O_5(OH)_4 + 4SiO_2 + 2K^{+1} + 2HCO_3$$

feldspar + carbonic acid + water →
kaolinite + silica + potassium + bicarbonate ion

This reaction is one of the most important reactions in chemical weathering. The product, kaolinite, is common in soils and is virtually insoluble in water. The other products, silica, potassium, and bicarbonate are typically dissolved in water and carried away during weathering.

Much of the material produced during chemical weathering is carried away in solution and deposited elsewhere, such as in the sea. The highest-temperature minerals are leached the most easily. Many minerals combine with oxygen in the atmosphere to form another mineral by oxidation. Iron is very easily oxidized from the Fe+2 state to the Fe+3 state, forming goethite or with the release of water, hematite.

$$2FeO + (OH)_2 \rightarrow Fe_2O_3 + H_2O$$

Different types of rock weather in distinct ways. For instance, granite contains potassium feldspar and weathers to form clays. Building stones are selected to resist weathering in different climates, but now, increasing acidic pollution is destroying many old landmarks. Chemical weathering results in the removal of unstable minerals and a consequent concentration of stable minerals. Included in the remains are quartz, clay, and other rare minerals such as gold and diamonds, which may be physically concentrated in placer deposits.

On many boulders, weathering penetrates only a fraction of the diameter of the boulder, resulting in a rind of the altered products of the core. The thickness of the rind itself is useful for knowing the age of the boulder, if rates of weathering are known. These types of weathering rinds are useful for determining the age of rockslides and rockfalls and the time interval between rockfalls in any specific area.

Exfoliation is a weathering process where rocks spall off in successive shells, like the skin of an onion. Exfoliation is caused by differential stresses within a rock formed during chemical weathering processes. For instance, feldspar weathers to clay minerals, which take up a larger volume than the original feldspar. When the feldspar minerals turn to clay, they exert considerable outward stress on the surrounding rock, which is able to form fractures parallel to the rock's surface. This need for increased space is accommodated by the minerals through the formation of these fractures, and the rocks on the hillslope or mountain are then detached from their base and more susceptible to sliding or falling in a mass-wasting event.

If weathering proceeds along two or more sets of joints in the subsurface, it may result in shells of weathered rock which surround unaltered rocks, looking like boulders. This is known as spheroidal weathering. The presence of the several sets of joint surfaces increases the effectiveness of chemical weathering, because the joints increase the available surface area to be acted on by chemical processes. The more subdivisions within a given volume, the greater the surface area.

Biological Weathering

Biological weathering is the least important of the different categories of weathering. In some places plants and microorganisms may derive nutrition from dissolving minerals in rocks and soil, thus contributing to their breakdown and weathering. There are enormous numbers of microorganisms and insects living in the soil horizon, and these contribute to the breakdown of organic material in the soils and also contribute their tests or bodies when they die. Biological weathering may also include some of the effects of roots pushing rocks apart or expanding cracks in the weathered rock horizon. These effects also move rock fragments, so they are discussed under the topic of mechanical weathering.

FACTORS THAT INFLUENCE WEATHERING

The effectiveness of weathering processes is dependent upon several different factors, explaining why some rocks weather one way at one location and a different way in another location. Rock type is an important factor in determining the weathering characteristics of a hillslope, because different minerals react differently to the same weathering conditions. For instance, quartz is resistant to weathering, and quartz-rich rocks typically form large mountain ridges. Conversely, shales readily weather to clay minerals, which are easily washed away by water, so shale-rich rocks often occupy the bottoms of valleys. Examples of topography being closely related to the underlying geology in this manner are abundant in the Appalachians, Rocky Mountains, and most other mountain belts of the world.

Rock texture and structure is important in determining the weathering characteristics of a rock mass. Joints and other weaknesses promote weathering by increasing the surface area for chemical reactions to take place on, as described above. They also allow water, roots, and mineral precipitates to penetrate deeply into a rock mass, exerting outward pressures that can break off pieces of the rock mass in catastrophic rockfalls and rockslides.

The slope of a hillside is important for determining what types of weathering and mass-wasting processes occur on that slope. Steep slopes let the products of weathering get washed away, whereas gentle slopes promote stagnation and the formation of deep weathered horizons.

Climate is one of the most important factors in determining how a site weathers. Moisture and heat promote chemical reactions, so chemical weathering processes are strong and fast and dominate over mechanical processes in hot, wet climates. In cold climates, chemical weathering is much less important. Mechanical weathering is very active during freezing and thawing, so mechanical processes such as ice wedging tend to dominate over chemical processes in cold climates. These differences are exemplified by two examples of weathering. In much of New England, a hike over mountain ridges will reveal fine, millimeter-thick striations that were formed by glaciers moving over the region more than 10,000 years ago. Chemical weathering has not removed even these one-millimeter-thick marks in 10,000 years. In contrast, new construction sites in the Tropics, such as roads cut through mountains, often expose fresh bedrock. In a matter of 10 years these road cuts will be so deeply eroded to a red soil-like material, called gruse, that the original rock will not be recognizable.

As in most things, time is important. It takes tens of thousands of years to wash away glacial grooves in cold climates, but in the Tropics, weathered horizons that extend to hundreds of meters may form over a few million years.

See also ATMOSPHERE; HYDROSPHERE; SOILS.

FURTHER READING

Birkland, P. W. *Soils and Geomorphology*. New York: Oxford University Press, 1984.

Wegener, Alfred Lothar (1880–1930) German *Meteorologist, Geologist* Alfred Lothar Wegener was born on November 1, 1880, in Berlin, Germany, and obtained a Ph.D. in astronomy from the University of Berlin in 1904. He is well known for his studies in meteorology and geophysics and is considered by many to be the father of the theory of continental drift. His is also known for his work on dynamics and thermodynamics of the atmosphere, atmospheric refraction and mirages, optical phenomena in clouds, acoustical waves, and the design of geophysical instruments. Wegener was an avid balloonist, and pioneered the use of weather balloons to monitor weather and air masses while he was working at the Royal Prussian Aeronautical Observatory near Berlin. Alfred and his brother, Kurt, broke a

world endurance record for hot-air balloons in 1906, staying aloft for more than 52 hours. Alfred married the daughter of famous Russian climatologist Vladdimir Koppen. On his fourth and last expedition to Greenland, Alfred and his companion Rasmus Villumsen became lost in a blizzard on November 2, 1930, and Wegener's body was not discovered until May 12, 1931.

Alfred Wegener's interest in meteorology and geology led him on a Danish expedition to the unmapped northeastern Greenland coast in 1906–08, mainly to study the circulation of polar air masses. This was the first of four Greenland expeditions he would make, and this area remained one of his dominant interests. In 1909 he took a position at the University of Marburg in Germany, where he lectured on meteorology, astronomy, and mapping. He authored a textbook in meteorology called *The Thermodynamics of the Atmosphere,* based on a series of lectures he gave at the university, and published it in 1911, when he was just 30 years old.

Wegener is most famous for being the first person to come up with the idea for continental drift. This interest was initially sparked while he was teaching at the University of Marburg in 1911 and noted the striking similarity of fossils from continents now separated by large oceans. The accepted theories for this similarity at the time were that land bridges between the continents occasionally rose up, allowing plants and animals to move from continent to continent. However, Wegener studied the apparent correspondence between the shapes of the coastlines of western Africa and eastern South America, and he hypothesized that the continents had themselves moved apart and that land bridges did not rise between stationary continents. Soon, he came up with a model in which most of the world's continents had drifted away from a former supercontinent starting around 180 million years ago. He continued to study the paleontological and geologic evidence, concluding that these similarities demanded a detailed explanation.

In 1912 Wegener returned to Greenland, on a perilous journey in which the team "narrowly escaped death" while climbing a tidewater glacier on the coast that suddenly began calving, and then he became the first person to spend the entire winter on the ice cap. In the spring the team broke another record, making the longest crossing of the ice sheet ever, walking 750 miles (1,200 km) across barren ice at elevations up to 10,000 feet (3,000 m). During this trip Wegener collected many scientific samples and data on glaciers and climate and became the first person to track storms over the polar ice cap.

Wegener continued to study many different features on the continents when he returned to Marburg and found that mountain ranges in South America

Alfred Wegener with pipe and parka *(Alfred-Wegener-Institut, Germany)*

and Africa lined up if the continents were once together, and belts of distinctive rock types, such as coal, also matched on different continents if they were restored to his hypothesized supercontinent of Pangaea (meaning all land). Alfred Wegener studied the fossils and found that narrow belts of distinctive fauna and flora, such as the reptiles *Mesosaurus* and *Lystrosaurus,* matched on the former supercontinent and that the *Glossopteris* flora from the southern continents also matched on his restored Pangaea map. Wegener wrote an extended account of his continental drift theory in his 1915 book *Die Entstehung der Kontinente und Ozeane (The Origin of the Continents and Oceans,* translated into English in 1966). In this book, he argued strongly against the land bridge hypothesis, stating that continents are made of different material (granite) that is less dense than the oceanic crust (made of basalt), and that the weaker material should be floating on the denser substratum much like an iceberg on water. He also provided evidence of the continents moving up and down relative to sea level, noting that the continents in northern latitudes rise up when glaciers retreat from an area, rising or rebounding in response to the reduction of weight on the crust. He reasoned that the continents must also be moving laterally, explain-

ing why coastlines of different continents could be restored so perfectly.

Wegener also argued against the then leading model for the origins of mountains. Others, such as the Tasmanian geologist S. Warren Carey, advocated that mountains formed by the cooling and shrinking of the Earth and that as the Earth shrunk mountains formed as wrinkles to accommodate the change in surface area. Wegener said that model did not account for the distribution and spacing of mountain ranges. He argued instead that they formed on the edges of continents that had drifted and collided with other continents, such as where India was currently crashing into Asia.

As a meteorologist he began to look at ancient climates, using paleoclimatic evidence he found to strengthen his theory of continental drift. One of his strongest arguments was that of the patterns of the Permian-Carboniferous glaciation on Pangaea. If the continents were restored, the paleoglaciers would have a pattern of all the ice moving away from a central core near the South Pole about 280 million years ago, whereas if the continents had their present distribution, then the glaciation patterns would be random and would require more water to make the amount of ice needed to explain the patterns than is available on Earth.

Wegener was by no means the first to think of the theory of continental drift. However, he was the first to go to great lengths to develop and establish the theory. His ideas were met for the large part by the geological community with strong opposition, anger, and hostility. Many geologists noted that Wegener was not trained as a geologist, so did not believe him, and published many disparaging comments on his theory and even his personality. Wegener had evidence for continental drift but could not provide a mechanism. Without a driving force he could do little to argue against those who said there was no driving mechanism, except to smoke his famous pipe and remain silent. He did argue, however, for the validity of his observations, especially that the continents move, that the movement causes deformation at their edges, and that earthquakes and volcanoes are associated with the edges of these great moving continents. It was not until 40–50 years later that proof of Wegener's theories came, in the form of the many proofs of plate tectonics that came with the geological revolution in the 1960s.

Wegener made his fourth and final trip to Greenland in the spring of 1930, never to return. The expedition started with 22 scientists and technicians, but bad weather plagued and delayed the initial setup of the base camp from May until mid-June, and the inland camp at Eismitte had only basic supplies. On September 21 Wegener led a team of 15 dogsleds to bring supplies to Eismitte, but bad weather convinced 12 of the 13 local Greenlanders to turn back after one week. After 40 days of traveling, Wegener and his two remaining companions reached Eismitte, delivered the supplies, and then celebrated Wegener's 50th birthday on November 1st. The next day Alfred and Rasmus Villumsen, his Greenlander companion, set out to return to base camp, but never arrived. In the spring, on May 12 a search party found Wegener's body, buried in the snow in sleeping bags, and they concluded he died of a heart attack in his sleep, from exertion of the trip. Villumsen was never found. His friends and companions erected an ice mausoleum over Wegener's body, all of which have since been covered deeply in the snow and ice of the glacier, and Wegener's body has become part of the glacial ice cap he devoted much of his life to studying.

See also GONDWANA, GONDWANALAND; HISTORICAL GEOLOGY; PLATE TECTONICS; SUPERCONTINENT CYCLES.

FURTHER READING

Koppen, Vladimir, and Alfred Wegener. *The Climates of the Geological Past.* London: D. Van Nostrand, 1863.

Wegener, Alfred. *The Origin of Continents and Oceans,* translated by John Biram. Mineola, N.Y.: Dover Publications, 1966.

———. *Thermodynamik der Atmosphare.* Leipzig: Verlag Von Johann Ambrosius Barth, 1911.

Wegener, Elsie, ed., with the assistance of Dr. Fritz Loewe. *Greenland Journey, The Story of Wegener's German Expedition to Greenland in 1930–31 As Told by Members of the Expedition and the Leader's Diary* (Translated from the seventh German edition by Winifred M. Deans). London: Blackie & Son Ltd, 1939.

Werner, Abraham Gottlob (1749–1817) Prussian, German *Geologist, Mineralogist* Abraham Werner was enormously influential in the field of geology. Werner developed techniques for identifying minerals using human senses, and this appealed to a broad audience interested in learning more about geology. Werner also proposed a new classification for certain geologic formations. In the 18th century, rocks were explained and were classified into three categories with accordance to the "biblical flood," including Primary for ancient rocks without fossils (believed to precede the flood), Secondary for rocks containing fossils (often attributed to the flood itself), and Tertiary for sediments believed to have been deposited after the flood. Werner did not dispute the commonly held belief of the biblical flood, but he did discover a different group of rocks that did not fit this classification—rocks with a few fossils that were younger than

primary rocks but older than secondary rocks. He called these "transition" rocks. Geologists of succeeding generations classified these rocks into the geologic periods still accepted today. Werner is the father of the discredited school of thought called Neptunism, that proposed that granitic rocks crystallized from mineralized fluids in the early Earth's oceans.

Abraham Gottlob Werner was born in Wehrau of Prussian Silesia in southeastern Germany on September 25, 1749. He was born into a mining family, and his father was a foreman at the foundry in Wehrau. Educated at Freiberg and Leipzig, Werner studied law and mining, then became an inspector and teacher of mining and mineralogy at the influential Freiberg Mining Academy in 1775. When he was studying at Leipzig in 1774, Werner wrote his first book on mineralogy, *Vonden äusserlichen Kennzeichen der Fossilien* (On the external characters of fossils, or of minerals), a book that became influential in its field. Werner did not publish many books or works after this, but instead was known through his lecturing, his mentoring of students, and many interactions with colleagues. Werner was plagued with health problems his entire life and rarely traveled; he spent his time quietly near Freiberg. When he was young he enjoyed mineral collecting, but as age and disease took their toll he abandoned field work. He died in Dresden on June 30, 1817.

Werner was one of the most important geologists of his time, in an era when the geologic and scientific community was actively presenting evidence that the geologic record preserves a history of the Earth that is very different from that advocated by the church at that time. Werner divided the strata of Europe and the world into five main series, including the Primitive, Transition, Secondary (or stratified), Alluvial (or Tertiary), and Volcanic (or Younger). His theory of neptunism, which was advocated by the church, was based on his idea that the early Primitive granites were crystallized from the sea before land had emerged from the worldwide ocean. Neptune was the Roman name for the ancient Greek god of the sea, Poseidon, and Werner's belief that nearly all rocks could be explained as precipitates from the ocean led to his theory's being called neptunism. Werner's Transition series included limestones, dikes, sills, and graywackes, which he suggested were universal formations that extended around the entire world. These were followed by the Secondary or Stratified series, including layered fossiliferous rocks and lava flows. Werner interpreted these to reflect the emergence of mountains from the primeval sea, depositing the products of their erosion on their flanks. These are followed by the Alluvial or Tertiary rocks consisting of poorly consolidated sands, gravels, and clays deposited as the oceans withdrew and receded from the continents. Finally, the Volcanic series consisted of younger lava flows associated with volcanic vents that Werner suggested were the product of subterranean coal fires.

Werner's ideas were debated, particularly his ideas on the origin of granites and basalts. Many other geologists noted that volcanic rocks occurred in places not known to have coal beds and thought they may have formed from melts of rock at depth. This alternate school of thought, called plutonism, pioneered by James Hutton, later became dominant and is accepted today along with the school of thought called uniformitarianism, whereby processes observed to create specific effects on the modern Earth may be assumed to be responsible for creating similar effects in the geological record. Werner was also criticized on the basis that the volume of water required by his theory for a universal ocean was enormous. He managed to avoid this question somewhat by suggesting many of the waters escaped or evaporated to space.

See also HISTORICAL GEOLOGY; HUTTON, JAMES; ORIGIN AND EVOLUTION OF THE EARTH AND SOLAR SYSTEM.

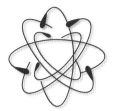

Appendix I
Chronology

ca. 5000 B.C.E.	Egyptians develop the balance and a standard unit of weight.
ca. 3000 B.C.E.	Egyptians develop a standard unit of length.
ca. 1450 B.C.E.	Egyptians develop the water clock.
ca. 1150 B.C.E.	The first geologic map, the Turin papyrus, is made in Egypt to help mine gold deposits.
ca. 550 B.C.E.	Pythagoras, in Greece, studies acoustics, relating the pitch of a tone to the length of the wind instrument or of the string producing it.
ca. 400 B.C.E.	Democritus, in Greece, states that all matter is made up of "atoms" and empty space.
ca. 370 B.C.E.	Aristotle, in Greece, describes free fall, but incorrectly claims that heavier bodies fall faster than lighter ones.
ca. 350 B.C.E.	Aristotle, in Greece, notes the slow rates of geologic processes such as erosion.
ca. 300 B.C.E.	Theophrastus, Greek scientist, publishes "On Stones," including one of the first systematic descriptions and classification of minerals, ores, and their behavior under specific tests such as burning.
ca. 270 B.C.E.	Ctesibius of Alexandria, Egypt, invents an accurate water clock, in use until the Renaissance.
ca. 260 B.C.E.	Archimedes, in Greece, studies floating bodies and states his principle of buoyancy. He also states the law of the lever.
ca. 130 B.C.E.	Hipparchus, from Greece, uses trigonometry to measure the sizes and distance to the Sun and Moon, and orbit of the Moon around Earth.
ca. 60 B.C.E.	Lucretius, in Greece, proposes the atomic nature of matter.

ca. 62 C.E.	Hero of Alexandria, Egypt, studies air pressure and vacuum.
ca. 50–70	Pliny the Elder publishes *Historia Naturalis* in 37 volumes. In these works he described many new minerals and ores, and defined the basis of crystallography.
ca. 110–50	In Greece, Ptolemy publishes *Almagest,* the first complete record of astronomy. This was followed by his *Geographia,* which discussed the geography of the Greco-Roman world, and *Tetrabiblos,* a discourse on astrology and natural philosophy.
ca. 700	I Hsing, in China, develops a mechanical clock.
ca. 1000	Jabir ibn Hayyan, Persian polymath, publishes on the geology of India, including the recognition that many of the rocks there were initially deposited deep under the seas.
ca. 1010	Ibn Sina (Avicenna in Latinized form), a Persian polymath, publishes *Kitab al-Shifa* (the Book of Cure, Healing or Remedy from ignorance), containing some of the early and influential works on Mineralogy and Meteorology, in six chapters: Formation of mountains; The advantages of mountains in the formation of clouds; Sources of water; Origin of earthquakes; Formation of minerals; and The diversity of earth's terrain. Many of these contributed to later theories of uniformitarianism, the law of superposition, and catastrophism.
ca. 1080	Shen Kuo (also known as Mengxi), Chinese polymath, formulates a theory of geomorphology, including deposition, uplift, erosion, and the

role of climate change, in studies of the Taihang Mountains of China.

ca. 1100 Abu'L-Fath 'Abd al-Rahman al-Khazini, in Persia, proposes that gravity acts toward the center of Earth.

ca. 1100 Maimonides (Rabbi Moshe ben Maimon) publishes The Guide for the Perplexed, including expositions on the Aristotelian geocentric models for the universe, but with a constantly changing universe.

ca. 1200 Jordanus de Nemore, in Germany, studies motion and explains the lever.

ca. 1235 Roger Bacon emphasizes the importance of experimentation.

1276 Roger Bacon, in England, proposes using lenses to correct vision.

1284 Witelo, in Poland, describes reflection and refraction of light.

1300 Rabbi Issac of Akko estimates the age of the universe to be 13.34 billion years old.

1305 Dietrich von Freiberg, in Germany, describes and explains rainbows.

1543 Nicolaus Copernicus publishes *De revolutionibus orbium coelestium*, recognizing that the Earth is not the center of the universe and that the solar system is heliocentric, with the Sun at the center.

1546 Niccolò Tartaglia, in Italy, studies projectile motion and describes the trajectory of a bullet.

1563 Tycho Brahe publishes his *Ephemeris*, tables of predictions of the locations of the stars, planets, and constellations.

1572 Tycho Brahe observes a supernova in the constellation Cassiopeia, changing the views that the heavens were unchanging.

1582 Galileo Galilei, in Italy, describes the motion of a pendulum, noticing that its period is constant and, for small amplitudes, independent of amplitude.

1583–86 Flemish scientist Simon Stevin investigates hydrostatics and free fall.

1590–95 Zacharias Janssen makes the first microscope.

1590–91 Galileo investigates falling bodies and free fall.

1596 Johannes Kepler publishes his work *Mysterium Cosmographicum*, a geometric model of a geocentric universe.

1600s Galileo develops the principle of inertia.

1600 English scientist William Gilbert studies magnetism and its relation to electricity and describes Earth as a magnet.

1601 Tycho Brahe suddenly dies. Some theories suggest he was murdered by Johannes Kepler, who took over his position and research.

1604 Johannes Kepler publishes observations of supernovas.

1609 Dutch lens maker Hans Lippershey invents the telescope. German astronomer Johannes Kepler presents his first and second laws of planetary motion.

1610 Galileo Galilei publishes his observations of several moons of Jupiter and uses this to argue for a Sun-centered model for the universe.

1617–21 Johannes Kepler publishes his *Epitome astronomia Copernicanae* (Epitome of Copernican astronomy), including the heliocentric model for the universe, the elliptical paths of planets, and all three laws of planetary motion.

1622 Willebrord Snell presents his law of refraction of light.

1636 French mathematician René Descartes advances understanding of rainbows.

1638 Galileo studies motion and friction.

1657 Christian Huygens publishes the first book on probability theory.

1660–62 Robert Boyle studies gases.

1666–1704 Sir Isaac Newton actively studies a wide range of natural phenomena.

1669 Danish anatomist and geologist Nicolaus Steno publishes *Prodromus*, the first work to show that fossils are the remains of formerly living organisms, as well as proposing the law of stratal superposition.

1705 Edmund Halley predicts the return of the comet named for him.

1714 Gottfried Leibniz proposes the conservation of energy.

1736   Carl Linnaeus publishes the first of 12 editions of *Systema naturae*, a book that outlined a system for classification of plants, animals, and minerals.

1736–65   Leonhard Euler studies theoretical mechanics using differential equations.

1738   Daniel Bernoulli investigates the theories of gases and of hydrodynamics.

1743–44   Jean d'Alembert studies energy in Newtonian mechanics and proposes a theory of fluid dynamics.

1751   American scientist Benjamin Franklin discovers that electricity can produce magnetism.

1754   Joseph Black discovers "fixed air" (carbon dioxide).

1772–88   French mathematician and physicist Joseph Lagrange investigates theoretical mechanics and proposes a new formulation of Newtonian mechanics.

1774   Abraham Gottlob Werner publishes his book "On the External Characters of Minerals" as a guide to identify minerals based on their characteristics. Werner also argued that granites crystallized from ocean waters, leading a school of thought called Neptunism.

1770–90   James Hutton pioneers the concept of uniformitarianism, that the natural processes that formed structures in old rocks are the same as the natural processes operating on Earth at present. This is often summarized as the quote "the present is the key to the past."

1776–84   French mathematician and physicist Pierre Laplace applies mathematical methods to theoretical physics, particularly to mechanics and electricity.

1785   James Hutton publishes his *Theory of the Earth*, becoming recognized as the father of modern geology. French physicist Charles Augustin de Coulomb proposes his law of electrostatics. French chemist Antoine Laurent Lavoisier develops the law of conservation of mass and names oxygen.

1790   The French Academy of Sciences establishes the metric system of measurement.

1796   British geologist and surveyor William Smith formulates the principle of fossil succession, laying the foundation for the science of stratigraphy.

1797   British physicists Benjamin Thompson and Benjamin Rumford study the heat generated by work. French chemist Joseph Proust develops the law of definite proportions.

1800   Italian physicist Alessandro Volta develops the electric battery.

1800–02   Jean-Baptiste Lamarck elaborates his theory of evolution based on the inheritance of modified traits.

1801   English physicist Thomas Young demonstrates that light is a wave phenomenon.

1802   French chemist Joseph Louis Gay-Lussac develops his gas law relating pressure and temperature.

1802–05   British chemist John Dalton develops his atomic theory.

1808   French scientist Étienne-Louis Malus discovers and investigates polarized light.

1809   English scientist Sir George Cayley publishes his theoretical studies of aerodynamics, laying the foundation for flight.

1811   Italian scientist Amedeo Avogadro describes gases in molecular terms.

1814   German physicist Joseph von Fraunhofer discovers and investigates optical spectra. French physicist Augustin-Jean Fresnel explains light polarization in terms of light's wave nature.

1818   The first national geologic map is produced, as a map of the United Kingdom by William Smith.

1820   Danish physicist Hans Christian Ørsted and French physicist André-Marie Ampère show that an electric current has a magnetic effect.

1821   English physicist Michael Faraday, who was studying electromagnetism, introduces the concept of magnetic field.

1824   French engineer Nicolas-Léonard-Sadi Carnot publishes his analysis of heat engines, which leads to the laws of thermodynamics.

1827   German physicist Georg Simon Ohm shows the proportionality of electric current and voltage, known as Ohm's law. English botanist Robert Brown discovers the motion of pollen grains suspended in a liquid, called Brownian motion.

1830–33   Charles Lyell publishes his work "Principles of Geology," in which he proposes that geological processes are very slow and the Earth is very old, and emphasizes the uniformitarian concepts of James Hutton.

1831   English physicist Michael Faraday discovers electromagnetic induction, that magnetism can produce electricity. American physicist Joseph Henry invents the electric motor.

1831–36   Charles Darwin collects evidence supporting his theory of evolution while traveling around the globe on the H.M.S. *Beagle.*

1833–60   Adam Sedgwick maps parts of Wales, and proposes divisions of Paleozoic time. He also identifies the origins of many geologic structures, such as folds and faults.

1837   James Dana publishes his "System of Mineralogy."

1840   English physicist James Prescott Joule develops the law of conservation of energy.

1842   Austrian scientist Christian Doppler explains the dependence of the observed frequency on the motion of the source or observer, called the Doppler effect.

1847   German scientist Hermann von Helmholtz expresses the conservation of energy in mathematical terms, the first law of thermodynamics.

1848   English scientist Sir William Thomson (Lord Kelvin) describes absolute zero temperature.

1850   German physicist Rudolf Clausius introduces the concept of entropy, which leads to the second law of thermodynamics. French physicist Jean-Bernard-Léon Foucault measures the speed of light in air, and later in water.

1851   Foucault uses a huge pendulum, known as a Foucault pendulum, to demonstrate the rotation of planet Earth.

1853   British geologist Henry Sorby pioneers the use of the polarizing microscope for petrography and publishes models for the origin of slaty cleavage in rocks.

1854   German mathematician Georg Riemann describes the geometry of curved spaces, applied later by modern physicists to relativity and other problems.

1858   Rudolf Virchow states that cells only arise from other cells. Darwin and Alfred Russel Wallace jointly propose the theory of natural selection.

1859   Charles Darwin publishes *On the Origin of Species by Means of Natural Selection, or the Preservation of Favored Races in the Struggle for Life.* Louis Pasteur disproves the notion of spontaneous generation.

1859   Scottish physicist James Clerk Maxwell develops the kinetic theory of gases, based on a statistical treatment of the gas particles.

1860   The famous debate on evolution by Thomas Henry Huxley and Bishop Samuel Wilberforce takes place at Oxford.

1862   German physicist Gustav Robert Kirchhoff introduces the concept of a blackbody, later study of which leads to the development of quantum mechanics.

1866   Haeckel coins the term *ecology* to mean the study of living organisms and their interactions with the environment and states his famous biogenetic law, that ontogeny recapitulates phylogeny.

1869   Russian physicist Dmitry Mendeleyev develops the periodic table based on atomic mass. American geologist and explorer John Wesley Powell leads a first expedition to the Grand Canyon, providing some of the first geological descriptions of the western United States.

1871   English physicist John William Strutt (Lord Rayleigh) mathematically relates the amount of scattering of light from particles, such as molecules, to the wavelength of the light, thus explaining the color of the sky.

1873 Scottish physicist James Clerk Maxwell presents his set of equations, known as Maxwell's equations, which form a theoretical framework for electromagnetism and predict the existence of electromagnetic waves. Dutch physicist Johannes van der Waals modifies the ideal gas equation to take into account weak attractive intermolecular forces, called van der Waals forces.

1875 Austrian geologist Eduard Suess publishes *Die Entstehung der Alpen,* a milestone paper in understanding the structural geology of the Alps of Europe and in relating his tectonic theories.

1877 English physicist John William Strutt (Lord Rayleigh) publishes his extensive work on acoustics, the science of sound waves. Grove Karl Gilbert publishes a monograph on the Henry Mountains, showing that intrusive plutons can deform host rock.

1879 Austrian physicist Josef Stefan shows experimentally that the rate of energy radiation from a body is proportional to the fourth power of the body's absolute temperature.

1883 Austrian physicist Ludwig Boltzmann explains Stefan's result as a property of blackbodies, known as the Stefan-Boltzmann law. This lays the foundation for the development of quantum mechanics.

1887 German physicist Heinrich Rudolf Hertz observes the photoelectric effect, the emission of electrons from a metal that is irradiated with light, and discovers radio waves. American scientists Albert Abraham Michelson and Edward Williams Morley attempt to measure the motion of Earth in the ether, a proposed medium for the propagation of electromagnetic waves, with a negative result, which leads to the special theory of relativity in the 20th century.

1890 The first monograph of the U.S. Geological Survey is published, by G. K. Gilbert on the origin of glacial Lake Bonneville.

1893 German physicist Wilhelm Wien shows experimentally that the wavelength at which a blackbody radiates at maximal intensity is inversely proportional to the absolute temperature, known as Wien's displacement law.

1895 Scottish geologist Andrew Lawson identifies and names the San Andreas fault and later (in 1908) authors a famous report on the 1906 earthquake. German physicist Wilhelm Conrad Röntgen discovers X-rays.

1896 French physicist Henri Becquerel discovers radioactivity in uranium.

1897 British physicist Sir J. J. Thomson discovers electrons and develops his "plum pudding model" of the atom, where the electrons are imbedded in a positively charged sphere.

1898 French chemist Marie Curie and physicist Pierre Curie isolate radium, which is highly radioactive.

1900 German physicist Max Planck proposes absorption and emission of radiation in discrete amounts, which introduces the quantum concept.

1904 Lord Ernest Rutherford performs the gold foil experiment that demonstrates the existence of the atomic nucleus.

1905 German/Swiss, later American, physicist Albert Einstein publishes his special theory of relativity. To explain the photoelectric effect, Einstein proposes that light consists of photons.

1910–40 Amadeus Grabau publishes a series of 10 books, outlining the stratigraphic and paleontological history of North America, and later China.

1910 Japanese geophysicist Motonori Matuyama recognizes magnetic reversals in Japanese basalts.

1911 Arthur Holmes becomes the first person to use the uranium-lead decay series to date rocks, showing that the Earth is more than a billion years old. Netherlands physicist Heike Kamerlingh Onnes discovers superconductivity. New Zealand–born British physicist Sir Ernest Rutherford discovers the structure of the atom. American physicist Robert A. Millikan uses his "oil drop method" to determine the charge of an electron.

1911–12 Using X-ray methods, British physicists William Henry Bragg and William Lawrence Bragg and German physicist Max von Laue discover the atomic structure of crystals.

1912 Alfred Wegener proposes a theory of continental drift.

1912–35 Victor Goldschmidt authors a series of papers that lay the foundation for modern geochemistry and proposes geochemical models for the composition of the Earth.

1913 Danish physicist Niels Bohr proposes a "planetary model" of the hydrogen atom to explain the hydrogen spectrum, in which the electrons rotate around the nucleus in orbits like planets around the sun. British physicist, Henry Moseley rearranges the periodic table based on atomic number.

1915 Albert Einstein proposes his general theory of relativity.

1916 American physicist Robert Millikan measures the value of the Planck constant, which characterizes all quantum phenomena.

1917 Joseph Grinnell coins the term *ecological niche*.

1919 English astronomer and physicist Arthur Eddington leads an expedition that measures the bending of starlight passing near the Sun during a total solar eclipse, which confirms the general theory of relativity.

1920 Pentti Eskola defines the concept of metamorphic facies.

1922 Soviet mathematician and meteorologist Aleksander Friedmann shows that the general theory of relativity predicts the Universe is expanding.

1923 French physicist Louis de Broglie proposes that matter possesses wave-like properties. American physicist Arthur Holly Compton demonstrates, through the Compton effect, that electromagnetic radiation consists of photons.

1925 German physicist Werner Heisenberg invents a matrix formulation of quantum mechanics. American astronomer Edwin Hubble identifies very distant stars outside the Milky Way Galaxy.

1926 Austrian physicist Erwin Schrödinger publishes his formulation of quantum mechanics in the form of the Schrödinger equation.

1927 Werner Heisenberg proposes his uncertainty principle. American physicists Clinton Davisson and Lester Germer show that electrons possess wave-like properties. Leading to the big bang theory, Belgian astronomer Georges Lemaître states that the universe began its expansion from a tiny, hot state.

1928 British geologist Arthur Holmes proposes that radioactive decay causes thermal convection in Earth's mantle and that this convection causes the continents to drift. British physicist Paul Dirac derives the Dirac equation, which predicts the existence of antiparticles.

1929 American astronomer Edwin Hubble demonstrates that all galaxies are receding from each other, indicating the expansion of the universe. Alexander du Toit begins to publish observations correlating fossils between South America and Africa in support of continental drift.

1931 Ernst Ruska and Max Knoll make the first electron microscope.

1932 Indian-born American astrophysicist Subrahmanyan Chandrasekhar proposes that when a sufficiently massive star reaches the end of its life, it will collapse to a black hole. American physicists Ernest O. Lawrence and M. Stanley Livingston invent the cyclotron, a particle accelerator for investigating nuclei and elementary particles. American physicist Carl D. Anderson discovers the positron, the electron's antiparticle. British physicist James Chadwick discovers the neutron.

1933 Swiss astronomer Fritz Zwicky studies the rotation of galaxies and shows that they must contain more mass than is visible, introducing the idea of dark matter. Georges Lemaître proposes the theory of the big bang, initially criticized as being too much like creationist accounts, but later endorsed by Albert Einstein.

1934 French physicists Irème Joliot-Curie and Frédéric Joliot-Curie produce the first artificial radioactive isotopes.

1935 Japanese physicist Hideki Yukawa proposes the theory of the nuclear force, binding protons and neutrons into nuclei, which predicts the existence of mesons.

1936 Hans Stille publishes his models for the tectonic pulses of Europe, correlating different events across Europe, Asia, and eventually the globe.

1937 Normal Levi Bowen proposes models of magmatic differentiation by partial melting and by fractional crystallization. American physicists Carl D. Anderson and Seth Neddermeyer discover the muon in cosmic rays.

1938 Soviet physicist Pyotr Kapitsa discovers that liquid helium exhibits superfluidity near 0 K. American physicist Hans Bethe explains the source of energy production in stars as nuclear fusion reactions.

1938–39 Austrian physicists Lise Meitner and Otto Frisch explain that the German chemists Otto Hahn and Fritz Strassmann achieved nuclear fission by bombarding uranium with neutrons.

1939 Linus Pauling publishes *The Nature of the Chemical Bond and the Structure of Molecules and Crystals: An Introduction to Modern Structural Chemistry.*

1941 Milutin Milankovitch publishes his model that changes in the orbital parameters of Earth change the amount of incoming solar radiation, leading to climate cycles on Earth, preserved as Milankovitch cycles in the rock record.

1942 A team led by Italian/American physicist Enrico Fermi produces the first controlled nuclear fission chain reaction. The United States initiates the Manhattan Project to construct a nuclear fission (atomic) bomb.

1943 As part of the Manhattan Project, the Los Alamos laboratory is built in New Mexico, directed by American physicist Robert Oppenheimer.

1945 Erwin Schrödinger publishes *What Is Life,* a book that inspires many biologists.

1946 American chemist Willard Frank Libby invents the carbon 14–dating technique for determining when living organisms died. The first programmable digital computer, the Electronic Numerical Integrator and Comparator (ENIAC) starts operation.

1947 A team directed by British astronomer Bernard Lovell completes construction of the first radio telescope. British physicist Cecil Frank Powell discovers the pion, predicted by Yukawa in 1935. American physicists John Bardeen, William Shockley, and Walter Brattain invent the transistor, a semiconductor device.

1948 George Gamow publishes a big bang model for the origin of the universe and predicts the presence of cosmic background radiation and the origin of the chemical elements.

1949 German-American physicist Maria Goeppert Mayer and German physicist Hans Jensen model the nucleus of atoms as consisting of shells of protons and neutrons. American geologist Francis Pettijohn publishes his book *Sedimentary Rocks,* which laid the foundation for modern sedimentology.

1950 Swedish astrophysicist Hannes Alfvén reaches an understanding of the physics of plasmas (ionized gases), with relevance to space science and, later, to nuclear fusion. The United States Congress creates the National Science Foundation for the funding of basic research and science education.

1951–52 American physicists Harold Ewen and Edward Mills Purcell observe the 21-cm radio signal from hydrogen atoms in space.

1954 American seismologist Hugo Benioff recognizes that earthquakes beneath island arcs are concentrated in narrow zones and suggests that the arcs are being thrust over sinking oceanic crust, a forerunner to the modern subduction zone model.

1956   American physicists Clyde Cowan, Frederick Reines, F. B. Harrison, H. W. Kruse, and A. D. McGuire discover the electron neutrino.

1956–57   Chinese-American physicist Chien-Shiung Wu experimentally confirms the proposal by the Chinese-American physicists Tsung-Dao Lee and Chen Ning Yang that the weak force might not obey reflection symmetry.

1957   American physicists John Bardeen, Leon Cooper, and Robert Schrieffer propose an explanation for superconductivity in terms of conduction by electron pairs. Bayer and General Electric develop polycarbonate plastics.

1958   Japanese physicist Leo Esaki invents the tunnel diode, which exploits quantum tunneling.

1959   Israeli physicist Yakir Aharonov and American physicist David Bohm predict that magnetic fields can affect particles in an observable, nonclassical way when the particles do not pass through the field. Austrian molecular biologist Max Perutz determines the structure of hemoglobin.

1960   Robert Dietz and Harry Hess propose the concept of seafloor spreading. The Aharonov-Bohm effect is observed. American physicist Theodore Maiman constructs the first laser from a ruby crystal.

1962   American physicists Leon M. Lederman, Melvin Schwartz, and Jack Steinberger discover the muon neutrino. Rachel Carson publishes *Silent Spring*, a book that stimulates the environmental movement.

1963   Dutch-American astronomer Maarten Schmidt discovers the first quasar (QUASi-stellAR radio source), a very distant object that appears similar to a star but radiates more than some galaxies.

1964   American physicists Murray Gell-Mann and George Zweig independently propose the existence of quarks as components of protons, neutrons, and other hadrons.

1965   American physicists Arno Penzias and Robert Wilson discover the cosmic microwave background. Canadian geologist J. Tuzo Wilson publishes "A New Class of Faults and Their Bearing on Continental Drift" in *Nature,* a paper widely held to be the start of the plate tectonic paradigm. Cambridge Instruments makes the first commercial scanning electron microscope.

1967   American physicists Steven Weinberg and Sheldon Glashow and Pakistani physicist Abdus Salam independently propose unifying the electromagnetic and weak forces to a single, electroweak force.

1967–68   British astronomers Jocelyn Bell and Anthony Hewish discover that certain stars, called pulsars, emit periodic radio pulses. American astrophysicist Thomas Gold explains pulsars as rotating neutron stars.

1969   A group of American physicists, including Jerome I. Friedman, Henry Kendall, and Richard E. Taylor, discover experimental evidence for the existence of quarks inside protons.

1970   John Dewey and John Bird apply the concept of plate tectonics to the ancient Appalachian mountain belt.

1970–73   Physicists develop the "standard model" of elementary particles, which includes the strong and electroweak forces.

1972   NASA launches the first Landsat satellite. Preston Cloud proposes models for the evolution of organisms and relationships between life, atmospheric chemistry, and geology for the early Earth.

1972   Stephen Jay Gould and Niles Eldridge propose the theory of punctuated equilibrium as a method for evolutionary events.

1974   English physicist Stephen Hawking proposes that black holes can radiate particles and eventually evaporate. American physicists Burton Richter and Samuel C. C. Ting and their groups independently discover the charm quark.

1974–77   American physicist Martin L. Perl and colleagues discover the tau particle.

1975–77 Polish-American mathematician Benoit B. Mandelbrot introduces the concept of fractals, patterns in systems that are similar to themselves at all scales.

1977 Robert Ballard and his team discover chemosynthetic communities surrounding hydrothermal vents. Carl Woese proposes a third domain of life, Archaea. American physicist Leon Lederman and colleagues discover the bottom quark.

1978 American astronomer Vera Rubin and others conclude from an analysis of the rotation of galaxies that the gravity from the visible stars is insufficient to keep them from flying apart, and they must contain invisible matter, called dark matter.

1979 French physicist Pierre-Gelles de Gennes publishes his work on the theories of polymers and liquid crystals.

1980 American physicist Alan Guth proposes adding inflation—a very brief period of extremely rapid expansion of the universe—to the big bang theory, in order to better explain observations.

1981 German physicist Gerd Binnig and Swiss physicist Heinrich Rohrer invent the scanning tunneling microscope, which can image surfaces to the detail of individual atoms.

1983 A team led by the Italian physicist Carlo Rubbia discovers the W and Z bosons, the carriers of the weak force.

1985 English chemist Sir Harry Kroto and American chemists Richard Smalley and Bob Curl discover the structure of the buckminsterfullerene molecule.

1986 Swiss physicist Karl Alexander Müller and German physicist Johannes Georg Bednorz discover high-temperature superconductors, materials that become superconducting at temperatures much farther above 0 K than were previously known.

1989 American astronomers Margaret Geller and John Huchra discover that the galaxies in the universe are located on thin sheets surrounding great voids that are empty of galaxies.

1989–92 The U.S. National Aeronautics and Space Administration (NASA) launches the Cosmic Background Explorer (COBE) satellite, which maps the radiation from the sky in all directions, the cosmic microwave background, a remnant from the big bang. The cosmic microwave background is found to be very uniform and to correspond to the radiation of a blackbody at a temperature of 2.725 K. Tiny angular fluctuations in the radiation's generally uniform distribution are detected, reflecting on some nonuniformity in the universe at a very early age.

1990 NASA launches the *Hubble Space Telescope* as a satellite above Earth's atmosphere to study the universe at high resolution.

1993 The U.S. Air Force completes the Global Positioning System (GPS), allowing users on Earth to locate themselves and navigate around the world.

1994 The top quark is discovered at Fermilab. Hubble telescope.

1995 American physicists Eric Cornell and Carl Wieman produce a Bose-Einstein condensate of 2,000 atoms at a temperature lower than 10^{-6} K, thus confirming a prediction of Bose-Einstein statistics. American geophysicists Xiaodong Song and Paul Richards demonstrate that Earth's solid inner core, with a diameter of 1,500 miles (2,400 km), rotates a little faster than the rest of the planet. The top quark is discovered by a group at Fermilab. Anti-hydrogen atoms, consisting of an antiproton and a positron, are created at the European Organization for Nuclear Research (CERN).

1998 Observations of supernovas indicate that the expansion of the universe is accelerating.

2000 A collaboration at Fermilab announces the detection of the tau neutrino.

2001 The first complete Archean ophiolite is discovered in China by American geologist T. Kusky and Chinese geologist J. H. Li.

2003 The Human Genome project, a collaborative group of scientists from many nations, complete the sequence of the human genome.

2006 President George W. Bush announces the Advanced Energy Initiative (AEI) to increase research on technology to reduce oil use for transportation, including hybrid-vehicle batteries, ethanol, and hydrogen–fuel cell vehicles and fueling stations. AEI also supports research into electricity production from clean coal, wind, and solar power.

2008 The U.S. Senate introduces *The Green Chemistry Research and Development Act of 2008* for the advancement of research into environmentally friendly chemicals by the National Science Foundation (NSF), National Institute of Standards and Technology (NIST), Environmental Protection Agency (EPA), and Department of Energy.

APPENDIX II
GLOSSARY

abrasion a process that occurs when particles of sand and other sizes are blown by the wind and impact each other

absorption in physics and astronomy, the process where energy from a photon is taken up by matter, typically electrons of an atom, and converted into some other form of energy (typically heat), causing reduction of light from a distant source. An absorption line is a dark line in an otherwise continuous spectrum, where the light from one narrow frequency range has been removed by some process, typically interaction with interstellar medium.

abyssal plains stable, flat parts of the deep oceanic floor, typically covered with fine-grained sedimentary deposits called deep sea oozes, derived from the small skeletons of siliceous organisms that fell to the seafloor

accreted terranes coherent regional-scale blocks of rocks with distinctive stratigraphic and structural histories that have been added to the edges of continents at convergent plate margins

accretion the transfer of material, such as sedimentary rocks, from an oceanic plate to an overriding continental plate at a convergent plate margin. Less commonly, the term may be applied to the addition of magma that crystallizes into rocks and is added to the extensional plate boundaries in mid-ocean ridges. The term can also be applied to the gravitational accumulation of dust, gas, and rocky bodies such as asteroids to a larger body such as a planet.

accretionary beaches wide summer beaches characterized by the movement of sediment from offshore to onshore

accretionary prism, accretionary wedge a structurally complex belt of rocks formed just above a subduction zone, characterized by strongly folded and faulted rocks scraped off from the downgoing oceanic plate

accretion disk a diffuse collection of material in orbit around a central celestial body

achondrite a stony meteorite that does not contain chondrules

acid rain rain or other forms of precipitation that have acidic components leading to a pH of less than 7. The extra acidity in acid rain comes from the reaction of air pollutants, mostly sulfur and nitrogen oxides, with water to form strong acids such as sulfuric and nitric acid

aerosol microscopic droplets or airborne particles that remain in the atmosphere for at least several hours

age elements elements that tend to accumulate in tissues as the tissues get older

aggradation the deposition of sediments along a river bed, causing the bed to rise

aggrading barriers barrier beaches that are growing upward in place as sea levels rise

alluvial fans coarse-grained deposits of alluvium that accumulate at the fronts of mountain canyons

Amors asteroids that orbit between Earth and Mars but do not cross Earth's orbit

amphibolite a metamorphic rock consisting mainly of minerals of the amphibole group such as hornblende, typically derived from a mafic igneous rock

amplification a process of transferring energy along the wave crest as the waves are forced to become shorter lengthwise, causing the wave height to increase

amplitude one-half of the wave height in a wave train, measured from the bottom of the trough to the top of the crest

andesite an intermediate composition, dark, fine-grained volcanic rock consisting mostly of plagioclase feldspar, with lesser amounts of biotite and hornblende. It is the extrusive equivalent of diorite and found mostly in continental margin magmatic arcs.

Andes Mountains a 5,000-mile- (8,000-km-) long mountain range in western South America, running generally parallel to the coast, between the

Caribbean coast of Venezuela in the north and Tierra del Fuego in the south

angle of repose the steepest angle at which a loose material such as sand, gravel, or boulders can be stacked

angular velocity a measure in degrees of the change in angle per unit time as an object spins about a pivot point. If an object is rotated about a pivot point, the angular velocity stays the same with increasing distance from the pivot point, but the linear velocity (speed) will increase with distance from the pivot point.

anorogenic granite a granite that intrudes a continental area at a time not associated with an orogeny or convergent margin tectonism, typically in the middle parts of supercontinents; may signal heating from below by mantle upwelling and thermal insulation by the large continent

antecedent stream a stream that has maintained its course across topography that is being uplifted by tectonic forces; these cross high ridges

anthropogenic an adjective used to describe changes caused by humans

anticline a fold in which the beds dip outward and the older rocks are in the center of the fold structure

anticlinorium a large regional scale anticline that may have smaller-scale anticlines and synclines superimposed on the larger-scale structure

antimatter the counterpart of matter, made of antiparticles the same way that matter is made of particles

antiparticle most kinds of particles have an associated antiparticle that has the same mass and opposite charge as the particle. For instance, electrons are associated with their antiparticle counterparts, the positrons.

aphelion the point in an orbit that is farthest from the Sun

Apollo asteroids asteroids that orbit less than one astronomical unit from the sun with periods longer than one year

aquicludes rock or soil units that stop the movement of water

aquifer any body of permeable rock or regolith saturated with water through which groundwater moves

aquitard rock or soil units that restrict the flow of water

Archaea a domain of life consisting of simple organisms that appeared on Earth by 3.85 billion years ago

Archean the oldest eon of geological time, ranging from 4.5 billion years ago until 2.5 billion years ago

arête a sharp-edged ridge that forms where two cirques intersect in a mountain range

argillite a clay-rich, fine-grained sedimentary rock

artesian pressure (also, artesian spring, artesian system) a permeable layer, typically a sandstone, that is confined between two impermeable beds, creating a confined aquifer. In these systems, water enters the system only in a small recharge area, and if this is in the mountains, then the aquifer may be under considerable pressure. This is known as an artesian system.

assimilation in geology, the process whereby a crystal or piece of rock is incorporated into an igneous melt, and its chemical constituents become part of the melt

asteroid a rocky or metallic body in space orbiting the Sun

asteroid belt area of the solar system between the orbits of Mars and Jupiter that contains millions of asteroids

asthenosphere weak, partially molten layer in the Earth beneath the lithosphere, extending to about 155 miles (250 km) depth. The lithosphere slides on top of this weak layer, enabling plate tectonics to happen.

astronomical unit (AU) the distance from the Sun to the Earth, or 93 million miles (150 million km)

astronomy the study of celestial objects and phenomena that originate outside the Earth's atmosphere

astrophysics the branch of astronomy that examines the behavior, physical properties, and dynamic processes of celestial objects and phenomena

Aten asteroids asteroids that orbit at less than one astronomical unit from the sun and have an orbital period of less than one year. Some of these cross Earth's orbit.

atmosphere the envelope of air that surrounds the Earth, held in place by gravity. The most abundant gas in the atmosphere is nitrogen (78 percent), followed by oxygen (21 percent), argon (0.9 percent), carbon dioxide (0.036 percent), and minor amounts of helium, krypton, neon, and xenon.

atmospheric pressure the force per unit area (similar to weight) that the air above a certain point exerts on any object below it

atolls geological structures that form circular, elliptical, or semicircular shaped islands made of coral reefs that rise from deep water. Atolls surround central lagoons, typically with no internal landmass

auriferous gold-bearing

aurora glowing arches or streamers of light sometimes visible in high latitudes around both the North and South Poles. Auroras are formed by interaction of the solar wind with the Earth's magnetic field.

avalanche a moving mixture of rock, soil, or regolith that moves rapidly, perhaps by riding on a cushion of air trapped as the material fell off a nearby mountain

avulsion the process by which a river breaks through a river levee and flows onto the floodplain

back arc the part of a magmatic arc that is on the continent side of an arc, or on the side furthest from the trench; includes back arc basins

backshore the area extending from the top of the beach ridge in the foreshore to the next feature (dune, seawall, forest, lagoon) closer to the mainland

banded-iron formation a distinctive type of sedimentary rock consisting of thin layers of iron oxides (typically magnetite or hematite) alternating with layers of iron-poor shale or chert. Banded-iron formations are common in some Precambrian rock sequences.

barrier islands narrow, linear, mobile strips of sand that are up to about 30–50 feet (10–15 m) above sea level and typically form chains located a few to tens of miles offshore along many passive margins. They are separated from the mainland by the back-barrier region, which is typically occupied by lagoons, shallow bays, estuaries, or marshes.

barrier spits narrow ridges of sand attached to the mainland at one end and terminating in a bay or the open ocean on the other end

basalt the most common igneous rock of the oceanic crust. Its subvolcanic or plutonic equivalent is called gabbro. The density of basalt is 3.0 g/cm³; its mineralogy includes plagioclase, clinopyroxene, and olivine.

basin, sedimentary basin a depression in the surface of the Earth or other celestial body is known as a basin. When this depression becomes filled with sediments, it is known as a sedimentary basin.

bayhead delta a delta deposited by a river in a bay or estuary, not in the open ocean

beach accumulations of sediment exposed to wave action along a coastline

beachface the side of a coast or barrier island facing the open ocean

bed load the coarse particles that move along or close to the bottom of the stream bed

bench a flat bedrock platform that borders many coastal cliffs, formed by erosion or beveling by waves

Benioff zone a narrow zone of earthquakes along a convergent boundary that marks the plane of slip between a subducting plate and an overriding plate, named after its discoverer, Hugo Benioff

benthic, benthos an environment that includes the ocean floor. Benthos are those organisms that dwell on or near the seafloor.

berm a small ridge and change in slope at the top of the foreshore

big bang theory a theory explaining the origin of the universe, in which the universe originated 10–20 billion years ago in a single explosive event during which the entire universe suddenly exploded out of nothing, reaching a pea-sized supercondensed state with a temperature of 10 billion million million degrees Celsius in (10^{-36}) of a second after the big bang

binary star systems two stars that rotate in orbit around each other

biofuels fuels such as methane that are produced from renewable biological resources including recently living organisms, such as plants, and their metabolic byproducts, such as manure

biological weathering the breaking down of rocks and minerals by biological agents such as roots, burrowing organisms, and bacteria

biome a climatically and biologically defined area with a distinctive community of plants, animals, and soil organisms

biosphere collection of all living organisms on the Earth and the environments in which they live

biostratigraphy the branch of stratigraphy that uses fossil assemblages and index fossils to correlate and assign relative ages to strata

bioturbation the process by which organisms burrow and dig into soft sediment and destroy fine-scale layering

blackbody a body that absorbs all radiation that falls on it when it is cold and emits at all wavelengths as it is progressively heated

black hole a super dense collection of matter that has collapsed from a giant star or stars and has such a strong gravity field that nothing can escape from it, not even light

black smoker chimneys hydrothermal vent systems that typically form near active magmatic systems along the mid-ocean ridge system, approximately two miles (3 km) below sea level. Black smokers form by seawater percolating into fractures in the seafloor rocks near the active spreading ridge, where the water gets heated to several hundred degrees Celsius. This hot pressurized water leaches minerals from the oceanic crust and extracts other elements from the nearby magma. The superheated water and brines then rise above the magma chamber in a hydrothermal circulation system and escape at vents on the seafloor, forming the black smoker hydrothermal vents.

blazar a very dense and variable source of energy associated with supermassive black holes at the center of a galaxy

body waves seismic waves that travel through the whole body of the Earth and move faster than surface waves. Body waves are of two types—P, or compressional waves, and S, for shear or secondary waves. P-waves or compressional waves deform material through a change in volume and density, and these can pass through solids, liquids and gases. The kind of movement associated with passage of a P-wave is a back and forth type of motion. S or secondary waves change the shape of a material but not its volume. Only solids can transmit shear waves, whereas liquids can not. Shear waves move material at right angles to the direction of wave travel, and thus they consist of an alternating series of sideways motions.

bolide a name for any unidentified object entering the planet's atmosphere

bore a wave that forms in some bays and estuaries, where the shape of the coastline tends to funnel water from the rising tides into narrower and narrower places, causing the water to pile up. When this happens, the volume of water entering the bay forms a wave called a tidal bore that moves inland, typically growing in height and forward velocity as the bay becomes narrower and narrower.

boudin a sausage-shaped lozenge of hard rock enclosed in a matrix of softer rock, typically occurring as a string or train of similar boudins. They form by a hard layer's being pulled apart into small sausage-shaped blocks during deformation, indicating extension parallel to the train of boudins

brachiopod a two-shelled marine invertebrate organism

brackish a term used to describe water that is mixed in salt content between salty and fresh

braid bars gravel and sand bars that separate the interconnected channels in a braided stream

breccia a fragmented rock produced by the breaking of preexisting rocks

bryozoans colonial organisms with hard skeletons of calcium carbonate, resembling coral

calc-alkaline a rock series, typically including basalt, andesite, dacite, and rhyolite, in which calcium oxide (CaO) generally declines with increasing silicon dioxide (SiO_2). The calc-alkaline rock series is typical of that found in island arc settings.

calcium-aluminum inclusions a group of very high-temperature minerals found inside some

chondrules, typically exhibiting textures like concentric skins of an onion

caldera a roughly circular or elliptical depression, often occupied by a lake, that forms when the rocks above a subterranean magma mass collapse into the magma during a cataclysmic eruption

calving the process whereby large blocks of ice plunge off the front of a tidewater glacier and fall into the sea, making icebergs

Cambrian the first geologic period of the Paleozoic era and the Phanerozoic eon. The Cambrian began 544 million years ago and ended 505 million years ago

capacity in hydrology, the potential load a stream can carry, measured in the amount (volume) of sediment passing a given point in a set amount of time

carbonaceous chondrites chondritic meteorites that contain organic material, such as hydrocarbons in rings and chains, and amino acids

carbonate a sediment or sedimentary rock containing the carbonate (CO_3^{2-}) ion. Typical carbonates include limestone and dolostone.

carbonatite an intrusive or extrusive igneous rock that has more than 50 percent carbonate minerals. Many carbonatites resemble marble in appearance, but are igneous in origin. Most carbonatites are associated with continental rift settings.

carbon cycle a complex series of processes where the element carbon makes a continuous and complex exchange between the atmosphere, hydrosphere, lithosphere and solid Earth, and biosphere

Carboniferous a late Paleozoic geologic period in which the Carboniferous system of rocks was deposited, between 360 and 286 million years ago

carbon sequestration a group of processes that enable long-term storage of carbon in the terrestrial biosphere, in the oceans, or deep underground, effectively isolating that carbon from the atmosphere

cave systems, caves underground openings or passageways in rock that are larger than individual spaces between the constituent grains of the rock. The term is often reserved for spaces that are large enough for people to enter, whereas other definitions reserve the term to describe any rock shelter, including overhanging cliffs.

cavitation a process that occurs when a stream's velocity is so high that the vapor pressure of water is exceeded and bubbles begin to form on rigid surfaces. These bubbles alternately form and then collapse with tremendous pressure, and they act as an extremely effective erosive agent

Cenozoic the most recent era of geologic time, marking the emergence of the modern Earth, starting at 66 million years ago and continuing until the present

Centaurs a group of asteroids that have highly eccentric orbits that extend beyond yet cross the orbits of Jupiter and Saturn, thus can potentially collide with these planets

centrifugal force a force, related to the spinning of the planet, that acts perpendicular to the axis of rotation of the Earth and affects the tides

charnockite an orthopyroxene-bearing granite consisting of quartz, perthite, and orthopyroxene, found in many high-grade Precambrian metamorphic terranes

chemical weathering decomposition of rocks through the alteration of individual mineral grains

chert a fine-grained almost glassy siliceous chemical sedimentary rock. Many cherts are made from the siliceous tests or outer skeletons of small organisms called radiolarians. Other chert is derived from volcanic activity, and still other is derived by chemical precipitation from seawater.

chlorofluorocarbons (CFCs) a group of inert, nontoxic, and easily liquefied chemicals that were once widely used in refrigeration. When released to the atmosphere they become long-lived greenhouse gases that contain carbon, hydrogen, fluorine, and chlorine, increasing in atmospheric concentration as a result of human activity. Chlorofluorocarbons are thought to cause depletion of the atmospheric ozone layer.

chondrite a stony meteorite containing chondrules

chondrules small lumps in chondritic meteorites, thought to represent melt droplets that formed before the meteorite fragments were accreted to asteroids, thus representing some of the oldest material in the solar system

chromite an iron-magnesium oxide mineral belonging to the spinel group and typically found in mantle peridotite and in some layered igneous intrusions

chromosphere the lower part of the solar atmosphere, resting directly above the photosphere

cirques bowl-shaped hollows that open downstream and are bounded upstream by a steep wall

clastic rocks composed of fragments, or clasts, of preexisting rock

cleavage in minerals, the way certain minerals break along specific crystallographic directions; in structural geology, where the preferred alignment of phyllosilicate minerals causes parallel planes of weakness to develop in a rock

climate the average weather of a place or region

climate change the phenomenon characterized by change in global temperatures, patterns of precipitation, wind, and ocean currents in response to human and natural causes

clouds visible masses of water droplets or ice crystals suspended in the lower atmosphere, generally confined to the troposphere

coastal plain relatively flat area along the coast, that was formerly below sea level during past high sea-level stand

coastal zone the region on the land that is influenced in some way by humidity, tides, winds, salinity, or biota from the sea. A more restrictive definition is the area between the highest point that tides influence the land and the point at which the first breakers form offshore.

collision zone a zone of uplifted mountains where two continental plates or magmatic arcs have collided as a result of subduction along a convergent plate boundary

coma the gaseous rim of a comet, from which the tail extends

comet an icy or mixed icy and rocky body that orbits the Sun and typically emits a long tail on its close approach to the Sun

compaction a phenomenon in which the pore spaces of a material are gradually reduced, condensing the material and causing the surface to subside

competence the maximum size of particles that can be entrained and transported by a stream under a given set of hydraulic conditions, measured in diameter of largest bed load

compressional waves *See* body waves.

conodonts extinct chordate, toothlike fossils that were part of a larger organism of the class Conodonta

consequent stream a stream whose course is determined by the direction of the slope of the land

constellation human groupings of stars in the sky into patterns, even though they may be far apart and lined up only visibly, and not close at all in space

continental crust the rock material that makes up the continents of the planet, extending to about 20 miles (35 km) depth and covering about 34.7 percent of the Earth's surface. Continental crust includes a wide variety of rock types and ages of rocks, but in general, the crust has an average composition of granodiorite.

continental drift a theory that was a precursor to plate tectonics, stating that the continents are relatively light objects that are floating and moving freely across a substratum of oceanic crust

continental margin the transition zone between thick buoyant continental crust and the thin dense submerged oceanic crust

continental shelf a submarine plain that forms the border of a continent, underlain by continental crust and having shallow water. Sedimentary deposits on continental shelves include muds, sands, and carbonates.

convection and the Earth's mantle a description of the main heat transfer mechanism in the Earth's mantle. Convection is a thermally driven process where heating at depth causes material to expand and become less dense, causing it to rise while being replaced by complementary cool material that sinks.

convergent boundaries places where two plates move toward each other, resulting in one plate sliding beneath the other when a dense oceanic plate is involved, or collision and deformation when continental plates are involved. These types of plate boundaries may experience the largest of all earthquakes.

coral invertebrate marine fossils of the phylum Cnidaria characterized by radial symmetry and a lack of cells organized into organs. They are related to jellyfish, hydroids, and sea anemones, all of which possess stinging cells. Corals are the best preserved of this phylum because they secrete a hard calcareous skeleton that typically forms hard reefs and gets preserved in the geological record.

Coriolis effect an apparent deflection of a freely moving body toward the right in the northern hemisphere and toward the left in the southern hemisphere

Coriolis force an apparent force that arises because the surface of the Earth at the equator is rotating through space faster than points on and near the poles. When water or air moves toward the pole it is moving faster than the new solid ground beneath it, causing objects to be deflected to the right in the northern hemisphere, and to the left in the southern hemisphere.

corrosiveness in soil, a measure of the ability to corrode or chemically decompose buried objects, such as pipes, wires, tanks, and posts

cosmic background radiation faint electromagnetic radiation that fills the universe, discovered in 1965, radiating as a thermal blackbody at 2.75 Kelvin. It is thought to be a remnant of the big bang.

cosmic rays extremely energetic particles that move through space at close to the speed of light. Most cosmic rays are made of atomic nuclei, about 90 percent of which are bare hydrogen nuclei (protons), 9 percent are helium nuclei (alpha particles), and about 1 percent are electrons (beta minus particles), but cosmic rays include a whole range of particles that spans the entire periodic table of the elements.

cosmology the study of the structure and evolution of the universe

covalent bond a type of chemical bonding characterized by the sharing of electrons between two or more atoms

cratons very old and stable portions of the continents that have been inactive for billions of years and typically have subdued topography including gentle arches and basins

creep the barely noticeable slow downslope flowing movement of regolith involving the slow plastic deformation of the regolith, as well as repeated microfracturing of bedrock at nearly imperceptible rates

Cretaceous the youngest of three periods of the Mesozoic, during which rocks of the Cretaceous system were deposited. The Cretaceous ranges from 144 million years ago (Ma) until 66.4 Ma and is divided into the Early and Late epochs and 12 ages.

cryosphere the portion of the planet where temperatures are so low that water exists primarily in the frozen state

crystal lattice the orderly, regular, and symmetric three-dimensional arrangement of atoms in a crystal

cumulates igneous rocks that form by the accumulation of crystals in a magma chamber, either from settling to the bottom and sides of the chamber, or floating to the top

cyclone a tropical storm equivalent to a hurricane, that forms in the Indian Ocean

dacite intermediate composition volcanic rock with high iron content, having a composition between andesite and rhyolite

dark matter a hypothetical form of matter of unknown composition but probably consisting of elemental particles. Because dark matter does not emit or reflect electromagnetic radiation, it can be detected only by observing its gravitational effects

debris avalanche a granular flow that moves at high velocity and covers large distances

debris flow the downslope movement of unconsolidated regolith, most of which is coarser than sand and typically chaotically mixed between fragments of different sizes. Some debris flows begin as slumps, but then continue to flow downhill as debris flows. They typically fan out and come to rest when they emerge out of steeply sloping mountain valleys onto lower-sloping plains.

decadal varying on scale of a decade, or 10-year time frame

deflation a process whereby wind picks up and removes material from an area, resulting in a reduction in the land surface

deformation of rocks a term, often used informally, to describe structural and geometric changes to rocks or rock masses. Deformation of rocks is measured by three components: strain, rotation, and translation. Strain measures the change in shape and size of a rock, rotation measures the change in orientation of a reference frame in the rock, and translation measures how far the reference frame has moved between the initial and final states of deformation.

delta low, flat deposits of alluvium at the mouths of streams and rivers that form broad triangular or irregular shaped areas that extend into bays, oceans, or lakes. They are typically crossed by many distributaries from the main river and may extend for a considerable distance underwater.

delta front an extremely sensitive environment located on the seaward edge of the delta. It is strongly affected by waves, tides, changing sea level, and changes to the flux or amount of sediment delivered to the delta front. Many delta fronts have an offshore sandbar known as a distributary mouth bar, or barrier island system, parallel to the coast along the delta front.

delta plain a coastal extension of the river system composed of river and overbank sedimentary deposits, in a flat meandering stream type of setting. These environments are at or near (or in some cases below) sea level, and it is essential that the overbank regions receive repeated deposits of muds and silts during flood stages to continuously build up the land surface as the entire delta subsides below sea level by tectonic processes.

dendritic drainage randomly branching stream patterns that form on horizontal strata or on rocks with uniform erosional resistance

depositional coast a coastline that is dominated by depositional landforms such as deltas or broad carbonate platforms

desalination any number of individual processes that remove salt from water and make it potable, or fit for drinking

desert an area characterized by receiving less than 1 inch (2.5 cm) of rain each year over an extended period of time

desertification the conversion of previously productive lands to desert

desert pavement a long-term stable surface in deserts characterized by pebbles concentrated along the surface layer

Devonian the fourth geological period in the Paleozoic era, spanning the interval from 408 to 360 million years ago

dextral a term used to describe the sense of motion along a fault where the block on the opposite side of the fault moves to the right

diagenesis a group of physical and chemical processes that affects sediments after they are deposited but before they undergo deformation and metamorphism

diamictite a poorly or nonsorted conglomerate containing a wide variety of rock fragments. Diamictites can be produced from glacial ice sheets, volcanic lahars, or debris flows, or by tectonic and erosional processes

diapir an igneous intrusion emplaced by buoyancy forces and differential pressure between the magma and surrounding country rock, and by the heat energy transferred from the magma to the country rock

diatreme a breccia-filled volcanic pipe formed by a gaseous explosion

differentiation in geology, the separation of one composition crystal or liquid from another, by processes that may include melting, crystallization, and removal of the phase from other parts of the system

diffraction a process that occurs when energy moves or is leaked sideways along a wave crest, enabling the wave to grow along the wave front to fill the available area inside a wide bay that the wave has entered through a narrow opening

dike any tabular, parallel-sided igneous intrusion that generally cuts across layering in the surrounding country rocks

dilation expansion of a rock caused by the development of numerous minor cracks or fractures in the rocks that form in response to the stresses concentrated along the fault zone

diorite gray-colored intermediate composition intrusive igneous rock composed of plagioclase feldspar, biotite, hornblende, and rarely pyroxene, small amounts of quartz, and other accessory minerals

dirty ice cometary ice that is mixed with dust or other material

discharge the amount of water moving past a point in a stream, usually measured in cubic feet or cubic meters per second

discharge areas the point where water leaves the groundwater system

dislocation a crystallographic defect in a crystal lattice, typically an extra row or wall of atoms, or a twisted misarrangement of the atomic structure

dispersion a process that moves energy from the initially high wave crest sideways, and in doing so takes energy away from the central area, decreasing the height of the wave

dissolution the process in which water reacts chemically with rocks and carries elements away in solution, gradually dissolving the rock. This often results in the formation of caves and other karst features, and it may result in underground collapse and sinking of the land.

dissolved load dissolved chemicals, such as bicarbonate, calcium, sulfate, chloride, sodium, magnesium, and potassium. The dissolved load tends to be high in streams fed by groundwater.

divergent boundaries places where two plates move apart, creating a void that is typically filled by new oceanic crust that wells up to fill the progressively opening hole; also called divergent margins

Doppler shift a shift in the frequency and wavelength of a wave for an observer moving relative to the source of the waves

downslope flow movement of a sediment/water/rock mixture downhill

drainage basin the total area that contributes water to a stream. The line that divides different drainage basins is known as a divide

drawdown the rapid retreat of water from a coastline immediately before a tsunami hits. The drawdown may resemble a rapidly retreating tide.

dropstones isolated pebbles or boulders in marine sediments, deposited when rocks trapped in floating icebergs melted out of the ice and dropped to seafloor

drought a period when the yearly rainfall for a region is significantly less than normal

drumlins teardrop-shaped accumulations of till that are up to about 150 feet (50 m) in height, and tend to occur in groups. These have a steep side that faces in the direction from which the glacier advanced and a back side with a more gentle slope. Drumlins are thought to form beneath ice sheets and record the direction of movement of the glacier.

dune low, wind-blown mounds of sand or granular material, with variable size and shape depending on sand supply, vegetation, and wind strength

dunite a plutonic igneous rock of ultramafic composition, composed almost entirely of the mineral olivine, with minor pyroxene, chromite, and pyrope

dwarf planet a celestial body that is in orbit around the Sun, has sufficient mass for its self-gravity to overcome rigid body forces so that it assumes a hydrostatic equilibrium (nearly round) shape, has not cleared the neighborhood around its orbit, and is not a satellite

dwarfs (stars) stars that have a size that is normal for their mass, that lie on the main sequence curve, and that are in the process of converting hydrogen to helium by nuclear fusion in their cores. Dwarfs are also classified as any star with a radius comparable to or smaller than the Earth's Sun, which is classified as a yellow dwarf.

Earth the third planet from the center of our solar system, located between Venus and Mars at a distance of 93 million miles (150×10^6 km) from the Sun

earthflow a generally slow-moving downslope flow that forms on moderate slopes with adequate moisture and develops preferentially in fine-grained deformable soils such as clays, as well as rocky soils that have a silt or clay matrix. Earthflows can contribute large amounts of sediment to streams and rivers and typically move in short periods of episodic movement or relatively steady movement in response to heavy rainfall events, earthquakes, irrigations, or other disturbances. Most earthflows move along a basal shear surface, are characterized by internal deformation of the sliding material, and do not fail catastrophically but do cause significant damage to infrastructure.

earthquake a sudden release of energy from slip on a fault, an explosion, or other event that causes the ground to shake and vibrate, associated with the passage of waves of energy released at its source. An earthquake originates in one place and then spreads out in all directions along the fault plane.

ebb tide outgoing tide

ecliptic plane the relatively flat plane that contains most of the planetary orbits around the Sun

eclogite a coarse grained mafic (basaltic in composition) metamorphic rock, containing red garnet and a green sodium-rich amphibole called omphacite. Eclogite is a high-pressure metamorphic rock, formed at cool temperatures and high pressures, so it is typically taken as an indicator of subduction zone metamorphism of a basaltic precursor.

economic geology the science of the study of earth materials that can be used for economic or industrial purposes

ecosystem a collection of the organisms and surrounding physical elements that together function as an ecological unit

edge wave a secondary wave, associated with tsunamis and other waves, that forms as the water from one wave crest retreats, moves back to sea, and interacts with the next incoming wave. Some of this water moves quickly sideways along the coast, setting up a new independent set of waves that oscillates up and down in amplitude along the coast, typically with a wavelength that is double that of the original tsunami.

ejecta material thrown out of an impact crater during an impact event

elastic rebound theory this theory states that recoverable (also known as elastic) stresses build up in a material until a specific level or breaking point is reached. When the breaking point or level is attained, the material suddenly breaks, releasing energy and stresses, causing an earthquake.

electromagnetic radiation self-propagating waves that move through matter or a vacuum, consisting of both electric and magnetic field components that vibrate perpendicular to each other and perpendicular to the direction of energy propagation. Electromagnetic radiation is classified into types according to frequency of the wave, in order of increasing frequency from radio waves, to microwaves, terahertz radiation, infrared, visible, and ultraviolet radiation, X-rays, and gamma rays.

electron-degenerate matter a form of degenerate matter in collapsed stars created where the pressure of overlying matter forces all of the electrons surrounding the nucleus into lowest energy quantum states

El Niño one of the better-known variations in global atmospheric circulation patterns that causes a warm current to move from the western Pacific to the eastern Pacific and that has global consequences in terms of changes in weather patterns. Full name is El Niño–Southern Oscillation (ENSO)

emission nebulae hot glowing clouds of ionized interstellar gas

enderbite granulite-facies metamorphic rock containing orthopyroxene, quartz, and plagioclase

entrainment the picking up of particles from the bed load of a stream and the erosion of material from the banks

environmental geology an applied interdisciplinary science focused on describing and understanding human interactions with natural geologic systems, where different systems of the lithosphere, biosphere, hydrosphere, and atmosphere interact, and changes in one system are seen to influence the other systems

Eocene the middle epoch of the Paleogene (Lower Tertiary) period, ranging from 57.8 million years ago to 36.6 million years ago, and divided from base to top into the Ypresian, Lutetian, Bartonian, and Priabonian ages

eolian sediments deposited by wind

epeirogenic a term referring to the vertical movements of continents

ephemerides predictions of when certain stars and planets would be found in specific locations

epicenter the point on the Earth's surface that lies vertically above the focus of an earthquake

equations of state thermodynamic equations that describe the state of matter under a given set of physical conditions such as temperature, pressure, volume, or internal energy

equinox the time, twice a year, that the tilt of the Earth places the Sun directly over the equator, and when the length of the day and night at the equator are equally long

erosion a group of processes that cause Earth material to be loosened, dissolved, abraded, or worn away and moved from one place to another

erosional coast a coastline that is actively eroding, typically with erosional cliffs or headlands

escape tectonic escape is process where a large block of rock slides sideways and moves out of a tectonic collision zone, generally moving toward an open ocean basin

estuary embayment along the coast that is open to the sea and influenced by tides and waves. Estuaries also have significant freshwater influence derived from river systems that drain into the head of the bay. The fresh and salt water mix within the estuary.

eukaryote an organism consisting of one or more cells possessing nuclei and membrane-bound organelles

eustatic sea-level change a sea-level change shown to be global in scale. Sea levels may also rise or fall for local (noneustatic) reasons, such as tectonic subsidence, or global reasons, such as melting of glaciers.

evaporation the conversion of a liquid to a vapor, such as water to vapor

evaporites sedimentary rocks made of water-soluble minerals that formed during the evaporation of a body of surface water, such as a lake or small ocean basin

evolution the cumulative and irreversible change of organisms through time. Results of this process explain the distribution and diversity of life throughout Earth history.

exfoliation a weathering process where rocks spall off in successive shells, like the skin of an onion. Exfoliation is caused by differential stresses within a rock formed during chemical weathering processes.

expansive clay and soils soils that add layers of water molecules between the plates of clay minerals (made of silica, aluminum, and oxygen), loosely bonding the water in the mineral, and that are capable of expanding by up to 50 percent more than their dry volume

extensional collapse a late-stage of orogenic evolution in which the crust is extended along numerous normal faults, initiated by the crust becoming too thick to be supported by weak rock layers at depth

extensional plate boundaries *See* divergent boundaries.

fabric a general term referring to the orientation of minerals in a rock, forming cleavage, foliation, lineation, gneissosity, or other structural elements

facies a distinctive set of igneous, metamorphic, or sedimentary characteristics of a group of rocks that separate and differentiate them from surrounding groups of rocks

falls in mass wasting, when regolith moves freely through the air and lands at the base of the slope or escarpment

fanglomerate a type of sedimentary deposit in which conglomerates are deposited as an alluvial fan derived from nearby uplifted mountains

fetch the distance over which the wind blows over water, forming waves of a particular wave set

fireball streak of light that moves across sky, produced by a meteorite or comet burning up in the atmosphere

firn frozen water that is transitional in density between snow and ice

fjords glacially carved, steep-sided valleys that are open to the sea

flash floods floods that rise suddenly, typically as a wall of water in a narrow canyon

flood abnormally high water, typically occurs when a river flows over its banks

flood basalt thick sequences of basaltic lava that cover very large areas of the continents or oceans, also known as traps or large igneous provinces

floodplain flat areas adjacent to rivers that naturally flood and are generally composed of unconsolidated sediments deposited by the river

flood tide incoming tide

flower structure a distinctive geometry of faults where a major nearly vertical strike-slip fault in the center of a mountain range is succeeded outward on both sides by more shallow dipping faults with strike-slip and thrust components of motion. A cross section drawn through these types of structures resembles a flower or a palm tree

flows in mass wasting, movements of regolith, rock, water, and air in which the moving mass breaks into many pieces that flow in a chaotic mass movement

fluvial deposits and landforms created by the action of flowing rivers and streams, and also the processes that occur in these rivers and streams

flysch a thick group of turbidites and shale deposited at the same time as mountain building, typically in a deep sea–basin environment or a forearc basin on a continent

focus the point in the Earth where the earthquake energy is first released, representing the area on one side of a fault that actually moves relative to the rocks on the other side of the fault plane. After the first slip event, the area surrounding the focus experiences many smaller earthquakes as the surrounding rocks also slip past each other to even out the deformation caused by the initial earthquake shock.

fold a geological structure in which deformation of the rock unit bends strata into inward or outward dipping layers. Inward dipping layers form synclines, and outward dipping layers form anticlines.

fold interference pattern a complexly shaped structure produced when a layer is affected by more than one generation of fold, typically producing strange three-dimensional warps and bends of the rock layers

forearc the part of a magmatic arc on the oceanward side of the arc, including the forearc basin, accretionary wedge, and trench. Many tsunamis are generated by thrust type earthquakes that uplift large sections of the forearc, displacing huge masses of water.

foredune ridge a linear ridge of sand that marks the boundary between the beachface and backbeach area

foreland basin a tectonic depression that forms adjacent to a mountain belt. Foreland basins form because the weight of the adjacent mountain belt causes the crust and entire lithosphere to bend in a process called flexure to accommodate the weight of the mountain belt.

foreshore a flat, seaward-sloping surface that grades seaward into the ridge and runnel. The foreshore is also known as the beachface.

fossil any remains, traces, or imprints of any plants or animals that lived on the Earth

fractionation in geochemistry, a separation process in which one phase (melt, solid, or fluid) of a mixture is removed in small quantities, changing the bulk composition of the remaining mixture. An example is when a crystal phase separates out of a magma by sinking to the bottom of the magma chamber, removing the chemical elements in those crystals from the overall melt composition, which then attains a different composition than before the crystals settled.

fracture any break in a rock or other body that may or may not have any observable displacement. Fractures include joints, faults, and cracks formed under brittle deformation conditions and are a form of permanent (nonelastic) strain.

fracture zone in oceanic crust, the zone that appears to be an extension of the transform fault but is not a plate boundary. Fracture zones represent the places where two parts of the same plate with different ages are juxtaposed by seafloor spreading.

fracture zone aquifer water that is stored in fractures in crystalline bedrock systems and is extractable for use by humans. Many buried granite and other bedrock bodies are cut by many fractures, faults, and other cracks, some of which may have open spaces along them. Fractures at various scales represent zones of increased porosity and permeability. They may form interconnected networks, and therefore, are able to store and carry vast amounts of water, forming fracture zone aquifers.

fusilinids fossils of a group of single-celled animals belonging to the phylum Protozoa. They grew in coral-like colonies, forming columnar, snail-like, or cigar-like shapes

gabbro a coarse-grained, dark, igneous intrusive rock consisting of the minerals pyroxene, plagioclase, amphibole, and olivine. Gabbro is the plutonic equivalent of basalt and is common in oceanic crust.

Gaia hypothesis a hypothesis suggesting that Earth's atmosphere, hydrosphere, geosphere, and biosphere interact as a self-regulating system that maintains conditions necessary for life to survive

galaxies gravitationally bound assemblages of stars, dust, gas, radiation, black holes, and dark matter. Most contain vast numbers of star systems and are located at enormous distances from the Milky Way Galaxy, such that the light reaching Earth from these galaxies was generated billions of years ago.

galaxy cluster a group of 50 or more galaxies held together by mutual gravitational attraction

gas hydrates solid, ice-like, water-gas mixtures that form at temperatures between 40° and 43°F (4–6°C) and pressures above 50 atmospheres. As a type of clathrate, gas hydrates consist of lattices that enclose or trap other molecules—typically methane gas, but also ethane, butane, propane, carbon dioxide, and hydrogen sulfides.

geochemical cycles the transportation, cycling, and transformation of the different chemical elements through various reservoirs or spheres in the Earth system, including the atmosphere, lithosphere, hydrosphere, and biosphere

geochemistry the study of the distribution and amounts of elements in minerals, rocks, ore bodies, rock units, soils, the Earth, atmosphere, and by some accounts, other celestial bodies, and the principles that govern the distribution and migration of these elements

geochronology the study of time with respect to Earth history including both absolute and relative dating systems as well as correlation methods

geodesy the study of the size and shape of the Earth and its gravitational field, and the determination of the precise locations of points on the surface

geodynamics the branch of geophysical science that deals with forces and physical processes in the interior of the Earth

geographic information systems (GIS) computer application programs that organize and link geographic-related information in a way that enables the user to manipulate that information in a constructive way. They typically integrate a database management system with a graphics display that shows links between different types of data.

geoid an imaginary surface near the surface of the Earth, along which the force of gravity is the same and equivalent to that at sea level

geological hazards natural geologic processes ranging from earthquakes and volcanic eruptions to the slow downhill creep of material on a hillside and the expansion of clay minerals in wet seasons. These processes are considered hazardous when they go to extremes and interfere with the normal activities of society.

geomagnetism the study of the Earth's magnetic field, including its generation, strength, and changes over time

geomorphology the study of the surface features and landforms on Earth

geophysics the study of the Earth by quantitative physical methods, with different divisions including solid-Earth geophysics, atmospheric and hydrospheric geophysics, and solar-terrestrial physics

geosyncline an obsolete term for a subsiding linear trough or basin formed by the accumulation of sedimentary rocks deposited in a basin and subsequently compressed, deformed, and uplifted into a mountain range, with attendant volcanism and plutonism caused by the melting of the deep sediments. Geosyncline theory was replaced by the theory of plate tectonics to explain features of basins and mountain ranges.

geothermal gradient a measure of how the temperature changes with increasing depth in the Earth

geyser springs in which hot water or steam sporadically or episodically erupts as jets from an opening in the surface, in some cases creating a tower of water hundreds of feet (tens of meters) high

glacial drift a general term for all sediment deposited directly by glaciers or by glacial meltwater in streams, lakes, and the sea

glacial erratic glacially deposited rock fragments with compositions different from underlying rocks

glacial marine drift sediment deposited on the seafloor from floating ice shelves or bergs, including any isolated pebbles or boulders that were initially trapped in glaciers on land, then floated in icebergs that calved off from tidewater glaciers

glacial moraine piles of sand, gravel, and boulders deposited by a glacier

glacial outwash plain a generally flat plain of gravel, sand, and other sediments deposited by meltwater in front of the terminus of a glacier

glacial rebound a process during which an ice sheet melts, leading to the upward rise (rebound) of a continent in response to the reduced weight load upon it

glacial striations scratches on the surface of bedrock, formed from a glacier dragging boulders across the bedrock surface

glacier any permanent body of ice (recrystallized snow) that shows evidence of gravitational movement

global warming a trend in climate characterized by progressive increases in the global average yearly temperature over many years

gneiss high-grade, regional metamorphic rock, with a medium to coarse grain size, showing a pronounced layering called gneissic foliation

Gondwana a late Proterozoic to Paleozoic supercontinent of the southern hemisphere, that included present day Africa, South America, Australia, India, Arabia, Antarctica, and many small fragments

graded stream a stream that has gradually adjusted its gradient to reach an equilibrium between sedimentary load, slope, and discharge

gradient in geology, a measure of the vertical drop in elevation over a given horizontal distance

granite, granodiorite a common igneous rock type found in the continental crust. The density of granodiorite is 2.6 g/cm³; its mineralogy includes quartz, plagioclase, biotite, and some potassium feldspar. Granite has more quartz than granodiorite. The volcanic or extrusive equivalent of granite is rhyolite, and of granodiorite, andesite.

granular flow a downslope flow in which the full weight of the flowing sediment is supported by grain-to-grain contact between individual grains

granulite a high-grade metamorphic rock formed at high pressures and temperatures, and characteristic of the middle to deep crust

gravity the attraction between any body in the universe and all other bodies

gravity anomaly geologically significant variations in gravity

gravity wave in fluid dynamics, waves generated in a fluid medium or at the interface between two fluids, such as air and water. In astrophysics and in general relativity theory, gravity waves can be thought of as gravitational radiation that results from a change in the strength of a gravitational field. Any mass that accelerates through space will produce a small distortion in the space through which it is traveling, producing a change in the gravitational field, or a gravity wave, that should be emitted at the speed of light.

great earthquakes earthquakes with magnitude greater than 8.0 on the Richter scale that often cause a vast amount of destruction and release a huge amount of energy. Most great earthquakes occur at convergent plate margins.

Great Ice Ages a term referring to the late Pleistocene glaciation

greenhouse effect a phenomenon characterized by abnormal warmth on Earth in response to the atmosphere's trapping incoming solar radiation by greenhouse gases

greenhouse gases gases, such as carbon dioxide (CO_2), that when built up in the atmosphere tend to keep solar heat in the atmosphere, resulting in global warming

greenschist a metamorphic rock that is generally derived from mafic rocks, altered to mineral assemblages of low regional metamorphic grade, called greenschist facies

greenschist facies a low to medium grade metamorphic zone

greenstone belts elongate accumulations of generally mafic volcanic and plutonic rocks, typically associated with immature graywacke types of sedimentary rocks, banded-iron formations, and less commonly carbonates and mature sedimentary rocks. Most greenstone belts are Archean or at least Precambrian in age, although similar sequences are known from orogenic belts of all ages. Most greenstone belts are metamorphosed to greenschist through amphibolite facies and are intruded by a variety of granitoid rocks.

Grenville province and Rodinia the youngest region of the Canadian shield, located outboard of the Labrador, New Quebec, Superior, Penokean, and Yavapai-Mazatzal provinces. It is the last part of the Canadian shield to experience a major deformational event, this being the Grenville orogeny at about 1.0 billion years ago, which was responsible for forming many folds and faults throughout the entire region during the amalgamation of numerous continents to form the supercontinent of Rodinia.

greywacke a sandy sedimentary rock with mud particles between the sand grains

groin walls of rock, concrete, or wood built at right angles to the shoreline that are designed to trap sand from longshore drift and replenish a beach

ground breaks fissures or ruptures that form where a fault cuts the surface and may be associated with mass wasting, or the movements of large blocks of land downhill. These ground breaks may have horizontal, vertical, or combined displacements across them and may cause considerable damage.

ground level the elevation of the land surface above sea level

ground motion shaking and other motion of the ground associated with the passage of seismic waves. The amount of ground motion associated with an earthquake generally increases with the magnitude of the earthquake, but depends also on the nature of the substratum. Ground motions are measured as accelerations, which indicate the rate of change of motion.

groundwater all the water contained within spaces in bedrock, soil, and regolith

Hadean Era the earliest time in Earth history, ranging from accretion at 4.56 billion years ago, until the time of the oldest rocks preserved at 3.96 billion years ago

Hadley cell belt of air that encircles the Earth, rising along the equator, dropping moisture as it rises in the tropics. As the air moves away from the equator at high elevations, it cools, becomes drier, and then descends at 15°–30°N and S latitude where it either returns to the equator or moves toward the poles. Similar circulation cells called mid-latitude cells dominate the air circulation between 30° and 60°N and S latitude, and polar cells occupy latitudes above 60°.

harzburgite an ultramafic igneous rock of the peridotite family, consisting of olivine and pyroxene. Harzburgite is a common rock in the mantle section of ophiolitic rock sequences.

heat capacity a measure of how much energy can be absorbed by a body without its changing temperature

heat of vaporization amount of energy to change water from a liquid state to a vapor

Heinrich Events specific intervals in the sedimentary record showing ice-rafted debris in the North Atlantic

helioseismology the science of studying the oscillations of the Sun

helium flash an uncontrolled explosive burning of helium that lasts for a few hours in the core of a red giant star as its core undergoes the conversion to helium fusion

heterotrophs organisms that cannot manufacture their own food but must intake carbon and nitrogen from their environment in the form of complex organic molecules

high-constructive deltas deltas that form where the fluvial transport dominates the energy balance on the delta. These deltas dominated by riverine processes are typically elongate, such as the modern delta at the mouth of the Mississippi.

high-destructive deltas deltas that form where the tidal and wave energy is high and much of the fluvial sediment gets reworked before it is finally deposited. In wave-dominated high-destructive deltas, sediment typically accumulates as curved barriers near the mouth of the river. Examples of wave-dominated deltas include the Nile and the Rhone deltas.

historical geology the science that uses the principles of geology to reconstruct and interpret the history of the Earth

horn peak in mountains that forms where three cirques meet

hothouse a time when the global climate was characterized by very hot conditions for an extended period

hot spot an area of unusually active magmatic activity that is not associated with a plate boundary. Hot spots are thought to form above a plume of magma rising from deep in the mantle.

Hubble's constant a constant of proportionality that gives the relationship between recessional velocity and distance in Hubble's Law

Hubble's law the law that states that the recessional velocity of a galaxy is proportional to its distance from the observer

hurricanes intense tropical storms with sustained winds of more than 74 miles per hour (119 km/hr) characterized by a central eye with calm or light winds and clear skies, surrounded by an eye wall, which is a ring of very tall and intense thunderstorms that spin around the eye, with some of the most intense winds and rain of the entire storm system. The eye is surrounded by spiral rain bands that spin counterclockwise in the Northern Hemisphere (clockwise in the Southern Hemisphere) in toward the eye wall, moving faster and generating huge waves as they approach the center. These storms are known as hurricanes if they form in the northern Atlantic or eastern Pacific Oceans, cyclones if they form in the Indian Ocean or near Australia, or typhoons if they form in the western Pacific Ocean.

hydraulic gradient the pressure difference between the elevation of the source area and the discharge area of an aquifer

hydraulic piping where water finds a weak passage through a levee

hydrocarbons and fossil fuels gaseous, liquid, or solid organic compounds consisting of hydrogen and carbon. Petroleum is a mixture of different types of hydrocarbons derived from the decomposed remains of plants and animals.

hydrologic (water) cycle sequential changes, both long- and short-term, in the Earth's hydrosphere, involving the movement and physical changes of water in the atmosphere and on the Earth's surface, involving processes such as precipitation, evaporation, transpiration, freezing, and melting

hydrosphere a dynamic mass of liquid, continuously on the move between the different reservoirs on land and in the oceans and atmosphere. The hydrosphere includes all the water in oceans, lakes, streams, glaciers, atmosphere, and groundwater, although most water is in the oceans.

hydrothermal a term used to describe any process that involves high temperature fluids, often in reference to hot springs, geysers, and related activities

ice ages times when the global climate was cold and large masses of ice covered many continents

igneous rocks rocks that have crystallized from a molten state known as magma. These include plutonic rocks, crystallized below the surface, and volcanic rocks, that have crystallized at the surface.

imbrication an arrangement of planar bodies in a regular stacked way that resembles a tipped over line of dominoes

impact crater a generally bowl-shaped depression excavated by the impact of a meteorite or comet with the Earth

impact glass glass produced by the sudden melting and homogenization and sudden cooling of rock during a meteorite impact event

impactor a general name for an object that strikes and creates a crater in another object

index fossil a fossil that is used to identify and define geological periods, or faunal stages. Ideal index fossils are short-lived, have a broad distribution, and are easy to identify.

inflationary theory a modification of the big bang theory, suggesting that the universe underwent a period of rapid expansion immediately after the big bang

infrared a type of electromagnetic radiation with a wavelength between 750 nanometers and 1 mm, longer than visible light but shorter than terahertz and microwave radiation

inselbergs steep-sided mountains or ridges that rise abruptly out of adjacent monotonously flat plains in deserts

interglacial period a period such as the present climate epoch, characterized as being between advances and retreats of major continental ice sheets. The glaciers could return, or global warming could melt the remaining polar ice, taking the planet out of the glacial epoch.

Intergovernmental Panel on Climate Change (IPCC) a scientific intergovernmental body set up by the World Meteorological Organization (WMO) and by the United Nations Environment Program (UNEP). The IPCC is open to all member countries of WMO and UNEP. Governments participate in plenary sessions of the IPCC where main decisions about the IPCC work program are taken and reports are accepted, adopted, and approved. They also participate in the review of IPCC reports. The IPCC includes hundreds of scientists from all over the world who contribute to the work of the IPCC as authors, contributors, and reviewers. As a United Nations body, the IPCC work aims at the promotion of the United Nations human development goals.

interstellar medium the areas or voids between the stars and galaxies, representing a nearly perfect vacuum, with a density a trillion trillion times less than that of typical stars

intertidal flat a flat area within the tidal range that sheltered from waves, dominated by mud, devoid of vegetation, and accumulating sediment; also called tidal flat

intertidal zone the area between the high-tide line and the low-tide line

intraplate earthquake a rare, occasionally strong type of earthquake that occurs in the interior of plates far from plate boundaries. The origins of these earthquakes are not well understood.

ionization the process of converting an atom or molecule into a charged ion by adding or removing charged particles such as electrons

iron meteorites meteorites that are composed mostly of metallic material

island arc *See* magmatic arc.

isoclinal folds folds in which relatively flat limbs on either side of the curved hinge are parallel to each other

isostasy a principle that states that the elevation of any large segment of crust is directly proportional to the thickness of the crust

isotopes forms of an element that have different atomic masses; all isotopes of an element have the same number of protons in the nucleus but differ in the number of neutrons

jet stream high-level, narrow, fast-moving currents of air that are typically thousands of miles (km) long, hundreds of miles (km) wide, and several miles (km) deep

Jupiter The fifth planet from the Sun, Jupiter is a gaseous giant of the solar system, with more than twice the mass of all the other planets combined, estimated at 1.9×10^{27} kilograms, or 318 Earth masses, and a radius of 44,268 miles (71,400 km), or 11.2 Earth radii

karst a type of landscape formed by the dissolution of underlying bedrock, typically limestone, characterized by caves, sinkholes, collapsed caves, and isolated towers

karst terrains areas that are affected by groundwater dissolution, cave complexes, and sinkhole development

katabatic wind a strong downhill wind of high-density air that moves downhill from a mountain, typically in a valley, and often at high velocity

Kepler's third law a law of planetary motion that states that the squares of the orbital periods of planets are directly proportional to the cubes of the axes, so that as the length of the orbit increases with distance, the orbital speed decreases

kerogen organic chemical compounds found in sedimentary rocks

khondalite granulite facies metamorphic rock containing quartz, sillimanite, garnet, K-feldspar, plagioclase, and graphite

kimberlite an unusual volcanic rock containing mixtures of material derived from the upper mantle and complex water-rich magmas. Many kimberlites intrude explosively, forming large pipes that fill with the magma, and many contain diamonds.

kinematic indicators features in deformed rocks that give clues about the sense of shearing in the deformational history of the rock. Typical kinematic indicators include deformed mineral grains, asymmetric foliations, and rotated grains.

kinematics the scientific study of motions without consideration of forces or stresses involved. Most of plate tectonics and structural geology are kinematic models.

Kuiper belt a distant part of the solar system, extending from 30–49 astronomical units, and containing many comets and asteroids

lagoon a rare class of restricted coastal bay that is separated from the ocean by an efficient barrier that blocks any tidal influx and that does not have significant freshwater influx from the mainland

lahar a mudflow formed by the mixture of volcanic ash and water. Lahars are common on volcanoes, both during and for years after major eruptions.

laminar flow in a stream, where paths of water particles are parallel and smooth and the flow is not very erosive. Resistance to flow in laminar systems is provided by internal friction between individual water molecules, and the resistance is proportional to flow velocity.

landslide a general name for any downslope movement of a mass of bedrock, regolith, or a mixture of rock and soil, commonly used to indicate any mass-wasting process

large igneous provinces, flood basalt deposits that include vast plateaus of basalts, covering large areas of some continents, exhibiting a tholeiitic basalt composition, but some show chemical evidence of minor contamination by continental crust. They are similar to anomalously thick and topographically high seafloor known as oceanic plateaus and to some volcanic rifted passive margins. Also known as continental flood basalts, plateau basalts, and traps

latent heat the amount of energy, in the form of heat, that is absorbed or released by a substance during a change in state such as from a liquid to a solid

laterite a red-colored soil rich in iron and aluminum that forms by intense weathering of underlying bedrock in tropical regions. The most common minerals in laterite are kaolinite, goethite, hematite, and gibbsite.

lava magma, or molten rock, that flows at the surface of the Earth

leeward the side of a mountain facing away from oncoming winds

leptynite granulite facies metamorphic rock containing quartz, garnet, K-feldspar, plagioclase, amphibole, sillimanite, and biotite

lherzolite an ultramafic igneous rock of the peridotite family consisting of olivine, orthopyroxene, and clinopyroxene, thought to be a major component of the mantle. Partial melt of lherzolite mantle in mid-ocean ridges forms a residual peridotite known as harzburgite, plus oceanic crust.

limestone a sedimentary carbonate rock made predominantly of the mineral calcite ($CaCO_3$)

lineation a penetrative linear element in a rock, typically formed by the alignment of minerals or deformed grains, pebbles, fossils, or other structural elements

liquefaction a process in which sudden shaking of certain types of water-saturated sands and muds turn these once-solid sediments into a slurry, a substance with a liquid-like consistency. Liquefaction occurs when individual grains move apart, and then water moves up in between the individual grains making the whole water/sediment mixture behave like a fluid.

lithosphere the rigid outer shell of the Earth that is about 75 miles (125 km) thick under continents and about 45 miles (75 km) thick under oceans. The basic theorem of plate tectonics is that the lithosphere of the Earth is broken into about 12 large rigid blocks or plates that are all moving relative to one another.

loess a deposit of fine-grained, wind-blown dust

longshore current the movement of water parallel to the coast along the beachface, caused by the oblique approach of waves

longshore drift the gradual transport of sand along a beach, caused by waves washing sand diagonally up the beachface, and gravity pulling the sand back down the beachface perpendicular to the shoreline

mafic an adjective describing generally dark-colored igneous rocks that are composed chiefly of iron-magnesium rich minerals and typically have silica contents from 45–52 percent, as determined by chemical analysis. Common mafic igneous rocks include basalt and gabbro.

magma molten rock, at high temperature. When magma flows on a surface, it is known as lava.

magmatic arc a line of volcanoes and igneous intrusions that forms above a subducting oceanic plate along a convergent margin. Island arcs are built on oceanic crust, and continental margin magmatic arcs are built on continental crust.

magnetic field, magnetosphere Earth's magnetic field is generated within the core of the planet. The field is generally approximated as a dipole, with north and south poles, and magnetic field lines that emerge at the magnetic south pole and reenter at the magnetic north pole. The field is characterized at each place on the planet by an inclination and a declination of the magnetic flux lines.

magnetic isogonic lines lines of constant magnetic declination

main sequence a continuous band of stars that appears on a plot of luminosity versus temperature for stars, indicating a stable relationship between luminosity, temperature, and star mass

mangal a dense, coastal, mangrove-dominated ecosystem

mantle the layer of the Earth between the crust and the core, containing dense, highly viscous mafic rocks that flow in convection cells

mantle plumes linear plumes of hot material that upwell from deep within the mantle, perhaps even from the core-mantle boundary

mantle root a very thick (up to 200 miles, or > 300 km) section of lithospheric mantle beneath many Archean cratons, containing cold material from which a melt has been removed, adding stability to the craton

marine terraces wave-cut benches or platforms that are uplifted above sea level, typically occurring in groups at different levels reflecting different stages of uplift

Mars the fourth planet from the Sun. Mars is only 11 percent of the mass of Earth and has an average density of 3.9 grams per cubic centimeter and a diameter of 4,222 miles (6,794 km).

mass extinction a time when large numbers of species and individuals within a species die off.

Mass extinction events are thought to represent major environmental catastrophes on a global scale. In some cases these mass extinction events can be tied to specific likely causes, such as meteorite impact or massive volcanism, but in others their cause is unknown. The Earth's biosphere has experienced five major and numerous less-significant mass extinctions in the past 500 million years (in the Phanerozoic era). These events occurred at the end of the Ordovician, in the late Devonian, at the Permian/Triassic boundary, at the Triassic/Jurassic boundary, and at the Cretaceous/Tertiary (K/T) boundary. Many mass extinctions can be shown to have occurred in less than a million or two years.

mass wasting the movement of material downhill without the direct involvement of water

meander a gentle bend in the trace of a river

mechanical weathering the disintegration of rocks, generally by abrasion

medical geology an emerging science that studies the effects of geological materials, processes, and trace elements in the environment on human and animal health

megaton a mass equal to 1,000,000 tons, at 2,000 pounds per ton in the United States. One ton is equal to 0.907 metric tonnes.

mélange a complexly mixed assemblage of rocks, characteristically formed at subduction zones

Mercalli intensity scale a scale that measures the amount of vibration people remember feeling for low-magnitude earthquakes and measures the amount of damage to buildings in high-magnitude events

Mercury the closest planet to the Sun, with a mass of only 5.5 percent of the Earth's, a diameter of 3,031 miles (4,878 km), and an average density of 5.4 grams per cubic centimeter

mesosphere region of the atmosphere that lies above the stratosphere, extending between 31 and 53 miles (85 km)

Mesozoic the fourth of five main geologic eras. The Mesozoic falls between the Paleozoic and Cenozoic, and includes the Triassic, Jurassic, and Cretaceous Periods. The era begins at 248 million years ago at the end of the Permian-Triassic extinction event and continues to 66.4 million years ago at the Cretaceous-Tertiary (K-T) extinction event.

metamorphic aureole a thin band of metamorphic rocks that are transposed from their initial state into their metamorphic equivalents in response to heating by an adjacent geologic body. Contact aureoles form next to igneous intrusions, and dynamothermal aureoles form beneath hot thrust sheets such as ophiolitic complexes.

metamorphism changes in the texture and mineralogy of a rock due to changes in the pressure, temperature, and composition of fluids, typically from plate margin processes

metasomatic the group of metamorphic processes responsible for changing a rock's composition or mineralogy by the gradual replacement of one component by another through the movement and reaction of fluids and gases in the pore spaces of a rock

meteor a streak of light in the sky formed by the burning of a meteorite or comet as it moves through the Earth's atmosphere

meteoric Water that has recently come from the Earth's atmosphere is called meteoric water. The term is usually used in studies of groundwater, to distinguish water that has resided in ground for extended periods of time versus water that has recently infiltrated the system from rain, snow melt, or stream infiltration.

meteorite rocky or metallic body that has fallen to Earth from space

meteorology the study of the Earth's atmosphere, along with its movements, energy, and interactions with other systems, and weather forecasting

microplate Most tectonic plates are large, typically thousands of miles (km) across. In some situations, such as during rifting and continental collision, relatively small areas of the continental lithosphere break up into many much smaller plates, typically tens to hundreds of miles across, known as microplates.

mid-ocean ridge system a 40,000-mile- (65,000-km-) long mountain ridge that runs through all the major oceans on the planet. The mid-ocean ridge system includes vast outpourings of young lava on the ocean floor and represents places where new oceanic crust is being generated by plate tectonics.

migmatite a metamorphic rock that contains layers or lenses of partially melted country rock interlayered or mixed with the original rock, which is typically strongly deformed

Milankovitch cycles variations in the Earth's climate that are caused by variations in the amount of incoming solar energy, induced by changes in the Earth's orbital parameters including tilt, eccentricity, and wobble

mineral any naturally occurring inorganic solid that has a characteristic chemical composition, a highly ordered atomic structure, and specific physical properties

mineralogy the branch of geology that deals with the classification and properties of minerals

Mohorovicic discontinuity the transition from crust to mantle, generally marked by a dramatic increase in the velocity of compressional seismic waves, from typical values of less than 4.7 miles per second to values greater than 4.85 miles per second (7.6–7.8 km/s)

molasse nonmarine, irregularly stratified conglomerate, sandstone, shale, and coal deposited during the late stages of mountain building

molecular attraction the force that makes thin films of water stick to things, instead of being forced to the ground by gravity

monsoon a wind system that influences large regions and has seasonally persistent patterns with pronounced changes from wet to dry seasons

monzonite intrusive igneous rock of intermediate composition, including roughly equal amounts of plagioclase and orthoclase feldspars, with minor hornblende, biotite, and other minerals. A rock with more than 10 percent quartz is classified as a quartz monzonite.

moraine ridgelike accumulations of glacial drift deposited at the edges of a glacier. Terminal moraines mark the farthest point of travel of a glacier, whereas lateral moraines form along the edges of a glacier.

morphodynamics the study of the effects of local coastal features on tsunami characteristics

mudflow a downslope flow that resembles a debris flow, except it has a higher concentration of water (up to 30 percent), which makes it more fluid, with a consistency ranging from soup to wet concrete. A mudflow often starts as a muddy stream in a dry mountain canyon, and as it moves it picks up more and more mud and sand, until eventually the front of the stream is a wall of moving mud and rock.

mylonite a strongly deformed rock that forms in ductile shear zones under dynamic recrystallization, forming a strongly foliated and lineated rock that is finer in grain size than the original rock

nappe a large sheetlike body of rock that has been moved by tectonic transport, such as thrusting, significantly far from its place of origin

neap tides tides that occur during the first and third quarters of the Moon and are characterized by lower than average tidal ranges

nebula an interstellar cloud of dust, gas, and plasma

Neogene the second of three periods of the Cenozoic, including the Paleogene, Neogene, and Quaternary, and the second of two subperiods of the Tertiary, younger than Paleogene. Its base is at 23.8 million years ago, and its top is at 1.8 million years ago, followed by the Quaternary period.

Neolithic in archeology, a term for the last division of the Stone Age, during which time humans developed agriculture and domesticated animals.

The transition from hunter-gatherer and nomadic types of existence to the development of farming took place about 10,000–8,000 years ago in the Middle East.

Neptune the eighth and farthest planet from the center of the solar system. Neptune orbits the Sun at a distance of 2.5 billion miles (4.1 billion km, or 30.1 astronomical units), completing each circuit every 165 years, and rotating about its axis every 18 hours. Neptune has a diameter of 31,400 miles (50,530 km) and a mass of more than 17.21 times that of Earth. Its density of 1.7 grams per cubic centimeter shows that the planet has a dense rocky core surrounded by metallic, molecular, and gaseous hydrogen, helium, and methane, giving the planet its blue color.

neutrino elementary particle that moves at nearly the speed of light, has no electrical charge, and passes through matter nearly undisturbed

neutron degeneracy pressure a strong force generated in collapsing giant stars by the repulsive forces between neutrons when they are forced closely together

nova a general name for a type of star that vastly increases in brightness (by up to 10,000 times) over very short periods of time, typically days or weeks

nucleosynthesis the production of heavier elements by the nuclear fusion of lighter elements in stars, novas, and supernovas

nucleus in the astronomical sense, the inner part of a comet, consisting of rock and ice. Also, the inner part of an atom consisting of protons and neutrons

nuée ardente a fast-moving glowing hot cloud of ash that can move down the flanks of volcanoes at hundreds of miles (km) per hour during eruptions. Nuée ardentes have been responsible for tens of thousands of deaths during eruptions.

nunatak an isolated mountain peak protruding from the top of a continental ice sheet or large mountain glacier

ocean basin submarine topographic depressions underlain by oceanic (simatic) crust

ocean currents the movement paths of water in regular courses, driven by the wind and thermohaline forces across the ocean basins

oceanic plateau regions of anomalously thick oceanic crust and topographically high seafloor. Many have oceanic crust that is 12.5–25 miles (20–40 km) thick and rise thousands of meters above surrounding oceanic crust of normal thickness.

oceanography the study of the physical, chemical, biological, and geological aspects of the ocean basins

olistostromes chaotic sedimentary deposits consisting of blocks of one rock type mixed with another. Most olistostromes are thought to be formed by the downslope movement of sedimentary packages.

ontogeny the development of an individual organism from young to old age, as from a tadpole to a frog

oolite a sedimentary deposit consisting of many sand-sized carbonate grains with an onion skin–like texture that form by rolling on the seafloor in shallow agitated waters, continuously being precipitated around a hard nucleus. Oolites form in waters that are saturated with calcium carbonate, typically in areas of high evaporation and agitation, which releases carbon dioxide.

Oort cloud a roughly spherical region containing many comets and other objects, extending from about 60 astronomical units beyond 50,000 AU, or about 1,000 times the distance from the Sun to Pluto, or about a light-year

ophiolite a group of mafic and ultramafic rocks including pillow lavas and plutonic and mantle rocks, generally interpreted to represent pieces of oceanic crust and lithosphere

Ordovician the second period of the Paleozoic Era, falling between the Cambrian and the Silurian, commonly referred to as the age of marine invertebrates. The base of the Ordovician is defined on the Geological Society of America time scale (1999) as 505 million years ago, and the top or end of the Ordovician is defined at 438 million years ago.

orogenic belts linear chains of mountains, largely on the continents, that contain highly deformed, contorted rocks that represent places where lithospheric plates have collided or slid past one another

orogeny the process of building mountains

orographic effect the phenomenon occurring when clouds move over a mountain range and cool, which decreases the capacity of the air to hold water, resulting in precipitation falling on the windward side of the range. As the air mass moves down the leeward side of the mountain, it warms and is able to hold more moisture than is present, so the leeward sides of mountains tend to be dry.

ostracods shrimplike fossils belonging to the class Crustacea

outwash plain a broad plain in front of a melting glacier, where glacial streams deposit gravels and sand

overland flow the movement of runoff in broad sheets

oxbow lake an elongate and curved lake formed by an abandoned meandering stream channel on a floodplain

oxidize to remove one or more electrons from an atom, molecule, or ion, or at least to increase the proportion of the electronegative part of a compound. Weathering can be an oxidizing process, for instance, the oxidation of carbon to yield carbon dioxide.

ozone hole Ozone (O_3) is a poisonous gas that is present in trace amounts in much of the atmosphere, but reaches a maximum concentration in a stratospheric layer between 9 and 25 miles (15–40 km) above the Earth, with a peak at 15.5 miles (25 km). The presence of ozone in the stratosphere is essential for most life on Earth, since it absorbs the most carcinogenic part of the solar spectrum. Since the middle 1980s, a large hole marked by large depletions of ozone in the stratosphere has been observed above Antarctica every spring, its growth aided by the polar vortex circulation. The hole has continued to grow, but the relative contributions to the destruction of ozone by CFCs, other chemicals (such as supersonic jet and space shuttle fuel), volcanic gases, and natural fluctuations is uncertain.

paired metamorphic belt the name for a type of metamorphic pattern characteristic of convergent margin arcs, where rocks near the trench experience high-pressure/low temperature metamorphism, and the rocks near the magmatic arc experience high-temperature/low-pressure metamorphism

paleoclimatology the study of past and ancient climates, their distribution and variation in space and time, and the mechanisms of long term climate variations

Paleolithic the first division of the Stone Age in archeological time, marked by the first appearance of humans and their associated tools and workings. The time of the Paleolithic corresponds generally with the Pleistocene (from 1.8 million years ago until 10,000 years ago) of the geological time scale but varies somewhat from place to place.

paleomagnetism the study of the record of Earth's magnetic field as preserved in lava flows, volcanic rocks, and sedimentary rocks, and the use of these data to place constraints on plate tectonic motions

paleontology the study of past life based on fossil evidence, with a focus on the lines of descent of organisms and the relationships between life and other geologic phenomena

paleoseismicity the study of past earthquakes. Most paleoseismicity studies rely on techniques such as digging trenches across faults to determine the recurrence intervals of fault segments. Some paleoseismicity studies combine archeology with geology and look for ruins of ancient civilizations that show signs of earthquake damage, then use isotopic or historical dating methods to determine the time of the ancient destructive earthquake

Paleozoic the era of geologic time between 544 and 250 million years ago, including seven geological periods and systems of rocks: the Cambrian, Ordovician, Silurian, Devonian, Carboniferous (Mississippian and Pennsylvanian), and Permian

Pangaea a supercontinent that formed in the Late Paleozoic and held together from 300 to 200 million years ago. Pangaea contained most of the planet's continental land masses.

parallax the apparent displacement or difference in orientation of an object when viewed along two different lines of sight

parsec a distance that a star must lie so that its parallax is equal to 1 arc second, equivalent to 206,000 AU (one astronomical unit equals 1.496 × 10^8 km), equivalent to 1.9 × 10^{13} miles, 3.09 × 10^{13} km, or 3.3 light years. A kiloparsec equals one thousand parsecs, a megaparsec equals one million parsecs, and a gigaparsec equals a billion parsecs, or 3.262 billion light-years, which is about one-fourteenth of the distance to the "horizon" of the presently observable universe.

partial melting the process in which a rock melts by only a small percentage, leaving mostly hot solid rock with a few (up to 10 or 20) percent liquid. These melts may stay in place, or move away to form larger magma bodies

passive continental margin a boundary between continental and oceanic crust that is not a plate boundary, characterized by thick deposits of sedimentary rocks. These margins typically have a flat, shallow-water shelf, then a steep drop off to deep ocean–floor rocks away from the continent.

pediments desert surfaces that slope away from the base of a highland and are covered by a thin or discontinuous layer of alluvium and rock fragments

pegmatite a coarse-grained igneous rock, typically composed of quartz, feldspar, mica, and other accessory minerals, sometimes gemstones. Most pegmatites are late-stage fluids associated with granitic plutons.

pelagic relating to the open sea, not along the shore or the bottom. Pelagic organisms are free-floating and do not need to be attached to the bottom or shoreline

pelite a clastic sedimentary rock with a grain size of less than 1/16 mm, including originally sandy or silty rocks

peneplain a nearly flat erosional plain produced by fluvial erosion, often recognized as a flat surface in uplifted mountain ranges

penumbra   part of a shadow where the occulting body obscures only part of the light source

perched water table   domed pockets of water at elevations higher than the main water table, resting on top of the impermeable layer

peridotite   a common rock of the mantle of the Earth. The average upper mantle composition is equivalent to peridotite. The density of peridotite is 3.3 g/cm³; its mineralogy includes olivine, clinopyroxene, and orthopyroxene.

perihelion   the point in an orbit closest to the Sun

permafrost   permanently frozen subsoil that occurs in polar regions and at some high altitudes

permeability   a body's capacity to transmit fluids or to allow the fluids to move through its open pore spaces

Permian   the last period in the Paleozoic era, lasting from 290–248 million years ago

petrography   the science that describes the minerals and textures in rock bodies

petroleum   a mixture of different types of hydrocarbons (fossil fuels) derived from the decomposed remains of plants and animals that are trapped in sediment, and which can be used as fuel

petrology   the branch of geology that attempts to describe and understand the origin, occurrence, structure, and evolution of rocks

Phanerozoic   the eon of geological time since the base of the Cambrian at 544 million years ago and extending to the present

phonon   a quantized mode of vibration of a crystal lattice

photic zone   zone through which light penetrates in the oceans

photoelectric effect   a quantum electronic phenomenon where electrons are emitted from an object after it absorbs energy from electromagnetic radiation such as visible light or X-rays

photosphere   the visible surface of the Sun

photosynthesis   the process carried out in green plants, algae, and some bacteria that involves trapping solar energy and using it to drive a series of chemical reactions that result in the production of carbohydrates such as glucose or sugar

phylogeny   the evolutionary history of an organism

pillow lava   a form of lava flow with bulblike or pillowlike shapes, generally basaltic in composition, that forms beneath water and is common on the ocean floor

placer   a deposit, typically found in streams, that contains high concentrations of gold (or another valuable mineral) in flakes or nuggets

plane strain   a type of strain or deformation of a rock in which the rock is shortened in one direction, elongated in a perpendicular direction, and has no change in the third perpendicular direction

planet   a celestial body that is in orbit around the Sun, has sufficient mass for its self-gravity to overcome rigid body forces so that it assumes a hydrostatic equilibrium (nearly round) shape, and has cleared the neighborhood around its orbit

planetary nebula   a type of emission nebula consisting of a glowing shell of gas and plasma surrounding a central core of a star that went through its hydrogen- and helium-burning stages, ejecting the outer layers of the dying giant star in one of the last stages of its evolution

planetesimal   a protoplanet growing in the solar nebula

plasma   a partially ionized gas in which some electrons are free rather than being bound to atoms or molecules

plate tectonics   a model that describes the process related to the slow motions of more than a dozen rigid plates of solid rock around on the surface of the Earth. The plates ride on a deeper layer of partially molten material that is found at depths starting at 60–200 miles (100–320 km) beneath the surface of the continents, and 1–100 miles (1–160 km) beneath the oceans

playa   a dry lake bed in a desert environment

Pleistocene   the older of two epochs of the Quaternary Period, lasting from 1.8 million years ago until 10,000 years ago, at the beginning of the Holocene Epoch

Plinian   a type of volcanic eruption characterized by a large and tall eruption column, typically reaching tens of thousands of feet (thousands of meters) into the air. Named after Pliny the Elder, from his description of Vesuvius

plume (radio galaxy)   a relatively narrow trail of high-energy radiation emitted from a radio galaxy

Pluto   The solar system has long been considered to have nine planets, with Pluto as the most-distant planet. In 2006 the International Astronomical Union met, decided that Pluto does not meet the formal criteria of being a planet, and demoted the status of the object to that of a "dwarf planet." Pluto is very small compared to all other planets, orbits far from the plane of the ecliptic, and resides in the Kuiper Belt, along with objects that are about the same order of magnitude in size and mass as Pluto. Pluto is thought to be composed of 98 percent nitrogen ice, along with methane and carbon monoxide, similar to other objects, including comets, in the Kuiper belt and in the scattered disk that overlaps with the outer edge of the Kuiper belt. If Pluto were to orbit close to the Sun, it would develop a long cometary tail, and would be classified as a comet, not a planet.

plutonic an igneous rock that has crystallized below the surface, typically characterized by coarse grain size and forming large intrusive bodies known as plutons

pluvial an extended period of abundant rainfall, particularly as applied to the Pleistocene epoch

podiform chromitite a concentration of chromite in a lens or pod, in an envelope or dike of dunite, in deformed mantle harzburgite tectonite. Podiform chromitites are typically found in ophiolites.

point bar deposits of sand and gravel along the inner bends of meandering streams

point defect a crystal defect in which something is wrong with the lattice at one point, such as an extra atom in one place, a missing atom at one place, or one atom substituting for another at one place

pole of rotation When a plate moves on the globe, its motion can be uniquely described by a rotation about a pole that goes through the center of the Earth, and exits at two places on the globe, known as poles of rotation.

polymorphs solids that have the same chemical formula but have different crystal lattice forms. Examples are carbon and diamond.

pore pressure In piles of sediments, the weight of the overlying material places the fluids under significant pore pressure, known as hydrostatic pressure, which keeps the pressure between individual grains in the regolith at a minimum. This in turn helps prevent the grains from becoming closely packed or compacted.

porosity the percentage of total volume of a body that consists of open spaces

positron the antiparticle, or antimatter counterpart, to the electron, containing an electric charge of +1 and the same mass as an electron. When a positron collides with an electron they annihilate each other, emitting two gamma ray photons.

Precambrian the oldest broad-grouping of geological time, stretching from the formation of the Earth at 4.5 billion years ago, including the Archean and Proterozoic eons, and ending at 540 million years ago

precipitation water that falls to the surface from the atmosphere in liquid, solid, or fluid form

prograding barriers barrier islands that are building themselves seaward with time, generally through a large sediment supply

prokaryote a single-celled organism lacking a cell nucleus, such as bacteria

Proterozoic the second eon of geological time, stretching from the end of the Archean eon at 2.5 billion years ago until the start of the Paleozoic at 540 million years ago

protoplanet a large body in the solar nebula that is in the process of growing by accretion into a planet

protosphere a surface on a star below which the material is opaque to the radiation it emits

pseudotachylytes glassy melt rocks that form along faults and from impact craters

pull-apart basin a rift-type basin formed along a bend in a strike-slip fault system or in an extensional area between two close and related strike-slip faults

pulsar strongly magnetized neutron stars that rotate with a regular period between a fraction of a second and 8.5 seconds, emitting a narrow beam of radio waves

pyroclastic a general term for rocks and material that are thrown from a volcano, including the explosive ash, bombs, and parts of the volcano ripped off the slopes during eruptions

quartzite a metamorphosed clean sandstone, in which fractures break through the individual grains as opposed to around them

quasar extremely powerful and distant galactic nucleus

Quaternary the last 1.8 million years of Earth history are known as the Quaternary period, divided into the older Pleistocene and the younger Holocene

radiation energy in the form of waves

radiative forcing the net change in downward minus the upward irradiance at the tropopause, caused by a change in an external driver such as a change in greenhouse gas concentration

radioactive decay the process whereby nuclei of unstable radioactive elements spontaneously break down to become more stable, emitting radiation as alpha particles, beta particles, or gamma rays

radio galaxies types of active galaxies that emit large amounts of electromagnetic radiation as radio waves

radiolarians small amoeboid protozoa that occur as free floating zooplankton in the open ocean, producing intricate exoskeletons that sink to form radiolarian ooze deposits on the sea floor. Many radiolarians make distinctive index fossils, so have been used to date marine stratigraphic units for much of the Phanerozoic.

radio waves electromagnetic waves with the longest wavelengths in the spectrum, between 1 mm and 100,000 km, commonly used to transport information through the atmosphere and space without wires

radon a heavy, poisonous gas produced as a product of radioactive decay in the uranium decay series. Radon presents a serious indoor health hazard in every part of the country.

rain shadow the area on the leeward side of a mountain where the air is descending, as a dry air mass, causing little rain to fall

rapakivi granite a hornblende-biotite granite with large rounded crystals of orthoclase feldspar surrounded by rims of oligoclase feldspar. Most rapakivi granites intruded orogens and cratons after the major deformation events, and are considered to be anorogenic.

recharge areas places where water enters the groundwater system

recumbent fold a flat-lying fold, with a subhorizontal axial plane and hinge

recurrence interval the average repeat time for earthquakes along a specific segment of a fault, based on the statistics of how frequently earthquakes of specific magnitude occur along individual segments of faults

recurved spit a sand spit that has ridges of sand that curve around the end of the spit that terminates in the sea, reflecting its growth

redshift a displacement of electromagnetic radiation emitted from a celestial body toward the lower energy red end of the visible spectrum, associated with an increase in the wavelength of the radiation since it was emitted. Redshifts are associated with objects moving away from the observer.

reef wave-resistant, framework-supported carbonate or organic mounds generally built by carbonate-secreting organisms; or, any shallow ridge of rock lying near the surface of the water

refraction the process of bending a wave front around an object, typically caused by friction on the base of the wave slowing its progress as it encounters shallow water, while parts of the wave still in deeper water continue to move quickly

regolith the outer surface layer of the Earth, consisting of a mixture of soil, organic material, and partially weathered bedrock

regression retreat of the sea from the shoreline, caused by eustatic sea-level fall or local effects

remote sensing the acquisition of information about an object by recording devices that are not in physical contact with the object. Different types include airborne or space-based techniques and sensors that measure different properties of earth materials, ground-based sensors that measure properties of distant objects, and techniques that penetrate the ground to map subsurface properties. Common use of the term *remote sensing* refers to the airborne and space-based observation systems, with ground-based systems more commonly referred to as geophysical techniques.

residence time average length of time that a substance remains in a geochemical system

resonances stable orbits where asteroids can remain for long periods of time, produced by an effect of the gravity and different orbital periods of the larger planets in the inner solar system. In physics, and in atmospheric and ocean dynamics, resonance can mean the tendency of a system to oscillate at a maximum amplitude at specific frequencies, known as the resonance frequencies for that system.

resonate oscillate back and forth. Many bodies have a specific rate or frequency at which they naturally resonate when excited with energy, known as a natural resonance. If a wave such as a tsunami excites the natural resonance of a harbor, then the height of a wave can be dramatically increased.

retrograding barriers barrier islands that are moving onshore with time

rhyolite an extrusive igneous rock that is the volcanic equivalent of granite, containing about 69 percent silicon dioxide (SiO_2), and typically containing the minerals quartz, alkali feldspar, and plagioclase

Richter scale an open-ended scale that gives an idea of the amount of energy released during an earthquake and is based on the amplitudes (half the height from wave-base to wave-crest) of seismic waves at a distance of 61 miles (100 km) from the epicenter. The Richter scale is logarithmic, where each step of 1 corresponds to a tenfold increase in amplitude. Because larger earthquakes produce more waves, an increase of 1 on the Richter scale corresponds to a 30-fold increase in energy released. Named after Frank Richter, an American seismologist

ridge and runnel the most seaward part of the beach, characterized by a small sandbar called a ridge, and a flat-bottom trough called the runnel, typically less than 30 feet (10 m) wide

rifts elongate topographic depressions, typically with faults along their margins, where the entire thickness of the lithosphere has ruptured in extension. These are places where the continents are beginning to break apart, and if successful, may form new ocean basins

rip current a strong current that moves perpendicularly away from the shore, typically localized by the presence of a jetty, sea floor topography, or other obstacle. Rip currents are dangerous as they can carry swimmers far out to sea.

river system the entire fluvial system, including all the tributaries of a river, its floodplain, water, and entrained and suspended sediment carried by the river

rockfall free falling of detached bodies of bedrock from a cliff or steep slope

rockslide the sudden downslope movement of newly detached masses of bedrock (or debris slides, if the rocks are mixed with other material or regolith)

Rossby waves dips and bends in the jet stream path

runoff water that fell as rain or snow and flows across the surface into streams or rivers, instead of being absorbed into the groundwater system

run-up the height of the tsunami above sea level at the farthest point it reaches on the shore

Saffir-Simpson scale a hurricane intensity scale that measures the damage potential of a storm, considering such factors as the central barometric pressure, maximum sustained wind speeds, and the potential height of the storm surge

saltation the movement of a particle by short intermittent jumps, such as caused by a water current lifting the particles

salt marshes coastal wetlands that form on the upper part of the intertidal zone where organic rich sediments are rarely disturbed by tides, providing a stable environment for grasses to take root. The low marsh area is defined as the part of the marsh that ranges from the beginning of vegetation to the least mean high tide. The high marsh extends from the mean high tide up to the limit of tidal influence.

sapping a process in which groundwater moves along the gravel and sand layers, seeping out along the river bank. This movement of groundwater can carry sediment away from the bank into the stream.

Saturn the sixth planet from the Sun, residing between Jupiter and Uranus, orbiting at 9.54 astronomical units (888 million miles, or 1,430 million kilometers) from the Sun, twice the distance from the center of the solar system as Jupiter, and with an orbital period of 29.5 Earth years. The mass of Saturn is 95 times that of Earth, yet it rotates at more than twice the rate of Earth.

schist medium-grade metamorphic rock that contains more than 50 percent platy and elongate minerals, aligned to produce a shiny luster. Most schists contain quartz, micas, and other accessory minerals such as garnet.

Schwarzschild radius a characteristic radius for a given mass at which, if the mass were compressed to fit inside that radius, no known force could stop the mass from collapsing to a singularity. It is also the radius for that mass at which no matter or energy could escape if the entire mass were compressed inside that radius.

seafloor spreading the process of producing new oceanic crust, as volcanic basalt pours out of the depths of the Earth, filling the gaps generated by diverging plates. Beneath the mid-oceanic ridges, magma rises from depth in the mantle and forms chambers filled with magma just below the crest of the ridges. The magma in these chambers erupts out through cracks in the roofs of the chambers, and forms extensive lava flows on the surface. As the two different plates on either side of the magma chamber move apart, these lava flows continuously fill in the gap between the diverging plates, creating new oceanic crust.

sea ice ice that has broken off an ice cap, from polar sea ice, or calved off a glacier and is floating in open water

sea-level rise the gradual increase in average height of the mean water mark with respect to the land

seasons variations in the average weather at different times of the year

sea stacks isolated columns of rock left by retreating cliffs, with the most famous being the Twelve Apostles along the southern coast of Australia

seawater intrusion the encroachment of seawater into drinking and irrigation wells, generally caused by overpumping of water from groundwater wells along the coast

sedimentary rock rocks that have consolidated from accumulations of loose sediment produced by physical, chemical, or biological processes

sedimentary structures the organized arrangement of sedimentary particles that form repeating patterns reflecting their origin. Types of sedimentary structures include sedimentary layers; ripples produced by currents moving the sedimentary particles as sets of small waves; megaripples, which are large ripples formed by unusually strong currents; mudcracks, produced by muddy sediments being dried by the Sun and shrinking and cracking; and other structures produced by organisms, such as burrows from worms, bivalves, and other organisms, trails, and footprints.

sediment flow a moving mass or deposit produced by mass-wasting processes that involve flows that are transitional between grain flows and stream-type flows in the relative amounts of sediment and water, and in velocity. Many different types of sediment flows include slurry flows, mudflows, debris flows, debris avalanches, earthflow, and loess flow.

seiche waves waves generated by the back-and-forth motion associated with earthquakes, causing a body of water (usually lakes or bays) to rock back and forth, gaining amplitude and splashing up to higher levels than normally associated with that body of water

seismic gaps places along large fault zones that have little or no seismic activity compared to adjacent parts of the same fault. Seismic gaps are generally interpreted as places where the fault zone is stuck, and where adjacent parts of the fault are gradually slipping along, slowly releasing seismic energy. Since the areas of the seismic gaps are not slipping, the energy gradually builds up in these sections, until it is released in a relatively large earthquake.

seismic reflection profile a geophysical cross section of part of the upper layers of the Earth produced by measuring the time it takes for a seismic wave (often artificially produced) to travel from the surface, bounce off a layer at depth, and travel back to the surface

seismic sea wave warning system a tsunami warning system in the Pacific Ocean that operates by monitoring seismograms to detect potentially seismogenic earthquakes, then monitors tide gauges to determine if a tsunami has been generated

seismograph a device built to measure the amount and direction of shaking associated with earthquakes

seismology the study of the propagation of seismic waves through the Earth, including analysis of earthquake sources, mechanisms, and the determination of the structure of the Earth through variations in the properties of seismic waves

sensitivity in soil, a measure of how the strength changes with shaking or with other disturbances such as those associated with excavation or construction

sequence stratigraphy the study of the large-scale three-dimensional arrangement of sedimentary strata and the factors that influence the geometry of these sedimentary packages. Sequences are groups of strata bounded above and below by identifiable surfaces that are at least partly unconformities.

shatter cone a group of fractures in a rock that forms a cone shape, and points towards the source of an explosion, such as a meteorite impact

shear waves *See* body waves.

sheeted dikes igneous dikes where each successive dike intrudes up the middle of the previously intruded dike, indicating intrusion in a region of extension, such as along a mid-ocean ridge

shrink/swell potential in soil, a measure of a soil's ability to add or lose water at a molecular level

Silurian the third period of Paleozoic time ranging from 443 Ma to 415 Ma, falling between the Ordovician and Devonian Periods. From base to top it is divided into the Llandoverian and Wenlockian ages or series (comprising the Early Silurian) and the Ludlovian and Pridolian ages or series (comprising the Late Silurian).

sinistral a term used to describe the sense of motion along a fault where the block on the opposite side of the fault moves to the left

sinkhole a large, generally circular depression that is caused by collapse of the surface into an underground open space

sinuosity ratio of the stream length to valley length

slaty cleavage a pervasive parallel foliation defined by fine-grained platy minerals such as chlorite

slickenside a smoothly polished surface marking a fault, typically containing striations or fibers oriented parallel to the movement direction on the fault

slides in mass wasting, when rock, soil, water, and debris move over and in contact with the underlying surface

slump a type of mass wasting where a large mass of rock or sediment moves downward and outward along an upward curving fault surface. Slumps may occur undersea or on the land surface.

slurry flow a moving mass of sediment saturated in water that is transported with the flowing mass. The mixture is so dense that it can suspend large boulders or roll them along the base.

SNC meteorites a class of meteorites that includes shergottites, nakhlites, and chassignites, believed to have originated from Mars and the Moon, being ejected by impact events on those planets, and landing on Earth

snowball Earth a time in Earth history when nearly all the water in the planet was frozen and glaciers existed at low latitudes

soil all the unconsolidated material resting above bedrock and serving as the natural medium for plant growth

soil profile a succession of distinctive horizons in a soil from the surface downward to unaltered bedrock

solar constant the amount of solar energy that reaches the Earth per unit time

solar luminosity a measure of the total energy emitted by the Sun, defined as the total energy reaching an imaginary sphere with a radius of 1 astronomical unit (AU, the distance from the Sun to the Earth)

solar mass the mass of the Sun, equivalent to 332,946 Earth masses, or 1.98892×10^{30} kg, commonly used as a standard reference to astronomical mass units

solar nebula the cloud of gases and solids in the early solar system from which the Sun and planets condensed and accreted

solar prominence a large bright feature that extends from the solar surface, often forming an arc-like shape

solar system The Earth's solar system represents the remnants of a solar nebula that formed in one of the spiral arms of the Milky Way Galaxy. After the condensation of the nebula, the solar system consisted of eight major planets, the moons of these planets, and many smaller bodies in interplanetary space. From the Sun outward, these planetary bodies include Mercury, Venus, Earth, Mars, the asteroids, Jupiter, Saturn, Uranus, and Neptune.

solifluction the slow viscous downslope movement of water-logged soil and debris. Solifluction is most common in polar latitudes where the top layer of permafrost melts, resulting in a water-saturated mixture resting on a frozen base.

solstice the time, occurring twice a year, when the Earth's axis is oriented the most toward or the most away from the Sun. The summer solstice happens on June 21 or 22, whereas the winter solstice happens every December 21 or December 22.

Spaceguard a term that refers to a number of different efforts to search for and monitor near-Earth objects

spectrum The electromagnetic spectrum is the range of all possible electromagnetic radiation. In astronomy, the word is typically used to refer to the spectrum of an object, such as a star, and in these cases refers to the characteristic distribution of electromagnetic radiation from that object.

spheroidal weathering a weathering process that proceeds along two or more sets of joints in the subsurface, resulting in shells of weathered rock which surround unaltered rocks, looking like boulders

spicules a several hundred mile (~500 km) diameter dynamic jet of material on the solar photosphere that shoots upward at 10–15 miles (~20 km) per second, and exists for only 5–10 minutes before fading away

spit a low tongue or embankment of land, typically consisting of sand and gravel, that terminates in the open water

springs places where groundwater flows out at the ground surface

spring tides tides that occur near the full and new Moons

stalactites pipelike formations that hang from the roof of a cave, formed by dripping water, typically composed of calcium carbonate

stalagmites pipelike formations that grow upward from the floor of caves, formed by accumulation of calcium carbonate from dripping water

standard model a model for the origin of the universe, stating that it formed 14 billion years ago in the big bang. This model suggests that galaxies and stars form 4.8 percent of the universe, 22.7 percent of the universe consists of dark matter, and 72.5 percent of the universe is nonmatter.

star a massive luminous ball of plasma in space

stony meteorites meteorites made of material typical of igneous rocks, such as minerals of olivine and pyroxene

stoping a mechanism of igneous intrusion whereby big blocks of country rock above a magma body get thermally shattered, drop off the top of the magma chamber, and fall into the chamber, and the magma then moves upward to take the place of the sunken blocks

storm beach thin strips of sand along a shoreline, formed typically in winter by strong winter storms that move sediment offshore from the beach

storm surge water that is pushed ahead of a storm and typically moves on land as an exceptionally high tide in front of a severe ocean storm such as a hurricane

stratigraphy the study of rock strata or layers, especially with concern for their succession, age relationships, lithologic composition, geometry, distribution, correlation, fossil content, and other aspects of the strata

stratosphere region of the atmosphere above the troposphere that continues to a height of about 31 miles (50 km)

stream capture an event that occurs when headland erosion diverts one stream and its drainage into another drainage basin

stream flow the flow of surface water in well-defined channels

strike-slip fault a vertical or nearly vertical fault that has horizontal or nearly horizontal motion along the fault

stromatolite a layered accretionary structure, formed by cyanobacteria, that traps, binds, and cements sedimentary grains, causing the structure to grow progressively outward from a starting nucleus

strong nuclear force one of the four basic forces in nature, responsible for holding the nucleus of atoms together, representing the interactions between quarks and gluons

structural geology the study of the deformation of the Earth's crust or lithosphere

subduction the destruction of oceanic crust and lithosphere by sinking back into the mantle at the deep ocean trenches. As the oceanic slabs go down, they experience higher temperatures that cause rock-melts or magmas to be generated, which then move upward to intrude the overlying plate. Since subduction zones are long narrow zones where large plates are being subducted into the mantle, the melting produces a long line of volcanoes above the down-going plate and forms a volcanic arc. Depending on what the overriding plate is made of, this arc may be built on either a continental or on an oceanic plate.

subduction erosion a process occurring at convergent margins in which material is eroded and scraped off the overriding plate and dragged down into the subduction zone

subduction zones long, narrow zones where large oceanic plates are sliding into the mantle. The down-going oceanic plates are associated with a zone of partial melting at about 60 miles (100 km) depth, generating magmas that rise to the surface, forming a long line of volcanoes above the down-going plate, and producing a volcanic arc. Depending on what the overriding plate is made of, this arc may be built on either a continental or on an oceanic plate.

subsequent stream a stream whose course has become adjusted so that it occupies a belt of weak rock or another geologic structure

subsidence sinking of one surface, such as the land, relative to another surface, such as sea level

Sun an average star, consisting of a glowing ball of gas held together by its own gravity and powered by nuclear fusion in its core, that sits 8 light minutes away from Earth

sun halos, sundogs, and sun pillars unusual optical phenomena related to the interaction of the Sun's rays with ice crystals in the upper atmosphere

supercontinent one large continent formed when plate tectonics and continental drift bring many or most of the planet's land masses into continuity. Examples of supercontinents include Pangaea that existed from about 300–200 million years ago and Gondwana that formed about 600 million years ago.

supercontinent cycle the semiregular grouping of the planet's landmasses into a single or several large continents that remain stable for a long period of time, then disperse, and eventually come back together as new amalgamated landmasses with a different distribution

supernova extremely luminous stellar explosion associated with large bursts of radiation that can exceed brightness of the entire galaxy it is associated with for a period of weeks or months

superposed streams streams whose courses were laid down in overlying strata onto unlike strata below

surface waves waves that travel along the surface, producing complicated types of twisting and circular motions, much like the circular motions exhibited by waves beyond the surf zone at the beach. Surface waves travel slower than either type of body waves, but they often cause the most damage due to their complicated types of motion.

suspended load the fine particles of silt and clay suspended in the stream that make it muddy and that move at the same or slightly lower velocity as the stream

suture zone a belt of highly deformed rocks in a modern or ancient mountain belt, typically containing fragments of ophiolites, mélanges, and blocks of high-grade metamorphic rock, marking the place where two continents or terranes have collided and an ocean has closed

swales the low areas in between dunes

synchrotron process the acceleration of charged particles to close to the speed of light in a magnetic field generating what is known as synchrotron radiation, governed by relativistic effects described by quantum mechanics

syncline a fold structure in which the associated beds dip inward and the youngest rocks occupy the core of the fold structure

synclinorium a large, regional-scale syncline that may have smaller-scale anticlines and synclines superimposed on the larger-scale structure

talus the entire body of rock waste sloping away from the mountains. The sediment composing talus is known as sliderock. This rock debris accumulates at the bases of mountain slopes, deposited there by rockfalls, rockslides, and other downslope movements

tectosphere the region of the Earth's crust occupied by tectonic plates

telescope any of a wide range of scientific instruments that observe remote objects by collecting electromagnetic radiation from them and enhancing this radiation by different processes in different types of telescopes

tephra a general term for all ash and rock fragments strewn from a volcano

terrace flat abandoned floodplain, sitting above the present floodplain level, that formed when a stream flowed above its present channel and floodplain level

terrane a fault-bounded block of rock that has a geologic history different from that of neighboring rocks, and likely was transported from far away by plate tectonic processes

Tertiary the first period of the Cenozoic era, extending from the end of the Cretaceous of the Mesozoic at 66 million years ago until the beginning of the Quaternary 1.6 million years ago. The Tertiary is divided into two periods, the older Paleogene (66–23.8 Ma) and the younger Neogene (23.8–1.8 Ma), and further divided into five epochs, the Paleocene (66–54.8 Ma), Eocene (54.8–33.7 Ma), Oligocene (38.7–23.8 Ma), Miocene (23.8–5.3 Ma), and Pliocene (5.3–1.6 Ma).

thalweg a line connecting the deepest parts of the channel of a stream or river

thermodynamics the study of the transformation of heat into and from other forms of energy, particularly mechanical, chemical, and electrical energy

thermohaline circulation vertical mixing of seawater driven by density differences caused by variations in temperature and salinity

thermosphere region of the atmosphere above the mesosphere that thins upward and extends to about 311 miles (500 km) above the surface

thixotropic a property of material such as mud that causes it to be fairly rigid when it is held or is still, but when it is shaken or disturbed, it rapidly turns into a fluid

tholeiitic relating to igneous rock dominated by plagioclase, and chemically containing less alkali elements (Na_2O plus K_2O) at similar SiO_2 than alkali basalt

thrust a contractional fault, or a reverse fault generally with shallow dips

thunderstorms convective storm systems that contain lightning and thunder and that form in unstable rising warm and humid air currents

tidal bore a breaking wave that migrates up the bay as the tide floods the bay, formed where the shape of the bay constricts the water flow, causing the wave to grow in height as it moves into smaller and smaller areas

tidal flat a flat area along the coast within the tidal range that is sheltered from waves, dominated by mud, devoid of vegetation, and accumulating sediment

tidal gauge a sensitive pressure meter placed on the seafloor that can accurately measure changes in the height of the sea surface and can be used for the detection of tides, storm surges, and tsunamis

tidal inlet break in barrier island system that allows water, nutrients, organisms, ships, and people easy access and exchange between the high-energy open ocean and the low-energy back-barrier environment consisting of bays, lagoons, tidal marshes, and creeks. Most tidal inlets are within barrier island systems, but others may separate barrier islands from rocky or glacial headlands.

tidal range the range in sea surface height between the high and low tide

tides the periodic rise and fall of the ocean surface, and alternate submersion and exposure of the intertidal zone along coasts

tidewater glaciers glaciers that are partly floating on the ocean, often in steep-walled fjords

till glacial drift that was deposited directly by the ice

tilt an astronomical measure of how much the Earth's rotational axis is inclined relative to the perpendicular to the plane of orbit

tombolo a spit that connects an offshore island with the mainland

tonalite an igneous rock that is similar to granite but is composed of the minerals plagioclase and quartz plus minor dark minerals such as amphibole, pyroxene, or biotite

tornado a rapidly circulating column of air with a central zone of intense low pressure that reaches the ground

trace fossil geological record of biological activity, such as footprints, worm burrows, or other markings preserved as fossils

transform boundaries places where two tectonic plates slide past each other, such as along the San Andreas fault in California, that often experience large earthquakes

transgression advance of the sea on the shore, caused by either a eustatic sea level rise or local effects

translational slide a variation of a slump in which the sliding mass moves not on a curved surface, but moves downslope on a preexisting plane, such as a weak bedding plane or a joint. Translational slides may remain relatively coherent or break into small blocks, forming a debris slide.

transmissivity the ability of a substance to allow water to move through it

transpiration the evaporation of water from the exposed parts of plants

trellis drainage parallel mainstream channels intersected at nearly right angles by tributaries

triple junction places where three plate boundaries meet

trondhjemite a light-colored intrusive igneous rock, consisting of plagioclase and quartz, common in the sheeted dike sections of ophiolites and in Archean greenstone belts

troposphere the lower 36,000 feet (10,972.8 m) of the atmosphere

tsunami a giant harbor or deepwater wave with long wavelengths, initiated by submarine landslides, earthquakes, volcanic eruptions, or another cause, that suddenly displaces large amounts of water. Tsunamis can be much larger than normal waves when they strike the shore, and they can cause great damage and destruction.

tsunami earthquakes a special category of earthquakes that generate tsunamis that are unusually large for the earthquake's magnitude

T-Tauri stage a stage of stellar evolution where a young (< 10 million years old) small star, less than three solar masses, is still undergoing gravitational contraction. This is an intermediate stage in stellar evolution between a protostar and a main sequence star (such as the Sun).

tube worm marine invertebrate belonging to the phylum Annelida. Tube worms are found near hot springs on the ocean floor. Some of these are thermophilic (heat-loving) and all are primitive organisms.

tuff a rock comprised of a cooled and crystallized volcanic ash, characterized by fine grain size and commonly having fine-scale layers and indications of high-temperature flow

turbidite a sedimentary unit characterized by being deposited by a turbidity current. Most turbidites show a sequence of coarse-grained sands at their base and fine silts or mud on their tops, indicting deposition during slowing of the current.

turbidity current subaqueous downslope flow of water-saturated sand and mud, that leaves behind a turbidite deposit of graded sand and shale

turbulent flow in a flowing stream, water that moves in different directions and often forms zones of sideways or short backward flows called eddies. These significantly increase the resistance to flow. In turbulent flows, the resistance is proportional to the square of the flow velocity.

ultramafic relating to dark-colored igneous rocks composed of iron-magnesium minerals, with silica content of less than 45 percent as determined by chemical analysis. Some ultramafic igneous rocks include peridotite and komatiite.

umbra the darkest part of a shadow, where the source of light is completely obscured by the occulting body

unconformity a buried erosion surface that separates two rock masses of different ages. Angular unconformities form when a deformation and tilting event separates the deposition of the two rock groups, nonconformities represent nondeposition, and disconformities form where a young rock sequence is deposited on an older igneous or metamorphic terrane.

underseepage the process of water seeping under a levee or other structure, often resulting in catastrophic failure of the levee

universe all matter and energy everywhere, including the Earth, solar system, galaxies, interstellar matter and space, regarded as a whole

Uranus the seventh planet from the Sun. Uranus is a giant gaseous sphere with a mass 15 times that of the Earth and a diameter four times as large as Earth's (51,100 km). The equatorial plane is circled by a system of rings, some associated with the smaller of the 15 known moons circling the planet. Uranus orbits the Sun at a distance of 19.2 astronomical units (Earth-Sun distance) with a period of 84 Earth years and has a retrograde rotation of 0.69 Earth days. Its density is only 1.2 grams per cubic centimeter.

urban heat island an effect where cities tend to hold heat more than the countryside

urbanization the process of building up and populating a natural habitat or environment, such that the habitat or environment no longer responds to input the way it did before being altered by humans

Vendian the last period of the Proterozoic in the Precambrian, lasting from 650–543 million years ago. The Vendian has a wonderful soft-bodied faunal assemblage preserved in it, but many of these organisms died off before the Phanerozoic era began.

Venus the second planet from the Sun. Venus has a planetary radius of 3,761 miles (6,053 km), or 95 percent of that of Earth's radius, orbits the Sun in a nearly circular path at 0.72 astronomical units (Earth-Sun distance), has a mass equal to 81 percent of Earth's, and has density of 5.2 grams per cubic centimeter.

vicariance biogeography the initially wide distribution of primitive groups that later were broken up by processes such as rifting and divergent plate tectonics, leading to evolution in individual isolated groups

viscosity a measure of the resistance to flow. The more viscous a fluid is, the more resistant it is to flow.

volcanic arc a line of volcanoes that forms above a subducting oceanic plate at a convergent boundary; *See also* magmatic arc.

volcano a mountain or other constructive landform built by a singular or a sequence of volcanic eruptions of molten lava and pyroclastic material from a volcanic vent

wadi a dry stream bed. This term is commonly used in Arabic countries.

water resources any sources of water that are potentially available for human use, including lakes, rivers, rainfall, reservoirs, and the groundwater system

water table the boundary between the saturated and unsaturated zones

wave base the depth at which all motion associated with the passage of the wave stops, and the water beneath this point experiences no effect from the passage of the wave above. In deepwater ocean waves, the particle motion follows roughly circular paths, where particles move approximately in a circle, and return back to their starting position after the wave passes. The amount of circular motion decreases gradually with depth, to a depth that equals one-half of the wavelength, where all motion associated with the wave passage stops.

wave fronts   imaginary lines drawn parallel to the wave crests. A wave moves perpendicular to the wave fronts.

wave height   the vertical distance from the crest to the bottom of the trough of a wave

wavelength   the distance between successive troughs or crests on a wave train

wave period   the time (in seconds) that it takes successive wave crests to pass a point

wave refraction   a phenomenon that occurs when a straight wave front approaches a shoreline obliquely. The part of the wave front that first feels shallow water (with a depth of less than one half of the wavelength, known as the wave base) begins to slow down while the rest of the wave continues at its previous velocity. This causes the wave front to bend, or be refracted.

wave train   waves of a certain character in a series, moving across the ocean or body of water

wave trap   areas such as some bays and other places along some shorelines that amplify the effects of waves coming in from a certain direction, making run-ups higher than average

weathering   the process of mechanical and chemical alteration marked by the interaction of the lithosphere, atmosphere, hydrosphere, and biosphere

welded barriers   barrier islands that have grown completely across a bay and sealed the water inside off from the ocean

Widmanstaetten texture   a criss-cross texture best shown on polished metallic surfaces of iron meteorites, produced by intergrown blades of iron and nickel minerals, with the size of the blades related to the cooling rate of the minerals

windward   the side of a mountain facing the oncoming prevailing winds, typically the wet side of a mountain range

wobble   an astronomical measure of the rotation axis describing a motion much like a top's rapidly spinning and rotating with a wobbling motion, such that the direction of tilt toward or away from the Sun changes, even though the tilt amount stays the same. This wobbling phenomenon is known as precession of the equinoxes.

xenolith   a piece of rock, typically derived from the mantle, caught and preserved in a crystallized magma or in a kimberlite pipe

yardangs   elongate streamlined wind-eroded ridges that resemble an overturned ship's hull sticking out of the water

zero-age main sequence   the time at which the stellar properties of an young star become stable and the star enters a steady period of burning or fusion

APPENDIX III
FURTHER RESOURCES

BOOKS

Abbott, Patrick L. *Natural Disasters*. 3rd ed. Boston: McGraw-Hill, 2002. A college freshman–level book about natural disasters, listing causes and examples.

Abrahams, A. D., and A. J. Parsons. *Geomorphology of Desert Environments*. Norwell, Massachusetts: Kluwer Academic Publishers for Chapman and Hall, 1994. This is a comprehensive textbook, describing the wide range of landforms and processes in desert environments.

Ahrens, C. D. *Meteorology Today, An Introduction to Weather, Climate, and the Environment*. 7th ed. Pacific Grove, Calif.: Brooks/Cole, 2003. An introductory text for freshman college levels on meteorology, weather, and climate.

Alley, W. M., T. E. Reilly, and O. L. Franke. *Sustainability of Ground-Water Resources*. United States Geological Survey Circular 1186, 1999. This book discusses the use and misuse of groundwater, including the effects of contamination and pollution.

Angelo, Joseph A. *Encyclopedia of Space and Astronomy*. New York: Facts On File, 2006. This is a comprehensive, high school–to-college level encyclopedia covering thousands of topics in astronomy.

Armstrong, B. R., and K. Williams. *The Avalanche Book*. Armstrong, Colo.: Fulcrum Publishing, 1992. This is a comprehensive yet readable book about avalanches and their hazards.

Ashworth, William, and Charles E. Little. *Encyclopedia of Environmental Studies*. New Ed. New York: Facts On File, 2001. A comprehensive encyclopedia for high school students covering diverse aspects of the environment.

Ball, Philip. *The Elements: A Very Short Introduction*. New York: Oxford University Press, 2005. A simple introduction to each of the groups of chemical elements.

Birkland, P. W. *Soils and Geomorphology*. New York: Oxford University Press, 1984. This is a general, college-level textbook on soil science and geomorphology.

Blong, Russell J. *Volcanic Hazards, A Sourcebook on the Effects of Eruptions*. New York: Academic Press, 1984. This book discusses the geological hazards associated with volcanic eruptions.

Botkin, D., and E. Keller. *Environmental Science*. Hoboken, N.J.: John Wiley and Sons, 2003. This is an introductory college-level book that discusses many issues of environmental sciences.

Bromley, D. Allan. *A Century of Physics*. New York: Springer, 2002. A tour from the last century of physics growth, impact, and directions. Numerous photos and illustrations.

Chaisson, Eric, and Steve McMillan. *Astronomy Today*. 2nd ed. Upper Saddle River, N.J.: 2007. A college-level textbook on astronomy.

Chapple, Michael. *Schaum's A to Z Physics*. New York: McGraw-Hill, 2003. Defines 650 key concepts with diagrams and graphs, intended for high school students and college freshmen.

Charap, John M. *Explaining the Universe: The New Age of Physics*. Princeton, N.J.: Princeton University Press, 2002. A description of the field of physics at the beginning of the 21st century.

Considine, Glenn D., ed. *Van Nostrand's Scientific Encyclopedia*. 10th ed. 3 vols. New York: John Wiley & Sons, 2008. Comprehensive general reference containing more than 10,000 entries on topics ranging from all scientific disciplines including the life sciences, Earth and atmospheric sciences, physical sciences, medicine, and mathematics, as well as many areas of engineering and technology.

Davis, R., and D. Fitzgerald. *Beaches and Coasts*. Malden, Mass.: Blackwell Publishing, 2004. This is a comprehensive, undergraduate-to-graduate–level text on processes and environments on beaches and coasts.

Dawson, A. G. *Ice Age Earth,* London: Routledge, 1992. This book describes environmental and geological conditions on the Pleistocene Earth during the ice ages.

Dennis, Johnnie T. *The Complete Idiot's Guide to Physics*. Indianapolis, Ind.: Alpha Books, 2003.

A friendly review of high school–level classical physics.

Dewick, Paul M. *Essentials of Organic Chemistry*. Hoboken, N.J.: John Wiley & Sons, 2006. An introductory textbook of organic chemistry.

The Diagram Group. *The Facts On File Physics Handbook*. Rev. ed. New York: Facts On File, 2006. Convenient resource containing a glossary of terms, short biographical profiles of celebrated physicists, a chronology of events and discoveries, and useful charts, tables, and diagrams.

Drew, D. *Karst Processes and Landforms*. New York: MacMillan Education Press, 1985. This is a comprehensive review of the geological conditions that lead to the development of karst terrains.

Elkins-Tanton, Linda T. *Asteroids, Meteorites, and Comets*. New York: Facts On File, 2006. This is a good high school book that covers the characteristics, formation, and evolution of asteroids, meteorites, and comets.

Falk, Dan. *Universe on a T-Shirt: The Quest for the Theory of Everything*. New York: Arcade Publishing, 2002. A story outlining developments in the search for the theory that will unify all four natural forces.

Fleisher, Paul. *Relativity and Quantum Mechanics: Principles of Modern Physics*. Minneapolis, Minn.: Lerner Publications, 2002. An introduction to the concepts of relativity and quantum mechanics, written for middle school students.

Francis, Peter. *Volcanoes, A Planetary Perspective*. Oxford, England: Oxford University Press, 1993. This book discusses the role of volcanoes in planetary processes such as magma budget, lithosphere-asthenosphere interactions, and variations among volcanoes on the planet.

Gillispie, Charles C., ed. *Dictionary of Scientific Biography*. 18 vols. New York: Charles Scribner's Sons, 1970–81. *New Dictionary of Scientific Biography*. 8 additional vols., 2007. More than 5,000 biographies of scientists and mathematicians from around the world.

Gordon, N. D., T. A. McMahon, and B. L. Finlayson. *Stream Hydrology—an Introduction for Ecologists*. New York: John Wiley and Sons, 1992. This is an elementary book for non-specialists on the hydrology and dynamics of streams.

Griffith, W. Thomas. *The Physics of Everyday Phenomena*. 4th ed. Boston: WCB/McGraw-Hill, 2004. A conceptual text for nonscience college students.

Hamblin, Jacob Darwin. *Science in the Early Twentieth Century: An Encyclopedia*. Santa Barbara, Calif.: ABC-CLIO, 2005. Alphabetical entries examining science from 1900–50.

Holton, Gerald James, and Stephen G. Brush. *Physics, the Human Adventure: From Copernicus to Einstein and Beyond*. New Brunswick, N.J.: Rutgers University Press, 2001. Comprehensive introduction intended for nonscience college students. Difficult reading but covers a lot of material.

Intergovernmental Panel on Climate Change. *Climate Change 2007: The Physical Science Basis. Contributions of Working Group I to the Fourth Assessment Report of the Intergovernmental Panel on Climate Change*, edited by S. Solomon, D. Qin, M. Manning, Z. Chen, M. Marquis, K. B. Averyt, M. Tignor, and H. L. Miller. Cambridge: Cambridge University Press, 2007. This is the most comprehensive and up-to-date scientific assessment of past, present, and future climate change.

Intergovernmental Panel on Climate Change. *Climate Change 2007: Impacts, Adaptation, and Vulnerability. Contributions of Working Group II to the Fourth Assessment Report of the Intergovernmental Panel on Climate Change*, edited by M. Parry, O. Canziani, J. Palutikof, P. van der Linden, and C. Hanson. Cambridge: Cambridge University Press, 2007. This is the most comprehensive and up-to-date scientific assessment of the impacts of climate change, the vulnerability of natural and human environments, and the potential for response through adaptation.

Intergovernmental Panel on Climate Change. *Climate Change 2007: Mitigation. Contributions of Working Group III to the Fourth Assessment Report of the Intergovernmental Panel on Climate Change*, edited by B. Metz, O. R. Davidson, P. R. Bosch, R. Dave, and L. A. Meyer. Cambridge: Cambridge University Press, 2007. This is the most comprehensive and up-to-date assessment of mitigation of future climate change.

Interrante, Leonard V., Lawrence A. Casper, and Arthur B. Ellis. *Materials Chemistry: An Emerging Discipline*. Washington, D.C.: American Chemical Society, 1995. A college-level introduction to material science.

James, Ioan. *Remarkable Physicists: From Galileo to Yukawa*. New York: Cambridge University Press, 2004. Contains brief biographies of 50 physicists spanning a period of 250 years, focusing on the lives rather than the science.

Kusky, T. M. *Asteroids and Meteorites: Catastrophic Collisions with Earth, The Hazardous Earth Set*. New York: Facts On File, 2009. A comprehensive account of asteroids, meteorites, and the effects of their collisions with Earth, illustrated with many examples for high school and college students.

———. *Climate Change: Shifting Glaciers, Deserts, and Climate Belts, The Hazardous Earth Set*.

New York: Facts On File, 2008. A comprehensive account of short-, medium-, and long-term climate change and what drives changes in climate, illustrated with many examples for high school and college students.

———. *The Coast: Hazardous Interactions within the Coastal Environment, The Hazardous Earth Set.* New York: Facts On File, 2008. A comprehensive account of hurricanes, coastal erosion, land subsidence, and rising sea levels, illustrated with many examples for high school and college students.

———. *Earthquakes: Plate Tectonics and Earthquake Hazards, The Hazardous Earth Set.* New York: Facts on File, 2008. A comprehensive account of earthquakes and their causes, illustrated with many examples for high school and college students.

———. *Encyclopedia of Earth and Space Science.* 2 vols. New York: Facts On File, 2009. Contains more than 200 entries on topics related to the NSES content standards for grades 9–12, a chronology, glossary, and further resources.

———. *Floods: Hazards of Surface and Groundwater Systems, The Hazardous Earth Set.* New York: Facts On File, 2008. A comprehensive account of the Earth's hydrosphere, including floods, groundwater contamination, and other water issues, illustrated with many examples for high school and college students.

———. *Geologic Hazards, A Sourcebook.* Westport, Conn.: Greenwood Press, 2002.

———. *Landslides: Mass Wasting, Soil, and Mineral Hazards, The Hazardous Earth Set.* New York: Facts On File, 2008. A comprehensive account of landslides and hazards including contaminants in soils, illustrated with many examples for high school and college students.

———. *Tsunamis: Giant Waves from the Sea, The Hazardous Earth Set.* New York: Facts On File, 2008. A comprehensive account of tsunamis, including the 2004 Indian Ocean tsunami, illustrated with many examples for high school and college students.

———. *Volcanoes: Eruptions and Other Volcanic Hazards, The Hazardous Earth Set.* New York: Facts On File, 2008. A comprehensive account of volcanic eruptions and their consequences, illustrated with many examples for high school and college students.

Leiter, Darryl J. *A to Z of Physicists.* New York: Facts On File, 2003. Profiles more than 150 physicists, discussing their research and contributions. Includes bibliography, cross-references, and chronology.

Leopold, L. B. *A View of the River.* Cambridge, Mass.: Harvard University Press, 1994. This is a layman's description of river systems.

Lerner, K. Lee, and Brenda Wilmoth Lerner, eds. *Gale Encyclopedia of Science.* 4th ed. 6 vols. Farmington Hills, Mich.: Gale Group, 2007. Provides an overview of current knowledge in all major areas of science, engineering, technology, mathematics, and the medical and health sciences, consisting of alphabetical entries of scientific concepts and terms.

Longshore, D. *Encyclopedia of Hurricanes, Typhoons, and Cyclones.* New Ed. New York: Facts On File, 1998. A comprehensive encyclopedia of hurricanes written for college and high school audiences and the general public.

Nemeh, Katherine H., ed. *American Men and Women of Science: A Biographical Dictionary of Today's Leaders in Physical, Biological, and Related Sciences.* 25th ed. 8 vols. Farmington Hills, Mich.: Thomson Gale, 2008. Brief profiles of nearly 135,000 living scientists.

Oakes, Elizabeth H. *Encyclopedia of World Scientists.* Rev. ed. 2 vols. New York: Facts On File, 2007. Profiles nearly 1,000 scientists from around the world.

Ritter, D. F., R. C. Kochel, and J. R. Miller. *Process Geomorphology.* 3rd ed. Dubuque, Iowa: W.C. Brown, 1995. This is a comprehensive book describing modern views on geomorphology and river system dynamics.

Rosen, Joe, and Lisa Q. Gothard. *Encyclopedia of Physical Science.* 2 vols. New York: Facts On File, 2009. Contains more than 200 entries on topics related to the NSES content standards for grades 9–12, a chronology, glossary, and further resources.

Trefil, James. *From Atoms to Quarks: An Introduction to the Strange World of Particle Physics.* Rev. ed. New York: Anchor Books, 1994. A primer on this complex subject written for general readers.

United States Environmental Protection Agency and Centers for Disease Control. *A Citizen's Guide to Radon: The Guide to Protecting Yourself and Your Family from Radon.* 2nd ed. EPA 402-K92-001, 1992. A general interest booklet on how to check for and reduce the risk of radon in homes and the environment.

INTERNET RESOURCES

The ABCs of Nuclear Science. Nuclear Science Division, Lawrence Berkeley National Laboratory. Available online. URL: http://www.lbl.gov/abc/. Accessed July 22, 2008. Introduces the basics of nuclear science—nuclear structure, radioactivity, cosmic rays, antimatter, and more.

American Chemical Society. Available online. URL: http://portal.acs.org/portal/acs/corg/content. Accessed July 22, 2008. Home page of ACS. Includes useful resources under education and information about new areas in chemistry.

American Geological Institute Government Affairs Program. Available online. URL: http://www.agiweb.org/gap/index.html. Accessed February 4, 2009. The AGI Government Affairs Program (GAP), established in 1992, serves as an important link between the federal government and the geoscience community. Through Congressional workshops, testimony, letters, and meetings, GAP ensures that the voices of the AGI Member Societies are heard on Capitol Hill and in the executive branch. At the same time, GAP is working to improve the flow of geoscience information to policy-makers. Equally important is the program's mission of providing federal science-policy information back to the member societies and the geoscience community at large.

American Institute of Physics: Center for History of Physics. AIP, 2004. Available online. URL: http://www.aip.org/history/. Accessed July 22, 2008. Visit the "Exhibit Hall" to learn about events such as the discovery of the electron or read selected papers of great American physicists.

American Museum of Natural History home page. Available online. URL: http://www.amnh.org/. Accessed February 6, 2008. Contains links for research conducted by the museum, and updates on scientific topics.

American Physical Society. A Century of Physics. Available online. URL: http://timeline.aps.org/. Accessed July 22, 2008. Wonderful, interactive timeline describing major events in the development of modern physics.

———. Physics Central. Available online. URL: http://www.physicscentral.com/. Accessed July 22, 2008. Updated daily with information on physics in the news, current research, and people in physics.

Astronomy Today. Available online. URL: http://www.astronomytoday.com/. Accessed February 5, 2009. Web site on news and interesting topics in astronomy.

Dinosaur Extinction Page. Available online. URL: http://web.ukonline.co.uk/a.buckley/dino.htm Accessed February 4, 2009. Web site offers short summaries of some theories of dinosaur extinction, including meteorite impacts and volcanic eruptions.

Fear of Physics. Available online. URL: http://www.fearofphysics.com/. Accessed July 22, 2008. Entertaining way to review physics concepts.

Federal Emergency Management Agency. Available online. URL: http://www.fema.gov. Accessed February 4, 2009. FEMA is the nation's premier agency that deals with emergency management and preparation and issues warnings and evacuation orders when disasters appear imminent. FEMA maintains a Web site that is updated at least daily, including information on hurricanes, floods, fires, national flood insurance, and information on disaster prevention, preparation, and emergency management. Divided into national and regional sites. Also contains information on costs of disasters, maps, and directions on how to do business with FEMA. FEMA, 500 C Street, SW, Washington, D.C. 20472.

Geology.com. Available online. URL: http://geology.com/. A Web site with geological news, educational links, photos, maps, and links to careers.

Global Volcanism Network, Museum of Natural History E-421, Smithsonian Institution. Available online. URL: http://www.volcano.si.edu/. Accessed February 4, 2009. The Global Volcanism Program (GVP) seeks better understanding of all volcanoes through documenting their eruptions—small as well as large—during the last 10,000 years. The range of volcanic behavior is great enough, and volcano lifetimes are long enough, that we must integrate observations of contemporary activity with historical and geological records of the recent past in order to prepare wisely for the future. By building a global framework of volcanism over thousands of years, and by stimulating documentation of current activity, the GVN attempts to provide a context in which any individual volcano's benefits and dangers can be usefully assessed. GVP also plays a central role in the rapid dissemination of information about ongoing volcanic activity on Earth by publishing eruption reports from local observers in the monthly *Bulletin* of the Global Volcanism Network.

Intergovernmental Panel on Climate Change. Available online. URL: http://www.ipcc.ch/index.htm. Accessed January 30, 2008. The IPCC is a scientific intergovernmental body set up by the World Meteorological Organization (WMO) and by the United Nations Environment Program (UNEP). The IPCC is open to all member countries of WMO and UNEP. Governments participate in plenary sessions of the IPCC where main decisions about the IPCC work program are taken and reports are accepted, adopted, and approved. They also participate in the review of IPCC Reports. The IPCC includes hundreds of scientists from all over the world who contribute to the work of the IPCC as authors, contributors, and reviewers. As a United Nations body, the IPCC aims at the promotion of the United Nations human development goals. The IPCC was established to provide the decision makers and others interested in climate change with an objective source of information about climate change. The IPCC does not conduct any research nor does it monitor climate-related data or parameters. Its role is to assess on a comprehensive, objective, open and transparent basis the latest scientific, technical and socioeconomic literature produced worldwide relevant to the understanding of the risk of human-induced climate change, its observed and projected impacts, and options for adaptation and mitigation. IPCC reports should be neutral with respect to policy, although they need to deal objectively with policy-relevant scientific, technical, and socioeconomic factors. They should be of high scientific and technical standards and aim to reflect a range of views, expertise, and wide geographical coverage.

Jones, Andrew Zimmerman. "Physics." About, Inc., 2004. Available online. URL: http://physics.about. com. Accessed July 22, 2008. Contains regular feature articles and much additional information.

Los Alamos National Laboratory, Tsunami Society. Available online. URL: http://library.lanl.gov/tsunami/. Accessed March 28, 2007. Site publishes an online journal in pdf format available for download, called the International Journal of the Tsunami Society. The journal comes out between two and five times per year.

National Aeronautical and Space Administration (NASA). Earth Observatory. URL: http://earthobservatory.nasa.gov/NaturalHazards/. Accessed August 26, 2006. Earth scientists around the world use NASA satellite imagery to better understand the causes and effects of natural hazards. This site posts many public domain images to help people visualize where and when natural hazards occur and to help mitigate their effects. All images in this section are freely available to the public for reuse or republication.

National Aeronautical and Space Administration (NASA). Near-Earth Object Program. Jet Propulsion Laboratory, 4800 Oak Grove Drive, Pasadena, California 91109, (818) 354-4321. Available Online. URL: http://neo.jpl.nasa.gov/. Accessed February 4, 2009. In 1998 NASA initiated a program called the "Near-Earth Object Program," whose aim is to catalog potentially hazardous asteroids that could present a hazard

to Earth. This program uses five large telescopes to search the skies for asteroids that pose a threat to Earth and to calculate their mass and orbits. So far, the largest potential threat known is from asteroid 99AN10, which has a mass of 2.2 billion tons and may pass within the orbit of the moon, at 7:10 A.M., August 7, 2027. NASA has another related program called "Deep Impact," designed to collect data on the composition of a comet named Tempel 1, which will be passing beyond the orbit of Mars. The comet is roughly the size of mid-town Manhattan, and the spacecraft will be shooting an object at the comet to determine its density by observing the characteristics of the impact.

National Oceanographic and Atmospheric Administration, Hazards research. Available online. http://ngdc.noaa.gov/seg/hazard/tsu.html Accessed March 28, 2007. Web site about hazards, including tsunamis, volcanoes, hurricanes, and droughts.

National Oceanic and Atmospheric Administration. Home page. Available online. URL: http://www. noaa.gov. Accessed July 22, 2008. A useful site for all areas of environmental research.

National Weather Service. Available Online. URL: http://www.nws.noaa.gov/om/brochures/ffbro. htm. Accessed December 10, 2007. The National Weather Service, FEMA, and the Red Cross maintain a Web site dedicated to describing how to prepare for floods, describing floods of various types, with in-depth descriptions of warnings and types of emergency kits that families should keep in their homes.

Natural Hazards Observer. Available online. URL: http://www.colorado.edu/hazards/o/. Accessed February 4, 2009. This Web site is the online version of the periodical, *The Natural Hazards Observer*. The *Observer* is the bimonthly periodical of the Natural Hazards Center. It covers current disaster issues; new international, national, and local disaster management, mitigation, and education programs; hazards research; political and policy developments; new information sources and Web sites; upcoming conferences; and recent publications. Distributed to more than 15,000 subscribers in the United States and abroad via printed copies of their Web site, the *Observer* focuses on news regarding human adaptation and response to natural hazards and other catastrophic events and provides a forum for concerned individuals to express opinions and generate new ideas through invited personal articles.

The Particle Adventure: The Fundamentals of Matter and Force. The Particle Data Group of the

Lawrence Berkeley National Laboratory, 2002. Available online. URL: http://particleadventure. org/. Accessed July 22, 2008. Interactive tour of quarks, neutrinos, antimatter, extra dimensions, dark matter, accelerators, and particle detectors.

SciTechDaily Review. Available online. URL: http:// www.scitechdaily.com/. Accessed February 10, 2008. Regularly updated science and technology coverage.

United States Environmental Protection Agency. Available online. URL: http://www.epa.gov. Accessed December 10, 2007. The EPA works with other government agencies and private organizations to monitor groundwater quality and contamination, superfund sites, and subsidence.

U.S. Geological Survey. Available online. URL: U.S. Department of the Interior, 345 Middlefield Road, Menlo Park, CA 94025; also, offices in Reston, Virginia, and Denver, Colorado; Main Offices URL: http://www.usgs.gov. Accessed February 4, 2009. Earthquake Hazards Program monitors recent earthquakes worldwide. The USGS is responsible for making maps of many of the different types of earthquake hazards discussed in this book, including earthquake-related shaking hazards, tsunamis, landslides, and others. This site also provides answers to frequently asked questions about earthquakes URL: http://earthquake.usgs.gov/. Accessed February 4, 2009. U.S. Geological Survey, National Earthquake Information Center, Federal Center, Box 25046, MS 967, Denver, CO 80225-0046, U.S.A.

Volcanoworld. Available online. URL: http://volcano.oregonstate.edu/. Accessed February 4, 2009. Presents updated information about eruptions and volcanoes and has many interactive pages designed for different grade levels from kindergarten through college and professional levels. Volcanoworld is an award-winning Web site, designed as a collaborative Higher Education, K-12, and Public Outreach project of the North Dakota and Oregon Space Grant Consortia administered by the Department of Geosciences at Oregon State University.

Windows to the Universe team. Fundamental Physics. Boulder, Colo.: ©2000–04 University Corporation of Atmospheric Research (UCAR), ©1995–99, 2000 The Regents of the University of Michigan. Available online. URL: http://www.windows.ucar.edu/tour/link=/physical_science/physics/physics.html. Accessed July 22, 2008. Still under construction, this site will contain a broad overview of physics and already has many links to physics topics including mechanics, electricity and magnetism, thermal physics, and atomic and particle physics.

PERIODICALS

The American Naturalist
Published for the American Society of Naturalists by The University of Chicago Press
1427 East 60th Street
Chicago, IL 60637
www.journals.uchicago.edu/AN/home.html

American Scientist
Published by Sigma Xi, The Scientific Research Society
P.O. Box 13975
Research Triangle Park, NC 27709
Telephone: (919) 549-0097
www.americanscientist.org

Astronomy
Published by Kalmbach Publishing Company
21027 Crossroads Circle
P.O. Box 1612
Waukesha, WI 53187-1612
Telephone 1-(800)-533-6644
Astronomy.com

Discover
Published by Buena Vista Magazines
114 Fifth Avenue
New York, NY 10011
Telephone: (212) 633-4400
www.discover.com

Earth
Published by American Geological Institute
2000 Florida Avenue N.W.
Washington, DC 20009-1277
Telephone: (202) 462-6900
http://www.earthmagazine.org/

Environmental Science and Technology
Published by the American Chemical Society
1155 16th Street N.W.
Washington, DC 20036
Telephone: (202) 872-4582
www.acs.org/est

GSA Today
Published by the Geological Society of America
P.O. Box 9140
Boulder, CO 80301-9140
Telephone: (303) 357-1000
http://www.geosociety.org/pubs/gsatguid.htm

Issues in Science and Technology
Published by The University of Texas at Dallas
P.O. Box 830688
Mail Station J030

Richardson, TX 75083-0688
Telephone: (800) 345-8112
www.issues.org

Journal of the American Chemical Society
Published by the American Chemical Society
1155 16th St. N.W.
Washington, DC 20036
Telephone: (202) 872-4614
www.acs.org

Natural History
Published by Natural History Magazine, Inc.
   in affiliation with the American Museum of
   Natural History
P.O. Box 5000
Harlan, IA 51593-0257
www.naturalhistorymag.com

Nature
The Macmillan Building
4 Crinan Street
London N1 9XW
United Kingdom
Telephone: +44 (0)20 7833 4000
www.nature.com/nature

New Scientist
6277 Sea Harbor Drive
Orlando, FL 32887
Telephone: (888) 822-3242
www.newscientist.com

Oceanus
Published by Woods Hole Oceanographic
   Institution

WHOI
Mail Stop 40
Woods Hole, MA 02543
www.oceanusmag.com

Physics Today
Published by the American Institute of Physics
Circulation and Fulfillment Division
Suite 1NO1
2 Huntington Quadrangle
Melville, NY 11747
Telephone: (516) 576-2270
www.physicstoday.org

Science
Published by the American Association for the
   Advancement of Science
1200 New York Avenue N.W.

Washington, DC 20005
Telephone: (202) 326-6417
www.sciencemag.org

Science News
Published by the Society for Science & the Public
1719 N Street N.W.
Washington, DC 20036
Telephone: (202) 785-2255
www.sciencenews.org

Scientific American
415 Madison Avenue
New York, NY 10017
Telephone: (212) 754-0550
www.sciam.com

SOCIETIES AND ORGANIZATIONS

American Association for the Advancement of Science (www.aaas.org), 1200 New York Avenue N.W., Washington, DC 20005. Telephone: (202) 326-6400

American Chemical Society (www.acs.org), 1155 16th Street N.W., Washington, DC 20036. Telephone: (202) 872-4600

American Geological Institute (http://www.agiweb.org/), 4220 King Street, Alexandria, VA 22302-1502. Telephone: (703) 379-2480

American Geophysical Union (http://www.agu.org/), 2000 Florida Avenue N.W., Washington, DC 20009-1277. Telephone: (202) 462-6900

American Physical Society (www.aps.org), One Physics Ellipse, College Park, MD 20740-3844. Telephone: (301) 209-3200

Geological Society of America (http://www.geosociety.org/), P.O. Box 9140, Boulder, CO 80301-9140. Telephone: (303) 357-1000

The Minerals, Metals, & Materials Society (www.tms.org), 184 Thorn Hill Road, Warrendale, PA 15086-7514. Telephone: (724) 776-9000

National Science Foundation (www.nsf.gov), 4201 Wilson Boulevard, Arlington, VA 22230. Telephone: (703) 292-5111; FIRS: (800) 877-8339; TDD: (800) 281-8749

Society of Physics Students (www.spsnational.org), American Institute of Physics, One Physics Ellipse, College Park, MD 20740-3843. Telephone: (301) 209-3007

Woods Hole Oceanographic Institute (http://www.whoi.edu/), 266 Woods Hole Road, Woods Hole, MA 02543. Telephone: (508) 289-2252. WHOI is the world's largest private, nonprofit ocean research, engineering, and education organization.

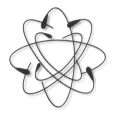

APPENDIX IV
THE GEOLOGIC TIMESCALE

THE GEOLOGIC TIMESCALE

Era	Period	Epoch	Age (millions of years)	First Life-forms	Geology
Cenozoic	Quaternary	Holocene	0.01		
		Pleistoscene	3	Humans	Ice age
	Tertiary	Pliocene	11	Mastodons	Cascades
		Neogene			
		Miocene	26	Saber-toothed tigers	Alps
		Oligocene	37		
		Paleogene			
		Ecocene	54	Whales	
		Paleocene	65	Horses, Alligators	Rockies
Mesozoic	Cretaceous		135		
	Jurassic		210	Birds, Mammals, Dinosaurs	Sierra Nevada, Atlantic
	Triassic		250		
Paleozoic	Permian		280	Reptiles	Appalachians
	Carboniferous	Pennsylvanian	310	Trees	Ice age
		Mississippian	345	Amphibians, Insects	Pangaea
	Devonian		400	Sharks	
	Silurian		435	Land plants	Laurasia
	Ordovician		500	Fish	
	Cambrian		544	Sea plants, Shelled animals	Gondwana
Proterozoic			700	Invertebrates	
			2500	Metazoans	
			3500	Earliest life	
Archean			4000		Oldest rocks
			4600		Meteorites

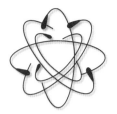

APPENDIX V
PERIODIC TABLE OF THE ELEMENTS

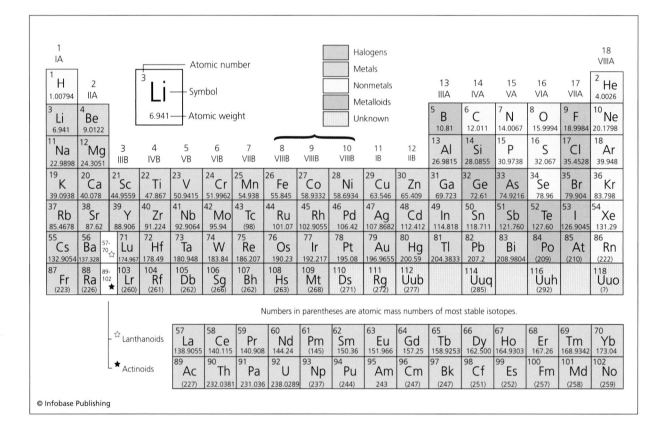

Numbers in parentheses are atomic mass numbers of most stable isotopes.

☆ Lanthanoids

★ Actinoids

© Infobase Publishing

(g) none
(c) nonmetallics

element	symbol	a.n.
carbon	C	6
hydrogen	H	1

(g) chalcogen
(c) nonmetallics

element	symbol	a.n.
oxygen	O	8
polonium	Po	84
selenium	Se	34
sulfur	S	16
tellurium	Te	52
ununhexium	Uuh	116

(g) alkali metal
(c) metallics

element	symbol	a.n.
cesium	Cs	55
francium	Fr	87
lithium	Li	3
potassium	K	19
rubidium	Rb	37
sodium	Na	11

(g) alkaline earth metal
(c) metallics

element	symbol	a.n.
barium	Ba	56
beryllium	Be	4
calcium	Ca	20
magnesium	Mg	12
radium	Ra	88
strontium	Sr	38

(g) none (c) metallics

element	symbol	a.n.	element	symbol	a.n.
aluminum	Al	13	scandium	Sc	21
bohrium	Bh	107	seaborgium	Sg	106
cadmium	Cd	48	silver	Ag***	47
chromium	Cr	24	tantalum	Ta	73
cobalt	Co	27	technetium	Tc	43
copper	Cu***	29	thallium	Tl	81
darmstadium	Ds	110	titanium	Ti	22
dubnium	Db	105	tin	Sn	50
gallium	Ga	31	tungsten	W	74
gold	Au***	79	ununbium	Uub	112
hafnium	Hf	72	ununtrium	Uut	113
hassium	Hs	108	ununquadium	Uuq	114
indium	In	49	vanadium	V	23
iridium	Ir ****	77	yttrium	Y	39
iron	Fe	26	zinc	Zn	30
lawrencium	Lr	103	zirconium	Zr	40
lead	Pb	82			
lutetium	Lu	71			
manganese	Mn	25			
meitnerium	Mt	109			
mercury	Hg	80			
molybdenum	Mo	42			
nickel	Ni	28			
niobium	Nb	41			
osmium	Os****	76			
palladium	Pd****	46			
platinum	Pt ****	78			
rhenium	Re	75			
rodium	Rh****	45			
roentgenium	Rg	111			
ruthenium	Ru****	44			
rutherfordium	Rf	104			

(g) pnictogen (c) metallics

element	symbol	a.n.
arsenic	As*	33
antimony	Sb*	51
bismuth	Bi	83
nitrogen	N	7
phosophorus	P**	15
ununpentium	Uup	115

(g) none (c) semimetallics

element	symbol	a.n.
boron	B	5
germanium	Ge	32
silicon	Si	14

(g) actinoid (c) metallics

element	symbol	a.n.
actinium	Ac	89
americium	Am	95
berkelium	Bk	97
californium	Cf	98
curium	Cm	96
einsteinium	Es	99
fermium	Fm	100
mendelevium	Md	101
neptunium	Np	93
nobelium	No	102
plutonium	Pu	94
protactinium	Pa	91
thorium	Th	90
uranium	U	92

(g) halogens (c) nonmetallics

element	symbol	a.n.
astatine	At*	85
bromine	Br	35
chlorine	Cl	17
fluorine	F	9
iodine	I	53
ununseptium	Uus*	117

(g) lanthanoid (c) metallics

element	symbol	a.n.
cerium	Ce	58
dysprosium	Dy	66
erbium	Er	68
europium	Eu	63
gadolinium	Gd	64
holmium	Ho	67
lanthanum	La	57
neodymium	Nd	60
praseodymium	Pr	59
promethium	Pm	61
samarium	Sm	62
terbium	Tb	65
thulium	Tm	69
ytterbium	Yb	70

(g) noble gases (c) nonmetallics

element	symbol	a.n.
argon	Ar	18
helium	He	2
krypton	Kr	36
neon	Ne	10
radon	Rn	86
xenon	Xe	54
ununoctium	Uuo	118

a.n. = atomic number
(g) = group
(c) = classification

* = semimetallics (c)
** = nonmetallics (c)
*** = coinage metal (g)
**** = precious metal (g)

© Infobase Publishing

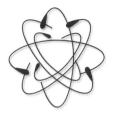

APPENDIX VI
SI UNITS AND DERIVED QUANTITIES

DERIVED QUANTITY	UNIT	SYMBOL
frequency	hertz	Hz
force	newton	N
pressure	pascal	Pa
energy	joule	J
power	watt	W
electric charge	coulomb	C
electric potential	volt	V
electric resistance	ohm	Ω
electric conductance	siemens	S
electric capacitance	farad	F
magnetic flux	weber	Wb
magnetic flux density	tesla	T
inductance	henry	H
luminous flux	lumen	lm
illuminance	lux	lx

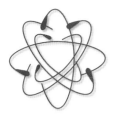

APPENDIX VII
MULTIPLIERS AND DIVIDERS FOR USE WITH SI UNITS

Multiplier	Prefix	Symbol	Divider	Prefix	Symbol
10^1	deca	da	10^{-1}	deci	d
10^2	hecto	h	10^{-2}	centi	c
10^3	kilo	k	10^{-3}	milli	m
10^6	mega	M	10^{-6}	micro	μ
10^9	giga	G	10^{-9}	nano	n
10^{12}	tera	T	10^{-12}	pico	p
10^{15}	peta	P	10^{-15}	femto	f
10^{18}	exa	E	10^{-18}	atto	a
10^{21}	zetta	Z	10^{-21}	zepto	z
10^{24}	yotta	Y	10^{-24}	yocto	y

APPENDIX VIII
ASTRONOMICAL DATA

Body	Mass (Kg)	Mean Radius (M)	Orbital Period (Years)	Mean Orbital Radius (M)
Sun	1.991×10^{30}	6.96×10^8	—	—
Mercury	3.18×10^{23}	2.43×10^6	0.241	5.79×10^{10}
Venus	4.88×10^{24}	6.06×10^6	0.615	1.08×10^{11}
Earth	5.98×10^{24}	6.37×10^6	1.00	1.50×10^{11}
Mars	6.42×10^{23}	3.37×10^6	1.88	2.28×10^{11}
Jupiter	1.90×10^{27}	6.99×10^7	11.9	7.78×10^{11}
Saturn	5.68×10^{26}	5.85×10^7	29.5	1.43×10^{12}
Uranus	8.68×10^{25}	2.33×10^7	84.0	2.87×10^{12}
Neptune	1.03×10^{26}	2.21×10^7	165	4.50×10^{12}
Moon	7.36×10^{22}	1.74×10^6	27.3 days	3.84×10^8

Note: Pluto, which had long been considered a planet, has been recategorized. It now belongs to the class of astronomical bodies called dwarf planets, which are smaller than planets.

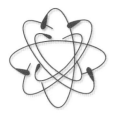

APPENDIX IX
ABBREVIATIONS AND SYMBOLS FOR PHYSICAL UNITS

Symbol	Unit	Symbol	Unit	Symbol	Unit
A	ampere	g	gram	min	minute
u	atomic mass unit	H	henry	mol	mole
atm	atmosphere	h	hour	N	Newton
Btu	British thermal unit	hp	horsepower	Pa	pascal
C	coulomb	Hz	hertz	rad	radian
cd	candela	in	inch	rev	revolution
°C	degree Celsius	J	joule	s	second
cal	calorie	K	kelvin	sr	steradian
d	day	kg	kilogram	T	tesla
eV	electron volt	L	liter	V	volt
F	farad	lb	pound	W	watt
°F	degree Fahrenheit	ly	light-year	Wb	weber
ft	foot	m	meter	yr	year
G	gauss	mi	mile	Ω	ohm

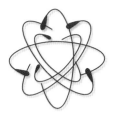

APPENDIX X
THE GREEK ALPHABET

Letter	Capital	Small	Letter	Capital	Small
Alpha	A	α	Nu	N	ν
Beta	B	β	Xi	Ξ	ξ
Gamma	Γ	γ	Omicron	O	o
Delta	Δ	δ	Pi	Π	π
Epsilon	E	ϵ	Rho	ρ	ρ
Zeta	Z	ζ	Sigma	Σ	σ, ς
Eta	H	η	Tau	T	τ
Theta	Θ	θ	Upsilon	Y	υ
Iota	I	ι	Phi	Φ	φ, ϕ
Kappa	K	κ	Chi	X	χ
Lambda	Λ	λ	Psi	Ψ	ψ
Mu	M	μ	Omega	Ω	ω

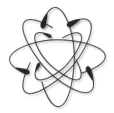

APPENDIX XI
COMMON CONVERSIONS
WITHIN THE METRIC SYSTEM

1 meter (m) = 100 centimeters (cm) = 1,000 millimeters (mm)
1 centimeter (cm) = 10 millimeters (mm) = 0.01 meters (m)
1 millimeter (mm) = 1000 micrometers (μm) = 1 micron (μ)

1 liter (L) = 1,000 milliliters (mL)
1 cubic centimeter (cc or cm^3) = 1 milliliter (mL)
1 milliliter (mL) = 1,000 microliters (μL)

1 kilogram (kg) = 1,000 grams (g)
1 gram (g) = 1,000 milligrams (mg)
1 milligram (mg) = 1,000 micrograms (μg)

APPENDIX XII
COMMON CONVERSIONS FROM U.S. CUSTOMARY TO METRIC UNIT VALUES

Quantity	To Convert From	To	Multiply by (Rounded to Nearest 1,000th)
mass	pounds (lb)	kilograms (kg)	0.454
	ounces (oz)	gram (g)	28.350
length	miles (mi)	kilometers (km)	1.609
	yards (yd)	meters (m)	0.914
	feet (ft)	meter (m)	0.305
	inches (in)	centimeters (cm)	2.540
area	square feet (ft^2)	square meter (m^2)	0.093
	square miles (mi^2)	square kilometers (km^2)	2.590
volume	gallon (gal)	liters (L)	3.785
	quarts (qt)	liters (L)	0.946
	fluid ounces (fl oz)	milliliters (mL)	29.574

APPENDIX XIII
TEMPERATURE CONVERSIONS

In the Celsius scale, 0°C is the freezing point of water and 100°C is the boiling point. In the Fahrenheit scale, 32°F is the freezing point of water and 212°F is the boiling point. In the Kelvin scale, 0 K is absolute zero temperature and the freezing point of water is 273.15 K. To convert temperature in degrees Celsius (T_C) to temperature in degrees Fahrenheit (T_F):

$$T_F = \frac{9}{5} T_C + 32$$

To convert temperature in degrees Fahrenheit (T_F) to temperature in degrees Celsius (T_C):

$$T_C = \frac{5}{9} (T_F - 32)$$

To convert temperature in degrees Celsius (T_C) to temperature in kelvins (T):

$$T = T_C + 273.15$$

To convert temperature in kelvins (T) to temperature in degrees Celsius (T_C):

$$T_C = T - 273.15$$

INDEX

Page numbers in **boldface** indicate main entries; *italic* page numbers indicate photographs and illustrations.

A

aa (lava flow) 488, 796, 800
Abbot, Henry 291
Abell 2218 galaxy cluster *304*
Abercromby, Ralph 141
Abhe, Lake 21
Abitibi belt 572
ablation zone 332
abrasion 206, 259, 333, 810
absolute humidity 553
abyssal plains 154, 225, 586–587
Acadian-Appalachian Orogeny 268
Acadian Orogeny 388, 575, 580–581
acapulcoites 62
Acasta gneisses 39, 177, 567–568
accelerations, in earthquakes 233–234
accretionary beaches 87
accretionary prism 454
accretionary processes 177–182
accretionary wedge **1–5**, *2*, 162
accumulation zone 332
Achankovil shear zone 446, 500
achondrites 60, 61–62, 543
acid rain 73
Aconcauga 22, 692
acritarchs 114
active arc 157, 160–161, 454
Adams, Mount 469
Adirondack dome 576
Adirondack Highlands *359*
Adirondack Highlands–Green Mountains block 363–364
Adirondack Lowlands 360, 362–363
Adirondack Mountains 581
Advanced Spaceborne Thermal Emission and Reflection Radiometer (ASTER) 655
Advanced Very High Resolution Radiometer (AVHRR) 655
advection 249
Afar Depression 21
Afar region 20–21, 196, 218, 512
Afghanistan *57*
Afif Terrane 31–32

African/Arabian boundary 218
African geology *5*, **5–22**, *6*
African/Somalian boundary 218
Agassiz, Louis 23, 694
age
    of amphibians 117, 120
    of coal 117
    of Earth 394–399, 652–653
    of fishes 683
    of invertebrates 112
    of mammals 741–742
    of marine invertebrates 598
    of reptiles 537–538
agnostic 272
Agulhas-Falkland fracture zone 695
Ahagger 208
Ahlmann, Hans Wilhelmsson 352
Ahmadi ridge 36, 37
A horizon, of soil 688, 689
Air Mountains 208
air pressure 71, 73–74
Airy model 328
Akhdar, Jabel 33, 35
Akhdar Group 33
Alaotra, Lake 505
Alaotra-Ankay rift 503–504
Alaotra Basin 505
Alaska
    Aleutian earthquake (1946) 236
    aurora borealis *78*
    Chugach terrane 2–3
    convergent plate margin of 157
    Glacier Bay National Park *335*
    gold deposits 238–239
    Lituya Bay tsunami (1958) 784
    Valdez earthquake (1964) 163–165
        ground breaks in 234
        ground level changes in 235–236, 670, 722, 725, 777
        magnitude of 231
        mass wasting in 520
        tsunami induced by 164, 755, 763, 776–779
    volcanoes of 804, 805
Alaskan Volcano Observatory 808
Albany-Fraser belt 364, 642
Albert, Lake 21, 724
Alboran Sea 586

Aldan River 667
Aldan shield 662, 663–665
Aldan Supergroup 663–664
Aldan-Timpton block 664
Aldrin, Edwin E. (Buzz) 441
Aleutian Islands earthquake (1946) 236
Aleutian Mountains 196
Alexander terrane 2
Algeria, flash floods in 278
Algoma-type banded-iron formations 638–639
Algonquin terrane 358–360, *359*
alkali series 457
Alleghenian (Appalachian) Orogeny 118, 267, 391, 575
Allende meteorites 61
allochthons
    Golconda 583
    Taconic 579–580
alluvial fans 206
alluvial terraces *326*
Almagro, Diego de 210, 694
Alpamayo Peak *22*
alpha decay 651
Alpine belt 663
Alpine fault 630, 749
Alps 264–265, *266*
    folded rock strata in *195*
    formation of 123, 265, 394
    ice caps of 334
    orogen of 153
    southern, flash floods in 278
Altaids 662, 664, 665–666
Altai Mountains 45, 118
Altiplano Plateau 22–23, 692, 694
altocumulus clouds 142
altostratus clouds 142
Altyn Tagh fault 749
Amagaon Orogeny 447
Amalia belt 7
Amar-Idsas belt, Al- 31
Amazon Basin 692, *693*
Amazonian craton 362, 364, 642, *693*, 695
Amazon River 23, *693*, 694
Ambilobe basin 503
Amboropotsy Group 502
AMCG suite 361, 363

Amor asteroids 65, 67
Ampanihy shear zone *501*
amphibians, age of 117, 120
amphibole 540, 560
amphibolite 540
Amu Darya River 58
Anabar Complex 665
Anabar shield 662, 665
Anai Mudi 442
Anak Krakatau 783
Ancient Gneiss Complex 5–6
Andaingo Heights 505
Andean-type arcs 158, *159*, 455–456, 719
Andean-type margin 361–362
Anderson, Don 511
Anderson, E. M. 296
Andersonian geometries 296
Andersson, Johan G. 695
Andes 22, **22–23**, 692–694
    ice caps of 334
    landslides in 523–524, 528
andesitic magma 434, 489, 509, 797
Andrew, Hurricane 322, 324, 409
Andriamena belt 502
Andriba Group 500
Andromeda Galaxy 303, 789
Angara (Siberian) craton 153, 662, 663–665, *664*
Angel Falls 695
Angolan abyssal plain 586
angrites 62
angular strain 718
angular unconformities *380, 381, 788, 789*
anhydrite 561, 678
Anialik River 567
Ankarafantsika Natural Reserve 262
Ankaratra Mountains 500
Ankaratra Plateau 504
Ankay-Alaotra rift 504, 505
Ankay Basin 505
annealing 188
anorthosite 361, 643
Antananarivo Block 500, *501*, 502–503
Antarctica **23–26**
    Dry Valleys 23, 25, 204, 210–211
    ice sheet/cap of *25*, 25–26, 136, 331, 334, 340, 342–343, 669, 673, 745
    meteorites in 60
    ozone hole above 73, 607–608, *608*
    sea ice around 335–336
Antarctic Circumpolar Current 587
Antarctic ice cores *132*
Antarctic Peninsula 24, 25, 26
Antarctic Treaty 23
antecedent stream 289
Antelope belt 13, 14, *17*
anthodites 121–122
anthracite 415
anticlines 195, 622

Antler Orogeny 117–118, 583
Antongil Block 500, *501*, 502
Anton terrane 177, 567–569, 571
Antwerp-Rossie suite 363
apatite 561
Apex chert 39, 43
aphanites 431–432, 506
Apollo 11 441
Apollo asteroids 65, 67, 549
Appalachian-Caledonide orogen 267, 567
Appalachian Mountains 153, 575–581, *580*
    Acadian Orogeny in 580–581
    as fold and thrust chain 196
    formation of 212–213, 575–576
    Penobscottian Orogeny in 268, 576–579
    Taconic Orogeny in 388, *389*, 575–576, 579–580
Appalachian (Alleghenian) Orogeny 118, 267, 391, 575
Appalachian-Ouachita orogen 575
apparent polar wandering (APW) 632
Appin Group 268
APW. *See* apparent polar wandering
Aqiq-Tuluhah Orogeny 29
aqueduct, California 423, *423*
aquifers 366, 367
    coastal, seawater intrusion in 371
    confined 367
    extraction from, and subsidence 722–724
    fracture zone 297–298
    horizontal 297, 368
    unconfined 367
Arabian Desert 208
Arabian folds 37
Arabian geology **26–39**, *27, 28*
Arabian-Nubian shield 6, 18–19, 26, 28, *30*, 32, 112, 642, *643*
Arabian platform 32
Arabian shield 26–29, *27*
    classification of rock units in 29
    geology of 28–29
    intrusive rocks of 29–30
    Phanerozoic cover of 32–33
    tectonic evolution of 31–32
    tectonic models of 26–28, *30*
Arabian/Somalian boundary 218
Aral Sea 58
Aravalli craton 443, 448–449, *450*
Aravalli-Delhi belt 448–449
Aravalli Mountains 449
Aravalli Supergroup 449
arc(s) 719–720
    active 157, 160–161, 454
    Andean-type 158, *159*, 455–456, 719
    compressional 161, 456–457
    convergent plate 157–161, *159, 160*, 455–457
    island **453–474**. *See also* island arc(s)

    Marianas-type 158, 455–456, 719
    Pacific-type *159*
    volcano types in 457
arc-continent collisions 161–162
Archean **39–44**, *40*, 382–383, 600, 636
Arctic, ozone hole above 73
Arctic Ocean
    basin of 586
    circulation patterns in 251–252, 587
    thermohaline circulation in 743–744
arc-trench migration 177, 180–181
Arduino, Giovanni 741
Argyll Group 268
Al Aridh Group 33
Ariel (moon) 791
Aristotle 306
Arizona Desert *205*
Armstrong, Neil A. 441
Aroan-Jaquie event 695
Aroostook-Matapedia trough 580, 581
Ar-Rayn terrane 31–32
Arrhenius, Svante 341, 351
ARS. *See* Attitude Reference System
arsenates 559
arsenic, in groundwater 370
artesian spring or well 367
arubites 62
Aruma Group 33
Arunter Inlier 78
Asama eruptions 804
asbestos 253–255, 323
Asbestos Hazard Emergency Response Act 255
ash, volcanic 796–797, 801–802, 805
Ash Shaqq 36
Ashwa fault zone 500
Asian geology 44, **44–59**, *46*
Asian monsoons 562–563
Asir terrane 31–32
assimilation 431
ASTER. *See* Advanced Spaceborne Thermal Emission and Reflection Radiometer
asteroid(s) **59–68**, 689–690
    composition and origin of 66–67
    as geological hazard 323
    impact with Earth 59, 65, 67, 122–123, 323, 393
        atmospheric shock waves of 543, 544–546
        craters/structures of 435–442
        mitigating dangers of 543, 547–549
        solid Earth shock waves of 547
        and tsunamis 543, 547, 760
        tsunamis caused by 543, 547
    inner solar system 63–64, 67
    Kuiper belt 66, 67–68
    locations in solar system 63–66

main belt *62,* 64–65
    meteorite origin as 543
    monitoring of 549–551
    near-collisions with Earth, recent
        549–551
    orbits of 60–66
    outer solar system 65–66, 67–68
asthenosphere **68–69,** 224, 328, 626,
    627
Aston, Francis William 398
astronomy **69–70**
astrophysics **70–71**
asymmetric rifting 215, *216*
Atacama Desert 22, *201,* 203, 204,
    210, 692, 694
ataxites 60, 63
Aten asteroids 65, 67
Athollian Orogeny 268
Atlantic Ocean
    basin of 586
    circulation patterns in 251
    opening of 123
    thermohaline circulation in 743–
        744, *744*
Atlantic-type margins 155
Atlantic-type ridges 218
Atlas Mountains 20
atmosphere **71–77,** 104, 225
    circulation patterns in 74–75, *76,*
        124–125, 553–554
    evolution of 75–77
    external energy-driven processes in
        251–252
    global climate effects of 74, 124–
        125, 126–127
    Precambrian 383
    Silurian and Devonian 388
    solar 728–729
    structure of *71,* 71–72
atmospheric geophysics 326
atmospheric pressure 71, 73–74
atmospheric shock waves
    of meteorite impact 543, 544–546
    of volcanoes 804
atolls 100, 170
atomic bomb 243
attitude 713
Attitude Reference System (ARS) 658
augmented GPS 318
aurora 77, **77–78,** *78*
aurora australis 72, **77–78**
aurora borealis 72, **77–78,** *78*
Austin Glen 576, 580
Australian geology **78–83,** *79, 80*
avalanches 523–524, 802. *See also*
    slides
Avalon Composite terrane 581
Avalonia 267
Avalon zone 576
AVHRR. *See* Advanced Very High
        Resolution Radiometer
avulsion of levee 289–290
Awash River 21
Axial Ranges 21

Ayres Rock 206
Azbine 208
Azores 586

B

back arc 157, 161, 454, 456
back arc basin accretion 177, 180
Back River area 570
backshore 86–87
backwash, tsunami 765
Bacon, Francis 606
Baggs Hill granite 579
Bag Pond Mountain 577
Bahah Group 29
Bahra anticline 36, 37
Baikal, Lake 215, 666–667
Bailadila Group 447
Bailey, Edward 499
Bain, Wadi Al- 36, 37
Baish Group 29
bajada (slope) 206
Balcones escarpment 277
Baldwin Hills Dam and Reservoir fail-
        ure 723–724
Baltic 267
Baltica 362
Baltic shield 265, 268–270, 662–663
Bam, Iran, earthquake (2003) 166
Bancroft terrane *359,* 360
Banded Gneiss complex 449
banded-iron formations (BIFs) 637–
        639, *638*
    Australian 79, *81,* 238
    early signs of life in 493–494
    Indian 443, 448
    iron deposits in 238
Bangemall basin 82
Bangladesh, storms and 276–277,
        407, 408
banner clouds 142
Baoule-Mossi domain 18
Barberton belt 7, 178–179, 354, 355
barchan dunes 207, 261
barred spiral galaxies 299, 300
barrier islands 88–91, *94*
barrier reefs 82–83, 99, 102, 170
barrier spits 90
Barringer, Daniel 437–438
Barringer meteor impact crater 436,
        437, 437–438
Bartholomew, John 23
Barzaman Formation 35
basalt 626–627
    flood. *See* flood basalts
    metamorphism of 539–540
basaltic magma 433, 489, 508–509,
        629, 797, 800–801
basin(s) **84–86.** *See also* specific
        basins
    back arc 161, 177, 180
    cratonic association with 42–43
    deflation 206, 259
    drainage 84, **220–221,** 289
    forearc 157, 160, 454, 456

ocean. *See* ocean basins
    pull-apart 85–86, 724–725, 750
Basin and Range Province 215, 394,
        583–584, 607
basin ranges 330
Bass basin 82
batholith *347,* **347–349,** 431, 508
bauxite 689
Bayer, Johann 151
Bay of Islands ophiolites 579, 595
beaches and shorelines **86–100,** *87,*
        90, 225
beach face 86
Beagle voyages 100, 192–193, 271,
        295, 677, 695
Beaufort gyre 251, 587
Beaulieu River belt 570
Becke, Friedrich 344
Beck Springs Dolomite 384
Becquerel, Henri 395, 652
bedding 678
Beechy Lake domain 571
Beehive Geyser 329
Beekmantown Group 576
Beforana-Alaotra belt 502
Bekily Block 500, *501,* 503
Belaya River 666
Belinda (moon) 791
Belingwe belt 13–15, *14, 15, 17,* 42,
        354
Belize Chamber 120, 479
Bellingshausen, Fabian Gottliev von
        23
Belomorian mobile belt 269
belts. *See* greenstone belt(s); orogenic
        belts; specific belts
Belt Supergroup 582–583
Bemarivo Block 500–501, *501*
benches, shoreline 99
benchmarks 316
Bend Formation 14
bending of rocks *195,* 195–196
Bengal, Bay of 52, 407, 408
Bengpal Group 447
Beniah Lake 570
Benioff, Hugo 632
Benioff zones 456, 632
benthic environment *101,* **101–102,**
        104
benthos *101,* **101–102**
Bergeron, Tor 124
Bering Sea 456
Bermuda platform 586
beta decay 651
Betsiboka estuary 264
Bhandara craton 443, 447
Bhilwara suite 449
Bhima basin 445
Bhola cyclone 276, 407
B horizon, of soil 688, 689
Bianca (moon) 791
BIFs. *See* banded-iron formations
big bang 175, 490–491, 602–604
Big Spring, Missouri 367

Big Thompson Canyon flood (1976) 277–278
Biligirirangan Hills 446
bimodal volcanoes 86
binary star systems **102–104**, *103*, 585
Bindura-Shamva belt 15
biodiversity 240–241
biogeochemistry 310
biogeography 382
biological weathering 810, 812
biomass fuels 622
biosphere **104–105**, 225
biostratigraphy 380
Bir Omq belt 31
Bishah-Rimmah Orogeny 29
bituminous coal 415
Black, Joseph 411
black dwarfs 222, 704–705
Blackfoot River *280*
black holes **105–107**
Black River Group 576
black smoker chimneys 104, **106–108**, *108*, 218, 329, 492–493
block lava 488, 800
Blue Ridge Mountains 579
body forces 716
body waves, earthquake 228–230, 679–680, *680*
Bogger, W. C. 344
Bohai basin 45, 55–56
Boil Mountain ophiolite 576–579
Boltwood, Bertram 396
Bonneville, Lake 197, 331
Bootlegger shale 235
Borah Peak 582
borates 559
Border Ranges fault 3–4
Border Ranges ultramafic-mafic complex (BRUMC) 4
Borrelly, Comet 144
Bouguer gravity anomaly 349, 567
Bouma, Arnold 294
Bouma sequence 294
Bowen, Norman Levi *109*, **109–110**, 435
Bowen basin 79
brachinites 62
Bradley, Dwight 618
Brahe, Tycho **110–111**, 482, 737
Brahmaputra River 52, 55, 276, 407, 408, 449
braid-delta 96, 200, 288
braided stream channels 287, *326*, 333, 661
Brasiliano events 695
Brazilian Orogeny 695
Brazilian shield 23, 692, 695
breaking of rocks 196
brittle ductile transition 215, *216*
Broken Hill Inlier 82
Brongniart, Alexandre 686
Bronson Hill–Boundary Mountain anticlinorium 581

Bronson Hill terrane 580
Bronze Age 609–610
Brooklands Formation 14
Brooks Range 582
Brown, Harrison 399
brown dwarfs 222, 700–701
Brownian motion 242
BRUMC. *See* Border Ranges ultramafic-mafic complex
Bubi belt 13, 14, *17*
Bubiyan 36
Buhwa belt 13, 14, 15–16, *17*
Bulawayan Group 11
Bulawayo belt 14, *17*
Bundelkhand massif 448, 449
Burbidge, Geoff 603
Burbidge, Margaret 603
Burgess shale 113–114, *114*, *494*
burial metamorphism 542
Burnell, Jocelyn Bell 646
Bushveld Complex 7–8, *8*
Byars, Carlos 439

C
CAIs. *See* calcium-aluminum inclusions
Calabrian arc 264
calc-alkaline series 457
calcite 561
calcium-aluminum inclusions (CAIs) 61
caldera *432*, 434, 757, 797
Caledonian belt 575
Caledonian Orogeny 268, 269–270
Caledonides 265, *266*, 266–270
California
earthquakes in 233, 319, 324, 681, 754. *See also* specific earthquakes
mass wasting in 523, 534
subsidence in 723–725
tsunamis of 763, 775, 779
water issue in 422–423, *423*
Callendar, Guy Stewart 342, 352
Callisto 478, 634
calving, of glaciers 334–336, *335*
Cambrian **112–115**, *115*, 385–388, 614–615
Cameron River belt 42, 354–355
Camille, Hurricane 406, 726
Canadian Rockies 582, 583
Canadian shield 357, 382, 385, 440
Canary Islands 785–786, 800
Canyon Mountain ophiolite 577
Cape Smith belt 355
Cape Verde Islands 525, 586, 759
Capricorn orogen 82
Capricornus Void 790
carbonaceous chondrites 61
carbonate(s) 559, 561
carbonate-iron formations 638
carbon cycle **115–117**, *116*, 309
carbon dating 312
carbon-detonation supernova 737

carbon dioxide
in atmosphere 71, 72–73, 74
in carbon cycle 116, 116–117, 309
as greenhouse gas 72–74, 125, 126–127, 341–342, 351–352
and origins of life 491–492
volcano release of 127–128
Carboniferous **117–120**, *118*, *119*, 391, 614–615
Caribbean seafloor 139, 487–488, 590
Carl Sagan Center for the Study of Life in the Universe 493
Carlsbad Cavern 120, 479
Carnegie Institution 309–310
Carter, Jimmy 254
Carthage-Colton mylonite 363
Cascade Mountains 196, 469–471, 607
Cascades Volcano Observatory 808
Cascadia subduction zone 681–682, 775
Caspian Sea 264, 265–266, *266*, 666
Cassini spacecraft 477
Cassiopeia supernova 110–111
catadioptric telescopes 740
catazonal plutons 348
Cathaysia block 46, 50
Cat's Eye Nebula 626
Catskill clastic wedge 388
Catskill Mountains 212–213, 575, 580–581
Caucasus Mountains 265
caves **120–122**, *121*, 479
cavitation 262–263, 420
Cayman fault zone 749
Cayman trough 750
Cenozoic *122*, **122–124**, 394, 564
Centaur asteroids 65–66
Centaurus Supercluster 304
Center for Education and Public Outreach 493
Center for SETI Research 493
Central African belt 17–18
Central American landslides 529–530
Central Asian orogenic belt 51
Central China orogen 51–52
central gneiss belt (CGB) 358–360, *359*
Central Gneissic Unit (Tokwe terrane) 12–14, 15–16, *17*
central granulite terrane (CGT) 358, *359*, 361
central metasedimentary belt (CMB) 358, *359*, 360–361
Central orogenic belt (China) 47–48, *49*
Central Plains orogen 575
Central Rand Group 7, 9–11
Central Siberian Plateau 488
CERCLA. *See* Comprehensive Environmental Response, Compensation, and Liability Act

1 Ceres asteroid 63
CERN 605
CFCs. *See* chlorofluorocarbons
CGB. *See* central gneiss belt
CGL. *See* Chibougamau-Gatineau
    Lineament
CGT. *See* central granulite terrane
Chadwick, George H. 625
Chain Lakes massif 576–579
chalk, Cretaceous 182, 184
Challenger voyage 593
Chamberlin, Mount 582
Chandrasekhar, Subramanyan 737
Chandrasekhar mass 737
Changbai Mountain 45
Changcheng system 47, 643
channelization, of rivers 421–426
channel patterns, stream 286–287,
    661–662
Charon 635
chassignites 62
Chegutu belt 14, *17*
chelogenic cycle 732
chemical sediment 677–678
chemical weathering 261, 810, 811–
    812, 813
Cherskogo Ranges 667
Chesapeake Bay 264
Cheshire Formation 14, 15
Cheyenne belt 641–642
Chibi granitic suite 14
Chibougamau-Gatineau Lineament
    (CGL) 361
Chi-Chi, Taiwan, earthquake (1999)
    165–166
Chicxulub asteroid impact 438–442,
    519, 543
    crater structure of 436, *436*,
        438–442
    and extinction 59, 122–123, 139,
        393, 438–440, 442, 519,
        543
    as geological hazard 323
    and life on Earth 145
    size of asteroid 65, 439–440
    and tsunamis 440, 760
Chile, earthquake and tsunami (1960)
    165, 231, 779–780
Chilimanzi suite 11, 15, 16
China 44–58
    flooding in 275, 281–282
    geomorphology of 44–46
    Haiyuan earthquake and landslide
        (1920) 527
    Pacific plate subduction in 56
    Shaanxi earthquake and landslide
        (1556) 526–527
    Sinchuan (Wenchuan) earthquake
        and landslide (2008) *234,
        235, 235, 319, 527*
Chingezi tonalite 13, 14
Chipinda batholith 13
Chipuriro belt 14, *17*
2060 Chiron asteroid 66

Chitradurga belt 443
chlorites 560–561
chlorofluorocarbons (CFCs) 73,
    607–608
chondrites 60–61, 67, 542–543
chondrules 60, 61, 67, 542–543
C horizon, of soil 688, *689*
Chotanagpur-Satpura belt 448
Chotanagpur terrane 448
chromates 559
chromite 239
chromosphere 727, 728
Chucuito, Lake 23
Chugach Mountains 334
Chugach–Prince William superterrane
    2–3
Chugach terrane 2–4
Chu Hsi 44
Churchill Province 574–575
cirque(s) 333
cirque glaciers 331
cirrocumulus clouds 142
cirrostratus clouds 142
cirrus clouds 141–142, *142*
cladistics 272–273
Clarke, F. W. 309
clastic sediment 677
clathrates 416
Claudette, Hurricane 277
clays 560–561
Clearwater Lake crater 436
cleavage
    mineral 561
    slaty 539
Cleveland Dyke 398
climate **124–125**
    astronomical forcing of 131–135,
        133, 134
    clouds and 143
    global, atmosphere and 74,
        124–125
    and weathering 813
climate change **125–140**, 336
    atmosphere and 126–127
    global warming **336–344**. *See also*
        global warming
    long-term
        natural 126–131
        and sea level 670, 671–673
    medium, natural 131–140
    ocean circulation and 131, 135–
        136, 589, 592
    orbital variations and 125–126,
        131–135, 555–558
    plate tectonics and 126, 127–130,
        129
    and seasonality 130–131
    short-term
        causes of 341
        v. medium-term paleoclimate
            record 342
        natural 131–140
        observed, effects of 338–341
    and sea level 670, 671

speed (pace) of transitions 140
supercontinents and 128–130
volcanoes and 127–130, 138–140,
    804–805
closed universe model 604
Closepet granite 443, 445
closing salts 678
cloud(s) **141–143**, *142, 143,* 553, 640,
    640
    molecular 453, *696,* 696–699
    thunderstorm 745–747, *747*
Cloud, Preston **140–141**
CMB. *See* central metasedimentary
    belt
coal 415
    age of 117
coal dust 255–256
coast (coastline) 86–100
coastal aquifers, seawater intrusion
    in 371
coastal deserts 204
coastal downwelling 589
coastal dunes 91, *92*
coastal lagoons 91–93, *94*
coastal living 88–89
coastal storms 276–277
coastal subsidence 323, 721, 725–726
coastal upwelling 588–589
coastal wetlands and marshes 95–96
Coast Range(s) 583
Coast Range batholith 347
coaxial deformation 718
Coble creep 188–189
Cocos plate 583
cold-based glaciers 263
cold dark matter 191
collapse sinkholes 481
Collins, Michael 441
collision(s) 157, 161–162
collisional foreland basins 85
Colombia, Nevada del Ruiz eruption
    (1985) 468–469
Colorado, Big Thompson Canyon
    flood (1976) 277–278
Colorado Plateau 583–584
Colorado River 202, 423, 636
Columbia River flood basalt 138, 139,
    485, *487,* 488
Columbus, Christopher 593
Coma cluster 303, 304, 789
comet(s) 66–67, **143–147**, *145,* 549,
    690, 691
    Halley's 372
    impact with Earth 145–147,
        435–442
    and origins of life 145
Comoros 504
compaction 189, 214, 721, 725–726
complex craters 436
Comprehensive Environmental
    Response, Compensation, and
    Liability Act (CERCLA) 254
compression, rock 195
compressional arcs 161, 456–457

compressional bends 750
compressional waves, earthquake 228, *229*, 679–680, *680*
concordant plutons 348, 431, 507–508
concordia 313, *313*
condensation 553
condensation theory, of solar system 599–600
conduction 743
confined aquifers 367
Congo craton 5, 6, 16–18, 346, 362, 364, 642
Congo volcano 724, 795, 801
Connecticut Valley–Gaspé trough 580–581
consequent stream 289
constellation **147–151**
continent(s) 224
continental area, and sea level 673
continental crust **151–153**, *153*, 184, 626, 714, *715. See also* craton(s)
continental divide 220, 582
continental drift **153–155**, *154*, 221, 224, 397, 607, 630, 813, 814
continental flood basalt 485. *See also* flood basalts
continental freeboard 185, 714
continental interior deserts 203
continental margin **155**. *See also* specific types
continental margin volcanoes 457–474
continental rifts 85, 214–218
continental shelves 225
continental shields 152, 196
continent-arc collisions 161–162
continent-continent collisions 162
Contwoyto terrane 567, 570, 571
convection 743
    and continental drift 155
    and Earth's mantle *156*, **156–157**, *157*, 249–250, 732–733
convection zone, of Sun 727
Convention on Biologic Diversity 240–241
convergent plate margin(s) 155, 629, *629*, 630
    arcs of 157–161, *159*, *160*, 455–457
    climate effects of 127–128
    deformation at 715–716
    earthquakes from 162–167, 225
    passive 155, **617–618**, 725–726
    subsidence at 725
    volcanoes of *160*, 160–161, 457–474, 799
convergent plate margin processes 1, **157–167**, 177–182
Cook, James 593
Copernicus, Nicolaus 69, **167–169**
Copper Age 610
copper deposits 239
Coppermine River basalts 642

coral *169*, **169–171**
coral reefs 82–83, 99–100, 102, 169–171
Cordelia (moon) 791
Cordilleran *22*, 22–23, 347, 394, 567, 582–583
CORE. *See Cosmic Background Explorer*
core, of Earth 714
core, solar 727, 728, 730–731
core collapse, in giant stars 736–737
core contraction, stellar 702–703
core-mantle boundary 680
Coriolis, Gustave **171**
Coriolis effect (force) 171, **172**
    and atmospheric circulation 71, 125, 172, 225, 251
    and hurricanes 405, 406
    and iceberg movement 336
    and meteorology 553
    and ocean currents 172, 587
    and trade winds 562
corona 727, 728–729, 730
correlation of rocks 712
corrosion 261
Cosiguana eruption (1835) 804
Cosmic Background Explorer (CORE) 172–173, 605
cosmic dust 437
cosmic microwave background radiation 172–173, 605
cosmic rays **173–174**, 649
cosmochemistry 310
cosmological principle 174
cosmology **174–176**
country rock 348, 431, 507–508
cover-collapse sinkholes 481
cover-subsidence sinkholes 481
Cox, Allen 632
Coy Pond ophiolite 578–579
Crab Nebula 737
Crater Lake 469, 797
craters, impact **435–442**
craton(s) 39, 151–154, **177–182**, 184, 196, 714. *See also* specific cratons
    African 5–22, *6*
    Antarctic 24
    Archean 40–42, *41*, 46–51, 79–82, 152–153, 177–182, 382–383
    Australian 78, 79–82
    Chinese 46–51
    greenstone belts of 352–357
    idealized cross section of *181*
    North American 183, 391, 566–575
    Precambrian 636, 695
    Russian 662–665
    South American 692, *693*, 695
cratonic basin association 42–43
cratonization 181–182
creep 188–189, 320–322, 332, 522, 532

Crescent City, California, tsunami (1964) 763, 779
Cressida (moon) 791
Cretaceous *182*, **182–184**, *183*, 393, 537
Cretaceous-Tertiary (K-T) Boundary 122–123
Cretaceous-Tertiary mass extinction 438–440, 517, 519
crocidolite 255
cross strata 678, *679*
crust **184–185**, 224, 714
    continental **151–153**, *153*, 626, 714, *715*
    deformation of 712–719
    oceanic 218–220, 596–597, 626–627, 629, 714, *715*
cryosphere 331
crystal(s) *185*, **185–189**, *186*, 561
crystal defects 186–188, *187*
crystal dislocations **185–189**
Cuddapah basin 445–446
Cuddapah Group 445
Cuizhangzi Complex 50
cumulonimbus clouds 142–143, 553, *554*, 745, 747
cumulus clouds 141–143, *143*, 553, 745–747
Cuquenan Falls 695
Curie, Marie 395, 652–653
Curie, Pierre 395
currents, ocean **587–589**, *588*
Cuvier, Georges 497, 686
Cybele asteroids 65
cyclones 403, 554. *See also* hurricane(s)
cyclothem 118, **711–712**
Cygnus X-1 star system 106
Cyprian arc 264

D

Dabie Shan belt 51–52
Dalma Group 448
Dalma thrust 448
Dalradian Supergroup 268
Daly, Reginald 607
Dalziel, Ian 362
Damara orogen 19
Dana, James Dwight **190**, 559
Danakil Horst 21
Dan River 426
Daraina-Milanoa Group 500–502
dark matter **191**
Darwin, Charles **191–194**
    and age of Earth 652
    atoll studies of 100, 170
    Lyell (Sir Charles) and 498–499
    Sedgwick (Adam) and 677
    South American exploration by 23, 694, 695
    theory of evolution 191–194, 271, 295, 652
Darwin's finches 193
Dasht-e-Kavir Desert *201*, 203, 208

DASI. *See* Degree Angular Scale Interferometer
dating systems 312–314, 394–399, 653
Davie Ridge 503–504
Dead Sea 724
Dead Sea Transform 35, 630, 749–750
debris avalanches 523–524
debris flows 522–523, 524–525, 758–759, 802
Deccan flood basalts 59, 123, 138–139, 184, 393, 442–443, 445, 485–486, 488, 513
Deccan Plateau 442
deep-sea oozes 101
Deep Space 1 spacecraft 144
deep subsidence 722
deflation 206, 259
deflation basins 206, 259
deforestation 689
deformation of rocks **194–197**, 712–719, *718*
Degree Angular Scale Interferometer (DASI) 605
Delhi craton 450
Delhi-Haridwar ridge 448
Delhi Supergroup 449
delta(s) 93, *94*, 96–97, **197–200**, 288–289, 725–726
delta front 199
delta plain 199
Denali fault 750
dendrochronology 312
dendroclimatology 312
denudation 357
deposition
    in fluvial systems 284–286
    in glacial systems 333–334
    in historical geology 377–378
derecho 746
Desaguadero River 23
desalination 427
Desdemona (moon) 791
desert(s) **200–211**
    drainage systems of 205–206
    formation of 203
    geomorphology of 326
    landforms of 201, 205, 260–261
    location of *201*, 203
    types of 203–204
    winds of 201, 206–207, 259–261
desertification 201–203, 322
Desgt-e Kavir desert 56
Desnoyers, Jules 648
detachment fault 215, *216*
detritus 677
deviatoric stress 716
Devils Tower, Wyoming 348, 431, 508
Devonian **211–213**, *212*, 388–391, 614–615
dewatering 214
Dewey, John F. **213–214**

dew point 553
Dharwar craton 443–446
diachronous boundaries 712
diagenesis **214**, 296, 538
Diamond, Marian 243
diamonds 178–179, 333, 445
diapirism 357, 431
Dibaya Complex 16
Dibdibba arch 36–37
Diego basin 503
Dietz, Robert Sinclair 397
differential GPS 317
differentiated nonchondrites 60, 67
dike(s) 348, 355, 431, *432*, 508
dike injection model 215
dilational bends 750–751
dinosaur extinction 59, 65, 67, 122–123, 323, 393, 438–440, 442, 519, 543
diogenites 62
Dione 668
dip 713
Dirac, Paul 243
disconformities *380*, *381*, 789
discordant plutons 431, 507–508
dislocation annihilation 188
dislocation job 188
dislocation pile-ups 188
dislocations, crystal **185–189**
dismembered ophiolites 597
dissipative beaches 87
dissolution 261, 368–369, 479
divergent plate margin(s) 1, 214, 629, *629*
    climate effects of 127–128
    in continents 215–217
    deformation at 715–716
    earthquakes from 225
    extension at, styles of 215, *216*
    in oceans 218–220
    subsidence at 724
divergent plate margin processes **214–220**
divide 220
Dneiper-Donets aulacogen 663
Doha 36
dolomite 561
Dolomite Mountains 558
Dominion Group 9–11
Dongargarh Supergroup 447
Dongwanzi ophiolite 493
Dongwanzi terrane 48
Don Quijote mission 548–549
Doppler shifts 103
double-line spectroscopic binaries 103
Douglas Harbour terrane 573
Doushantuo Formation 46
downbursts 746
downwelling, coastal 589
drainage basin 84, **220–221**, 289
drainage system **220–221**, 289
    desert 205–206
    rift 215–217
Drake, Sir Francis 593

Drake, Frank 493
Drake equation 493
Drake Passage 694
draperies, speleothem 121
drinking water standards 369
dripstone 121
drizzle 640
Dronning Maud Land 24
droughts 201, 202, 322
drumlins 98, 333
dry lake beds 206
Dry Valleys 23, 25, 204, 210–211
ductile deformation 195, 716
Dumbbell Nebula 626
dunes
    coastal 91, *92*
    desert 207–208, 259–261
Dunnage terrane 576
Durness sequence 268
dust
    interstellar 451–453, *452*
    windblown 207–208
dust clouds 452–453
Du Toit, Alex L. 154, **221–222**, 391, *392*, 607, 630–631
dwarf planets 634–635
dwarf stars **222–223**

E
Eads, James B. 291
early evolution **491–495**
early Mesozoic truncation event 583
Earth **224**, **224–225**, 689–690
    age of 394–399, 652–653
    Gaia hypothesis of **299**
    geochemistry of 310–311, *311*
    magnetic field of 324–325, *325*, **509–511**, 611–613, 631
    origin and evolution of **599–602**
    theory of Hutton (James) 410–414
Earth axis
    and climate *133*, *134*, 135, 557
    and external energy sources 250
    and sea-level rise 672–673
Earth orbit
    and climate change 125–126, 131–135, *133*, *134*, 555–558
    and external energy sources 250
    and Milankovitch cycles 555–558, *556*, 671–673
    and sea-level rise 671–673
earthquakes **225–237**, *226*
    convergent plate margin 162–167, 225
    deadliest 227
    divergent plate margin 225
    epicenter of 227, *228*
    ground breaks in 234
    ground level changes in 235–236
    ground motion in 232, 232–234, *233*, *234*, *235*, 722
    hazards in 232–236, 319
    liquefaction in 235, *235*, 321

magnitude of 230–232
measurement of 230–232, *231*
meteorite impact and 547
origins of 227–230
and sea level 670
subsidence in 722
transform plate margin 752–754
and tsunamis 755–756, *756*,
        769–781
volcano-induced 803–804
waves of 227–230, 327, 679–683,
        680
in Western U.S., prediction of
        681–682
*Earth Resources Technology Satellite
        (ERTS-1)* 655–656
Earth systems
definition of 248
energy in **248–252**
earth system science 252
East African orogen 6, 18–20, *19*,
        112, 500, 642
East African Orogeny 346
East African Rift 6, *20*, 20–21, 85,
        196
extension into Madagascar 503–
        504
oceanic linkage of 215, 218
subsidence at 724
volcanoes of 795
East Anatolian transform 749
East Antarctic craton 24
Eastern Atlantic tsunami (1755)
        772–773
Eastern block (China) 47, 50
Eastern Desert of Egypt 208
Eastern Dharwar craton 443, 445–446
Eastern Ghats of India 364, 442, 443,
        446–447, 642
Eastern Goldfields 80, 81
Eastern Rift Zone 401
East European craton *266*, 662–663,
        664
East Pacific Rise 104, *108*
East St. Louis, Missouri 321
ebb-tidal current 93
ebb-tidal delta 93
eccentricity, orbital 132–135, *133*,
        134, 250, 672
eclipsing binaries 103–104
economic geology **237–240**
ecosystem **240–241**
Eddington, Arthur 243
edge dislocations, in crystals *187*,
        187–188
Ediacaran 382, 384–385
Ediacaran fauna 614, 644
Ediacaran Hills 382, 384–385, 644
Edward, Lake 21, 724
Egypt, water supply for 424–426
Einstein, Albert 69, 106, 175, **241–
        244**, *242*, 305
Ekman spirals 587
elastic deformation 194–195

elastic lithosphere 495
Elbert, Mount 582
Elbruz Mountains 265
Eldredge, Niles 272, 295
electromagnetic spectrum **244–246**,
        245, 649–650
elements, supernovas and formation of
        737–739, *738*
Ellesmere Island 270
Ellet, Charles A. 291
elliptical galaxies 299, 300
Ellsworth Land 26
Ellsworth Mountains 24, 25
El Niño–Southern Oscillation (ENSO)
        131, 136–137, *137*, **247–248**,
        554
and Indian monsoon 563
La Niña v. 137, 247, 554
and sea-level rise 136–137, 247,
        343, 671
El Salvador, landslide in (1998)
        529–530
Elsasser, Walter M. 325, 510
Elsonian Orogeny 358
Eltanin fault zone 749
Elysium region, of Mars 514
Elzevirian batholith 360, 362
Elzevirian Orogeny 358, 360
Elzevir terrane *359*, 360, 362–363
Emba River 265
E = mc² 241, 242
emerald *561*
Emi Koussi 208
emission nebulae 626, 701
emplacement, pluton 349, 431–432
Empty Quarter (Rub'a Khali) desert
        27
Enceladus 668
Encke, Comet 146
Enderby Land 24
end moraines 97, 333
End of Greatness 789, 790
energy
in Earth systems **248–252**
external sources of 248, 250–252
geothermal 330
internal sources of 248–250
renewable sources of 622–623
energy-mass equation 241, 242
English River belt 572
Enriquillo-Plaintain Garden fault sys-
        tem 754
ENSO. *See* El Niño-Southern
        Oscillation
enstatite chondrites 61
entrainment 285
environmental geochemistry 310
environmental geology **252–259**
Environmental Protection Agency
        (EPA) 254, 255, 369, 370
Eocambrian (Ediacaran) 382, 384–
        385
Eocene 122, 123, **259**, 393, 394, 741
eolian **259–261**, *260*

EPA. *See* Environmental Protection
        Agency
epeirogenic movements 343
epicenter 227, *228*
Epipaleolithic 610
epizonal plutons 348
equilibrium line, glacier 332
equinoxes, precession of *133*, *134*,
        134–135
era or erathem 712
Erastosthenes 314
ergs 261
Eridanus Supervoid 790
Eriksson, Leif 593
erosion **261–263**, *262*
in fluvial systems 284–286
sea-level rise and 670
water as agent of 261–262,
        419–420
erratics, glacial 97–98, 333
*ERTS-1. See Earth Resources
        Technology Satellite*
eruptions, historical **453–474**. *See
        also* volcano(es)
Eskola, Pentti **263**, 540
estuary *263*, **263–264**
Etendeka flood basalts 139, 487
Ethiopia 20–21, 196, 218, 512
Ethiopian Plateau 21
Euburnean domain 18
eucrites 62, *624*
Eudoxus of Cnidus 149
15 Eunomia asteroid 63
Europa 478, 634
European geology **264–271**, *265*, *266*
European Laboratory for Particle
        Physics 605
European Space Agency 548–549
eustatic sea-level changes 343–344,
        378–379, 670, 671
evaporite sediment 678
event horizon 106–107
Events, Heinrich 135–136
Everest, Mount 45, *53*, *55*, 449
evolution **271–273**
early **491–495**
of Earth and solar system **599–
        602**, 690–691
in fossil record 272–273, 295
of humans 610–611, 648
of mammals 742
Mendel genetics and 272
processes driving 515–516
of stars 696–700, *697*, *698*, **701–
        705**, *702*
supercontinents and 733–734
theory of Darwin (Charles) 191–
        194, 271, 295, 652
theory of Haeckel (Ernst) 271–272
theory of Lamarck (Jean-Baptiste)
        271
theory of Lyell (Sir Charles)
        498–499
of universe **602–606**

Ewing, Maurice 374, 632
exfoliation 812
exfoliation domes 347–348
exosphere 72
Exploits subzone 578–579
exploration geochemistry 310
extension, at divergent plate margins 215, *216*
extensional arcs 161
extensional collapse, late 357
extensional foreland basins 85
extension joints 296
extinctions, mass **514–520**, *515*
    causes of *515*
    Chicxulub asteroid impact and 59, 122–123, 139, 393, 438–440, 442, 519, 543
    examples of 517–519
    history of life and 517
    Permian-Triassic 517, 518, 615, 620–621, 734
    processes driving 515–516
extraterrestrial life 492–493
extratropical cyclones 406–408
extrusive rocks 431, 506, 507
Eyre, Lake 79, 83

F
facies
    metamorphic 540–542
    sedimentary 377–378, *378*
Failaka 36
Faizabad ridge 448
Falkland Islands 694–695
Falkland Plateau 692, 694–695
falls (mass wasting) 521–524
fan-delta 96, 200, 288
Farallon plate 583
fault(s) 196, 295–296. *See also* specific faults
    detachment 215, *216*
    earthquakes on 225–230
    in greenstone belts 355–357
    ruptures of (ground breaks) 234
    transform 630, 749–754, *750*
fault-block mountains 153, 607
Federal Emergency Management Agency (FEMA) 806
feldspars 561
FEMA. *See* Federal Emergency Management Agency
Fennoscandian (Baltic) shield 268–270
Ferrel cells 76, 124
Fertile Crescent 565, *565*
Filabusi belt 13, 14, *17*
finches, Darwin's 193
fires, earthquakes and 236
firestorm, global 547
firths 267
fishes, age of 683
fission-track dating 314
fjord(s) 331, 333
fjord glaciers 331

flares, solar 78, *727, 729–730*
flash floods 274, 277–278, 282, 420
flats, intertidal 94–95
Flinders Ranges 644
Flinton Group 360
flood(s) **274–283**, 420
    coastal storms and storm surges 276–277
    flash 274, 277–278, 282
    frequency of 274–275, *275*
    as geological hazard 322
    regional disasters 278–282
    river 274–275, 659
    types of 275–278
    volcanoes and 802
flood basalts 123, 138–140, 184, 393, *432*, **485–488**
    continental 486–487
    geographic distribution of *486*
    mantle plumes and 512–513
    submarine 487–488
    volcanism of, environmental hazards of 488
floodplains 288, 420, 661
flood-tidal current 93
flood-tidal delta 93
Flora asteroids 65
Flores, Indonesia, tsunami (1992) 773–776
Florida sinkholes 479–480, 721
flow(s) 521–535. *See also* landslides
flower structure 750
flowstone 121
fluid dynamics 350
fluid mechanics 315–316
fluvial-dominated deltas 96, *198*
fluvial/fluvial systems **283–293**, 661–662
    depositional features of 288–289
    erosion, sediment transport, and deposition in 284–286
    geometry of 283–284
Flynn diagram *719, 719*
flysch 158, **293–294**, *294*, 454, 576, 666
focus, of earthquake 227
fog 640
fold and thrust mountains 196, 607
fold hinge 195
folding/folds, rock *195*, 195–196, 355–356
foliation 539
forearc basin 157, 160, 454, 456
foreland basins 84–85
foreshore 86
formation(s) 377, 711
Fortescue Group 79
Fort Victoria belt 14
Fossa Magna 196
fossil(s) **294–295**, 679
    index 380, 712
    study by Steno (Nicolaus) 705, 708–709
fossil fuels **414–416**, 621–624

fossil record 272–273, 295, 380
fossil succession 686
Fourier, Joseph 341, 351
Fournier ophiolites 579
Fowler, William 603
fractional crystallization 109–110
fracture 194–195, 196, **295–298**
fracture zone(s) 751–752. *See also* specific fracture zones
fracture zone aquifers 297–298, 367–368
Fram Strait 587
Frances, Hurricane *404*
Franciscan Complex 3, 454, 476, 583
Francistown Complex 13
Franklin, Sir John 335
Franklinian orogen 567
Franklin Mountains 582
Fraunhofer, Joseph von 69
Fredericton trough 581
free-air gravity anomaly 349
freeboard, continental 185, 714
freeze-thaw cycle 522
freezing rain 553, 640
freshwater, diminishing supply of 416, 420–421
Friesland Orogeny 269–270
fringing reefs 99, 170
Frisius, Gemma 315
front (weather) 554
Frontenac terrane 359, 360, 362–363
Front Ranges 583
frost heaving 522
frost wedging 333
fuel shortage stage, of star 702
Fuji, Mount 153, *161*, 476, 607, 797
Fujita, Theodore 748
Fujita tornado scale 748
fumaroles 330, 402

G
gabbro 219
Gabon-Chaillu block 16–18
Gaia hypothesis **299**
Galápagos fault zone 749
galaxies **299–303**, *300*, 789–790
    evolution of 302–303
    physical properties of 301–302
    radio **653–654**
    studies by Hubble (Edwin) 299–301, 402–403
    types of 299–301
galaxy clusters **303–304**, *304*, 789–790
Galilei, Galileo 69, 168, **304–307**, 481, 483, 740
Galileo spacecraft 64, 492
Galle, Johann Gottfried 634
Galveston, Texas, hurricanes 324, 409–410
Gama, Vasco da 593
gamma ray(s) *245*, 246, 649–650, 651
gamma-ray astronomy 70

gamma-ray telescopes 740–741
Gamow, George **307**, 603
Gander terrane 576
Ganges River 55, 449, 450–451
    delta of 96, 197, 200, 288–289,
        407, 408
    storm surge of 276–277
Ganymede 478, 634
Gareip belt 19
Garibaldi, Mount 469
garnet 560
Garnet Peak 582
gas
    interstellar 451–453, *452*
    natural 415–416, 621–624, 760
    volcanic 797, 802–803
Gascoyne Complex 82
gas hydrate(s) 416
gas hydrate eruptions, and tsunamis
    760
gas mains, earthquake damage to 236
951 Gaspra asteroid 64
Gatoma belt 14
Gawler craton 78, 79, 81
GCOS. *See* Global Climate Observing
    System
gelatinous plankton 619
Geller, Margaret 790
general relativity 69, 106, 175, 243,
    350, 490, 602–603
geochemical cycles **307–309**
geochemical cycling 310
geochemistry **309–311**, 344
geochronology **312–314**, 394–399,
    652–653
geodesy **314–315**
geodynamics **315–316**
geodynamo theory 325, 510
geographic information systems (GIS)
    316–318
geoid 314–315, *318*, **318–319**, 732–
    733, *733*
geological hazards **319–324**
geologic map, world's first 687
geologic time
    Archean **39–44**, *40*
    Cambrian **112–115**, *115*
    Carboniferous **117–120**, *118*, *119*
    Cenozoic *122*, **122–124**
    Cretaceous *182*, **182–183**, *183*
    Devonian **211–213**, *212*
    geochronology **312–314**, 394–399,
        652–653
    Mesozoic **537–538**
    Paleolithic **609–611**
    Paleozoic 391, **614–615**
    Phanerozoic 39, *40*, **625**
    Precambrian **636–639**, *637*
    Proterozoic *641*, **641–645**
    Silurian 388–391, **683–684**
    Tertiary **741–742**
geology
    African 5, **5–22**, *6*
    Antarctica **24–25**

Arabian **26–39**, *27*, *28*
Asian *44*, **44–59**, *46*
Australian **78–83**, *79*, *80*
economic **237–240**
environmental **252–259**
European **264–271**, *265*, *266*
historical **376–394**
Indian **442–445**, *443*, *444*
North American **566–585**, *567*
petroleum **621–624**
Russian **662–667**, *663*, *664*
South American *692*, **692–696**,
    693
structural **712–719**
geomagnetic reversal **324–325**
geomagnetism **324–325**, *325*
geomorphology **325–326**
Geophysical Laboratory 309–310
geophysics **326–328**
GeoProbe model *623*
George Creek Group 79
geostrophic currents 587–588
geosynclinal theory 710
geothermal energy 330, 622
geysers *329*, **329–330**, 402
GF. *See* Grenville front
GFTZ. *See* Grenville front tectonic zone
Gilbert, Grover K. 96, 197, **330–331**,
    438
Gippsland basin 82
GIS. *See* geographic information sys-
    tems
glacial drift 97, 197–199, 333
glacial erratics 97–98, 333
glacial landforms 332–333
glacial maximum 197–199, 633–634
glacial moraine 97, 333
glacial plucking 333
glacial retreat 131, 555
glacial striations 332–333
glacial systems **331–336**
glaciated coasts 97–98
glaciation 332–333, 429–431
glacier(s) **331–336**
    calving of 334–336, *335*
    erosion by 263
    formation of 332
    as geological hazard 322–323
    movement of 332
    transport by 333–334
Glacier Bay National Park *335*
Glen Canyon Dam 423
Glengarry basin 82
Global Climate Observing System
    (GCOS) 338
global firestorm 547
global positioning systems (GPS) 315,
    316–318, 808
global warming 25–26, 72–74, 125–
    127, **336–344**
    and glaciers 332
    and sea-level changes 337, 338–
        340, *339*, 342–344, 669,
        673

global winter 547
gneisses 539. *See also* specific gneisses
    Archean 39
    greenstone contact relationships
        with 354–355
Gobi Desert 45–46, *201*, 203, 207,
    208, 210, 261
Godavari rift 445, 446
Godavari River 442
Golconda allochthon 583
Gold, Thomas 646
gold deposits 238, *238*
    Alaska 238–239
    Indian 443, 444, 445
    Witwatersrand basin 8, 9, 11,
        238
Goldschmidt, Victor M. 310, **344–**
    345
Golmud River *326*
Gondwana/Gondwanaland *345*,
    345–346, 357, 362, 391–393,
    392
    in Carboniferous period 117–118,
        118
    climate effects of 128–130
    East Antarctica in 25
    formation of 19–20, 112–113
    in Precambria 385
"Gondwanas" 443
Gonu, Typhoon 278
Gorda plate 583
Gosses Bluff crater 436
Gott, J. Richard 790
Goudalie belt 573
Gould, John 193
Gould, Stephen Jay 272, 295
GPS. *See* global positioning systems
Grabau, Amadeus William 44,
    346–347
graded stream 288, 330
grain boundary diffusion 188–189
Grampian Group 268
Grampian Orogeny 268
Grand Banks, Newfoundland, tsuna-
    mis 786–787
Grand Canyon 636, 711
Grande, Lake 23
Grand Teton 582
granite **347–349**
    late-stage 181–182
granite-anorthosite association 643
granite batholith *347*, **347–349**
granite-greenstone terranes 40–42,
    177, 382–383
granitic magma 433–434, *434*, 509
granodiorites 40–41, 626
Grant, Ulysses S. 401
granular flows 523
granulite-gneiss belts 42, 382–383
gravitational tractor strategies, for
    asteroids 549
gravity **349–350**
    and black holes 105–107
    and mass wasting 520

and star formation 696
and subsidence 722
gravity anomalies 327–328, **349–350,** 567
gravity wave **350**
Great Attractor 304
Great Barrier Reef 82–83, 99, 102, 170
Great Basin 330
Great Bear arc 575
Great Boundary fault 448, 449
Great Dike (Zimbabwe) 12
"great" earthquakes 162
Greater Kingham Mountains 45
Great Glen Fault 268
Great Rift Valley. *See* East African Rift
Great Salt Lake 331
Great Sandy Desert (Arabia) 203
Great Sandy Desert (Australia) *201,* 203, 261
Great Wall (galaxies) 304, 790
Great Wall of China 45
Greece, Thera eruption 457–459, *458*
Greek constellations 147–151
greenhouse effect 72–74, 125–127, 143, 338, 341–342, **351–352**
Greenland, ice sheet/cap of 334, 340, 342–343, 669, 673
Green Mountains 579, 581
Greenough, George Bellas 687
Green River 636
greenschist 540
greenstone 540
greenstone belt(s) 39, **352–357,** 382–383
    African 5–22, *17*
    Archean 40–42
    European 269
    geometry of 353–354
    gneiss contact relationships of 354–355
    Indian 443–449
    Madagascar 502–503
    North American 567–575
    structural elements of *353,* 355–357
    structural v. stratigraphic thickness of 354
Grenville front (GF) 358, 361
Grenville front tectonic zone (GFTZ) 358, *359*
Grenville orogenic period 364, 642
Grenville (Ottawan) Orogeny 357–358, 360–364, 575–576, 642
Grenville Province 212, **357–364,** 383, 575–576, 642
    belts of, and Rodinia 364
    subprovinces of 358–361, *359*
    tectonic evolution of 361–364
Grenville Supergroup 360, 361
Grenvillian belts 357–358
Groth, Paul von 344
ground breaks 234

ground level changes, in earthquakes 235–236
ground motion, earthquake *232, 232–234, 233, 234, 235,* 722
groundwater **365–371,** 417–418
    contamination of 369–371
    dissolution by 368–369
    extraction of, and subsidence 722–724
    in fracture zone aquifers 297–298, 367–368
    movement of 365–366
    system of 366, *366*
    use of 367–368
groundwater basin 84
Guiana Highlands 695
Guiana shield 18, 23, 692, 695
Gulf Stream 140, 589
Gunflint Formation 384
Gunnedah basin 79
Guth, Alan 603, 606
guyots 374
Gwanda belt 13, 14, *17*
Gweru-Mvuma belt 13, *17*
gypsum 122, 561

H

Hackett River arc 567, 570–571
Hadean 39–40, 382, 600, 636
Hadhramaut Group 33–34
Hadley, George 124
Hadley cell(s) 71, 74–75, 76, 124, 136, 247, 251
Hadley cell deserts 203
Haeckel, Ernst 271–272
Haima Group 33
Haiti earthquake (2010) 753–754
Haiyuan, China, earthquake and landslide (1920) 527
Hajar (Oman) Mountains 32–34, *57,* 278
Hajar Supergroup 35
Hajar Unit 33
Hale-Bopp comet *145*
halides 559, 561
Hali schists 29, 32
halite 678
Hall, James 413
Halley, Edmond **372**
Halley's Comet 372
halos, sun **731–732**
Hamersley Basin 79, 238, 639
Hamersley Gorge *81*
Hamersley Group 43, 79
Hamrat Duru Group 33, 35
Harare belt 14, 15, *17*
harbor wave 754, 764. *See also* tsunami(s)
hardness, mineral 562
hardpans 206
hard water 369
Harrats 138
harzburgite 594–595, 629
Hasbani River 426

Haughton, Samuel 395
Hawaii
    hot spot 130, *400,* 400–402, *401,* 525, 673, 735
    submarine landslides of 759
    tsunamis of 755, 765–766, 779, 780, 784–785
    volcanoes of 400, *400,* 488, *488,* 506, 796, 797, *797,* 799, 800, *807*
Hawaiian Volcano Observatory 808
Hawasina nappes 33, 35
Hawkeye suite 364
Hawking, Stephen 305
hazard(s). *See also* specific hazards
    earthquake 232–236, 319
    elements, minerals, and materials 252–258, 323
    geological **319–324**
    meteorite impact 543–547
    volcano 319–320, *320,* 796, 799–805
    water as 420
Hearne Province 574
heat 743
heat transfer
    Earth 156–157, 248–250
    radiative 649
    Sun 728
heavy elements, supernovas and formation of 737–739, *738*
Hebrides Islands 267
Hecynian Orogeny 118
Heezen, Bruce 632
Heinrich events 745
Hekla volcano 220, 401, 795
Helana Formation *644*
Helan Mountains 45
helictites 121
heliocentric model 167–168, 304–305, 482–483
helioseismology 728
helium fusion stage, of star 703–704
Hellenic arc 264, 457–459, *458*
Henbury crater 435
Hengduan Mountains 45, 52
Hengshan belt 39, 42, 47–48
Henry the Navigator (prince of Portugal) 593
Heraclides, Ponticus 111
Herodotus 96, 197
Hertzsprung-Russell (H-R) diagram 696, *697,* 702
Hess, Harry **372–375,** 397, 632
Hess Rise 139, 487, 590
heterogeneous strain 717
heterosphere *71,* 72
Hevelius, Johannes 151
Hewish, Antony 646
hexahedrites 60, 63
high-constructive deltas 96, 199–200
high-destructive deltas 96, 200
high-grade granulite-gneiss assemblage 383

High Himalaya 449–450
Highland Boundary Fault 267, 268
high-mass stars 705, 736
Hijaz island arc 32
Hijaz terrane, Al- 31–32
Hilda asteroids 64, 65
Hildebrand, Alan 439
Hildebrandsson, H. Hildebrand 141
Hills Cloud 66
Hilo, Hawaii, tsunami
    1946 755, 765–766, 779
    1960 780
Himalaya Mountains 45, 52, *53*, 55,
        123, 449–450, *450*
    formation of 123, 162, 394, 443,
        449
    ice caps of 334
Himalayan orogen 362
Himalayan terrane 52–53, *54*
Hipparchus 149, **375–376**
historical eruptions **453–474**, 800, 806
historical geology **376–394**
    depositional environment in
        377–378
    facies in 377–378, *378*
    rock record in 376–377
    stratigraphic principles in 376–377
    unconformities and gaps in *380*,
        381, *788*, 788–789
Historical period 610
Hoffman, Axel 12
Hoffman, Paul 362
Hokonui Formation 14
Holmes, Arthur 155, **394–400**, 607,
        630
Holmes-Houtermans model 399
Holocene 122, 124, 393, 394, 648
Homo erectus 610–611
homogenous strain 717
Homo habilis 610
Homo sapiens sapiens 610–611
homosphere *71*, 72
Honduras, landslide in (1998) 529–
        530
Hood, Mount 469
Hooker Chemical 254
Hooke's law 194
Hoover Dam 423
horizontal aquifers 297, 368
horn(s) 333
Horn, Cape 694
hornfels rocks 348
Horsehead Nebula *698*
horse latitudes 562
hot dark matter 191
hot spots 130, **400–402**
    Hawaiian 130, *400*, 400–402,
        401, 525, 673, 735
    Iceland 219–220, 401, 513
    mantle plumes and 512–513, *513*
    Marion 504–505
    and sea-level changes 735
    submarine landslides at 525, 759
    volcanoes of 798–799

hot springs 329–330
Houtermans, Fiesel 399
Howard, Luke 141
howardites 62
Hoyle, Fred 603
H-R diagram. *See* Hertzsprung-
        Russell diagram
Hubble, Edwin 174, 299, 301, **402–
        403**, 490, 603
Hubble classification 299–301
Hubble's constant 174, 301–302, 403
Hubble's law 69, 174–175, 301–302,
        302, 403
Huchra, John 790
Hudson, Henry 593
Hugo, Hurricane 199
Huinaymarca, Lake 23
human evolution 610–611, 648
Humason, Milton 174
Humber zone 576, 578
Humboldt, Alexander von 497
humidity 553
Humphreys, A. A. 291
Hungaria objects 65
Hunter-Bowen Orogeny 82
Huqf Group 33
Huronian Supergroup 643
hurricane(s) **403–410**, 554. *See also*
        specific hurricanes
    categories of 405–406
    cross section of *405*
Hurricane Mountain mélange 576
Hussein (king of Jordan) 424
Hutton, James **410–414**, 497, 652,
        816
Huwaimiliyah dunes, Al- 36
Huxley, Thomas H. 272
Huygens, Christian 414
hydrocarbons and fossil fuels **414–
        416**
hydroelectricity 622–623
hydrogen shell burning, stellar 702–
        703
hydrologic cycle 308, 416, 418, *419*
hydrology 416
hydrolysis 811–812
hydrosphere 104–105, 225, **416–428**,
        417
hydrospheric geophysics 326
hydrothermal deposits 330
hydrothermal fluids 330
hydrothermal solutions 330
hydrothermal vents 104–105,
        106–108, 218, 319–320, *475*,
        492–493
hydroxides 559
10 Hygiea asteroid 63
hypabyssal rocks 431
hyposometric diagrams 714, *715*

I

Iapetus (moon) 668
Iapetus Ocean 113, 212, 267, 268,
        388–390, *390*, 575–576

ice ages 131, **429–431**
icebergs 334–336
ice caps 331, 334, 342–344
    Antarctic *25*, 25–26, 136, 334,
        340
    Greenland 334, 340
ice-dam floods 274, 276
Iceland
    divergent plate margin processes in
        219–220, 629
    hot spot 220, 401, 513
    volcanic activity in 220, 401, 795,
        800, 802
ice sheets 331, 334, 342–343
    Antarctic *25*, 25–26, 136, 331,
        334, 340, 342–343, 669,
        673, 745
    Greenland 334, 340, 342–343,
        669, 673
    ice age 429–431
    melting of, and sea level 334, 669,
        673
    and thermohaline circulation 745
243 Ida asteroid 64
igneous province, large **485–488**. *See
        also* flood basalts
igneous rock 109, **431–435**, 506–508
Ike, Hurricane 91, 199, 324
Illinoian glacial maximum 634
Illinois River
    flooding 278–281, *279*
    liquefaction of floodplain 321
impact basin 84
impact crater structures **435–442**, *436*
Incas 23
index fossils 380, 712
India
    flooding in 277
    monsoons of 562–563
Indian geology **442–445**, *443*, *444*
Indian Ocean
    basin of 586
    circulation pattern in 251
    tsunami (2004) 88–89, 91, 96,
        163, 236, 592, 755, 768–
        772, 771, 774–775
Indian shield 443–449
Indo-Gangetic Plain 443, 449,
        450–451
Indonesia
    Flores tsunami (1992) 773–776
    island arc system of *461*
    Krakatau eruption (1883) *463*,
        464–465, 592–593, 756,
        757, 782–783, 804
    Ruang eruption (1871) 757
    Tambora eruption (1815) 138,
        461–464, *463*, 781–782,
        805
Indus River 52, 55, 449, 450–451
Indus-Tsangpo suture 450, *450*
Indus Valley 451
inflationary theory, of universe 603–
        604

infrared astronomy 70
infrared radiation 244, 245, *245,* 649–650
inlets, tidal 93–94, *94*
Inner Mongolia Plateau 45
Innuitian belt 575
inselbergs 206
interfluve 220
interglacial periods 131, 555, 669
Intergovernmental Panel on Climate Change (IPCC) 125, 336, 337, 341, 352
International Astronomical Union 634–635
interstellar clouds *696,* 696–699, *698*
interstellar medium **451–453,** *452*
interstitials, crystal 186, *187,* 188
intertidal flats 94–95
intraplate volcanoes 796
intrusive rocks 431, *432,* 506, 507–508
Inukjuak terrane 573
invertebrates
    age of 112
    marine, age of 598
Io 478, 634
iodates 559
iodine 253
ionizing radiation 649
ionosphere 72
IPCC. *See* Intergovernmental Panel on Climate Change
Ipswich basin 79
Iran
    Bam earthquake (2003) 166
    mountains of 56–58, *57*
Iranian Plateau 56
Iraq, water supply for 426
Iron Age 609–610
iron meteorites 60, 62–63, *542, 543*
iron ore 237–238
Irrawaddy River 407
irregular galaxies 299, 300–301
irregular galaxy clusters 303–304
Irumide belt 364, 642
Irving, Earl 632
Isalo Group 503
island(s), barrier 88–91, *94*
island arc(s) **453–474,** 630, 632
    Arabian 31–32
    cross section of *454*
    types of 455–457
    volcano types in *457*
island arc accretion 177
isostacy 328
isostatic anomalies 328
isostatic compensation 218
isostatic correction 349
isostatic models 328
isotope geochemistry 310
isotopic dating 312–314, 394–399, 653
Israel, water supply for 426
Isua belt 39, 43, 177

Italy, Vaiont landslide (1963) 527–528, 532
Itremo Group 502

J
J'Alain, Jabal 33
Jal al-Zor hills 36
Jamestown ophiolite 178–179, 355
Janssen, Zacharias 69, 740
Japan **475–476**
    Asama eruptions of 804
    Kobe earthquake (1995) 166–167, 228, 236, 722
    tsunamis of 755, 757, 758, 765, 779, 780–781
Japanese Disease Control Research Center 767
Japan Sea 56
Jeddah Group 29
Jefferson, Mount 469
jet airplane trails 143
jet streams 72, 75, 251, 553
Jibalah Group 31
Johnson, Peter 28–29
Johnstone, Sir John Vanden Bempde 687
joints 295–296, 810–811
jökulhlaups 802
Jolmo Lungma. *See* Everest, Mount
Joly, John 395
Jordan, water supply for 426
Jordan River 426
Jovian asteroids 64
Jovian planets 690–691
Juan de Fuca plate 469
Juliet (moon) 791
Junggar basin 45
3 Juno asteroid 63
Jupiter **476–478,** *477,* 689–691
Jurassic 537
Juric, Mario 790
juvenile island arc accretion 177

K
Kaapvaal craton 5–11, *6, 8,* 43, 153, 177, 354, 355
Kahmmah Group 33
Kaieteur Falls 695
Kaladgi basin 445
Kalahari craton 5–11, *6, 8,* 346, 364, 642
Kalahari Desert *201,* 203, 210
Kamaishi (Sanriku), Japan (1896) 755, 779
Kama River 666
Kameni Islands 457
Kangdian rift 51
Kangkar Tesi Mountains 45
Kansan glacial maximum 634
Kaoko belt 19
Kapuskasing structure 572
Karakoram Mountains 45, 52
Kara Sea 666
Karelian craton 269

Karera fault 449
Karoo basin 8
Karoo flood basalts 695
Karoo sequence 8
karst 120–121, **479–481**
karst terranes 368–369
Kasai block 16
Kasila Group 18
Kaskaskia Sequence 388
katabatic winds 25
Kathadin, Mount 580
Katmai eruption (1912) 804
Katrina, Hurricane 88–89, 91, 199, 290, 323–324, 408–409, 721, 726
Kaveri River 442
Kawr, Jabal 35
Kawr Group 33
Kazakhstan microcontinent 666
Kearsarge-Central Maine basin 580, 581
Keck Mirror Array 604
Kelvin, William Thomson, Lord 395–396, 652–653
Kenai Mountains 334
Kenema-Man domain 18
Kepler, Johannes 69, 110, **481–484**
Kepler's supernova 110, *736*
kettle holes 333
Keweenawan rift 642
Khairagarh Orogeny 447
Khamis Mushayt Gneiss 32
Khanka belt 51
Khoshilat Maqandeli 120, 479
Kibalian block 18
Kibaran belt 364, 642
Kilauea volcano 400, *400,* 488, 506, 797, 799, *807*
Kilimanjaro, Mount 795
kimberlites 47, 178–179, 184, 445
Kimberly craton 78, 79, 82
kinematic models 713
kinetic impact strategies, for asteroids 548–549
King Survey 331
Kipchak arc 665–666
Kirchhoff, Gustav 69
Kirkwood, Daniel 64
Kirkwood Gaps 64
Kirwan, Richard 413
Kivu, Lake 21, 724
Klamath Mountains 476
klippe 369
Kobe, Japan, earthquake (1995) 166–167, 228, 236, 722
Kodaikanal massif 446
Kola Peninsula 269
Kolar belt 445
Kolbeinsky ridge 220, 401
Kolyma Ranges 667
Komatii Formation 7
komatiite 41–42, 639
Komati River 639
Koodoovale Formation 14

Koolyanobbing shear 81, 357
Koppen, Wladimir 124
Koronis asteroids 64, 65
Krakatau eruption (1883) *463,*
    464–465, 592–593, 756, 757,
    782–783, 804
Kresak, Lubor 146
Krishna River 442
K-T Boundary. *See* Cretaceous-
    Tertiary Boundary
Kubbar 36
Kuiper belt 66, 67–68, 144, 435,
    549
Kunlun Mountains 45, 52
Kura River 265
Kurnool Group 445
Kuskokwim River *660*
Kusky, Timothy 12, 363, 567
Kuunga Orogeny 20, 346
Kuwait 35–38, *37,* 206
Kuwait arch 36–37
Kuwait Bay 36

L

Laborde, Albert 395
Lacaille, Nicolas-Louis de 151
Lachlan orogen 78, *80,* 82
lagoons, coastal 91–93, *94*
Lagrange, Joseph-Louis 64
Lagrange orbits 64
lahar 802
Lahasa terrane 52
Lake Ambrose belt 578
lake beds, dry 206
Lallah Rookh belt 356
Lamarck, Jean-Baptiste 271
laminar flow, stream 284
Lamont-Doherty Geological
    Observatory 632
Lamyuka Complex 665
landfills 370
Landsat systems 655–656, 658
landslides 234–235, 320, 521–535
    active, monitoring of 533–534
    disasters, examples of 525–532
    mitigation of damages from
        534–535
    reducing hazards and dangers of
        532–535
    submarine 524–525, 758–760,
        783–787
    and tsunamis 758–760, 783–787
    volcanoes and 802
La Niña 137, 247, *554*
lapilli 796, 801
Lapworth, Charles 598
Laramide Orogeny 583
Laramide ranges 574
Large Hadron Collider (LHC) 604–
    605
large igneous province **485–488.** *See
    also* flood basalts
Larsen ice shelf 26
Lassen Peak 469

late extensional collapse 357
lateral blast, of volcano 757–758, 801
lateral continuity, principle of 705,
    709
laterite 689
Laurasia 117, 392
Laurentia 267, 358, 362
Laurentide ice sheet 25–26
lava 431, **488–490,** 506, *506,* 796–
    797, 800–802
lavaka *262*
Lavoisier, Antoine 309, 377
Lawson, Andrew Cooper **489–490**
Lawson, Robert 397
Lazarev, Mikhail 23
lead, primeval 398–399
lead contamination 256
lead dating systems 312–314, *313,*
    394, 396–399
lead deposits 238
leading plate margins 155
Leatherman Peak 582
Lebanon, water supply for 426
Lebombo sequence 5
Lemâitre, Georges **490–491,** 603
Lena River 667
lenticular clouds 142
Leonean Orogeny 18
Leonid showers 59, 542, 544
Lepelle terrane 573
Lesser Antilles arc 800–801
Lesser Kinghan Mountains 45
levees 289–293, 659
    avulsion of 289–290
    failure of, modes of 292
    Mississippi River 279–281, 283,
        289–293, *291,* 408–409,
        421
    Yellow River 281–282
Le Verrier, Urbain 634
Lewisian belt 39, 42
Leyte, Philippines, landslide (2006)
    530–532
LHC. *See* Large Hadron Collider
lherzolite 594–595
Libby, Montana, asbestos 255
Libby, William F. 312
Liberian domain 18
Liberian Orogeny 18
Libyan Desert 200, 203, 208, 209
life's origins 145, 383–384, **491–495,**
    602, 733–734
lightning 745, 747
lignite 415
limestone 540, *718*
Limpopo belt 5, 6, 7, 13, 39, 42
Limpopo Province *8, 9,* 16, 383
Linde, Andrei 606
linear dunes 208, 261
line defects, in crystals 186–188, *187*
Lippershey, Hans 69, 740
liquefaction 235, *235,* 321
Lisbon-Eastern Atlantic tsunami
    (1755) 772–773

lithosphere 104, 224–225, **495–496,**
    626, *627*
lithosphere deformation 712–719
"little ice age" 429
Little Salt Lake 331
Lituya Bay, Alaska, tsunami (1958) 784
Liyah ridge 36
Llandoverian Age or Series 683
Local Group (galaxy cluster) 303, 789
Lochaber Group 268
lochs 267
lodranites 62
loess 204, 208, 259
loess plateau, Asian 45
Logan, Sir William 360
Loma Prieta, California, earthquake
    (1989) 233, 681, 753
Long Beach, California 723
long-period comets 143
Long Range Peninsula 576
longshore currents 86, 94
longshore drift 86
Lonsdale, Peter 104
Los Angeles, water supply for 422–
    423, *423*
Lost River Range 582
Louisiana. *See* New Orleans
Love, William T. 254
Love Canal 254
Lovecock, James 299
Lowell, Percival 301, 634
Lower Greenstones 13–15, *17*
Lower Gwanda belt 13, 14, *17*
low-grade metamorphism 538
low-silica lava 488
low velocity zone 680
Luanyi gneiss 16
Luban Mountains 45
Lufilian arc 19
Luis Alves cratonic fragment *693*
Luria arc 19
Lurian belts 364, 642
luster, of minerals 561–562
Lut Desert 56, 203, 208
Lyell, Sir Charles 193, 259, 460,
    496–499, 537, 633, 652

M

Mackenzie Mountains 582
Mackenzie swarm 642
Mackeral sky 142
macrobursts 746
Madagascar **500–506,** *501,* 522–523
Madagascar craton 5
Maevatana belt 502
mafic lava 488
Magellan, Ferdinand 593, 694
Magellan, Strait of 694
Magellanic Clouds 300, 303, 789
Magellan Ranges 692, 694
magma **506–511**
    andesitic 434, 489, 509, 797
    basaltic 433, 508–509, 629, 797,
        800–801

composition of 506–507
granitic 433–434, *434, 509*
igneous rock formed from 431–432, 506–508
origin of 432–433, 508
rhyolitic 489, 797
solidification of 434–435, 509
surface flow. *See* lava
magmatic arc 128
magmatic differentiation
by fractional crystallization 109
by partial melting 109, 433, *433,* 508
magnetic field 324–325, *325,* **509–511,** 611–613, 631
magnetic stripes, seafloor 611–613, *612*
magnetosphere *77,* 77–78, **509–511**
Mahajana basin 503
main asteroid belt *62,* 64–65
Main Boundary Thrust 443, 449
Main Central Thrust 449
Main Ethiopian Rift 20–21
Main Frontal Thrust 449, 450
main sequence star 696–700, 701–702
Main Uralian fault 666
Majils Al Jinn Cave 120, 479
Majunga basin 503
Makran Mountains 56–58, *57*
Malafundi granites 16
Malakialana Group 502
Malawi, Lake 21, 724
Malawi orogen 19
mammals, age of 741–742
mammatus clouds 142–143
Mammoth Cave 120, 479
Mammoth hot springs 402
Manhattan Project 243
Manicougan crater 436, 440
Manihiki Plateau 139, 487, 590
Manjeri Formation 13–15, 16
Man shield 6, 18
mantle 224, **511–512,** 714
mantle convection *156,* **156–157,** *157,* 249–250, 732–733
mantle plumes **512–513,** *513*
map, geologic, world's first 687
Marampa Group 18
marble *539,* 540
marginal basins 161, 456
margins, continental **155.** *See also* specific types
Marianas-type arcs 158, 455–456, 719
Marie Byrd Land 24, 26
marine invertebrates, age of 598
Mariner 10 spacecraft *537*
marine regression 377–380, *379,* 388–390, 735
marine transgression 377–380, *379,* 735
Sauk 385–387, *386*
Tippecanoe 387–388

Marion hot spot 504–505
Mariotte, Edme 341, 351
Maromokotro, Mount 500
Mars 492, *513,* **513–514,** 689–690
marshes, coastal 95–96
Martinique, Mount Pelée eruption (1902) 465–468, *466–467,* 801
Mashaba tonalite 12–13, *14*
mass-energy equation 241, 242
mass extinctions **514–520,** *515*
causes of *515*
Chicxulub asteroid impact and 59, 122–123, 139, 393, 438–440, 442, 519, 543
examples of 517–519
history of life and 517
Permian-Triassic 517, 518, 615, 620–621, 734
processes driving 515–516
massifs 361
mass wasting 234–235, 263, 285, 292, **520–536**
disasters, examples of 525–532
driving forces of 520–521
mitigation of damages from 534–535
physical conditions controlling 521
prediction of 532–533
processes of 521–524
reducing hazards and dangers of 532–535
and tsunamis 758
Masvingo belt 14, 15, *17*
Matius, Jacob 69
matrix 431, 506
matrix theory, of universe 606
Matsitama belt 13
Matterhorn 265, 333
Matthews, Drummond H. 375, 632
Matuyama, Motonori 632
Mauna Loa volcano 796, 797, 799
mausim 562
Mayon volcano *320,* 799
Mazama, Mount 469–470
Mazatzal orogen 575
Mazinaw terrane *359,* 360
McHugh Complex 3
McKinley, Mount 750
McLelland, Jim 363
meandering streams 286–287, 659, *660,* 661
mechanical models 713–714
mechanical weathering 810–811, 813
Mediterranean Sea 264, 586
Mekong River 52
mélanges 1, 158, 268, 455, *536, 536*
melting, partial 109, 433, *433,* 508
membrane theory, of universe 606
Mendel, Gregor Johan 271
Mendocino fault zone 749
Mercalli, Giuseppe 231
Mercalli scale 231–232

Mercury *537,* **537,** 689–690
Merrimack trough 581
Mesolithic 610
mesopause *71, 72*
Mesoproterozoic 382
mesoscale convective systems 746–747
mesosiderites 60, 63
mesosphere *71, 72,* 76
Mesozoic 393, **537–538**
Mesozoic truncation event, early 583
mesozonal plutons 348
metallic ores 237–240
metamorphic facies 540–542
metamorphic rocks **538–542**
metamorphism **538–542**
changes in 539
grades of 538–539
plate tectonics and 540–542, *541*
metamorphosed ophiolites 597
metasomatic **542**
Metazoa 614
meteor(s) 59, **542–552**
Meteor Crater, Arizona 436
meteoric **552**
meteorites 59–60, **542–552**
composition of 60–63, 66–67
as geological hazard 323
impact with Earth
atmospheric shock waves of 543, 544–546
craters/structures of 435–442
hazards of 543–547
mitigating dangers of 543, 547–549
and sea level 670
solid Earth shock waves of 547
and tsunamis 543, 547, 760
tsunamis caused by 543, 547
near-collisions with Earth, recent 549–551
origin of 66–67, 543
and origin of Moon 600–601, *601*
meteorology **552–555**
meteor showers 59, 542, 544
methane, as greenhouse gas 73, 341
Mexico City
earthquake (1985) 233
subsidence in 723
mica 558, 560–561, *718*
microbursts 746
microfossils, Archean 43
microwave radiation 244, *245,* 649–650
microwave remote sensing 655
mid-Atlantic ridge 220
Mid-Devonian Acadian Orogeny 212–213
Middle East, water issue in 421–422, 424–426, *425*
Middle Hamersley Group 79
Middle Marker 355

Midlands belt 13, 14, 15, *17*
midlatitude deserts 203
midocean ridge basalt (MORB)-type
    ophiolites 178
midocean ridge system 214–215,
    218–220, 225, 632
    geological hazards in 319–320
    volcanoes of 798–799
    volume of, and sea-level changes
        673
Mid-Pacific Mountains 139, 487, 590
Midyan terrane 31–32
Milankovitch, Milutin M. 131, 250,
    555, 671
Milankovitch cycles 118, 131, *134,*
    134–135, 250, 555–558, *556,*
    557, 671–673
Milky Way Galaxy *603,* 789
    black hole in 106
    cosmic microwave background
        radiation in 172–173
    in galaxy cluster 303, 304
    as spiral galaxy 300
Miller, Stanley L. 384
Mimas 668
mineral(s) **558–562**
    composition of *559*
    deposits 237–240
    hazardous 252–258, 323
    properties of 561–562
    silicate tetrahedron of 559–561,
        560
    types of *559, 559*
mineralogy **558–562**
mine tailings *253*
mini-craton 12
Minnesota River Valley Province
    572
Minto block 573
Miocene 122, 123, 393, 394, 741
Miramichi massif 581
Miranda (moon) 791–792
mirror plane symmetry 185–186
Miskan 36
Mississippian Period 117–120, 391
Mississippi River 88–89
    delta of *96, 197,* 197–200, 288–
        289, 377
    flooding 278–281, *279,* 283, 322,
        421
    levee building on 279–281, 283,
        289–293, *291,* 408–409,
        421
    liquefaction in floodplain 321
Mississippi River Commission (MRC)
    279, 291–293
Missouri River
    flooding 278–281, *279,* 322
    floodplain 282, 321
Mitch, Hurricane 406, 529–530
Miyashiro, Akiho 475
modern constellations 150–151
Mohorovičić, Andrija 327, 595–596,
    680–682

Mohorovicic (Moho) discontinuity
    33, 151, 327, 495, 595–596,
    626, 682
Moh's hardness scale 562
Moine thrust 268
Moinian Assemblage 268
Mojave Desert *201,* 203–204
molasse 294, 388, 449, 666
molecular clouds 453, *696,* 696–699,
    698
molybdates 559
monoclines 195
Monoun, Lake 803
monsoon(s) **562–563**
monsoon deserts 204
Moon
    impact craters of 435–436, 441
    origin and evolution of 599–602,
        601
    Pluto comparison with 634
moon(s). *See* specific moons
moonmilk 122
Moores, Eldridge 362
moraine, glacial 97, 333
MORB. *See* midocean ridge basalt-
    type ophiolites
Morin massif 361
Morley, Lawrence 375
Morondova basin 503
Mosetse Complex 13
Motloutse Complex 13
Motosuko Lake *476*
Mo Tzu (Mozi) 306
mountain(s). *See also* specific moun-
    tains and ranges
    fault-block 153, 196–197
    fold and thrust 196
    orogeny of **606–607**
    volcanic 153
mountain belts 153, 196–197, 225,
    630
mountain glaciers 331
Mount Bruce Supergroup 79
Mount Darwin belt 14, *17*
Mount d'Or tonalite 13
Mount Isa Complex *80,* 82
Moyar-Bhavani-Attur shear zone 446
Mozambique belt 18–19, 31, 112
Mozambique Channel 500
Mozambique Ocean 345–346, 391,
    642
MRC. *See* Mississippi River
    Commission
M-theory, of universe 606
Mtshingwe Group 13, 14
mudflows 523, 802
mudstones 539
Mulangwane Range 14
Muldersdrif belt 7
Mulholland, William 423
Murchison, Roderick I. 211, 377, 497,
    620, 676–677
Murchison belt 7
Murchison Province 80, 81

Musgrave orogen 78, *80,* 82
Mushandike granitodiorite 13
Mutare belt 14, 15–16, *17*
Muti Formation 33
Mweza belt 13, 14, 15–16, *17*
Myanmar, storms and 407

N

Nabarro-Herring creep 186
Nabberuh basin 79
Nabitah belt 31
Najd fault system 31
nakhlites 62
Nallamalai Group 445
Namaqua-Natal belt 5, 364, 642
Namche Barwa 55, 449
Namib Desert *201,* 204, 261
Nandgaon Orogeny 447
Nanga Parbat 55, 449
Nanhua rift 51
Nanling Mountains 45
Napier Complex 24
Narryer Gneiss Complex 80
NASA. *See* National Aeronautics and
    Space Administration
Nasser, Lake 424–426
Natal'n, Boris 180
National Aeronautics and Space
    Administration (NASA) 144,
    492–493, 514, 547–548,
    657–658
National Oceanic and Atmospheric
    Administration (NOAA) 766,
    767, 774–775
native elements *559,* 561
natural gas 415–416, 621–624
natural gas eruption, and tsunamis
    760
natural selection 191–194, 272
Navstar Satellites 316–318
Nazca plate 692–693, 779
Neanderthals *272,* 610
near-Earth objects. *See* asteroid(s);
    meteorites
near-infrared radiation 650
Nebraskan glacial maximum 633–
    634
Nebraska Sand Hills 261
nebula 452, *452*
    emission 626, 701
    planetary 626, **626,** 704
nebular theory, of solar system 599,
    600
Negev Desert 209
Negro, Rio 693
nektonic **618–620**
Neogene 122, 393, **564,** 741
Neolithic **564–566,** 610
Neoproteroic 382
Neo-Tethys Ocean 265
Neptune 566, **566,** 689–690
neptunism 411–413, 815–816
Nereid 566
Netherlands, subsidence and 721

neutronization 736–737

Nevada del Ruiz eruption (1985) 468–469

Nevados Huascaran landslides 523–524, 528, 532

New Foundland tsunamis 786–787

New Orleans
    flood control for (levees) 279–281, 283, 289–293, 290–291, *291*, 408–409
    flooding 88–89, 281
    hurricanes and 409–410
    sea-level rise and 670
    subsidence and 323, 377, 721, *721*, 725–726

Newton, Sir Isaac 69, 481, 740

Ngezi Group 14, 15

Ngorongoro crater 21

Nicaragua
    Cosiguana eruption (1835) 804
    landslide in (1998) 529–530
    tsunami (1992) 776

nickel deposits 239

Nier, Alfred 398–399

Nile River 659
    delta of 96, 197, 200, 288–289
    modifications of 289
    water supply from 424–426

Nilgiri Hills 442, 446

nimbostratus clouds 142

nimbus clouds 141–143

Ningxia Autonomous Region 207, 260

Nipissing terrane 358, *359*

nitrates 559

nitrous oxide, as greenhouse gas 73, 341

NOAA. *See* National Oceanic and Atmospheric Administration

nonchondrites 60, 67

noncoaxial deformation 718

nonconformities *380, 381,* 789

nonionizing radiation 649

nontransform extension regions 752

normal faults 196

Normanskill formation 580

Norseman-Wiluna belt 353–355, 356

North American cratons 183, 391, 566–575

North American geology **566–585,** *567*

North Anatolian fault 264, 630, 749, 753

North Atlantic Igneous Province 138–139, 487

North-Central Afar Rift 20–21

North China craton 46, 47–50, 55, 108, 153, 179–180, 353, 597, 618, 642

northern lights (aurora borealis) 72, **77–78,** *78*

North Hebel orogen 49

North Qinling terrane 51

Northridge, California, earthquake (1994) 319, 324

Norway tsunamis 786–787

Notre Dame arc 579

nova 110–111, **585,** 735

Nubian Desert 208

nuclear attack, for deflecting asteroids 548

nuées ardentes 757, 797, 801, 802

Nuna 574–575

Nyiragongo volcano 724, 795, 801

Nymph Lake 402

Nyos, Lake 803

O

Oberon 791

Ob Gulf 666

Ob-Irtysh drainage system 666

oblate strain 719, *719*

obliquity, of Earth axis 132–135, *133, 134,* 250, 557, 672–673

Ob River 667

observational astronomy 69–70

Occidental Petroleum 254

ocean(s) 224–225. *See also* specific oceans
    divergent plate margins in 218–220
    exploration and study of 591–594
    external energy-driven processes in 251–252
    thermohaline circulation in 131, 135–136, 251–252, 587, **743–745,** *744*

ocean basins **586–587**
    exploration of 591–594
    spreading of 218, 372–375, 611–613

ocean crust accretion 177

ocean currents **587–589,** *588*

ocean floor *101,* 101–102, 104

oceanic crust 218–220, 596–597, 626–627, 629, 714, *715*

oceanic plateau 139, 487–488, **589–591,** *590*

oceanic plateau accretion 177, 180

oceanography *591,* **591–594,** *592*

octahedrites 60, 63

oil 414–415, 621–624

oil field 415

oil pool 414–415

Olber, Heinrich 174

Olber's paradox 174

Old Faithful 329, 330, 402

Ol Doinyo Lengai volcano 795

Old Red Sandstone 267, 268, 270, 377, 388, 581, 614

Oligocene 122, 123, 393, 394, 741

olistostromes 454

olivine 559, 560

Olondo Group 663–664

"Oman Exotics" 35

Oman Mountains 32–34, *57,* 278

Ontong-Java Plateau 139, 487–488, 590–591

Onverwacht Group 355

Oort Cloud 66, 67, 68, 144, 435, 549, 600, 691

Ophelia (moon) 791

ophiolite(s) **594–597.** *See also* specific ophiolites
    accretion 177–180
    cross section of *595*
    formation of, process of 596–597
    geographic distribution of *594*
    oceanic 219–220
    world's oldest 597

optical astronomy 70

optical telescopes 740

orbit(s)
    asteroid 60–66
    comet 143
    Earth
        and climate 125–126, 131–135, *133, 134,* 555–558
        and external energy sources 250
        and Milankovitch cycles 555–558, *556,* 671–673
        and sea-level rise 671–673

ordinary chondrites 61

Ordos basin 45

Ordos block 47

Ordovacian 112, 598, **598–599,** *599,* 614–615

ore deposits/reserves 237–240

Orellana, Francisco de 23

organic geochemistry 310

original horizonality, principle of 705, 709

origin and evolution
    of Earth and solar system **599–602,** 690–691
    of universe **602–606**

origins of life 145, 383–384, **491–495,** 602, 733–734

Orinocan-Nickerie event 695

Orion Nebula 698

Orkney Islands 267

Orlando, Florida, sinkholes 479–480

orogen(s) 151, 153, 184, 196, 630, 714. *See also* specific orogens
    accretionary 180–181
    Pan African 5
    Proterozoic 641–642

orogenic belts 153, 154, 184, 196, 225

orogeny **606–607.** *See also* specific orogenies

Ottawa (Grenville) Orogeny 357–358, 360–364, 575–576, 642

Otway basin 82

Ouachita Orogeny 118

outwash 333

Ouzzalian craton 208

Oweineat Mountains 208

overland flow 262

Owen, Richard 193

Owen Fracture Zone 35

Owens Valley 423

oxide(s) 559, 561
oxide-iron formations 638
ozone 73, 607
ozone hole 73, **607–608**, *608*

P
Pacific Northwest
  earthquake prediction for 681–
    682
  tsunami risk for 775
Pacific Ocean
  basin of 586
  circulation patterns in 251
  thermohaline circulation in *744,*
    744–745
Pacific plate subduction 56
Pacific Tsunami Warning Center 766,
    767, 774–775
Pacific-type arcs *159*
Pacific-type margins 155
Pacific-type ridges 218
Pacini, Franco 646
pack ice 335–336
Padbury basin 82
pahoehoe 488–489, 796, 800
Pakistan earthquake (2005) 165, *232*
Paleocene 122, 123, 393, 394, 741
paleoclimatology **609**
Paleogene 122, 259, 393, 741
paleogeography 382
paleolith(s) 609
Paleolithic **609–611**
paleomagnetism 375, 510, **611–613,**
    631
paleontology **613–614**
Paleoproterozoic 382
Paleozoic 112, 117, 211, 385, 391,
    614–615, 683
Palestinians, water supply for 426
Palghat-Cauvery shear zone 446, 500
2 Pallas asteroid 63
pallasites 60, 63
Palmer Land 26
palm tree structure 750
Pan African belts 18–20, *19,* 112, 642
Pan African origens 5
Paneth, Fritz 398
Pangaea 357, 391, **616–617,** 732
  breakup of 123, 182–183, 392–
    393, 394, 617
  in Carboniferous period 117
  climate effects of 128–130
  continental drift and *154,* 154–
    155
  formation of 345–346, 616
  landmass distribution in *616*
panspermia 492
parabolic dunes 207, 261
parallel strata 678
Paraná Basin *693*
Paraná flood basalts 139, 184, 487,
    695
Paraná River 23, *692, 693*

Parguazan event 695
Parnaiba Basin *693*
Parry Sound terrane *359,* 359–360
partial melting 109, 433, *433,* 508
partial ophiolites 597
particle accelerator 604–605
particle radiation 649
Partridgeberry Hills granite 579
passive margin 155, **617–618,** 725–
    726
Paterson orogen *80, 82*
Patterson, Clair 399
pavements, desert 206
peat 415, *688*
pediments, desert 206
pelagic **618–620**
Pelée, Mount, eruption (1902) 465–
    468, *466–467,* 801, 804
Pelmo massif *557*
Penfield, Glen 439
Peninsular Gneiss 445
Peninsular terrane 2
Pennsylvanian Period 117–120, 391
Penobscot-Exploits arc 579
Penobscottian Orogeny 268, 576–579
Penrose-type ophiolite 594–596
Pensacola Mountains 25
Penzias, Arno 173, 603
Pequeno, Lake 23
Percival, John 573, 604
peridot 560
peridotite 68–69, 626–627, 629
permeability 365, 367
Permian 614–615, *620,* **620–621**
Permian-Triassic mass extinction 517,
    518, 615, 620–621, 734
Perseid showers 59, 542, 544
Perth basin 79
Peru
  El Niño and 136–137, 246–247
  landslides in 523–524, 532
petrography **624**
petroleum 414–415, 621–624
petroleum geology **621–624**
petrologic Moho 495
petrology **624**
Pettijohn, Francis John **624–625,** *625*
phanerites 431–432, 506
Phanerozoic 39, *40,* 112, 385, **625**
phase rule 310
phenocrysts 431, 506
Philippines
  Leyte landslide (2006) 530–532
  volcanoes of 138, *320,* 471–473,
    488, 758, 799, 802–803,
    805–807
Phillips, John 686–687
Phlegraean Fields 459
phosphates 559, 561
photodisintegration 736
photosphere 727, 728
photosynthesis 104, 384, **625–626**
physical weathering 261

phytoplankton 619
piedmont glaciers 331
Pietersburg belt 7
Pikwitonei uplift 573
Pilbara craton *41, 43,* 78, 79, *80, 81,*
    354–356
Pilbara Supergroup 79
pillars, sun **731–732**
Pinatubo, Mount 138, 471–473, 488,
    799, 802–803, 805–807
Pinzon, Vincent 23, 694
Pioneer fault zone 749
Pipestone Pond ophiolite 578–579
Piscataquis arc 580, 581
Pitman, Walter 632
placer gold 238–239
plane strain 719, *719*
planet(s). *See also* specific planets
  definition of 634–635
  Jovian 690–691
  terrestrial 690
planetary motion, laws of 482–483
planetary nebula 626, **626,** 704
planktonic **618–620**
plateau basalt 485. *See also* flood
    basalts
plate tectonics 224, **626–633,** *627,*
    628, 629, 631. See also specific
    physical features and processes
  accretionary wedge 1–5, *2,* 162
  and climate 126, 127–130, *129*
  continental drift 153–155, *154*
  earthquakes 225–230, *226*
  and geological hazards 319–320
  historical development of 630–
    632
  horizontal v. vertical 356
  and metamorphism 540–542, *541*
  and sea-level rise 673–674
  and structural geology 712–713,
    714–716
  and subsidence 724–725
  and volcanoes 795–796, 798–799
platinum group elements 239
playas 206
Playfair, John 413
Pleistocene 122, 123–124, 393, 609,
    610, **633–634,** 648
plieus clouds 142
Plinian eruption column 461, 797
Pliny the Elder 461, 797
Pliny the Younger 461
Pliocene 122, 393, 741
plumose structures 296
Pluto 66, **634–635**
pluton(s) *348,* 348–349, 431, *432,*
    507–508
pluton emplacement mechanisms 349,
    431–432
plutonic rocks 431, 506, 507–508
plutonism 410, 816
point defects, in crystals 186–188,
    187

Point Lake belt 42, 354–355, 567–569
polar bears, global warming and *337*
Polar cells *76*, 124
polar deserts 204
polar glaciers 331–332
polar low 408
Pongola basin *8*
Pongola Supergroup 7, 43
Pontiac belt 572
Poopo, Lake 23
Popelogan arc 579
population growth 323, *422*
porosity 365, 367
porphyries 431, 506
Portia (moon) 791
potassium-argon dating 314
potential field studies 327–328
Potsdam sandstone 576
Powell, John Wesley 423, **635–636**
Powell, Lake 202
Powell Survey 331
Pratt model 328
Precambrian *113*, 382–385, **636–639**, *637*
Precambrian-Phanerozoic transition 385
precession of equinoxes *133*, *134*, 134–135
precipitation *553*, **640**
precipitation patterns, global warming and 340–341
pressure solution 188–189
Prévost, Constant 497
primary waves, earthquake 228, *229*, 679–680, *680*
primeval lead 398–399
primitive nonchondrites 60, 67
Prince Charles Mountains 24
prodelta 199
prograde 538–539
prolate strain *719*, *719*
prominences, solar 730
Protalus ramparts *633*
Proterozoic *39*, *40*, 382, 383, 636, *641*, **641–645**
protostar 698–699
Proxima Centauri 223
pryoclast(s) 796
pryoclastic flows 757
Ptolemy (Claudius Ptolemaeus) 111, 151, 593, **645–646**
Puck (moon) 791
pull-apart basin 85–86, 724–725, 750
pulsar **646**
pumice 797
punctuated equilibrium 272, 295
Purcell Supergroup 582–583
Purcell trench 582
pure shear model 215, *216*, 718
Purtuniq ophiolite 355
Puyehue, Mount 165
P-waves, earthquake 228, *229*, 679–680, *680*

pyroclastic flow 796, 800–802
pyroxene 560

Q
Qahlah Formation 33
Qalluviartuuq belt 573
Qaruh 36
Qatari Desert *260*
Qattara Depression 208
Qiadam basin 45
Qiangtang terrane 52, *54*
Qianxi-Taipingzhi terrane 48
Qinghai-Xizang Plateau. *See* Tibetan Plateau
Qinglong basin 48
Qinling-Dabie-Sulu belt 51–52
Qinling Mountains 45
QSC 502
Quadrilatero Ferrifero 695
quartz 558, 561, *718*
quartz dust 255
quartzite 540
quartzo-feldspathic gneiss terranes 177
quasars **647–648**
Quaternary 122, 393, 633, **648**
Queen Maud Land 26
Que Que belt 14
Quetico Province 572
Qurain, Al- 36

R
radar interferometry 808
radar remote sensing 655, 657–659
RADARSAT 658–659
radiation **649–651**, 743
    cosmic microwave background **172–173**
    electromagnetic 244–246, *245*, 649–650
radiation zone, of Sun 727
radiative forcing 341
radiative heat 649, 743
radioactive dating 312–314, 394–399, 653
radioactive decay **651–653**
radio astronomy 70
radio galaxies **653–654**
radiolarians 3
radio telescopes 740
radio waves 244–245, 649–650
radon 256–258, 651–652
Rae Province 574
Ragbah Orogeny 29
rain 553, 640
    acid 73
    freezing 553, 640
Rainier, Mount 469, *633*
rain-shadow deserts 203–204
Rajasthan Desert 204
Raleigh number 249
ramp valleys 750
Ranotsara fault zone 500, *501*, 503

Reagan, Ronald 255
recurved spits 90, 93
red dwarfs 222–223
red giants 703
Redoubt, Mount 805
Red River fault 749
Red Sea 215, 218, 617, 724
redshift 301–302, 304, 402–403, 789–790
reefs 82–83, 99–100, 102, 169–171
Rees, Martin 606
reflecting telescopes 740
reflective beaches 87
refracting telescopes 740
regional deformation of rocks 196
regional metamorphic rocks 540
regolith 205, 520
regression, marine 377–380, *379*, 388–390, 735
Reguibat shield *6*, 18
regular galaxy clusters 303
Reisner, M. 423
relative humidity 553
relative sea-level rise 670, 671
relativity, theory of 69, 106, 175, 241–244, 350, 490, 602–603
Reliance Formation 14, 15
remote sensing **654–659**, *657*
renewable energy 622–623
reptiles, age of 537–538
resonances 64
Réunion Island 525, 759
reverse faults 196
Reykjanes Ridge 219, 401, 513, 629
Reynolds, Doris L. 397–398
Rhea 668
Rheic Ocean 118
rheology 316
Rheticus, Georg Joachim 168
Rhodesdale Complex 11
Rhodesian craton. *See* Zimbabwe craton
Rhône delta 96, 200
rhyolitic magma 489, 797
Richardson, Benjamin 686
Richardson Mountains 582
Richter scale 230–232
ridge(s). *See also* specific ridges
    beach 86
    oceanic 218–220, *219*, 225
ridge push 715
Riedel fractures 751
Ries Crater 564
rifts 724. *See also* specific rifts
    continental 84, 85, 214–218
        evolution of *217*
        extension of 215, *216*
        oceanic 218–220
Rigolet Orogeny 364
Rimah, Wadi Ar- 37
Rimmah Orogeny 29
Ring of Fire 779, 795–796
Ringwood, Alfred E. (Ted) 511

Rio de la Plata craton 692, *693, 695*
Rio Grande Ridge 513
Rio Grande rift 724
Rita, Hurricane 88, 290, 323, 408, 721, 726
Ritter, Mount, eruption (1888) 757
river(s) 659–662. *See also* specific rivers
    channel patterns 286–287
    dynamics of flow 287–288
    erosion, sediment transport, and deposition in 284–286
    fluvial action of 283–293, 661–662
    geometry of 283–284
    modifications of
        to alleviate water shortages 421–426
        effects on dynamics 289–293
river floods 274–275, 659
river systems **659–662**
RLMZ. *See* Robertson Lake mylonite zone
Robert's Arm-Annieopsquotch belt 579
Robertson Lake mylonite zone (RLMZ) 361
Robson, Mount 582
rock(s). *See also* specific types and processes
    bending or folding of *195*, 195–196
    breaking of 196
    correlation of 712
    deformation of **194–197**, 712–719, *718*
    igneous **431–435**
    metamorphic **538–542**
    sedimentary **677–679**
rock cycle 307
rockfalls 523
rock flour 333
Rocknest Formation 558
Rockport-Hyde-School-Wellesley-Wells suite 363
rock record 376–377
rockslide 523
rocky coasts 98–99
Rocky Mountains 118, 153, 581–584
Rodinia 50–51, 81, 642
    climate effects of 128–130
    Greenville province and **357–364**
Rodinia-Sunsas belt 364, 642
Rokelides 18
Rokell River Group 18
Romanche fault zone 749
Rosalind (moon) 791
Rossby waves 75, 251
Ross Ice Shelf 26, 336, 587
rotation, and deformation of rocks 194
Roter Kamm 436
roto inversion 186

rouche moutonnées 333
rounding 678–679
Rover (Mars mission) 492
RRR triple junction 218
Ruang eruption (1871) 757
Rub'al Khali Desert *27, 201,* 203, 208, 261
Runcorn, Stanley K. 632
run-up, tsunami *764,* 764–765
Rush Lake 331
Russian craton *266,* 662–663, *664*
Russian geology **662–667,** *663, 664*
Rutherford, Ernest 395, 398, 653
Ryoke-Abukuma belt 476

S

Sabia Formation 29
Saffir-Simpson scale 405–406
Sagan, Carl 492
Sagan criteria for life 492
Saha, A. K. 448
Sahara Desert 203, 204, 208–210, 261
Sahara meta-craton 6
Sahel region *201,* 201–203, 322
Sahtan Group 33
Saih Hatat 33, 35
St. Elmo's fire 804
St. Helens, Mount 153, 469–471, 607, 757–758, 796, 799, 801–802, 806–807
St. Louis flooding 278–281, *279*
Sakamena Group 503
Sakoa Group 503
Sakoli Group 447
salinity, ocean 743–745
    and climate 131, 135–136
    and currents 587
    external energy-driven processes and 251–252
Salisbury (Harare) belt 14, *17*
Salle de la Verna 120, 479
salt(s) 678
saltation 260, 284
Salton Sea 724, 750
Salween River 52
Sambirano Group 500
San Andreas Fault *195,* 225–226, 583, 629, 630, 681, 712–713, 725, 742, 749–750, *750,* 753
Sanbagawa belt 475–476
sand, windblown 207–208, 259–261, *260*
sand dunes
    coastal 91, *92*
    desert 207–208, 259–261
sand sea 261
sandstone 540
San Francisco earthquake (1906) 228, 236, 752–753
San Francisco Peaks 797
San Joaquin Valley 723
Sanriku, Japan, tsunami (1896) 755, 779

Santori Islands 457–459
Santorini eruption and tsunami (1600) 756–757
São Francisco craton *693*
São Luis cratonic fragment *693*
SAR. *See* Synthetic Aperture Radar
Sarawak Chamber 120, 479
Sargurs 445
Sarkar, S. N. 448
Sarmatia block 663
satellite, global positioning 315, 316–318
satellite imagery 655–658
Satpura Orogeny 447, 449
Saturn **668,** *669,* 689–690
Sauk transgression 385–387, *386*
Sausar Group 447
Sayq Plateau 33
schist 539, *539*
schistosity 539
Schwarzschild, Karl 106
Schwarzschild radius 106
Schweig, Eugene 321
Scott, Robert F. 335
Scottish Highlands 265, 267–268
screw dislocations, in crystals *187,* 187–188
Scripps Institute of Oceanography 593–594, 632
seafloor magnetism 611–613, *612, 613*
seafloor spreading 218, 372–375, 611–613
sea ice 334–336
sea-level fall (regression) 377–380, *379,* 388–390, 735
sea-level rise 89, 128–130, 139, **669–675**
    causes of 669, 670–671
    climate change and
        long-term 670, 671–673
        short-term 670, 671
    continental area changes and 673
    eustatic (global) 343–344, 378–379, 670, 671
    and evolution/extinction 515–516
    global warming and 337, 338–340, *339,* 342–344, 669, 673
    as hazard 409
    ice age and 429
    melting of ice sheets and 334, 669, 673
    mid-ocean ridge volume and 673
    plate tectonics and 670, 673–674
    relative 670, 671
    subsidence and 721
    supercontinents and 673–674, *734,* 734–735
    transgression 377–380, *379,* 735
        Sauk 385–387, *386*
        Tippecanoe 387–388
    unconformity bound sequences and *381,* 381–382

Search for Extra-Terrestrial
    Intelligence (SETI) 492–493
Seasat Synthetic Aperture Radar 658
seasonality 130–131
sea stacks 99
Seattle area
    earthquake prediction for 681–
        682
    tsunami risk for 775
seawater **675–676**
    composition of 675–676
    intrusion in coastal aquifers 371
Sebakwian greenstones 15–16
secondary waves, earthquakes 228–
        230, *229*, 679–680, *680*
SEDEX deposit 238
Sedgwick, Adam 211, 377, 614,
        676–677
sediment
    chemical 677–678
    clastic 677
    evaporite 678
    transport in fluvial systems 284–
        286
sedimentary basin **84–86**
sedimentary facies 377–378, *378*
sedimentary rock **677–679**
    Milankovitch cycles and 134–135,
        557, *558*
    stratification of 678
    surface features of 679
sedimentary structures 678
sedimentation **677–679**
seiche waves 236
seismic discontinuity 680
seismic Moho 495
seismic waves 227–230, 327, 679–
        683, *680*
seismographs 230, *231*, 682–683
seismology 326, 327, **679–683**
selenium 253
Selukwe (Shurugwi) belt 13, *17*
Semail ophiolite *27*, 33–34, *34*, 123,
        577, 595
Sengör, A. M. Celal 180
Septentrional fault system 754
septic systems 369–370
sequence stratigraphy *381*, 381–382,
        683
Séries Quartzo-Schisto-Calcaire 502
SETI. *See* Search for Extra-Terrestrial
        Intelligence
SETI Institute 492–493
Sevier belt 583
Sevier Lake 331
Sevier Orogeny 583
sewage contamination 370–371
Shaanxi, China, earthquake and land-
        slide (1556) 526–527
Shabani gneiss 13, *14*
shale 539
shallow subsidence 722
Shams, Jabal 33
Shamvaian Group 11, 15, 16, *17*

Shandong Hills 45
Shangani belt 13
Shangani Complex 11
Shanxi Graben 55–56
Shapley, Harlow 490
Sharbot Lake terrane *359*, 360
Shasta, Mount 469
Shatsky Rise 139, 487, 590
shear 716, 718–719
shear joints 296
shear waves, earthquake 228–230,
        229, 679–680, *680*
shelves, continental 225
shergottites 62
Shetland Islands 267
shield(s) 39, 636. *See also* specific
        shields
    continental 152, 196
    speleothem 121
shield volcanoes 797, 799
Shimanto belt 3
shock waves
    of meteorite impact 543, 544–
        547
    of volcanoes 804
Shonbein, Christian F. 309
shorelines **86–100**, 225
shoshonite group 457
Shugat Al-Huwaimiliyah 36
Shurugwi belt 13, *17*
Shuttle Imaging Radar (SIR-A, SIR-B)
        657–658
Siberia 362, 667
    comet impact in 145–147, 435,
        543, 544–545, 546
    taiga forest of 667
Siberian craton 153, 662, 663–665,
        664
Siberian flood basalts 139–140, 488,
        621, 662, 666–667, 733–734
Sichuan, China, earthquake and land-
        slide (2008) 527
Sichuan basin 45
Sidr, Typhoon 277
Sierra Nevada 476, 583
Sierra Nevada batholith 347, *347*
silica 255–256, 432, 488, 507
silicate(s) 559, 561
silicate-iron formations 638
silicate tetrahedron 559–561, *560*
sills 431, *432*
Silobela belt 13, 14
Silurian 388–391, 614–615, **683–
        684**
silver deposits 238
simple craters 436
simple shear model 215, *216*, 718–
        719
simple visual binaries 103
Sinai Desert 209
Sinchuan, China, earthquake and
        landslide (2008) 230, *233*,
        234, 235, 235, 319, 527
Sind Desert 203

Singhbhum craton 443, 447–448
Singhbhum-Dhalbhum belt 448
Singhbhum Group 448
Singhbhum thrust 448
single-apparition comets 143
Sinian 382, 384–385
sinkholes 120–121, 323, 368–369,
        479–481, 721
Sipolilo belt 14
SIR-A/SIR-B. *See* Shuttle Imaging
        Radar
Sitter, William de 603
Siwalik molasse 449
Skeleton Coast 204, 210
skolithos 268
slate 539
slaty cleavage 539
Slave Province 153, 177, 179, 352,
        354–355, 567–571, *568, 569,*
        574–575
Sleat Group 268
Sleepy Dragon terrane 567, 570, 571
sleet 640
slides 521–535
    active, monitoring of 533–534
    disasters, examples of 525–532
    mitigation of damages from
        534–535
    reducing hazards and dangers of
        532–535
    submarine 524–525, 758–760,
        783–787
    and tsunamis 758–760, 783–787
    volcanoes and 802
Slipher, Vesto 301
slip systems, crystal 186–188
Sloan Great Wall 790
Sloss, Laurence L. 381
slump(s) 521–525, *533*, 758
slurry flow 522
small solar system bodies 635
Smith, William **684–688**
Snake River Plain flood basalts 402
SNC meteorites 62
snow 553, 640
    changes in cover *339*
"snowball Earth" 331, *332*
Soddy, Frederick 395
sodium cycle 308–309
soils **688–689**, *689*
solar core 727, 728, 730–731
solar cycle 729–730
solar flares 78, *727*, 729–730
solar mass, and stellar evolution
        700–701
solar power 623
solar prominences 730
solar radiation 727–728
    and climate change 131–135, *133,
        134*
    as external energy source 248,
        250
    Milankovitch cycles and 555–558,
        556

solar system **689–691**
  origin and evolution of **599–602,** *600,* 690–691
  physical properties of objects in 690
solar wind *77,* 77–78
Solger, F. 44
solid-Earth geophysics 326
solid Earth shock waves, of meteorite impact 547
solifluction 522
solution sinkholes 480–481
Somali plate 505
Songban-Ganzi terrane 52, *54*
Songliao basin 45
Sonoma Orogeny 583
Sonoran Desert 203–204
Sorby, Henry Clifton **691**
sorting 678
South Africa, Vredefort dome of 7, 435, 441
South American geology *692,* **692–696,** *693*
South China craton 46, 50–51
Southern Cross Province 80, 81
Southern Granulite craton 443, 446
Southern Hills (Asia) 45
southern lights (aurora australis) 72, **77–78**
Southern Oscillation. *See* El Niño-Southern Oscillation
Southern Volcanic Zone 220, 401
South Iceland Seismic Zone 220, 401
South Lake ophiolite 578
South Qinling terrane 51
Southwest, American 202, 421–423
space geodesy 315
spaceguard 548
special relativity 242
speleothems 121–122
spiral galaxies 299–300
Spitzbergen Island 269–270, 581
SPOT. *See* Systeme Pour l'Observation de la Terre
spreading centers, oceanic 218, 372–375
springs 366, *367*
spruce bark beetle *340*
squall lines 746–747
stalactites *121*
stalagmites *121*
standard model, of universe 604–606
Stanovoy Ranges 667
star(s)
  black holes **105–107**
  constellation **147–151**
  death of 704, 736
  dwarfs **222–223**
  evolution of 696–700, *697, 698,* **701–705,** *702*
  formation of *696,* **696–701**
  galaxies **299–303**
  high-mass 705, 736

main sequence 696–700, 701–702
nova 110–111, **585,** 735
pulsar **646**
Stardust Comet Sample Return Mission 144
star systems, binary **102–104,** *103,* 585
stationary fronts 554
steady state theory, of universe 604
Steamboat geyser 402
Steep Rock Lake belt 42, 354, 618
Steep Rock platform 43
Steinman, G. 594
stellar evolution 696–700, *697, 698,* **701–705,** *702*
Steno, Nicolaus **705–710**
Stille, Wilhelm Hans **710–711**
Stoer Group 268
Stone Age 609–610
stony-iron meteorites 60, 63, 543
stony meteorites 60, 542
stopping, magma 431
Storegga, Norway, tsunamis 786–787
storm(s) 276–277, 745–749
storm beaches 87–88
storm ridge 86
storm surges 276–277, 406, 407, 408, 758
Stowe, Clive W. 11–12, 13
straight channels, stream 286
straight line winds 746
strain, and deformation of rocks 194–195, 717–719
strata, relative ages of 380
stratification 678, **711–712**
stratigraphy 376–383, **711–712**
  sequence **683**
  Smith and birth of 686–687
  time 711–712
stratocumulus clouds 142
stratopause 71, 72
stratosphere *71, 72,* 76
stratovolcanoes 797
stratus clouds 141–142
stream(s) 659–662
  capacity of 285
  channel patterns 286–287
  drainage by 220–221, 289
  dynamics of flow 287–288
  erosion by 261–263, 419–420
  fluvial action of 283–293, 661–662
  geometry of 283–284
  graded 288, 330
  modifications of
    to alleviate water shortages 421–426
    effects on dynamics 289–293
  sediment transport and deposition in 284–286
stream capture 289
stream flow 262, 287–288

stream terraces 288
stress, and deformation of rocks 194–195, 716–717
strike 713
strike line 713
strike-slip faults 196
strike-slip plate margins 1. *See also* transform plate margin(s)
stromatolites 43, 384, 643–644, *644*
Stronghyle volcano 458
structural geology **712–719**
Strutt, Robert J. 395–396
stylolites 188–189, 214
subduction 157–162, **719–720**
  accretionary wedge in 1–5
  collisions in 161–162
  Pacific plate 56
subduction zone **719–720**
Subgan Group 663–664, 664–665
submarine flood basalts 487–488
submarine landslides 524–525, 758–760, 783–787
subsequent stream 289
subsidence **720–726**
  coastal 323, 721, 725–726
  sinkholes 120–121, 323, 368–369, 479–481, 721
Sudan, water supply for 425–426
Sudbury crater 436, *436*
Suess, Eduard 345, 606, 616
Suhaylah Formation 33
Sukinda thrust 448
Sukma Group 447
sulfates 559, 561
sulfide(s) 559, 561
sulfide-iron formations 638
sulfur dioxide, in atmosphere 73
Sulu belt 51
Sumatra earthquake (2004) 88–89, 163
  ground level changes in 235–236
  magnitude of 231
  tsunami induced by 88–89, 91, 95–96, 163, 236, 592, 755, 768–772, *771,* 774–775
Sun **726–731**
  atmosphere of 728–729
  as external energy source 248, 250
  interior of *727,* 728
  origin and evolution of 599–602
  physical properties of 727–728
sundogs **731–732**
sun halos **731–732**
sun pillars **731–732**
sunspots 78, 729–730
supercell thunderstorms 746, 749
supercluster 304, 789–790
supercontinent(s). *See also* Gondwana/ Gondwanaland; Nuna; Pangaea; Rodinia
  and climate 128–130, *129,* 131
  and evolution of life 733–734
  and mantle convection 732–733

and sea-level changes 673–674, 734, 734–735
and seasonality 131
supercontinent cycles **732–735**
Superfund Act 254
Superior Province 352, 354–356, 361, 568, 571–573, 573, 574–575
Superior-type banded-iron formations 638–639
supernova 110–111, **735–739**, 736
superposed stream 289
superposition, law of 705, 709, 711
surface forces 716
surface waves, earthquake 230
Surma fault zone 500
Sutherland belt 7
Sutlej River 52
Sutton, A. H. 732
Svalbard 266, 269–270
Svecofennian Orogeny 269
S-waves, earthquake 228–230, 229, 679–680, 680
Swaziland Supergroup 5–7
Sydney basin 79
synclines 195
Synthetic Aperture Radar (SAR) 658–659
Syr Darya River 58
Syria, water supply for 426
Syrian Desert 203
system, stratigraphic 712
Système pour l'Observation de la Terre (SPOT) 655, 656–657, 658
Szilard, Leo 243

T

Taal volcano 758
Taconic Orogeny 161, 212, 267, 268, 388, 389, 536, 575, 576, 579–580
TAG. See Trans-Atlantic Geotraverse hydrothermal mound
Taif island arc 32
taiga forest, Siberian 667
Taihang Mountains 45
Taihangshan Massif 56
Taishan Mountains 45
Taiwan, Chi-Chi earthquake (1999) 165–166
Taklimakan Desert 201, 203, 207, 208
Talang, Mount, eruption (2005) 771
Tally Pond volcanics 578–579
Tambora eruption (1815) 138, 461–464, 463, 781–782, 805
Tanganyika, Lake 21, 724
Tanglha Mountains 45
Tan-Lu fault system 51
Tanzania block 18
Tanzanian craton 153
Tar Desert 208
Tarib island arc 32
Tarim basin 9, 45
Tarim block 46, 47–50

tarns 333
Tasman orogen 78, 80, 82
Tati belt 13, 17
tectonic(s) 326. See also plate tectonics
tectonic mélanges 1, 158, 268, 455, 536, **536**
tectosphere 636
Teixeira, Pedro 23, 694
telescope(s) 304–306, **740–741**
television waves 649–650
Temek River 265
temperate glaciers 332
temperature
    changes in 126–127, 127, 132, 337–338, 338, 339. See also global warming
    as climate factor 125
    ocean 743–745
        and climate change 131, 135–136
        and currents 587
        external energy-driven processes and 251–252
    and star formation 696
tephra 796–797, 801–802
tequis 695
terahertz radiation 650
Terai belts 451
terminal moraines 97, 333
terranes 177. See also specific terranes
    Archean 39, 40–42
    Proterozoic 641
terrestrial planets 690
Tertiary 122, 123, 393, 394, 564, 741, **741–742**
Tethyan belt 594
Tethys (moon) 668
Tethys Ocean 118, 123, 184, 259, 264–265, 394
Teton Range 582
Texas
    flooding in 277
    subsidence in 721, 723
thalweg 661
Thaqab Formation 33
Thar Desert 203, 204
Tharsis region, of Mars 514
Theespruit area 354
Theespruit Formation 7
Thelon-Talston arc and orogen 574
Thematic Mapper (TM) 655–656, 658
theory of Earth 410–414
Thera eruption 457–459, 458
thermodynamics **742–743**
thermohaline circulation 131, 135–136, 251–252, 587, **743–745**, 744
thermoluminescence 314
thermosphere 71, 72, 76
Thetford Mines ophiolite 578
tholeiitic series 457
Thompson belt 573

thorium-lead dating 312, 313–314, 398–399
Three Sisters 469
thrust faults 196
Thule asteroids 64
thunderstorms 554, **745–749**, 747
Tibesti 208
Tibetan Himalaya 449–450
Tibetan Plateau 45, 52–54, 54, 162, 449
tidal delta 93
tidal flats 94–95
tidal inlets 93–94, 94
tidal power 623
tidal waves 754–755. See also tsunami(s)
tide-dominated deltas 96, 198, 200
tidewater glaciers 331
Tienshan Mountains 45
Tierra del Fuego 694
Tigris-Euphrates drainage basin 424–426
Tikkerutuk terrane 573
till 97–98, 333
time. See geologic time
timelines 712
Times Beach, Missouri 254
time stratigraphy 711–712
Tippecanoe Sequence 387–388
Tiquinia, Strait of 23
Titan (moon) 634, 668
Titania 791
Titanic 334
Titicaca, Lake 22–23, 694
Tjornes fracture zone 220, 401
TM. See Thematic Mapper
Tokwe River gneiss 13
Tokwe terrane 12–14, 15–16, 17
Tokyo tsunami (1923) 758
Tombaugh, Clyde 634
tombolos 90
Tomiko terrane 358, 359
tonalite(s) 40–41, 577–579
tonalite-trondhjemite-granodiorite group (TTG) 18
Torino Hazard Scale for Near-Earth Objects 549, 550
tornado(es) 748, 748–749
tornado alley 748–749
Tornquist Sea 267
Torridon Group 268
Toubkal, Jabel 20
4179 Toutatis asteroid 65
Townsend, Joseph 686
toxic materials 252–258
trade wind(s) 553, **562–563**
trade wind deserts 203
trailing margins 155, **617–618**
Transamazonian Orogeny 695
Transantarctic Mountains 23, 24, 25, 26, 334
Trans-Atlantic Geotraverse (TAG) hydrothermal mound 108
transcontinental arch 385

transcurrent plate margins 155. *See also* transform plate margin(s)
Trans-European suture zone 663
transform faults 630, 749–754, *750*
transform plate margin(s) 1, 155, 629, *629, 630*
  climate effects of 127–128
  in continents 749–751
  deformation at 715–716
  earthquakes from 752–754
  in oceans *751,* 751–752
  subsidence at 724–725
transform plate margin processes **749–754**
transgression, marine 377–380, *379,* 735
  Sauk 385–387, *386*
  Tippecanoe 387–388
Trans-Himalaya batholith 450
Trans-Hudson orogen 572, 573, 574
translation, and deformation of rocks 194
translational slide 521–522
translation gliding 186–188
Trans-Neptunian objects 66, 67–68, 635
transportation, in erosion 261
transportation network, earthquake damage to 236
Transvaal Supergroup 7–8, 43, 441
transverse dunes 207
Transverse Ranges, California 155, 750
traps. *See* flood basalts
travertine 122
trench(es) 155, 157–161, 177, 180–181, 225, 454
trench pull 715
Trenton Group 576
Triassic 537
Triassic-Jurassic mass extinction 517, 518–519
trilobites 114–115
Trinity River 423
triple junction 218
Triton 634, 635
Trojan asteroids 64, 65
tropopause *71,* 72
troposphere *71,* 71–72, 76, 251
Tsaidam basin 9
Tsarantana sheet 502
Tsarantana thrust zone 502
tsunami(s) 88–89, 91, 96, 163, 236. *See also* specific tsunamis
  earthquake-induced 755–756, *756,* 769–781
  force of impact and backwash of 765
  generation mechanisms **754–769**
  as geological hazard 319, *323*
  hazard zones for 766–767
  historical accounts **769–787**
  landslide-induced 758–760, 783–787

  meteorite impact and 543, 547, 760
  monitoring for 765–766, 774–775
  movement of
    encounter with shallow water 762–763
    in open ocean 761–762
    physics of 760–765
  natural gas and gas hydrate eruption-induced 760
  prediction of 766
  public education on 767–768
  reducing threat of 765–768
  risk mapping for 766–767
  run-up of *764,* 764–765
  volcanic eruption-induced 756–758, 781–783, 803–804
  warning systems for 767, 775
  weather-related 760
TTG. *See* tonalite-trondhjemite-granodiorite group
Tuareg shield 18
Tucker, Robert 502
Tuluhah Orogeny 29
tungstates 559
Tunguska, comet impact at 145–147, 435, 543, 544–545, 546
turbidites 293–294, 454–455
turbidity currents 525, 759
turbulent flow, stream 284
Turee Creek Group 79
Turkana, Lake 21
Turkey
  earthquake (1999) 753
  water supply for 426
Turkic-type orogeny accretion 177, 180–181
Tuwaiq Mountains 26
Twelve Apostles (sea stacks) 99
twinning, crystal 186
twist hackle 296
Tychonic System 111
Tycho's supernova 737
Tyndall, John 341, 351
typhoons 403. *See also* hurricane(s)

U

Uchi-Sachigo arc 572–573
Ukrainian shield 662, 663
ultraviolet astronomy 70
ultraviolet radiation 244, *245,* 245–246, 649–650
Umar Group 33
Umbriel 791
Umm Al-Maradim 36
Umm Al-Naml 36
Umm Al-Neqqa dunes 36
Umtali belt 14
Umtali line 16, 180
Uncompahgre rift 583
unconfined aquifers 367
unconformities *380, 381, 788,* **788–789**
unconformity bound sequences *381,* 381–382

Ungava orogen 355
uniformitarianism 193, 413, 497, 652, 816
U.S. Army Corps of Engineers 281, 321
U.S. Coast and Geodetic Survey 316
U.S. Geological Survey (USGS) 309, 316, 321, 370, 534, 766, 805, 806, 807
universe **789–790**
  big bang theory of 175, 490–491, 602–604
  closed model of 604
  cosmological models of 176
  heliocentric model of 167–168, 304–305, 482–483
  inflationary theory of 603–604
  origin and evolution of **602–606**
  standard model of 604–606
  steady state theory of 604
Unzen, Mount, eruption (1792) 757
Upper Greenstones 14–15, *17*
upwelling, coastal 588–589
Ural Mountains 266, 663, *664,* 666
Ural-Okhotsk belt 666
Ural River 265, 666
uranium 240, 256–257, 651–652
uranium-lead dating 312–313, *313,* 394, 397–399
Uranus 689–690, **790–792,** *791*
urbanization, and flash floods 282
urelites 62
Urey, Harold C. 384
Uruacuan event 695
USGS. *See* U.S. Geological Survey
utilities, earthquake damage to 236
Utsalik terrane 573
Uwaynat Mountains 208
Uyak Complex 3

V

vacancies, crystal 186, *187*
Vaiont, Italy, landslide (1963) 527–528, 532
Valdez, Alaska, earthquake (1964) 163–165
  ground breaks in 234
  ground level changes in 235–236, 670, 722, 725, 777
  magnitude of 231
  mass wasting in 520
  tsunami induced by 164, 755, 763, 776–779
valley(s)
  glaciated 333
  ramp 750
valley glaciers 331
vanadates 559
Van Allen radiation belts 77, 77–78
Variscan belt 663
varve 678
Vendian 382, 384–385
Vendoza fauna 614
Venice, Italy 377, 670, 672, 721, 723

Ventersdorp Supergroup 7, 43, 441
Venus 689–690, **793–795**, *794*
Verkhoyansk-Kolyma block 662, *664, 665*
Verkhoyansk Ranges 665, 667
vermiculite 255
Vernadsky Institute 310
Very Large Array 740
Very Long Baseline Interferometry (VLBI) 740
4 Vesta asteroid 62, 63, 66
Vestfold Hills 24
Vesuvius eruption 459–461, 757, 795, 797, 801, 804
Vikings 593
Vindhyan Supergroup 449
Vine, Fred J. 375, 632
Virgo cluster 304, 789
visible light 244, 245, 649–650
Vitim River 667
VLBI. *See* Very Long Baseline Interferometry
VMS. *See* volcanogenic massive sulfide deposits
voids, in universe 790
volcanic arcs 160–161, 453–474. *See also* island arc(s)
volcanic blocks 801
volcanic bombs 796, 801
volcanic landforms 795, 796–797
volcanic mountains 153, 607
volcanic necks 348, 431, 508
volcanic rocks 431, 506, 507
volcano(es) **795–809**
 bimodal 86
 climate effects of 127–130, 138–140, 804–805
 convergent margin 160–161, *161,* 457–474
 and earthquakes 803–804
 eruption column of *798*
 explosive eruptions of 796–797
 geographic distribution of *796*
 hazards of 319–320, *320,* 796, 799–805
 historical eruptions of **453–474,** 800, 806
 long-term effects of eruptions 805
 nonexplosive eruptions of 796
 oceanic (ridges) 218–220
 plate tectonics and 795–796, 798–799
 precursors of eruptions 806–807
 predicting and monitoring 805–808
 and subsidence 722
 and tsunamis 756–758, 781–783, 803–804
 types of eruptions 796–797
volcanogenic massive sulfide (VMS) deposits 108
Volga River 265
Volgo-Uralia block 662–663
Vorokafortra shear zone *501*

Voronezh uplift 663
Voyager 1 492
Voyager 2 634
Vredefort dome 7, 435, 441
vulcanists 410
Vulcan objects 64
Vumba belt 13, *17*
Vurney, Venetia 634

W
W. R. Grace and Company 255
Waagen, Wilhelm Heinrich 272
WAAS system 318
Wabigoon terrane 572
wadi(s) 297–298
Wahiba Sand Sea 261
Waisa Group 33
Walcott, Charles D. 113–114
Walther, Johannes 378
Walther's law 378
Walvis Ridge 513
Warba 36
warm-based glaciers 263
Warrawoona Group 79, 384
water
 contamination of 369–371
 desalination of 427
 erosion by 261–263, 419–420
 fracture zone aquifers 297–298, 367–368
 fresh, diminishing supply of 416, 420–421
 groundwater **365–371**, 417–418
 as hazard 420
 hydrosphere 104–105, 225, **416–428**, *417*
 meteoric 552
 properties of 417–418
 as resource 420
 shortages of, river modifications to alleviate 421–426
 vapor, in atmosphere 72, 74
water cycle 308, 416, 418, *419*
water wells 366, 367–368
Watts, William 395
wave(s)
 earthquake 227–230, 327, 679–683, *680*
 gravity **350**
 seiche 236
 tsunami. *See* tsunami(s)
wave cyclones 406–408
wave-dominated deltas 96, *198, 200*
wave-particle duality 414
Wawa terrane 572
weather 552–555
weather extremes 340–341
weathering 261, **810–813,** *811*
weather-related tsunamis 760
Weddell Sea 743–744
Wegener, Alfred Lothar **813–815,** *814*
 and Du Toit (Alexander) 154, 221, 391, 607
 and Hess (Harry) 373, 374

 and Holmes (Arthur) 397, 607
 on Pangaea 391, 616
 theory of continental drift 153–154, 391, 397, 607, 630, 813, 814
Weihe-Shanxi graben system 55–56
welded barriers 91
wells, water 366, 367–368
Wenchuan, China, earthquake and landslide (2008) 230, *233, 234,* 235, *235,* 319, 527
Wenlockian Age or Series 683
Werner, Abraham Gottlob 411, **815–816**
West African craton 5, 6, 18
West Antarctic Ice Sheet 24
Western block (China) 47, 50
Western Desert of Egypt 207, 260, 424–426, 657
Western Dharwar craton 443–445
Western Ghats of India 442–443
Western Gneiss Terrane 80
Western Rift Zone 220, 401
Western U.S.
 earthquake prediction for 681–682
 tsunami risk for 775
West Nile Complex 18
West Rand Group 7, 9–11
wetlands, coastal 95–96
Whangaehu River *320*
Wheeler, George M. 330
Wheeler Survey 330–331
Whim Creek belt 356
Whim Creek Group 79
Whin Sill 398
Whipple, F. J. W. 146
white dwarfs 222, 585, 704, 737
white smoker chimneys 104
Widmannstatten texture 63
Wiechert, E. 683
Wild 2 comet 144
Wilde, Simon 80
Wilkes Land 26
Wilkins ice shelf 26
Wilson, J. Tuzo 607, 630, 632
Wilson, John F. 11
Wilson, Robert 173, 603
wind(s) 553
 desert 201, 206–207, 259–261
 erosion by 263
 katabatic 25
 thunderstorm 746
 trade 553, **562–563**
windblown sand and dust 207–208, 259–261, *260*
Windemere Supergroup 582–583
wind power 623
Wind River Range 574, 582
winter, global 547
Winter Park sinkhole 480
Wisconsin glacial maximum 634
Wittenoom asbestos 255
Witwatersrand basin 7, 8, 8–11, *9,* 238

Witwatersrand Supergroup 9–11, 43, 441
WMO. *See* World Meteorological Organization
Woods Hole Oceanographic Institute 593–594, 632
Wopmay orogen 567–568, 575, 641
work hardening 187, 188
World Meteorological Organization (WMO) 338
Wrangellia superterrane 2–3
Wrangellia terrane 2
Wui Mountains 45
Wutai Shan ophiolite 493
Wutai-Taihang Mountains 48
Wynne-Edwards, H. R. 361
Wyoming Province 568, 574–575

X

X-ray(s) 244, *245*, *246*, 649–650
X-ray astronomy 70
X-ray telescopes 740–741

Y

Yablonovy Ranges 667
Yafikh-Ragbah Orogeny 29
Yanbu belt 31

Yangtze block 46, 50
Yangtze River 52, 281–282, 463
yardangs 206
Yarlung River 52
Yarlungzangbo suture 52, *54*
Yavapai orogen 575
"year without a summer" 461–464, 782, 805
yellow dwarfs 222
Yellowknife belt 42, 355, 567
Yellow River flooding 281–282, 463
Yellowstone Lake 402
Yellowstone National Park *329*, 329–330, 401–402, 434, 509, 513, 797
Yen Cheng-ching 44
Yerrida basin 82
Yilgarn belt 42, 352
Yilgarn craton 78, 79–81, *80*, 352, 353–357
Yinchuan-Hetao graben system 56
Yinshan Mountains 45
Yosemite National Park *811*
Yucatán asteroid impact 438–442, 519, 543
    crater structure of 436, *436*, 438–442

and extinction 59, 122–123, 139, 393, 438–440, 442, 519, 543
as geological hazard 323
and life on Earth 145
size of asteroid 65, 439–440
and tsunamis 440, 760
Yunnan-Guizhou Plateau 45

Z

Zagros Mountains 37, 56–58, *57*
Zambezi belt 19
Zangbo (Brahmaputra) River 52, 55, 276, 407, 408, 449
Zeederbergs Formation 14
zero-age main sequence star 700
Zimbabwe craton 5, 6, *8*, 11–16, *17*, 180, 352, 355, 618
zinc deposits 238
zooplankton 619, *619*
Zucchi, Niccolo 740
Zunhua belt 48–50
Zunhua ophiolite 493
Zuni Sequence 183
Zwankendaba arc 14